A Commemorative Issue in Honor of Professor Jose M. Kenny: Advances in Polymer Composites and Nanocomposites

A Commemorative Issue in Honor of Professor Jose M. Kenny: Advances in Polymer Composites and Nanocomposites

Guest Editors

Miguel Ángel López Manchado
Laura Peponi
Luigi Torre

MDPI

Basel • Beijing • Wuhan • Barcelona • Belgrade • Novi Sad • Cluj • Manchester

Guest Editors
Miguel Ángel López Manchado
Institute of Polymer Science and Technology (ICTP)
CSIC
Madrid
Spain

Laura Peponi
Institute of Polymer Science and Technology (ICTP)
CSIC
Madrid
Spain

Luigi Torre
Department of Civil and Environment Engineering
University of Perugia
Terni
Italy

Editorial Office
MDPI AG
Grosspeteranlage 5
4052 Basel, Switzerland

This is a reprint of the Special Issue, published open access by the journal *Polymers* (ISSN 2073-4360), freely accessible at: https://www.mdpi.com/journal/polymers/special_issues/00YP5LA2I4.

For citation purposes, cite each article independently as indicated on the article page online and as indicated below:

Lastname, A.A.; Lastname, B.B. Article Title. *Journal Name* **Year**, *Volume Number*, Page Range.

ISBN 978-3-7258-4023-6 (Hbk)
ISBN 978-3-7258-4024-3 (PDF)
https://doi.org/10.3390/books978-3-7258-4024-3

Contents

polymers

MDPI

Review

Bio-Based and Biodegradable Polymeric Materials for a Circular Economy

Víctor Oliver-Cuenca [1], Valentina Salaris [1], Pedro Francisco Muñoz-Gimena [1], Ángel Agüero [2,3], Mercedes A. Peltzer [4,5], Victoria Alcázar Montero [3,6], Marina P. Arrieta [3,6], Jaume Sempere-Torregrosa [7], Cristina Pavon [7], Maria Dolores Samper [7], Gema Rodríguez Crespo [1], Jose M. Kenny [8], Daniel López [1,*] and Laura Peponi [1,*]

1 Instituto de Ciencia y Tecnología de Polímeros, ICTP-CSIC, Calle Juan de la Cierva 3, 28006 Madrid, Spain; victor.oc@ictp.csic.es (V.O.-C.); v.salaris@ictp.csic.es (V.S.); pfmunoz@ictp.csic.es (P.F.M.-G.); gema@ictp.csic.es (G.R.C.)
2 Instituto Universitario de Tecnología de Materiales (IUTM), Universitat Politècnica de València (UPV), Plaza Ferrándiz y Carbonell 1, 03801 Alcoy, Spain; anagrod@epsa.upv.es
3 Departamento de Ingeniería Química Industrial y del Medio Ambiente, Escuela Técnica Superior de Ingenieros Industriales, Universidad Politécnica de Madrid (ETSII-UPM), Calle José Gutiérrez Abascal 2, 28006 Madrid, Spain; mariavictoria.alcazar@upm.es (V.A.M.); m.arrieta@upm.es (M.P.A.)
4 Laboratory of Obtention, Modification, Characterization, and Evaluation of Materials (LOMCEM), Department of Science and Technology, University of Quilmes, Bernal B1876BXD, Argentina; mercedes.peltzer@unq.edu.ar
5 National Scientific and Technical Research Council (CONICET), Buenos Aires C1425FQB, Argentina
6 Grupo de Investigación en Polímeros, Caracterización y Aplicaciones (POLCA), 28006 Madrid, Spain
7 Instituto de Tecnología de Materiales (ITM), Universitat Politècnica de València (UPV), Plaza Ferrándiz y Carbonell 1, 03801 Alcoy, Spain; jausemto@epsa.upv.es (J.S.-T.); cripava1@epsa.upv.es (C.P.); masammad@upv.es (M.D.S.)
8 STM Group, University of Perugia, Strada Pentima 4, 05100 Terni, Italy; jose.kenny@unipg.it
* Correspondence: daniel.l.g@csic.es (D.L.); lpeponi@ictp.csic.es (L.P.)

Abstract: Nowadays, plastic contamination worldwide is a concerning reality that can be addressed with appropriate society education as well as looking for innovative polymeric alternatives based on the reuse of waste and recycling with a circular economy point of view, thus taking into consideration that a future world without plastic is quite impossible to conceive. In this regard, in this review, we focus on sustainable polymeric materials, biodegradable and bio-based polymers, additives, and micro/nanoparticles to be used to obtain new environmentally friendly polymeric-based materials. Although biodegradable polymers possess poorer overall properties than traditional ones, they have gained a huge interest in many industrial sectors due to their inherent biodegradability in natural environments. Therefore, several strategies have been proposed to improve their properties and extend their industrial applications. Blending strategies, as well as the development of composites and nanocomposites, have shown promising perspectives for improving their performances, emphasizing biopolymeric blend formulations and bio-based micro and nanoparticles to produce fully sustainable polymeric-based materials. The Review also summarizes recent developments in polymeric blends, composites, and nanocomposite plasticization, with a particular focus on naturally derived plasticizers and their chemical modifications to increase their compatibility with the polymeric matrices. The current state of the art of the most important bio-based and biodegradable polymers is also reviewed, mainly focusing on their synthesis and processing methods scalable to the industrial sector, such as melt and solution blending approaches like melt-extrusion, injection molding, film forming as well as solution electrospinning, among others, without neglecting their degradation processes.

Keywords: sustainable polymers; bio-based polymers; biodegradable polymers; revalorization; circular economy; nanocomposites; plasticizers; natural polymers; nanoparticles; degradation; processing

Citation: Oliver-Cuenca, V.; Salaris, V.; Muñoz-Gimena, P.F.; Agüero, Á.; Peltzer, M.A.; Montero, V.A.; Arrieta, M.P.; Sempere-Torregrosa, J.; Pavon, C.; Samper, M.D.; et al. Bio-Based and Biodegradable Polymeric Materials for a Circular Economy. *Polymers* **2024**, *16*, 3015. https://doi.org/10.3390/polym16213015

Academic Editor: Piotr Dobrzynski

Received: 29 August 2024
Revised: 4 October 2024
Accepted: 17 October 2024
Published: 28 October 2024

1. Introduction

The use of plastics, which are cheap and versatile materials with a broad range of properties and applications, has strongly increased in the last decades. Since the beginning, traditional polymers have been used as "structural" and "single-use" materials. Moreover, fossil-based polymers such as polyethylene terephthalate (PET), polyethylene (PE), polyvinylchloride (PVC), polypropylene (PP), and polystyrene (PS) have been widely used in different sectors such as construction, electronics, packaging, automotive, sensors, etc. However, their long degradation rate and the large amount of waste generated represent critical points that cannot be neglected in their future development. For this reason, many efforts are now focused on the study of bio-based and biodegradable polymers and their potential industrial application. Moreover, it is important to consider both their processing methods and their end-of-life. Not all bio-based polymers are biodegradable, and some fossil-based polymers can be biodegradable (Figure 1). For example, bio-polyethylene is produced from bio-based sugarcane derivatives but is not biodegradable, whilst polycaprolactone (PCL) is a biodegradable non-bio-based polymer [1].

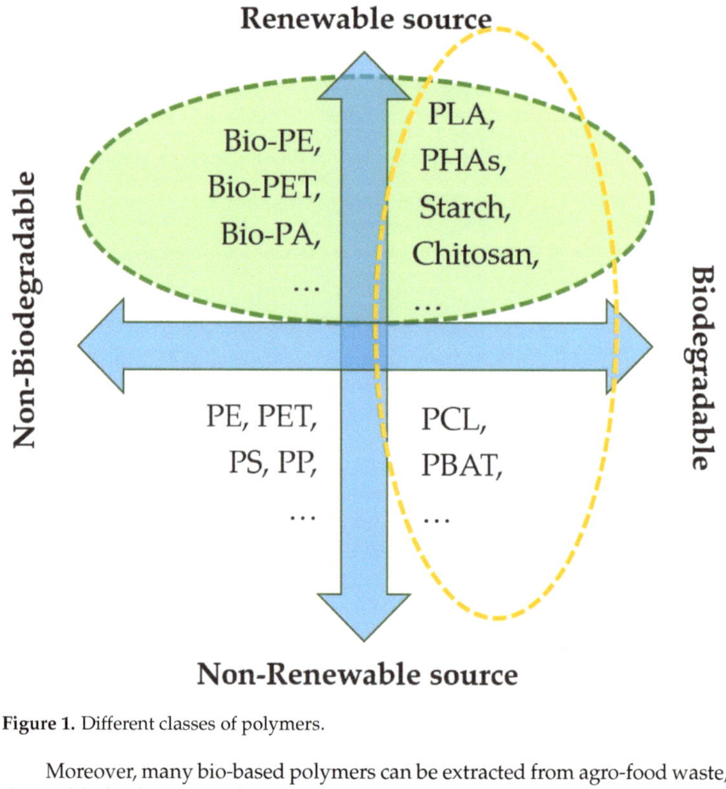

Figure 1. Different classes of polymers.

Moreover, many bio-based polymers can be extracted from agro-food waste, increasing their added values upscaling. In particular, plastic demand is expected to double before 2050, leading to an increase in global pollution caused by plastic waste and greenhouse gas emissions. Various options are taken into account to overcome this issue, including plastic recycling and the use of bio-based and biodegradable polymers [2].

Nowadays, two are the key points regarding plastics: the attention must be focused on designing multifunctional polymeric materials able to be used on demand, as well as we need to shift our effort from a linear to a circular economy, taking into account also the end-of-life of polymer-based materials, as the idea of having a world without plastic is quite impossible.

In this Review, we summarize the current state of the art of the most important bio-based and biodegradable polymer improvements reported in the literature, preferably in the last twenty-four years, as well as their main results in terms of blends and/or composites, nanocomposites polymeric materials. We highlight not only the importance of using a good polymeric matrix but also the effect of the addition of micro/nanoparticles and the use of vegetable oils as additives in terms of plasticizers and compatibilizers from a sustainable point of view. In particular, this Review is mainly focused on their synthesis and processing methods scalable to the industrial sector, such as melt blending approaches (melt-extrusion, injection molding, film forming) as well as solution electrospinning, among others, without neglecting to discuss the importance of their degradation process looking for a right balance between industrial interests and environmental issues, as summarized in Scheme 1.

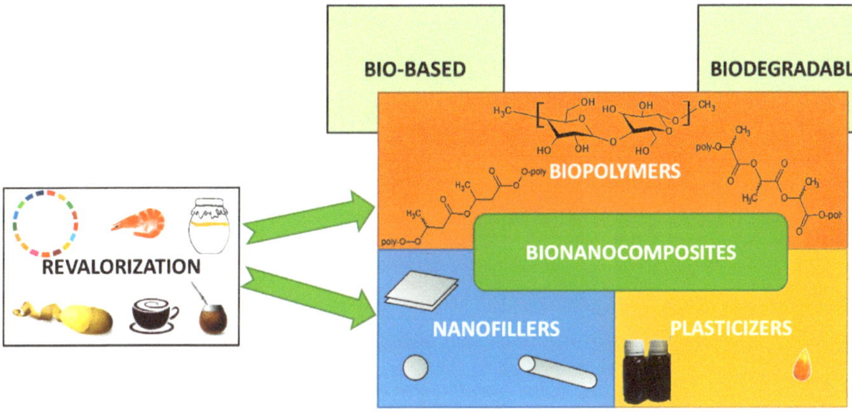

Scheme 1. Review contents.

Nowadays, the terms biopolymers, biomass, bio-based, biocompatible, biodegradable as well as bioplastic are widely used. However, it is important to note, as recommended by IUPAC [3], that biopolymers are macromolecules formed by living organisms. At the same time, the use of bioplastics is discouraged, preferably using the expression "bio-based" polymers [4]. Biodegradable polymers, which can be obtained either from renewable resources or from chemical synthesis, can be degraded by living organisms into non-toxic byproducts [5], which can be obtained either from renewable resources or from chemical synthesis. Degradation is carried out by microorganisms producing H_2O, CO_2, and methane at different rates according to polymer structure, crystallinity, morphology, and molecular weight [6]. However, to avoid confusion, we define sustainable polymers as all the bio-based and biodegradable polymeric materials, as reported in Figure 2. In particular, sustainable polymers, both bio-based and biodegradable, can be divided into three main groups according to their source, such as drop-in bioplastics derived from renewable or non-renewable sources [1,7] (i.e., cellulose, starch, and chitosan are biopolymers derived from natural sources such as plants and animals [8], biodegradable polymers as polyesters such as poly(lactic acid) (PLA) and PCL, polycarbonates, polyamides, and polyanhydrides, produced by chemical synthesis [9] and polymers produced by microorganisms including polyhydroxyalkanoates (PHAs), polyhydroxybutyrate (PHB), poly(3-hydroxybutyrate-co-3-hydroxyvalerate) (PHBV), bacterial cellulose, dextran, etc. [7].

Furthermore, bio-based polymers find applications as antioxidant and antimicrobial materials [10] in tissue engineering [11], medical and food applications [12], water treatment [13], etc.

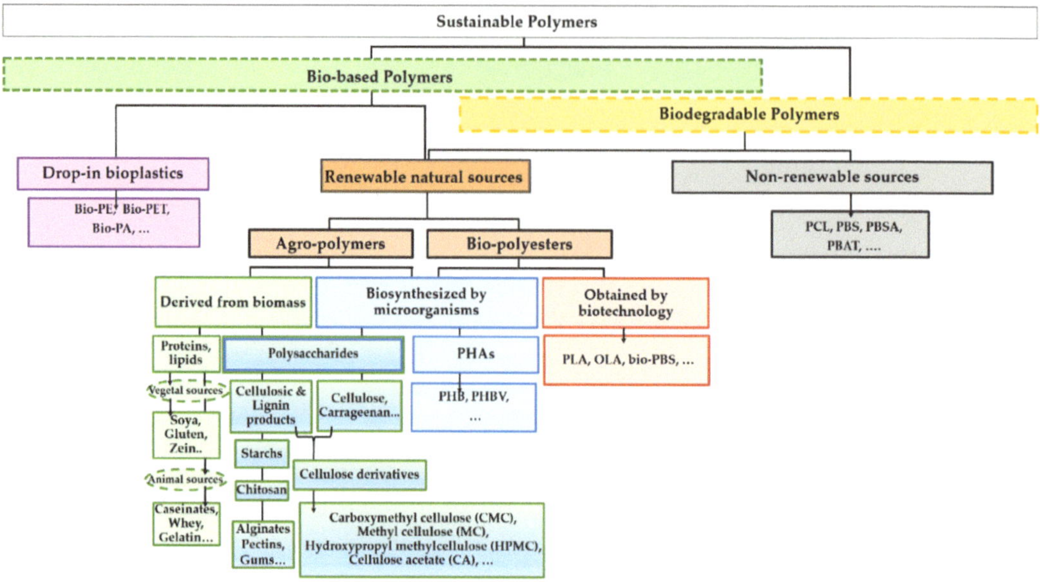

Figure 2. Classification of sustainable polymers according to their sources.

Nonetheless, bio-based polymers are characterized by poor mechanical properties, low chemical resistance, and reduced processing windows. To improve these features, biopolymeric composites, and nanocomposites are produced by adding nano or micro reinforcements, as particles or fibers, to obtain a synergistic effect and enhanced properties [14]. Another approach is the fabrication of blends with other bio-based polymers [15] or with synthetic polymers such as PE, polyamides (nylon), PS, synthetic rubber, PVC, etc., to produce materials with both increased processability, resistance, and impact strength but also with shorten degradation of the material [16].

The production of bioplastics has increased again in 2023 after stagnation in previous years, and a significant increase is expected in the coming years from 2.18 million tons in 2023 to approximately 7.43 million tons in 2028. Moreover, it is noteworthy that 52.1% of bioplastics are biodegradable, with PLA being the most produced, as can be seen in Figure 3 [17].

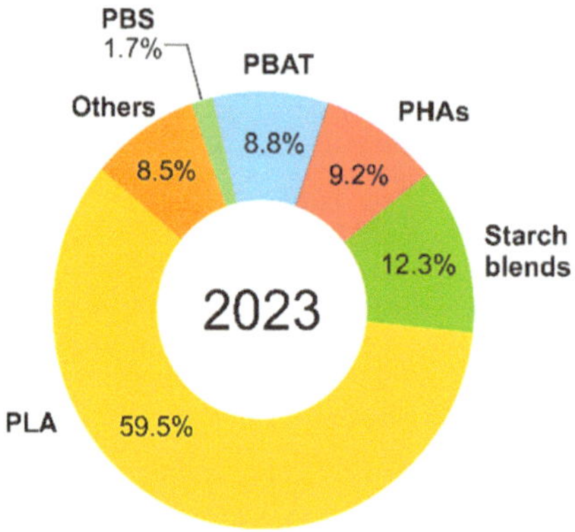

Figure 3. Global production capacities of biodegradable/compostable plastics by material type in 2023, adapted from European-Bioplastics-Association, 2023.

Numerous researchers have studied bio-based and/or biodegradable polymeric bends in recent years to modify different properties. A solution to improve the miscibility of these materials is the use of different types of plasticizers and compatibilizers to improve miscibility between the polymers of the blend and certain properties, such as mechanical properties, to obtain a more competitive material.

Moreover, another interesting approach to improve the polymer's performance to extend its industrial application is the development of bionanocomposites, which refer to composites based on a biopolymeric matrix loaded with nanofillers that present at least one dimension in the nanoscale range (below 100 nm) [18]. In this regard, if good nanofillers dispersion is achieved within the polymeric matrix, the nanosized fillers can considerably improve some specific properties of the biopolymeric matrices into the final bionanocomposites through strong and large polymer/nanofillers interactions. Bionanocomposite production offers a multitude of advantages as nanosized particles can enhance the mechanical properties, thermal stability, and barrier properties of polymers, as well as they can also provide the final bionanocomposite with specific functionalities as active (e.g., antimicrobial or antioxidant properties) or smart capability (e.g., shape memory, self-healing, or sensitivity to external stimuli) making the resulting bionanocomposites extremely versatile and suitable for several industrial applications at the same time that they are inherently more sustainable than nanocomposites based on traditional plastics derived from oil. Thus, bionanocomposites offer a promising pathway to achieving several of the Sustainable Development Goals by providing sustainable and high-performance polymeric materials that can replace conventional composites based on traditional plastics in many areas such as biomedical, healthcare, construction, electronics, packaging, agricultural applications, and water and air remediation. The introduction of bionanocomposites into the industrial sector not only supports environmental preservation but also promotes economic and social sustainability.

Nanofillers, Figure 4, are mainly classified into three categories according to their sizes and shapes on the dimensionality of the nanosized particles [18–20]: (i) two-dimensional nanofillers (2D) or nanofibrous with two dimensions beyond the nanoscale range that exhibit a diameter below 100 nm and have an aspect ratio of at least 100 (Figure 4a). The most well-known 2D nanoparticles are carbon nanotubes and nanocellulose substrates (cellulose nanofibers, nanocrystals, or nanorods); (ii) one-dimensional nanofillers (1D) with only one dimension in the nanometer scale, are layered materials (nanolayers) classically with a thickness on the order of 1 nm, but with an aspect ratio following their two remaining dimensions of at least 25 (Figure 4b). The most popular 1D nanofillers are layered silicates, graphene, and covalent organic frameworks; and (iii) zero-dimensional nanofillers (0D) or referred to as equiaxed (isodimensional) nanoparticles with all dimensions confined below 100 nm (Figure 4c). This category includes nanoparticles such as silica nanoparticles and nanosized metal oxides, buckyballs, and quantum dots.

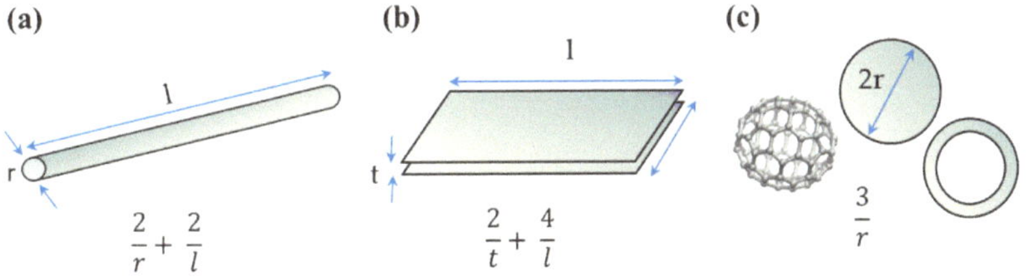

Figure 4. Nanofiller geometries and their respective surface area-to-volume ratios. (**a**) two-dimensional (2D); (**b**) one-dimensional (1D); and (**c**) zero-dimensional (0D) nanofillers.

In the case of 2D and 1D nanofillers, the surface area/volume is governed by the first term in the equation. The second term ($2/l$ and $4/l$, respectively) has a minimal influence and is commonly omitted compared to the first one. Therefore, changing the nanofiller diameter (changing either the layer thickness or fiber diameter) from the micrometer to the nanometer scale will affect the surface area-to-volume ratio by three orders of magnitude. Meanwhile, for 0D nanofillers, the smaller the nanofiller diameter, the greater the surface area per unit volume [18].

Nanofillers can additionally be classified depending on their pathway as (i) natural, produced through natural processes in the environment (e.g., minerals, volcanic dust, etc.), (ii) incidental, as a waste of industrial processes (e.g., coal combustion, welding fumes, diesel exhaust, etc.) and (iii) engineered nanoparticles, produced either by "top-down approach" mainly by milling a large sample to obtain nanosized particles of the starting material, or by "bottom-up" approach, assembling smaller subunits through crystal growth or chemical synthesis from atoms or molecules up to nanofillers, better-controlling sizes, shapes, and size ranges [19,20].

Many nanosized fillers based on bio-based polymers can be produced by "top-down approaches," starting from a polymer and reaching nanosized particles based on it. Similarly, bionanocomposites production are mainly produced by "top-down approaches" as the ingredients (polymer, nanoparticles, and other additives) are macroscopically mixed either at laboratory scale level by dissolving the polymer and dispersing the nanoparticle within the polymeric matrix and further allowing the solvent to evaporate (solvent casting method) or produced by industrial processing technologies based on melt blending approaches. In this regard, the major challenge in the development of nanocomposite materials is to achieve nanosized particle dispersion within the polymeric matrix [20]; this is why other additives are sometimes incorporated to facilitate the nanosized particles dispersion within the polymeric matrix, such as the case of plasticizers. As plasticizers increase the polymer chain mobility, they can easily increase the nanoparticle distribution along the polymeric matrix.

The most important bio-based and biodegradable polymeric matrices and nanosized particles extracted from them are summarized in this Review. Moreover, the strategy to improve polymers' performance through the development of bionanocomposites is reviewed, as well as the different strategies used to improve the dispersion of nanosized particles within the polymeric matrix and the top-down processing technologies to produce bionanocomposites. Finally, the potential applications of such bionanocomposites are discussed and highlighted in relation to both industrial interests and environmental issues.

In order to obtain optimization in the biopolymeric blends as well as on the bionanocomposites, however, we also have to consider the importance of plasticizers and/or compatibilizers in the composition of the materials and in their final properties. In fact, plasticizers are essential additives in polymer formulation, playing a fundamental role in improving their physical and mechanical properties. They are organic compounds characterized by their ability to intercalate between polymeric chains, reducing intermolecular forces of attraction and facilitating sliding between them [21]. For this reason, they increase the flexibility, malleability, and processability of polymers as they modify viscosity, decrease the glass transition temperature, and enhance the ductile properties of polymers [22].

Plasticizers can be classified as primary or secondary plasticizers, depending on their primary function in the polymer formulation. Primary plasticizers provide flexibility, impact resistance, and tear resistance by using a single plasticizing compound that is miscible with the polymer. However, they may pose challenges in terms of thermal stability, migration, and environmental compatibility. The most commonly used primary plasticizers are phthalates such as diisononyl phthalate (DINP) and diisodecyl phthalate (DIDP), although other examples include adipates, such as diocyl adipate, sebacates like dioctyl sebacate, azelates, trimellitates, phosphoric acid esters, or epoxy softeners [23]. On the other hand, secondary plasticizers are less polar than primary ones and have lower compatibility with polymers. Being less soluble and having low compatibility, they are usually employed in combination with primary plasticizers. The mixture of both plasticizers provides a more stable blend against migration [23,24].

The benefits of using plasticizers are copious. Firstly, they enable the production of polymers with a wide range of mechanical and thermal properties, making them suitable for a variety of industrial applications, from flexible packaging to medical devices. Additionally, plasticizers can enhance the durability and impact resistance of materials, contributing to a longer lifespan and better service performance. Finally, plasticizers can also facilitate the processing and manufacturing of polymers, reducing production costs and increasing efficiency in the supply chain.

However, it is important to consider that plasticizer migration can occur due to factors such as temperature, pressure, exposure to chemicals, or contact with food and bodily fluids. This migration can lead to environmental contamination and human exposure to these compounds, raising concerns about potential toxicity and long-term health effects [25]. Phthalate plasticizers are a concern due to their migration and toxicity. As a result, both the European Union and the United States, through the Environmental Protection Agency (EPA) and the Food and Drug Administration (FDA), respectively, have prohibited or restricted the use of several phthalates in various products, especially those intended for children or for storing food, cosmetics, or medications [24,26].

A promising alternative is the use of natural plasticizers, as they exhibit low toxicity and migration. In this group, glycerol [27], vegetable oils [28], sugars such as sorbitol [29], or citric acid [30], among others, can be found. The use of natural plasticizers has not only been explored as a potential substitution for phthalates but is also associated with the growing interest in the development of new polymeric materials.

In a wide range of applications, biodegradable polymers are recognized as excellent candidates due to their environmentally friendly decomposition process. To broaden their applicability, it is necessary to adapt some of their intrinsic properties. Enhancing flexibility and reducing brittleness address the material's limitations in various application contexts [31]. Similarly, modifying processing conditions, such as reducing the melt viscosity, facilitates easier molding and shaping, thereby expanding their usability across diverse manufacturing processes [32]. Furthermore, optimizing thermal properties to meet specific requirements while ensuring the polymers' biodegradability is preserved is crucial for their effective use in industries ranging from packaging and agriculture to medical devices, where their degradation process does not contribute to pollution [33]. These adaptations are pivotal for integrating biodegradable polymers into sectors that demand environmental sustainability alongside functional performance [34].

Therefore, in this Review, the most important bio-based and biodegradable polymeric matrices, plasticizers, and compatibilizers, as well as nanosized particles extracted from natural sources, are summarized. In particular, the strategy to improve polymer performance through the development of bionanocomposites is reviewed, as well as the different strategies used to improve the dispersion of nanosized particles within the polymeric matrix, as well as the importance of the use of vegetable oils as additives. In particular, 10 different bio-based and/or biodegradable polymers are discussed, such as PLA, PBAT, PHAs, Starch, Cellulose, Lignin, Chitosan, Gelatin, polybutylene succinate (PBS), poly(itaconic acid), summarizing their synthesis and processing approaches. Moreover, for every polymer, the last part of the discussion is dedicated to their degradation process, which is considered a very important aspect of their end-of-life from a circular economy perspective. Therefore, in the case of natural polymers, such as starch, cellulose, and lignin, we discuss the main results in terms of extraction of nanofillers from them, evidencing their double face to be used as polymeric matrices from one side and as nanofillers from the other one.

2. Polylactic Acid

Polylactic acid (PLA) is an aliphatic polyester (2-hydroxypropionic acid) that was synthesized for the first time in 1932 by Carothers [35]. PLA is a biodegradable, bio-based, and biocompatible polymer derived from natural and renewable resources, such as agricultural waste [36]. This polymer is constituted by monomers of lactic acid, which are present as D and L enantiomers. Indeed, three types of PLA are present: poly-L-lactic acid (PLLA), poly-D-lactic acid (PDLA), and poly-D,L-lactic acid (PDLLA) [37].

The content of enantiomers affects the properties of the polymer; PLLA and PDLA are semi-crystalline polyesters with a melting temperature (T_m) of around 175 °C and a glass transition temperature (T_g) of 55–60 °C, while higher content of D-isomer in PDLLA leads to the formation of amorphous materials with T_g of 55–60 °C [18,38].

The homopolymer presents four polymorphic crystalline structures, α, α′, β, and γ, which can be formed in different crystallization conditions [39]. In particular, the α-phase is the most common and stable crystalline structure and can be obtained from the melt or solution, while the α′-phase is produced at lower crystallization temperatures and can be converted to the more stable α-phase at temperatures above 120 °C [40]. The β-phase is formed by deforming the α-form at high temperature and high drawing ratio, whereas the γ-form is obtained through epitaxial crystallization [41].

The crystallinity of PLA influences its barrier properties. Indeed, PLA presents CO_2 and O_2 permeability coefficients similar to PS and PET but with lower barrier properties against water vapor [42]. The crystallinity of the polymer also affects its properties. PLA presents high modulus (3 GPa) and tensile strength (50–70 MPa). However, it has a brittle nature [43,44].

In order to improve the mechanical properties, in particular its toughness, blends of PLA with more flexible polymers have been produced, [45,46]. From 1988 to 2024, many works have been published based on PLA blends, as reported in Figure 5.

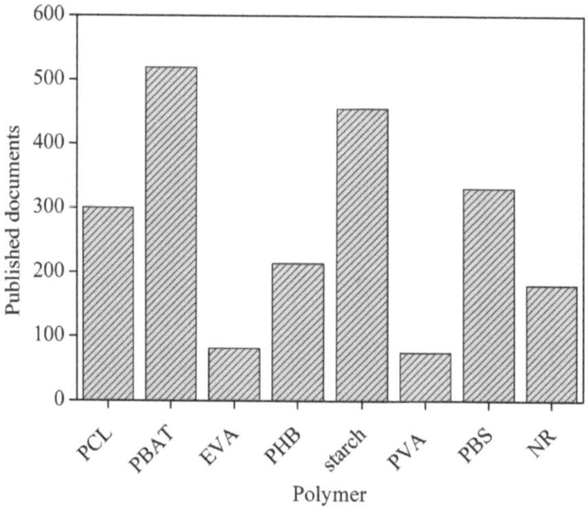

Figure 5. Number of publications of PLA blends (Scopus Source up to June 2024).

In addition, various organic and inorganic reinforcements such as carbon, metals, fibers, clays, and natural materials have been incorporated into the polymeric matrix to improve its permeability mechanical, antimicrobial, and thermal properties [47,48].

Three routes have been reported for the synthesis of PLA (Mw > 10,000), as shown in Figure 6, involving (a) direct condensation polymerization, (b) azeotropic dehydration condensation, and (c) lactic ring opening polymerization (ROP), which is the most common process for the production of PLA with high molecular weight [49].

Polycondensation leads to the fabrication of polymeric chains with low molecular weights due to the increased viscosity of the polymer melt and the presence of impurities and water as byproducts [50].

Azeotropic dehydration condensation permits the production of high molecular weight PLA without using chain extenders and esterification adjuvants. It involves a distillation of lactic acid for 2–3 h at 130 °C under reduced pressure. Subsequently, catalyst and diphenyl ether are added, then a molecular sieve is used, and the solvent is returned to the container for a further 30–40 h at 130 °C. Finally, the polymer is separated for further purification [51].

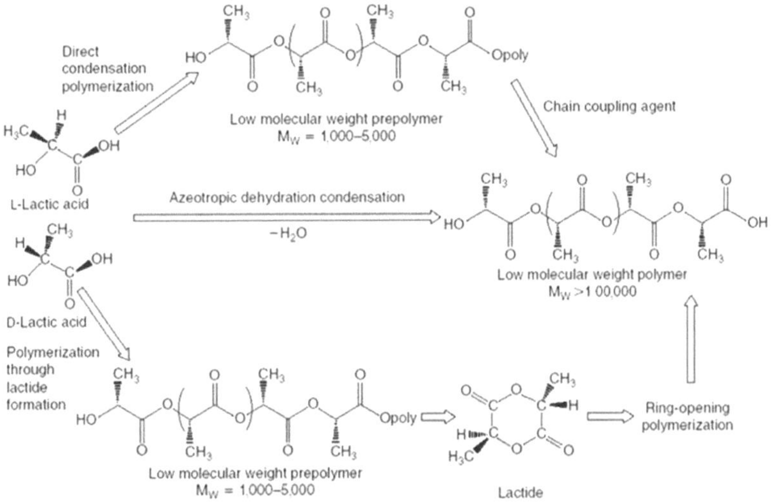

Figure 6. Synthetic routes of PLA synthesis, adapted from [49].

ROP implicates polycondensation, depolymerization, and ring opening of the LA lactide. Although this route needs several purification steps and requires the use of heavy metal catalysts (i.e., Tin(II) octoate, Sn(Oct)2)) [52], it is the most industrially used to obtain high molecular weight PLA [18].

On the other hand, the synthesis of shorter polymer chains of poly lactic acid from the monomer, changing the polymerization conditions to obtain compounds with different molar masses, is an alternative to get the oligomeric lactic (OLA), which acts as a plasticizer [53].

PLA is the most used bio-based polymer, and several processing methods are used to achieve different properties for numerous applications. The main manufacturing techniques include extrusion, cast films, injection molding, thermoforming, foaming, 3D printing, and electrospinning [54].

Melt blending approaches are the processing technologies most studied for producing PLA-based materials as they are used at the industrial level. PLA has been blended with many plasticizers (glycerol [55], citrate esters [55,56], polyethylene glycol (PEG) [55,56], OLA [55,57], vegetable oils [58], among others) to reach a flexible material, while it has also been blended with many nanoparticles to obtain PLA-based nanocomposites such as with carbon nanotubes [59], clays [60,61], nanocellulose [62], metal nanoparticles [63], etc.

Extrusion is one the most used techniques for PLA processing, typically by using conventional screws with L/D of 24–30 and compression ratios of 2–3. The polymer is heated 40–50 °C above its melting temperature due to the presence of heater bands and the friction of the molten polymer between the screw and the barrel, which permit the melting of crystalline regions [64].

Bourbigot et al. [65] prepared a PLA-based nanocomposite with multi-wall carbon nanotube (MWNT) through reactive extrusion, observing an improvement in flame retardancy. Galvez et al. [66] investigated the influence of the speed of the extrusion process for PLA plasticized with acetyl tributyl citrate (ATBC) between 10–30 wt%.

Extrusion is the step that precedes other processes, such as injection, thermoforming, or spinning.

In injection molding, the polymer is injected into a mold following extrusion. In this case, the screw not only rotates but also moves forward to introduce the molten polymer into the mold cavities. During the injection process, the mold is kept closed, and the screw is maintained in the forward position to compensate for the shrinkage of the material caused by the solidification of the material itself. The mold presents a cooling system, which regulates the cooling process. When the material is cooled down, the molded part is ejected from the mold.

To optimize PLA injection, a compression ratio of 2.5–3 and a melting temperature of around 200–205 °C are usually used. It is very important to control the shrinkage of the molded parts, which are generally brittle due to the fast physical aging caused by the low T_g of PLA [67].

Aliotta et al. [68] studied the effect of three nucleating agents, LAK, talc, and calcium carbonate ($CaCO_3$), on the mechanical and thermal behavior of PLA samples obtained through injection molding. LAK can increase PLA crystallinity but reduce flexural modulus and Charpy impact resistance, while an opposite effect was observed with talc and calcium carbonate.

Pawlak et al. [69] fabricated a composite constituted by PLA, maleinized linseed oil (MLO) as a plasticizer, and functionalized sheep wool fiber by extrusion and then injection molding, obtaining a hydrophobic material with improved mechanical properties.

The biocompatibility of PLA allows it to be used in food packaging, such as cups, lips, blister packaging, containers, and trays that can be produced through the thermoforming process. This technique consists of first heating to soften the polymer by irradiating it with infrared radiation (IR), then compressed air is applied, and the material is pressed against the mold, which is usually an aluminum mold. Specific storage conditions are required for PLA sheets since they are very fragile at room temperature. The ultimate product should be stored at 40 °C or below relative humidity (RH) under 50%, reducing moisture absorption and achieving unwinding resistance [70].

Barletta et al. [71] studied the effect of $CaCO_3$ as reinforcement of PLA and polybutylene succinate (PBS) blends obtained by thermoforming using molds specially designed for polyolefins. Wand et al. [72] prepared nanofibrillar PLA/PET through high-pressure microcellular injection molding (HPMIM) to improve PLA melt strength, crystallinity, and foaming ability.

PLA has also been used to produce thin films, usually of 0.076 mm in thickness and sheets of 0.25 mm, by cast extrusion. PLA is first extruded through a lip die, whose gap is fixed around 10% higher than the thickness required for films and sheets, then is quenched on polished chrome rollers refrigerated with circulating cooled water. PLA films are usually orientated between 2 to 10 times their original length at 60 to 80 °C, in particular 2–3 times in the machine direction stretch ratios and 2–4 times transverse stretch ratios, and at higher D-lactide contents, the stretch ratios can be increased. The rapid cooling of the material provides good optical properties due to the transparent appearance caused by low crystallinity degree; besides, PLA films present a high Young modulus [54,73].

Furthermore, PLA porous foam structures have been produced for biomedical scaffolds and tissue engineering, as well as for packaging and construction; the main techniques comprise extrusion foaming and foam injection molding. In the first case, the polymer and blowing agent are injected into the extruder to further emerge from the die; the pressure drop induces a thermodynamic instability causing foaming [74].

Meanwhile, foam injection molding uses supercritical nitrogen as a foaming agent; besides, it reduces the polymer melt viscosity and permits the use of lower temperatures and pressures, reducing processing costs. This represents an advantage when biodegradable polymers such as PLA are used since they are thermally sensitive and present a narrow processing temperature range. Nevertheless, the foaming of PLA with this technique needs to be highly monitored due to its low melt strength and slow crystallization kinetics [75].

In the last decade, additive manufacturing (AM) has become one of the most important techniques to develop three-dimensional (3D) objects. Among the various techniques, fused deposition modeling (FDM) was revealed to be one the most used due to its low cost and low maintenance. Furthermore, thermoplastic polymers with low melting points, like PLA, are required [76]. The FDM method presents one or two extruder head nozzles from where the melted polymeric filament is ejected onto a platform layer by layer. Then, the printed polymer solidifies to produce the final 3D object [77].

PLA is one of the most common polymers used in FDM. However, it presents some disadvantages, such as low toughness and elongation at break, low thermal stability, and a small processing window. Moreover, PLA presents low mechanical strength compared to acrylonitrile butadiene styrene (ABS) and polycarbonate (PC). For this reason, the use of PLA composites produced by the incorporation of reinforcement is one of the main approaches used to improve its mechanical properties [78].

Hong et al. [79] incorporated lignin and COOH-lignin into the PLA matrix, achieving a better adhesion between functionalized lignin and PLA and aiming to reduce the cost of printing PLA 3D filaments. Zoumaki et al. [80] covered 3D-printed PLA filament with maize starch-based conventional and nanocomposite films. Starch-based skins enhanced the mechanical properties of neat PLA in terms of strength and stiffness.

Another technique to produce 3D PLA-based scaffolds is electrospinning. This versatile and low-cost technique enables the fabrication of fibers with diameters in the range of micro and nanometers through the application of a high potential field between a needle and a collector where the fibers are collected [81]. Different types of collector permit the fabrication of both random and orientated fibers (Figure 7), which present different properties such as mechanical behavior and crystallinity depending on their orientation.

(a) (b)

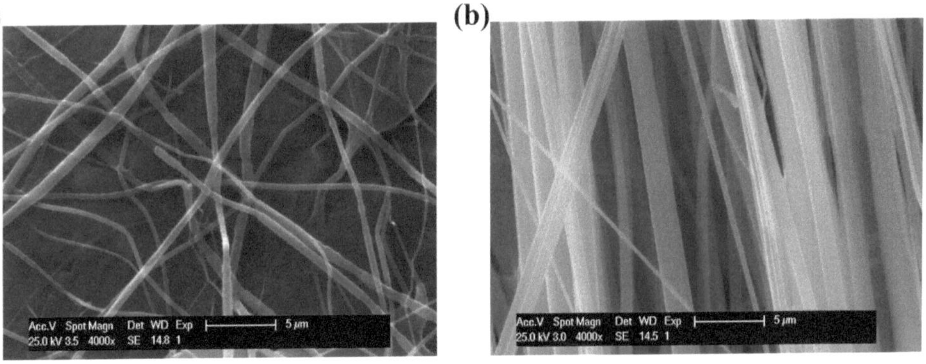

Figure 7. SEM images of (**a**) PLA random nanofibers and (**b**) PLA-aligned nanofibers.

In particular, nanoscale fibers present high porosity and high aspect ratio, which allows the incorporation of bioactive substances, which is ideal for biomedical applications [82]. Moreover, PLA fibers find applications in food packaging for their biodegradability and hydrophobicity, improving the water barrier properties of the package [83] and as a membrane due to the interconnectivity of pores, which is useful in the separation process [84].

Duygulu et al. [85] blended ceftriaxone disodium into PLA solution and tested different processing parameters such as flow rates and voltages. The produced electrospun nanofibers showed antimicrobial behavior against *Escherichia coli* (*E. coli*). Meanwhile, Zhu et al. [86] developed a porous membrane of PLA and HKUST-1 by electrospinning, increasing the hydrophobicity and antibacterial activity against *E. coli* and *Staphylococcus aureus (S. aureus)*, with an antibacterial rate of 99.99% with *S. aureus*.

PLA is the most produced biodegradable polymer worldwide and is highly valued in everything from packaging to biomedicine. However, its inherent rigidity and fragility limit its direct application. To mitigate this drawback, the use of plasticizers is a viable solution, as they allow the improvement of flexibility and toughness without compromising their biodegradability.

One way to improve the flexibility of PLA is the use of its own monomers: L-lactide, D-lactide, D,L-lactide monomers, and oligomeric lactic acid (OLA). In this regard, Ara and Fiori patented the process for the plasticization of lactic acid polymers, focusing on the use of oligomeric lactic acid as it is the same molecule with lower molecular weight and thus ensuring good compatibility with PLA polymeric matrix [87]. These plasticizers, used in concentrations between 10 and 25 wt%, improve ductile properties and lower the glass transition temperature. Additionally, they seem to have good storage stability for applications in contact with food [53,88–90].

The use of vegetable oils and their derivatives has shown broad interest, as they demonstrate good plasticizing results using low amounts, around 5 wt%, improving ductility, decreasing the glass transition temperature, and showing good compatibility with PLA. These results have been obtained with different modified vegetable oils (MVO), such as epoxidized soybean oil (ESBO) [91], epoxidized linseed oil (ELO) [92], maleinized linseed oil (MLO) [93], epoxidized jatropha oil (EJO) [94] among many others. Other molecules derived from vegetable oils that have been studied as plasticizers for PLA include glycerol, tributyrin, trilaurin, and tristearin, with tributyrin showing the best plasticizing effect, followed by glycerol with just 10 wt% [95]. To improve the plasticizing effect of glycerol, it has been combined with other molecules like succinic anhydride and alcohols for applications in flexible food packaging [96] or modified with different carbonates [97].

PLA has also been plasticized with different polymers, such as polycaprolactone [98], which significantly reduces the Tg, increasing the elongation at break and decreasing the crystallinity of the PLA. Polyamide 6 and polyamide 11 have also been used to improve the ductile properties of PLA; however, better ductile properties are achieved if they are combined with other additives, such as chain extenders [99,100]. Poly(ethylene glycol) (PEG) of different molecular weights, with lower molecular weight PEGs, for example, PEG 200 and PEG 400, being the best candidates for use as a PLA plasticizer, significantly reducing the Tg by up to 30 wt% and increasing the elongation at break [101–103]. Thermoplastic starch (TPS) has also been used as a plasticizer for PLA despite its immiscibility. However, TPS-rich domains lead to an improvement in the ductile properties of PLA up to 30 wt% TPS percentages [104,105]. To improve the miscibility of this system, compatibilizing agents such as MLO can be incorporated [106].

Other molecules, such as acid esters, like citric acid ester [107], cinnamic acid ester [108,109], or tartaric acid ester [110,111], among others, seem to be very interesting for use as plasticizers, in addition to incorporating other functionalities such as antioxidant agents, UV light stability, compatibilizers, or improvement of barrier properties that make them ideal for packaging applications, structural applications, or long-lasting products.

PLA can degrade in particular conditions of temperature, moisture, oxygen, and microorganisms to water, carbon dioxide, and other non-toxic products. The main degradation mechanisms include hydrolytic, enzymatic, microbial, thermal, and photodegradative mechanisms [112].

Hydrolytic degradation of PLA occurs in the presence of water and is mainly by bulk erosion, which happens when the rate of water diffusion is higher than the hydrolysis rate, and the material tends to shrink. The cleavage of the ester bond takes place initially in the amorphous regions, leading to a first increase in crystallinity. Hydrolysis leads to the formation of monomers and oligomers with hydroxyl and carboxylic acid end groups that can catalyze the hydrolysis reaction, increasing the degradation speed [113].

Vaid et al. [114] studied the hydrolytic degradation of electrospun PLA nanofibers in different conditions of pH and exposure time and compared them to molecular dynamics (MD) simulations and density functional theory (DFT) calculations. A faster degradation was found at higher pH (10) after 25 days, while slower degradation was observed at lower pH values (7.4, 2, and 3). Furthermore, they observed that PLA nanofibers degrade through a surface erosion mechanism, in contrast to the typical bulk erosion observed in PLA, caused by the increased crystallinity due to the electrospinning process.

Qu et al. [115] studied the accelerating effect of zinc oxide nanoparticles (ZnO NPs) on the hydrolytic degradation of PLA in water, with an activation energy of 38% lower than neat PLA, while Zanella et al. [116] studied the effect of the amount of silver NPs in the in vitro degradation of the nanocomposite in artificial saliva observing that the degradation time of the materials decreased by increasing the amount of NPs.

The cleavage of the ester bond also occurs through enzymatic degradation. Indeed, enzymes produced by microorganisms such as *lipases*, *esterases*, *kutinases*, and *proteases* can catalyze the hydrolysis of PLA chains in particular conditions of pH, temperature, and crystallinity of the polymer. Hegyesi et al. [117] studied the degradation of PLA with *lipase* and *proteinase K*, finding that the second one is more effective.

Donate et al. [118] fabricated composite scaffolds of PLA and $CaCO_3$ and β-tricalcium phosphate ($Ca_3(PO_4)_2$, β-TCP) in order to increase the osteoconductivity of the material and studied the enzymatic degradation by using *proteinase K* and compared with neat PLA, observing an accelerating effect in degradation caused by the presence of the additives.

Microbial degradation starts after PLA chains with high molecular weight undergo hydrolysis and smaller molecules are formed. Thus, microorganisms first release extracellular depolymerizes to further reduce the molecular weight of PLA with the presence of inducers such as gelatin, silk fibroin, and elastin. Subsequently, oligomers are absorbed by microbial membranes where intracellular enzymes convert PLA to water or CH_4 and CO_2 [112].

Moon et al. [119] studied the biodegradability of PLA in anaerobic digester sludge by evaluating the extent of the gas production, in particular, and the formation of PLA degradation by-product, observing an increased gas liberation as the surface area of the material increased.

Bubpachat et al. [120] isolated two bacteria from soils and wastewater capable of degrading PLA, *Stenotrophomonas pavanii* and *Pseudomonas geniculata*, respectively. These can produce *proteases*, and we studied the optimal pH for both *proteases* produced by the two types of bacteria, finding 7.5 and 8.0 as optimum values, respectively.

Microbial populations are also present in compost, which is an organic-rich, biological environment where biodegradable polymers are transformed by fungi, bacteria, and actinomycetes into H_2O, CO_2, inorganic compounds, and biomass in aerobic conditions. The first step involves the conversion of the polymer into small molecules, followed by a thermophilic phase where the temperature is increased due to microbial activity. Subsequently, in the cooling phase, the temperature is decreased, and humus-like substances are produced. Since plastic recycling machines are not well-equipped to recycle food-contaminated plastics, bags, and containers, compostable plastics such as PLA are a possible solution to deal with this issue [121].

Thermal degradation of PLA can be easily studied through thermogravimetric analysis. Some researchers have studied the degradation mechanism and have suggested that above 200 °C the mechanism involves different reactions, such as intra-molecular transesterification leading to the formation of oligomeric rings, which is the main degradation pathway, cis-elimination leading to acrylic acid and fragmentation producing CO_2 and acetaldehyde, causing the reduction in molecular weight of the polymeric chains and loss of mechanical performances [122].

Furthermore, PLA is sensitive to ultraviolet (UV) and slightly sensitive to sunlight. The degradation caused by the exposition of materials to radiation is called photodegradation. When aliphatic polyesters such as PLA are exposed to UV radiation, carbon-carbon double bonds and carboxyl end groups are formed. Another mechanism suggested for the photodegradation of PLA entails photolysis of the ester bond of the main chain and the production of carboxylic acid and diketone end groups [123].

3. Poly(Butylene Adipate-*co*-Terephthalate) (PBAT)

Poly(butylene adipate-*co*-terephthalate), known as PBAT, is a synthetic polymer that is biodegradable under composting conditions and, therefore, regarded as an environmentally friendly alternative to replace unbiodegradable thermoplastic films for packaging or agricultural applications [124]. PBAT combines an aromatic polyester and an aliphatic polyester, usually obtained by reacting butylene glycol (BDO), adipic acid (AA), and terephthalic acid (TPA) [125], as seen in Figure 8. In this copolymer, the aromatic monomer contributes to good thermal stability and mechanical properties, and the aliphatic monomer provides flexibility and biodegradability [126]. PBAT is more flexible than most biodegradable polyesters, including PLA and PBS, with a Young modulus of 20–35 MPa, a tensile strength of 32–36 MPa, and high elongation at break (almost 700%) [127]. The resulting high flexibility is ideal for food packaging applications and is frequently compared to low-density polyethylene (LDPE) because of its high molecular weight and branched molecular structure [128].

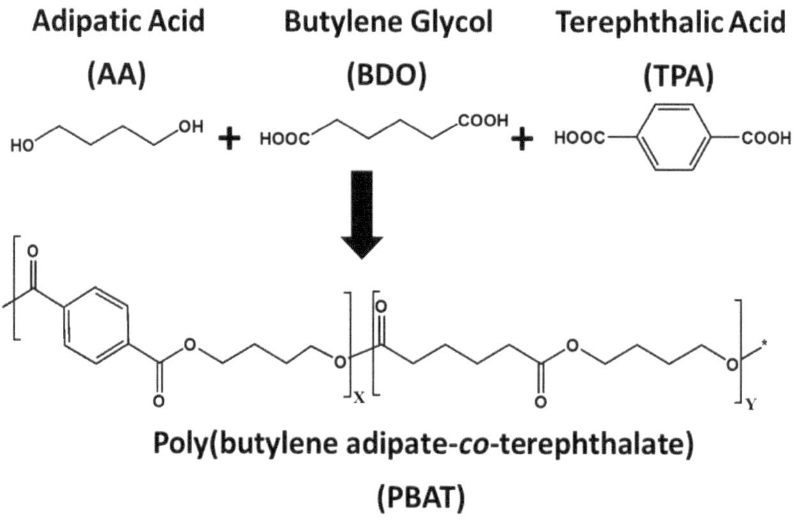

Figure 8. Synthesis of PBAT.

PBAT has been marketed under the trading name Ecoflex®, initially introduced by BASF towards the close of the 1990s [129], and PBAT-related copolymeric product (Biomax®) produced by DuPont [127]. Ecoflex® is widely used to make bags, mulching films, laminated sheets, plastics for wrapping food, toothbrushes, etc. [130]. PBAT's inferior thermal-mechanical properties compared with polyolefins and high cost (~4 USD/Kg) compared with polyethylene (PE), polypropylene (PP), and polystyrene (PS) (<1 USD/Kg) [131], encourages forming composite products in combination with other inexpensive polymeric materials. Moreover, it is worth noting that the relatively low tensile strength, melt strength, and heat resistance greatly limit its industrial applications and development [132]. Additionally, unmodified PBAT is easily broken in outdoor applications such as mulch films after a short time, and its O_2 and CO_2 barrier properties are also clearly lower than other common packaging materials like PET [133]. Thus, enhancing the mechanical and thermal properties of PBAT by methods such as blending, composites, and chain extension is necessary to increase its application scope.

Typically, the synthesis of PBAT is carried out in two steps: esterification, during which comparatively low-MW oligomers form, and a final polycondensation at high temperature and vacuum to get the required Mn values of about 50 kDa [134]. These conditions are necessary to eliminate the smaller, lighter molecules, like water, which can cause hydrolyzation. The esterification reaction between AA and TPA, usually in a 50:50 ratio and excess BDO, to produce the PBAT oligomer is carried out at 200–220 °C [135]. The polycondensation reaction to increase the molecular weight of PBAT polyester is carried out at 250–260 °C, with a catalyzer and low pressure, eliminating the excess BDO. A molar ratio of 50:50 (AA:TPA) produces a PBAT with good processability, optimized mechanical properties, and good compostability for packaging film applications [136]. Organometallic compounds based on zinc, tin, and titanium are normally used as catalysts during polycondensation [137]. Nucleating agents, such as talc, chalk, mica, or silicon oxides, are frequently used to enhance the crystallization ability of PBAT, improving its mechanical and physical properties [137]. Phosphoric compounds such as phosphoric acid and phosphorous acid help stabilize color in PBAT but slow down the condensation rate [125].

On an industrial scale, PBAT can be produced by batch or continuous processes utilizing traditional polyester manufacturing technologies. However, batch processing can produce partially heterogeneous reactions with various intermediate products and struggle to suppress possible hydrolysis reactions during poly-condensation, decaying the molecular weight of PBAT [138]. Therefore, a continuous process with sequential reactors is preferred, as it offers a higher control over product quality and molecular weight. Despite being mainly derived from fossil sources, PBAT is potentially obtainable from bio-derived raw materials [125]. Alternatively, PBAT can also be produced from the low molecular weight degradation product of post-consumer PET waste since a significant portion of PET's structural backbone is TPA [139].

Wu et al. [132] studied the influence of the branching effect on PBAT using dicumyl peroxide. Generally, samples displayed higher elastic modulus and higher crystallization with increasing branching levels. Although the branching structures promoted nucleation, they reduced the diffusion rate of PBAT chain segments during the transformation of the PBAT. Moreover, melt crystallization was easier due to the higher crystallization temperature as the branching level increased.

PBAT-based films can be manufactured through multiple processing techniques, including solution casting, compression, molding, blown film extrusion, and hot pressing. Solvent casting is a simple and cost-effective method to prepare PBAT-based films, suitable for small laboratory-scale production. Chloroform is frequently the preferred solvent [140], with process conditions varying from stirring at 60 °C for 24 h and casting at room temperature for 48 h [141] to stirring at 60 °C for 1 h and drying at 70 °C for 5 h [131]. Solvent casting has also been used to prepare films by blending PBAT with other polymers, including PLA [142], PBS [143], and ternary blends such as PBAT/PLA/Babassu [126]. However, solvent casting is a less efficient method to elaborate PBAT composites when compared to other solvent-free techniques like melt processing [127].

Extrusion and blown methods are commonly used for large-scale film productions, requiring advanced equipment and large batch samples. Extrusion has several benefits, such as better control over feed, residence time and mixing degree, ease of handling high-viscosity polymers, and a wide range of processing conditions [144]. Blown film extrusion is one of the most important industrial manufacturing processes for polymeric films and is used to elaborate PBAT-based films in packaging, bags, and coverings. The process involves extruding a molten polymer into a tube shape, drawing it in both extrusion and transverse directions, and then blowing it by injecting air into a larger tube ratio [145]. The orientation of macromolecules and the final morphology of blown films depend on the chosen process parameters. PBAT/TPS (70/30) mulch films prepared by blow extrusion and compared to PE mulch films exhibited higher elongation at break and a seemingly low WVP but lower maximum tensile strength [146]. Other works have used blown processing to develop PBAT/TPS films with α-tocopherol as an antioxidant [147], PBAT/lignin films for enhanced photo-stability [148], or PBAT/TPS films loaded with polyhexamethylene guanidine hydrochloride (PHMG) for antimicrobial properties [149].

Recently, new applications for PBAT using less common PBAT processing techniques have been developed, including electrospun PBAT scaffolds with nanohydroxyapatite (nHAp) to improve bone regeneration [150], hollow PBAT electrospun microfibers loaded porous MFs (HPMFs) for controlled and sustained release of polypeptides [151], and 3D printing PBAT powder for selective laser sintering additive manufacturing [152].

PBAT is a biodegradable synthetic polymer based on fossil resources, with high elongation at break and very flexible. However, PBAT's high cost has restricted its application; as a result, research frequently focuses on combining it with other inexpensive polymeric materials with acceptable mechanical qualities [153]. PBAT is particularly indicated for the production of flexible films blended with PLA and TPS, as seen in Figure 9, or mineral fillers; the first study in this regard dates from 2001.

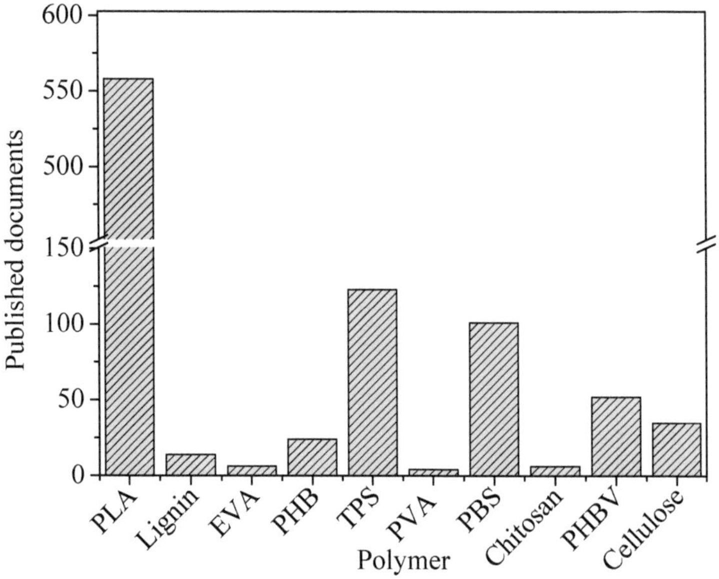

Figure 9. Number of publications of PBAT blends (Scopus Source up to June 2024).

Poly(lactic acid) (PLA) is a bio-based and biodegradable polyester with good processability, high transparency, and complementary mechanical properties to PBAT, exhibiting high stiffness but also high brittleness [154]. Nevertheless, because of the poor compatibility between PBAT and PLA, improving the properties of PBAT by direct melt blending these polymers can be challenging [155]. The compatibility can be improved using physical compatibilization, usually through synthesizing block/graft copolymers or reactive compatibilization strategies, such as employing dicumyl peroxide or multifunctional epoxides (ADR) [156].

Biodegradable plastic based on blends of polyester and thermoplastic starch (TPS) dominates the market of biodegradable polymers, representing more than 50% of the market share [157]. Therefore, it is unsurprising to see starch, mainly with glycerol in its thermoplastic form, has been recurrently applied as a filler for PBAT. Starch is highly regarded because of its biomass origin, high availability, quick biodegradation, affordable pricing, and low carbon footprint. However, the mechanical properties and WVP of films prepared with TPS/PBAT blends are inferior to those of conventional films, further declining as the amount of starch increases [158]. Moreover, TPS and PBAT are incompatible due to their poor interfacial affinity as a consequence of the hydrophilic nature of TPS and the relative hydrophobicity of PBAT [159]. PBAT/TPS blends require high shear forces, thus, higher screw speeds in the extrusion compounding process to obtain a homogeneous dispersion of TPS inside the PBAT matrix [135].

An effective way to improve the compatibility between TPS and PBAT is to functionalize the PBAT matrix by grafting highly reactive functions. These grafted functions can react with the hydroxyl groups of starch to form covalent bonds, providing better control of the phase size and strong interfacial adhesion [160]. Alternatively, compatibilizers including citric, maleic, and tartaric acids [159], soybean oil [158], or sodium hypophosphite as a co-agent with citric acid [161], reduce the interfacial tension caused by the difference in polarity between PBAT and TPS. A method to address the poor mechanical and water sensitivity of the TPS phase without sacrificing biodegradability is to reinforce it with the addition of nanofillers. Multiple nanofillers have been proposed in PBAT/TPS blends, including halloysite clay nanotubes (HNTs) [162], needle-like sepiolite [163], bentonite (BT) [164], and cellulose nanofibrils (CNFs) [157].

Nanocomposites offer an attractive method to improve pristine PBAT properties. A wide range of nanomaterials have been added as reinforcements or additives to PBAT films derived from natural organic sources, such as acetylated cellulose nanocrystals (CNFs) [131] and CNFs modified with octadecyl isocyanate [165] and inorganic resources such ZnO [141], TiO_2 [166] or Ag [140] nanoparticles, and layered silicate nanoclays including MMT [167], octadecylamine-modified MMT [168] or MMT/Cloisite/Bentonite [169].

A comparison between PBAT/MMT and PBAT/CTAB-modified MMT (CMMT) nanocomposites was carried out by Mondal et al. [170]. PBAT/CMMT showed greater property enhancement, decreasing WVP, and stronger antimicrobial activity, especially against gram-positive bacteria. However, PBAT/CMMT displayed less biodegradation after 100 days. Han and Park [171] proposed the elaboration of a PBAT/PE blend to improve the mechanical properties of PBAT mulching films via blown film extrusion and extend its decomposition period. The blend was compounded with a PE mixture (LDPE:LLDPE = 15:85) and added peroxide di-(2-tert-butly-peroxyisopropyl) benzene (BIBP) as a crosslinking agent. The compatibilization effect of BIBP increased the storage modulus of the blend, provided higher bubble stability during extrusion, and improved the film's surface morphology.

The degradation of PBAT waste can be divided into abiotic degradation (physical and chemical degradation) and biotic degradation. Biodegradability in composting conditions and soil burial conditions is affected by both biotic and abiotic factors of the environment, such as temperature, moisture, pH, bio-surfactant, and enzymes, and by polymer characteristics, such as chain flexibility, crystallinity, regularity and heterogeneity, functional groups, and Mw [172]. Biotic degradation is a slow process that uses microorganisms or other organisms to degrade PBAT into carbon dioxide and water. In contrast, the fast abiotic process degrades PBAT into smaller molecules. Moreover, abiotic methods preferentially target the chemical bonds with higher energy (especially C-O and C=O) of PBAT, eventually shortening the polymer chain and then reducing the molecular weight [173].

PBAT can theoretically be disposed directly into the soil without recycling, entirely decomposing into water, carbon dioxide, and microbial biomass [174]. Nevertheless, PBAT has a very low biodegradation rate compared to those of other biodegradable polymers such as PHAs and PCL, caused by the high resistance to enzymes to the TPA ester bonds and the small number of PBAT-degraders in the environment [175]. Therefore, the biodegradation rate is highly influenced by the aromatic segment of TPA, with faster biodegradation occurring with decreased TPA content in the structural backbone [172]. Ma et al. [176] explained how biodegraded PBAT films in soil exhibited an increase of O elements and a decrease of C elements in XPS, indicating that the degradation process involved the oxidation of molecular chains. In addition, the hydrolysis of ester bonds in PBAT films led to the breakdown of molecular chains and subsequent degradation into smaller molecules.

The biodegradation rate under mesophilic (37 °C) and thermophilic (55 °C) anaerobic conditions was studied by Svoboda et al. [177] by measuring biogas production. Under mesophilic conditions, the biodegradation after 126 days was only 2.2%, Mw had changed from 93,000 to 25,500 g/mol, and crystallization had experienced slight changes. On the other hand, under thermophilic conditions, biodegradation reached 8.3% after 126 days, Mw decreased from 93,000 to 9430 g/mol, and the crystallization behavior changed significantly.

Nomadolo et al. [178] compared the biodegradation of PBAT/PLA and PBAT/PBS blend films under industrial and home composting conditions. PBAT-PBS reached 82% and PBAT-PLA 90% mineralization in 120 days in industrial conditions, and under home composting for 200 days, samples reached 72% and 50% degradation, respectively. Dammack et al. [179] examined the use of maleated poly(butylene adipate-*co*-terephthalate) (PBATg-MA) and maleic anhydride (MA) as compatibilizers in PBAT/TPS blends and their effect on biodegradation. In the presence of MA, the degradation profile of the blends was close to that of neat PBAT and TPS. On the contrary, in the presence of PBATg-MA, biodegradation of the blends was 70–75% after 80 days, while it exceeded more than 85% without any compatibilizer. The authors explained that PBATg-MA might have created a chemical continuity between the PBAT and the TPS phase, generating an interfacial domain that is harder to degrade.

The photodegradation and thermophile biodegradation of two commercial mulching films based on PBAT and a PBAT/PLA blend were studied by Morro et al. [180]. Photodegradation was efficient in both films, reducing the Mw by 16% for PBAT and 31% in the blended film. The photodegraded samples then presented higher susceptibility to biodegradation as the reduction in molecular weights facilitated bacterial action. On a similar note, Zhang et al. [181] examined the effect of light stabilizers on the weathering resistance of PBAT/PLA blend films. The addition of a light stabilizer prevented the chain scission of polyesters by absorbing damaging UV, preventing a significant decrease in Mw.

Yang et al. [182] employed *Thermobifida fusca* cutinases, a type of enzyme, to break PBAT films into fragments in 4 h, with further digestion into small particles in 12 h and a complete decomposition in 48 h. Another alternative approach was proposed by Pang et al. [183], employing a 1,4-butanediol-alkali combined method for the hydrolysis of PBAT. Subsequently, a two-step synthesis produces a foam nickel-supported NiOOH catalyst for the electrocatalytic reformation of the PBAT hydrolysate into succinate salts, a higher-value chemical.

4. Polyhydroxyalkanoates

Polyhydroxyalkanoates (PHAs) are a natural, biodegradable, and biocompatible family of aliphatic polyesters that find applications in food packaging and the biomedical field [184].

This biopolymer is produced by more than 300 bacteria, both Gram-positive and Gram-negative, under particular conditions [185].

The general structure of PHAs is reported in Figure 10.

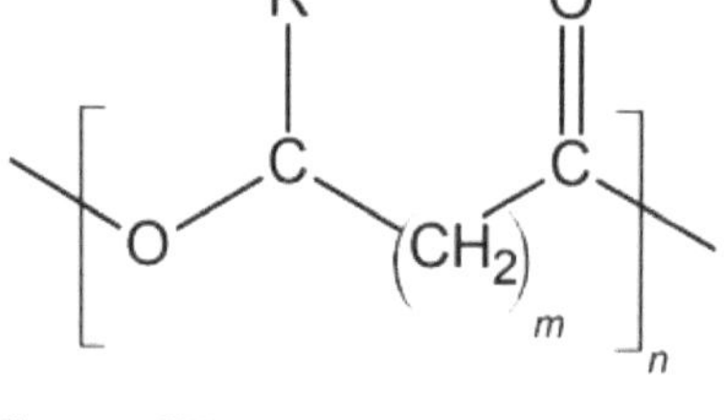

Figure 10. PHAs general structure.

The –R group can be a saturated, unsaturated, or branched alkyl group, and according to the chain length, can be divided into three categories [186,187]: (i) short chain length PHA *(scl-PHA)* that consists of monomers of 3–5 carbons and present a high degree of crystallinity, (ii) medium chain length PHA *(mcl-PHA)* with monomers of 6–14 carbons that can be aliphatic or aromatic. They present a low degree of crystallinity and are elastomeric. (iii) long chain length PHA with 15 or more carbons in the building block.

PHAs are generally UV stable and present a variable degree of crystallinity. Indeed, they could be completely amorphous or reach a crystallinity of 70%, depending on the monomeric building block [188].

Their mechanical properties are similar to those of polypropylene (PP) and polyethylene (PE) but more brittle and with lower elongation at break. However, the mechanical properties depend on the length of the polymeric chain and the crystallinity of the material [189]. In general, PHAs present a Young modulus between 0.008 and 3.5 GPa and tensile strength between 6.6 and 50 MPa [190].

Concerning the thermal properties, T_m decreases by increasing the side chain of the carbon length while T_g increases. Indeed, T_m can vary from 53 to 180 °C and T_g from 1 to 50 °C [191].

Due to their biodegradability and barrier properties, PHAs find applications in particular in industrial applications such as food packaging [192]. Genovesi et al. [193] prepared blends of poly(3-hydroxybutyrate-co-3-hydroxyhexanoate) (PHBH) and poly(3-hydroxybutyrate-4-hydroxybutyrate) P(3HB)(4HB) through reactive extrusion obtaining a film suitable as packaging material for vegetables and fruits, with increased applications.

Their biocompatibility and biodegradability allow the use of these kinds of polymers for biomedical applications such as medical devices and implants [194].

Ye et al. [195] developed a material for bone tissue engineering by using PHA as a biodegradable polymeric matrix reinforced with β-tricalcium phosphate (β-TCP) at different percentages from 0 to 20 wt% to improve its mechanical properties and osteoconductivity and finding that this scaffold can promote the proliferation, adhesion, and migration MC3T3-E1.

PHAs have many advantages, but their poor mechanical properties and low thermal stability represent a shortcoming for their processing and applications, which are overcome with the production of blends or nanocomposites [196].

Polyhydroxybutyrate (PHB) belongs to the family of PHAs and is characterized by a $-CH_3$ group in the side chain, which makes it highly crystalline, thermoplastic, and hydrophobic [197].

Bio-based synthesized PHB presents an isotactic structure with the carbons in the R(-)-configuration. Indeed, only this configuration can be recognized by microorganisms and degraded, while synthetic PHB presents more irregularities and cannot be degraded to a high extent [198].

The high crystallinity (around 70%) is the main reason for its outstanding mechanical properties, with an elastic modulus around 2.5–3 GPa and tensile strength of 25–40 MPa, comparable with PP and PE [199].

PHB, which presents less permeability than PE and PP [200], finds application in food packaging as a replacement for common PP-based packaging [201]. Furthermore, as bio-based and biodegradable polymer it is widely used in the biomedical field [202], tissue regeneration [203], and drug delivery [204].

However, PHB presents low thermal stability, and the narrow range between its Tm and Td, which is close to 180 °C, represents a limitation in the processability of this polymer [205].

In order to overcome these problems, PHB was blended with various natural and synthetic polymers since 1998, as reported in Figure 11 [206]. The addition of reinforcement as fibers or both organic and inorganic nanoparticles also permits the improvement of mechanical and conductive properties and increases the thermal stability of PHB [207].

Poly(3-hydroxybutyrate-co-3-hydroxyvalerate), PHBV, is obtained by the copolymerization of 3-hydroxybutanoic acid and 3-hydroxypentanoic acid or 3-hydroxyvalerate (HV) and can be produced by both gram-positive and gram-negative bacteria, as *Ralstonia* or *Alcaligenes eutrophus* [208].

The physical and mechanical properties of PHBV depend on the 3HV content; indeed, the crystallinity and melting point decrease by increasing 3HV content while the degradation rate increases [209].

Since the presence of HV leads to the formation of a less crystalline structure, the reduced chain packing increases flexibility and toughness with respect to PHB [210].

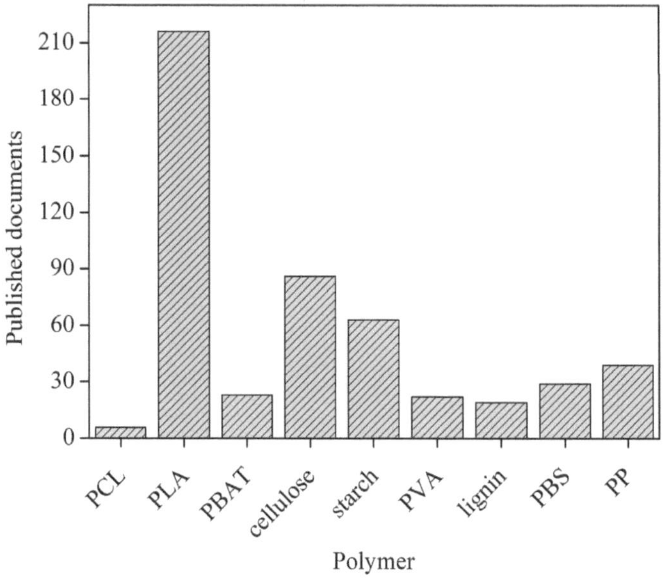

Figure 11. Number of publications of PHB blends (Scopus Source up to June 2024).

Besides its biodegradability and biocompatibility, it presents some shortcomings, such as hydrophobicity, poor mechanical properties, and low thermal stability, which can be improved by fabricating nanocomposites with natural fibers, clays, silicates, etc. [211]. Moreover, blending PHBV with other polymers such as PCL, PLA, PBS, or cellulose and starch, as shown in Figure 12, permits the overcoming of these issues [212].

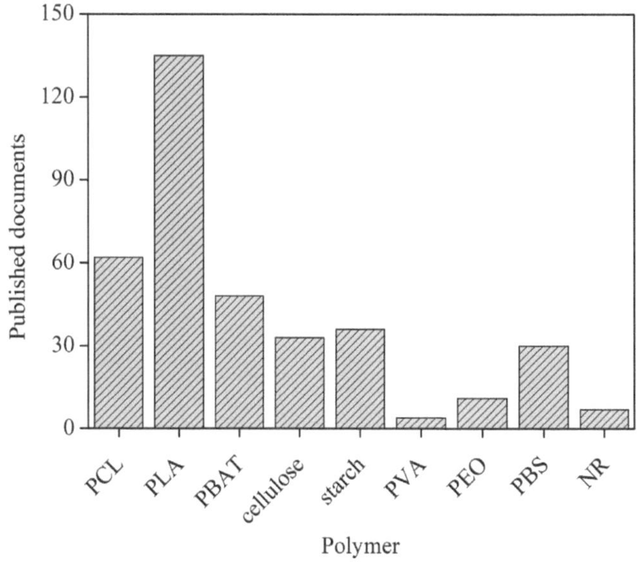

Figure 12. Number of publications of PHBV blends (Scopus Source up to June 2024).

– Natural biosynthesis

PHAs are usually produced through the induction of enzymatic reactions in microorganisms from diverse feedstocks [213].

When there is an excess in carbon source and a lack of nutrients such as nitrogen, sulfur, and phosphates, bacteria start to accumulate PHA, stored as granules in the cytoplasm, which undergo depolymerization and are metabolized by bacteria when the external carbon source is consumed [187].

In general, the biosynthesis of PHAs can be summarized in three different pathways: direct synthesis pathway, β-Oxidation pathway, and fatty acid synthesis pathway [214].

– Direct synthesis pathway

The first route, which is more typical for the synthesis of scl-PHA, is carried out by *Ralstonia eutropha* and *Azotobacter beijerinckii*. The synthesis involves the conversion of Acetyl-CoA to Acetoacetyl-CoA, which is subsequently transformed into (R)-3-Hydroxyacyl-CoA, catalyzed by *PhaA* and *PhaB* respectively. Finally, (R)-3-Hydroxyacyl-CoA is converted into scl-PHA [215].

– β-Oxidation pathway

This pathway is one of the main routes for the synthesis of mcl-PHA, which occurs in *Pseudomonas* strains by using fatty acids [216].

Acyl-CoA is used as substrate and is converted into Enoyl-CoA, which is subsequently transformed into (R)-3-hydroxyoxycil-CoA catalyzed by (R)-specific-enoyl-CoA hydrase. Lastly, (R)-3-hydroxyoxycil-CoA is used as a substrate by PHA *synthase* [217].

– Fatty acid synthesis pathway

mcl-PHA can be produced either through the β-Oxidation pathway or through the fatty acid synthesis pathway, both using fatty acids as starting material [218].

In this route, fatty acid synthesis leads to the production of 3-ketoacyl-ACP, then transformed into (R)-3-hydroxyacyl-ACP. The enzyme *PhaG* catalyzes the transformation of (R)-3-hydroxyacyl-ACP into (R)-3-Hydroxyacyl-CoA, which is finally converted into mcl-PHA [219].

– Bioengineering strategies

As a result of the advancement of bioengineering, new methods have been developed in order to increase the bio-catalytic activity of PHA synthase, the utilization of new HA monomers, and the genetic modifications of bacteria [220,221].

Thomas et al. [222] carried out the characterization of *Halomonas* sp. SF2003, which is a marine bacterium able to produce PHA by using two genes identified as PhaC1 and PhaC2, and the study of potential carbon substrates for PHA production and the production itself. Furthermore, PhaC1 and PhaC2 *synthases* were expressed in *Cupriavidus necator* (*C. necator*) H16 $PHB^{-}4$ (DSM 541). *Halomonas* sp. SF2003 and *C. necator* showed preferably glucose and fructose as substrates, respectively.

Cha et al. [223] produced four types of scl-PHA through the modification of the levulinic acid (LA) metabolic pathway in *Pseudomonas putida* EM42 and expressing different types of PHA *synthase* to produce PHA with different monomer fractions, reaching 40% of cell dry mass under non-optimized flask culture.

Another method used for PHA synthesis is the use of waste-activated sludge (WAS), which also permits a sustainable treatment of wastewater and represents a source of mixed microbial culture (MMC). The process entails a first enrichment of *bacteria* in a sequencing batch reactor (SBR) and a second accumulation of PHA in *bacteria* through the aerobic dynamic feeding method (ADF), where *bacteria* are submitted to alternating feat and famine periods [224]. PHA accumulation has reached 90% gPHA/gVSS on a laboratory scale and a value between 40 and 80% gPHA/gVSS on an industrial scale [225].

Estévez-Alonso et al. [226] carried out the production of PHA through activated sludge from municipal wastewater treatment and demonstrated that the addition of calcium stimulates the selective enrichment of PHA-accumulating bacteria and improves PHA production three times more compared to control without calcium after 48 h.

Zhang et al. [227] investigated the production of PHA using volatile fatty acids (VFAs) from acetate to valerate in SBRs inoculated with activated sludge in ADF at nitrogen-limiting conditions with high yields.

As with all types of PHAs, PHB as well is synthesized by microalgae and in bacterial cells in carbon-excess and critical environment conditions as limited nutrients [187].

Vroman individuated three main routes to synthesize PHB [197]. ROP of β-butyrolactone (BL) is one of them [228]. Yang et al. [229] carried out the synthesis of PHB through ROP by using different bifunctional organoboron catalysts in order to avoid the use of metallic catalysts and obtain a polymer selectivity above 99% with the optimized catalyst. Jian et al. [230] used Zirconium compounds as initiators and obtained a PHB with an average molecular weight of 12,000 g/mol.

The utilization of natural/transgenic plants, which contain acetyl-CoA that is used as a substrate, is another approach [231,232]. Lössl et al. [233] used transplastomic tobacco to obtain PHB, and Parveez et al. [234] realized PHB biosynthesis in oil palm by using three bacterial genes, *bktB*, *phaB*, and *phaC*.

The third and more common method is through bacterial fermentation [235]. Acetyl-CoA is the main substrate; two molecules are condensed by *PhaA* to obtain acetoacetyl-CoA, which is subsequently reduced by *PhaB* to 3-hydroxybutyryl-CoA. Lastly, *PhaC* catalyzes the polymerization of (R)-3-hydroxybutyryl-CoA [236].

El-Kadi et al. [237] produced PHB using three different carbon sources: dried whey, sugar beet molasses, and date molasses. Thirty isolated bacteria were tested, and *Bacillus paramycoides* MCCC 1A04098, *Azotobacter salinestris* NBRC 102611, and *Brevundimonas naejangsanensis* BIO-TAS2-2 were identified as the best bacteria to obtain PHB.

C. necator is widely used for PHB production. Lambauer et al. [238] carried out first the cultivation of *C. necator* H16 in the presence of an explosive gas mixture, H_2, O_2, and CO_2 under O_2 in a bioreactor; 98% of PHB was obtained from CO_2. However, in order to increase PHB production, metabolic engineering is used to modify the native metabolic pathways of this bacterium [239].

The biosynthesis of PHB in bacteria involves two parallel pathways, one for the production of (3-hydroxybutyrate) monomer and the other for the production of (3-hydroxyvalerate) monomer [210].

In the first step, two molecules of acetyl-CoA, derived from glycolysis and tricarboxylic acid (TCA) cycle, are condensed by *phaA* into acetoacetyl-CoA, which is further converted into 3-hydroxybutyryl-CoA by *phaB*. Other substrates, such as propionic acid, which is the precursor of 3HV, are metabolized to obtain propionyl-CoA that subsequently reacts with acetyl-CoA to form 3-ketovaleryl-CoA. *PhaB* catalyzes the reduction of 3-ketovaleryl-CoA to 3-hydroxyvaleryl-CoA. Finally, the polymerization of the two monomers is catalyzed by *phaC* [240].

There are different bacteria able to produce PHBV; Zhao et al. [241] studied the production of PHBV in *Halogranum amylolyticum* (PHBV-H) and *Ralstonia eutropha* (PHBV-B) by using different carbon sources (Figure 13).

The two copolymers presented similar content of 3HV monomer, almost 21%, but exhibited different mechanical and thermal properties, crystallinity, surface energy, and wettability.

Meng et al. [242] carried out the production of PHBV in *E. coli* by expressing two genes, namely *phaCp* and *phaABp*, cloned from *Propylenella binzhouense* L72T. The highest yield (1.06 g/L) was obtained when glucose and propionic acid were used as substrates. The obtained PHBV showed improved thermal stability (298 °C) and high crystallinity.

Recently, few works have been focused on the reutilization of waste as carbon sources. Indeed, Altun et al. [243] generated PHBV from waste cooking oil in a lab-scale bioreactor and reinforced it with pistachio shell flour to fabricate a biocomposite through melt mixing. Suhazsini et al. [244] obtained PHBV from *Bacillus megaterium* of cheese whey permeate by co-feeding propionic acid used as a plasticizer. Parameters such as time, C/N ratio, and propionic acid content were optimized. Indeed, they obtained the highest yield (3.64 g/L) after 72 h, 0.1% of propionic acid, and a C/N ratio of 15.

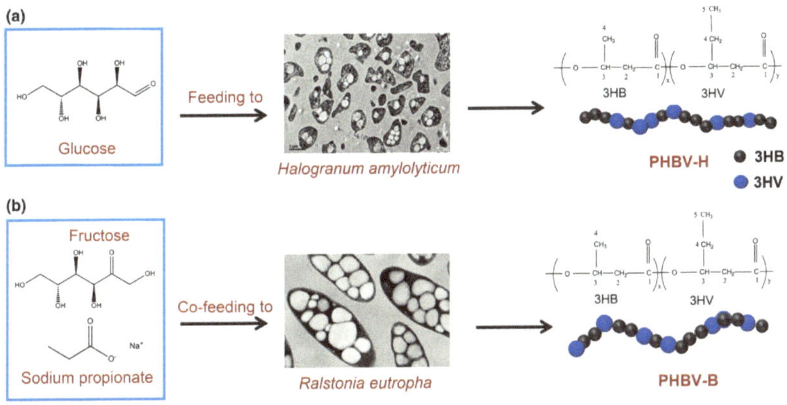

Figure 13. Schematic representation of PHBV production (**a**) in *H. amylolyticum* and (**b**) *R. eutropha* [241].

Furthermore, PHBV production was also studied in activated sludge. Chen et al. [245] used primary sludge (PS) and waste-activated sludge (WAS) as carbon sources, achieving 65.6% of the content, and identified *Thauera* as the predominant strain.

However, fed-batch fermentation, which consists of the continuous addition of substrates to microorganisms and incessant removal of products, is the most used method to obtain a higher amount of PHBV [240].

Chang et al. [246] utilized genetically modified *Methylorubrum extorquens* AM1 to over-express PHBV. Fed-batch fermentation implemented with multiple cycles was used to remove the excess of accumulated sodium, obtaining 2.76 g/L of copolymer.

Hathi et al. [247] reutilized food waste from restaurants to produce PHBV, using *C. necator* as a PHBV-producing strain through a fed-batch fermentation strategy. They obtained 5.32 g/L of PHBV with a 50% molar 3HV fraction.

The recovery process, which involves the separation of PHA from biomass after its production, is the first step before the processing of PHA. Several approaches have been adopted in order to extract polymeric chains from cells; the downstream processing depends on various factors such as microbial strain, type of PHA, and content [248].

The typical processing techniques adopted for polymer manufacturing are used for PHAs as well, as solvent casting, fiber spinning, injection molding and film extrusion, phase separation, freeze drying, foaming, 3D printing, and the synthesis of NPs [249,250].

Mármol et al. [251] produced films of PHA by solvent casting, reinforced with graphene nanoplatelets (GNPs) at different concentrations and studied the mechanical, electrical, and piezoelectric response of this material that found potential applications as sensors.

Chen et al. [252] fabricated PHA/polyvinyl alcohol (PHA/PVA) membranes by combining electrospinning and spin-coating methods. The membrane is used as a drug delivery system for levofloxacin (LEV), which was added at 0.05 and 0.1 wt% and showed antibacterial activity against *E. coli* and *S. aureus*.

Liu et al. [253] integrated graphene-decorated gold NPs (RGO/Au) into poly-3 hydroxybutyrate-co-12 mol.% (P3HB-co-12 mol.%) hydroxyhexanoate (HHx) fibers through electrospinning. They produced a scaffold suitable for nerve tissue regeneration by combining the conductivity of graphene with nanofibers. Indeed, the electroactive scaffold promoted Schwann cell (SC) proliferation and migration.

Ivorra-Martinez et al. [254] reinforced poly(3-hydroxybutyrate-co-3-hydroxyhexanoate) [P(3HB-co-3HHx)] with hydroxyapatite nanoparticles (nHA) in a range between 2.5 and 20 wt%. The nanocomposite produced through injection molding exhibited improved mechanical properties, in particular with concentrations above 10 wt% of NPs, which presented ductility and strength similar to native bone.

PHA has also been found to be applicable in soft tissue engineering. Chen et al. [255] developed a scaffold via a phase separation method based on 3-hydroxybutyrate (3HB) and 4-hydroxybutyrate (4HB) (P34HB) (or PAP34HB) reinforced with ZnO NPs for oral soft tissue regeneration and antibacterial activity against bacteria as *Porphyromonas gingivalis* and *Streptococcus mutans.*

Thermally induced phase separation followed by freeze drying was used by Zeinali et al. [256] to produce a 3D microporous scaffold of poly(hydroxybutyrate-co-hydroxyvalerate) (PHBV) copolymers. The scaffold promotes cell adhesion and proliferation, which makes this scaffold suitable for tissue regeneration.

Lopresti et al. [257] blended PLLA with PHA at 30 wt% in order to produce a foam suitable as a scaffold for tissue engineering. Blending PLLA with PHA improved the wettability, biocompatibility, and proliferation capacity of the scaffold.

Zhang et al. [258] developed a 3D-printed scaffold based on poly(3-hydroxybutyrate-co-4-hydroxybutyrate) (P34HB) by fused deposition modeling (FDM). PHA scaffold was integrated with bone morphogenetic protein 2 (*BMP2*) and immobilized through a polydopamine (PDA) coating, improving its osteogenic bioactivity and cell adhesion and proliferation.

As a thermoplastic polymer, PHB can be processed with typical processing techniques such as film casting and blowing, injection molding, electrospinning, extrusion, thermoforming, etc. [259].

However, it is extremely brittle and presents a narrow processability range of temperature due to its T_d close to its T_m (180 °C), limiting the residence time of some techniques to a few minutes [260]. To overcome this issue, Uzun et al. [261] fabricated different composites of PHB and three different bio-fillers as lignin, alpha-cellulose, and cellulose nanofibrils by solvent casting, observing an increased thermal performance in terms of T_c and T_m, particularly when lignin was added.

In order to expand PHA application fields, Anbukarasu et al. [262] developed a nanocomposite constituted by PHB and GNP through solvent casting, suitable as an actuator or sensor. The resistivity was studied at different casting temperatures, and observing a decrease in resistivity by increasing the casting temperature, from 42.3 Ω cm for 80 °C to 3.01 Ω cm for 110 °C and 1.5 Ω cm for 140 °C. Furthermore, the resistivity decreases with the physical aging due to an increase in the crystallinity and density of the material, resulting in a smaller separation of GNP reinforcement, improving the mobility of electrons between the nanoplatelets.

With the aim of replacing petroleum-based plastics with biodegradable polymers, Hernandez et al. [263] added microspheres, carbon fibers, or polyethylene glycol (PEG) of different MW: 2000, 10,000, and 20,000 through injection molding. They studied the influence of the reinforcement in thermal stability, observing a slight increase only when PEG at higher MW was added; however, all the samples presented a reduction in hardness and compressibility.

Bucci et al. [264] fabricated material for food packaging through injection molding; PHB material was compared with PP-based packaging in terms of dimensional and mechanical properties. PHB resulted in a more rigid material with respect to PP, showing a lower deformation of about 50%.

PHB-based material for packaging applications was studied also by Olejnik et al. [265] by blending PHB with PLA at different mass ratios, 100/0, 50/10, 50/20, 40/30, 50/50, 30/40, 20/50, 10/50 and 0/100 mass ratio through extrusion, identifying PLA/PHB 50/10 the blend with the optimal properties.

Weinmann et al. [198] realized a chemical modification of PHB by reactive extrusion in order to get either branched or crosslinked material with improved thermal and mechanical properties, also with the addition of rapeseed oil and PEG 1500 and PEG 4000 as plasticizers. The addition of plasticizers increased the melt stability and melt viscosity. Furthermore, the materials presented higher thermal stability by retarding degradation reactions.

Electrospinning is another technique used for PHB manufacturing. Arrieta et al. [266] fabricated PLA-PHB nanofibers in different proportions (100:0; 75:25; 50:50, 25:75, and 0:100) and studied their thermal, mechanical, and morphological properties. The proportion PLA-PHB 75:25 showed the highest interactions between the two polymers, and PHB reduced the presence of beads. Furthermore, they added acetyl tri-n-butyl citrate (ATBC) as a plasticizer to the best formulation, obtaining a more flexible and ductile material with improved thermal stability.

Kundrat et al. [267] fabricated PHB meshes for drug delivery applications. Levofloxacin was incorporated into the nanofibers depending on the viscosity of the PHB solution and morphology of nanofibers in a range of 14.4–81.8%. The sponge model incorporated the highest amount of levofloxacin, while nanofibers with and without beads entrapped 22.3% and 14.4%, respectively. Moreover, drug release was studied in PBS, and all samples were able to release between 30.4 and 32.5% of the incorporated drug.

PHB has also been processed by thermoforming. Sánchez-Safont et al. [268] produced thermoformed trays of PHB, highly purified commercial cellulose fibers (TC90) were used as filler and hexamethylene diisocyanate (HMDI) as compatibilizer. Wei et al. [269] obtained a cross-linked PHB with dicumyl peroxide (DCP) (0.25–1 wt%), obtaining an improvement in melt strength, which is an important property for processing techniques such as thermoforming.

The common processing techniques used for PHB can be used to process PHBV as well. PHBV presents higher flexibility with respect to PHB, and the presence of HV expands the processing range of temperatures with higher melt stability [270].

Erceg et al. [271] fabricated films based on PHBV and PCL-diol by solvent casting, obtaining the optimal composition in terms of film morphology with 60 wt% of PCL.

Perez-Martinez et al. [272] fabricated PHBV/PLA sheets for food packaging in order to replace PP, which is commonly used in this field. The films were developed by solvent casting and extrusion in different proportions: 50/50, 60/40, 80/20, and 100/0 (%w/w). PHBV presents exceptional barrier properties, while PLA improves the thermal stability of the material. The extruded films showed better mechanical properties, and the formulation PHBV/PLA 50/50 presented properties similar to those of PP.

With the same purpose, Samaniego-Aguilar et al. [273] obtained a binary blend of PHBV and thermoplastic polyurethane (TPU) by reactive extrusion and using hexamethylene diisocyanate (HDMI) as a reactive agent and compatibilizer. The material presented improved ductility and impact toughness, due in particular to the presence of HDMI.

Frącz et al. [274] developed PHBV-based reinforced with 30% of three types of fibers (1 mm length): flax, hemp, and wood. The material was processed first by extrusion and then by injection molding. Only hemp and flax fibers improved the mechanical properties of neat PHBV.

Mallegni et al. [275] reutilized agro-food residue as coffee silverskin (CS) to produce a PHBV/CS biocomposite, utilizing ATBC as a plasticizer and $CaCO_3$ as reinforcement. Furthermore, they produced coffee capsule bodies through injection molding and studied the migration properties of the biocomposite with simulants at 100 °C, which was below the limit (10 mg/dm^2) needed for materials apt for food-packaging applications.

Electrospinning is a versatile technique used to produce nanofibers, and PHBV electrospun nanofibers find applications in the biomedical field and industrial applications [276].

Rabbani et al. [277] fabricated PHBV beaded nanofibers and beadless microfibers. In order to increase the porosity and roughness of nanofibers, polydimethylsiloxane (PDMS) treated silica was added to the solution and subsequently treated with stearic acid at 80 °C. The incorporation of NPs and stearic acid decreases the crystallinity of the material.

Furthermore, they obtained a superhydrophobic material with high static contact angle WCAs $\geq 155^{\circ}$ and sliding angles $\leq 5^{\circ}$ suitable for biomedical applications but also water filtration, oil-water separation, and food packaging.

Karbowniczek et al. [278] developed PHBV-based electrospun fibers reinforced with HA for bone tissue engineering. The hybrid material improved filopodia formation, cell proliferation, and anchoring within the scaffold.

Another system widely used for the production of scaffolds for tissue engineering is 3D printing. Kontogianni et al. [279] blended PHBV with PLA and PCL in a ratio PLLA/PCL/PHBV 90/5/5 wt% through twin-screw extrusion. Furthermore, they added nano-HA and strontium-substituted nano-HA (Sr-nano-HA) to produce the filament to be printed by FDM. The presence of reinforcement and the porous structure of the scaffold facilitated cell attachment and proliferation.

de Carvalho et al. [280] used the same technique to fabricate a filament of PHBV reinforced with $ZrO_2 \cdot nH_2O$ particles between 1 and 10% wt/wt. The presence of particles decreases the crystallinity of PHBV and increases the relative density.

There are numerous studies on the compatibilizing effect of modified oils in blends of immiscible polymers. They observed how the separate incorporation of epoxidized soybean oil (ESO), acrylate soybean oil (AESO), and maleinized soybean oil (MSO) improved compatibility in immiscible blends of PLA/PHB/PCL (60/10/30), obtaining formulations with improved ductile properties in all cases [281]. Other blends with different percentages of PLA (75%)—PHB (25%) obtained an interesting result after the incorporation of modified oils, epoxidized (ECO) and maleinized (MCO) corn oil showed an effect on the mechanical properties, obtaining an increase on ductility for both plasticizer, thermal properties were slightly reduced, but the incorporation does not affect the thermal stability of initial polymers. Also, the degradability of samples showed an increase in the degradability grade compared with the initial blend, PLA—PHB (Sempere-Torregrosa et al., 2022 [282]).

Adding the OLA plasticizer to a blend of PLA (85%) and PHB (15%) reduces tensile strength but increases ductility, indicating a more flexible material and a decrease in Tg was observed. Although other thermal properties decrease, thermal stability remains stable up to 20% by weight of OLA. With 30% OLA, degradation intensifies. Also, water vapor and oxygen permeability increase, but crystallinity and hydrophobic behavior are suitable for food packaging applications [88].

The incorporation of a polyester plasticizer (Lapol 108) at two different concentrations, 5 and 7% by weight per 100 parts of the blend PLA/PHB, was studied. Thermal values show only one Tg value, indicating polymers of the blend are miscible. The rest of the results showed that the plasticizer does not affect the thermal stability of PLA and PHB. Also, the addition of plasticizer improved ductile properties, obtaining an increase of elongation at break values [283].

PHAs are biodegradable polymers obtained through biosynthesis and biopolymerization by microorganisms. The most studied and employed ones are poly(3-hydroxybutyrate) (PHB) and the copolymer poly(3-hydroxybutyrate-co-3-hydroxyvalerate) P(3HB-co-3HV) or PHBV. Additionally, their mechanical and thermal properties are like those of common polymers obtained from fossil sources. Although their properties vary considerably depending on their structure, they are generally characterized by a high crystallinity of up to 80%, exhibiting high rigidity and brittleness, as well as low thermal stability. These properties can be modified by the incorporation of plasticizers, resulting in significant variation.

The plasticizers used to modify the properties of PHAs are based on the use of vegetable oils and their derivatives, such as glycerol, citrates, and glycol esters. Seoane et al. studied the effect of epoxidized soybean oil (ESBO) on the properties of PHB. The results demonstrated how the incorporation of 5% by weight of oil increased the elongation at the break of PHB by 45% and slightly reduced its tensile strength and elastic modulus. As for thermal properties, the incorporation of 5% by weight of ESO slightly increased the onset temperature of PHB degradation without affecting the melting temperature, thus maintaining its thermal stability [284]. The effect of vegetable oils such as epoxidized linseed oil and ESBO have also been studied as compatibilizers of PHBV, both showing good plasticizing properties incorporating 10 wt% since they reduce tensile strength and elastic modulus and increase elongation at break. In addition, a decrease of between 4 and 5 °C in the melting temperature was also observed [285].

Glycerol (G) and its derivatives, such as glycerol tributyrate (GTB), glycerol triacetate (GTA), and polyglycerols (PGs), have also been studied as plasticizers for PHB. Among them, only the additives GTB and GTA, by acting as plasticizers, improved the mobility of the polymer chains, which reduced the glass transition temperature and modified the mechanical properties, decreasing tensile strength and Young's modulus while increasing elongation at break. On the other hand, the additives G and PGs had a negative effect due to thermal degradation, which decreased the strength and elasticity of the material [286].

Some authors have also studied the compatibility and effectiveness of acetyl tributyl citrate (ATBC) as a plasticizer for PHB. They investigated the effect of incorporating different amounts of ATBC (10, 20, 25, and 30% by weight) on the mechanical and thermal properties of PHB, observing how the plasticized samples exhibited lower rigidity and greater ductility than pure PHB, resulting in decreased tensile strength, reduced elastic modulus, and increased elongation at break of the plasticized samples. Also, the sample plasticized with 30 wt% of ATBC decreased the melting temperature by more than 10 °C and the Tg by more than 35 °C. However, this plasticizer considerably reduced the thermal stability of PHB, lowering its onset degradation temperature to 189 °C for the sample plasticized with 30% by weight of ATBC [287]. Parra et al. studied the effect of different amounts (2, 5, 10, 20, and 40 wt%) of PEG in PHB films obtained through dissolution in chloroform. A considerable increase in elongation was achieved at breakup to 32% with 20 wt% by weight of PEG. Furthermore, the sample with 10% by weight of PEG reduces the melting temperature by 21 °C and reduces the crystallinity of the material, increasing the permeability to water vapor and also increasing the degradation rate [288].

PHAs, as polyesters, undergo aerobic and anaerobic degradation following the mechanisms described previously.

In particular, since this kind of polymers are synthesized by bacteria in excess of the carbon source when the external carbon source is exhausted, the microorganisms themselves release enzymes called PHA depolymerase, which are able to hydrolyze PHA chains into smaller products soluble in water that can be absorbed by cells and metabolized [289].

The degradation of PHAs has been studied in soil, river and seawater, sludge, and compost [290,291] but also in phosphate buffer, simulated body fluids, and muscular tissue [292,293].

Ong et al. [294] studied the degradation of films of four different types of PHAs, poly(3-hydroxybutyrate) [P(3HB)], poly(3HB-co-8 mol% 3-hydroxyhexanoate) [P(3HB-co-8 mol% 3HHx)], P(3HB-co-12 mol% 3HHx) and P(3HB-co-21 mol% 3HHx) with three different thickness. The films were buried in a secondary forest for 8 weeks, and the soil bacterial population was studied and directly correlated with the degradation of PHA, observing a richness in PHA-degrading bacteria close to the sampling sites.

Borelbach et al. [295] investigated the degradation of PHA blended PLA nanofibers in soil for 12 weeks and also the degradation in compost at 58 °C for 4 weeks and evaluated the degradation by studying the variation in morphological, mechanical, and thermal properties.

Komiyama et al. [296] analyzed the degradation of poly[(R)-3-hydroxybutyrate-co-(R)-3-hydroxyvalerate] (P(3HB-co-3HV)) powder and processed as cast film, undrawn fiber and fivefold-drawn fibers in seawater and freshwater. They observed that powder showed faster degradation with respect to the other materials, and fivefold-drawn fibers showed slower degradation with respect to undrawn fibers due to the higher crystallinity.

Furthermore, they observed that the biodegradation of the materials depends on the amount of bacteria present in the degradation medium.

Few works have been focused on tuning the degradation rate of PHAs by blending with other polymers. Kim et al. [297] achieved to accelerate degradation rate of poly(3-

hydroxyundecenoate) (PHU), which is a mcl-PHA with a long side chain that makes it highly hydrophobic, by blending with poly(lactide-co-glycolide), PLGA. The higher compatibility of PHU with blood proteins was combined with the better mechanical properties and higher degradation rate of PLGA, obtaining a material suitable for biomedical applications.

PHB is the most common PHA and can be degraded by microorganisms such as *Alcaligenes*, *Comamonas*, and *Pseudomonas* [298].

PHB degradation has been studied in different environments, such as freshwater, soil, composting, landfills, and sludge, and different strains able to degrade PHB have been studied [299,300]. PHB degradation was also studied in other conditions, such as hydrolytic and thermal degradation [301,302].

Sabapathy et al. [303] studied the degradation of PHB-sugar cane bagasse (SCB) in soil and water, which showed a degradation between 7 and 18 weeks.

Rech et al. [304] prepared films of PHB loaded with eugenol by solvent casting and studied the degradation for 60 days in agricultural, sandy, and landfill soil, observing a higher degradation rate in agricultural soil due to the higher presence of microorganisms and phosphorus availability.

Pei et al. [305] evaluated PHB degradation in municipal waste-activated sludge over two weeks at different pH, 5.5, 7.0, and 10 at different temperatures (25, 37, and 55 °C).

Brzezinka et al. [306] incorporated three types of polyhexamethylene guanidine (PHMG) derivatives in PHB and studied biodegradability in compost and river water. They were able to isolate the strains present in the film; *Bacillus toyonensis* HW1 and *Variovorax boronicumulans* HK3 were identified.

Cho et al. [307] studied six bacteria present in marine soil, in particular *Bacillus* sp. JY14. After 14 days of incubation, *Bacillus* sp. JY14 reached almost 98% of PHB degradation in solid and liquid conditions.

Furthermore, they studied the microbial degradation of plasticized PHB with 10% and 20% tributyl citrate by using *Microbulbifer* sp. SOL66 [308] (Figure 14).

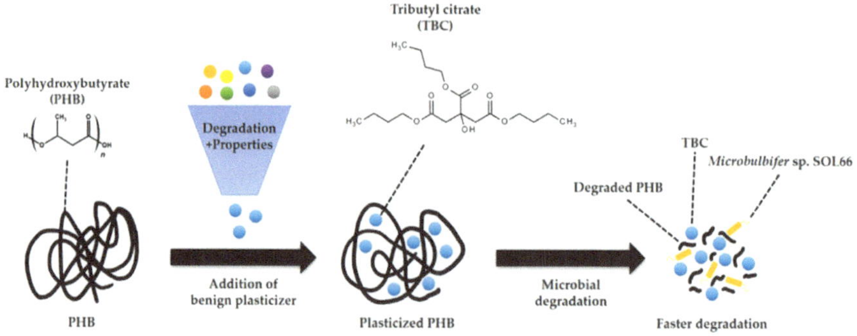

Figure 14. PHB degradation by *Microbulbifer* sp. SOL66 [308].

The presence of plasticizers increased the degradation rate of PHB by 88% after 1 day. However, the samples with and without tributyl citrate presented similar degradation after 3 days.

ATBC as a plasticizer was also used by Mosnáčková et al. [309] in order to study the hydrolytic degradation of a blend constituted by PLA and PHB reinforced with hydrolyzed keratin showing faster degradation at 60 °C.

Zhou et al. [310] fabricated an artificial peripheral nerve conduit (PNC) through electrospinning of PHB/chitosan (PHB/CS) composite polymers and studied the hydrolytic degradation at different pH, observing a decreasing degradation rate from acidic > alkaline > PBS.

Râpă et al. [311] studied the in vitro degradation of PHB/bacterial cellulose (BC) in PBS of 7.4 for 20 days. The reduction in crystallinity caused by the presence of BC increased the degradation rate of PHB.

The degradation of PHBV was studied in different environments, such as soil, compost, freshwater, seawater, hydrolytic degradation, enzymatic, and aerobic composting [312].

Since PHBV is widely used in food-packaging applications, a lot of work has been focused on soil degradation.

Kim et al. [313] studied the degradation of PHBV in soil with 100% and 40% RH. PHBV was totally degraded in 100% RH soil after two weeks, showing a reduction in mechanical properties already after 3 days, while showing small changes in mechanical and thermal properties, crystallinity, and MW over six weeks in 40% RH soil.

Furgier et al. [314] studied the degradation in the soil of PHBV reinforced with three different reinforcements derived from organic waste, human hair waste (HHW), sawdust (SD), and chitin from shrimp shells for five months. The organic reinforcement is able to increase the biodegradation of the composite, showing 35% weight loss with HHW, with respect to neat PHBV (only 5% weight loss).

Furthermore, Hernández-García et al. [315] studied the degradation of films based on PLA/PHBV 75:25 and 2% (*w*/*w*) of ferulic, p-coumaric or protocatechuic acid and 15 wt% of polyethylene glycol (PEG 1000), in composting conditions during 35 and 45 days. PLA-PHBV films were degraded already after 20 days, while films with p-coumaric and protocatechuic after 21 and 26 days retarding the biodegradation.

Bonnenfant et al. [316] studied the biodegradability of PHBV reinforced with acid and quercetin. The degradation was studied in both culture media of *Streptomyces exfoliates* and in compost. Both polyphenols are able to inhibit bacterial growth, however, without preventing the mineralization of the material within the time established by the NF T51 800 standard [317].

Meereboer et al. [318] studied for the first time the degradation in the simulated marine environment of PHBV reinforced with *Miscanthus* (Misc) fibers and distillers' dried grains with soluble (DDGS) using ASTM standard D7991-15 [319]. PHBV/Misc (85/15) and (75/25) showed 15 and 25%, while PHBV/DDGS (85/15) and (75/25) presented 17 and 40% more biodegradation compared to PHBV, respectively. The two biocomposites were totally degraded in 412 and 295 days, respectively.

Hammiche et al. [320] studied the effect of alfa fiber as reinforcement in the PHBV matrix. Alkali treatment of the alfa fiber surface improved the compatibility between the polymeric matrix and reinforcement. Moreover, treated composites showed higher degradation with respect to neat polymeric matrix and reinforced material without treatment in distilled and seawater.

Furthermore, hydrolytic degradation of PHBV was carried out by Patel et al. [321]. They developed a porous bone scaffold based on PHBV by utilizing Selective Laser Sintering (SLS), which was incubated in PBS for over 20 weeks. A bulk degradation mechanism was observed, and the crystallinity of the scaffold increased by around 7–15% due to a first degradation of the amorphous region and a further recrystallization of lower MW chains.

Zhu et al. [322] studied the hydrolytic degradation of PHBV reinforced with three types of cellulose nanoparticles (CNPs). The hydrogen bonds between CNPs and PHBV slow down the penetration of PBS and, therefore, the degradation of the material, in particular with 5 wt% of CNPs.

Kovalcik et al. [323] synthesized two types of PHAs through *C. necator* by using fructose and grape sugar extract to obtain PHBV-1 and PHBV-2, respectively. The enzymatic degradation of both PHAs was studied in synthetic gastric juice and PBS for 81 days with lipase. PHBV-2, which showed a lower crystallinity of almost 50% with respect to PHBV-1, also showed faster degradation, demonstrating the crystallinity-dependence of degradation.

5. Starch

Starch is a natural biopolymer extracted from various botanical sources. Worldwide production is mainly based on four common sources: corn, which represents 80% of the world's manufactory for bioplastics [324], cassava, wheat, and potato. However, starch has been extracted from diverse natural sources to obtain different properties of the extracted starch and diversification of agricultural resources [325].

The structure of starch is formed by the succession of α-D-glucose linked to each other through (1 → 4) glycosidic bonds. This biopolymer presents two different structures in its matrix [326]: the slightly branched named amylose (20–30% of the structure) and the highly branched denominated amylopectin. The predominant linear structure of amylose is mainly linked through (1 → 4) glycosidic bonds with a slight presence of (1 → 6) glycosidic bonds, which allows for its helicoidal 3D structure. In the case of amylopectin, although being linked through (1 → 4) glycosidic bonds, it is also bonded through (1 → 6) glycosidic bonds, resulting in a highly branched structure with 24 to 30 residues of glucose units. Due to the presence of many end-chain points onto which enzymes can attack, this structure has a higher solubility and can be degraded more quickly than amylose [327]. The ratio of amylose and amylopectin depends on the source of extraction of the starch [328], as well as its characteristics and properties, as seen in Table 1. This ratio varies in most cases from 20 to 30% amylose and 70 to 80% amylopectin [329].

Table 1. Corn, potato, and rice starch content.

	Corn (Maize) Starch	
Content	**Amount**	**Units**
Density	1.356–1.4029	g/cm^3
Amylose	24.64–29.4	g/100 g
Amylopectin	72–75.36	g/100 g
Crude fats	0.32–7.13	g/100 g
Crude proteins	0.38–7.70	g/100 g
Moisture content	10.45–12.7	%
	Potato Starch	
Content	**Amount**	**Units**
Amylose	17–38.8	g/100 g
Amylopectin	73.42	g/100 g
Crude fats	0.06–0.20	g/100 g
Crude proteins	0.01–0.09	g/100 g
Moisture content	9–11	%
	Rice Starch	
Content	**Amount**	**Units**
Amylose	32–4.1	g/100 g
Amylopectin	74.56	g/100 g
Crude fats	1.9–28.1	g/100 g
Crude proteins	0–3.0	g/100 g
Moisture content	13	%

Several methods have been developed and used for the extraction of native starch from a wide variety of botanical sources; due to this reason, the ratio between amylose and amylopectin may vary from each extraction source. Of all the extraction methods reported for starch obtention, two are the most widely used.

In the first method, the botanical source is grounded, followed by its treatment with a dissolution of hydrogen sulfite 0.1 M at 50 °C for 24 h. After this treatment, the substrate obtained is grounded again with distillate water to extract all the starch present in the substrate, followed by several screenings through different mesh sizes to separate the residues from the starch. Finally, the starch slurry is centrifuged several times and dried in an oven at 40 °C for 12 h [330].

In the second method, the ground from the botanical source is treated with an aqueous solution of sodium metabisulfite 0.1 M for 30 min. This homogenate is screened through different mesh sizes, keeping the filtrate until the complete deposition of starch. This starch is recovered and treated with an aqueous solution of NaOH 0.2 M for 1–2 h and finally dried at 40 °C for 12 [331].

Starch can present different granule shapes and sizes, depending on the botanical source, as seen in Table 2. For example, some starch granules have angular shapes like corn starch, while potato granules exhibit irregular and oval shapes, and rice display pentagonal granules.

Table 2. Size and shape of starch granules from different botanical sources.

Plant Source	Size (μm)	Shape
Maize	2–30	Round and polyhedral
Waxy maize	2–30	Round and polyhedral
Wheat	10 and 10–30	Discs
Waxy wheat	>10	Spherical and lenticular
Rice	<20 and 2–8	Polygonal and angular
Tapioca	5–45	Spherical/lenticular
Cassava	30	Kettle-drum truncated
Potato	<110	Oval and irregular
Sweet potato	1–100	Oval, spherical, round
Pea	2–40	Oval, spherical, round
Lotus	14–31	Oblong

Due to limitations compared to conventional plastic materials, such as a high permeability to water vapor and solubility in water, a high gas permeability, and proximity to its melting and degradation temperatures, the native starch form cannot be used as a material in various industrial applications, especially the food packaging industry [332], for this reason the formation of blends based on starch are a good alternative to overcome these limitations as shown in Figure 15 using a data range from 1956 up to June 2024

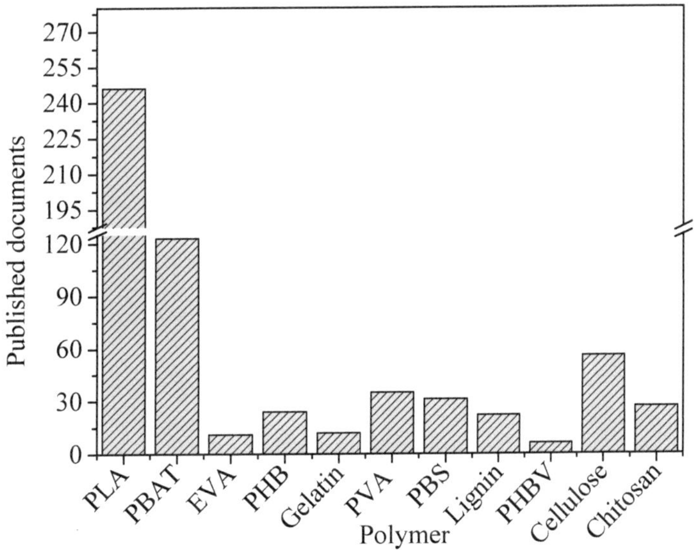

Figure 15. Number of publications of TPS blends (Scopus Source up to June 2024).

However, starch can be processed to obtain a continuous polymeric entangled phase denominated thermoplastic starch (TPS), which can have industrial applications. To obtain TPS, it is necessary to add water and/or plasticizers [333] and have elevated temperatures and shear. The process involves various chemical and/or physical interactions such as water diffusion, starch granules' expansion, gelatinization, and polymer melting [334]. The presence of a plasticizer changes the molecular structure of starch by destroying the inter and intra-molecular hydrogen bonding of the starch granule [244]. Meanwhile, in the gelatinization process, heat and pressure destroy the crystalline structure of starch, which transforms into amorphous structures with the presence of water and/or plasticizers [335], as shown in Figure 16.

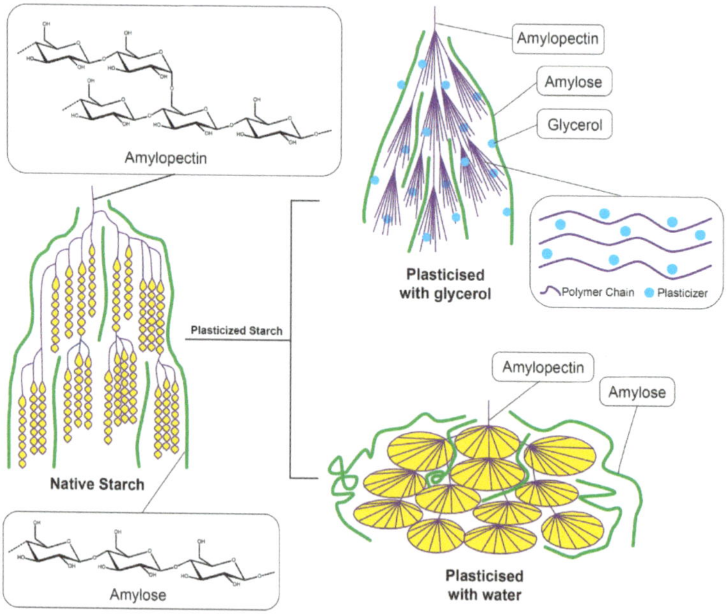

Figure 16. Starch microstructural changes under treatment with plasticizers or water [332].

Solvent casting is the most widespread processing method on a laboratory scale, as it does not require long processing times or high amounts of materials [336]. In this method, the starch matrix is heated above its gelatinization temperature, generally between 80 and 90 °C, for 15–20 min under constant mechanical stirring or ultrasonication. After this process, filler materials can be added and mixed, as well as through mechanical stirring or ultrasonication. Finally, the solution is poured into a glass or petri dish and dried in an oven to remove the solvent for 48 h until constant weight.

Despite its simplicity, practicality, and its cost-effectiveness. The properties obtained from materials processed through this technique can vary significantly between batches, influenced by environmental conditions, which can also slow the production process [337].

Extrusion is one of the common processing techniques for melting thermoplastic matrixes. A rotating extruder is utilized to turn solid pellets of a particular polymer into liquid, with a motor that assists in rotating the screw and a metal barrel that may be heated electrically. The polymer melts because of the heater's heat energy, the frictional heat produced by the screw's friction, and the immense quantity of heat emitted by the barrel. The L/D controls the residence time, and shear melts is the proportion of the screw's flight length towards its external diameter. Higher L/D ratio screws allow for better mixing, longer melt residence times, and greater shear heating extruder [338].

Thermoplastic starch, TPS, is obtained after mixing starch with a plasticizer (often glycerol or water) to allow for the melting of the material at a temperature lower than the decomposition temperature of starch [339]. TPS is produced during the extrusion process as a result of the initial starch structure expanding, breaking, and melting in conditions of high temperature and shear. The partially crystalline structure of starch can be destroyed and transformed into an amorphous state. A decrease in the degree of crystallinity has been observed in various types of starch after the extrusion process as well as the granule morphology and the molecular structure of starch [340].

Several conditions of extrusion have been tested on starch matrices, and several research studies have been done on this topic. In the study carried out by González-Seligra et al., the morphology of the granules of a matrix of starch/glycerol was studied with different screw speeds. After characterizing the starch obtained, they concluded that the effect of different screw speeds was determinant on the final structure and physicochemical properties of the starch, with 80 rpm being the ideal speed for obtaining a homogeneous TPS [341]. Pushpadass et al. [342] conducted a similar experiment but used different temperatures and different plasticizers, from 110 °C to 120 °C, concluding that the temperature needed for the correct formation of TPS is intrinsically connected to the plasticizer used. Logié et al. [343] compared the mechanism of extrusion with the same conditions for different types of starch. In this case, potato and pea starch did not show significantly different results; the only essential difference was that potato starch presented larger macromolecular chain splitting.

Three-dimensional printing (3D printing) is a technology that uses computer softwares to design and print 3D models, with the material being added together (such as liquid molecules or powder granules being fused), typically layer by layer. This technique offers the advantage of producing complex shapes that would be impossible to construct by hand, including hollow parts or parts with internal truss structures to reduce weight. This technology has revolutionized the manufacturing industry.

Starch materials printed by 3D printing are characterized by a smooth surface, which can be compared to the surface observed in the filaments of oil-based polymers [344]. The suitability of starch-based materials for 3D printing applications is determined by their printability, gelling behavior, mechanical properties, and heat resistance. Often, the printability of starch-based materials can be enhanced by combining them with other materials [345].

Electrospinning is a method of producing fine fibers with diameters in the nanometer and micron range from polymer solutions to melts by electrostatic interaction. It is widely used in the food and biomedical sectors. In this process, a solution of the polymer is forced through a needle with a high voltage power supply, forming a conical droplet named Taylor cone due to this potential. When the electrostatic force overcomes the surface tension, the polymer solution is ejected from the Taylor cone towards the grounded collector. The solvent evaporates before the jet reaches the collector, and the polymer chains in the jet tend to stretch and orient themselves [346].

Starch has promising applications through this technique due to its high bioavailability and biocompatibility. Several studies have demonstrated the potential use of this polymer with other polymers. Hanaa et al. [347] successfully processed these various mixtures of starch and polyvinyl alcohol to improve the spinning ability of neat starch, concluding that the one that presented better characteristics was 40:60 PVA:Starch.

Malafatti et al. [348] obtained nanofibers with various ratios of PLA:Starch, from 10% to 50% (*w*/*w*) loaded with manganese carbonate as a source of Mg^{2+} ions used as a model fertilizer and source of Mn^{2+} ions required for plant metabolism. As a result, they found out that the addition of 20% of starch was the optimal proportion, but in general, the addition of starch into the PLA matrix showed a better affinity for water molecules, allowing better diffusional performance for application in fertilizer release, also showing characteristics capable of packaging and releasing particles.

Asl et al. [349] designed scaffolds based on a blend of PHB and starch (from 5 to 15 wt%) for bone tissue engineering. They observed that the addition of starch decreased the fiber diameters, decreased the brittleness and crystallinity of PHB, and enhanced thermal stability. The addition of up to 10 wt% starch also resulted in an improvement in mechanical and tensile strength and toughness. Also, scaffolds showed higher hydrophilic nature and higher degradation compared with neat PHB.

Preparations of TPS blends with conventional polymers were made to create novel materials that are low-cost and highly biodegradable. Much work has been done on creating PLA/starch mixes to lower the overall cost of raw materials and improve their degradability. However, the poor interfacial contact between hydrophilic starch granules and hydrophobic PLA is the main issue with this blend system [350].

Different adipate and citrate esters were used as plasticizers to produce PLA/TPS sheets by a calendering-extrusion process. The sheets were made using 30-weight percent TPS and 70-weight percent PLA plus plasticizer (*w*/*w*). The plasticizer corresponds to 10% of the PLA component. It was found that the mechanical characteristics and processability of PLA/TPS sheets are enhanced by the addition of adipate or citrate esters. Better plasticization was determined, and diethyl adipate produced lower tensile strength and young modulus and higher elongation at break among the plasticizers tested. Additionally, diethyl adipate increases the water vapor permeability (Shirai et al., 2015).

It was discovered that substituting formamide and water for glycerol as TPS plasticizers significantly increases the fluidity of TPS/PLA, which enhances the plasticization of the starch, raising the mixture's dispersion and lowering the interfacial tension to produce homogenous blends. Additionally, the TPS/PLA blend's qualities were enhanced by formamide and water. This was due to the PLA and TPS plasticized with formamide and water becoming more compatible and dispersed, resulting in an increase in elongation at a break of 4% while maintaining the tensile strength at 31 MPa, which was comparable to neat PLA [350].

The composition and characteristics of TPS/PLA blends where sorbitol, glycerol, and combinations of glycerol and sorbitol plasticize the TPS phase were examined. It was discovered that even in the absence of compatibilization, the sorbitol-plasticized TPS phase could be more evenly and finely distributed inside the PLA matrix. The TPS/PLA mixes that were plasticized with sorbitol showed significantly greater modulus and tensile strength. It was also demonstrated by thermogravimetric studies to be less volatile. Moreover, when compared to virgin PLA, the blends, including TPS plasticized with glycerol alone, exhibited the lowest shear viscosity. The viscosity of the blend rose when sorbitol gradually replaced glycerol. The blend viscosity for the TPS-only sorbitol plasticized material was comparable to that of virgin PLA [351].

Starch has received much attention in recent decades to produce biodegradable polymers as a potential replacement for plastics derived from petrochemicals [352]. Starch in its native state does not exhibit thermoplastic behavior. Therefore, the structure of granular starch must be broken down to employ starch in the plastics sector [353]. The molecular structure of starch granules changes to thermoplastic through plasticization, which requires (1) temperatures greater than 70 °C, (2) shear stress, and (3) the presence of plasticizers [354]. Plasticizers increase the free volume and decrease the molecular contacts between the polymer chains, increasing the polymers' flexibility, plasticity, processability, and elongation [355].

Plasticizers with polar groups—such as OH, COOH, and NH2—are required to disrupt molecular connections between starch chains and generate hydrogen bonds [356]. Even though water is a good starch plasticizer, it may volatilize during the starch plasticization process. Therefore, the plasticizers used to obtain TPS from starch are polyols, amines, amides, and monosaccharides. Also, the combination of different plasticizers was investigated.

Glycerol is the most used polyol for manufacturing TPS because it has greater plasticizing and greater diffusion in the starch chains [357]. On the other hand, sorbitol has a lower volatility than glycerol and forms stronger bonds with starch chains, resulting in a material with high tensile stress and low elongation at break [351,358]. However, the tendency of sorbitol to migrate to the component surface and its eventual recrystallization limit its usage as a plasticizer for starch [351]. Studies using the combinations of the two plasticizers have been conducted to achieve a synergy between their respective qualities. The results indicate that a 2:1 ratio of sorbitol to glycerol in a formulation containing 42 weight percent plasticizer yields a formulation with a Tg below room temperature, adequate tensile strength, and significant flexibility [359]. Additionally, it was found that a 1:1 ratio keeps the plasticizers more tightly bonded, hindering them from migrating out of the material [360]. A promising polyol that is in the study as a TPS plasticizer isosorbide, which is non-toxic and seems to prevent TPS retrogradation. Moreover, it would allow energy savings because the temperature profile needed with this plasticizer is 20 °C lower than the employed glycerol [361].

The effectiveness of plasticizers with amine groups in plasticizing starch has been demonstrated to be greater than that of alcohols. TPS plasticized with formamide presents the highest elongation at breakage and the lowest tensile strength compared to sorbitol, glycerol, ethylene glycol, and urea, which points to a higher plasticization [362]. Urea, which is an amine, prevents starch from retrograding, but because it is a solid with no internal flexibility, the TPS that was plasticized with urea became brittle and stiff [363]. It was found that a 1:1 ratio of urea and ethanolamine in a formulation containing 30% plasticizer displayed higher mechanical properties than TPS plasticized by glycerol in the same content [364]. However, the utilization of amines and amides is limited due to their toxicity or unsuitability for interaction with food [357].

Fructose was studied as a plasticizer to produce TPS films. It was determined that in comparison to the other films—plasticized with urea, tri-ethylene glycol, and tri-ethanolamine- the fructose-plasticized film absorbed less water (110 for 30% F at 90 min). Additionally, according to the FT-IR results, fructose was able to generate stronger and more stable hydrogen bonds with the hydroxyl group of starch molecules than other plasticizers [365].

Like cellulose, starch has also been used to produce nanoparticles, namely starch nanocrystals (SNCs) through acid hydrolysis. The major challenge in the production of SNC is completely removing the amorphous region of starch granules but not destroying the starch crystalline structure [366]. The classical extraction procedure of SNCs from their natural starch granules consists of the degradation of native starch granules through strong, controlled acid hydrolysis conditions, mainly using H_2SO_4 [367]. HCl hydrolysis is also common but requires a longer time and yields less amount of individual starch nanoplatelets [366]. The acid concentration and time of the acid hydrolysis process depend on the relative proportion of amylose and amylopectin as well as the granule morphology (i.e., size, shape, and surface state), mainly related to the botanical origin of the starch [367]. Firstly, regions of low lateral order are dissolved, and hydrolysis of the amorphous domains takes place while the highly crystalline water-insoluble lamellae remain intact [366]. The acid hydrolysis reaction takes at least 5 days. Then, the final suspension is washed by successive centrifugations and can also be dialyzed in distilled water until reaching neutral pH [336,366]. A sonication or Ultra-Turrax step is generally applied to avoid agglomeration and to improve the SNC dispersion in the final suspension [336,366]. The obtained SNCs through the acid hydrolysis process are of diverse shapes and sizes depending on the starchy plant sources as well as the preparation condition used [366,367]. Starches are categorized into three crystalline types known as A (closely packed with water molecules between each double helical structure), B (more open and water molecules are located in the central cavity formed by six double helices), and C (a mixture of both A-type and B-type) that could lead in different nanocrystals morphology [336]. Figure 17 shows a schematic representation of a classical process to obtain SNCs from native starch granules using an acid hydrolysis process.

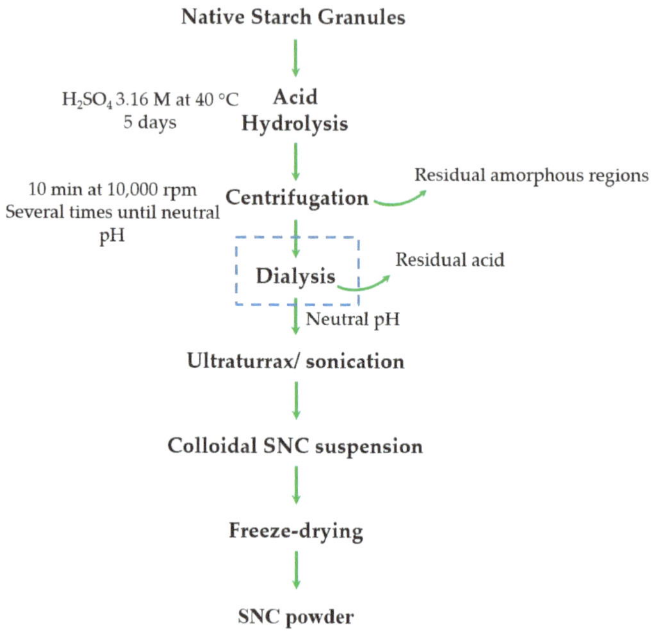

Figure 17. Schematic representation of an acid hydrolysis process to obtain SNCs.

SNCs have been used to reinforce starch polymeric matrix, improving the overall mechanical properties of the material [336]. Moreover, SNCs have also been used to reinforce biopolyester matrices, such as to promote the PHB spherulite growth [368] as well as to improve the water and oxygen barrier properties of PLA [369]. In Figure 18, the number of publications of starch nanocrystals with different polymeric matrixes is reported.

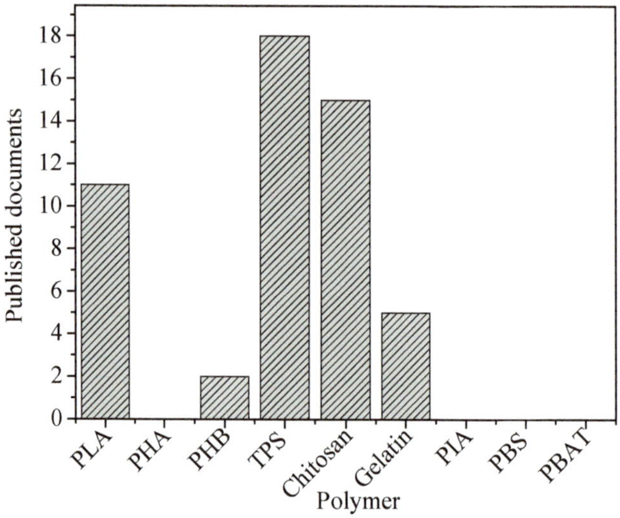

Figure 18. Number of publications of starch nanocrystals (Scopus Source up to June 2024).

The main mechanisms in starch degradation can be divided into two: enzymes and thermal degradation.

The term 'starch-degrading enzymes' is often imprecise. Enzymes that act on starch granules can also utilize solubilized α-glucans. Additionally, the structural changes on the surface of native starch may facilitate the accessibility of degrading enzymes, but these processes are not yet fully understood. For this reason, we can differentiate different types of starch-degrading enzymes; most of them cleave inter-glucose bonds, although there are other enzymes that can act on inter-glucose bonds, they are functionally related to their cleavage for starch degradation [370]. The first step in the starch degradation pathway must, therefore, be catalyzed by an enzyme capable of metabolizing polymers at the surface of a semicrystalline granule rather than in a soluble form. Although several different types of enzymes are capable of releasing soluble glucans from purified starch granules in vitro experiments, no enzyme has been shown to be able to do this [371]. Among these enzymes, we can divide them into several groups in relation to the mechanism of action of the enzyme: exo-hydrolases (β-Amylases), phosphorylases, endo-hydrolases (α-Amylases), debranching Enzymes (DBEs) and disproportionating enzymes (DPE1 and DPE2).

Characteristically, this type of material presents a multi-step thermal degradation, which shows mainly two stages of its degradation. The first one is typically located at 80–120 °C, depending on its temperature with the botanical source of starch. This first loss of weight is attributed to the evaporation or dehydration of loosely bound water and low molecular weight molecules, and the second step represents around 70% of the total thermal degradation, at around 270–400 °C which corresponds with the degradation of saccharide rings [372].

6. Cellulose

Cellulose consists of a semi-crystalline and high-molecular-weight homopolymer of β-1,4-D-anhydroglucopyranosyl [373]. Cellulose is the most widely distributed bio-based and biodegradable polymer in nature since it is part of higher plants and available all over the world with around 8×10^{10} tons of annual production. It is mainly extracted from wood (50% of cellulose content) and also from cotton (90% of cellulose content), where individual crystalline cellulose chains (20–50 nm) form cellulose fibers arranged in layers bounded by a cross-linking polymer (lignin) and by strong hydrogen bonds interactions [373–375]. Therefore, crystalline regions of cellulose are embedded in an amorphous matrix of hemicelluloses and lignin [376], where the high amount of hydroxyl group interactions plays a major role in the cellulose crystalline packing, which governs its physical properties [377]. There are a number of cellulose derivatives that can be prepared from cellulose, including microcrystalline cellulose (MCC), a purified pulp of cellulose that is obtained by acid treatment of the alpha cellulose (type I_{β}) coming from fibrous plants; it is partially depolymerized cellulose [373] with dimensions ranging between 10 and 50 μm [378]. Depending on the wood sources, MCC can exhibit significant variations in structural organization, such as crystalline region percentage or chemical composition, varying the proportions of cellulose, hemicelluloses, and lignin [373]. The degree of polymerization of the cellulose depends on the starting material, resulting in approximately 10,000 glucopyranose units in the case of wood cellulose and 15,000 glucopyranose units for cotton cellulose [374,379]. For nanocomposite production reinforced with nanocellulose (cellulosic material with at least one dimension in the nanometer scale), there are two main structures of cellulose extracted by top-down method from plants that lead to nanocellulose, nanofibrilated cellulose (NFCs) and cellulose nanocrystals (CNCs). While plants are the main source of cellulose and nanocellulose, it is noteworthy that cellulose can alternatively be biosynthesized through a bottom-up strategy from glucose with the aid of acetic acid-producing bacteria, known as bacterial cellulose (BC) [379]. Although cellulose, BC is structurally different with respect to plant-sourced cellulose, presenting a more compact network within the nanometer due to the predominant I_{α} allomorph content in its structure [380]. Cellulose is likewise found in several microorganisms, such as fungi, bacteria, and algae [375,381]. Although plant-derived cellulose is a renewable polymer, environmental concerns associated with both intensive deforestation and the chemical processing extraction methods are currently driving forces to produce cellulose in a more sustainable way, such as from microorganisms [382]. Particular attention is being paid to the obtention of BC from food waste as well as food by-products, such as from the kombucha probiotic beverage production, since the side product BC produced is considered a discarding by-product [380,383,384]. While plant-derived cellulose also contains lignin and hemicellulose that should be eliminated through strong basic and/or acid conditions, BC is mostly composed of pure cellulose, showing some advantages including higher purity, crystallinity, mechanical strength, and hydrophilicity [380], at the same time it is easily produced without the need of strong basic and/or acid extraction procedures that are questioned from an environmental point of view [383,384]. In this regard, BC can be obtained from the revalorization of food waste, introducing the process in the circular economy concept.

The classical extraction procedure of cellulose from its natural plant matrix consists of the removal of lignin, hemicelluloses, and other resins throughout an alkali treatment (2% NaOH at 80 °C) that removes the soluble polysaccharides, followed by a bleaching treatment (i.e., $NaClO_2$/sodium acetate buffer pH = 4.8, 80 °C) to eliminate the majority of the matrix materials, i.e., lignin, hemicellulose or polyphenols while leaving nanocellulose intact [377,385]. Then, the bleached fibers could either be mechanical or enzymatic and disintegrated, leading to NFC, or subjected to acid hydrolysis to obtain CNC [378].

- Nanofibrillated cellulose extraction

The mechanical (i.e., refining, grinding, high-pressure homogenization, etc.) or enzymatic disintegration of wood or cotton pulp fibers leads to a fine network of alternating crystalline and amorphous and flexible cellulose fibrils with less than 100 nm. The denominations most often used to describe such fibrils is Nano/Microfibrillated Cellulose (NFC/MFC) [386,387]. The mechanical disintegration process can also be combined with chemical or enzymatic pretreatments [374], and then the previously cut softwood fibers (0.6–0.7 mm) are homogenized under controlled temperature (70–80 °C) and high pressure (55 MPa). Then, the mechanical process itself is conducted in such a way that the applied high shear forces irreversible separate the highly ordered cellulose microfibrils, leading to a viscous aqueous suspension. This suspension is further dried with carbon dioxide, yielding isolated fibril and microfibril aggregates (diameters of about 100 nm and 2–10 nm thick), while 10–50 microfibrils compose the NFC [388,389]. Figure 19 shows a schematic representation of a classical process to obtain NFC from wood or other annual plants using a mechanical top-down approach leading to the obtention of NFC gel-like suspension [387].

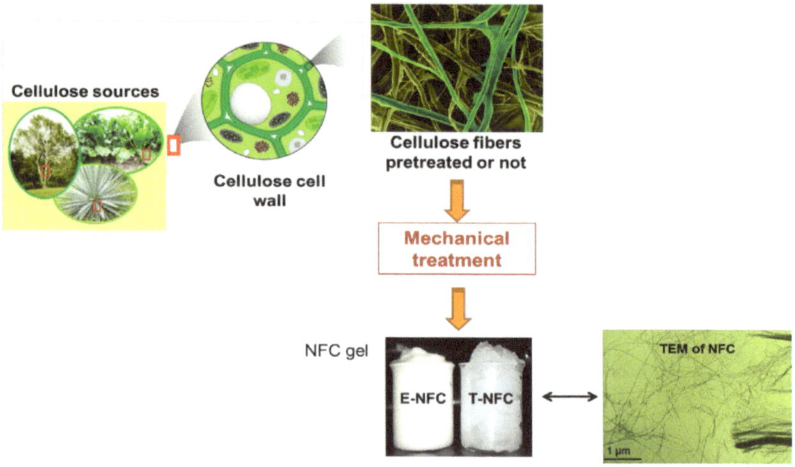

Figure 19. Classical process to obtain nanofibrillated cellulose (NFC). Reprinted from Missoum et al., 2013 [387] under Creative Commons CC BY license.

The main mechanisms of extraction of nanocellulose can be summarized as the extraction of cellulose nanocrystals (CNC) by acid hydrolysis (Figure 20) and bacterial cellulose obtention.

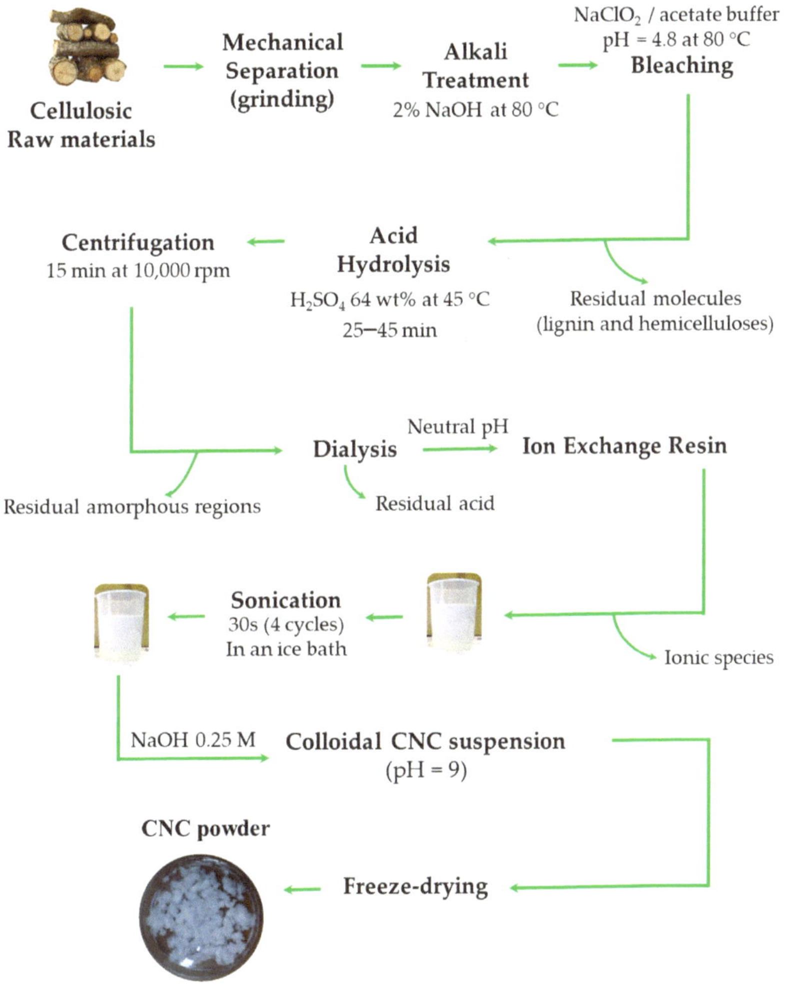

Figure 20. Schematic representation of pre-treatment, purification steps of the raw materials and CNC. Adapted from. Arrieta et al. [378].

The amorphous regions of cellulose are susceptible to hydrolysis; thus, typical procedures to produce CNC starting from natural fibers, waste, or commercial MCC, are conducted through strong and strictly controlled acid hydrolysis conditions (H_2SO_4 64 wt% at 45 °C for 25–45 min) [18], yielding shorter and more crystalline fragments of MCC which, in turn, lead to the formation of highly crystalline (between 54% and 88%) and high-purity single cellulose nanocrystals (CNC) [20,373,390]. Strong acid hydrolysis conditions promote a transverse cleavage of cellulose, disrupting the amorphous regions between cellulose microfibrils, leaving the crystalline regions intact [374]. The acid is subsequently removed with deionized water through centrifugation, followed by dialysis until a neutral pH is attained. Sulfuric acid is currently chosen among other studied acids (i.e., hydrochloric acid, phosphoric, and hydrobromic acids) since it yields in nanoparticles, CNC, with negatively charged surfaces, which forms a stable colloidal suspension due to the esterification of hydroxyl groups in C6-position by sulfate ions [391]. The ionic species are then removed through an ion exchange resin with the exception of the H^+ counter ions associated with the sulfate groups on the CNC surfaces [392,393]. Then, a controlled sonication step is generally applied to improve the CNC dispersion in the final suspension while preventing the desulfation of the sulfate groups on the cellulose surface due to overheating [377,394]. The CNC colloidal suspension is raised up to pH 9 (0.25 M NaOH), promoting the interchange of H^+ counter ions associated with the sulfate groups with Na^+ on the nanocrystal surface, affording CNC with improved thermal stability and with dimensions ranging from 100 to 300 nm in length and between 5 and 10 nm in width [392]. Finally, the CNC suspension undergoes a freeze-drying process to produce the nanoparticles in the form of a powder able to be processed by industrial melt processing techniques [377,378]. However, during the freeze-drying process, the intrinsic hydrophilic nature of cellulose combined with the nanoparticle's large specific surface area and/or high surface energy promote irreversible agglomeration, leading to the formation of strong additional hydrogen bond interactions between CNC [388,395].

As was already commented cellulose can also be obtained through a bottom-up method by biosynthesis from glucose through acetic acid-producing bacteria, i.e., *Acetobacter xylinus, Acetobacter hansenii and Acetobacter pasteurianus* [375,396] in a culture medium, containing carbon and nitrogen sources under controlled temperature and pH conditions for several days, leading to microfibrils based on aggregates of relatively straight, continuous and dimensionally uniform (2 to 4 nm in diameter) glucan chains, which finally bundle to form microbial cellulose ribbons, known as bacterial nanocellulose (BNC) with a diameter less than 100 nm [388].

Concerning the revalorization of food waste, the production of bacterial cellulose from the kombucha beverage industry has gained considerable attention during the last few years due to the kombucha's relatively high cellulose-generating capacity [383,384,397,398]. Kombucha beverage results from the fermentation of sugared infusions by a symbiotic community of bacteria and yeast (SCOBY). These aerobic microbial strains, mainly from the *Acetobacter Xylinum* family, in the kombucha culture produce the broth fermentation of a cellulosic membrane as a by-product during the infusion fermentation that keeps growing while sucrose is still available, often leading to the formation of multiple separate layers [380].

Many reports focus on the study of culture methods, fermentation techniques, media formulations, and bioreactor designs. Another strategy is the design of culture media using sugared wastes added with alcohols or organic acids at low concentrations to activate cellulose synthesis [398,399]. In addition, it was reported that there is an increase in cellulose yield production due to a synergic metabolism of yeast and bacteria, and this effect is present during kombucha fermentation. The kombucha symbiotic community of bacteria and yeast (SCOBY) ecosystem involves cooperative and competitive interactions but predominates the symbioses between microorganisms [399]. Regarding bioreactor selection, static cultures with large surfaces promote cellulose production and prevent shear stress to cells during nanofiber exportation. The surface area is directly linked to the availability of oxygen in the growth medium as dissolved oxygen, and the surface-to-volume ratio (S/V) determines the aeration conditions [398,399]. In this regard, Ramirez Tapia et al., 2023 studied the optimization of BC production during the fermentation of kombucha in yerba mate infusion. During the study, the evaluated variables were culture volume (mL): 50, 150, and 250; and initial sucrose concentration (g/L) 20, 100, and 200. Response variables were broth pH, film production (g/L), soluble solids (% d.m.), total sugar consumption (g/L), and yield conversion (g/g). results demonstrated that a lower level of volume and higher sucrose content favored the formation of the film [397].

With the aim to improve the mechanical, thermal, and barrier properties of other polymers, cellulose, and its derivatives have been widely blended with other biopolymeric matrices, producing polymer blends, composites, and nanocomposites with enhanced properties shown in Figure 21 using a range of years from 1943 up to June 2024.

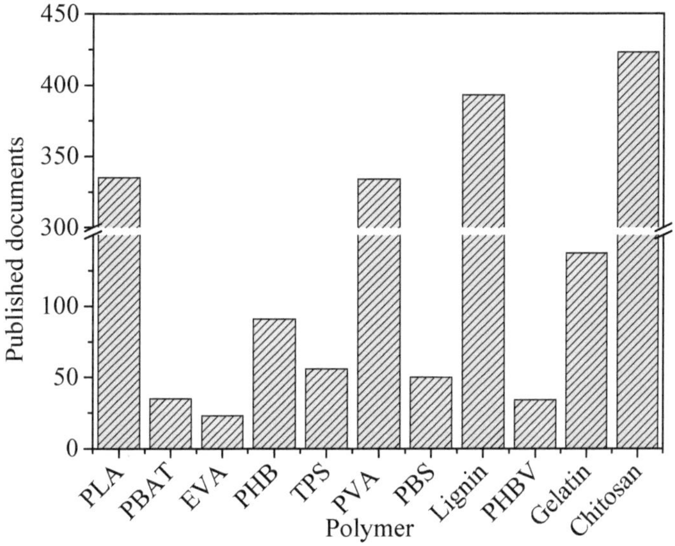

Figure 21. Number of publications of cellulose blends (Scopus Source up to June 2024).

At the industrial level, as already mentioned, cellulose-based blends, composites, and nanocomposites are mainly processed by melt-processing approaches to improve the physical properties of the final material formulation over a wide range through a readily available in the plastic manufacturing industry, simple and cost-effective processing technology. The melt processing of a polymeric matrix reinforced with cellulose tends to produce cellulose particle aggregation, which in turn leads to their phase-separation from the matrix, with a consequent reduction of the overall properties of the composite and/or nanocomposite [20,400]. As the major challenge in the production of nanocomposites is to achieve a homogeneous dispersion of the nanoparticle into the polymeric matrix, to promote the good dispersion of cellulose particles into the melt-processed polymeric matrix, lowering the surface energy of the cellulose particles is required and thus, cellulose micro and/or nanoparticles are frequently physically or chemically modified [388,395].

As lignocellulosic materials start to degrade close to 220 °C are ideal candidates to reinforce biopolymeric matrices, including biopolyesters such as semicrystalline PLA (with melting temperature between 130 and 180 °C [18,401]) and PHAs (with T_m between 165 and 190 °C [402]. Thus, CNC has been successfully introduced in PLA [18,392,403] and PHAs (PHB-CNC [404] and PHBV-CNC [405]) matrices leading to nanocomposites with improved thermal and mechanical properties, while it has also been used as a compatibilization strategy in PLA-PHB based blends [406–408]. In this regard, PLA-PHB blends reinforced with CNC have also been processed by the electrospinning technique, avoiding the PHA matrix thermal degradation [407]. Additionally, other typically used bioplastics at the industrial level, such as TPS (with Tm between 80 and 140 °C [409], have been reinforced with CNC, leading to materials with improved oxygen barrier and mechanical properties [410].

Agüero et al., 2023, used bacterial cellulose from kombucha fermented in wasted yerba mate (KMW) and new yerba mate (KMN) [384] as well as in spent coffee grounds (SCG), leading to spent coffee kombucha (SCK) and coffee kombucha (CK) [383] to reinforce a mechanically recycled (reprocessed 3 times, r3) PLA (r-PLA) and neat PLA matrix and further plasticized with a citrate ester (acetyl tributyl citrate) or maleinized linseed oil

(MLO). The plasticized r3-PLA-ATBC matrix was then loaded with KMW and KMN in 1 and 3 wt%, while PLA was loaded in 3 and 5 wt% with SCK and CK. The strategy used to obtain bacterial cellulose in the form of powder is described in Figure 22.

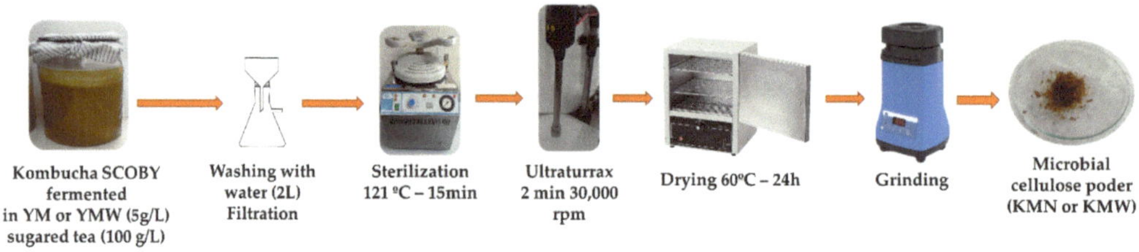

Figure 22. The classical process to obtain kombucha microbial cellulose. Reprinted from Agüero et al. [384] under Creative Commons CC BY license.

The results obtained in the study suggested the potential interest in recycled and bio-based materials using food waste, which means the revalorization of food residues and their industrial application in food packaging and/or agricultural films [383,384].

Finally, the obtention of nanocomposites by a solution casting process is briefly commented on. As it has been shown, the main drawback of cellulose for its use as reinforcement in polymer matrices is its strong hydrophilic character, which means that common solvents cannot be used in the casting method. Extensive research has been conducted trying to find the more appropriate solvent, and several solvent systems have been studied: LiCl/N,N-dimethylacetamide(DMAc) [411]; $Ca(SCN)_2$/water [412], LiOH/urea/water [413], chloroform/N, N-dimethylformamide (DMF) [414] and ionic liquids. Although solution casting is, in general, a simple process, the incompatibility of cellulose with polymer matrices requires the use of these non-common solvents, making the obtention of composites a major challenge. Besides, some of these solvents are not recommended due to their toxicity.

Cellulose is widely chemically modified by functionalization of its hydroxyl groups, which typically involves etherification and esterification, aimed at creating new desired characteristics, tuning the solubility, or in order to improve properties. The majority of cellulose derivatives are commercially available, and some typical examples of chemically modified cellulose are cellulose acetate (CA), methyl cellulose, carboxymethyl cellulose (CMC), cellulose nitrate, and cellulose sulfate [381]. Likewise, to improve the compatibility of nanocellulose with other polymeric matrices when producing polymeric nanocomposites, nanocellulose is typically grafted with specific groups, including the formation of esters, ethers, carbamates, and siloxanes [415]. The relatively reactive surface of nanocellulose allows for "grafting onto" and "grafting from" strategies (Figure 23). The "grafting onto" strategies are based on the attachment of previously synthesized polymer chains with reactive end-groups onto modified hydroxyl groups of the nanocellulose surface [374]. But nanocellulose is more frequently chemically modified by using the "grafting from" strategy, in which a polymer is directly grown in situ at the nanocellulose surface from the reactive hydroxyl groups, which act as initiating sites for ring-opening polymerization [374]. Navarro-Baena et al. studied the thermally-activated shape memory behavior of polyurethanes reinforced with modified cellulose nanocrystals by grafting PLLA onto their surface [416]. Peltzer et at. modified the CNC surface by grafting from L-lactide by ring-opening polymerization with the aim of improving the compatibility of nanocrystals and hydrophobic polymer matrices [417]. Sessini et al. studied the effect of the addition of polyester-grafted-cellulose nanocrystals on the shape memory properties of biodegradable PLA/PCL nanocomposites [418]. Mujica-García et al. studied the effect of the addition of functionalized cellulose nanocrystals on PLA melt-spun fibers [419]. Peponi et al. functionalized CNC by grafting from both PCL and PLA on the surface of the CNC by bio-catalysis by using an enzyme as catalyzer [420].

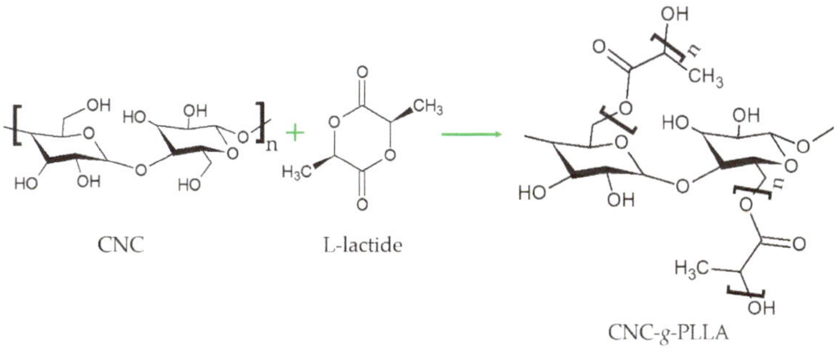

Figure 23. CNC functionalization with PLLA by "grafting from" reaction.

Additionally, CNC can be easily physically modified by simply adding surfactants to the CNC suspension that offer both hydrophobic and hydrophilic end-groups. The CNC adsorbs the hydrophilic tails of the surfactant on its surface, and the hydrophobic head offers a non-polar surface able to lower the surface tension of the CNC, which in turn provides a better nanoparticle with non-polar polymeric matrices. Recently, Bovi et al. used bacterial cellulose and rice husk cellulose as reinforcement of PLA by means of the application of three strategies for the incorporation of the filler: direct mixing of dried milled nanocellulose and PLA, masterbatch by solvent-casting of native nanocelluloses followed by melt compounding and finally, masterbatch by solvent-casting of acetylated nanocelluloses followed by melt compounding. Masterbatches of PLA/nano celluloses were useful for the incorporation of the fillers into processing equipment due to a reduction of the hornification of cellulose during drying. Indeed, acetylation of nanocelluloses improved the dispersion of the filler. This study demonstrated that filler loading played an important role in mechanical properties, such as increasing the stiffness of the films while reducing the translucency of the materials [421]. In Figure 24, the number of publications of cellulose nanocrystals with different polymeric matrixes is reported.

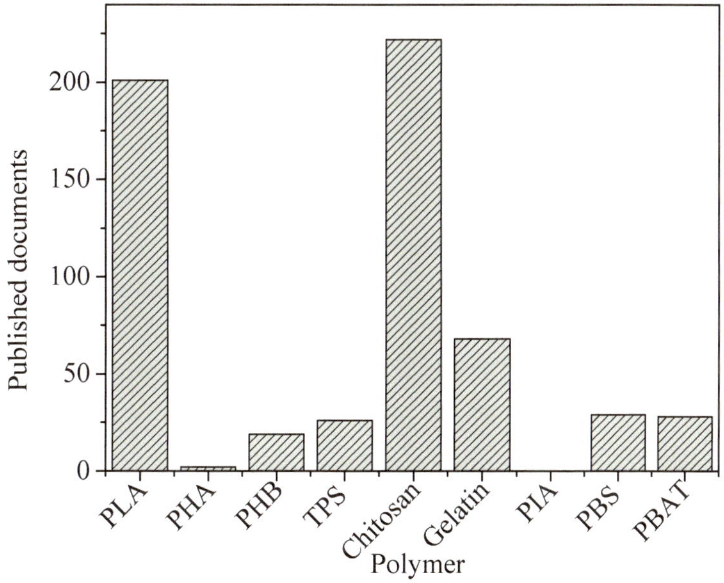

Figure 24. Number of publications of cellulose nanocrystals (Scopus Source up to June 2024).

7. Lignin

Lignin, together with cellulose and hemicellulose, is the major component of the secondary cell walls present in vascular plants. Lignin usually represents between 15% and 30% of the total dry weight and provides rigidity, water-impermeability, and resistance against microbial attack. However, lignin is usually present as a by-product of the bioethanol and wood pulp processing industries [422]. It is either discarded as waste or used as fuel [423].

Lignin is a complex aromatic polymer derived from three monoligols, namely p-cumaryl, coniferyl, and sinapyl alcohols. In the polymeric form, the monolignols become individual units of p-hydroxyphenyl (H), guaiacyl (G), and syringyl (S), which are present in varying amounts and connected through several different possible C–O or C–C bonds forming a highly crosslinked and branched structure, Figure 25. The proportion of H, G, and S depends on the plant species and origin [424]. However, a small percentage of lignin building blocks are not derived from traditional monolignols, such as hydroxycinnamyl alcohols. These non-traditional building blocks include dihydroconiferyl alcohol units, as well as units with aryl conjugated carbonyl/carboxyl groups, such as vanillin, cinnamic aldehydes (conifer and sinapaldehyde), vanillic acid, and the cinnamic acids (ferulic and sinapic acid). In some lignin, acylation occurs with p-hydroxybenzoate (poplar species) or with p-coumarate (grasses) [425].

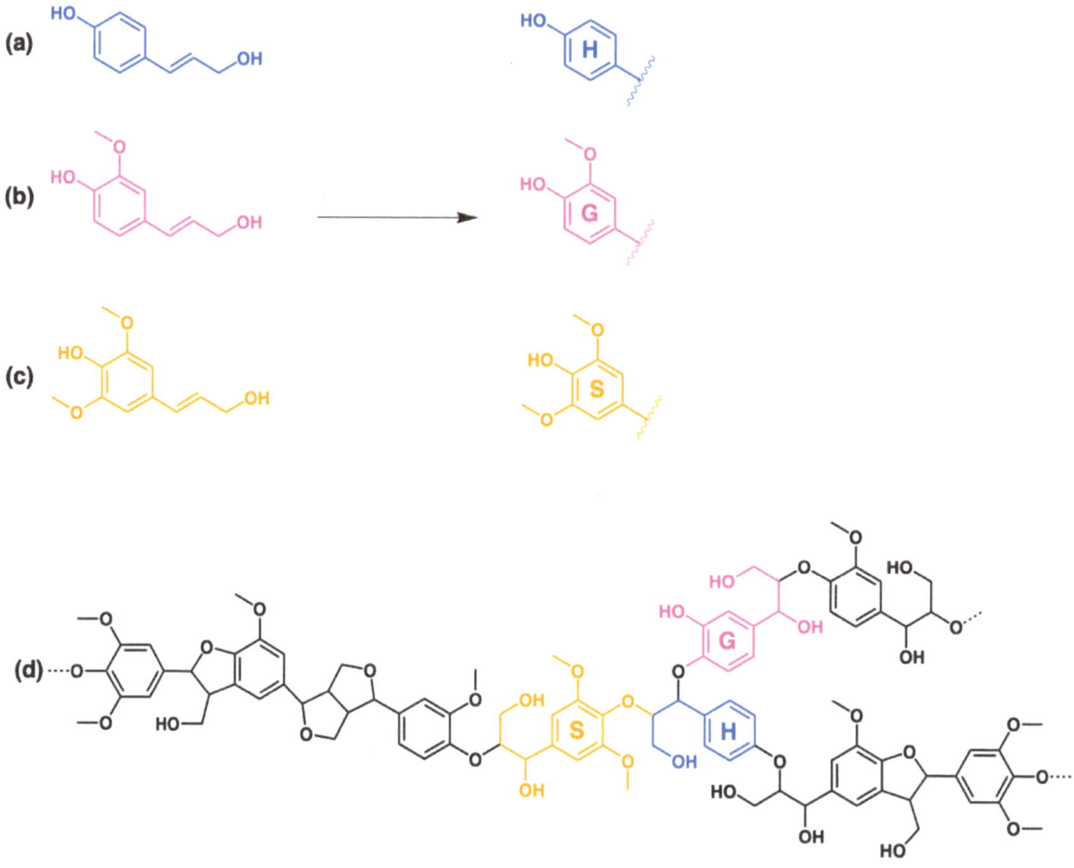

Figure 25. Monoligols present on the structure of lignin: (**a**) p-cumaryl alcohol and p-hydroxyphenyl; (**b**) coniferyl alcohol and guaiacyl; (**c**) sinapyl alcohol and syringyl and (**d**) lignin structure.

Lignification is a dynamic, flexible process reinforcing cell walls according to the different needs (water conduction, mechanical support, defense) of the plant during its life [426].

During the S1 and S2 developmental stages of the secondary wall, monolignols are initially integrated into the middle lamella of the primary cell wall. Subsequently, incorporation into the secondary wall primarily occurs after the S3 stage.

Lignification of the vessel wall, cell corner, and middle lamella begins in the early stages of xylem differentiation. In contrast, lignification of fiber secondary walls is most active during the middle and late stages.

High-resolution microautoradiography of developing hardwood xylem has revealed that the three monolignol units are incorporated at different stages of cell wall formation. During the earliest stages of S1 layer formation, H (p-hydroxyphenyl) units are primarily incorporated in the cell corner and middle lamella. The deposition of G lignin continues from early to late stages, while S lignin is mainly deposited in the middle and late developmental stages, particularly in the secondary fiber wall, after the formation of the inner layer of the secondary wall has begun [427].

The main sources of technical lignin are pulp and paper mills and emerging cellulosic biorefineries, as well as agricultural and forestry residues. The annual production of industrial lignin is about 1.8×10^9–2×10^{10} tonnes, which is mainly obtained as a by-product from the pulp and paper industry and emerging cellulosic biorefineries, while a cellulosic ethanol plant can produce 100,000–200,000 tonnes of lignin/year [428].

Many extraction methods have been used to isolate lignin from the vegetal, but not one allows for obtaining it in its native form (protolignin). Therefore, pretreatment is necessary for the valorization of the different compounds present in the matrix. The traditional pretreatment of lignocellulose aims to eliminate the hemicellulosic barrier surrounding cellulose and disrupt the lignocellulosic structure. This enhances cellulose accessibility and enzymatic hydrolysis, primarily for bioethanol production. To effectively use lignocellulosic compounds, it is crucial to select the optimal pretreatment method, considering the specific requirements of each compound and its corresponding final product. The main pretreatment methods can be divided into three major groups: physical, physicochemical, chemical, and biological, as shown in Figure 26 [429].

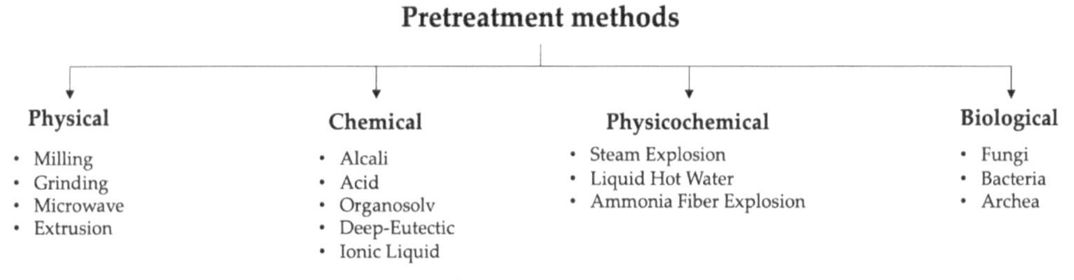

Figure 26. Different pretreatment methods for lignocellulosic biomass.

These methods have their advantages even though. For example, alkaline processes attack or disrupt ether bonds, like β-O-4′. The most relevant technical lignins for material applications are lignosulfonates (LS), kraft lignin (KL), organosolv lignin (OSL), and soda lignin (SL) [430]. For this reason, the combination of different techniques is a good alternative in order to extract lignin in a similar way as proto lignin can and to make the process more sustainable. Nunes da Silva et al. used a combination of an acid extraction with two different acids, in this case, phosphoric acid and sulfuric acid, followed by an alkaline delignification of the cellulignin resulting from the acid treatment [431].

Liu et al. used microwave-assisted with a ternary deep eutectic solvent composed of formic acid, lactic acid, and choline chloride. This method showed better yield and purity of the recovered lignin as well as its molecular weight, dispersity, thermal stability, and antioxidant capacity [432].

– Lignocelulosic nanoparticles extraction

Lignocellulosic nanoreinforcements, have gained huge interest mainly because of their high availability. In this regard, Arrieta et al. [433] developed a simple and green method to produce lignocellulosic nanoparticles from yerba mate (Ilex paraguariensis, a typical South American infusion) waste. The lignocellulosic nanoparticles were produced by an aqueous extraction process, avoiding the use of acids and/or organic solvents. In brief, the yerba mate waste was dried after its consumption and further subjected to an aqueous extraction process (in a flask with distilled water, heated up to 100 °C under reflux for 60 min and under vigorous magnetic stirring). The obtained solution was filtered off twice (filter paper Whatman Grade 541: 22 μm particle retention), frozen, and then freeze-dried to obtain a powder. Figure 27 shows a schematic representation of the procedure used to obtain lignocellulose nanoparticles from yerba mate waste granules using an aqueous extraction process.

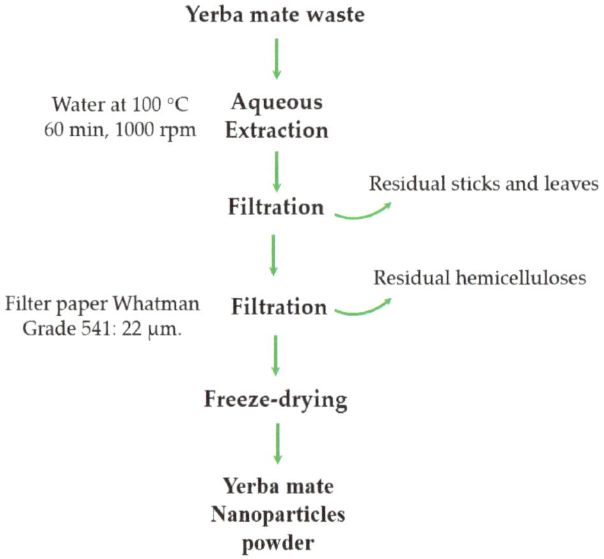

Figure 27. Schematic representation of the production of lignocellulosic nanoparticles from yerba mate wastes through an aqueous extraction procedure.

These lignocellulosic nanoparticles have been used to reinforce the PLA matrix, allowing the obtained materials to have improved thermal resistance and mechanical performance [433].

– Lignin nanoparticles extraction

Lignin nanoparticles (LNPs) can be obtained from various resources following different chemical/physical strategies [434]. In this sense, Gupta et al. [435] have synthesized LNPs by means of the acid precipitation method, leading to individual and randomly distributed nanoparticles. For that, lignin is suspended in distilled water and stirred for 2 h at room temperature, and then the temperature is varied between 50 and 95 °C for 4 h. Once the lignin suspension reaches the desired temperature, a 50% sodium hydroxide solution and/or an ammonia hydroxide (25%) solution are added and maintained under stirring for 2 h (the pH varies in the range of 7.5–12). Then, a formaldehyde solution (37%) is added to the reaction mixture, and the obtained LNPs are further precipitated at pH 2 (HCl, 2%), separated by centrifugation, washed, and finally dried. Two alternative methods to obtain lignin nanoparticles with very different stability upon change of pH have been compared by Frangville et al. [436]. One of them leads to lignin nanoparticles that are stable at low pH and based on the acidic precipitation of lignin from a high pH aqueous solution. The other one, based on the precipitation of low sulfonated lignin from an ethylene glycol solution through a diluted acidic aqueous solution, leads to very stable lignin nanoparticles over a wide range of pH. Lignin nanoparticles have widely been used to reinforce biopolyesters such as PLA, showing that low amounts of nanolignin (1 wt%) can be processed by melt-extrusion achieving an improvement of the mechanical properties while higher amounts lead to nanolignin aggregation [437]. Although nanolignin was less used to reinforce PHAs, there are some works in which it is evident that it is a useful nanoparticle to improve the thermal stability of PHB [438].

Lignin is a promising material for additive manufacturing due to its good mechanical properties, which gives lignin-based products good strength and durability. However, lignin has a complex structure, high molecular weight, huge steric hindrance, and low reactivity, which severely limits its applications in 3D printing materials. Researchers have focused on the depolymerization or modification of lignin to improve its reactivity [439].

Even though lignin has some interesting properties as its use in additive processing in photo-curing 3D printing material could be a photosensitive prepolymer, as can be the case of the study carried out by Nisha et al. in which lignin extracted from biomass sources was used for the synthesis of an ionic liquid-lignin epoxy prepolymer [440]. The most used 3D printing techniques for lignin processing are those based on fusion. In this study, Prakash et al. developed lignin-acrylonitrile butadiene styrene through extrusion, which was 3D printed through the FDM technique [441]. In the study by Nguyen et al., a blend of nylon and lignin was obtained through extrusion with different weight ratios of 40, 50, and 60% [442]. Or the study carried out by Tanase-Opedal et al. in which a blend of PLA/lignin with different ratios of lignin was produced by extrusion and then 3D printed through the FDM technique [443].

Due to the variance in the physicochemical properties depending on the method chosen for the extraction, several parameters have to be taken into consideration in the extrusion process:

- Structure: syringyl lignin melts easier compared to guanicyl lignin, this is due to the availability of guanicyl on the second one which can easily form condensed structures with C_5-position of p-hydroxyphenyl. This forms β-5′ and 5-5′ linkages, which limits molecular rotation [444].
- Aryl-ether bonds (β-O-4′): It has been shown that a higher ratio of aryl-ether linkages, specifically β-O-4′, increases thermal mobility [445].
- Decomposition temperature: The greater its value, the more volatile the compound formed as a consequence of degradation during the extrusion process, which can compromise the final structure of the product [444].
- Glass transition temperature (T_g): Glass transition temperature is related to free polymer chain movements; for this reason, lower glass transition temperatures facilitate polymer chain movement in extrusion. In the case of lignin, a higher concentration of condensation products has, as a consequence, a reduction of free volume, increasing glass transition temperature [445].

The increase in the compatibility of lignin with other polymer (Figure 28) matrixes is a good approach in order to enhance its mechanical and thermal properties. This strategy has been under research since 1976. In order to achieve this, several chemical modifications have been researched in the last few years using different strategies.

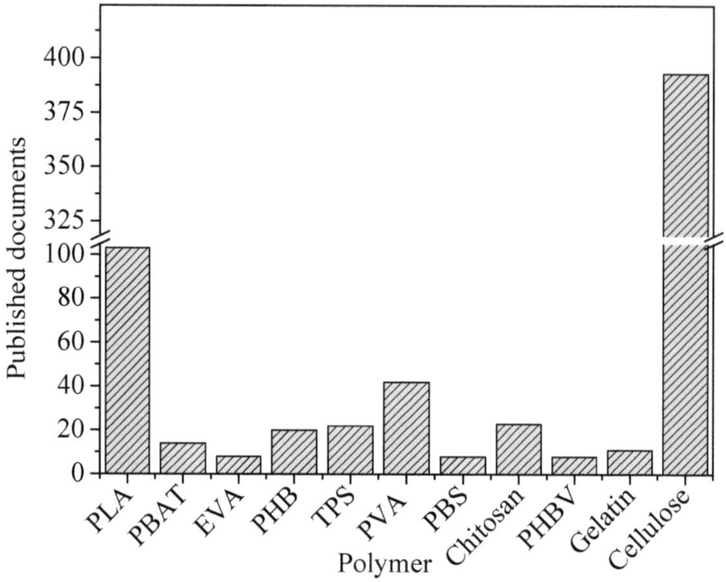

Figure 28. Number of publications of lignin blends (Scopus Source up to June 2024).

(a) Fragmentation of lignin. Due to the multiple low molecular weights in which lignin can be degraded, the use of lignin as raw material due to this fragmentation of this structure to obtain molecules with high added value like vanillin or quinines is an interesting approach for the valorization of lignin residues. Diverse techniques have been used in order to achieve this purpose, such as pyrolysis, which is based on the thermochemical decomposition of the lignin by heating biomass at 500 °C, oxidation which aims to obtain phenolic derivates using different oxidant molecules as can be metallic oxides, air, and oxygen which also permits to conserve the aromatic rings giving aldehydes as a result or gasification which is a process that converts lignin into gases, the gasification of lignin involves a series of reactions, including hydrolysis, polymerization, hydrogenation, water-gas shift, steam reforming, and methanation. These reactions are influenced by various factors such as temperature, reaction time, pressure, concentration, and catalysts [446].

(b) Synthesis of new chemical active sites. The insertion of new reactive sites can solve some of the limitations and increase the range of applications of the material. Lignin is a complex biopolymer with a variety of functional groups. Due to this reason, different strategies have been implemented, such as sulfomethylation/sulfonation, amination, hydroxylation, alkylation/dealkylation, nitration, and halogenation [447]. These mechanisms are shown in Figure 29.

(c) Functionalization of hydroxyl groups: Due to the high presence of hydroxyl groups, which are the most reactive functional groups and can affect the chemical reactivity of the newly formed material. Modifications on hydroxyl groups can lead to the formation of polyol derivatives of lignin; the most common modifications are shown in Figure 30 [446].

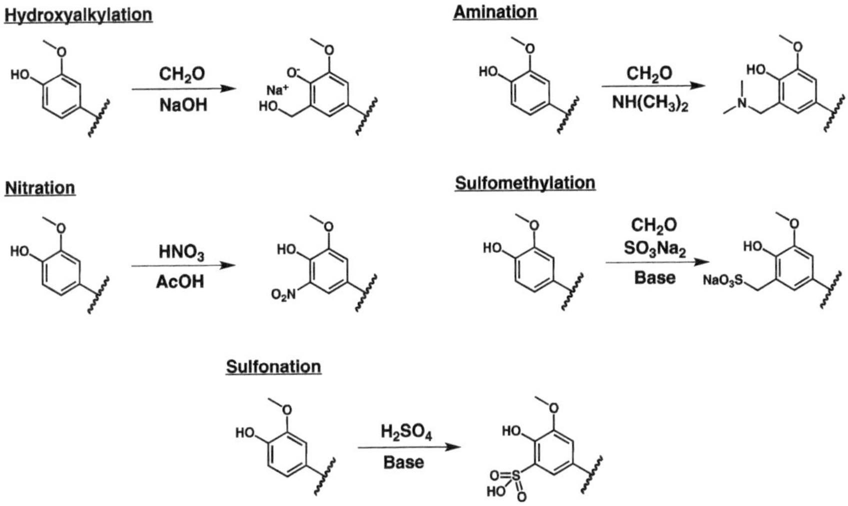

Figure 29. Different mechanisms for the insertion of new chemical active sites on lignin.

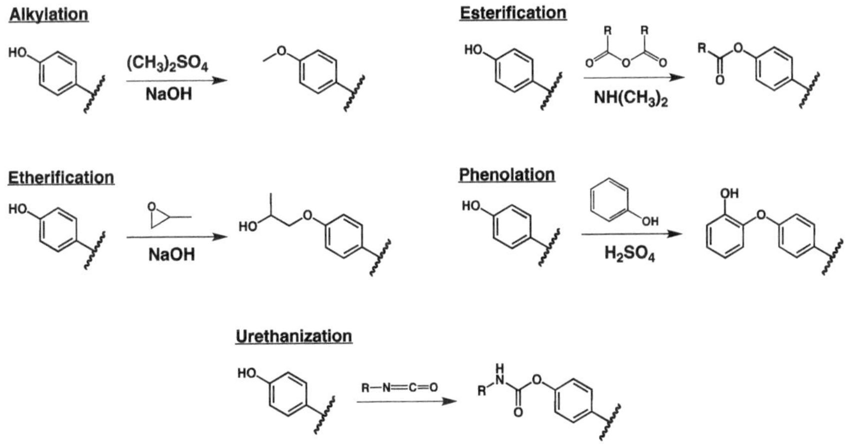

Figure 30. Different mechanisms for the functionalization of hydroxyl groups on lignin.

Lignin is the most abundant source of carbon in the soil after cellulose. Lignin degradation can play a significant role in enhancing Earth's biofuel resources and serve as an alternative to the harsh technologies used in the paper and pulp industry [448]. Several mechanisms have been studied for lignin degradation, with bacterial and fungal degradation being the most significant.

– Fungal degradation

Several fungal species are widely spread in hardwood forests, and the principal example of a wood-degradation fungal species is white-rot fungi. They are called this because when wood decays, they cause the tree to appear bleached and light in color [449]. These microorganisms are capable of degrading lignin along with other components present in wood, such as cellulose.

White-rot fungi produce several types of extracellular oxidative enzymes. These enzymes include laccases and high redox potential ligninolytic peroxidases such as lignin peroxidase, manganese peroxidase, and versatile peroxidase. Their production is triggered by secondary metabolism conditions in response to nutrient depletion [450].

Various species of brown-rot fungi have been reported to have lignin degradation properties, although these properties are inferior to the ones observed in white-rot fungi [451]. This is because, unlike white-rot fungi, brown-rot fungi primarily target cellulose for degradation, but at the same time, they also modify the chemical structure of lignin via demethylation [452].

– Microbial degradation

Even though fungal degradation of lignin is the most studied process for lignin degradation due to the high bioavailability of fungi, it is well-known that several types of bacteria, such as *Streptomyces viridosporus* which was confirmed to degrade lignin in the 1980s by the report of Ramachandra et al. [453], are able to produce CO_2 from lignocellulosic residues.

It was not until 2011 that Ahmad et al. characterized two DNA strains from Rhodococcus jostii RHA1 with peroxidase activity and sequence similarity to open reading frames in other lignin-degrading microbes [454]. Further dye-decolorizing peroxidases active for lignin degradation have also been identified in *Amycolatopsis sp. 75iv2* [455], *Pseudomonas fluorescens* [456] and *Thermobifida fusca* [457].

Most of the microbial mechanisms that can degrade lignin are aerobic organisms, even though there have been reported anaerobic organisms [458,459] and facultative anaerobic bacteria [460].

– Enzymes involved in lignin degradation

(a) Fungal peroxidases. All are members of class II peroxidases or plant peroxidases within the superfamily of heme peroxidases. In this class, extracellular lignin peroxidase, manganese peroxidase, and versatile peroxidase are used.

(b) Dye-decoloring peroxidases (DyP-type peroxidase). This type of heme-containing peroxidases has been recently discovered [461]; these peroxidases are inherent to certain types of bacteria, which are capable of degrading lignin along with other compounds [462].

Due to its complex structure and composition, lignin degradation is strongly affected by its nature, reaction temperature, heating rate, and degradation atmosphere, which also affects the temperature domain of degradation, conversion, and product yields [463].

Lignin decomposes at a slower rate than cellulose and hemicellulose components of biomass over a broader temperature range of 200–500 °C. This decomposition presents 3 different peaks, which correspond with β-O-4 linkages at a range between 180 and 310 °C [464], C-C linkages at a range between 310 and 415 °C [465], and β-β linkages at a range from 415 to 500 °C [466].

The degradation of the polymer structure in lignin starts at relatively low temperatures, 200–275 °C, with the main process occurring at around 400 °C, with the formation of aromatic hydrocarbons, phenolics, hydroxyphenolics and guaiacyl/syringyl-type compounds, most of the products having phenolic -OH groups [463].

8. Chitosan

Chitosan is a natural polymer widely used in environmentally friendly applications due to its naturally occurring characteristics and properties. It is a polymer obtained through the extensive deacetylation of chitinous materials, which serves as a fundamental structural component found in the outer shells of marine invertebrates: crabs, prawns, shrimps, krills, lobsters, squids and crawfish, insects: cockroaches, mosquitoes, silkworms, honeybees and microorganism: fungus, yeast and algae [467–470]. In Figure 31, the number of scientific papers from 1983 up to date on chitosan blends is shown.

Chitosan is a cationic polysaccharide with a chemical structure similar to cellulose and chitin. It is primarily composed of β(1-4) linked glucosamine units, with a certain percentage of N-acetyl-glucosamine units (Figure 32).

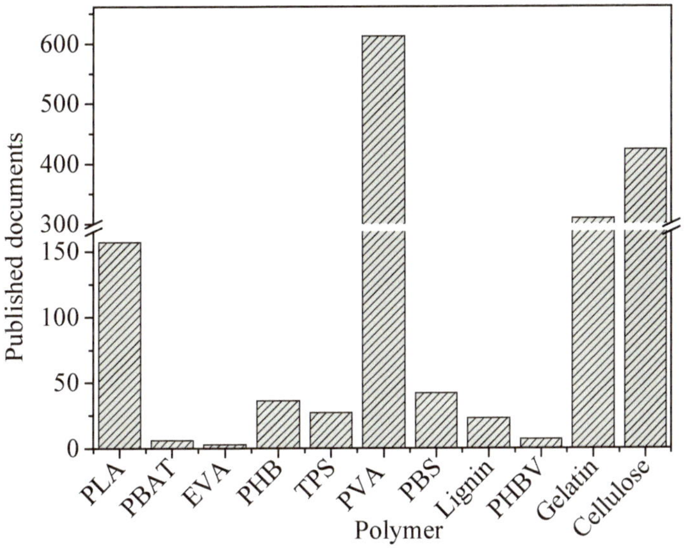

Figure 31. Number of publications of chitosan blends (Scopus Source up to June 2024).

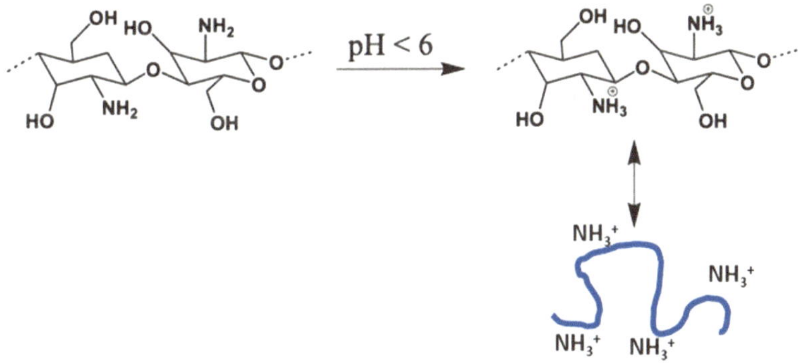

Figure 32. Chemical structure of Chitosan and process of transformation into a polycationic electrolyte.

Chitin is made of acetamido-2-deoxy-β-D-glucose groups connected by β (1 → 4), analogous to the structure of cellulose [471]. Furthermore, it possesses acetamide units at C2. Chitosan is achieved by removing an adequate number of acetyl groups in chitin, which makes the molecule become more soluble in nearly all diluted acids [472,473]. This process is called deacetylation and leads to the discharge of amine groups to form chitosan, which is a heteropolymer with cationic characteristics, with a direct chain comprising β (1 → 4) connected 2-acetamido-2-deoxy-β-D-glucopyranose and 2-amino-2-deoxy-β-D-glucopyranose. It is important to highlight that not all of the acetamide groups in the chitosan structure are replaced by amine groups. In fact, the degree of deacetylation can range from 30% to 95%, depending on the source and production method. This deacetylation degree has a critical influence on solubility, chemical reactivity, and biodegradability [474,475].

Chitosan is soluble in dilute acidic aqueous solvents due to the addition of H+ to -NH_2 groups. Its oligomers also dissolve under physiological conditions. However, because of hydrogen bonding and hydrophobic interactions, even completely protonated chitosan tends to aggregate in acidic conditions. The hydrophobic activity is based on the collaboration between the core polysaccharide backbone and N-acetyl groups at C2 [476,477]. The transformation of -NH_2 groups into protonated amine groups (NH_3^+) within the chitosan polymer chain renders it polycationic and highly alkaline, which is responsible for its exceptional properties, including metal chelation, solubility in various media, the ability to bind with microbial cells, film-forming capacity, viscosity, optical properties, and structural characteristics [478–480].

The main source of Chitosan production is the extraction from shells of marine invertebrates, which are constituted by about 15–40% chitin, 20–40% protein, and 20–50% calcium and magnesium carbonate accompanied by other minor components such as lipids and astaxanthin [473]. The traditional way to produce Chitosan from Chitin using chemical methods includes 3 steps: deproteinization, demineralization, and deacetylation. Furthermore, the industrial production of Chitosan includes a fourth step, which is named decolorization (bleaching). Some authors have described that the order of the different steps, as well as the raw material used for Chitosan production, could affect extraction results. For example, for Chitosan obtained from shells with a high content of protein (i.e., shrimp), the extraction should start with deproteinization followed by demineralization [481], but if it is obtained from crab shells, demineralization with acid treatment should precede deproteinization step [482]. Other authors have reported that using both procedures indistinctly there can be obtained good results [483–485].

In the chemical isolation method, minerals and proteins can be eliminated from the structure of chitin by using strong acid and alkali, respectively, resulting in a huge amount of destructive acidic or basic wastewater and severe ecological glitches. Therefore, biological methods can be performed to overcome the disadvantages of the chemical isolation process. The biological methods can be divided into enzymatic reactions and microbial fermentation. The chitin produced by the biological method has a higher molecular weight and crystallinity compared to the chemical method [486,487]. Nevertheless, longer fermentation cycles and expensive enzymes are referenced as the main bottlenecks of the biological process [488].

– Deproteinization

The most common method for this is to use strong bases such as NaOH and high temperatures to render the proteins water-soluble and wash them from the chitin. Sodium hydroxide (NaOH) is the most common base used for deproteination due to its low cost. Reaction temperatures are typically between 65 °C and 100 °C and in treatment methods for 0.5 to 72 h [489]. Concerns about polluting the environment with high concentrations of sodium ions have directed researchers to alternative bases with less toxic cations such as KOH, $NaHCO_3$, Na_2CO_3, K_2CO_3, Na_3PO_4, $Ca(OH)_2$, Na_2SO_3, $CaHSO_3$, and Na_2S or other deproteination methodologies [490].

– Demineralization

This step typically involves exposing crustacean shells to acidic conditions to dissolve shell minerals that are then separated from the residual shell solids. Both mineral and organic acids have been employed for demineralization such as 0.5 to 5 M HCl, H_2SO_4, HNO_3 or CH_3COOH solutions at varying temperatures, from room temperature to 90 °C, and different treatment periods, from 0.5 to 36 h. Hydrochloric acid (HCl) is the most used, though concerns about environmental toxicity have prompted a turn toward organic acids such as citric, acetic, and lactic acids [491].

– Decolorization

This step aims to remove pigments and colors associated with flexible fragments of chitinous materials. Chitin is treated with various reagents such as potassium permanganate and oxalic acid, acetone, ethanol and hydrogen peroxide [492]

– Deacetylation

The degree of deacetylation plays a pivotal role in solubility, chemical reactivity, and biodegradability of chitosan. Typically, to achieve deacetylation, chitin is subjected to treatment with a concentrated alkaline solution, such as 40% to 70% NaOH or KOH, at different temperatures (ranging from room temperature to 70–120 °C) for a duration spanning from 40 min to three days. This process aims to remove acetyl groups from the chitin polymer, ultimately yielding chitosan. The caustic conditions of chemical deacetylation can lead to reductions in the molecular weight of products. Deacetylation reactions tend to last between 30 min and 5 h; some reactions last up to 24 h Longer reaction times, higher temperatures, and higher alkali concentrations correspond to higher degrees of deacetylation at the cost of molecular weight [489,490,493].

Recently, it is forcefully emerging a wide range of processing technologies involving greener solvents that are characterized by lower vapor pressure, higher thermal stability, better bonding abilities and lower toxicity compared with conventionally used solvents [494]. Ionic liquids (ILs), which are organic salts in the liquid state at room temperature (or close), can outperform the disadvantages of chemical and biological-based methods for the efficient extraction of chitin and chitosan [495].

ILs have been widely adopted in chitosan processing for pharmaceutical applications due to their ability to dissolve hydrogen bonding among chains, not affecting the molecular weight [496]. However, they have some disadvantages such as an increased cost, though some of the cost could be mitigated in the recyclability, and their hazardous toxicity and very low biodegradability [497].

As an alternative, deep eutectic solvents (DESs), recognized as the new green solvents, own similar features to ILs [498].

Deep eutectic solvents (DESs) are composed of a hydrogen bond donor and acceptor duo, having solvation characteristics similar to ionic liquids. Initially, DESs featured metal ions. However, the pursuit of reduced toxicity led to the development of natural deep eutectic solvents (NADESs), comprising primary metabolites. NADESs present a more economical and biodegradable alternative for biomass extraction. Within NADESs utilized for chitin extraction, combinations of choline chloride (ChCl) and organic acids predominate due to their minimal toxicity and capacity to eliminate calcium carbonate. Betaine serves as another prevalent hydrogen bond acceptor, complemented by various organic molecules like glycerol, ethylene glycol, and urea, acting as hydrogen bond donors [499,500].

Chitosan has found widespread application in medicine, cosmetics industry, agriculture, food processing, and paper industry, among others [501–504]. This is attributed to its remarkable properties, including biocompatibility, biodegradability, non-toxicity, antimicrobial properties, hydrophilicity, and its potential as an anticholesterolemic agent [505,506]. Furthermore, chitosan possesses distinct physicochemical characteristics, including its insolubility in water and typical organic solvents and solubility in weak acids, it has poor mechanical properties and limited water resistance. Despite these limitations, chitosan offers an excellent oxygen barrier, is readily available on a large scale, and features functional groups that facilitate chemical modifications [507].

Chitosan films have been widely studied for food-related applications for the quality preservation of a variety of food since it possesses antimicrobial activity.

Solution-casting method has been widely used to prepare chitosan-based coatings for many food applications with particular interest in fruits as it is an edible polymer. However, chitosan films have relatively high water vapor permeability and this is why lipid materials (i.e.,: fatty acids, fatty acid esters, and essential oils) are frequently added to chitosan-based formulations [508,509]. In this context, Vargas et al. have developed antimicrobial chitosan coatings with oleic acid in 1, 2 and 4% (v/v) for strawberry protection and found that chitosan-based coating with increasing amounts of oleic acid leads to better preservation of the fruit color, water resistance and mechanical performance during storage, while protected strawberries against fungal infection [508]. Similarly, Perdonés et al. developed chitosan-lemon essential oil edible coatings to extend strawberry shelf-life and observed that when using less amount of 3 wt% of lemon essential oils the chitosan coatings were able to induce fungal decay, while although showing changes in the sensory attributes maintaining pleasant changes [510].

Chitosan has been also widely used in blend formulations or as a reinforcing filler. In this regard, Bonilla et al. have developed PLA-Chitosan composites by melt-extrusion. They used different chitosan particle sizes (the starting one of 715 μm and then grinding this one to a 180 μm) and blended PLA with different amounts of both chitosan particles of 5 and 10 wt%, and observed that the chitosan presence leads to PLA-chitosan films that were able to protect minced meat samples from coliform microorganisms (about 2 log CFU/g) [511].

Arrieta et al. have prepared electrospun plasticized PLA-PHB blends with ATBC and further reinforced them with chitosan particles (chitosan particles between 50 and 100 μm) in 1 and 5 wt%. Although chitosan-loaded electrospun PLA-PHB-ATBC fibers presented some structural defect such as beads, the chitosan-loaded mats showed increased mechanical resistance at the expense of the reduction of the elongation at break [512].

9. Gelatin

Gelatin is an affordable and abundant water-soluble protein obtained from partial or full hydrolysis of collagen. It's a colorless and tasteless substance whose properties vary based on the extraction process [513].

Bovine and pig skins, bones, and demineralized hooves are the primary sources of gelatin extraction. Among these, one of the primary sources of gelatin is pig skin. The entire amount of gelatin produced is made up of 46% porcine, 29.4% bovine hides, and 23.1% pork and cattle skeletons [514]. While pig skin is the most often used raw material for commercial gelatin production. Fish, mammals, or birds killed using the halal method can also be used as an alternate raw material [515]. Fish and poultry by-products are becoming more and more popular as alternative gelatin sources. Due to their high collagenous protein content, poultry wastes including blood, viscera, feet, and bone are increasingly widely utilized in the food industry [515].

Structure of gelatin is shown in Figure 33. Gelatin approximately contains moisture 5–8%, ash 1.9–8.9%, total lipids 1.6–4.5%, and total protein 80–91% [516].

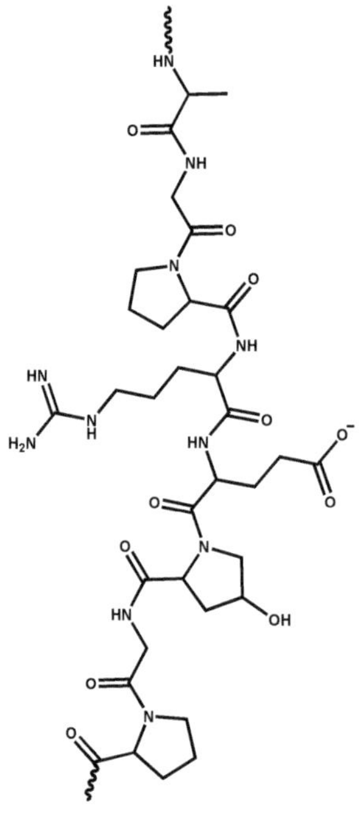

Figure 33. Chemical structure of gelatin.

Protein gelatin is transparent, brittle, and odorless. Its color can range from dull to slightly yellow. It appears as a tasteless sheet, flake, or powdered form. Glycerol, hot water, and acetic acid can dissolve it, but organic solvents cannot [515]. Gelatin is an amphoteric substance. Depending on the pre-treatment procedure, under acid and alkaline pre-treatment conditions, two types of gelatin are obtainable that are commercially known as type-A (isoelectric point at pH~8–9) and type-B (isoelectric point at pH~4–5) [517].

Gelatin produced from various animal byproducts has strong mechanical and barrier properties. The triple helical structure of the gelatin protein provides physical strength, and the different amino acids present in the gelatin structure easily absorb UV radiation and protect packaged food products from oxidative damage as well as water-binding ability, film- and foam-forming capacity, and emulsifying tendency [518].

Even though gelatin presents interesting properties for its use as material for the packaging industry, it also has some limitations for its use as raw material for this type of application, these films become fragile under drying conditions, thereby becoming rigid, brittle, and exhibit poor elongation at break. Due to the presence of polar groups on its structure, these films are highly hydrophilic, for this reason, these films present poor protection against highly moisture environments, in which they may dissolve, swell or disintegrate [519]. To overcome these drawbacks, the blending of gelatin with other matrixes has been widely explored from 1968 up to date as shown in Figure 34.

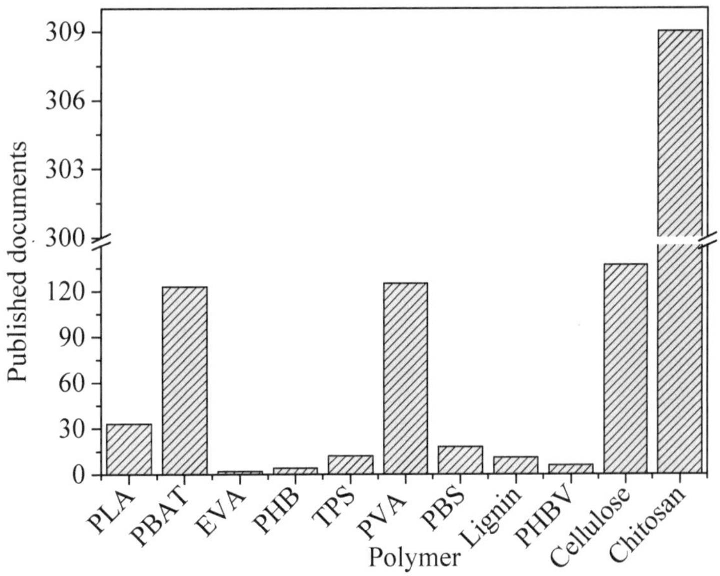

Figure 34. Number of publications of gelatin blends (Scopus Source up June 2024) (1968–2024).

– Solvent casting

As previously mentioned, solvent casting is a technique that allows us to obtain polymeric films on a laboratory scale. In the case of gelatin, several studies have been performed in order to evaluate the properties of this polymeric matrix for its use in the packaging sector usually mixed with other bio-based polymers as PLA.

In this study, Hosseini et al. [520] successfully processed a multi-layer film based on PLA and gelatin, first the PLA layer was formed through this processing technique after which the gelatin film was cast above this first PLA film, the film-forming solution was formed dissolving 3 g of gelatin in 100 mL of water for 30 min and then heated at 45 °C for 30 min while stirring and 30% *w*/*w* glycerol.

Kurek et al. [521] produced a blend of chitosan and gelatin through this technique adding to the film-forming solution gallic acid as an antioxidant agent. Their goal was to try to produce an antioxidant material capable of storing olive oil, as a result, they obtained a material capable of reducing the amount of volatiles observed as markers of olive oil degenerative oxidation.

Zong et al. [522] developed a binary blend of starch and gelatin loaded with sweet potato anthocyanins for its use in packaging as well as a monitor of the freshness of mushrooms. They observed a change on the color of the films in function of the state of oxidation of the mushrooms as well as an enhancement of the mechanical properties of the film.

- Electrospinning

Electrospinning is an interesting technique for its use in the development of nanofiber for the packaging sector, the nanometric scale of the fibers permits the films to strongly attach to the microbial cells, destroy their cell structure, and show antimicrobial activity in the polymeric system [518].

Duan et al. [523] produced gelatin/chitosan nanofibers loaded with curcumin as a multifunctional additive, the effect of the addition of curcumin into the matrix was evaluated observing that an addition of 0.2% m/m of curcumin presented high sensitivity of colorimetric behavior to ammonia (within 3 min) improving also the antimicrobial and antioxidant activity if the matrix.

Thermal degradation is influenced by the fact that gelatin has relatively small amounts of cysteine and cystine, which contain sulfhydryl and disulfide groups, and cannot form disulfide bonds during heating; therefore, heating results in shorter polypeptide chain lengths and a decrease in particle size. An 80% decrease in gelatin particle size was observed after 1 h of heating, indicating that a high degree of degradation of gelatin had occurred [524].

At high temperatures, gelatin presents a single-step degradation in the range of temperatures from 280 °C to 400 °C [525].

10. Poly(Butylene Succinate)

Poly(butylene succinate) (PBS) (Figure 35) is a biodegradable aliphatic polyester constituted by succinic acid and 1,4-butanediol that can be obtained both from petroleum and renewable sources as starch, cellulose and glycerol [526]. PBS production involves a first esterification of succinic acid and 1,4-butanediol obtaining oligomers, followed by the polycondensation of produced oligomers to obtain higher molecular weights [527].

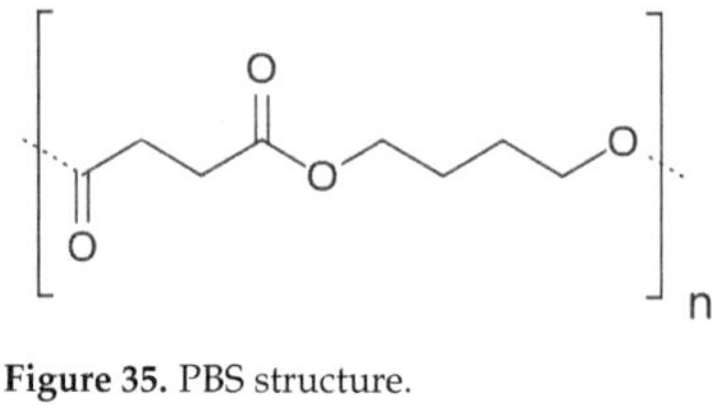

Figure 35. PBS structure.

PBS is characterized by a Tm around 110 and 115 °C and Tg between −40 and −10 °C and good mechanical properties (elastic modulus 300 MPa and tensile strength of 20 to 60 MPa). Furthermore, PBS shows higher ductility respect to PLA and PHAs [528].

Besides the good thermo-mechanical properties, PBS can be processed with common processing techniques and is susceptible to biodegradation [529].

Degradation of PBS was studied in activated sludge, compost but also other methods as hydrolytic and enzymatic degradation [527]. Mao et al. [530] isolated *Fusarium* sp. FS1301, which is a PBS-degrading strain, by screening oil-polluted soil, and investigated the behavior of PBS, observing the degradation of both crystalline and amorphous regions.

Nelson et al. [531] studied the degradation of (13C-) labeled PBS in soil in order to evaluate the formation of 13CO2 derived from PBS mineralization quantifying almost 65% of added 13C. Lyyra et al. [532] studied the hydrolytic degradation of PLA/PBS blends reinforced with bioactive glass, observing a fast loss of mechanical properties and molecular weight. Due to its biodegradability, PBS finds applications such as active packaging [533], potential biomedical applications [534], drug delivery [535], electronic devices [536], etc. In order to obtain PBS-based materials suitable for different applications, Ostheller et al. [537] fabricated PBS nanofibers through melt electrospinning, which can be used as stents, sutures and in tissue engineering. With the aim of developing a material for food packaging applications, Xu et al. [538] produced PBS-based films reinforced with chitin whiskers (CHW) and nanocrystalline cellulose (NCC) by melt extrusion, observing an increase in mechanical and barrier properties, with a reduction in oxygen transmission rate and water vapor transmission from 737.7 to 280 cc/m^2/day and from 83.8 to 49.4 g/m^2/day, respectively, by adding 3% of NCC. While Rennert et al. [539] produced a fully bio-based composite constituted by bio-based PBS (bioPBS) reinforced with coffee parchment (PMT) with increased antioxidant properties through injection molding.

PBS presents low flexibility. Indeed, copolymers are produced to improve this feature. Copolymerization of succinic acid and 1,4-butanediol with adipic acid leads to the production of poly(butylene succinate-co-adipate) or PBSA [540]. However, PBSA is characterized by poor mechanical properties in terms of Young modulus and tensile strength, but its lower crystallinity and higher flexibility lead to a high degradation rate with respect to PBS [541,542]. PBSA finds application in food packaging as biodegradable active packaging, blended with other polymers such as PLA [543] and poly(propylene carbonate) (PCC) [544], or by fabricating nanocomposites using PBSA usually reinforced with layered double hydroxides (LDHs) to improve the gas barrier performance [545]. A summary of the number of publications from 1971 up to date of works related to PBS blends with other polymeric matrices, Figure 36.

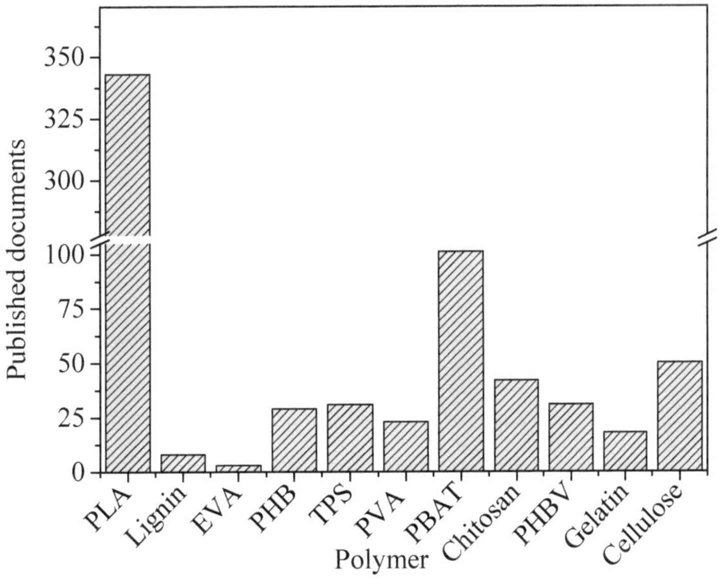

Figure 36. Number of publications of PBS blends (Scopus Source up to June 2024).

11. Poly(Itaconic Acid)

As previously mentioned, nowadays, there is a need for bio-based and bio-degradable polymers with improved mechanical and functional properties; due to this reason, the use of alternative polymers is necessary. In this context, alternative bio-based polymers like polyitaconic acid have attracted the attention of numerous researchers over the past decade. Itaconic acid is the only one that provides the possibility of substituting acrylic acid and methacrylic acid. Furthermore, it can be used as a building block in producing essential polymers, such as resins, fibers, and rubber [546].

Itaconic acid (2-methylenebutane-1,4-dicarboxylic acid), as shown in Figure 37, is a bio-based monomer structurally related to acrylic and methacrylic acid but polymerizes at a significantly slower rate than these monomers. Monomers derived from itaconic acid are asymmetric, having one carboxyl group on the α-carbon atom and another on the β-carbon atom relative to the double bond.

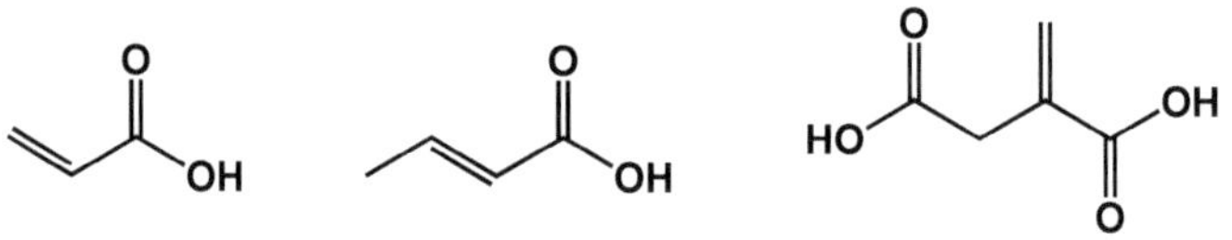

Figure 37. Acrylic, methacrylic, and itaconic acid structure, respectively.

The identification of the biosynthetic pathway of itaconic acid has been known since 1936, when the synthesis of itaconic acid in Aspergillus itaconicus was reported, postulating that the decarboxylation reaction of cis-aconitate was the key process in the biosynthesis of itaconic acid. Later on, Bentley and Thiessen [547] identified the cis-aconitate dehydrogenase (CadA) in *Aspergillus terreus*, confirming that the citric acid from the Krebs cycle is the precursor in the formation of itaconic acid.

Nowadays, *A. terreus* is the strain used for the industrial production of itaconic acid, with glucose and sucrose as the main carbon source [548].

There have been attempts to homopolymerize itaconic acid via free radical polymerization since the late 50 s using potassium peroxodisulfate, dibenzoyl peroxide (DBPO), and azoisobutyric acid nitrile (AIBN) as initiators [549,550]. As a result, polymers with low molecular masses, broad PDIs, uncontrolled branched architecture and low to medium conversion were obtained. Several alternatives have been used in order to solve these downsides as can be the use of cyclic derivatives of itaconic acid [551], diaconate [552], or other derivatives can be singly substituted derivatives of itaconic [553], disubstituted itaconic acid derivatives [554,555].

As mentioned before, the synthesis of poly(itaconic acid) has severe disadvantages which have to be solved, due to this reason most of the processing of poly(itaconic acid) is produced via binary or ternary blends or modification of the polymer, those

- Electrospinning

Diverse combinations of poly(itaconic acid) with different matrixes have been studied for their processing through the electrospinning technique.

Baskan et al. [556] synthesized a copolymer of poly(itaconic acid) and polyacrylonitrile by emulsion polymerization, this copolymer was then dissolved and added a certain amount of $AgNO_3$ as a source of silver nanoparticles. This mixture was then processed through an electrospinning technique using as parameters 0.8 mm needle diameter, 15 kV voltage, 1 mL/h flow rate and 10 m distance between needle tip and collector. The fibers obtained for this copolymer presented a higher diameter than the observed in polyacrylonitrile fibers and presented a reduced cyclization temperature.

Rzayev et al. [557] used a colloidal nanofibre using solution blends of polyvinyl alcohol/octadecylamine montmorillonite, an alternating copolymer of itaconic anhydride with 2-vinyl-1,3-dioxalane and their Ag-carrying complexes for its use as semiconductor electrolytes. In this case, they used the following conditions, voltage of 20 kV, flow rate of 0.7 mL/h and distance of 17 cm between needle tip and grounded collector.

Wilke et al. [558] synthesized a microgel based on poly(N-vinylcaprolactam-co-itaconic acid) for the controlled release of zinc ions in biomedical applications. Subsequently, the microgel was suspended in tetrahydrofuran, and a fixed amount of zinc chloride was dissolved in it. It was then treated with ultrasound for a few minutes and stirred at room temperature for one hour. Subsequently, 2.2 equivalents of sodium hydroxide were added, and the solvent was eliminated by evaporation. The product was washed, centrifuged and freeze-dried. The microgel was combined with PCL in a 7:3 ratio, PCL:composite. The electrospinning parameters employed were a flow rate of 0.3–0.5 mL/h, a distance between the spinneret and the target of 15–20 cm, and an acceleration voltage of 24–26 kV. These fibers presented a regulated zinc ion release during a period of up to five days, it also shows high biocompatibility proving that microfibers do not affect the activity of the human dermal fibroblasts.

- Solvent casting

Nesic et al. [559] prepared a series of films comprising chitosan and poly(itaconic acid) at varying ratios for use as a novel biosensor. These films were prepared by mixing solutions of both polymers and then cast into a petri dish, which was subsequently dried until a constant mass was achieved at 40 °C. The films exhibited superior proton conductivity in comparison to the synthetic sensor Nafion 117, as well as superior water stability.

Işıklan et al. [559] grafted poly(itaconic acid) onto poly(vinyl alcohol) with cerium(IV) ammonium nitrate as an initiator at 45 °C as a separator for acid aqueous solutions. Membranes were prepared by casting technique at 90 °C for 12 h and the cast on a petri dish. The addition of poly(itaconic acid) into the matrix improved the performance of the polymer and showed good permeation results.

Poly(itaconic acid) is well known as a biodegradable polymer, at low molecular weights poly(itaconic acid) is readily degradable in an aqueous medium, even some derivatives from poly(itaconic acid) can biodegrade faster than cellulose-based materials under compost conditions [560].

In relation to thermal degradation, a first degradation is usually observed at 100 °C as a consequence of the loss of moisture, at 170 °C a first degradation step is observed and continues up to 200 °C due to the formation of the polyanhidre, a second decomposition step started at 250 °C due to decarboxylation step of anhydrides formed in high molecular weight poly(itaconic acids) and a last decomposition step at ~450 °C correspond to the scission of the polymeric backbone [561].

12. Vegetable Oils as Plasticizers

Vegetable oils, which can come from oleaginous plants and trees, are found in abundance in all areas of the planet, making them ideal as an alternative raw material for different applications, including their use as a plasticizer, since they are readily available, biodegradable and have low toxicity. These oils are primarily composed of triglycerides, whose structure can be seen in Figure 38. Triglycerides are made up of three fatty acids attached to a glycerol backbone. Most common oils contain fatty acids whose length varies between 14 and 22 carbon atoms, with between 0 and 3 double bonds per fatty acid. The distribution of fatty acids in some common oils can be observed in Table 3 [562].

Triglycerides have active sites that are susceptible to modification through chemical reactions; these sites include the double bonds, allylic carbons, the ester group, and alpha carbon atoms with respect to the ester group. These active sites can be used to introduce other groups into the triglyceride, thus obtaining molecules with greater reactivity. On one hand, the ester group composed of glycerol and fatty acids can be esterified and used as plasticizers [563]. On the other hand, a double bond can be used to modify the triglycerides and obtain molecules with greater reactivity that increase the compatibility of vegetable oils with polymers, suitable for use as plasticizing additive, as can be seen in Figure 38. From the virgin triglyceride (1), maleates (2) can be attached, or the unsaturation can be transformed into epoxy (3). Also, from the epoxidized triglyceride (3), acrylates can be incorporated using acrylic acid (4) [562].

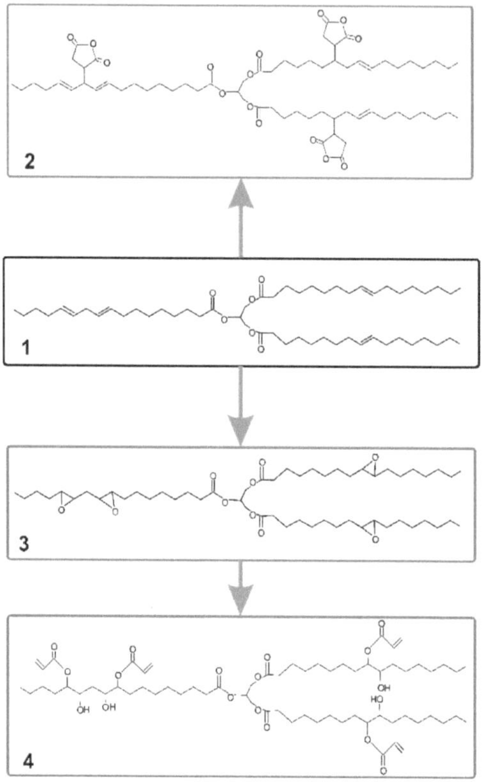

Figure 38. Schematic representation of the triglyceride molecule (**1**) and different chemical modifications: maleinization (**2**), epoxidation (**3**), and acrylation (**4**).

Table 3. Distribution of fatty acids in various vegetable oils, adapted from Khot S. N. et al [562].

Fatty Acid	C:DB [(a)]	Canola	Corn	Cottonseed	Linseed	Olive	Palm	Rapeseed	Soybean	High Oleic [(b)]
Myristic	14:0	0.1	0.1	0.7	0.0	0.0	1.0	0.1	0.1	0.0
Myristoleic	14:1	0.0	0.0	0.0	0.0	0.0	0.0	0.0	0.0	0.0
Palmitic	16:0	4.1	10.9	21.6	5.5	13.7	44.4	3.0	11.0	6.4
Palmitoleic	16:1	0.3	0.2	0.6	0.0	1.2	0.2	0.2	0.1	0.1
Margaric	17:0	0.1	0.1	0.1	0.0	0.0	0.1	0.0	0.0	0.0
Margaroleic	17:1	0.0	0.0	0.1	0.0	0.0	0.0	0.0	0.0	0.0
Stearic	18:0	1.8	2.0	2.6	3.5	2.5	4.1	1.0	4.0	3.1
Oleic	18:1	60.9	25.4	18.6	19.1	71.1	39.3	13.2	23.4	82.6
Linoleic	18:2	21.0	59.6	54.4	15.3	10.0	10.0	13.2	53.2	2.3
Linolenic	18:3	8.8	1.2	0.7	56.6	0.6	0.4	9.0	7.8	3.7
Arachidic	20:0	0.7	0.4	0.3	0.0	0.9	0.3	0.5	0.3	0.2
Gadoleic	20:1	1.0	0.0	0.0	0.0	0.0	0.0	9.0	0.0	0.4
Eicosadienoic	20:0	0.0	0.0	0.0	0.0	0.0	0.0	0.7	0.0	0.0
Behenic	22:0	0.3	0.1	0.2	0.0	0.0	0.1	0.5	0.1	0.3
Erucic	22:1	0.7	0.0	0.0	0.0	0.0	0.0	49.2	0.0	0.1
Lignoceric	24:0	0.2	0.0	0.0	0.0	0.0	0.0	1.2	0.0	0.0
Average DB/ triglyceride		3.9	4.5	3.9	6.6	2.8	1.8	3.8	4.6	3.0

[(a)] Carbon number (C): Double bonds number (DB); [(b)] Genetically modified soybean oil with high oleic acid content (DuPont).

Epoxidized soybean oil (ESBO) is one of the most studied oils as a plasticizer in various bioplastics. For example, when used as a plasticizer for PLA, it increases the elongation at break by 63% using 9 wt% of ESBO [564]. However, migration of ESBO in PLA films is observed after 30 days of storage under ambient conditions, leading to a slight increase in the glass transition temperature (T_g) and a decrease in elongation at break and break energy. Despite this, the film exhibited suitable behavior for flexible packaging applications even after an extended storage period [91]. Another alternative is the use of epoxidized soybean oil methyl ester (ESOME), which manages to increase the elongation at break to over 700% using 40 wt%, although the maximum strength is significantly reduced. A balance between ductile and strong properties is achieved with 10 wt% of ESOME [565]. This oil has also been studied as a plasticizer in other polymers such as PBS, where using 5 wt% of ESBO results in a fifteenfold increase in elongation at break and also reduces the T_g [566] or for the production of plasticized ethyl cellulose films, with concentrations ranging from 15–25 wt% of ESBO, achieving a balance in mechanical properties and better thermal stability, as well as non-flammability [567]. In addition to its plasticizing effect, it also has a compatibilizing effect in blends, as is the case with PLA/PBAT blends [568] or ternary blends of PLA/PCL/PHB [281].

Another chemical modification that can be applied to soybean oil is acrylation through peroxidation. Acrylated epoxidized soybean oil (AESO) also shows a plasticizing effect on PLA, as there is an improvement in ductile properties such as elongation at break and impact energy absorption, along with a reduction in the glass transition temperature (T_g) with contents ranging from 2.5 to 7.5 wt% in PLA. These properties favor the use of these combinations for applications in rigid materials with energy absorption capacity [569]. Like ESBO, AESO also exhibits a compatibilizing effect, in this case between PLA filled with orange peel flour. This helps to enhance the interfacial adhesion between the filler and the polymer matrix, improving the overall properties of the composite material [570]. Ozonized soybean oil (OSBO) has also been explored. This modification involves applying an ozone stream to a mixture of soybean oil and ethylene glycol, which causes the double bonds to react, thereby increasing the oxygen content in the triglycerides. Although OSBO enhances the ductile properties of PLA, such as impact resistance and elongation at break [571], its plasticizing effect is less significant compared to using ESBO or AESO.

Epoxidized linseed oil (ELO) has also shown significant interest in its plasticizing properties. Comparing the plasticizing properties that ELO and ESBO provide to PLA, it is observed that both improve ductile properties such as elongation at break and impact resistance, and both oils decrease the glass transition temperature (T_g). Additionally, they also exhibit a compatibilizing effect between PLA and sisal fibers, as they improve the adhesion between the fibers and the PLA thanks to the epoxy groups. This compatibilizing effect is more pronounced when using ELO, as it contains more epoxy groups per triglyceride than ESBO [572]. The plasticizing and compatibilizing effect of epoxidized linseed oil (ELO) has been demonstrated with various cellulosic fillers and PLA, including hazelnut shell flour or kraft pulp fiber. This use of ELO enhances the mechanical properties of PLA composites by improving the integration and dispersion of the cellulosic material within the PLA matrix. The presence of ELO aids in bridging the interaction between the bio-based polymer and the natural fibers, thereby improving the overall structural integrity and performance of the composite materials [92,573,574].

Maleinized linseed oil (MLO) also shows a plasticizing effect on PLA, increasing the elongation at break and reducing the glass transition temperature. With just 5 phr (parts per hundred resin, equivalent to 4.8 wt%) of MLO, the elongation at break of PLA increases by 40.5%, and this increase grows with higher concentrations of MLO. This indicates that MLO can significantly enhance the flexibility and workability of PLA, making it more suitable for applications requiring more ductile properties [93]. Maleinized linseed oil (MLO) also exhibits a compatibilizing effect, for example, as a compatibilizing/plasticizing agent in blends of PLA with TPS (30 wt%). This combination shows a notable increase in elongation at break and impact resistance, with an increase of over 500% compared to unblended

PLA, when using 6 phr (5.7 wt%) of MLO. This significant enhancement highlights MLO's effectiveness in improving the mechanical properties of PLA/TPS blends, making it a valuable additive for producing more durable and flexible bioplastic materials [106]. The compatibilizing effect of epoxidized linseed oil (ELO) and maleinized linseed oil (MLO) is also evident in lignin/natural fiber composites, such as those found in Arboform®. Both oils improve impact resistance, tensile and flexural strength, and elongation at break. The best properties are observed when using 5 wt% of MLO. This enhancement can be attributed to the functional groups in MLO and ELO, which improve the interaction between the lignin matrix and natural fibers, leading to composites with better mechanical properties and more effective stress transfer across the materials [575].

Other modified vegetable oils studied as potential plasticizers in bioplastics include epoxidized cottonseed oil (ECSO). By using up to 10 wt% of ECSO as a plasticizer in PLA (polylactic acid), the elongation at break can be increased by more than 1100% compared to pure PLA. Additionally, this increases impact resistance and lowers the glass transition temperature (T_g), showing very low migration in n-hexane [576]. Maleinized cottonseed oil (MCSO) also has a good plasticizing effect on PLA, increasing elongation at break. Its maximum plasticizing effect is achieved with a 7.5 wt% addition, and it also improves the barrier properties of PLA, making it ideal for food packaging applications [58]. Carbonell-Verdu et al. reported better plasticizing properties of ECSO compared to MLO, possibly due to the control of the CSO modification process since it was carried out in a laboratory, while MLO was an industrial product.

Epoxidized sunflower oil (ESO) has also been investigated as a plasticizer for PLA, showing that the ductile properties of PLA are improved but without achieving as good ductile properties as with ESBO. Moreover, while optimal properties are achieved with 9 wt% of ESBO, with ESO, it is necessary to increase above 15 wt% to achieve an optimum increase in properties [577]. Chia seed oil (CO), which has a high iodine value (above 190 g I2/100 g oil), also appears to be an interesting oil. For this reason, Dominguez-Candela et al. studied epoxidized chia seed oil (ECO) as a plasticizer in PLA, achieving an improvement in ductile properties, with a 700% increase in elongation at break compared to PLA [578]. Maleinized oils from Brazil nut seed (MBNO) and hemp seed (MHO) have also been studied as plasticizers for PLA, improving elongation at break by 643% with 7.2 phr of MBNO and 771% with 10 phr of MHO compared to PLA [579]. Epoxidized corn oil (ECornO) and maleinized corn oil (McornO) have also been used as plasticizers/compatibilizers in blends of PLA (75 wt%) and PHB (25 wt%), with both showing an improvement in the ductile properties of the system. The best ductile properties were achieved using 10 wt% of ECornO or 5 wt% of McornO, improving the elongation at break of the PLA-PHB blend by more than 400% and by more than 800% for neat PLA [282]. Other modified oils with interesting characteristics for use as plasticizers include epoxidized karanja oil (EKO) and epoxidized jatropha oil (EJO). Using 5 wt% of EKO relative to PLA achieves a balance between tensile strength (only a 9% reduction), elongation at break (improved by 77%), and impact resistance (improved by 42%) [580]. Using just 3 wt% of epoxidized jatropha oil (EJO) relative to PLA results in a 7000% increase in elongation at break and a decrease in the glass transition temperature (T_g) by 3 °C [94]. MVOs appear to be promising candidates as plasticizing agents in bio-based polymers, particularly with PLA, as can be seen in Table 4.

In the other hand, MVOs also have a compatibilizing effect between PLA blends and PLA composites with cellulosic fibers. It is noteworthy that both effects are enhanced when the oils are composed of fatty acids with more double bonds, and the compatibilizing effect is better with maleinized oils than with epoxidized oils.

Table 4. Comparative properties of PLA plasticized with different modified vegetable oils.

MVO	Blending Ratio		Properties				Ref.
	PLA Grade	Plasticizer (wt%)	Tensile Strength (MPa)	Elongation at Break (%)	Impact Energy (kJ/m^2)	Tg (°C)	
ESBO	N.I.	- 9	42.0 57.7	4.0 6.5	- -	- -	[564]
ESBO	2003D	- 10	≈66 ≈49	≈7.5 ≈20	16 ≈32	55.5 51.6	[572]
ESBO	2002D	- 20	13 10	5 34	- -	64 59	[577]
ESOME	2003D	- 10	51 29	2 368	- -	71.3 51.2	[565]
AESO	6201D	- 10	66 60	4.1 10.6	19.5 35.1	62.8 59.4	[569]
OSBO	4043D	- 10	≈55 ≈40	≈5 ≈10	≈2 ≈2.2	59.8 50.1	[571]
ELO	2003D	- 10	≈66 ≈51	≈7.5 ≈15	≈16 ≈35	55.5 49.0	[572]
MLO	6201D	- 13	≈64 ≈40	≈8 ≈74	30.9 50.3	65.4 60.8	[93]
ECSO	6201D	- 10	63.7 49.5	9 110	30.8 38.4	66.7 60.7	[576]
MCSO	6210D	- 7.5	≈47 ≈38	≈4 ≈16	- -	63.0 61.6	[58]
ESO	2002D	- 20	13 11	5 16	- -	64 60	[577]
ECO	2003D	- 10	45 ≈34	8.0 64.5	37.1 68.3	62 56.8	[578]
MBNO	N.I.	- 7	35.8 ≈48	7.4 52	35.5 ≈46	61 58.7	[579]
MHO	N.I.	- 9	35.8 ≈48	7.4 61	35.5 ≈49	61 56.9	[579]
EKO	6201D	- 5	46.5 43.3	8.3 14.7	36.6 51.9	61.2 58.3	[580]
EJO	4042D	- 3	58 ≈42	5.4 388	- -	62.8 59.9	[94]

N.I.: Not Indicated. ≈: approximate value, obtained from figures.

13. Future Perspectives

Nowadays, the environmental society concern and the European strategies to reduce the consumption of single-use plastics will open new opportunities to introduce biodegradable plastic formulations into the market.

In fact, according to the last report of Plastics Europe [581] the production of plastics from bio-based feedstock still accounts for only 1% of the plastics produced in Europe. However, bio-based feedstock availability is steadily increasing and plastics production from bio-based feedstock has significant potential growth. Moreover, increasing efforts in research and in the managing of plastic wastes are required to increase the circularity of polymer-based materials. From the research point of view there is a need to improve the effectiveness of the different polymerization processes to produce bio-based polymers including the wide variety of the potential biomasses available and to develop greener processing strategies. On the other hand, the management of plastic wastes must increase the efforts for the collection and differentiation of the different plastic fractions with particular attention to the separation of the biodegradable fraction to be used in compost production. In fact, following Plastic Europe recommendations [581], with mounting concerns about food waste's environmental impact, compostable plastics present a promising solution. By supporting the quality of collected bio-waste, certified compostable products mitigate contamination of this type of waste by contributing to more effective waste management.

Bio-waste collection bags, fruit labels, coffee capsules, and food-soiled packaging are all suitable applications for compostable plastics. Standardization plays a pivotal role in validating biodegradability claims. There is an increasing call for uniform requirements for industrial composting procedures, testing, and certifications. Such standards will assist plastic producers and waste managers to achieve a better sorting and recycling of bio-waste and align with shared sustainability objectives [581].

Not only materials but also processing technologies should be considered from an environmentally-friendly point of view. For example, if we want to analyze the electrospinning technique in a circular economy point of view, we have to remark that from one side, normally, no temperature is needed to electrospun the polymeric solutions with a consequence reduction of the energy consumption and from the other side we have to focus the attention on solvents. In this regard, two main approaches can be considered as significant challenges for the near future: the use of more environmentally-friendly solvents and the intensification of solvent recovery and reuse.

The current state of the art of the most used bio-based and biodegradable polymeric matrices has been reviewed here from a circular economy perspective. Their chemical structure, processing methods and their main properties as well as their degradation processes have been analyzed. Then, the most important strategies to produce blend formulations, composites and nanocomposites as well as the introduction of natural plasticizers to improve their performance for extending their industrial applications is summarized.

Author Contributions: Conceptualization, L.P.; methodology, L.P., D.L., M.P.A. and M.D.S.; data curation, V.O.-C.; V.S., L.P., D.L., M.P.A. and M.D.S.; writing—original draft preparation, V.O.-C.; V.S., P.F.M.-G., Á.A., M.A.P., V.A.M., M.P.A., J.S.-T., C.P., M.D.S., G.R.C., J.M.K., D.L., and L.P.; writing—review and editing, V.O.-C.; V.S., J.M.K., D.L., M.P.A., M.D.S. and L.P.; supervision, L.P., D.L., M.P.A. and M.D.S.; project administration, L.P., D.L., M.P.A. and M.D.S.; funding acquisition, L.P., D.L., M.P.A. and M.D.S. All authors have read and agreed to the published version of the manuscript.

Funding: This research was funded by the Agencia Estatal de Investigación (AEI, MICINN, Spain), and Fondo Europeo de Desarrollo Regional (FEDER, EU), PID2021-123753NB-C31, PID2021-123753NA-C32, PID2021-123753NA-C33, and NextGenerationEU MRR TED2021-129335B-C21.

Institutional Review Board Statement: Not applicable.

Data Availability Statement: Not applicable.

Acknowledgments: V.O.-C. acknowledges Spanish government for the grant PRE2020-103097 funded by MCIN/AEI/10.13039/501100011033 and "ESF Investing in your future". The authors want to know MICIN for PID2021-123753NB-C31 and TED2021-129335B-C21 projects and for RED2022-134806-T.

Conflicts of Interest: The authors declare no conflict of interest.

References

1. Balart, R.; Garcia-Garcia, D.; Fombuena, V.; Quiles-Carrillo, L.; Arrieta, M.P. Biopolymers from Natural Resources. *Polymers* **2021**, *13*, 2532. [CrossRef] [PubMed]
2. Bachmann, M.; Zibunas, C.; Hartmann, J.; Tulus, V.; Suh, S.; Guillén-Gosálbez, G.; Bardow, A. Towards Circular Plastics within Planetary Boundaries. *Nat. Sustain.* **2023**, *6*, 599–610. [CrossRef]
3. Nagel, B.; Dellweg, H.; Gierasch, L.M. Glossary for Chemists of Terms Used in Biotechnology (IUPAC Recommendations 1992). *Pure Appl. Chem.* **1992**, *64*, 143–168. [CrossRef]
4. Vert, M.; Doi, Y.; Hellwich, K.-H.; Hess, M.; Hodge, P.; Kubisa, P.; Rinaudo, M.; Schué, F. Terminology for Biorelated Polymers and Applications (IUPAC Recommendations 2012). *Pure Appl. Chem.* **2012**, *84*, 377–410. [CrossRef]
5. Biswas, M.C.; Jony, B.; Nandy, P.K.; Chowdhury, R.A.; Halder, S.; Kumar, D.; Ramakrishna, S.; Hassan, M.; Ahsan, M.A.; Hoque, M.E.; et al. Recent Advancement of Biopolymers and Their Potential Biomedical Applications. *J. Polym. Environ.* **2022**, *30*, 51–74. [CrossRef]
6. Samir, A.; Ashour, F.H.; Hakim, A.A.A.; Bassyouni, M. Recent Advances in Biodegradable Polymers for Sustainable Applications. *npj Mater. Degrad.* **2022**, *6*, 68. [CrossRef]

7. Udayakumar, G.P.; Muthusamy, S.; Selvaganesh, B.; Sivarajasekar, N.; Rambabu, K.; Banat, F.; Sivamani, S.; Sivakumar, N.; Hosseini-Bandegharaei, A.; Show, P.L. Biopolymers and Composites: Properties, Characterization and Their Applications in Food, Medical and Pharmaceutical Industries. *J. Environ. Chem. Eng.* **2021**, *9*, 105322. [CrossRef]
8. Zhong, Y.; Godwin, P.; Jin, Y.; Xiao, H. Biodegradable Polymers and Green-Based Antimicrobial Packaging Materials: A Mini-Review. *Adv. Ind. Eng. Polym. Res.* **2020**, *3*, 27–35. [CrossRef]
9. George, A.; Sanjay, M.R.; Srisuk, R.; Parameswaranpillai, J.; Siengchin, S. A Comprehensive Review on Chemical Properties and Applications of Biopolymers and Their Composites. *Int. J. Biol. Macromol.* **2020**, *154*, 329–338. [CrossRef]
10. Sivakanthan, S.; Rajendran, S.; Gamage, A.; Madhujith, T.; Mani, S. Antioxidant and Antimicrobial Applications of Biopolymers: A Review. *Food Res. Int.* **2020**, *136*, 109327. [CrossRef]
11. Biswal, T. Biopolymers for Tissue Engineering Applications: A Review. *Mater. Today Proc.* **2019**, *41*, 397–402. [CrossRef]
12. Jaya, B.; Brajesh, B.; Antonella, F.; Giovanna, L.D.; Amit, K. Biopolymer: A Sustainable Material for Food and Medical Applications. *Polymers* **2022**, *14*, 983. [CrossRef] [PubMed]
13. Udayakumar, G.P.; Muthusamy, S.; Selvaganesh, B.; Sivarajasekar, N.; Rambabu, K.; Sivamani, S.; Sivakumar, N.; Maran, J.P.; Hosseini-Bandegharaei, A. Ecofriendly Biopolymers and Composites: Preparation and Their Applications in Water-Treatment. *Biotechnol. Adv.* **2021**, *52*, 107815. [CrossRef] [PubMed]
14. Díez-Pascual, A.M. Synthesis and Applications of Biopolymer Composites. *Int. J. Mol. Sci.* **2019**, *20*, 2321. [CrossRef]
15. Aydemir, D.; Gardner, D.J. Biopolymer Blends of Polyhydroxybutyrate and Polylactic Acid Reinforced with Cellulose Nanofibrils. *Carbohydr. Polym.* **2020**, *250*, 116867. [CrossRef]
16. Patel, R.; Trivedi, R.; Raj, M.; Raj, L. *A Brief Review of Polymeric Blends Based on Natural Polymers and Synthetic Thermoplastics Polymers*; Springer International Publishing: Berlin/Heidelberg, Germany, 2024; Volume 78, ISBN 0123456789.
17. European-Bioplastics-Association. *Bioplastics Market Development Update 2023*; European Bioplastics e.V.: Berlin, Germany, 2023.
18. Raquez, J.-M.; Habibi, Y.; Murariu, M.; Dubois, P. Polylactide (PLA)-Based Nanocomposites. *Prog. Polym. Sci.* **2013**, *38*, 1504–1542. [CrossRef]
19. Kumar, A.P.; Depan, D.; Singh Tomer, N.; Singh, R.P. Nanoscale Particles for Polymer Degradation and Stabilization—Trends and Future Perspectives. *Prog. Polym. Sci.* **2009**, *34*, 479–515. [CrossRef]
20. Peponi, L.; Puglia, D.; Torre, L.; Valentini, L.; Kenny, J.M. Processing of Nanostructured Polymers and Advanced Polymeric Based Nanocomposites. *Mater. Sci. Eng. R Rep.* **2014**, *85*, 1–46. [CrossRef]
21. Carraher, C.E., Jr. *Seymour/Carraher's Polymer Chemistry*; CRC Press: Boca Raton, FL, USA, 2003; ISBN 020391130X.
22. Bartolome, M.J.; Bischof, S.; Pellis, A.; Konnerth, J.; Wimmer, R.; Weber, H.; Schwaiger, N.; Guebitz, G.M.; Nyanhongo, G.S. Enzymatic Synthesis and Tailoring Lignin Properties: A Systematic Study on the Effects of Plasticizers. *Polymer* **2020**, *202*, 122725. [CrossRef]
23. Thakur, K.; Thakur, N.; Sharma, N.; Mahajan, K. A Comprehensive Review On Plasticizing Agent. *J. Emerg. Technol. Innov. Res.* **2021**, *8*, a79–a85.
24. Vieira, M.G.A.; Da Silva, M.A.; Dos Santos, L.O.; Beppu, M.M. Natural-Based Plasticizers and Biopolymer Films: A Review. *Eur. Polym. J.* **2011**, *47*, 254–263. [CrossRef]
25. Meeker, J.D.; Sathyanarayana, S.; Swan, S.H. Phthalates and Other Additives in Plastics: Human Exposure and Associated Health Outcomes. *Philos. Trans. R. Soc. B Biol. Sci.* **2009**, *364*, 2097–2113. [CrossRef] [PubMed]
26. Jagadeeswaran, I.; Sriram, H. EU 1907/2006–Registration, Evaluation, Authorisation and Restriction of Chemicals. In *Medical Device Guidelines and Regulations Handbook*; Springer: Berlin/Heidelberg, Germany, 2022; pp. 237–260.
27. Bilck, A.P.; Müller, C.M.O.; Olivato, J.B.; Mali, S.; Grossmann, M.V.E.; Yamashita, F. Using Glycerol Produced from Biodiesel as a Plasticiser in Extruded Biodegradable Films. *Polímeros* **2015**, *25*, 331–335. [CrossRef]
28. Bocqué, M.; Voirin, C.; Lapinte, V.; Caillol, S.; Robin, J. Petro-based and Bio-based Plasticizers: Chemical Structures to Plasticizing Properties. *J. Polym. Sci. A Polym. Chem.* **2016**, *54*, 11–33. [CrossRef]
29. Tian, H.; Liu, D.; Yao, Y.; Ma, S.; Zhang, X.; Xiang, A. Effect of Sorbitol Plasticizer on the Structure and Properties of Melt Processed Polyvinyl Alcohol Films. *J. Food Sci.* **2017**, *82*, 2926–2932. [CrossRef]
30. Feng, G.; Hu, L.; Ma, Y.; Jia, P.; Hu, Y.; Zhang, M.; Liu, C.; Zhou, Y. An Efficient Bio-Based Plasticizer for Poly (Vinyl Chloride) from Waste Cooking Oil and Citric Acid: Synthesis and Evaluation in PVC Films. *J. Clean. Prod.* **2018**, *189*, 334–343. [CrossRef]
31. Mitrus, M.; Wojtowicz, A.; Moscicki, L. Biodegradable Polymers and Their Practical Utility. In *Thermoplastic Starch: A Green Material for Various Industries*; Wiley: Hoboken, NJ, USA, 2009; pp. 1–33.
32. Telis, V.R.N. *Biopolymer Engineering in Food Processing*; CRC Press: Boca Raton, FL, USA, 2012; ISBN 1439844941.
33. Songtipya, L.; Kalkornsurapranee, E.; Songtipya, P.; Sengsuk, T.; Promsung, R.; Chuaybamrung, A.; Johns, J. Enhancing Thermo-Mechanical Properties of Thermoplastic Starch/Natural Rubber Blends Through the Synergistic Combination of PEG and Modified Natural Rubber. *J. Polym. Environ.* **2023**, *32*, 1868–1878. [CrossRef]
34. Mukherjee, C.; Varghese, D.; Krishna, J.S.; Boominathan, T.; Rakeshkumar, R.; Dineshkumar, S.; Rao, C.V.S.B.; Sivaramakrishna, A. Recent Advances in Biodegradable Polymers–Properties, Applications and Future Prospects. *Eur. Polym. J.* **2023**, *192*, 112068. [CrossRef]
35. Hamad, K.; Kaseem, M.; Yang, H.W.; Deri, F.; Ko, Y.G. Properties and Medical Applications of Polylactic Acid: A Review. *Express Polym. Lett.* **2015**, *9*, 435–455. [CrossRef]

36. Ilyas, R.A.; Sapuan, S.M.; Harussani, M.M.; Hakimi, M.Y.A.Y.; Haziq, M.Z.M.; Atikah, M.S.N. Polylactic Acid (PLA) Biocomposite: Processing, Additive manufacturing and advanced applications. *Polymers* **2021**, *13*, 1326. [CrossRef]
37. Zhao, X.; Liu, J.; Li, J.; Liang, X.; Zhou, W.; Peng, S. Strategies and Techniques for Improving Heat Resistance and Mechanical Performances of Poly(Lactic Acid) (PLA) Biodegradable Materials. *Int. J. Biol. Macromol.* **2022**, *218*, 115–134. [CrossRef] [PubMed]
38. Jamshidi, K.; Hyon, S.H.; Ikada, Y. Thermal Characterization of Polylactides. *Polymer* **1988**, *29*, 2229–2234. [CrossRef]
39. Kalish, J.P.; Aou, K.; Yang, X.; Hsu, S.L. Spectroscopic and Thermal Analyses of A′ and α Crystalline Forms of Poly(l-Lactic Acid). *Polymer* **2011**, *52*, 814–821. [CrossRef]
40. Chen, X.; Kalish, J.; Hsu, S.L. Structure Evolution of A′-Phase Poly(Lactic Acid). *J. Polym. Sci. B Polym. Phys.* **2011**, *49*, 1446–1454. [CrossRef]
41. Puchalski, M.; Kwolek, S.; Szparaga, G.; Chrzanowski, M.; Krucinska, I. Investigation of the Influence of PLA Molecular Structure on the Crystalline Forms (α″ and α) and Mechanical Properties OfWet Spinning Fibres. *Polymers* **2017**, *9*, 18. [CrossRef]
42. Colomines, G.; Ducruet, V.; Courgneau, Ć.; Guinault, A.; Domenek, S. Barrier Properties of Poly(Lactic Acid) and Its Morphological Changes Induced by Aroma Compound Sorption. *Polym. Int.* **2010**, *59*, 818–826. [CrossRef]
43. Barkhad, M.S.; Abu-jdayil, B.; Mourad, A.H.I.; Iqbal, M.Z. Thermal Insulation and Mechanical Properties of Polylactic Acid (PLA) at Different Processing Conditions. *Polymers* **2020**, *12*, 2091. [CrossRef]
44. Singha, S.; Hedenqvist, M.S. A Review on Barrier Properties of Poly(Lactic Acid)/Clay Nanocomposites. *Polymers* **2020**, *12*, 1095. [CrossRef]
45. Sangeetha, V.H.; Deka, H.; Varghese, T.O.; Nayak, S.K. State of the Art and Future Prospectives of Poly(Lactic Acid) Based Blends and Composites. *Polym. Compos.* **2018**, *39*, 81–101. [CrossRef]
46. Li, X.; Lin, Y.; Liu, M.; Meng, L.; Li, C. A Review of Research and Application of Polylactic Acid Composites. *J. Appl. Polym. Sci.* **2023**, *140*, e53477. [CrossRef]
47. Bikiaris, N.D.; Koumentakou, I.; Samiotaki, C.; Meimaroglou, D.; Varytimidou, D.; Karatza, A.; Kalantzis, Z.; Roussou, M.; Bikiaris, R.D.; Papageorgiou, G.Z. Recent Advances in the Investigation of Poly(Lactic Acid) (PLA) Nanocomposites: Incorporation of Various Nanofillers and Their Properties and Applications. *Polymers* **2023**, *15*, 1196. [CrossRef] [PubMed]
48. Mulla, M.Z.; Rahman, M.R.T.; Marcos, B.; Tiwari, B.; Pathania, S. Poly Lactic Acid (Pla) Nanocomposites: Effect of Inorganic Nanoparticles Reinforcement on Its Performance and Food Packaging Applications. *Molecules* **2021**, *26*, 1967. [CrossRef] [PubMed]
49. Li, G.; Zhao, M.; Xu, F.; Yang, B.; Li, X.; Meng, X.; Teng, L. Synthesis and Biological Application of Polylactic Acid. *Molecules* **2020**, *25*, 5023. [CrossRef] [PubMed]
50. de Albuquerque, T.L.; Marques Júnior, J.E.; de Queiroz, L.P.; Ricardo, A.D.S.; Rocha, M.V.P. Polylactic Acid Production from Biotechnological Routes: A Review. *Int. J. Biol. Macromol.* **2021**, *186*, 933–951. [CrossRef] [PubMed]
51. Garlotta, D. A Literature Review of Poly (Lactic Acid). *J. Polym. Environ.* **2019**, *9*, 63–84. [CrossRef]
52. Singhvi, M.S.; Zinjarde, S.S.; Gokhale, D.V. Polylactic Acid: Synthesis and Biomedical Applications. *J. Appl. Microbiol.* **2019**, *127*, 1612–1626. [CrossRef]
53. Burgos, N.; Tolaguera, D.; Fiori, S.; Jiménez, A. Synthesis and Characterization of Lactic Acid Oligomers: Evaluation of Performance as Poly(Lactic Acid) Plasticizers. *J. Polym. Environ.* **2014**, *22*, 227–235. [CrossRef]
54. Castro-Aguirre, E.; Iñiguez-Franco, F.; Samsudin, H.; Fang, X.; Auras, R. Poly(Lactic Acid)—Mass Production, Processing, Industrial Applications, and End of Life. *Adv. Drug Deliv. Rev.* **2016**, *107*, 333–366. [CrossRef]
55. Martin, O.; Avérous, L. Poly(Lactic Acid): Plasticization and Properties of Biodegradable Multiphase Systems. *Polymer* **2001**, *42*, 6209–6219. [CrossRef]
56. Courgneau, C.; Domenek, S.; Guinault, A.; Avérous, L.; Ducruet, V. Analysis of the Structure-Properties Relationships of Different Multiphase Systems Based on Plasticized Poly(Lactic Acid). *J. Polym. Environ.* **2011**, *19*, 362–371. [CrossRef]
57. Burgos, N.; Martino, V.P.; Jiménez, A. Characterization and Ageing Study of Poly(Lactic Acid) Films Plasticized with Oligomeric Lactic Acid. *Polym. Degrad. Stab.* **2013**, *98*, 651–658. [CrossRef]
58. Carbonell-Verdu, A.; Garcia-Garcia, D.; Dominici, F.; Torre, L.; Sanchez-Nacher, L.; Balart, R. PLA Films with Improved Flexibility Properties by Using Maleinized Cottonseed Oil. *Eur. Polym. J.* **2017**, *91*, 248–259. [CrossRef]
59. Mai, F.; Habibi, Y.; Raquez, J.-M.; Dubois, P.; Feller, J.-F.; Peijs, T.; Bilotti, E. Poly(Lactic Acid)/Carbon Nanotube Nanocomposites with Integrated Degradation Sensing. *Polymer* **2013**, *54*, 6818–6823. [CrossRef]
60. Fukushima, K.; Tabuani, D.; Dottori, M.; Armentano, I.; Kenny, J.M.; Camino, G. Effect of Temperature and Nanoparticle Type on Hydrolytic Degradation of Poly(Lactic Acid) Nanocomposites. *Polym. Degrad. Stab.* **2011**, *96*, 2120–2129. [CrossRef]
61. Zaldua, N.; Mugica, A.; Zubitur, M.; Iturrospe, A.; Arbe, A.; Re, G.L.; Raquez, J.-M.; Dubois, P.; Müller, A.J. The Role of PLLA-g-Montmorillonite Nanohybrids in the Acceleration of the Crystallization Rate of a Commercial PLA. *CrystEngComm* **2016**, *18*, 9334–9344. [CrossRef]
62. Fortunati, E.; Armentano, I.; Zhou, Q.; Iannoni, A.; Saino, E.; Visai, L.; Berglund, L.A.; Kenny, J.M. Multifunctional Bionanocomposite Films of Poly(Lactic Acid), Cellulose Nanocrystals and Silver Nanoparticles. *Carbohydr. Polym.* **2012**, *87*, 1596–1605. [CrossRef]
63. Leonés, A.; Salaris, V.; Mujica-Garcia, A.; Arrieta, M.P.; Lopez, D.; Lieblich, M.; Kenny, J.M.; Peponi, L. PLA Electrospun Fibers Reinforced with Organic and Inorganic Nanoparticles: A Comparative Study. *Molecules* **2021**, *26*, 4925. [CrossRef]
64. Lim, L.T.; Auras, R.; Rubino, M. Processing Technologies for Poly(Lactic Acid). *Prog. Polym. Sci.* **2008**, *33*, 820–852. [CrossRef]

65. Bourbigot, S.; Fontaine, G.; Gallos, A.; Bellayer, S. Reactive Extrusion of PLA and of PLA/Carbon Nanotubes Nanocomposite: Processing, Characterization and Flame Retardancy. *Polym. Adv. Technol.* **2011**, *22*, 30–37. [CrossRef]
66. Gálvez, J.; Correa Aguirre, J.P.; Hidalgo Salazar, M.A.; Mondragón, B.V.; Wagner, E.; Caicedo, C. Effect of Extrusion Screw Speed and Plasticizer Proportions on the Rheological, Thermal, Mechanical, Morphological and Superficial Properties of PLA. *Polymers* **2020**, *12*, 2111. [CrossRef]
67. Kühnert, I.; Spörer, Y.; Brünig, H.; Tran, N.H.A.; Rudolph, N. *Industrial Applications of Poly(Lactic Acid)*; Springer: Berlin/Heidelberg, Germany, 2018; Volume 282, ISBN 9783319754581.
68. Aliotta, L.; Sciara, L.M.; Cinelli, P.; Canesi, I.; Lazzeri, A. Improvement of the PLA Crystallinity and Heat Distortion Temperature Optimizing the Content of Nucleating Agents and the Injection Molding Cycle Time. *Polymers* **2022**, *14*, 977. [CrossRef] [PubMed]
69. Pawlak, F.; Aldas, M.; Parres, F.; López-Martínez, J.; Arrieta, M.P. Silane-Functionalized Sheep Wool Fibers from Dairy Industry Waste for the Development of Plasticized Pla Composites with Maleinized Linseed Oil for Injection-Molded Parts. *Polymers* **2020**, *12*, 2523. [CrossRef] [PubMed]
70. Swetha, T.A.; Bora, A.; Mohanrasu, K.; Balaji, P.; Raja, R.; Ponnuchamy, K.; Muthusamy, G.; Arun, A. A Comprehensive Review on Polylactic Acid (PLA)—Synthesis, Processing and Application in Food Packaging. *Int. J. Biol. Macromol.* **2023**, *234*, 123715. [CrossRef] [PubMed]
71. Barletta, M.; Puopolo, M. Thermoforming of Compostable PLA/PBS Blends Reinforced with Highly Hygroscopic Calcium Carbonate. *J. Manuf. Process* **2020**, *56*, 1185–1192. [CrossRef]
72. Wang, G.; Zhao, J.; Wang, G.; Zhao, H.; Lin, J.; Zhao, G.; Park, C.B. Strong and Super Thermally Insulating In-Situ Nanofibrillar PLA/PET Composite Foam Fabricated by High-Pressure Microcellular Injection Molding. *Chem. Eng. J.* **2020**, *390*, 124520. [CrossRef]
73. Jamshidian, M.; Tehrany, E.A.; Imran, M.; Jacquot, M.; Desobry, S. Poly-Lactic Acid: Production, Applications, Nanocomposites, and Release Studies. *Compr. Rev. Food Sci. Food Saf.* **2010**, *9*, 552–571. [CrossRef]
74. Nofar, M.; Park, C.B. Poly (Lactic Acid) Foaming. *Prog. Polym. Sci.* **2014**, *39*, 1721–1741. [CrossRef]
75. Volpe, V.; Pantani, R. Foam Injection Molding of Poly(Lactic) Acid: Effect of Back Pressure on Morphology and Mechanical Properties. *J. Appl. Polym. Sci.* **2015**, *132*, 42612. [CrossRef]
76. Liu, Z.; Wang, Y.; Wu, B.; Cui, C.; Guo, Y.; Yan, C. A Critical Review of Fused Deposition Modeling 3D Printing Technology in Manufacturing Polylactic Acid Parts. *Int. J. Adv. Manuf. Technol.* **2019**, *102*, 2877–2889. [CrossRef]
77. Valino, A.D.; Dizon, J.R.C.; Espera, A.H.; Chen, Q.; Messman, J.; Advincula, R.C. Advances in 3D Printing of Thermoplastic Polymer Composites and Nanocomposites. *Prog. Polym. Sci.* **2019**, *98*, 101162. [CrossRef]
78. Tümer, E.H.; Erbil, H.Y. Extrusion-Based 3D Printing Applications of PLA Composites: A Review. *Coatings* **2021**, *11*, 390. [CrossRef]
79. Hong, S.H.; Park, J.H.; Kim, O.Y.; Hwang, S.H. Preparation of Chemically Modified Lignin-Reinforced Pla Biocomposites and Their 3d Printing Performance. *Polymers* **2021**, *13*, 667. [CrossRef]
80. Zoumaki, M.; Mansour, M.T.; Tsongas, K.; Tzetzis, D.; Mansour, G. Mechanical Characterization and Finite Element Analysis of Hierarchical Sandwich Structures with PLA 3D-Printed Core and Composite Maize Starch Biodegradable Skins. *J. Compos. Sci.* **2022**, *6*, 118. [CrossRef]
81. Salaris, V.; Leonés, A.; Lopez, D.; Kenny, J.M.; Peponi, L. Shape-Memory Materials via Electrospinning: A Review. *Polymers* **2022**, *14*, 995. [CrossRef] [PubMed]
82. Gao, X.; Han, S.; Zhang, R.; Liu, G.; Wu, J. Progress in Electrospun Composite Nanofibers: Composition, Performance and Applications for Tissue Engineering. *J. Mater. Chem. B* **2019**, *7*, 7075–7089. [CrossRef] [PubMed]
83. Wu, J.H.; Hu, T.G.; Wang, H.; Zong, M.H.; Wu, H.; Wen, P. Electrospinning of PLA Nanofibers: Recent Advances and Its Potential Application for Food Packaging. *J. Agric. Food Chem.* **2022**, *70*, 8207–8221. [CrossRef]
84. Ahmed, F.E.; Lalia, B.S.; Hashaikeh, R. A Review on Electrospinning for Membrane Fabrication: Challenges and Applications. *Desalination* **2015**, *356*, 15–30. [CrossRef]
85. Duygulu, N.E.; Ciftci, F.; Ustundag, C.B. Electrospun Drug Blended Poly(Lactic Acid) (PLA) Nanofibers and Their Antimicrobial Activities. *J. Polym. Res.* **2020**, *27*, 12–16. [CrossRef]
86. Zhu, Y.; Yang, D.; Li, J.; Yue, Z.; Zhou, J.; Wang, X. The Preparation of Ultrathin and Porous Electrospinning Membranes of HKUST-1/PLA with Good Antibacterial and Filtration Performances. *J. Porous Mater.* **2023**, *30*, 1011–1019. [CrossRef]
87. Paolo, A.R.A.; Stefano, F. Method for the Plasticization of Lactic Acid Polymers. WO2009092825A1, 30 July 2009.
88. Armentano, I.; Fortunati, E.; Burgos, N.; Dominici, F.; Luzi, F.; Fiori, S.; Jiménez, A.; Yoon, K.; Ahn, J.; Kang, S.; et al. Processing and Characterization of Plasticized PLA/PHB Blends for Biodegradable Multiphase Systems. *Express Polym. Lett.* **2015**, *9*, 583–596. [CrossRef]
89. López-Rodríguez, N.; Sarasua, J.R. Plasticization of Poly-l-lactide with L-lactide, D-lactide, and d, L-lactide Monomers. *Polym. Eng. Sci.* **2013**, *53*, 2073–2080. [CrossRef]
90. Srithep, Y.; Pholharn, D. Plasticizer Effect on Melt Blending of Polylactide Stereocomplex. *e-Polymers* **2017**, *17*, 409–416. [CrossRef]
91. Xing, C.; Matuana, L.M. Epoxidized Soybean Oil-plasticized Poly (Lactic Acid) Films Performance as Impacted by Storage. *J. Appl. Polym. Sci.* **2016**, *133*, 43201. [CrossRef]

92. Balart, J.F.; Fombuena, V.; Fenollar, O.; Boronat, T.; Sánchez-Nacher, L. Processing and Characterization of High Environmental Efficiency Composites Based on PLA and Hazelnut Shell Flour (HSF) with Biobased Plasticizers Derived from Epoxidized Linseed Oil (ELO). *Compos. B Eng.* **2016**, *86*, 168–177. [CrossRef]
93. Ferri, J.M.; Garcia-Garcia, D.; Montanes, N.; Fenollar, O.; Balart, R. The Effect of Maleinized Linseed Oil as Biobased Plasticizer in Poly (Lactic Acid)-based Formulations. *Polym. Int.* **2017**, *66*, 882–891. [CrossRef]
94. Chieng, B.W.; Ibrahim, N.A.; Then, Y.Y.; Loo, Y.Y. Epoxidized Jatropha Oil as a Sustainable Plasticizer to Poly (Lactic Acid). *Polymers* **2017**, *9*, 204. [CrossRef] [PubMed]
95. Petchwattana, N.; Sanetuntikul, J.; Narupai, B. Plasticization of Biodegradable Poly (Lactic Acid) by Different Triglyceride Molecular Sizes: A Comparative Study with Glycerol. *J. Polym. Environ.* **2018**, *26*, 1160–1168. [CrossRef]
96. Halloran, M.W.; Danielczak, L.; Nicell, J.A.; Leask, R.L.; Marić, M. Highly Flexible Polylactide Food Packaging Plasticized with Nontoxic, Biosourced Glycerol Plasticizers. *ACS Appl. Polym. Mater.* **2022**, *4*, 3608–3617. [CrossRef]
97. Seo, H.J.; Seo, Y.H.; Park, S.U.; Lee, H.J.; Lee, M.R.; Park, J.H.; Cho, W.Y.; Lee, P.C.; Lee, B.Y. Glycerol-Derived Organic Carbonates: Environmentally Friendly Plasticizers for PLA. *RSC Adv.* **2024**, *14*, 4702–4716. [CrossRef]
98. Wachirahuttapong, S.; Thongpin, C.; Sombatsompop, N. Effect of PCL and Compatibility Contents on the Morphology, Crystallization and Mechanical Properties of PLA/PCL Blends. *Energy Procedia* **2016**, *89*, 198–206. [CrossRef]
99. Khankrua, R.; Pivsa-Art, S.; Hiroyuki, H.; Suttiruengwong, S. Effect of Chain Extenders on Thermal and Mechanical Properties of Poly (Lactic Acid) at High Processing Temperatures: Potential Application in PLA/Polyamide 6 Blend. *Polym. Degrad. Stab.* **2014**, *108*, 232–240. [CrossRef]
100. Nam, B.; Son, Y. Enhanced Impact Strength of Compatibilized Poly (Lactic Acid)/Polyamide 11 Blends by a Crosslinking Agent. *J. Appl. Polym. Sci.* **2020**, *137*, 49011. [CrossRef]
101. Chieng, B.W.; Ibrahim, N.A.; Yunus, W.M.Z.W.; Hussein, M.Z. Plasticized Poly (Lactic Acid) with Low Molecular Weight Poly (Ethylene Glycol): Mechanical, Thermal, and Morphology Properties. *J. Appl. Polym. Sci.* **2013**, *130*, 4576–4580. [CrossRef]
102. Kodal, M.; Sirin, H.; Ozkoc, G. Long-and Short-Term Stability of Plasticized Poly (Lactic Acid): Effects of Plasticizers Type on Thermal, Mechanical and Morphological Properties. *Polym. Bull.* **2019**, *76*, 423–445. [CrossRef]
103. Wang, B.; Hina, K.; Zou, H.; Zuo, D.; Yi, C. Thermal, Crystallization, Mechanical and Decomposition Properties of Poly (Lactic Acid) Plasticized with Poly (Ethylene Glycol). *J. Vinyl Addit. Technol.* **2018**, *24*, E154–E163. [CrossRef]
104. Ferri, J.M.; Garcia-Garcia, D.; Carbonell-Verdu, A.; Fenollar, O.; Balart, R. Poly (Lactic Acid) Formulations with Improved Toughness by Physical Blending with Thermoplastic Starch. *J. Appl. Polym. Sci.* **2018**, *135*, 45751. [CrossRef]
105. Müller, P.; Bere, J.; Fekete, E.; Móczó, J.; Nagy, B.; Kállay, M.; Gyarmati, B.; Pukánszky, B. Interactions, Structure and Properties in PLA/Plasticized Starch Blends. *Polymer* **2016**, *103*, 9–18. [CrossRef]
106. Ferri, J.M.; Garcia-Garcia, D.; Sánchez-Nacher, L.; Fenollar, O.; Balart, R. The Effect of Maleinized Linseed Oil (MLO) on Mechanical Performance of Poly (Lactic Acid)-Thermoplastic Starch (PLA-TPS) Blends. *Carbohydr. Polym.* **2016**, *147*, 60–68. [CrossRef]
107. Labrecque, L.V.; Kumar, R.A.; Dave, V.; Gross, R.A.; McCarthy, S.P. Citrate Esters as Plasticizers for Poly (Lactic Acid). *J. Appl. Polym. Sci.* **1997**, *66*, 1507–1513. [CrossRef]
108. Barandiaran, A.; Gomez-Caturla, J.; Ivorra-Martinez, J.; Lascano, D.; Selles, M.A.; Moreno, V.; Fenollar, O. Esters of Cinnamic Acid as Green Plasticizers for Polylactide Formulations with Improved Ductility. *Macromol. Mater. Eng.* **2023**, *308*, 2300022. [CrossRef]
109. Ordoñez, R.; Atarés, L.; Chiralt, A. Antibacterial Properties of Cinnamic and Ferulic Acids Incorporated to Starch and PLA Monolayer and Multilayer Films. *Food Control* **2022**, *136*, 108878. [CrossRef]
110. Gomez-Caturla, J.; Dominguez-Candela, I.; Medina-Casas, M.P.; Ivorra-Martinez, J.; Moreno, V.; Balart, R.; Garcia-Garcia, D. Improvement of Poly (Lactide) Ductile Properties by Plasticization with Biobased Tartaric Acid Ester. *Macromol. Mater. Eng.* **2023**, *308*, 2200694. [CrossRef]
111. Zawada, K.; Plichta, A.; Jańczewski, D.; Hajmowicz, H.; Florjanczyk, Z.; Stępień, M.; Sobiecka, A.; Synoradzki, L. Esters of Tartaric Acid, a New Class of Potential "Double Green" Plasticizers. *ACS Sustain. Chem. Eng.* **2017**, *5*, 5999–6007. [CrossRef]
112. Zaaba, N.F.; Jaafar, M. A Review on Degradation Mechanisms of Polylactic Acid: Hydrolytic, Photodegradative, Microbial, and Enzymatic Degradation. *Polym. Eng. Sci.* **2020**, *60*, 2061–2075. [CrossRef]
113. Martín-Alonso, M.D.; Salaris, V.; Leonés, A.; Hevilla, V.; Muñoz-Bonilla, A.; Echeverría, C.; Fernández-García, M.; Peponi, L.; López, D. Centrifugal Force-Spinning to Obtain Multifunctional Fibers of PLA Reinforced with Functionalized Silver Nanoparticles. *Polymers* **2023**, *15*, 1240. [CrossRef]
114. Vaid, R.; Yildirim, E.; Pasquinelli, M.A.; King, M.W. Hydrolytic Degradation of Polylactic Acid Fibers as a Function of Ph and Exposure Time. *Molecules* **2021**, *26*, 7554. [CrossRef]
115. Qu, M.; Tu, H.; Amarante, M.; Song, Y.Q.; Zhu, S.S. Zinc Oxide Nanoparticles Catalyze Rapid Hydrolysis of Poly(Lactic Acid) at Low Temperatures. *J. Appl. Polym. Sci.* **2014**, *131*, 1–7. [CrossRef]
116. Zanella, G.S.; Becker, D.; dos Santos Schneider, A.L.; Pezzin, A.P.T.; Silva, D.A.K.; Nogueira, A.L. PLLA–Silver Nanoparticles Bionanocomposite Membranes: Preparation, Antibacterial Activity, and in Vitro Hydrolytic Degradation Assessment. *J. Appl. Polym. Sci.* **2019**, *136*, 47998. [CrossRef]
117. Hegyesi, N.; Zhang, Y.; Kohári, A.; Polyák, P.; Sui, X.; Pukánszky, B. Enzymatic Degradation of PLA/Cellulose Nanocrystal Composites. *Ind. Crops Prod.* **2019**, *141*, 111799. [CrossRef]

118. Donate, R.; Ortega, Z. Enzymatic Degradation Study of PLA-Based Composite Scaffolds. *Rev. Adv. Mater. Sci.* **2020**, *59*, 170–175. [CrossRef]
119. Moon, J.; Kim, M.Y.; Kim, B.M.; Lee, J.C.; Choi, M.C.; Kim, J.R. Estimation of the Microbial Degradation of Biodegradable Polymer, Poly(Lactic Acid) (PLA) with a Specific Gas Production Rate. *Macromol. Res.* **2016**, *24*, 415–421. [CrossRef]
120. Bubpachat, T.; Sombatsompop, N.; Prapagdee, B. Isolation and Role of Polylactic Acid-Degrading Bacteria on Degrading Enzymes Productions and PLA Biodegradability at Mesophilic Conditions. *Polym. Degrad. Stab.* **2018**, *152*, 75–85. [CrossRef]
121. Karamanlioglu, M.; Preziosi, R.; Robson, G.D. Abiotic and Biotic Environmental Degradation of the Bioplastic Polymer Poly(Lactic Acid): A Review. *Polym. Degrad. Stab.* **2017**, *137*, 122–130. [CrossRef]
122. Li, J.; Zheng, W.; Li, L.; Zheng, Y.; Lou, X. Thermal Degradation Kinetics of G-HA/PLA Composite. *Thermochim. Acta* **2009**, *493*, 90–95. [CrossRef]
123. Santonja-Blasco, L.; Ribes-Greus, A.; Alamo, R.G. Comparative Thermal, Biological and Photodegradation Kinetics of Polylactide and Effect on Crystallization Rates. *Polym. Degrad. Stab.* **2013**, *98*, 771–784. [CrossRef]
124. Li, M.; Jia, Y.; Shen, X.; Shen, T.; Tan, Z.; Zhuang, W.; Zhao, G.; Zhu, C.; Ying, H. Investigation into Lignin Modified PBAT/Thermoplastic Starch Composites: Thermal, Mechanical, Rheological and Water Absorption Properties. *Ind. Crops Prod.* **2021**, *171*, 113916. [CrossRef]
125. Jiao, J.; Zeng, X.; Huang, X. An Overview on Synthesis, Properties and Applications of Poly(Butylene-Adipate-Co-Terephthalate)–PBAT. *Adv. Ind. Eng. Polym. Res.* **2020**, *3*, 19–26. [CrossRef]
126. França, D.C.; Almeida, T.G.; Abels, G.; Canedo, E.L.; Carvalho, L.H.; Wellen, R.M.R.; Haag, K.; Koschek, K. Tailoring PBAT/PLA/Babassu Films for Suitability of Agriculture Mulch Application. *J. Nat. Fibers* **2019**, *16*, 933–943. [CrossRef]
127. Ferreira, F.V.; Cividanes, L.S.; Gouveia, R.F.; Lona, L.M.F. An Overview on Properties and Applications of Poly(Butylene Adipate-Co-Terephthalate)–PBAT Based Composites. *Polym. Eng. Sci.* **2019**, *59*, E7–E15. [CrossRef]
128. Roy, S.; Ghosh, T.; Zhang, W.; Rhim, J.W. Recent Progress in PBAT-Based Films and Food Packaging Applications: A Mini-Review. *Food Chem.* **2024**, *437*, 137822. [CrossRef]
129. Lin, Q. Synthetic Biodegradable Polymers. In *Introduction to Biomaterials*; Tsinghua University Press: Beijing, China, 2005; pp. 187–196; ISBN 9783642271533.
130. Kanwal, A.; Zhang, M.; Sharaf, F.; Li, C. Polymer Pollution and Its Solutions with Special Emphasis on Poly (Butylene Adipate Terephthalate (PBAT)). *Polym. Bull.* **2022**, *79*, 9303–9330. [CrossRef]
131. Francisco, A.B.F.D.P.; Lorevice, M.V.; Claro, P.I.C.; Gouveia, R.F. Comprehensive Study of Cellulose Nanocrystals Acetylation Effects on Poly (Butylene Adipate-Co-Terephthalate) Nanocomposite Films Obtained by Solvent Casting and Heat Pressing. *Ind. Crops Prod.* **2022**, *177*, 114459. [CrossRef]
132. Wu, D.; Xu, R.; Zhang, Y.; Shi, S. Influence of Branching on the Mechanical, Rheological and Crystallization Properties of Poly(Butylene Adipate-Co-Terephthalate). *J. Therm. Anal. Calorim.* **2022**, *147*, 9539–9557. [CrossRef]
133. Huang, F.; Wu, L.; Li, B.G. Sulfonated Biodegradable PBAT Copolyesters with Improved Gas Barrier Properties and Excellent Water Dispersibility: From Synthesis to Structure-Property. *Polym. Degrad. Stab.* **2020**, *182*, 21–26. [CrossRef]
134. Nifant'ev, I.E.; Bagrov, V.V.; Komarov, P.D.; Ilyin, S.O.; Ivchenko, P.V. The Use of Branching Agents in the Synthesis of PBAT. *Polymers* **2022**, *14*, 1720. [CrossRef]
135. Aversa, C.; Barletta, M.; Cappiello, G.; Gisario, A. Compatibilization Strategies and Analysis of Morphological Features of Poly(Butylene Adipate-Co-Terephthalate) (PBAT)/Poly(Lactic Acid) PLA Blends: A State-of-Art Review. *Eur. Polym. J.* **2022**, *173*, 111304. [CrossRef]
136. Mahata, D.; Karthikeyan, S.; Godse, R.; Gupta, V.K. Poly(Butylene Adipate-Co-Terephthalate) Polyester Synthesis Process and Product Development. *Polym. Sci. Ser. C* **2021**, *63*, 102–111. [CrossRef]
137. Burford, T.; Rieg, W.; Madbouly, S. Biodegradable Poly(Butylene Adipate-Co-Terephthalate) (PBAT). *Phys. Sci. Rev.* **2023**, *8*, 1127–1156. [CrossRef]
138. Hu, L.; Wu, L.; Song, F.; Li, B.G. Kinetics and Modeling of Melt Polycondensation for Synthesis of Poly[(Butylene Succinate)-Co-(Butylene Terephthalate)], 1–Esterification. *Macromol. React. Eng.* **2010**, *4*, 621–632. [CrossRef]
139. Austin, H.P.; Allen, M.D.; Donohoe, B.S.; Rorrer, N.A.; Kearns, F.L.; Silveira, R.L.; Pollard, B.C.; Dominick, G.; Duman, R.; El Omari, K.; et al. Characterization and Engineering of a Plastic-Degrading Aromatic Polyesterase. *Proc. Natl. Acad. Sci. USA* **2018**, *115*, E4350–E4357. [CrossRef]
140. Shankar, S.; Rhim, J.W. Tocopherol-Mediated Synthesis of Silver Nanoparticles and Preparation of Antimicrobial PBAT/Silver Nanoparticles Composite Films. *LWT* **2016**, *72*, 149–156. [CrossRef]
141. Venkatesan, R.; Rajeswari, N. ZnO/PBAT Nanocomposite Films: Investigation on the Mechanical and Biological Activity for Food Packaging. *Polym. Adv. Technol.* **2017**, *28*, 20–27. [CrossRef]
142. Thiyagu, T.T.; Sai Prasanna Kumar, J.V.; Gurusamy, P.; Sathiyamoorthy, V.; Maridurai, T.; Vr, A.P. Effect of Cashew Shell Biomass Synthesized Cardanol Oil Green Compatibilizer on Flexibility, Barrier, Thermal, and Wettability of PLA/PBAT Biocomposite Films. *Biomass Convers. Biorefinery* **2023**, *13*, 11841–11851. [CrossRef]
143. Pulikkalparambil, H.; Phothisarattana, D.; Promhuad, K.; Harnkarnsujarit, N. Effect of Silicon Dioxide Nanoparticle on Microstructure, Mechanical and Barrier Properties of Biodegradable PBAT/PBS Food Packaging. *Food Biosci.* **2023**, *55*, 103023. [CrossRef]

144. Gupta, V.; Biswas, D.; Roy, S. A Comprehensive Review of Biodegradable Polymer-Based Films and Coatings and Their Food Packaging Applications. *Materials* **2022**, *15*, 5899. [CrossRef]
145. Azevedo, J.V.C.; Ramakers-Van Dorp, E.; Grimmig, R.; Hausnerova, B.; Möginger, B. Process-Induced Morphology of Poly(Butylene Adipate Terephthalate)/Poly(Lactic Acid) Blown Extrusion Films Modified with Chain-Extending Cross-Linkers. *Polymers* **2022**, *14*, 1939. [CrossRef]
146. Bilck, A.P.; Grossmann, M.V.E.; Yamashita, F. Biodegradable Mulch Films for Strawberry Production. *Polym. Test.* **2010**, *29*, 471–476. [CrossRef]
147. Lopes, A.C.; Barcia, M.K.; Veiga, T.B.; Yamashita, F.; Grossmann, M.V.E.; Olivato, J.B. Eco-Friendly Materials Produced by Blown-Film Extrusion as Potential Active Food Packaging. *Polym. Adv. Technol.* **2021**, *32*, 779–788. [CrossRef]
148. Botta, L.; Titone, V.; Teresi, R.; Scarlata, M.C.; Lo Re, G.; La Mantia, F.P.; Lopresti, F. Biocomposite PBAT/Lignin Blown Films with Enhanced Photo-Stability. *Int. J. Biol. Macromol.* **2022**, *217*, 161–170. [CrossRef]
149. Liu, W.; Yu, W.; Wang, J.; Gao, J.; Ding, Y.; Zhang, S.; Zheng, Q. Enhanced Mechanical and Long-Lasting Antibacterial Properties of Starch/PBAT Blown Films via Designing of Reactive Micro-Crosslinked Starch. *Int. J. Biol. Macromol.* **2024**, *266*, 131366. [CrossRef]
150. Santana-Melo, G.F.; Rodrigues, B.V.M.; da Silva, E.; Ricci, R.; Marciano, F.R.; Webster, T.J.; Vasconcellos, L.M.R.; Lobo, A.O. Electrospun Ultrathin PBAT/NHAp Fibers Influenced the in Vitro and in Vivo Osteogenesis and Improved the Mechanical Properties of Neoformed Bone. *Colloids Surf. B Biointerfaces* **2017**, *155*, 544–552. [CrossRef] [PubMed]
151. Liu, Y.; Yang, L.; Chen, G.; Liu, Z.; Lu, T.; Yang, Y.; Yu, J.; Kang, D.; Yan, W.; He, M.; et al. PBAT Hollow Porous Microfibers Prepared via Electrospinning and Their Functionalization for Potential Peptide Release. *Mater. Des.* **2021**, *207*, 109880. [CrossRef]
152. Colucci, G.; Piano, M.; Lupone, F.; Badini, C.; Bondioli, F.; Messori, M. Preparation and 3D Printability Study of Bio-Based PBAT Powder for Selective Laser Sintering Additive Manufacturing. *Mater. Today Chem.* **2023**, *33*, 101687. [CrossRef]
153. Akhir, M.A.M.; Mustapha, M. Formulation of Biodegradable Plastic Mulch Film for Agriculture Crop Protection: A Review. *Polym. Rev.* **2022**, *62*, 890–918. [CrossRef]
154. Pietrosanto, A.; Scarfato, P.; Maio, L. Di Development of Eco-Sustainable PBAT-Based Blown Packaging Applications. *Materials* **2020**, *13*, 5395. [CrossRef]
155. Liu, W.; Wang, Y.; Xiang, S.; Liu, H. Unveiling the Effect of Enhanced Interfacial Compatibility on the Mechanical Properties of PLA/PBAT Blends. *Polymer* **2024**, *296*, 126815. [CrossRef]
156. Sui, X.; Zhao, X.; Wang, Z.; Sun, S. Super-Ductile and Stiff PBAT/PLA Biodegradable Composites Balanced with Random PMMA-Co-GMA Copolymer as Compatibilizer. *Polym. Int.* **2022**, *72*, 333–341. [CrossRef]
157. Fourati, Y.; Tarrés, Q.; Delgado-Aguilar, M.; Mutjé, P.; Boufi, S. Cellulose Nanofibrils Reinforced PBAT/TPS Blends: Mechanical and Rheological Properties. *Int. J. Biol. Macromol.* **2021**, *183*, 267–275. [CrossRef]
158. Brandelero, R.P.H.; Grossmann, M.V.; Yamashita, F. Films of Starch and Poly(Butylene Adipate Co-Terephthalate) Added of Soybean Oil (SO) and Tween 80. *Carbohydr. Polym.* **2012**, *90*, 1452–1460. [CrossRef]
159. Nayak, S.K. Biodegradable PBAT/Starch Nanocomposites. *Polym.-Plast. Technol. Eng.* **2010**, *49*, 1406–1418. [CrossRef]
160. Nabar, Y.; Raquez, J.M.; Dubois, P.; Narayan, R. Production of Starch Foams by Twin-Screw Extrusion: Effect of Maleated Poly(Butylene Adipate-Co-Terephthalate) as a Compatibilizer. *Biomacromolecules* **2005**, *6*, 807–817. [CrossRef] [PubMed]
161. Garcia, P.S.; Grossmann, M.V.E.; Shirai, M.A.; Lazaretti, M.M.; Yamashita, F.; Muller, C.M.O.; Mali, S. Improving Action of Citric Acid as Compatibiliser in Starch/Polyester Blown Films. *Ind. Crops Prod.* **2014**, *52*, 305–312. [CrossRef]
162. Dang, K.M.; Yoksan, R.; Pollet, E.; Avérous, L. Morphology and Properties of Thermoplastic Starch Blended with Biodegradable Polyester and Filled with Halloysite Nanoclay. *Carbohydr. Polym.* **2020**, *242*, 116392. [CrossRef] [PubMed]
163. Olivato, J.B.; Marini, J.; Yamashita, F.; Pollet, E.; Grossmann, M.V.E.; Avérous, L. Sepiolite as a Promising Nanoclay for Nano-Biocomposites Based on Starch and Biodegradable Polyester. *Mater. Sci. Eng. C* **2017**, *70*, 296–302. [CrossRef]
164. Lendvai, L.; Apostolov, A.; Karger-Kocsis, J. Characterization of Layered Silicate-Reinforced Blends of Thermoplastic Starch (TPS) and Poly(Butylene Adipate-Co-Terephthalate). *Carbohydr. Polym.* **2017**, *173*, 566–572. [CrossRef]
165. Pinheiro, I.F.; Ferreira, F.V.; Souza, D.H.S.; Gouveia, R.F.; Lona, L.M.F.; Morales, A.R.; Mei, L.H.I. Mechanical, Rheological and Degradation Properties of PBAT Nanocomposites Reinforced by Functionalized Cellulose Nanocrystals. *Eur. Polym. J.* **2017**, *97*, 356–365. [CrossRef]
166. Phothisarattana, D.; Wongphan, P.; Promhuad, K.; Promsorn, J.; Harnkarnsujarit, N. Biodegradable Poly(Butylene Adipate-Co-Terephthalate) and Thermoplastic Starch-Blended Tio2 Nanocomposite Blown Films as Functional Active Packaging of Fresh Fruit. *Polymers* **2021**, *13*, 4192. [CrossRef]
167. Someya, Y.; Sugahara, Y.; Shibata, M. Nanocomposites Based on Poly(Butylene Adipate-Co-Terephthalate) and Montmorillonite. *J. Appl. Polym. Sci.* **2005**, *95*, 386–392. [CrossRef]
168. Chen, J.H.; Chen, C.C.; Yang, M.C. Characterization of Nanocomposites of Poly(Butylene Adipate-Co- Terephthalate) Blending with Organoclay. *J. Polym. Res.* **2011**, *18*, 2151–2159. [CrossRef]
169. Mohanty, S.; Nayak, S.K. Aromatic-aliphatic Poly(Butylene Adipate- Co -terephthalate) Bionanocomposite: Influence of Organic Modification on Structure and Properties. *Polym. Compos.* **2010**, *31*, 1194–1204. [CrossRef]
170. Mondal, D.; Bhowmick, B.; Mollick, M.M.R.; Maity, D.; Ranjan Saha, N.; Rangarajan, V.; Rana, D.; Sen, R.; Chattopadhyay, D. Antimicrobial Activity and Biodegradation Behavior of Poly(Butylene Adipate-Co-Terephthalate)/Clay Nanocomposites. *J. Appl. Polym. Sci.* **2014**, *131*, 40079. [CrossRef]

171. Han, J.G.; Park, S.J. Fabrication of PBAT/Polyethylene Blends Mulching Film via Blown Film Extrusion Process. *Korea Aust. Rheol. J.* **2020**, *32*, 79–86. [CrossRef]
172. Kijchavengkul, T.; Auras, R.; Rubino, M.; Selke, S.; Ngouajio, M.; Fernandez, R.T. Biodegradation and Hydrolysis Rate of Aliphatic Aromatic Polyester. *Polym. Degrad. Stab.* **2010**, *95*, 2641–2647. [CrossRef]
173. Ye, H.; Li, Q.; Li, J.; Li, D.; Ao, Z. Review on the Abiotic Degradation of Biodegradable Plastic Poly(Butylene Adipate-Terephthalate): Mechanisms and Main Factors of the Degradation. *Chin. Chem. Lett.* **2024**, *113*, 109861. [CrossRef]
174. Zumstein, M.T.; Schintlmeister, A.; Nelson, T.F.; Baumgartner, R.; Woebken, D.; Wagner, M.; Kohler, H.P.E.; McNeill, K.; Sander, M. Biodegradation of Synthetic Polymers in Soils: Tracking Carbon into CO2 and Microbial Biomass. *Sci. Adv.* **2018**, *4*, eaas9024. [CrossRef]
175. Soulenthone, P.; Tachibana, Y.; Muroi, F.; Suzuki, M.; Ishii, N.; Ohta, Y.; Kasuya, K. ichi Characterization of a Mesophilic Actinobacteria That Degrades Poly(Butylene Adipate-Co-Terephthalate). *Polym. Degrad. Stab.* **2020**, *181*, 109335. [CrossRef]
176. Ma, J.; Cao, Y.; Fan, L.; Xie, Y.; Zhou, X.; Ren, Q.; Yang, X.; Gao, X.; Feng, Y. Degradation Characteristics of Polybutylene Adipate Terephthalic Acid (PBAT) and Its Effect on Soil Physicochemical Properties: A Comparative Study with Several Polyethylene (PE) Mulch Films. *J. Hazard. Mater.* **2023**, *456*, 131661. [CrossRef]
177. Svoboda, P.; Dvorackova, M.; Svobodova, D. Influence of Biodegradation on Crystallization of Poly (Butylene Adipate-Co-Terephthalate). *Polym. Adv. Technol.* **2019**, *30*, 552–562. [CrossRef]
178. Nomadolo, N.; Dada, O.E.; Swanepoel, A.; Mokhena, T.; Muniyasamy, S. A Comparative Study on the Aerobic Biodegradation of the Biopolymer Blends of Poly(Butylene Succinate), Poly(Butylene Adipate Terephthalate) and Poly(Lactic Acid). *Polymers* **2022**, *14*, 1894. [CrossRef]
179. Dammak, M.; Fourati, Y.; Tarrés, Q.; Delgado-Aguilar, M.; Mutjé, P.; Boufi, S. Blends of PBAT with Plasticized Starch for Packaging Applications: Mechanical Properties, Rheological Behaviour and Biodegradability. *Ind. Crops Prod.* **2020**, *144*, 112061. [CrossRef]
180. Morro, A.; Catalina, F.; Sanchez-León, E.; Abrusci, C. Photodegradation and Biodegradation Under Thermophile Conditions of Mulching Films Based on Poly(Butylene Adipate-Co-Terephthalate) and Its Blend with Poly(Lactic Acid). *J. Polym. Environ.* **2019**, *27*, 352–363. [CrossRef]
181. Zhang, T.; Han, W.; Zhang, C.; Weng, Y. Effect of Chain Extender and Light Stabilizer on the Weathering Resistance of PBAT/PLA Blend Films Prepared by Extrusion Blowing. *Polym. Degrad. Stab.* **2021**, *183*, 109455. [CrossRef]
182. Yang, Y.; Min, J.; Xue, T.; Jiang, P.; Liu, X.; Peng, R.; Huang, J.W.; Qu, Y.; Li, X.; Ma, N.; et al. Complete Bio-Degradation of Poly(Butylene Adipate-Co-Terephthalate) via Engineered Cutinases. *Nat. Commun.* **2023**, *14*, 1645. [CrossRef] [PubMed]
183. Pang, W.; Li, B.; Wu, Y.; Zeng, Q.; Yang, J.; Zhang, Y.; Tian, S. Upgraded Recycling of Biodegradable PBAT Plastic: Efficient Hydrolysis and Electrocatalytic Conversion. *Chem. Eng. J.* **2024**, *486*, 150342. [CrossRef]
184. Jo, J.; Kim, H.; Jeong, S.Y.; Park, C.; Hwang, H.S.; Koo, B. Changes in Mechanical Properties of Polyhydroxyalkanoate with Double Silanized Cellulose Nanocrystals Using Different Organosiloxanes. *Nanomaterials* **2021**, *11*, 1542. [CrossRef]
185. Najah, M.I.; Aisyah, R.; Adzila, S.; Haq, R.H.A. Mechanical Properties of Calcium Phosphate Reinforced Polyhydroxyalkanoate (PHA) Biocomposite. *J. Thermoplast. Compos. Mater.* **2023**, *36*, 3294–3311. [CrossRef]
186. Sharma, V.; Sehgal, R.; Gupta, R. Polyhydroxyalkanoate (PHA): Properties and Modifications. *Polymer* **2021**, *212*, 123161. [CrossRef]
187. Prados, E.; Maicas, S. Bacterial Production of Hydroxyalkanoates (PHA). *Univers. J. Microbiol. Res.* **2016**, *4*, 23–30. [CrossRef]
188. Tarrahi, R.; Fathi, Z.; Seydibeyoğlu, M.Ö.; Doustkhah, E.; Khataee, A. Polyhydroxyalkanoates (PHA): From Production to Nanoarchitecture. *Int. J. Biol. Macromol.* **2020**, *146*, 596–619. [CrossRef]
189. Laycock, B.; Halley, P.; Pratt, S.; Werker, A.; Lant, P. The Chemomechanical Properties of Microbial Polyhydroxyalkanoates. *Prog. Polym. Sci.* **2013**, *38*, 536–583. [CrossRef]
190. Nigmatullin, R.; Thomas, P.; Lukasiewicz, B.; Puthussery, H.; Roy, I. Polyhydroxyalkanoates, a Family of Natural Polymers, and Their Applications in Drug Delivery. *J. Chem. Technol. Biotechnol.* **2015**, *90*, 1209–1221. [CrossRef]
191. Guo, W.; Yang, K.; Qin, X.; Luo, R.; Wang, H.; Huang, R. Polyhydroxyalkanoates in Tissue Repair and Regeneration. *Eng. Regen.* **2022**, *3*, 24–40. [CrossRef]
192. Philip, S.; Keshavarz, T.; Roy, I. Polyhydroxyalkanoates: Biodegradable Polymers with a Range of Applications. *J. Chem. Technol. Biotechnol.* **2007**, *82*, 233–247. [CrossRef]
193. Genovesi, A.; Aversa, C.; Barletta, M. Polyhydroxyalkanoates-Based Cast Film as Bio-Based Packaging for Fresh Fruit and Vegetables: Manufacturing and Characterization. *J. Polym. Environ.* **2023**, *31*, 4522–4532. [CrossRef]
194. Kalia, V.C.; Patel, S.K.S.; Lee, J.-K. Exploiting Polyhydroxyalkanoates for Biomedical Applications. *Polymers* **2023**, *15*, 1937. [CrossRef]
195. Ye, X.; Zhang, Y.; Liu, T.; Chen, Z.; Chen, W.; Wu, Z.; Wang, Y.; Li, J.; Li, C.; Jiang, T.; et al. Beta-Tricalcium Phosphate Enhanced Mechanical and Biological Properties of 3D-Printed Polyhydroxyalkanoates Scaffold for Bone Tissue Engineering. *Int. J. Biol. Macromol.* **2022**, *209*, 1553–1561. [CrossRef]
196. Kumar, V.; Sehgal, R.; Gupta, R. Blends and Composites of Polyhydroxyalkanoates (PHAs) and Their Applications. *Eur. Polym. J.* **2021**, *161*, 110824. [CrossRef]
197. McAdam, B.; Fournet, M.B.; McDonald, P.; Mojicevic, M. Production of Polyhydroxybutyrate (PHB) and Factors Impacting Its Chemical and Mechanical Characteristics. *Polymers* **2020**, *12*, 2908. [CrossRef]

198. Weinmann, S.; Bonten, C. Thermal and Rheological Properties of Modified Polyhydroxybutyrate (PHB). *Polym. Eng. Sci.* **2019**, *59*, 1057–1064. [CrossRef]
199. Yeo, J.C.C.; Muiruri, J.K.; Thitsartarn, W.; Li, Z.; He, C. Recent Advances in the Development of Biodegradable PHB-Based Toughening Materials: Approaches, Advantages and Applications. *Mater. Sci. Eng. C* **2018**, *92*, 1092–1116. [CrossRef]
200. Sharma, N. Polyhydroxybutyrate (PHB) Production by Bacteria and Its Application as Biodegradable Plastic in Various Industries. *Acad. J. Polym. Sci.* **2019**, *2*, 6–9. [CrossRef]
201. Markl, E. PHB-Bio Based and Biodegradable Replacement for PP: A Review. *Nov. Tech. Nutr. Food Sci.* **2018**, *2*, 206–209. [CrossRef]
202. El-hadi, A.M.; Abd Elbary, A.M. Design of the Electrically Conductive PHB Blends for Biomedical Applications. *J. Mater. Sci. Mater. Electron.* **2018**, *29*, 16496–16506. [CrossRef]
203. Movahedi, M.; Karbasi, S. A Core-Shell Electrospun Scaffold of Polyhydroxybutyrate-Starch/Halloysite Nanotubes Containing Extracellular Matrix and Chitosan for Articular Cartilage Tissue Engineering Application. *J. Polym. Environ.* **2023**, *31*, 3052–3069. [CrossRef]
204. Fernandes, J.G.; Correia, D.M.; Botelho, G.; Padrão, J.; Dourado, F.; Ribeiro, C.; Lanceros-Méndez, S.; Sencadas, V. PHB-PEO Electrospun Fiber Membranes Containing Chlorhexidine for Drug Delivery Applications. *Polym. Test.* **2014**, *34*, 64–71. [CrossRef]
205. Hong, S.G.; Hsu, H.W.; Ye, M.T. Thermal Properties and Applications of Low Molecular Weight Polyhydroxybutyrate. *J. Therm. Anal. Calorim.* **2013**, *111*, 1243–1250. [CrossRef]
206. Moreira, J.B.; Kuntzler, S.G.; da Silva Vaz, B.; da Silva, C.K.; Costa, J.A.V.; de Morais, M.G. Polyhydroxybutyrate (PHB)-Based Blends and Composites. In *Biodegradable Polymers, Blends and Composites*; Elsevier Ltd.: Amsterdam, The Netherlands, 2021; ISBN 9780128237915.
207. Briassoulis, D.; Tserotas, P.; Athanasoulia, I.G. Alternative Optimization Routes for Improving the Performance of Poly(3-Hydroxybutyrate) (PHB) Based Plastics. *J. Clean. Prod.* **2021**, *318*, 128555. [CrossRef]
208. Policastro, G.; Panico, A.; Fabbricino, M. *Improving Biological Production of Poly(3-Hydroxybutyrate-Co-3-Hydroxyvalerate) (PHBV) Co-Polymer: A Critical Review*; Springer: Dordrecht, The Netherlands, 2021; Volume 20, ISBN 1115702109575.
209. Rivera-Briso, A.L.; Serrano-Aroca, Á. Poly(3-Hydroxybutyrate-Co-3-Hydroxyvalerate): Enhancement Strategies for Advanced Applications. *Polymers* **2018**, *10*, 732. [CrossRef]
210. Tebaldi, M.L.; Maia, A.L.C.; Poletto, F.; de Andrade, F.V.; Soares, D.C.F. Poly(-3-Hydroxybutyrate-Co-3-Hydroxyvalerate) (PHBV): Current Advances in Synthesis Methodologies, Antitumor Applications and Biocompatibility. *J. Drug Deliv. Sci. Technol.* **2019**, *51*, 115–126. [CrossRef]
211. Ibrahim, M.I.; Alsafadi, D.; Alamry, K.A.; Hussein, M.A. Properties and Applications of Poly(3-Hydroxybutyrate-Co-3-Hydroxyvalerate) Biocomposites. *J. Polym. Environ.* **2021**, *29*, 1010–1030. [CrossRef]
212. Kiruthika, A.V. PHBV Based Blends and Composites. In *Biodegradable Polymers, Blends and Composites*; Elsevier Ltd.: Amsterdam, The Netherlands, 2021; ISBN 9780128237915.
213. Behera, S.; Priyadarshanee, M.; Vandana; Das, S. Polyhydroxyalkanoates, the Bioplastics of Microbial Origin: Properties, Biochemical Synthesis, and Their Applications. *Chemosphere* **2022**, *294*, 133723. [CrossRef] [PubMed]
214. Che, L.; Jin, W.; Zhou, X.; Han, W.; Chen, Y.; Chen, C.; Jiang, G. Current Status and Future Perspectives on the Biological Production of Polyhydroxyalkanoates. *Asia-Pac. J. Chem. Eng.* **2023**, *18*, e2899. [CrossRef]
215. Zhou, W.; Bergsma, S.; Colpa, D.I.; Euverink, G.J.W.; Krooneman, J. Polyhydroxyalkanoates (PHAs) Synthesis and Degradation by Microbes and Applications towards a Circular Economy. *J. Environ. Manag.* **2023**, *341*, 118033. [CrossRef] [PubMed]
216. Liu, S.; Narancic, T.; Tham, J.L.; O'Connor, K.E. β-Oxidation–Polyhydroxyalkanoates Synthesis Relationship in Pseudomonas Putida KT2440 Revisited. *Appl. Microbiol. Biotechnol.* **2023**, *107*, 1863–1874. [CrossRef] [PubMed]
217. Diniz, M.S.D.F.; Mourão, M.M.; Xavier, L.P.; Santos, A.V. Recent Biotechnological Applications of Polyhydroxyalkanoates (PHA) in the Biomedical Sector—A Review. *Polymers* **2023**, *15*, 4405. [CrossRef]
218. Alves, A.A.; Siqueira, E.C.; Barros, M.P.S.; Silva, P.E.C.; Houllou, L.M. Polyhydroxyalkanoates: A Review of Microbial Production and Technology Application. *Int. J. Environ. Sci. Technol.* **2023**, *20*, 3409–3420. [CrossRef]
219. Zhang, X.; Lin, Y.; Wu, Q.; Wang, Y.; Chen, G.Q. Synthetic Biology and Genome-Editing Tools for Improving PHA Metabolic Engineering. *Trends Biotechnol.* **2020**, *38*, 689–700. [CrossRef]
220. Saravanan, K.; Umesh, M.; Kathirvel, P. *Microbial Polyhydroxyalkanoates (PHAs): A Review on Biosynthesis, Properties, Fermentation Strategies and Its Prospective Applications for Sustainable Future*; Springer: New York, NY, USA, 2022; Volume 30, ISBN 0123456789.
221. Zou, H.; Shi, M.; Zhang, T.; Li, L.; Li, L.; Xian, M. Natural and Engineered Polyhydroxyalkanoate (PHA) Synthase: Key Enzyme in Biopolyester Production. *Appl. Microbiol. Biotechnol.* **2017**, *101*, 7417–7426. [CrossRef]
222. Thomas, T.; Sudesh, K.; Bazire, A.; Elain, A.; Tan, H.T.; Lim, H.; Bruzaud, S. PHA Production and Pha Synthases of the Halophilic *Bacterium halomonas* sp. SF2003. *Bioengineering* **2020**, *7*, 29. [CrossRef]
223. Cha, D.; Ha, H.S.; Lee, S.K. Metabolic Engineering of Pseudomonas Putida for the Production of Various Types of Short-Chain-Length Polyhydroxyalkanoates from Levulinic Acid. *Bioresour. Technol.* **2020**, *309*, 123332. [CrossRef]
224. Inoue, D.; Fukuyama, A.; Ren, Y.; Ike, M. Optimization of Aerobic Dynamic Discharge Process for Very Rapid Enrichment of Polyhydroxyalkanoates-Accumulating Bacteria from Activated Sludge. *Bioresour. Technol.* **2021**, *336*, 125314. [CrossRef]
225. Pei, R.; Vicente-Venegas, G.; Tomaszewska-Porada, A.; Van Loosdrecht, M.C.M.; Kleerebezem, R.; Werker, A. Visualization of Polyhydroxyalkanoate Accumulated in Waste Activated Sludge. *Environ. Sci. Technol.* **2023**, *57*, 11108–11121. [CrossRef] [PubMed]

226. Estévez-Alonso, Á.; Arias-Buendía, M.; Pei, R.; van Veelen, H.P.J.; van Loosdrecht, M.C.M.; Kleerebezem, R.; Werker, A. Calcium Enhances Polyhydroxyalkanoate Production and Promotes Selective Growth of the Polyhydroxyalkanoate-Storing Biomass in Municipal Activated Sludge. *Water Res.* **2022**, *226*, 119259. [CrossRef] [PubMed]
227. Zhang, Z.; Lin, Y.; Wu, S.; Li, X.; Cheng, J.J.; Yang, C. Effect of Composition of Volatile Fatty Acids on Yield of Polyhydroxyalkanoates and Mechanisms of Bioconversion from Activated Sludge. *Bioresour. Technol.* **2023**, *385*, 129445. [CrossRef]
228. Vroman, I.; Tighzert, L. Biodegradable Polymers. *Materials* **2009**, *2*, 307–344. [CrossRef]
229. Yang, L.; Zhang, Y.; Yang, G.; Xie, R.; Wu, G. Controlled Ring-Opening Polymerization of β-ButyrolactoneViaBifunctional Organoboron Catalysts. *Macromolecules* **2021**, *54*, 5509–5517. [CrossRef]
230. Jiang, J.; Choi, J.; Yoon, S. Living Ring-Opening Polymerization of β-Butyrolactone Initiated by Mononuclear Zirconium Compounds Containing Sterically Hindered N,O-Chelate and Anionic Dimethylamide Ligands. *RSC Adv.* **2023**, *13*, 10379–10383. [CrossRef]
231. Tsuda, H.; Shiraki, M.; Inoue, E.; Saito, T. Generation of Poly-β-Hydroxybutyrate from Acetate in Higher Plants: Detection of Acetoacetyl CoA Reductase- and PHB Synthase-Activities in Rice. *J. Plant Physiol.* **2016**, *201*, 9–16. [CrossRef]
232. Snell, K.D.; Peoples, O.P. Polyhydroxyalkanoate Polymers and Their Production in Transgenic Plants. *Metab. Eng.* **2002**, *4*, 29–40. [CrossRef]
233. Lössl, A.; Eibl, C.; Harloff, H.J.; Jung, C.; Koop, H.U. Polyester Synthesis in Transplastomic Tobacco (*Nicotiana tabacum* L.): Significant Contents of Polyhydroxybutyrate Are Associated with Growth Reduction. *Plant Cell Rep.* **2003**, *21*, 891–899. [CrossRef]
234. Parveez, G.K.A.; Bahariah, B.; Ayub, N.H.; Masani, M.Y.A.; Rasid, O.A.; Tarmizi, A.H.; Ishak, Z. Production of Polyhydroxybutyrate in Oil Palm (Elaeis Guineensis Jacq.) Mediated by Microprojectile Bombardment of PHB Biosynthesis Genes into Embryogenic Calli. *Front. Plant Sci.* **2015**, *6*, 598. [CrossRef] [PubMed]
235. Raju, N.N.; Sankaranarayanan, M. Enhancing the Production of Polyhydroxybutyrate, a Biodegradable Polymer by an Optimized Process Using a Novel Klebsiella Pneumonia Strain. *Top. Catal.* **2024**, *67*, 181–191. [CrossRef]
236. Batista, M.B.; Teixeira, C.S.; Sfeir, M.Z.T.; Alves, L.P.S.; Valdameri, G.; Pedrosa, F.D.O.; Sassaki, G.L.; Steffens, M.B.R.; de Souza, E.M.; Dixon, R.; et al. PHB Biosynthesis Counteracts Redox Stress in Herbaspirillum Seropedicae. *Front. Microbiol.* **2018**, *9*, 472. [CrossRef]
237. El-kadi, S.M.; Elbagory, M.; El-zawawy, H.A.H.; El-shaer, H.F.A. Biosynthesis of Poly-ß-Hydroxybutyrate (PHB) from Different Bacterial Strains Grown on Alternative Cheap Carbon Sources. *Polymers* **2021**, *13*, 3801. [CrossRef] [PubMed]
238. Lambauer, V.; Kratzer, R. Lab-Scale Cultivation of Cupriavidus Necator on Explosive Gas Mixtures: Carbon Dioxide Fixation into Polyhydroxybutyrate. *Bioengineering* **2022**, *9*, 204. [CrossRef]
239. Zhang, L.; Jiang, Z.; Tsui, T.H.; Loh, K.C.; Dai, Y.; Tong, Y.W. A Review on Enhancing Cupriavidus Necator Fermentation for Poly(3-Hydroxybutyrate) (PHB) Production From Low-Cost Carbon Sources. *Front. Bioeng. Biotechnol.* **2022**, *10*, 946085. [CrossRef]
240. Jin, A.; del Valle, L.J.; Puiggalí, J. Copolymers and Blends Based on 3-Hydroxybutyrate and 3-Hydroxyvalerate Units. *Int. J. Mol. Sci.* **2023**, *24*, 17250. [CrossRef]
241. Zhao, Y.; Rao, Z.; Xue, Y.; Gong, P.; Ji, Y.; Ma, Y. Biosynthesis, Property Comparison, and Hemocompatibility of Bacterial and Haloarchaeal Poly(3-Hydroxybutyrate-Co-3-Hydroxyvalerate). *Sci. Bull.* **2015**, *60*, 1901–1910. [CrossRef]
242. Meng, D.; Miao, C.; Liu, Y.; Wang, F.; Chen, L.; Huang, Z.; Fan, X.; Gu, P.; Li, Q. Metabolic Engineering for Biosynthesis of Poly(3-Hydroxybutyrate-Co-3-Hydroxyvalerate) from Glucose and Propionic Acid in Recombinant Escherichia Coli. *Bioresour. Technol.* **2022**, *348*, 126786. [CrossRef]
243. Altun, M.; Çelebi, M.; Şen, S.; Kökpınar, Ö.; Kaçoğlu, H.S. Characterization of Polyhydroxyalkanoate-Based Composites Derived from Waste Cooking Oil and Agricultural Surplus. *Polym. Compos.* **2023**, *44*, 4055–4068. [CrossRef]
244. Suhazsini, P.; Keshav, R.; Narayanan, S.; Chaudhuri, A.; Radha, P. A Study on the Synthesis of Poly (3-Hydroxybutyrate-Co-3-Hydroxyvalerate) by Bacillus Megaterium Utilizing Cheese Whey Permeate. *J. Polym. Environ.* **2020**, *28*, 1390–1405. [CrossRef]
245. Chen, Y.; Li, M.; Meng, F.; Yang, W.; Chen, L.; Huo, M. Optimal Poly (3-Hydroxybutyrate/3-Hydroxyvalerate) Biosynthesis by Fermentation Liquid from Primary and Waste Activated Sludge. *Environ. Technol.* **2014**, *35*, 1791–1801. [CrossRef] [PubMed]
246. Chang, W.; Yoon, J.; Oh, M.K. Production of Polyhydroxyalkanoates with the Fermentation of Methylorubrum Extorquens Using Formate as a Carbon Substrate. *Biotechnol. Bioprocess. Eng.* **2022**, *27*, 268–275. [CrossRef]
247. Hathi, Z.J.; Haque, M.A.; Priya, A.; Qin, Z.H.; Huang, S.; Lam, C.H.; Ladakis, D.; Pateraki, C.; Mettu, S.; Koutinas, A.; et al. Fermentative Bioconversion of Food Waste into Biopolymer Poly(3-Hydroxybutyrate-Co-3-Hydroxyvalerate) Using Cupriavidus Necator. *Environ. Res.* **2022**, *215*, 114323. [CrossRef] [PubMed]
248. Koller, M.; Niebelschütz, H.; Braunegg, G. Strategies for Recovery and Purification of Poly[(R)-3-Hydroxyalkanoates] (PHA) Biopolyesters from Surrounding Biomass. *Eng. Life Sci.* **2013**, *13*, 549–562. [CrossRef]
249. Puppi, D.; Pecorini, G.; Chiellini, F. Biomedical Processing of Polyhydroxyalkanoates. *Bioengineering* **2019**, *6*, 108. [CrossRef]
250. Koller, M. Biodegradable and Biocompatible Polyhydroxy-Alkanoates (PHA): Auspicious Microbial Macromolecules for Pharmaceutical and Therapeutic Applications. *Molecules* **2018**, *23*, 362. [CrossRef]
251. Mármol, G.; Sanivada, U.K.; Fangueiro, R. Effect of GNPs on the Piezoresistive, Electrical and Mechanical Properties of Pha and Pla Films. *Fibers* **2021**, *9*, 86. [CrossRef]

252. Chen, S.; Gao, J.; Yan, E.; Wang, Y.; Li, Y.; Lu, H.; Fan, L.; Wang, D.; An, Q. A Novel Porous Composite Membrane of PHA/PVA via Coupling of Electrospinning and Spin Coating for Antibacterial Applications. *Mater. Lett.* **2021**, *301*, 130279. [CrossRef]
253. Liu, Q.; Liu, G.; Liu, X.; Yang, M.; Xing, S.; Du, Y.; Xiong, X. Synthesis of an Electrospun PHA/RGO/Au Scaffold for Peripheral Nerve Regeneration: An in Vitro Study. *Appl. Nanosci.* **2020**, *10*, 687–694. [CrossRef]
254. Ivorra-martinez, J.; Quiles-carrillo, L.; Boronat, T. Assessment of the Mechanical and Thermal Properties of Injection-Molded Poly(3-Hydroxybutyrate-Co-3-Hydroxyhexanoate)/Hydroxyapatite Nanoparticles Parts for Use in Bone Tissue Engineering. *Polymers* **2020**, *12*, 1389. [CrossRef] [PubMed]
255. Chen, F.; Liu, X.; Ge, X.; Wang, Y.; Zhao, Z.; Zhang, X.; Chen, G.Q.; Sun, Y. Porous Polydroxyalkanoates (PHA) Scaffolds with Antibacterial Property for Oral Soft Tissue Regeneration. *Chem. Eng. J.* **2023**, *451*, 138899. [CrossRef]
256. Zeinali, R.; Khorasani, M.T.; Behnamghader, A.; Atai, M.; Del Valle, L.; Puiggalí, J. Poly(Hydroxybutyrate-Co-Hydroxyvalerate) Porous Matrices from Thermally Induced Phase Separation. *Polymers* **2020**, *12*, 2787. [CrossRef] [PubMed]
257. Lopresti, F.; Liga, A.; Capuana, E.; Gulfi, D.; Zanca, C.; Inguanta, R.; Brucato, V.; La Carrubba, V.; Carfì Pavia, F. Effect of Polyhydroxyalkanoate (PHA) Concentration on Polymeric Scaffolds Based on Blends of Poly-L-Lactic Acid (PLLA) and PHA Prepared via Thermally Induced Phase Separation (TIPS). *Polymers* **2022**, *14*, 2494. [CrossRef] [PubMed]
258. Zhang, X.; Li, J.; Chen, J.; Peng, Z.X.; Chen, J.N.; Liu, X.; Wu, F.; Zhang, P.; Chen, G.Q. Enhanced Bone Regeneration via PHA Scaffolds Coated with Polydopamine-Captured BMP2. *J. Mater. Chem. B* **2022**, *10*, 6214–6227. [CrossRef] [PubMed]
259. Raza, Z.A.; Abid, S.; Banat, I.M. Polyhydroxyalkanoates: Characteristics, Production, Recent Developments and Applications. *Int. Biodeterior. Biodegrad.* **2018**, *126*, 45–56. [CrossRef]
260. Turco, R.; Santagata, G.; Corrado, I.; Pezzella, C.; Di Serio, M. In Vivo and Post-Synthesis Strategies to Enhance the Properties of PHB-Based Materials: A Review. *Front. Bioeng. Biotechnol.* **2021**, *8*, 619266. [CrossRef]
261. Uzun, G.; Aydemir, D. Biocomposites from Polyhydroxybutyrate and Bio-Fillers by Solvent Casting Method. *Bull. Mater. Sci.* **2017**, *40*, 383–393. [CrossRef]
262. Anbukarasu, P.; Sauvageau, D.; Elias, A. The Effects of Solvent Casting Temperature and Physical Aging on Polyhydroxybutyrate-Graphene Nanoplatelet Composites. *Polym. Compos.* **2021**, *42*, 1451–1461. [CrossRef]
263. Hernandez, M.M.; Gupta, N.S.; Lee, K.S.; Dumont, J.H.; Pital, A.C.; Marrone, B.L.; Iverson, C.N. Characterization of Polyhydroxybutyrate-Based Composites Prepared by Injection Molding. *Polymers* **2021**, *13*, 4444. [CrossRef]
264. Bucci, D.Z.; Tavares, L.B.B.; Sell, I. PHB Packaging for the Storage of Food Products. *Polym. Test.* **2005**, *24*, 564–571. [CrossRef]
265. Olejnik, O.; Masek, A.; Zawadziłło, J. Processability and Mechanical Properties of Thermoplastic Polylactide/Polyhydroxybutyrate (PLA/PHB) Bioblends. *Materials* **2021**, *14*, 898. [CrossRef] [PubMed]
266. Arrieta, M.P.; López, J.; López, D.; Kenny, J.M.; Peponi, L. Development of Flexible Materials Based on Plasticized Electrospun PLA-PHB Blends: Structural, Thermal, Mechanical and Disintegration Properties. *Eur. Polym. J.* **2015**, *73*, 433–446. [CrossRef]
267. Kundrat, V.; Cernekova, N.; Kovalcik, A.; Enev, V.; Marova, I. Drug Release Kinetics of Electrospun PHB Meshes. *Materials* **2019**, *12*, 1924. [CrossRef]
268. Sánchez-Safont, E.L.; Aldureid, A.; Lagarón, J.M.; Cabedo, L.; Gámez-Pérez, J. Study of the Compatibilization Effect of Different Reactive Agents in Phb/Natural Fiber-Based Composites. *Polymers* **2020**, *12*, 1967. [CrossRef]
269. Wei, L.; McDonald, A.G. Peroxide Induced Cross-Linking by Reactive Melt Processing of Two Biopolyesters: Poly(3-Hydroxybutyrate) and Poly(l-Lactic Acid) to Improve Their Melting Processability. *J. Appl. Polym. Sci.* **2015**, *132*, 41724. [CrossRef]
270. Bugnicourt, E.; Cinelli, P.; Lazzeri, A.; Alvarez, V. Polyhydroxyalkanoate (PHA): Review of Synthesis, Characteristics, Processing and Potential Applications in Packaging. *Express Polym. Lett.* **2014**, *8*, 791–808. [CrossRef]
271. Erceg, T.; Rackov, S.; Terek, P.; Pilić, B. Preparation and Characterization of PHBV/PCL-Diol Blend Films. *Polymers* **2023**, *15*, 4694. [CrossRef]
272. Perez-Martinez, V.; Bello-Rocha, L.; Rodríguez-Rodriguez, C.; Sierra, C.A.; Castellanos, D.A. Obtention and Characterization of PLA/PHBV Thin Sheets by Solvent Casting and Extrusion with Application in Food Packaging. *Bull. Mater. Sci.* **2024**, *47*, 47. [CrossRef]
273. Samaniego-Aguilar, K.; Sánchez-Safont, E.; Arrillaga, A.; Anakabe, J.; Gamez-Perez, J.; Cabedo, L. In Service Performance of Toughened PHBV/TPU Blends Obtained by Reactive Extrusion for Injected Parts. *Polymers* **2022**, *14*, 2337. [CrossRef]
274. Frącz, W.; Janowski, G.; Smusz, R.; Szumski, M. The Influence of Chosen Plant Fillers in PHBV Composites on the Processing Conditions, Mechanical Properties and Quality of Molded Pieces. *Polymers* **2021**, *13*, 3934. [CrossRef] [PubMed]
275. Mallegni, N.; Signori, F.; Gigante, V.; Amato, G.; Navarini, L.; Seggiani, M.; Cinelli, P.; Lazzeri, A. Poly(3-Hydroxybutyrate-Co-3-Hydroxyvalerate) (PHBV) Based Biocomposites Containing Coffee Silverskin: Injection Moulding and Migration Performances Assessment. *Chem. Eng. Trans.* **2022**, *92*, 643–648. [CrossRef]
276. Kaniuk, Ł.; Stachewicz, U. Development and Advantages of Biodegradable PHA Polymers Based on Electrospun PHBV Fibers for Tissue Engineering and Other Biomedical Applications. *ACS Biomater. Sci. Eng.* **2021**, *7*, 5339–5362. [CrossRef] [PubMed]
277. Rabbani, S.; Jafari, R.; Momen, G. Superhydrophobic Micro-Nanofibers from PHBV-SiO_2 Biopolymer Composites Produced by Electrospinning. *Funct. Compos. Mater.* **2022**, *3*, 1. [CrossRef]
278. Karbowniczek, J.E.; Kaniuk, Ł.; Berniak, K.; Gruszczyński, A.; Stachewicz, U. Enhanced Cells Anchoring to Electrospun Hybrid Scaffolds With PHBV and HA Particles for Bone Tissue Regeneration. *Front. Bioeng. Biotechnol.* **2021**, *9*, 632029. [CrossRef]

279. Kontogianni, G.I.; Bonatti, A.F.; De Maria, C.; Naseem, R.; Melo, P.; Coelho, C.; Vozzi, G.; Dalgarno, K.; Quadros, P.; Vitale-Brovarone, C.; et al. Promotion of In Vitro Osteogenic Activity by Melt Extrusion-Based PLLA/PCL/PHBV Scaffolds Enriched with Nano-Hydroxyapatite and Strontium Substituted Nano-Hydroxyapatite. *Polymers* **2023**, *15*, 1052. [CrossRef]
280. de Carvalho, J.G.; Zanini, N.C.; Claro, A.M.; do Amaral, N.C.; Barud, H.S.; Mulinari, D.R. Composite Filaments OF PHBV Reinforced with ZrO2·nH2O Particles for 3D Printing. *Polym. Bull.* **2022**, *79*, 2113–2132. [CrossRef]
281. Garcia-Campo, M.J.; Quiles-Carrillo, L.; Masia, J.; Reig-Pérez, M.J.; Montanes, N.; Balart, R. Environmentally Friendly Compatibilizers from Soybean Oil for Ternary Blends of Poly(Lactic Acid)-PLA, Poly(ε-Caprolactone)-PCL and Poly(3-Hydroxybutyrate)-PHB. *Materials* **2017**, *10*, 1339. [CrossRef]
282. Sempere-Torregrosa, J.; Ferri, J.M.; de la Rosa-Ramírez, H.; Pavon, C.; Samper, M.D. Effect of Epoxidized and Maleinized Corn Oil on Properties of Polylactic Acid (PLA) and Polyhydroxybutyrate (PHB) Blend. *Polymers* **2022**, *14*, 4205. [CrossRef]
283. Abdelwahab, M.A.; Flynn, A.; Chiou, B.-S.; Imam, S.; Orts, W.; Chiellini, E. Thermal, Mechanical and Morphological Characterization of Plasticized PLA-PHB Blends. *Polym. Degrad. Stab.* **2012**, *97*, 1822–1828. [CrossRef]
284. Seoane, I.T.; Manfredi, L.B.; Cyras, V.P. Effect of Two Different Plasticizers on the Properties of Poly (3-hydroxybutyrate) Binary and Ternary Blends. *J. Appl. Polym. Sci.* **2018**, *135*, 46016. [CrossRef]
285. Seydibeyoğlu, M.Ö.; Misra, M.; Mohanty, A. Synergistic Improvements in the Impact Strength And% Elongation of Polyhydroxybutyrate-Co-Valerate Copolymers with Functionalized Soybean Oils and POSS. *Int. J. Plast. Technol.* **2010**, *14*, 1–16. [CrossRef]
286. Quispe, M.M.; Lopez, O.V.; Boina, D.A.; Stumbé, J.-F.; Villar, M.A. Glycerol-Based Additives of Poly (3-Hydroxybutyrate) Films. *Polym. Test.* **2021**, *93*, 107005. [CrossRef]
287. Wang, L.; Zhu, W.; Wang, X.; Chen, X.; Chen, G.; Xu, K. Processability Modifications of Poly (3-hydroxybutyrate) by Plasticizing, Blending, and Stabilizing. *J. Appl. Polym. Sci.* **2008**, *107*, 166–173. [CrossRef]
288. Parra, D.F.; Fusaro, J.; Gaboardi, F.; Rosa, D.S. Influence of Poly (Ethylene Glycol) on the Thermal, Mechanical, Morphological, Physical–Chemical and Biodegradation Properties of Poly (3-Hydroxybutyrate). *Polym. Degrad. Stab.* **2006**, *91*, 1954–1959. [CrossRef]
289. Numata, K.; Abe, H.; Iwata, T. Biodegradability of Poly(Hydroxyalkanoate) Materials. *Materials* **2009**, *2*, 1104–1126. [CrossRef]
290. White, E.M.; Horn, J.; Wang, S.; Crawford, B.; Ritchie, B.W.; Carraway, D.; Locklin, J. Comparative Study of the Biological Degradation of Poly(3-Hydroxybutyrate-Co-3-Hydroxyhexanoate) Microbeads in Municipal Wastewater in Environmental and Controlled Laboratory Conditions. *Environ. Sci. Technol.* **2021**, *55*, 11646–11656. [CrossRef]
291. Boey, J.Y.; Mohamad, L.; Khok, Y.S.; Tay, G.S.; Baidurah, S. A Review of the Applications and Biodegradation of Polyhydroxyalkanoates and Poly(Lactic Acid) and Its Composites. *Polymers* **2021**, *13*, 1544. [CrossRef]
292. Vodicka, J.; Wikarska, M.; Trudicova, M.; Juglova, Z.; Pospisilova, A.; Kalina, M.; Slaninova, E.; Obruca, S.; Sedlacek, P. Degradation of P(3HB-Co-4HB) Films in Simulated Body Fluids. *Polymers* **2022**, *14*, 1990. [CrossRef]
293. Ansari, S.; Sami, N.; Yasin, D.; Ahmad, N.; Fatma, T. Biomedical Applications of Environmental Friendly Poly-Hydroxyalkanoates. *Int. J. Biol. Macromol.* **2021**, *183*, 549–563. [CrossRef]
294. Ong, S.Y.; Sudesh, K. Effects of Polyhydroxyalkanoate Degradation on Soil Microbial Community. *Polym. Degrad. Stab.* **2016**, *131*, 9–19. [CrossRef]
295. Borelbach, P.; Kopitzky, R.; Dahringer, J.; Gutmann, P. Degradation Behavior of Biodegradable Man-Made Fibers in Natural Soil and in Compost. *Polymers* **2023**, *15*, 2959. [CrossRef] [PubMed]
296. Komiyama, K.; Omura, T.; Iwata, T. Effect of Morphology and Molecular Orientation on Environmental Water Biodegradability of Poly[(R)-3-Hydroxybutyrate-Co-(R)-3-Hydroxyvalerate]. *Polym. Degrad. Stab.* **2021**, *193*, 109719. [CrossRef]
297. Kim, H.W.; Chung, C.W.; Kim, Y.B.; Rhee, Y.H. Preparation and Hydrolytic Degradation of Semi-Interpenetrating Networks of Poly(3-Hydroxyundecenoate) and Poly(Lactide-Co-Glycolide). *Int. J. Biol. Macromol.* **2005**, *37*, 221–226. [CrossRef] [PubMed]
298. Roohi; Zaheer, M.R.; Kuddus, M. PHB (Poly-β-Hydroxybutyrate) and Its Enzymatic Degradation. *Polym. Adv. Technol.* **2018**, *29*, 30–40. [CrossRef]
299. Kliem, S.; Kreutzbruck, M.; Bonten, C. Review on the Biological Degradation of Polymers in Various Environments. *Materials* **2020**, *13*, 4586. [CrossRef]
300. Park, S.L.; Cho, J.Y.; Choi, T.R.; Song, H.S.; Bhatia, S.K.; Gurav, R.; Park, S.H.; Park, K.; Joo, J.C.; Hwang, S.Y.; et al. Improvement of Polyhydroxybutyrate (PHB) Plate-Based Screening Method for PHB Degrading Bacteria Using Cell-Grown Amorphous PHB and Recovered by Sodium Dodecyl Sulfate (SDS). *Int. J. Biol. Macromol.* **2021**, *177*, 413–421. [CrossRef]
301. Balogová, A.F.; Trebuňová, M.; Ižaríková, G.; Kaščák, L.; Mitrík, L.; Klímová, J.; Feranc, J.; Modrák, M.; Hudák, R.; Živčák, J. In Vitro Degradation of Specimens Produced from Pla/Phb by Additive Manufacturing in Simulated Conditions. *Polymers* **2021**, *13*, 1542. [CrossRef]
302. Kervran, M.; Vagner, C.; Cochez, M.; Ponçot, M.; Saeb, M.R.; Vahabi, H. Thermal Degradation of Polylactic Acid (PLA)/Polyhydroxybutyrate (PHB) Blends: A Systematic Review. *Polym. Degrad. Stab.* **2022**, *201*, 109995. [CrossRef]
303. Sabapathy, P.C.; Devaraj, S.; Anburajan, P.; Parvez, A.; Kathirvel, P.; Qi, X. Active Polyhydroxybutyrate (PHB)/Sugarcane Bagasse Fiber-Based Anti-Microbial Green Composite: Material Characterization and Degradation Studies. *Appl. Nanosci.* **2023**, *13*, 1187–1199. [CrossRef]

304. Rech, C.R.; da Silva Brabes, K.C.; Bagnara e Silva, B.E.; Bittencourt, P.R.S.; Koschevic, M.T.; da Silveira, T.F.S.; Martines, M.A.U.; Caon, T.; Martelli, S.M. Biodegradation of Eugenol-Loaded Polyhydroxybutyrate Films in Different Soil Types. *Case Stud. Chem. Environ. Eng.* **2020**, *2*, 100014. [CrossRef]
305. Pei, R.; Tarek-Bahgat, N.; Van Loosdrecht, M.C.M.; Kleerebezem, R.; Werker, A.G. Influence of Environmental Conditions on Accumulated Polyhydroxybutyrate in Municipal Activated Sludge. *Water Res.* **2023**, *232*, 119653. [CrossRef] [PubMed]
306. Swiontek Brzezinka, M.; Richert, A.; Kalwasińska, A.; Świątczak, J.; Deja-Sikora, E.; Walczak, M.; Michalska-Sionkowska, M.; Piekarska, K.; Kaczmarek-Szczepańska, B. Microbial Degradation of Polyhydroxybutyrate with Embedded Polyhexamethylene Guanidine Derivatives. *Int. J. Biol. Macromol.* **2021**, *187*, 309–318. [CrossRef] [PubMed]
307. Cho, J.Y.; Lee Park, S.; Lee, H.J.; Kim, S.H.; Suh, M.J.; Ham, S.; Bhatia, S.K.; Gurav, R.; Park, S.H.; Park, K.; et al. Polyhydroxyalkanoates (PHAs) Degradation by the Newly Isolated marine *Bacillus* sp. JY14. *Chemosphere* **2021**, *283*, 131172. [CrossRef] [PubMed]
308. Cho, J.Y.; Kim, S.H.; Jung, H.J.; Cho, D.H.; Kim, B.C.; Bhatia, S.K.; Ahn, J.; Jeon, J.M.; Yoon, J.J.; Lee, J.; et al. Finding a Benign Plasticizer to Enhance the Microbial Degradation of Polyhydroxybutyrate (PHB) Evaluated by PHB Degrader *Microbulbifer* sp. SOL66. *Polymers* **2022**, *14*, 3625. [CrossRef]
309. Mosnáčková, K.; Šišková, A.O.; Kleinová, A.; Danko, M.; Mosnáček, J. Properties and Degradation of Novel Fully Biodegradable Pla/Phb Blends Filled with Keratin. *Int. J. Mol. Sci.* **2020**, *21*, 9678. [CrossRef]
310. Zhou, Y.; Li, Y.; Li, D.; Yin, Y.; Zhou, F.L. Electrospun PHB/Chitosan Composite Fibrous Membrane and Its Degradation Behaviours in Different PH Conditions. *J. Funct. Biomater.* **2022**, *13*, 58. [CrossRef]
311. Râpă, M.; Zaharia, C.; Stănescu, P.O.; Căărică, A.; Matei, E.; Predescu, A.M.; Pantilimon, M.C.; Vidu, R.; Predescu, C.; Cioflan, H. In Vitro Degradation of PHB/Bacterial Cellulose Biocomposite Scaffolds. *Int. J. Polym. Sci.* **2021**, *2021*, 3820364. [CrossRef]
312. Lyshtva, P.; Voronova, V.; Barbir, J.; Leal Filho, W.; Kröger, S.D.; Witt, G.; Miksch, L.; Sabowski, R.; Gutow, L.; Frank, C.; et al. Degradation of a Poly(3-Hydroxybutyrate-Co-3-Hydroxyvalerate) (PHBV) Compound in Different Environments. *Heliyon* **2024**, *10*, e24770. [CrossRef]
313. Kim, J.; Gupta, N.S.; Bezek, L.B.; Linn, J.; Bejagam, K.K.; Banerjee, S.; Dumont, J.H.; Nam, S.Y.; Kang, H.W.; Park, C.H.; et al. Biodegradation Studies of Polyhydroxybutyrate and Polyhydroxybutyrate-Co-Polyhydroxyvalerate Films in Soil. *Int. J. Mol. Sci.* **2023**, *24*, 7638. [CrossRef]
314. Furgier, V.; Root, A.; Heinmaa, I.; Zamani, A.; Åkesson, D. Development and Characterisation of Composites Prepared from PHBV Compounded with Organic Waste Reinforcements, and Their Soil Biodegradation. *Materials* **2024**, *17*, 768. [CrossRef] [PubMed]
315. Hernández-García, E.; Vargas, M.; Chiralt, A.; González-Martínez, C. Biodegradation of PLA-PHBV Blend Films as Affected by the Incorporation of Different Phenolic Acids. *Foods* **2022**, *11*, 243. [CrossRef] [PubMed]
316. Bonnenfant, C.; Chatellard, L.; Gontard, N.; Aouf, C. Effect of Quercetin and Gallic Acid on the Microbial Degradation of Poly(3-Hydroxybutyrate-Co-3-Hydroxyvalerate) (PHBV) Materials. *J. Polym. Environ.* **2023**, *31*, 1478–1488. [CrossRef]
317. *NF T 51-800:2015*; Plastics—Specifications for Plastics Suitable for Home Composting. Association Française de Normalisation: La Plaine Saint-Denis, France, 2015.
318. Meereboer, K.W.; Pal, A.K.; Cisneros-López, E.O.; Misra, M.; Mohanty, A.K. The Effect of Natural Fillers on the Marine Biodegradation Behaviour of Poly(3-Hydroxybutyrate-Co-3-Hydroxyvalerate) (PHBV). *Sci. Rep.* **2021**, *11*, 911. [CrossRef]
319. *ASTM D7991-15*; Standard Test Method for Determining Aerobic Biodegradation of Plastics Buried in Sandy Marine Sediment under Controlled Laboratory Conditions. American Society for Testing and Materials: Philadelphia, PA, USA, 2015. [CrossRef]
320. Hammiche, D.; Boukerrou, A.; Grohens, Y.; Guermazi, N.; Arrakhiz, F.E. Mechanical Properties and Biodegradation of Biocomposites Based on Poly (Hydroxybutyrate-Co-Valerate) and Alfa Fibers. *J. Polym. Res.* **2020**, *27*, 308. [CrossRef]
321. Patel, R.; Monticone, D.; Lu, M.; Grøndahl, L.; Huang, H. Hydrolytic Degradation of Porous Poly(Hydroxybutyrate-Co-Hydroxyvalerate) Scaffolds Manufactured Using Selective Laser Sintering. *Polym. Degrad. Stab.* **2021**, *187*, 109545. [CrossRef]
322. Zhu, J.; Chen, Y.; Yu, H.Y.; Guan, Y.; Zhou, Y.; Yang, X.; Zou, Z.Y.; Tam, K.C. Comprehensive Insight into Degradation Mechanism of Green Biopolyester Nanocomposites Using Functionalized Cellulose Nanocrystals. *ACS Sustain. Chem. Eng.* **2019**, *7*, 15537–15547. [CrossRef]
323. Kovalcik, A.; Obruca, S.; Kalina, M.; Machovsky, M.; Enev, V.; Jakesova, M.; Sobkova, M.; Marova, I. Enzymatic Hydrolysis of Poly(3-Hydroxybutyrate-Co-3-Hydroxyvalerate) Scaffolds. *Materials* **2020**, *13*, 2992. [CrossRef]
324. Henning, F.G.; Ito, V.C.; Demiate, I.M.; Lacerda, L.G. Non-Conventional Starches for Biodegradable Films: A Review Focussing on Characterisation and Recent Applications in Food Packaging. *Carbohydr. Polym. Technol. Appl.* **2022**, *4*, 100157.
325. Zhu, X.; Dai, Q.; Silveira Hornung, P.; Lopes Teixeira, G.; Nkhata Malunga, L.; Ames, N.; Beta, T. Valorizing Canadian Oat Cultivars to Obtain Non-Conventional Starch: Pasting, Physicochemical, and Morphological Properties. *Starch-Stärke* **2023**, *75*, 2200142. [CrossRef]
326. González-Mendoza, M.; Martínez-Bustos, F.; Castaño-Tostado, E.; Amaya-Llano, S.L. Effect of Microwave Irradiation on Acid Hydrolysis of Faba Bean Starch: Physicochemical Changes of the Starch Granules. *Molecules* **2022**, *27*, 3528. [CrossRef] [PubMed]
327. Niranjana Prabhu, T.; Prashantha, K. A Review on Present Status and Future Challenges of Starch Based Polymer Films and Their Composites in Food Packaging Applications. *Polym. Compos.* **2018**, *39*, 2499–2522. [CrossRef]
328. Fredrikssona, H.; Silveriob, J.; Andemon', R.; Eliassonb, A.-C.; Amanta, P. The Influence of Amylose and Amylopectin Characteristics on Gelatinization and Retrogradation Properties of Different Starches. *Carbohydr. Polym.* **1998**, *35*, 119–134.

329. Jane, J.; Chen, Y.Y.; Lee, L.F.; McPherson, A.E.; Wong, K.S.; Radosavljevic, M.; Kasemsuwan, T. Effects of Amylopectin Branch Chain Length and Amylose Content on the Gelatinization and Pasting Properties of Starch. *Cereal Chem.* **1999**, *76*, 629–637. [CrossRef]
330. Singh Sandhu, K.; Singh, N. Relationships between Selected Properties of Starches from Different Corn Lines. *Int. J. Food Prop.* **2005**, *8*, 481–491. [CrossRef]
331. Muñoz-Gimena, P.F.; Oliver-Cuenca, V.; Peponi, L.; López, D. A Review on Reinforcements and Additives in Starch-Based Composites for Food Packaging. *Polymers* **2023**, *15*, 2972. [CrossRef]
332. Cataño, F.A.; Moreno-Serna, V.; Cament, A.; Loyo, C.; Yáñez-S, M.; Ortiz, J.A.; Zapata, P.A. Green Composites Based on Thermoplastic Starch Reinforced with Micro- and Nano-Cellulose by Melt Blending—A Review. *Int. J. Biol. Macromol.* **2023**, *248*, 125939.
333. García, N.L.; Ribba, L.; Dufresne, A.; Aranguren, M.; Goyanes, S. Effect of Glycerol on the Morphology of Nanocomposites Made from Thermoplastic Starch and Starch Nanocrystals. *Carbohydr. Polym.* **2011**, *84*, 203–210. [CrossRef]
334. Surendren, A.; Mohanty, A.K.; Liu, Q.; Misra, M. A Review of Biodegradable Thermoplastic Starches, Their Blends and Composites: Recent Developments and Opportunities for Single-Use Plastic Packaging Alternatives. *Green. Chem.* **2022**, *19*, 8606–8636.
335. Abdullah, Z.W.; Dong, Y. Recent Advances and Perspectives on Starch Nanocomposites for Packaging Applications. *J. Mater. Sci.* **2018**, *53*, 15319–15339. [CrossRef]
336. Sessini, V.; Arrieta, M.P.; Kenny, J.M.; Peponi, L. Processing of Edible Films Based on Nanoreinforced Gelatinized Starch. *Polym. Degrad. Stab.* **2016**, *132*, 157–168. [CrossRef]
337. Borbolla-Jiménez, F.V.; Peña-Corona, S.I.; Farah, S.J.; Jiménez-Valdés, M.T.; Pineda-Pérez, E.; Romero-Montero, A.; Del Prado-Audelo, M.L.; Bernal-Chávez, S.A.; Magaña, J.J.; Leyva-Gómez, G. Films for Wound Healing Fabricated Using a Solvent Casting Technique. *Pharmaceutics* **2023**, *15*, 1914. [CrossRef] [PubMed]
338. Singh, B.; Sharma, C.; Sharma, S. Fundamentals of extrusion processing. In *Novel Food Processing Technologies*; New India Publishing Agency: New Delhi, India, 2017.
339. Combrzyński, M.; Wójtowicz, A.; Mitrus, M.; Oniszczuk, T.; Matwijczuk, A.; Pawelczyk, P.; Mościcki, L. Effect of Starch Type and Screw Speed on Mechanical Properties of Extrusion-Cooked Starch-Based Foams. *Int. Agrophys.* **2019**, *33*, 233–240. [CrossRef]
340. Ye, J.; Hu, X.; Luo, S.; Liu, W.; Chen, J.; Zeng, Z.; Liu, C. Properties of Starch after Extrusion: A Review. *Starch-Stärke* **2018**, *70*, 1700110. [CrossRef]
341. González-Seligra, P.; Guz, L.; Ochoa-Yepes, O.; Goyanes, S.; Famá, L. Influence of Extrusion Process Conditions on Starch Film Morphology. *LWT* **2017**, *84*, 520–528. [CrossRef]
342. Pushpadass, H.A.; Marx, D.B.; Hanna, M.A. Effects of Extrusion Temperature and Plasticizers on the Physical and Functional Properties of Starch Films. *Starch-Stärke* **2008**, *60*, 527–538. [CrossRef]
343. Logié, N.; Della Valle, G.; Rolland-Sabaté, A.; Descamps, N.; Soulestin, J. How Does Temperature Govern Mechanisms of Starch Changes during Extrusion? *Carbohydr. Polym.* **2018**, *184*, 57–65. [CrossRef] [PubMed]
344. Adewale, P.; Yancheshmeh, M.S.; Lam, E. Starch Modification for Non-Food, Industrial Applications: Market Intelligence and Critical Review. *Carbohydr. Polym.* **2022**, *291*, 119590. [CrossRef]
345. Cheng, H.; Chen, L.; McClements, D.J.; Yang, T.; Zhang, Z.; Ren, F.; Miao, M.; Tian, Y.; Jin, Z. Starch-Based Biodegradable Packaging Materials: A Review of Their Preparation, Characterization and Diverse Applications in the Food Industry. *Trends Food Sci. Technol.* **2021**, *114*, 70–82. [CrossRef]
346. Liu, G.; Gu, Z.; Hong, Y.; Cheng, L.; Li, C. Electrospun Starch Nanofibers: Recent Advances, Challenges, and Strategies for Potential Pharmaceutical Applications. *J. Control. Release* **2017**, *252*, 95–107. [CrossRef]
347. Hanaa, J.; Abd Alradha, R.; Alzubaidy, A. Preparation and Characterization of Polyvinyl Alcohol/Starch/Bio Oil Extraction Blends Coating for Food Packaging Using Electrospinning Technique. *Egypt. J. Chem.* **2022**, *66*, 141–149. [CrossRef]
348. Malafatti, J.O.D.; de Oliveira Ruellas, T.M.; Sciena, C.R.; Paris, E.C. PLA/Starch Biodegradable Fibers Obtained by the Electrospinning Method for Micronutrient Mineral Release. *AIMS Mater. Sci.* **2023**, *10*, 200–212. [CrossRef]
349. Asl, M.A.; Karbasi, S.; Beigi-Boroujeni, S.; Zamanlui Benisi, S.; Saeed, M. Evaluation of the Effects of Starch on Polyhydroxybutyrate Electrospun Scaffolds for Bone Tissue Engineering Applications. *Int. J. Biol. Macromol.* **2021**, *191*, 500–513. [CrossRef] [PubMed]
350. Wang, N.; Yu, J.; Chang, P.R.; Ma, X. Influence of Formamide and Water on the Properties of Thermoplastic Starch/Poly(Lactic Acid) Blends. *Carbohydr. Polym.* **2008**, *71*, 109–118. [CrossRef]
351. Li, H.; Huneault, M.A. Comparison of Sorbitol and Glycerol as Plasticizers for Thermoplastic Starch in TPS/PLA Blends. *J. Appl. Polym. Sci.* **2011**, *119*, 2439–2448. [CrossRef]
352. Lauer, M.K.; Smith, R.C. Recent Advances in Starch-based Films toward Food Packaging Applications: Physicochemical, Mechanical, and Functional Properties. *Compr. Rev. Food Sci. Food Saf.* **2020**, *19*, 3031–3083. [CrossRef]
353. Angellier, H.; Molina-Boisseau, S.; Dole, P.; Dufresne, A. Thermoplastic Starch– Waxy Maize Starch Nanocrystals Nanocomposites. *Biomacromolecules* **2006**, *7*, 531–539. [CrossRef]
354. Jiugao, Y.; Ning, W.; Xiaofei, M. The Effects of Citric Acid on the Properties of Thermoplastic Starch Plasticized by Glycerol. *Starch-Stärke* **2005**, *57*, 494–504. [CrossRef]
355. Jia, P.; Xia, H.; Tang, K.; Zhou, Y. Plasticizers Derived from Biomass Resources: A Short Review. *Polymers* **2018**, *10*, 1303. [CrossRef]

356. Zdanowicz, M.; Johansson, C. Mechanical and Barrier Properties of Starch-Based Films Plasticized with Two-or Three Component Deep Eutectic Solvents. *Carbohydr. Polym.* **2016**, *151*, 103–112. [CrossRef]
357. Montilla-Buitrago, C.E.; Gómez-López, R.A.; Solanilla-Duque, J.F.; Serna-Cock, L.; Villada-Castillo, H.S. Effect of Plasticizers on Properties, Retrogradation, and Processing of Extrusion-obtained Thermoplastic Starch: A Review. *Starch-Stärke* **2021**, *73*, 2100060. [CrossRef]
358. Yang, J.-H.; Yu, J.-G.; Ma, X.-F. Study on the Properties of Ethylenebisformamide and Sorbitol Plasticized Corn Starch (ESPTPS). *Carbohydr. Polym.* **2006**, *66*, 110–116. [CrossRef]
359. Esmaeili, M.; Pircheraghi, G.; Bagheri, R. Optimizing the Mechanical and Physical Properties of Thermoplastic Starch via Tuning the Molecular Microstructure through Co-plasticization by Sorbitol and Glycerol. *Polym. Int.* **2017**, *66*, 809–819. [CrossRef]
360. Krogars, K.; Heinämäki, J.; Karjalainen, M.; Niskanen, A.; Leskelä, M.; Yliruusi, J. Enhanced Stability of Rubbery Amylose-Rich Maize Starch Films Plasticized with a Combination of Sorbitol and Glycerol. *Int. J. Pharm.* **2003**, *251*, 205–208. [CrossRef]
361. Battegazzore, D.; Bocchini, S.; Nicola, G.; Martini, E.; Frache, A. Isosorbide, a Green Plasticizer for Thermoplastic Starch That Does Not Retrogradate. *Carbohydr. Polym.* **2015**, *119*, 78–84. [CrossRef]
362. Zuo, Y.; Gu, J.; Tan, H.; Zhang, Y. Thermoplastic Starch Prepared with Different Plasticizers: Relation between Degree of Plasticization and Properties. *J. Wuhan Univ. Technol. Mater. Sci. Ed.* **2015**, *30*, 423–428. [CrossRef]
363. Thomas, M.S.; Richard, V.G. Amino Acids as Plasticizers for Starch-Based Plastics. *Starch-Stärke* **1997**, *49*, 245–249.
364. Ma, X.F.; Yu, J.G.; Wan, J.J. Urea and Ethanolamine as a Mixed Plasticizer for Thermoplastic Starch. *Carbohydr. Polym.* **2006**, *64*, 267–273. [CrossRef]
365. Edhirej, A.; Sapuan, S.M.; Jawaid, M.; Zahari, N.I. Effect of Various Plasticizers and Concentration on the Physical, Thermal, Mechanical, and Structural Properties of Cassava-Starch-Based Films. *Starch-Stärke* **2017**, *69*, 1500366. [CrossRef]
366. Lin, N.; Huang, J.; Chang, P.R.; Anderson, D.P.; Yu, J. Preparation, Modification, and Application of Starch Nanocrystals in Nanomaterials: A Review. *J. Nanomater.* **2011**, *2011*, 573687. [CrossRef]
367. Dufresne, A. Polysaccharide Nano Crystal Reinforced Nanocomposites. *Can. J. Chem.* **2008**, *86*, 484–494. [CrossRef]
368. Zhang, G.; Wu, D.; Xie, W.; Wang, Z.; Xu, C. Green Poly(β-Hydroxybutyrate)/Starch Nanocrystal Composites: Tuning the Nucleation and Spherulite Morphology through Surface Acetylation of Starch Nanocrystal. *Carbohydr. Polym.* **2018**, *195*, 79–88. [CrossRef] [PubMed]
369. Sharafi Zamir, S.; Fathi, B.; Ajji, A.; Robert, M.; Elkoun, S. Crystallinity and Gas Permeability of Poly (Lactic Acid)/Starch Nanocrystal Nanocomposite. *Polymers* **2022**, *14*, 2802. [CrossRef] [PubMed]
370. Nakamura, Y. (Ed.) *Starch*; Springer: Tokyo, Japan, 2015; ISBN 978-4-431-55494-3.
371. Smith, A.M.; Zeeman, S.C.; Smith, S.M. STARCH DEGRADATION. *Annu. Rev. Plant Biol.* **2005**, *56*, 73–98. [CrossRef]
372. Sanyang, M.L.; Sapuan, S.M.; Jawaid, M.; Ishak, M.R.; Sahari, J. Development and Characterization of Sugar Palm Starch and Poly(Lactic Acid) Bilayer Films. *Carbohydr. Polym.* **2016**, *146*, 36–45. [CrossRef]
373. Thoorens, G.; Krier, F.; Leclercq, B.; Carlin, B.; Evrard, B. Microcrystalline Cellulose, a Direct Compression Binder in a Quality by Design Environment—A Review. *Int. J. Pharm.* **2014**, *473*, 64–72. [CrossRef]
374. Habibi, Y. Key Advances in the Chemical Modification of Nanocelluloses. *Chem. Soc. Rev.* **2014**, *43*, 1519–1542. [CrossRef]
375. Wang, J.; Tavakoli, J.; Tang, Y. Bacterial Cellulose Production, Properties and Applications with Different Culture Methods—A Review. *Carbohydr. Polym.* **2019**, *219*, 63–76. [CrossRef]
376. Siqueira, G.; Bras, J.; Dufresne, A. Cellulosic Bionanocomposites: A Review of Preparation, Properties and Applications. *Polymers* **2010**, *2*, 728–765. [CrossRef]
377. Brinchi, L.; Cotana, F.; Fortunati, E.; Kenny, J.M. Production of Nanocrystalline Cellulose from Lignocellulosic Biomass: Technology and Applications. *Carbohydr. Polym.* **2013**, *94*, 154–169. [CrossRef]
378. Arrieta, M.P.; Peltzer, M.A.; López, J.; Peponi, L. PLA-Based Nanocomposites Reinforced with CNC for Food Packaging Applications: From Synthesis to Biodegradation. In *Industrial Applications of Renewable Biomass Products*; Springer International Publishing: Cham, Switzerland, 2017; pp. 265–300.
379. Lavoine, N.; Desloges, I.; Bras, J. Microfibrillated Cellulose Coatings as New Release Systems for Active Packaging. *Carbohydr. Polym.* **2014**, *103*, 528–537. [CrossRef] [PubMed]
380. Ramírez Tapias, Y.A.; Di Monte, M.V.; Peltzer, M.A.; Salvay, A.G. Bacterial Cellulose Films Production by Kombucha Symbiotic Community Cultured on Different Herbal Infusions. *Food Chem.* **2022**, *372*, 131346. [CrossRef] [PubMed]
381. Machado, B.; Costa, S.M.; Costa, I.; Fangueiro, R.; Ferreira, D.P. The Potential of Algae as a Source of Cellulose and Its Derivatives for Biomedical Applications. *Cellulose* **2024**, *31*, 3353–3376. [CrossRef]
382. Liu, Z.; Wang, Y.; Guo, S.; Liu, J.; Zhu, P. Preparation and Characterization of Bacterial Cellulose Synthesized by Kombucha from Vinegar Residue. *Int. J. Biol. Macromol.* **2024**, *258*, 128939. [CrossRef]
383. Agüero, A.; Lascano, D.; Ivorra-Martinez, J.; Gómez-Caturla, J.; Arrieta, M.P.; Balart, R. Use of Bacterial Cellulose Obtained from Kombucha Fermentation in Spent Coffee Grounds for Active Composites Based on PLA and Maleinized Linseed Oil. *Ind. Crops Prod.* **2023**, *202*, 116971. [CrossRef]
384. Agüero, Á.; Corral Perianes, E.; Abarca de las Muelas, S.S.; Lascano, D.; de la Fuente García-Soto, M.D.M.; Peltzer, M.A.; Balart, R.; Arrieta, M.P. Plasticized Mechanical Recycled PLA Films Reinforced with Microbial Cellulose Particles Obtained from Kombucha Fermented in Yerba Mate Waste. *Polymers* **2023**, *15*, 285. [CrossRef]

385. Siqueira, G.; Bras, J.; Dufresne, A. Cellulose Whiskers versus Microfibrils: Influence of the Nature of the Nanoparticle and Its Surface Functionalization on the Thermal and Mechanical Properties of Nanocomposites. *Biomacromolecules* **2009**, *10*, 425–432. [CrossRef]
386. Jonoobi, M.; Oladi, R.; Davoudpour, Y.; Oksman, K.; Dufresne, A.; Hamzeh, Y.; Davoodi, R. Different Preparation Methods and Properties of Nanostructured Cellulose from Various Natural Resources and Residues: A Review. *Cellulose* **2015**, *22*, 935–969. [CrossRef]
387. Missoum, K.; Belgacem, M.N.; Bras, J. Nanofibrillated Cellulose Surface Modification: A Review. *Materials* **2013**, *6*, 1745–1766. [CrossRef]
388. Miao, C.; Hamad, W.Y. Cellulose Reinforced Polymer Composites and Nanocomposites: A Critical Review. *Cellulose* **2013**, *20*, 2221–2262. [CrossRef]
389. Abdul Khalil, H.P.S.; Bhat, A.H.; Ireana Yusra, A.F. Green Composites from Sustainable Cellulose Nanofibrils: A Review. *Carbohydr. Polym.* **2012**, *87*, 963–979. [CrossRef]
390. Pei, A.; Zhou, Q.; Berglund, L.A. Functionalized Cellulose Nanocrystals as Biobased Nucleation Agents in Poly(l-Lactide) (PLLA)—Crystallization and Mechanical Property Effects. *Compos. Sci. Technol.* **2010**, *70*, 815–821. [CrossRef]
391. Bondeson, D.; Mathew, A.; Oksman, K. Optimization of the Isolation of Nanocrystals from Microcrystalline Cellulose by Acid Hydrolysis. *Cellulose* **2006**, *13*, 171–180. [CrossRef]
392. Fortunati, E.; Peltzer, M.; Armentano, I.; Torre, L.; Jiménez, A.; Kenny, J.M. Effects of Modified Cellulose Nanocrystals on the Barrier and Migration Properties of PLA Nano-Biocomposites. *Carbohydr. Polym.* **2012**, *90*, 948–956. [CrossRef] [PubMed]
393. Arrieta, M.P.; Fortunati, E.; Burgos, N.; Peltzer, M.A.; López, J.; Peponi, L. Nanocellulose-Based Polymeric Blends for Food Packaging Applications. In *Multifunctional Polymeric Nanocomposites Based on Cellulosic Reinforcements*; William Andrew Publishing: Norwich, NY, USA, 2016; pp. 205–252.
394. Bondeson, D.; Oksman, K. Dispersion and Characteristics of Surfactant Modified Cellulose Whiskers Nanocomposites. *Compos. Interfaces* **2007**, *14*, 617–630. [CrossRef]
395. Mariano, M.; El Kissi, N.; Dufresne, A. Cellulose Nanocrystals and Related Nanocomposites: Review of Some Properties and Challenges. *J. Polym. Sci. B Polym. Phys.* **2014**, *52*, 791–806. [CrossRef]
396. Shah, N.; Ul-Islam, M.; Khattak, W.A.; Park, J.K. Overview of Bacterial Cellulose Composites: A Multipurpose Advanced Material. *Carbohydr. Polym.* **2013**, *98*, 1585–1598. [CrossRef]
397. Tapias, Y.A.R.; Di Monte, M.V.; Peltzer, M.A.; Salvay, A.G. Kombucha Fermentation in Yerba Mate: Cellulose Production, Films Formulation and Its Characterisation. *Carbohydr. Polym. Technol. Appl.* **2023**, *5*, 100310. [CrossRef]
398. Blanco Parte, F.G.; Santoso, S.P.; Chou, C.-C.; Verma, V.; Wang, H.-T.; Ismadji, S.; Cheng, K.-C. Current Progress on the Production, Modification, and Applications of Bacterial Cellulose. *Crit. Rev. Biotechnol.* **2020**, *40*, 397–414. [CrossRef]
399. Gullo, M.; La China, S.; Falcone, P.M.; Giudici, P. Biotechnological Production of Cellulose by Acetic Acid Bacteria: Current State and Perspectives. *Appl. Microbiol. Biotechnol.* **2018**, *102*, 6885–6898. [CrossRef]
400. Spinella, S.; Lo Re, G.; Liu, B.; Dorgan, J.; Habibi, Y.; Leclère, P.; Raquez, J.-M.; Dubois, P.; Gross, R.A. Polylactide/Cellulose Nanocrystal Nanocomposites: Efficient Routes for Nanofiber Modification and Effects of Nanofiber Chemistry on PLA Reinforcement. *Polymer* **2015**, *65*, 9–17. [CrossRef]
401. Auras, R.; Harte, B.; Selke, S. An Overview of Polylactides as Packaging Materials. *Macromol. Biosci.* **2004**, *4*, 835–864. [CrossRef] [PubMed]
402. Garcia-Garcia, D.; Quiles-Carrillo, L.; Balart, R.; Torres-Giner, S.; Arrieta, M.P. Innovative Solutions and Challenges to Increase the Use of Poly(3-Hydroxybutyrate) in Food Packaging and Disposables. *Eur. Polym. J.* **2022**, *178*, 111505. [CrossRef]
403. Oksman, K.; Mathew, A.P.; Bondeson, D.; Kvien, I. Manufacturing Process of Cellulose Whiskers/Polylactic Acid Nanocomposites. *Compos. Sci. Technol.* **2006**, *66*, 2776–2784. [CrossRef]
404. S. de O. Patrício, P.; Pereira, F.V.; dos Santos, M.C.; de Souza, P.P.; Roa, J.P.; Orefice, R.L. Increasing the Elongation at Break of Polyhydroxybutyrate Biopolymer: Effect of Cellulose Nanowhiskers on Mechanical and Thermal Properties. *J. Appl. Polym. Sci.* **2013**, *127*, 3613–3621. [CrossRef]
405. Chen, J.; Deng, L.; Chen, Q. Influence of Different Dimensional Nanoparticles on the Properties of Poly(β-Hydroxybutyrate-Co-Valerate) Nanocomposites. *J. Polym. Eng.* **2024**, *44*, 162–172. [CrossRef]
406. Arrieta, M.P.; Fortunati, E.; Dominici, F.; López, J.; Kenny, J.M. Bionanocomposite Films Based on Plasticized PLA-PHB/Cellulose Nanocrystal Blends. *Carbohydr. Polym.* **2015**, *121*, 265–275. [CrossRef]
407. Arrieta, M.P.; López, J.; López, D.; Kenny, J.M.; Peponi, L. Biodegradable Electrospun Bionanocomposite Fibers Based on Plasticized PLA–PHB Blends Reinforced with Cellulose Nanocrystals. *Ind. Crops Prod.* **2016**, *93*, 290–301. [CrossRef]
408. Arrieta, M.P.; Fortunati, E.; Dominici, F.; Rayón, E.; López, J.; Kenny, J.M. PLA-PHB/Cellulose Based Films: Mechanical, Barrier and Disintegration Properties. *Polym. Degrad. Stab.* **2014**, *107*, 139–149. [CrossRef]
409. Xie, F.; Pollet, E.; Halley, P.J.; Avérous, L. Starch-Based Nano-Biocomposites. *Prog. Polym. Sci.* **2013**, *38*, 1590–1628. [CrossRef]
410. González, K.; Iturriaga, L.; González, A.; Eceiza, A.; Gabilondo, N. Improving Mechanical and Barrier Properties of Thermoplastic Starch and Polysaccharide Nanocrystals Nanocomposites. *Eur. Polym. J.* **2020**, *123*, 109415. [CrossRef]
411. McCormick, C.L.; Callais, P.A.; Hutchinson, B.H., Jr. Solution Studies of Cellulose in Lithium Chloride and N,N-Dimethylacetamide. *Macromolecules* **1985**, *18*, 2394–2401. [CrossRef]

412. Jeong, M.J.; Lee, S.; Yang, B.S.; Potthast, A.; Kang, K.Y. Cellulose Degradation by Calcium Thiocyanate. *Polymers* **2019**, *11*, 1494. [CrossRef] [PubMed]
413. Cai, J.; Zhang, L. Rapid Dissolution of Cellulose in LiOH/Urea and NaOH/Urea Aqueous Solutions. *Macromol. Biosci.* **2005**, *5*, 539–548. [CrossRef] [PubMed]
414. Benini, K.C.; Cioffi, M.O.H.; Voorwald, H.J.C. PHBV/Cellulose Nanofibrils Composites Obtained by Solution Casting and Electrospinning Process. *Matéria* **2017**, *22*, e11837. [CrossRef]
415. Paquet, O.; Krouit, M.; Bras, J.; Thielemans, W.; Belgacem, M.N. Surface Modification of Cellulose by PCL Grafts. *Acta Mater.* **2010**, *58*, 792–801. [CrossRef]
416. Navarro-Baena, I.; Kenny, J.M.; Peponi, L. Thermally-Activated Shape Memory Behaviour of Bionanocomposites Reinforced with Cellulose Nanocrystals. *Cellulose* **2014**, *21*, 4231–4246. [CrossRef]
417. Peltzer, M.; Pei, A.; Zhou, Q.; Berglund, L.; Jiménez, A. Surface Modification of Cellulose Nanocrystals by Grafting with Poly(Lactic Acid). *Polym. Int.* **2014**, *63*, 1056–1062. [CrossRef]
418. Sessini, V.; Navarro-Baena, I.; Arrieta, M.P.; Dominici, F.; López, D.; Torre, L.; Kenny, J.M.; Dubois, P.; Raquez, J.-M.; Peponi, L. Effect of the Addition of Polyester-Grafted-Cellulose Nanocrystals on the Shape Memory Properties of Biodegradable PLA/PCL Nanocomposites. *Polym. Degrad. Stab.* **2018**, *152*, 126–138. [CrossRef]
419. Mujica-Garcia, A.; Hooshmand, S.; Skrifvars, M.; Kenny, J.M.; Oksman, K.; Peponi, L. Poly(Lactic Acid) Melt-Spun Fibers Reinforced with Functionalized Cellulose Nanocrystals. *RSC Adv.* **2016**, *6*, 9221–9231. [CrossRef]
420. Peponi, L.; Barrera-Rivera, K.A.; Kenny, J.M.; Marcos-Fernandez, Á.; Martinez-Richa, A.; López, D. Bio-Catalysis for the Functionalization of Cellulose Nanocrystals. *Nanomaterials* **2022**, *12*, 4064. [CrossRef] [PubMed]
421. Bovi, J.; Delgado, J.F.; de la Osa, O.; Peltzer, M.A.; Bernal, C.R.; Foresti, M.L. Biobased Composites of Poly(Lactic Acid) Melt Compounded with Bacterial and Vegetal Nanocelluloses Incorporated through Different Strategies. *Polymers* **2024**, *16*, 898. [CrossRef] [PubMed]
422. He, Z.-W.; Yang, J.; Lü, Q.-F.; Lin, Q. Effect of Structure on the Electrochemical Performance of Nitrogen- and Oxygen-Containing Carbon Micro/Nanospheres Prepared from Lignin-Based Composites. *ACS Sustain. Chem. Eng.* **2013**, *1*, 334–340. [CrossRef]
423. He, Z.; Wang, M.; Ma, S. Porous Lignin-Based Composites for Oil/Water Separation: A Review. *Int. J. Biol. Macromol.* **2024**, *260*, 129569. [CrossRef] [PubMed]
424. Das, A.K.; Mitra, K.; Conte, A.J.; Sarker, A.; Chowdhury, A.; Ragauskas, A.J. Lignin-A Green Material for Antibacterial Application—A Review. *Int. J. Biol. Macromol.* **2024**, *261*, 129753. [CrossRef]
425. Bujanovic, B.; Ralph, S.; Reiner, R.; Hirth, K.; Atalla, R. Polyoxometalates in Oxidative Delignification of Chemical Pulps: Effect on Lignin. *Materials* **2010**, *3*, 1888–1903. [CrossRef]
426. Wang, Y.; Chantreau, M.; Sibout, R.; Hawkins, S. Plant Cell Wall Lignification and Monolignol Metabolism. *Front. Plant Sci.* **2013**, *4*, 220. [CrossRef]
427. Dixon, R.A.; Chen, F.; Guo, D.; Parvathi, K. The Biosynthesis of Monolignols: A "Metabolic Grid", or Independent Pathways to Guaiacyl and Syringyl Units? *Phytochemistry* **2001**, *57*, 1069–1084. [CrossRef]
428. Vasile, C.; Baican, M. Lignins as Promising Renewable Biopolymers and Bioactive Compounds for High-Performance Materials. *Polymers* **2023**, *15*, 3177. [CrossRef]
429. Tanis, M.H.; Wallberg, O.; Galbe, M.; Al-Rudainy, B. Lignin Extraction by Using Two-Step Fractionation: A Review. *Molecules* **2023**, *29*, 98. [CrossRef]
430. Moreno, A.; Sipponen, M.H. Lignin-Based Smart Materials: A Roadmap to Processing and Synthesis for Current and Future Applications. *Mater. Horiz.* **2020**, *7*, 2237–2257. [CrossRef]
431. Nunes da Silva, V.F.; Farias de Menezes, F.; Gonçalves, A.R.; Martín, C.; de Moraes Rocha, G.J. Modulating the Properties and Structure of Lignins Produced by Alkaline Delignification of Sugarcane Bagasse Pretreated with Two Different Mineral Acids at Pilot-Scale. *Int. J. Biol. Macromol.* **2024**, *263*, 130111. [CrossRef]
432. Liu, W.; Li, Z.; Ren, Q.; Jiang, C.; Feng, J.; Hou, Q. Upgrading Lignin Macromolecular by Green and Recyclable Ternary Deep Eutectic Solvents. *Bioresour. Technol.* **2024**, *394*, 130230. [CrossRef] [PubMed]
433. Arrieta, M.P.; Peponi, L.; López, D.; Fernández-García, M. Recovery of Yerba Mate (Ilex Paraguariensis) Residue for the Development of PLA-Based Bionanocomposite Films. *Ind. Crops Prod.* **2018**, *111*, 317–328. [CrossRef]
434. Fortunati, E.; Yang, W.; Luzi, F.; Kenny, J.; Torre, L.; Puglia, D. Lignocellulosic Nanostructures as Reinforcement in Extruded and Solvent Casted Polymeric Nanocomposites: An Overview. *Eur. Polym. J.* **2016**, *80*, 295–316. [CrossRef]
435. Gupta, A.K.; Mohanty, S.; Nayak, S.K. Synthesis, Characterization and Application of Lignin Nanoparticles (LNPs). *Mater. Focus.* **2014**, *3*, 444–454. [CrossRef]
436. Frangville, C.; Rutkevičius, M.; Richter, A.P.; Velev, O.D.; Stoyanov, S.D.; Paunov, V.N. Fabrication of Environmentally Biodegradable Lignin Nanoparticles. *ChemPhysChem* **2012**, *13*, 4235–4243. [CrossRef]
437. Yang, W.; Fortunati, E.; Dominici, F.; Kenny, J.M.; Puglia, D. Effect of Processing Conditions and Lignin Content on Thermal, Mechanical and Degradative Behavior of Lignin Nanoparticles/Polylactic (Acid) Bionanocomposites Prepared by Melt Extrusion and Solvent Casting. *Eur. Polym. J.* **2015**, *71*, 126–139. [CrossRef]
438. Lugoloobi, I.; Li, X.; Zhang, Y.; Mao, Z.; Wang, B.; Sui, X.; Feng, X. Fabrication of Lignin/Poly(3-Hydroxybutyrate) Nanocomposites with Enhanced Properties via a Pickering Emulsion Approach. *Int. J. Biol. Macromol.* **2020**, *165*, 3078–3087. [CrossRef]

439. Wan, Z.; Zhang, H.; Guo, Y.; Li, H. Advances in Catalytic Depolymerization of Lignin. *ChemistrySelect* **2022**, *7*, e202202582. [CrossRef]
440. Nisha, S.S.; Nikzad, M.; Al Kobaisi, M.; Truong, V.K.; Sbarski, I. The Role of Ionic-Liquid Extracted Lignin Micro/Nanoparticles for Functionalisation of an Epoxy-Based Composite Matrix. *Compos. Sci. Technol.* **2019**, *174*, 11–19. [CrossRef]
441. Prakash, V.R.A.; Bourchak, M.; Alshahrani, H.; Juhany, K.A. Development of Cashew Nut Shell Lignin-Acrylonitrile Butadiene Styrene 3D Printed Core and Industrial Hemp/Aluminized Glass Fiber Epoxy Biocomposite for Morphing Wing and Unmanned Aerial Vehicle Applications. *Int. J. Biol. Macromol.* **2023**, *253*, 127068. [CrossRef] [PubMed]
442. Nguyen, N.A.; Barnes, S.H.; Bowland, C.C.; Meek, K.M.; Littrell, K.C.; Keum, J.K.; Naskar, A.K. A Path for Lignin Valorization via Additive Manufacturing of High-Performance Sustainable Composites with Enhanced 3D Printability. *Sci. Adv.* **2018**, *4*, eaat4967. [CrossRef] [PubMed]
443. Tanase-Opedal, M.; Espinosa, E.; Rodríguez, A.; Chinga-Carrasco, G. Lignin: A Biopolymer from Forestry Biomass for Biocomposites and 3D Printing. *Materials* **2019**, *12*, 3006. [CrossRef] [PubMed]
444. Souto, F.; Calado, V.; Pereira, N. Lignin-Based Carbon Fiber: A Current Overview. *Mater. Res. Express* **2018**, *5*, 072001. [CrossRef]
445. Balakshin, M.Y.; Capanema, E.A. Comprehensive Structural Analysis of Biorefinery Lignins with a Quantitative ^{13}C NMR Approach. *RSC Adv.* **2015**, *5*, 87187–87199. [CrossRef]
446. Figueiredo, P.; Lintinen, K.; Hirvonen, J.T.; Kostiainen, M.A.; Santos, H.A. Properties and Chemical Modifications of Lignin: Towards Lignin-Based Nanomaterials for Biomedical Applications. *Prog. Mater. Sci.* **2018**, *93*, 233–269. [CrossRef]
447. Laurichesse, S.; Avérous, L. Chemical Modification of Lignins: Towards Biobased Polymers. *Prog. Polym. Sci.* **2014**, *39*, 1266–1290. [CrossRef]
448. Datta, R.; Kelkar, A.; Baraniya, D.; Molaei, A.; Moulick, A.; Meena, R.; Formanek, P. Enzymatic Degradation of Lignin in Soil: A Review. *Sustainability* **2017**, *9*, 1163. [CrossRef]
449. Martínez, A.T.; Speranza, M.; Ruiz-Dueñas, F.J.; Ferreira, P.; Camarero, S.; Guillén, F.; Martínez, M.J.; Gutiérrez, A.; del Río, J.C. Biodegradation of Lignocellulosics: Microbial, Chemical, and Enzymatic Aspects of the Fungal Attack of Lignin. *Int. Microbiol.* **2005**, *8*, 195–204.
450. Bumpus, J.A.; Aust, S.D. Biodegradation of Environmental Pollutants by the White Rot Fungus *Phanerochaete Chrysosporium*: Involvement of the Lignin Degrading System. *BioEssays* **1987**, *6*, 166–170. [CrossRef]
451. Bugg, T.D.H.; Ahmad, M.; Hardiman, E.M.; Rahmanpour, R. Pathways for Degradation of Lignin in Bacteria and Fungi. *Nat. Prod. Rep.* **2011**, *28*, 1883. [CrossRef] [PubMed]
452. Atiwesh, G.; Parrish, C.C.; Banoub, J.; Le, T.T. Lignin Degradation by Microorganisms: A Review. *Biotechnol. Prog.* **2022**, *38*, e3226. [CrossRef] [PubMed]
453. Ramachandra, M.; Crawford, D.L.; Hertel, G. Characterization of an Extracellular Lignin Peroxidase of the Lignocellulolytic Actinomycete Streptomyces Viridosporus. *Appl. Environ. Microbiol.* **1988**, *54*, 3057–3063. [CrossRef]
454. Ahmad, M.; Roberts, J.N.; Hardiman, E.M.; Singh, R.; Eltis, L.D.; Bugg, T.D.H. Identification of DypB from *Rhodococcus Jostii* RHA1 as a Lignin Peroxidase. *Biochemistry* **2011**, *50*, 5096–5107. [CrossRef]
455. Brown, M.E.; Barros, T.; Chang, M.C.Y. Identification and Characterization of a Multifunctional Dye Peroxidase from a Lignin-Reactive Bacterium. *ACS Chem. Biol.* **2012**, *7*, 2074–2081. [CrossRef]
456. Rahmanpour, R.; Bugg, T.D.H. Characterisation of Dyp-Type Peroxidases from Pseudomonas Fluorescens Pf-5: Oxidation of Mn(II) and Polymeric Lignin by Dyp1B. *Arch. Biochem. Biophys.* **2015**, *574*, 93–98. [CrossRef]
457. Rahmanpour, R.; Rea, D.; Jamshidi, S.; Fülöp, V.; Bugg, T.D.H. Structure of Thermobifida Fusca DyP-Type Peroxidase and Activity towards Kraft Lignin and Lignin Model Compounds. *Arch. Biochem. Biophys.* **2016**, *594*, 54–60. [CrossRef]
458. DeAngelis, K.M.; Allgaier, M.; Chavarria, Y.; Fortney, J.L.; Hugenholtz, P.; Simmons, B.; Sublette, K.; Silver, W.L.; Hazen, T.C. Characterization of Trapped Lignin-Degrading Microbes in Tropical Forest Soil. *PLoS ONE* **2011**, *6*, e19306. [CrossRef]
459. Duan, J.; Liang, J.; Wang, Y.; Du, W.; Wang, D. Kraft Lignin Biodegradation by *Dysgonomonas* sp. WJDL-Y1, a New Anaerobic Bacterial Strain Isolated from Sludge of a Pulp and Paper Mill. *J. Microbiol. Biotechnol.* **2016**, *26*, 1765–1773. [CrossRef]
460. Rashid, G.M.M.; Durán-Peña, M.J.; Rahmanpour, R.; Sapsford, D.; Bugg, T.D.H. Delignification and Enhanced Gas Release from Soil Containing Lignocellulose by Treatment with Bacterial Lignin Degraders. *J. Appl. Microbiol.* **2017**, *123*, 159–171. [CrossRef] [PubMed]
461. Kim, S.J.; Shoda, M. Purification and Characterization of a Novel Peroxidase from *Geotrichum candidum* Dec 1 Involved in Decolorization of Dyes. *Appl. Environ. Microbiol.* **1999**, *65*, 1029–1035. [CrossRef] [PubMed]
462. Yoshida, T.; Sugano, Y. A Structural and Functional Perspective of DyP-Type Peroxidase Family. *Arch. Biochem. Biophys.* **2015**, *574*, 49–55. [CrossRef] [PubMed]
463. Brebu, M.; Vasile, C. Thermal degradation of lignin—A review. *Cellul. Chem. Technol.* **2010**, *44*, 353.
464. Chu, S.; Subrahmanyam, A.V.; Huber, G.W. The Pyrolysis Chemistry of a β-O-4 Type Oligomeric Lignin Model Compound. *Green. Chem.* **2013**, *15*, 125–136. [CrossRef]
465. Kim, J.-Y.; Shin, E.-J.; Eom, I.-Y.; Won, K.; Kim, Y.H.; Choi, D.; Choi, I.-G.; Choi, J.W. Structural Features of Lignin Macromolecules Extracted with Ionic Liquid from Poplar Wood. *Bioresour. Technol.* **2011**, *102*, 9020–9025. [CrossRef]
466. El Moustaqim, M.; El Kaihal, A.; El Marouani, M.; Men-La-Yakhaf, S.; Taibi, M.; Sebbahi, S.; El Hajjaji, S.; Kifani-Sahban, F. Thermal and Thermomechanical Analyses of Lignin. *Sustain. Chem. Pharm.* **2018**, *9*, 63–68. [CrossRef]

467. Issahaku, I.; Tetteh, I.K.; Tetteh, A.Y. Chitosan and Chitosan Derivatives: Recent Advancements in Production and Applications in Environmental Remediation. *Environ. Adv.* **2023**, *11*, 100351. [CrossRef]
468. Mikušová, V.; Mikuš, P.; Piozzi, A. Advances in Chitosan-Based Nanoparticles for Drug Delivery. *Int. J. Mol. Sci.* **2021**, *22*, 9652. [CrossRef]
469. Haider, A.; Khan, S.; Iqbal, D.N.; Shrahili, M.; Haider, S.; Mohammad, K.; Mohammad, A.; Rizwan, M.; Kanwal, Q.; Mustafa, G. Advances in Chitosan-Based Drug Delivery Systems: A Comprehensive Review for Therapeutic Applications. *Eur. Polym. J.* **2024**, *210*, 112983. [CrossRef]
470. Lv, Y.; Zhai, C.; Sun, G.; He, Y. Chitosan as a Promising Materials for the Construction of Nanocarriers for Diabetic Retinopathy: An Updated Review. *J. Biol. Eng.* **2024**, *18*, 18. [CrossRef] [PubMed]
471. Dohendou, M.; Pakzad, K.; Nezafat, Z.; Nasrollahzadeh, M.; Dekamin, M.G. Progresses in Chitin, Chitosan, Starch, Cellulose, Pectin, Alginate, Gelatin and Gum Based (Nano)Catalysts for the Heck Coupling Reactions: A Review. *Int. J. Biol. Macromol.* **2021**, *192*, 771–819. [CrossRef] [PubMed]
472. Pérez-Álvarez, L.; Ruiz-Rubio, L.; Vilas-Vilela, J.L. Determining the Deacetylation Degree of Chitosan: Opportunities To Learn Instrumental Techniques. *J. Chem. Educ.* **2018**, *95*, 1022–1028. [CrossRef]
473. Pellis, A.; Guebitz, G.M.; Nyanhongo, G.S. Chitosan: Sources, Processing and Modification Techniques. *Gels* **2022**, *8*, 393. [CrossRef] [PubMed]
474. Foster, L.J.R.; Ho, S.; Hook, J.; Basuki, M.; Marçal, H. Chitosan as a Biomaterial: Influence of Degree of Deacetylation on Its Physiochemical, Material and Biological Properties. *PLoS ONE* **2015**, *10*, e0135153. [CrossRef] [PubMed]
475. Yu, M.; Zhang, K.; Guo, X.; Qian, L. Effects of the Degree of Deacetylation on the Single-Molecule Mechanics of Chitosans. *J. Phys. Chem. B* **2023**, *127*, 4261–4267. [CrossRef]
476. Kono, H.; Kato, T. Elucidation of Substituent Distribution States for Carboxymethyl Chitosan by Detailed NMR Analysis. *Carbohydr. Polym. Technol. Appl.* **2021**, *2*, 100175. [CrossRef]
477. Khan, A.; Alamry, K.A. Recent Advances of Emerging Green Chitosan-Based Biomaterials with Potential Biomedical Applications: A Review. *Carbohydr. Res.* **2021**, *506*, 108368. [CrossRef]
478. El-Araby, A.; Janati, W.; Ullah, R.; Ercisli, S.; Errachidi, F. Chitosan, Chitosan Derivatives, and Chitosan-Based Nanocomposites: Eco-Friendly Materials for Advanced Applications (a Review). *Front. Chem.* **2023**, *11*, 1327426. [CrossRef]
479. Lingait, D.; Rahagude, R.; Gaharwar, S.S.; Das, R.S.; Verma, M.G.; Srivastava, N.; Kumar, A.; Mandavgane, S. A Review on Versatile Applications of Biomaterial/Polycationic Chitosan: An Insight into the Structure-Property Relationship. *Int. J. Biol. Macromol.* **2024**, *257*, 128676. [CrossRef]
480. Mahmoud, M.G.; El Kady, E.M.; Asker, M.S. Chitosan and Glucan, Properties and Applications. *World J. Agric. Soil. Sci.* **2019**, *3*, 1–19. [CrossRef]
481. Kandile, N.G.; Zaky, H.T.; Mohamed, M.I.; Nasr, A.S.; Ali, Y.G. Extraction and Characterization of Chitosan from Shrimp Shells. *Open J. Org. Polym. Mater.* **2018**, *8*, 33–42. [CrossRef]
482. Ali, M.; Shakeel, M.; Mehmood, K. Extraction and Characterization of High Purity Chitosan by Rapid and Simple Techniques from Mud Crabs Taken from Abbottabad. *Pak. J. Pharm. Sci.* **2019**, *32*, 171–175. [PubMed]
483. Zuorro, A.; Moreno-Sader, K.A.; González-Delgad, Á.D. Inherent Safety Analysis and Sustainability Evaluation of Chitosan Production from Shrimp Exoskeleton in Colombia. *Water* **2021**, *13*, 553. [CrossRef]
484. Hosney, A.; Ullah, S.; Barčauskaitė, K. A Review of the Chemical Extraction of Chitosan from Shrimp Wastes and Prediction of Factors Affecting Chitosan Yield by Using an Artificial Neural Network. *Mar. Drugs* **2022**, *20*, 675. [CrossRef]
485. Bonfante-Alvarez, H.; De Avila-Montiel, G.; Herrera-Barros, A.; Torrenegra-Alarcón, M.; González-Delgado, Á.D. Evaluation of Five Chitosan Production Routes with Astaxanthin Recovery from Shrimp Exoskeletons. *Chem. Eng. Trans.* **2018**, *70*, 1969–1974. [CrossRef]
486. Hajji, S.; Ghorbel-Bellaaj, O.; Younes, I.; Jellouli, K.; Nasri, M. Chitin Extraction from Crab Shells by Bacillus Bacteria. Biological Activities of Fermented Crab Supernatants. *Int. J. Biol. Macromol.* **2015**, *79*, 167–173. [CrossRef]
487. Pakizeh, M.; Moradi, A.; Ghassemi, T. Chemical Extraction and Modification of Chitin and Chitosan from Shrimp Shells. *Eur. Polym. J.* **2021**, *159*, 110709. [CrossRef]
488. Giraldo, J.D.; García, Y.; Vera, M.; Garrido-Miranda, K.A.; Andrade-Acuña, D.; Marrugo, K.P.; Rivas, B.L.; Schoebitz, M. Alternative Processes to Produce Chitin, Chitosan, and Their Oligomers. *Carbohydr. Polym.* **2024**, *332*, 121924. [CrossRef]
489. Kozma, M.; Acharya, B.; Bissessur, R. Chitin, Chitosan, and Nanochitin: Extraction, Synthesis, and Applications. *Polymers* **2022**, *14*, 3989. [CrossRef]
490. Kou, S.G.; Peters, L.M.; Mucalo, M.R. Chitosan: A Review of Sources and Preparation Methods. *Int. J. Biol. Macromol.* **2021**, *169*, 85–94. [CrossRef] [PubMed]
491. El Knidri, H.; Belaabed, R.; Addaou, A.; Laajeb, A.; Lahsini, A. Extraction, Chemical Modification and Characterization of Chitin and Chitosan. *Int. J. Biol. Macromol.* **2018**, *120*, 1181–1189. [CrossRef] [PubMed]
492. Said Al Hoqani, H.A.; AL-Shaqsi, N.; Hossain, M.A.; Al Sibani, M.A. Isolation and Optimization of the Method for Industrial Production of Chitin and Chitosan from Omani Shrimp Shell. *Carbohydr. Res.* **2020**, *492*, 108001. [CrossRef] [PubMed]
493. Agarwal, M.; Agarwal, M.K.; Shrivastav, N.; Pandey, S.; Gaur, P. A Simple and Effective Method for Preparation of Chitosan from Chitin. *Int. J. Life-Sci. Sci. Res.* **2018**, *4*, 1721–1728. [CrossRef]

494. Serri, C.; Cruz-Maya, I.; Bonadies, I.; Rassu, G.; Giunchedi, P.; Gavini, E.; Guarino, V. Green Routes for Bio-Fabrication in Biomedical and Pharmaceutical Applications. *Pharmaceutics* **2023**, *15*, 1744. [CrossRef]
495. Yavir, K.; Konieczna, K.; Marcinkowski, Ł.; Kloskowski, A. Ionic Liquids in the Microextraction Techniques: The Influence of ILs Structure and Properties. *TrAC Trends Anal. Chem.* **2020**, *130*, 115994. [CrossRef]
496. Morais, E.S.; Da Costa Lopes, A.M.; Freire, M.G.; Freire, C.S.R.; Coutinho, J.A.P.; Silvestre, A.J.D. Use of Ionic Liquids and Deep Eutectic Solvents in Polysaccharides Dissolution and Extraction Processes towards Sustainable Biomass Valorization. *Molecules* **2020**, *25*, 3652. [CrossRef]
497. Sun, Y.; Jing, X.; Ma, X.; Feng, Y.; Hu, H. Versatile Types of Polysaccharide-Based Drug Delivery Systems: From Strategic Design to Cancer Therapy. *Int. J. Mol. Sci.* **2020**, *21*, 9159. [CrossRef]
498. Khajavian, M.; Vatanpour, V.; Castro-Muñoz, R.; Boczkaj, G. Chitin and Derivative Chitosan-Based Structures—Preparation Strategies Aided by Deep Eutectic Solvents: A Review. *Carbohydr. Polym.* **2022**, *275*, 118702. [CrossRef]
499. Bisht, M.; Macário, I.P.E.; Neves, M.C.; Pereira, J.L.; Pandey, S.; Rogers, R.D.; Coutinho, J.A.P.; Ventura, S.P.M. Enhanced Dissolution of Chitin Using Acidic Deep Eutectic Solvents: A Sustainable and Simple Approach to Extract Chitin from Crayfish Shell Wastes as Alternative Feedstocks. *ACS Sustain. Chem. Eng.* **2021**, *9*, 16073–16081. [CrossRef]
500. Rodrigues, L.A.; Radojčić Redovniković, I.; Duarte, A.R.C.; Matias, A.A.; Paiva, A. Low-Phytotoxic Deep Eutectic Systems as Alternative Extraction Media for the Recovery of Chitin from Brown Crab Shells. *ACS Omega* **2021**, *6*, 28729–28741. [CrossRef] [PubMed]
501. Liu, Y.; Yuan, Y.; Duan, S.; Li, C.; Hu, B.; Liu, A.; Wu, D.; Cui, H.; Lin, L.; He, J.; et al. Preparation and Characterization of Chitosan Films with Three Kinds of Molecular Weight for Food Packaging. *Int. J. Biol. Macromol.* **2020**, *155*, 249–259. [CrossRef] [PubMed]
502. Kulka, K.; Sionkowska, A. Chitosan Based Materials in Cosmetic Applications: A Review. *Molecules* **2023**, *28*, 1817. [CrossRef] [PubMed]
503. Rezaei, F.S.; Sharifianjazi, F.; Esmaeilkhanian, A.; Salehi, E. Chitosan Films and Scaffolds for Regenerative Medicine Applications: A Review. *Carbohydr. Polym.* **2021**, *273*, 118631. [CrossRef] [PubMed]
504. Ali, M.; Cybulska, J.; Frąc, M.; Zdunek, A. Application of Polysaccharides for the Encapsulation of Beneficial Microorganisms for Agricultural Purposes: A Review. *Int. J. Biol. Macromol.* **2023**, *244*, 125366. [CrossRef]
505. Ambaye, T.G.; Vaccari, M.; Prasad, S.; van Hullebusch, E.D.; Rtimi, S. Preparation and Applications of Chitosan and Cellulose Composite Materials. *J. Environ. Manag.* **2022**, *301*, 113850. [CrossRef]
506. Abd El-Hack, M.E.; El-Saadony, M.T.; Shafi, M.E.; Zabermawi, N.M.; Arif, M.; Batiha, G.E.; Khafaga, A.F.; Abd El-Hakim, Y.M.; Al-Sagheer, A.A. Antimicrobial and Antioxidant Properties of Chitosan and Its Derivatives and Their Applications: A Review. *Int. J. Biol. Macromol.* **2020**, *164*, 2726–2744. [CrossRef]
507. Zena, Y.; Periyasamy, S.; Tesfaye, M.; Tumsa, Z.; Jayakumar, M.; Mohamed, B.A.; Asaithambi, P.; Aminabhavi, T.M. Essential Characteristics Improvement of Metallic Nanoparticles Loaded Carbohydrate Polymeric Films—A Review. *Int. J. Biol. Macromol.* **2023**, *242*, 124803. [CrossRef]
508. Vargas, M.; Albors, A.; Chiralt, A.; González-Martínez, C. Quality of Cold-Stored Strawberries as Affected by Chitosan–Oleic Acid Edible Coatings. *Postharvest Biol. Technol.* **2006**, *41*, 164–171. [CrossRef]
509. Dutta, P.K.; Tripathi, S.; Mehrotra, G.K.; Dutta, J. Perspectives for Chitosan Based Antimicrobial Films in Food Applications. *Food Chem.* **2009**, *114*, 1173–1182. [CrossRef]
510. Perdones, A.; Sánchez-González, L.; Chiralt, A.; Vargas, M. Effect of Chitosan–Lemon Essential Oil Coatings on Storage-Keeping Quality of Strawberry. *Postharvest Biol. Technol.* **2012**, *70*, 32–41. [CrossRef]
511. Bonilla, J.; Fortunati, E.; Vargas, M.; Chiralt, A.; Kenny, J.M. Effects of Chitosan on the Physicochemical and Antimicrobial Properties of PLA Films. *J. Food Eng.* **2013**, *119*, 236–243. [CrossRef]
512. Arrieta, M.P.; López, J.; López, D.; Kenny, J.M.; Peponi, L. Effect of Chitosan and Catechin Addition on the Structural, Thermal, Mechanical and Disintegration Properties of Plasticized Electrospun PLA-PHB Biocomposites. *Polym. Degrad. Stab.* **2016**, *132*, 145–156. [CrossRef]
513. Zhang, W.; Hedayati, S.; Tarahi, M.; Can Karaca, A.; Hadidi, M.; Assadpour, E.; Jafari, S.M. Advances in Transglutaminase Cross-Linked Protein-Based Food Packaging Films; a Review. *Int. J. Biol. Macromol.* **2023**, *253*, 127399. [CrossRef] [PubMed]
514. Alipal, J.; Mohd Pu'ad, N.A.S.; Lee, T.C.; Nayan, N.H.M.; Sahari, N.; Basri, H.; Idris, M.I.; Abdullah, H.Z. A Review of Gelatin: Properties, Sources, Process, Applications, and Commercialisation. *Mater. Today Proc.* **2021**, *42*, 240–250. [CrossRef]
515. Rather, J.A.; Akhter, N.; Ashraf, Q.S.; Mir, S.A.; Makroo, H.A.; Majid, D.; Barba, F.J.; Khaneghah, A.M.; Dar, B.N. A Comprehensive Review on Gelatin: Understanding Impact of the Sources, Extraction Methods, and Modifications on Potential Packaging Applications. *Food Packag. Shelf Life* **2022**, *34*, 100945. [CrossRef]
516. Valcarcel, J.; Fraguas, J.; Hermida-Merino, C.; Hermida-Merino, D.; Piñeiro, M.M.; Vázquez, J.A. Production and Physicochemical Characterization of Gelatin and Collagen Hydrolysates from Turbot Skin Waste Generated by Aquaculture Activities. *Mar. Drugs* **2021**, *19*, 491. [CrossRef]
517. Abedinia, A.; Mohammadi Nafchi, A.; Sharifi, M.; Ghalambor, P.; Oladzadabbasabadi, N.; Ariffin, F.; Huda, N. Poultry Gelatin: Characteristics, Developments, Challenges, and Future Outlooks as a Sustainable Alternative for Mammalian Gelatin. *Trends Food Sci. Technol.* **2020**, *104*, 14–26. [CrossRef]
518. Luo, Q.; Hossen, M.A.; Zeng, Y.; Dai, J.; Li, S.; Qin, W.; Liu, Y. Gelatin-Based Composite Films and Their Application in Food Packaging: A Review. *J. Food Eng.* **2022**, *313*, 110762. [CrossRef]

519. Andreuccetti, C.; Galicia-García, T.; González-Nuñez, R.; Martínez-Bustos, F.; Grosso, C.R.F. Native and Modified Gelatin Films Produced by Casting, Extrusion, and Blowing Extrusion Processes. *Polym. Renew. Resour.* **2017**, *8*, 11–26. [CrossRef]
520. Hosseini, S.F.; Javidi, Z.; Rezaei, M. Efficient Gas Barrier Properties of Multi-Layer Films Based on Poly(Lactic Acid) and Fish Gelatin. *Int. J. Biol. Macromol.* **2016**, *92*, 1205–1214. [CrossRef] [PubMed]
521. Kurek, M.; Descours, E.; Poldan, P.; Julou, A.; Pitois, A.; Klepac, D.; Vallet, N.; Galić, K. Possibility of Storing Olive Oil in Antioxidant Biobased Pouches Made of Chitosan and Gelatin. *Food Hydrocoll.* **2024**, *151*, 109835. [CrossRef]
522. Zong, Z.; Liu, M.; Chen, H.; Farag, M.A.; Wu, W.; Fang, X.; Niu, B.; Gao, H. Preparation and Characterization of a Novel Intelligent Starch/Gelatin Binary Film Containing Purple Sweet Potato Anthocyanins for Flammulina Velutipes Mushroom Freshness Monitoring. *Food Chem.* **2023**, *405*, 134839. [CrossRef] [PubMed]
523. Duan, M.; Sun, J.; Huang, Y.; Jiang, H.; Hu, Y.; Pang, J.; Wu, C. Electrospun Gelatin/Chitosan Nanofibers Containing Curcumin for Multifunctional Food Packaging. *Food Sci. Hum. Wellness* **2023**, *12*, 614–621. [CrossRef]
524. Qi, J.; Zhang, W.; Feng, X.; Yu, J.; Han, M.; Deng, S.; Zhou, G.; Wang, H.; Xu, X. Thermal Degradation of Gelatin Enhances Its Ability to Bind Aroma Compounds: Investigation of Underlying Mechanisms. *Food Hydrocoll.* **2018**, *83*, 497–510. [CrossRef]
525. Bertoldo, M.; Bronco, S.; Gragnoli, T.; Ciardelli, F. Modification of Gelatin by Reaction with 1,6-Diisocyanatohexane. *Macromol. Biosci.* **2007**, *7*, 328–338. [CrossRef]
526. Puchalski, M.; Szparaga, G.; Biela, T.; Gutowska, A.; Sztajnowski, S.; Krucińska, I. Molecular and Supramolecular Changes in Polybutylene Succinate (PBS) and Polybutylene Succinate Adipate (PBSA) Copolymer during Degradation in Various Environmental Conditions. *Polymers* **2018**, *10*, 251. [CrossRef]
527. Aliotta, L.; Seggiani, M.; Lazzeri, A.; Gigante, V.; Cinelli, P. A Brief Review of Poly (Butylene Succinate) (PBS) and Its Main and Applications. *Polymer* **2022**, *14*, 844. [CrossRef]
528. Platnieks, O.; Gaidukovs, S.; Kumar Thakur, V.; Barkane, A.; Beluns, S. Bio-Based Poly (Butylene Succinate): Recent Progress, Challenges and Future Opportunities. *Eur. Polym. J.* **2021**, *161*, 110855. [CrossRef]
529. Barletta, M.; Aversa, C.; Ayyoob, M.; Gisario, A.; Hamad, K.; Mehrpouya, M.; Vahabi, H. Poly(Butylene Succinate) (PBS): Materials, Processing, and Industrial Applications. *Prog. Polym. Sci.* **2022**, *132*, 101579. [CrossRef]
530. Mao, H.; Liu, H.; Gao, Z.; Su, T.; Wang, Z. Biodegradation of Poly(Butylene Succinate) by *Fusarium* sp. FS1301 and Purification and Characterization of Poly(Butylene Succinate) Depolymerase. *Polym. Degrad. Stab.* **2015**, *114*, 1–7. [CrossRef]
531. Nelson, T.F.; Baumgartner, R.; Jaggi, M.; Bernasconi, S.M.; Battagliarin, G.; Sinkel, C.; Künkel, A.; Kohler, H.P.E.; McNeill, K.; Sander, M. Biodegradation of Poly(Butylene Succinate) in Soil Laboratory Incubations Assessed by Stable Carbon Isotope Labelling. *Nat. Commun.* **2022**, *13*, 5691. [CrossRef] [PubMed]
532. Lyyra, I.; Sandberg, N.; Parihar, V.S.; Hannula, M.; Huhtala, H.; Hyttinen, J.; Massera, J.; Kellomäki, M. Hydrolytic Degradation of Polylactide/Polybutylene Succinate Blends with Bioactive Glass. *Mater. Today Commun.* **2023**, *37*, 107242. [CrossRef]
533. Łopusiewicz, Ł.; Zdanowicz, M.; Macieja, S.; Kowalczyk, K.; Bartkowiak, A. Development and Characterization of Bioactive Poly(Butylene-Succinate) Films Modified with Quercetin for Food Packaging Applications. *Polymers* **2021**, *13*, 1798. [CrossRef]
534. Domínguez-Robles, J.; Larrañeta, E.; Fong, M.L.; Martin, N.K.; Irwin, N.J.; Mutjé, P.; Tarrés, Q.; Delgado-Aguilar, M. Lignin/Poly(Butylene Succinate) Composites with Antioxidant and Antibacterial Properties for Potential Biomedical Applications. *Int. J. Biol. Macromol.* **2020**, *145*, 92–99. [CrossRef]
535. Guidotti, G.; Soccio, M.; Bondi, E.; Posati, T.; Sotgiu, G.; Zamboni, R.; Torreggiani, A.; Corticelli, F.; Lotti, N.; Aluigi, A. Effects of the Blending Ratio on the Design of Keratin/Poly (Butylene Succinate) Nanofibers for Drug Delivery Applications. *Biomolecules* **2021**, *11*, 1194. [CrossRef]
536. Hsieh, H.C.; Wu, N.; Chuang, T.H.; Lee, W.Y.; Chen, J.Y.; Chen, W.C. Eco-Friendly Polyfluorene/Poly(Butylene Succinate) Blends and Their Electronic Device Application on Biodegradable Substrates. *ACS Appl. Polym. Mater.* **2020**, *2*, 2469–2476. [CrossRef]
537. Ostheller, M.E.; Balakrishnan, N.K.; Groten, R.; Seide, G. Detailed Process Analysis of Biobased Polybutylene Succinate Microfibers Produced by Laboratory-Scale Melt Electrospinning. *Polymers* **2021**, *13*, 1024. [CrossRef]
538. Xu, J.; Manepalli, P.H.; Zhu, L.; Narayan-Sarathy, S.; Alavi, S. Morphological, Barrier and Mechanical Properties of Films from Poly (Butylene Succinate) Reinforced with Nanocrystalline Cellulose and Chitin Whiskers Using Melt Extrusion. *J. Polym. Res.* **2019**, *26*, 188. [CrossRef]
539. Rennert, M.; Hiller, B.T. Influence of Coffee Variety and Processing on the Properties of Parchments as Functional Bioadditives for Biobased Poly(Butylene Succinate) Composites. *Polymers* **2023**, *15*, 2985. [CrossRef]
540. Jariyasakoolroj, P.; Chongcharoenyanon, B.; Wadaugsorn, K. Kinetic Migration of PBS and PBSA Biopolymers Prepared by Cast Film Extrusion and Biaxial Stretching: A Combined Experimental and Modeling Approach. *J. Appl. Polym. Sci.* **2024**, *141*, e55323. [CrossRef]
541. Altieri, R.; Seggiani, M.; Esposito, A.; Cinelli, P.; Stanzione, V. Thermoplastic Blends Based on Poly(Butylene Succinate-Co-Adipate) and Different Collagen Hydrolysates from Tanning Industry—II: Aerobic Biodegradation in Composting Medium. *J. Polym. Environ.* **2021**, *29*, 3375–3388. [CrossRef]
542. Puekpoonpoal, N.; Phattarateera, S.; Kerddonfag, N.; Aht-Ong, D. Morphology Development of PLAs with Different Stereo-Regularities in Ternary Blend PBSA/PBS/PLA Films. *Polym.-Plast. Technol. Mater.* **2021**, *60*, 1672–1685. [CrossRef]
543. Yang, C.; Tang, H.; Wang, Y.; Liu, Y.; Wang, J.; Shi, W.; Li, L. Development of PLA-PBSA Based Biodegradable Active Film and Its Application to Salmon Slices. *Food Packag. Shelf Life* **2019**, *22*, 100393. [CrossRef]

544. Jiang, G.; Yu, L. High Strength and Barrier Properties of Biodegradable PPC/PBSA Blends Prepared by Reaction Compatibilization for Promising Application in Packaging. *Macromol. Mater. Eng.* **2021**, *306*, 2000723. [CrossRef]
545. Delorme, A.E.; Radusin, T.; Myllytie, P.; Verney, V.; Askanian, H. Enhancement of Gas Barrier Properties and Durability of Poly(Butylene Succinate-Co-Butylene Adipate)-Based Nanocomposites for Food Packaging Applications. *Nanomaterials* **2022**, *12*, 978. [CrossRef]
546. Okabe, M.; Lies, D.; Kanamasa, S.; Park, E.Y. Biotechnological Production of Itaconic Acid and Its Biosynthesis in Aspergillus Terreus. *Appl. Microbiol. Biotechnol.* **2009**, *84*, 597–606. [CrossRef]
547. Bentley, R.; Thiessen, C.P. Biosynthesis of itaconic acid in aspergillus terreus. *J. Biol. Chem.* **1957**, *226*, 673–687. [CrossRef]
548. Zhang, R.; Liu, H.; Ning, Y.; Yu, Y.; Deng, L.; Wang, F. Recent Advances on the Production of Itaconic Acid via the Fermentation and Metabolic Engineering. *Fermentation* **2023**, *9*, 71. [CrossRef]
549. MARVEL, C.S.; SHEPHERD, T.H. Polymerization Reactions of Itaconic Acid and Some of Its Derivatives. *J. Org. Chem.* **1959**, *24*, 599–605. [CrossRef]
550. Nakamoto, H.; Ogo, Y.; Imoto, T. Polymerization of Itaconic Acid under High Pressure. *Die Makromol. Chem.* **1968**, *111*, 104–114. [CrossRef]
551. Osman, S.M.; El-Newehy, M.H.; Al-Deyab, S.S.; El-Faham, A. Microwave Synthesis and Thermal Properties of Polyacrylate Derivatives Containing Itaconic Anhydride Moieties. *Chem. Cent. J.* **2012**, *6*, 85. [CrossRef] [PubMed]
552. Katsikas, L.; Milovanović, M.; Popović, I.G. Hindered, 1,1-Disubstituted Monomers. Chain Transfer to Benzene in the Radical Polymerisation of Di-n-Butyl Itaconate. *Eur. Polym. J.* **2008**, *44*, 3028–3031. [CrossRef]
553. Schneider-Chaabane, A.; Bleicher, V.; Rau, S.; Al-Ahmad, A.; Lienkamp, K. Stimulus-Responsive Polyzwitterionic Surfaces Made from Itaconic Acid: Self-Triggered Antimicrobial Activity, Protein Repellency, and Cell Compatibility. *ACS Appl. Mater. Interfaces* **2020**, *12*, 21242–21253. [CrossRef] [PubMed]
554. Sollka, L.; Lienkamp, K. Progress in the Free and Controlled Radical Homo- and Co-Polymerization of Itaconic Acid Derivatives: Toward Functional Polymers with Controlled Molar Mass Distribution and Architecture. *Macromol. Rapid Commun.* **2021**, *42*, 2000546. [CrossRef]
555. Chiloeches, A.; Funes, A.; Cuervo-Rodríguez, R.; López-Fabal, F.; Fernández-García, M.; Echeverría, C.; Muñoz-Bonilla, A. Biobased Polymers Derived from Itaconic Acid Bearing Clickable Groups with Potent Antibacterial Activity and Negligible Hemolytic Activity. *Polym. Chem.* **2021**, *12*, 3190–3200. [CrossRef]
556. Baskan, H.; Esentürk, I.; Dösler, S.; Sarac, A.S.; Karakas, H. Electrospun Nanofibers of Poly (Acrylonitrile-Co-Itaconic Acid)/Silver and Polyacrylonitrile/Silver: In Situ Preparation, Characterization, and Antimicrobial Activity. *J. Ind. Text.* **2021**, *50*, 1594–1624. [CrossRef]
557. Rzayev, Z.M.O.; Şimşek, M.; Bunyatova, U.; Salamov, B. Novel Colloidal Nanofiber Semiconductor Electrolytes from Solution Blends of PVA/ODA–MMT, Poly (Itaconic Anhydride- Alt -2-Vinyl-1,3-Dioxalan) and Its Ag-Carrying Polymer Complex by Reactive Electrospinning. *Colloids Surf. A Physicochem. Eng. Asp.* **2016**, *492*, 26–37. [CrossRef]
558. Wilke, P.; Coger, V.; Nachev, M.; Schachschal, S.; Million, N.; Barcikowski, S.; Sures, B.; Reimers, K.; Vogt, P.M.; Pich, A. Biocompatible Microgel-Modified Electrospun Fibers for Zinc Ion Release. *Polymer* **2015**, *61*, 163–173. [CrossRef]
559. Nesic, A.R.; Onjia, A.; Ostojic, S.B.; Micic, D.M.; Velickovic, S.J.; Antonovic, D.G. Novel Biosensor Films Based on Chitosan. *Mater. Lett.* **2016**, *167*, 47–49. [CrossRef]
560. Chiloeches, A.; Cuervo-Rodríguez, R.; López-Fabal, F.; Fernández-García, M.; Echeverría, C.; Muñoz-Bonilla, A. Antibacterial and Compostable Polymers Derived from Biobased Itaconic Acid as Environmentally Friendly Additives for Biopolymers. *Polym. Test.* **2022**, *109*, 107541. [CrossRef]
561. Huerta-Ángeles, G.; Kanizsová, L.; Mielczarek, K.; Konefał, M.; Konefał, R.; Hodan, J.; Kočková, O.; Bednarz, S.; Beneš, H. Sustainable Aerogels Based on Biobased Poly (Itaconic Acid) for Adsorption of Cationic Dyes. *Int. J. Biol. Macromol.* **2024**, *259*, 129727. [CrossRef] [PubMed]
562. Khot, S.N.; Lascala, J.J.; Can, E.; Morye, S.S.; Williams, G.I.; Palmese, G.R.; Kusefoglu, S.H.; Wool, R.P. Development and Application of Triglyceride-Based Polymers and Composites. *J. Appl. Polym. Sci.* **2001**, *82*, 703–723. [CrossRef]
563. Satriana; Arpi, N.; Lubis, Y.M.; Adisalamun; Supardan, M.D.; Mustapha, W.A.W. Diacylglycerol-Enriched Oil Production Using Chemical Glycerolysis. *Eur. J. Lipid Sci. Technol.* **2016**, *118*, 1880–1890. [CrossRef]
564. Xu, Y.-Q.; Qu, J.-P. Mechanical and Rheological Properties of Epoxidized Soybean Oil Plasticized Poly(Lactic Acid). *J. Appl. Polym. Sci.* **2009**, *112*, 3185–3191. [CrossRef]
565. Zych, A.; Perotto, G.; Trojanowska, D.; Tedeschi, G.; Bertolacci, L.; Francini, N.; Athanassiou, A. Super Tough Polylactic Acid Plasticized with Epoxidized Soybean Oil Methyl Ester for Flexible Food Packaging. *ACS Appl. Polym. Mater.* **2021**, *3*, 5087–5095. [CrossRef]
566. Zhao, Y.; Qu, J.; Feng, Y.; Wu, Z.; Chen, F.; Tang, H. Mechanical and Thermal Properties of Epoxidized Soybean Oil Plasticized Polybutylene Succinate Blends. *Polym. Adv. Technol.* **2012**, *23*, 632–638. [CrossRef]
567. Yang, D.; Peng, X.; Zhong, L.; Cao, X.; Chen, W.; Zhang, X.; Liu, S.; Sun, R. Green Films from Renewable Resources: Properties of Epoxidized Soybean Oil Plasticized Ethyl Cellulose Films. *Carbohydr. Polym.* **2014**, *103*, 198–206. [CrossRef]
568. Han, Y.; Shi, J.; Mao, L.; Wang, Z.; Zhang, L. Improvement of Compatibility and Mechanical Performances of PLA/PBAT Composites with Epoxidized Soybean Oil as Compatibilizer. *Ind. Eng. Chem. Res.* **2020**, *59*, 21779–21790. [CrossRef]

569. Quiles-Carrillo, L.; Duart, S.; Montanes, N.; Torres-Giner, S.; Balart, R. Enhancement of the Mechanical and Thermal Properties of Injection-Molded Polylactide Parts by the Addition of Acrylated Epoxidized Soybean Oil. *Mater. Des.* **2018**, *140*, 54–63. [CrossRef]
570. Quiles-Carrillo, L.; Montanes, N.; Lagaron, J.M.; Balart, R.; Torres-Giner, S. On the Use of Acrylated Epoxidized Soybean Oil as a Reactive Compatibilizer in Injection-molded Compostable Pieces Consisting of Polylactide Filled with Orange Peel Flour. *Polym. Int.* **2018**, *67*, 1341–1351. [CrossRef]
571. Chaochanchaikul, K.; Pongmuksuwan, P. Influence of Ozonized Soybean Oil as a Biobased Plasticizer on the Toughness of Polylactic Acid. *J. Polym. Environ.* **2021**, *30*, 1095–1105. [CrossRef]
572. Orue, A.; Eceiza, A.; Arbelaiz, A. Preparation and Characterization of Poly (Lactic Acid) Plasticized with Vegetable Oils and Reinforced with Sisal Fibers. *Ind. Crops Prod.* **2018**, *112*, 170–180. [CrossRef]
573. Paunonen, S.; Berthold, F.; Immonen, K. Poly (Lactic Acid)/Pulp Fiber Composites: The Effect of Fiber Surface Modification and Hydrothermal Aging on Viscoelastic and Strength Properties. *J. Appl. Polym. Sci.* **2020**, *137*, 49617. [CrossRef]
574. Valente, B.F.A.; Karamysheva, A.; Silvestre, A.J.D.; Neto, C.P.; Vilela, C.; Freire, C.S.R. Epoxidized Linseed Oil as a Plasticizer for All-Cellulose Composites Based on Cellulose Acetate Butyrate and Micronized Pulp Fibers. *Ind. Crops Prod.* **2023**, *202*, 116980. [CrossRef]
575. Dominici, F.; Samper, M.D.; Carbonell-Verdu, A.; Luzi, F.; López-Martínez, J.; Torre, L.; Puglia, D. Improved Toughness in Lignin/Natural Fiber Composites Plasticized with Epoxidized and Maleinized Linseed Oils. *Materials* **2020**, *13*, 600. [CrossRef]
576. Dolores, M. Plasticization Effect of Epoxidized Cottonseed Oil (ECSO) on Poly (Lactic Acid). *Ind. Crops Prod.* **2017**, *104*, 278–286.
577. Bouti, M.; Irinislimane, R.; Belhaneche-Bensemra, N. Properties Investigation of Epoxidized Sunflower Oil as Bioplasticizer for Poly (Lactic Acid). *J. Polym. Environ.* **2022**, *30*, 232–245. [CrossRef]
578. Dominguez-Candela, I.; Ferri, J.M.; Cardona, S.C.; Lora, J.; Fombuena, V. Dual Plasticizer/Thermal Stabilizer Effect of Epoxidized Chia Seed Oil (*Salvia hispanica* L.) to Improve Ductility and Thermal Properties of Poly (Lactic Acid). *Polymers* **2021**, *13*, 1283. [CrossRef]
579. Perez-Nakai, A.; Lerma-Canto, A.; Domingez-Candela, I.; Garcia-Garcia, D.; Ferri, J.M.; Fombuena, V. Comparative Study of the Properties of Plasticized Polylactic Acid with Maleinized Hemp Seed Oil and a Novel Maleinized Brazil Nut Seed Oil. *Polymers* **2021**, *13*, 2376. [CrossRef]
580. Garcia-Garcia, D.; Carbonell-Verdu, A.; Arrieta, M.P.; López-Martínez, J.; Samper, M.D. Improvement of PLA Film Ductility by Plasticization with Epoxidized Karanja Oil. *Polym. Degrad. Stab.* **2020**, *179*, 109259. [CrossRef]
581. The Circular Economy for Plastics, A European Analysis, Plastics Europe, MARCH 2024. Available online: https://plasticseurope.org/wp-content/uploads/2024/05/Circular_Economy_report_Digital_EXEC_light_FINAL.pdf (accessed on 28 August 2024).

polymers

MDPI

Article

Biobased Composites of Poly(Lactic Acid) Melt Compounded with Bacterial and Vegetal Nanocelluloses Incorporated through Different Strategies

Jimena Bovi [1,2], Juan Francisco Delgado [1,2,3], Orlando de la Osa [3], Mercedes Ana Peltzer [3,4], Celina Raquel Bernal [1,2] and María Laura Foresti [1,2,*]

[1] Universidad de Buenos Aires, Facultad de Ingeniería, Buenos Aires C1127AAR, Argentina; jbovi@fi.uba.ar (J.B.); jfdelgado@fi.uba.ar (J.F.D.); cbernal@fi.uba.ar (C.R.B.)
[2] CONICET—Universidad de Buenos Aires, Instituto de Tecnología en Polímeros y Nanotecnología (ITPN), Buenos Aires C1127AAR, Argentina
[3] Universidad Nacional de Quilmes, Departamento de Ciencia y Tecnología, Laboratorio de Obtención, Modificación, Caracterización y Evaluación de Materiales (LOMCEM), Quilmes B1876BXD, Argentina; odelaosa@unq.edu.ar (O.d.l.O.); mercedes.peltzer@unq.edu.ar (M.A.P.)
[4] Consejo Nacional de Investigaciones Científicas y Técnicas (CONICET), Buenos Aires C1425FQB, Argentina
* Correspondence: mforesti@fi.uba.ar

Abstract: In the current contribution, bacterial nanocellulose obtained from a by-product of Kombucha tea production and vegetal nanocellulose isolated from milled rice husks were employed as fillers of PLA-based composites prepared by intensive mixing followed by compression molding. Given the challenges associated with the incorporation of nanocelluloses—initially obtained as aqueous suspensions—into melt compounding processes, and also with achieving a proper dispersion of the hydrophilic nanofillers within PLA, three different nanofibrils incorporation strategies were studied: i.e., direct mixing of dried milled nanocelluloses and PLA; masterbatching by solvent casting of native nanocelluloses followed by melt compounding; and masterbatching by solvent casting of acetylated nanocelluloses followed by melt compounding. Composites with varying filler content (from 0.5 wt.% to 7 wt.%) were characterized in terms of morphology, optical properties, and mechanical performance. Results revealed the relative suitability of each strategy employed to promote nanocelluloses dispersion within the PLA matrix. PLA/nanocellulose masterbatches prepared by solvent casting proved to be particularly useful for feeding the nanocelluloses into the processing equipment in a dry state with limited hornification. Acetylation also contributed to a better dispersion of both nanocelluloses within the PLA matrix, although no clear positive impact on the mechanical properties of the films was observed. Finally, filler loading played an important role in the films' properties by increasing their stiffness while reducing their translucency.

Keywords: poly(lactic acid); composites; bacterial nanocellulose; rice husk nanocellulose; filler incorporation strategies

Citation: Bovi, J.; Delgado, J.F.; de la Osa, O.; Peltzer, M.A.; Bernal, C.R.; Foresti, M.L. Biobased Composites of Poly(Lactic Acid) Melt Compounded with Bacterial and Vegetal Nanocelluloses Incorporated through Different Strategies. *Polymers* **2024**, *16*, 898. https://doi.org/10.3390/polym16070898

Academic Editor: Andrea Ehrmann

Received: 16 February 2024
Revised: 13 March 2024
Accepted: 18 March 2024
Published: 25 March 2024

1. Introduction

In the recent decades, the persistence of conventional non-biodegradable plastics in the environment has emerged as a global concern; and biodegradable polymers have garnered considerable attention. Currently, one of the most popular biodegradable polymers commercially available is poly(lactic acid) (PLA), a compostable thermoplastic polyester derived from biomass whose price has significantly declined over the last two decades [1,2].

Despite the relatively high mechanical strength and stiffness of poly(lactic acid), its gloss, and UV stability, PLA exhibits some disadvantages such as brittleness, a low crystallization rate, and relatively low barrier properties. In this regard, a well-established strategy to improve the performance of PLA is the development of composites with different types and amounts of fillers. Among them, in recent years, different biofillers have been assayed, with those of cellulosic nature being the most studied ones [3–5]. In particular, a huge variety of cellulosic nanomaterials has been employed [6,7]. Nanocelluloses are recognized for their numerous attractive attributes such as biodegradability, their non-toxic nature, the wide availability and renewability of the cellulosic sources, their lightweight nature, and impressive stiffness and strength, among others [6].

Over the last two decades, great advances in the development of PLA/nanocellulose composites prepared by solution casting have been achieved [8,9]. However, even if casting/evaporation-based processing techniques facilitate good dispersion of the nanofillers in the matrix, the amount of solvent and the time required usually limit their industrial scaling, so conventional melt processing techniques are preferred for large-volume production [10]. Nevertheless, achieving a proper nanofiller dispersion through melt compounding of nanocelluloses (especially cellulose nanofibrils) and PLA is much more challenging. Furthermore, cellulose nanofibrils are obtained and commonly commercialized as diluted aqueous suspensions, so their dehydration is highly desirable to avoid PLA degradation during thermal melting. However, upon water removal, extensive hydrogen bonding between cellulose nanofibrils is promoted, leading to their irreversible aggregation in a process known as hornification [11]. In the absence of superficial cellulose–water interaction, additional hydrogen bonds are created between adjacent cellulose nanofibrils, preventing further dispersion in aqueous or organic media [10]. In particular, oven-drying of cellulose nanofibrils is a great challenge since the hornification extent caused by this technique is usually greater than that induced by other drying methods [12,13]. Accordingly, direct oven-drying of cellulose nanofibrils prior to melt compounding with PLA usually hinders the later dispersion of nanocellulosic fillers within the polymeric matrix [14,15]. Moreover, the poor interfacial compatibility between the hydrophilic surface of nanocelluloses and the hydrophobic PLA further hampers the dispersion of cellulose nanofibrils.

In this context, in recent years, a few contributions have proposed the elaboration of masterbatches by solvent casting as a middle step in the development of PLA/nanocelluloses melt-compounded composites with suitable levels of nanofiller dispersion [16,17]. In those masterbatches, PLA acts as a carrier to feed the nanocellulose into the processing equipment in a dry form, with PLA chains placed between cellulose nanofibrils, limiting hornification. On the other hand, aiming to improve the chemical compatibility between PLA and nanocelluloses, different strategies for the surface hydrophobization of the nanofibrils have been proposed, including polymer grafting, silanization, esterification (mainly acetylation), etc. [18].

In the current contribution, bacterial nanocellulose obtained from a by-product of Kombucha tea production (BNC) and vegetal nanocellulose isolated from milled rice husks (VNC) were employed as fillers of PLA-based composites prepared by intensive mixing followed by compression molding. Given the challenges associated with achieving a proper dispersion of cellulose nanofibrils—initially obtained as aqueous suspensions—within the PLA matrix by melt processing, three different filler incorporation strategies were assayed. In the first place, oven-dried milled bacterial and vegetal nanocelluloses were directly mixed with PLA. Secondly, masterbatches with native BNC and VNC were prepared by solvent casting, diluted with PLA pellets, and melt compounded. Finally, the masterbatches were elaborated by the same protocol, but using acetylated bacterial and vegetal nanocelluloses (AcBNC and AcVNC, respectively). Composites with varying filler contents were characterized and compared in terms of morphology, optical properties, and mechanical performance.

2. Materials and Methods

2.1. Materials

Commercial PLA (Nature Works® trade name 4043D) consisting of 6% D-lactide and 94% L-lactide, with a density of 1.24 g/cm^3 and a melt flow index of 6 g/10 min (210 °C, 2.16 kg), was employed as the matrix of the composites. Before melt compounding, PLA pellets were vacuum dried at 80 °C for 4 h, as recommended by the supplier. Black tea (Green Hills, Buenos Aires, Argentina) and glucose (Xantana, Buenos Aires, Argentina) were used in the formulation of culture media for the production of bacterial nanocellulose. The SCOBY (Symbiotic Culture of Bacterial and Yeast) was kindly supplied by Aloha Fermentos (Buenos Aires, Argentina). Rice husks remaining from rice production (Chajarí, Entre Ríos, Argentina) were washed, micronized, and sieved to retain particles smaller than 125 µm. Hydrochloric acid (Biopack, Buenos Aires, Argentina), sodium hydroxide (Anedra, Buenos Aires, Argentina), and sodium chlorite solution (Anedra, Buenos Aires, Argentina) were employed for the isolation of vegetal cellulose, whereas potassium hydroxide (Cicarelli, Santa Fe, Argentina) was used in the purification of bacterial nanocellulose. Acetic acid, acetic anhydride, acetone, and chloroform (all from Cicarelli, Santa Fe, Argentina) were used for nanocellulose acetylation and for the preparation of the masterbatches. All chemicals used were of analytical grade and employed as received.

2.2. Bacterial Nanocellulose Production

Bacterial nanocellulose was isolated from the by-product of Kombucha tea production, as described previously [19]. Briefly, a black tea infusion (10 g/L in boiling water) was sweetened with 60 g/L of glucose. After cooling, the inoculum was added at a concentration of 10% *v*/*v*, and covered containers were kept in static conditions at 28–30 °C for 14 days. After this period, the floating nanocellulose membranes developed on the surface of the broth were harvested, thoroughly rinsed with distilled water to remove the culture medium, broken down for 4 min using a kitchen blender in a KOH 5% (*w*/*v*) solution, left in alkali for 16 h at room temperature, and successively rinsed with distilled water.

2.3. Vegetal Nanocellulose Production

Sieved milled rice husks (particle size < 125 µm) were put in contact (0.05 g/mL) with a sodium hydroxide solution (4 wt.%), refluxed for 2 h with magnetic stirring, exhaustively washed with distilled water, and filtered. The solid obtained was then treated (0.03 g/mL, dry weight) with a sodium chlorite solution (1.7% *w*/*v*, in acetic acid/sodium acetate/distilled water buffer, pH 4) at 90 °C for 2 h under reflux, followed by extensive washing with distilled water. The resulting cellulosic fraction was treated (0.04 g/mL) with a hydrochloric acid solution (1 M) at 90 °C for 45 min, and then washed again with distilled water until the washing waters reached the pH of distilled water [20]. The cellulose suspension (1 wt.%) was subsequently passed through a high-pressure homogenizer NS1001L (Niro-Soavi, Parma, Italy) operated at 100 MPa for 1 h.

2.4. Acetylation of Nanocelluloses

Aqueous suspensions BNC and VNC (1 wt.%) were solvent exchanged through acetic acid into acetic anhydride (twice each, using 200 mL). Vacuum-filtrated BNC and VNC (2.5 g dry weight) were then transferred into a glass flask containing 250 mL of acetic anhydride. The system, equipped with a reflux condenser, was heated at 120 °C for 1 h in a thermostatized oil bath under continuous magnetic stirring. After this period of time, the obtained AcBNC and AcVNC were vacuum-filtrated and thoroughly washed with ethanol and distilled water. The attained degrees of substitution (DS) of both modified nanocelluloses were quantified by heterogeneous saponification of dried samples according to Ávila Ramírez et al. (2014) [21].

2.5. Masterbatches and Composites Preparation

In the first strategy assayed, homogenized nanocelluloses (18,000 rpm, 5 min, 5 times) were dried (110 °C, 24 h), milled, and directly introduced into a Brabender internal mixer (170 °C, 50 rpm, 8 min) with the required amount of PLA pellets to reach filler concentrations of 0.5, 1, 1.5, 3, 5, and 7 wt.%. The composites obtained with this strategy were named PLA/BNC-D and PLA/VNC-D. For comparison, neat PLA was also melt compounded using the same processing conditions. As a second strategy, masterbatches (MBs) were prepared by solvent casting. For this purpose, aqueous suspensions of native bacterial and vegetal nanocelluloses were solvent-exchanged through acetone (twice) into chloroform (twice) with homogenization (18,000 rpm, 5 min). Subsequently, nanocellulose suspensions were homogenized in chloroform (65 mL) containing dissolved PLA at proper concentrations to achieve masterbatch filler contents of 16.7 wt.%, cast onto glass vessels covered with Teflon tape, and dried (room temperature, 48 h; followed by 60 °C in a vacuum oven, 2 weeks). Then, the masterbatches were milled, dried under the same conditions as additional PLA pellets (80 °C, under vacuum, 4 h), and melt compounded in a Brabender internal mixer (170 °C, 50 rpm, 8 min) to reach final filler concentrations of 0.5, 1, 1.5, 3, 5, and 7 wt.%. The composites obtained using this strategy were named PLA/BNC-MB and PLA/VNC-MB. The third strategy was the same as the second one, but acetylated bacterial nanocellulose and acetylated vegetal nanocellulose instead of their native counterparts were used as fillers. The resulting composites from this last strategy were named PLA/AcBNC-MB and PLA/AcVNC-MB.

To obtain the films, 5.5 g of the melt-compounded samples (PLA and each of the described composites) previously dried at 80 °C under vacuum overnight were compression molded using a hydraulic press model 3853 (Carver, Wabash, IN, USA) at 170 °C without any pressure, 8 min, followed by 170 °C at 2 MPa for 5 min, and cooling at room temperature. The thickness of all composites ranged from 0.6 to 0.9 mm, and their formulations are summarized in Table 1.

Table 1. PLA/nanocelluloses composite formulations.

Composite	PLA (wt.%)	BNC (wt.%)	VNC (wt.%)	AcBNC (wt.%)	AcVNC (wt.%)
PLA	100	-	-	-	-
PLA/BNC-D 0.5%	99.5	0.5	-	-	-
PLA/BNC-D 1%	99	1	-	-	-
PLA/BNC-D 1.5%	98.5	1.5	-	-	-
PLA/BNC-D 3%	97	3	-	-	-
PLA/BNC-D 5%	95	5	-	-	-
PLA/BNC-D 7%	93	7	-	-	-
PLA/VNC-D 0.5%	99.5	-	0.5	-	-
PLA/VNC-D 1%	99	-	1	-	-
PLA/VNC-D 1.5%	98.5	-	1.5	-	-
PLA/VNC-D 3%	97	-	3	-	-
PLA/VNC-D 5%	95	-	5	-	-
PLA/VNC-D 7%	93	-	7	-	-
PLA/BNC-MB 0.5%	99.5	0.5	-	-	-
PLA/BNC-MB 1%	99	1	-	-	-
PLA/BNC-MB 1.5%	98.5	1.5	-	-	-
PLA/BNC-MB 3%	97	3	-	-	-
PLA/BNC-MB 5%	95	5	-	-	-
PLA/BNC-MB 7%	93	7	-	-	-
PLA/VNC-MB 0.5%	99.5	-	0.5	-	-
PLA/VNC-MB 1%	99	-	1	-	-
PLA/VNC-MB 1.5%	98.5	-	1.5	-	-
PLA/VNC-MB 3%	97	-	3	-	-
PLA/VNC-MB 5%	95	-	5	-	-
PLA/VNC-MB 7%	93	-	7	-	-

Table 1. *Cont.*

Composite	PLA (wt.%)	BNC (wt.%)	VNC (wt.%)	AcBNC (wt.%)	AcVNC (wt.%)
PLA/AcBNC-MB 0.5%	99.5	-	-	0.5	-
PLA/AcBNC-MB 1%	99	-	-	1	-
PLA/AcBNC-MB 1.5%	98.5	-	-	1.5	-
PLA/AcBNC-MB 3%	97	-	-	3	-
PLA/AcBNC-MB 5%	95	-	-	5	-
PLA/AcBNC-MB 7%	93	-	-	7	-
PLA/AcVNC-MB 0.5%	99.5	-	-	-	0.5
PLA/AcVNC-MB 1%	99	-	-	-	1
PLA/AcVNC-MB 1.5%	98.5	-	-	-	1.5
PLA/AcVNC-MB 3%	97	-	-	-	3
PLA/AcVNC-MB 5%	95	-	-	-	5
PLA/AcVNC-MB 7%	93	-	-	-	7

2.6. *Characterization of Nanocelluloses*

2.6.1. Field Emission Scanning Electron Microscopy (FESEM)

Drops of aqueous suspensions (0.01 wt.%) of BNC, VNC, AcBNC, and AcVNC were placed on microscope slides, dried at 110 °C for 5 min, sputter coated with gold, and observed using a Zeiss Crossbeam 340 FESEM microscope (Carl Zeiss Microscopy GmbH, Oberkochen, Germany) operated at an accelerating voltage of 3 kV. Quantitative image analysis was performed with the help of the image processing software Image J 1.53e. In order to guarantee the statistical validity of the analysis, at least 200 measurements were performed.

2.6.2. Atomic Force Microscopy (AFM)

Drops of aqueous suspensions (0.2 wt.%) of BNC, VNC, AcBNC, and AcVNC were placed on a mica support and analyzed using a Nanoscope V-Multimode atomic force microscope (Veeco, Santa Barbara, CA, USA) in tapping mode (intermittent contact) using a silicon nitride probe Arrow-NCR-50 (Nano World, Neuchâtel, Switzerland). The resonance frequency of the cantilever was 258 kHz, with a constant force of 42 N m^{-1} and a tip radius ranging from 5 to 10 nm. Scanning frequencies typically ranged from 1.0 to 1.5 Hz.

2.6.3. Fourier Transform Infrared Spectroscopy (FTIR)

Fourier transform infrared spectra of BNC, VNC, AcBNC, and AcVNC were acquired with an IR Affinity-1 Shimadzu Fourier Transform Infrared Spectrophotometer (Kyoto, Japan) in absorbance mode. Dried milled samples and KBr (110 °C, overnight) were pressed at a 1:20 ratio into discs at 8 kg/cm^2 and the IR spectra were recorded in the range of 3800 to 900 cm^{-1} with a total of 32 scans at a resolution of 4 cm^{-1}. Baselines were corrected and normalized against the intensity of the absorption at 1165 cm^{-1}, attributed to the (C–O–C) linkage of cellulose [22].

2.6.4. X-ray Diffraction Analysis (XRD)

A Rigaku D/Max-C wide-angle automated X-ray diffractometer with vertical goniometer was employed to collect the X-ray diffraction patterns of BNC, VNC, AcBNC, and AcVNC (Cu/Kα, 0.154 nm, 40 kV, 30 mA). Diffractograms were recorded at a step size of 0.02° in the 2θ range of 10–50°. For the calculation of the crystallinity index of all samples, Segal's empirical Equation (1) was applied [23]:

$$CrI\ (\%) = \frac{I_{002} - I_{am}}{I_{002}} * 100\% \quad (1)$$

where I_{002} is the maximum intensity value of the 002 lattice diffraction and represents crystalline and amorphous material, and I_{am} is the intensity at 2θ = 18° and represents amorphous material only.

2.7. Characterization of PLA/Nanocelluloses Composites

2.7.1. Environmental Scanning Electron Microscopy (ESEM)

Cryo-fractured surfaces of PLA and PLA/nanocelluloses composites obtained at liquid nitrogen temperature were analyzed using an FEI Quanta 200 microscope operated at an accelerating voltage of 15 kV. All samples were sputter-coated with a thin layer of gold before observations.

2.7.2. Optical Properties

Light transmittance curves of PLA and PLA/nanocelluloses composites were obtained in the range of 200–800 nm using a Shimadzu UV-visible spectrophotometer (Model UV-1650pc). The color of the films was evaluated using a Minolta colorimeter (CR-20, Konica Minolta, Inc., Tokyo, Japan) previously calibrated with a white reflector plate. Measurements were expressed as colorimetric coordinates in the CIELAB scale: L* (lightness, from 0: black to 100: white), a* (from green (−) to red (+)), and b* (from blue (−) to yellow (+)). Color differences (ΔE) were calculated using Equation (2):

$$\Delta E = \sqrt{\Delta L*^2 + \Delta a*^2 + \Delta b*^2} \quad (2)$$

2.7.3. Mechanical Properties

Uniaxial tensile tests were performed on PLA and PLA/nanocellulose composites using dumbbell samples (at least five per system) in an INSTRON dynamometer 5985 (Norwood, MA, USA) at 5 mm/min, using a load cell of 1 kN [24]. Young's modulus, tensile strength, and strain-at-break values were obtained from the stress–strain curves. Average values of tensile parameters were reported along with their deviations.

3. Results and Discussion

3.1. Nanocelluloses

Figure 1 shows collected photographs of the raw materials used for nanocellulose isolation, as well as the purified bacterial and vegetal celluloses (i.e., BNC and VNC, respectively). FESEM micrographs of BNC, VNC, AcBNC, and AcVNC are also shown. Figure 1A is a photograph of the floating pellicle developed during Kombucha production, while Figure 1B shows the purified BNC isolated according to Section 2.2. Melanoidins present in the Kombucha tea give the pellicle a brownish color. Blending in an alkali solution promotes the removal of these components, resulting in the gel-like material with a whitish color shown in Figure 1B. Figure 1C,D present the original and the milled rice husks, respectively. According to the literature, rice husks are composed of cellulose (25–35 wt.%), hemicellulose (18–21 wt.%), lignin (26–31 wt.%), silica (15–17 wt.%), and other soluble compounds (2–5 wt.%) [25,26]. Figure 1E shows the appearance of the VNC obtained after purification and fibrillation of cellulose from rice husks as described in Section 2.3.

FESEM images collected in Figure 1F–I illustrate the morphology of the native and acetylated nanocelluloses obtained. In BNC and AcBNC images, the typical intertwisted bacterial cellulose nanoribbons with an average width of 40 ± 10 nm and several micrometers in length can be distinguished, in agreement with values reported in the literature [27,28]. Nanocelluloses derived from rice husks showed an average width of 27 ± 6 nm, also in accordance with previous reports [17,29]. Regarding VNC and AcVNC length, even if micrometric nanofibrils seem to prevail, some shorter fibrils can be distinguished in less concentrated zones.

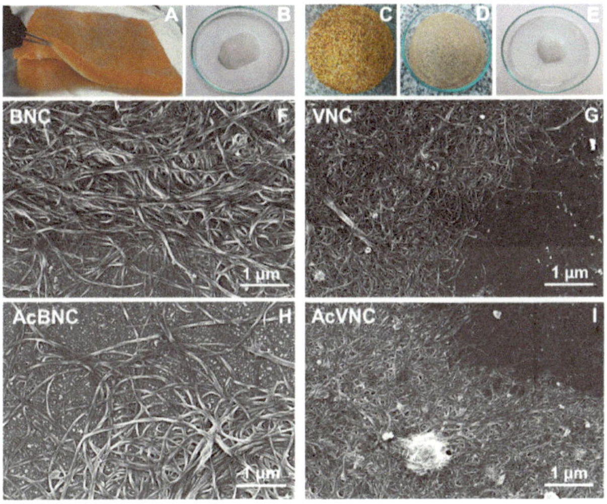

Figure 1. Photographs of the unpurified floating pellicle developed during Kombucha tea production (**A**), bacterial nanocellulose recovered after purification of the pellicle (**B**), rice husks washed (**C**) and milled (**D**), and vegetal nanocellulose isolated from rice husks (**E**). FESEM micrographs of diluted aqueous suspensions of BNC (**F**), VNC (**G**), AcBNC (**H**), and AcVNC (**I**).

Topographical images of nanocelluloses obtained by atomic force microscopy allowed better estimating their length. Whereas native and acetylated bacterial nanocelluloses (Figures 2A and 2C, respectively) show nanofibrils with lengths between 3–15 µm, in native and acetylated vegetal nanocelluloses images (Figures 2B and 2D, respectively) micrometric-in-length nanofibrils seem to coexist with a significant fraction of much shorter fibrils (up to 200 nm long). The latter observations suggest that the final step during VNC isolation involving hydrochloric acid might have promoted the reduction in the length of a fraction of the fibrils. Similar results were reported in the isolation of cellulose nanofibrils from other vegetal sources using protocols that involved treatments with hydrochloric acid solutions under comparable conditions (usually aimed at removing compounds other than cellulose [30] and/or improving fiber exfoliation [31]).

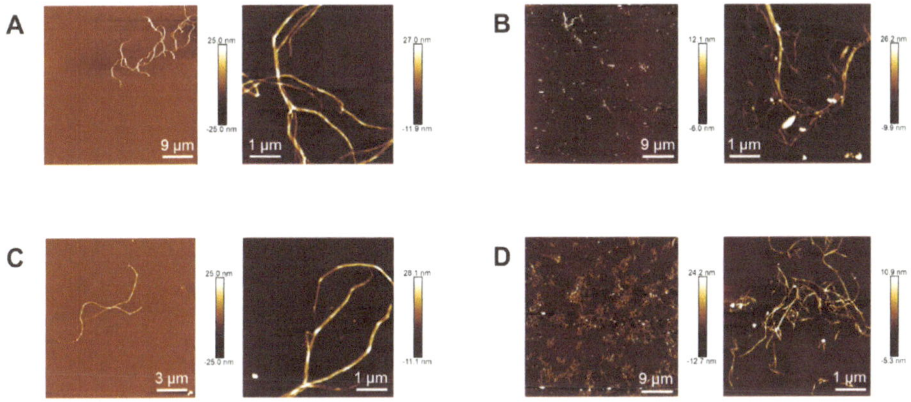

Figure 2. Topographical images of BNC (**A**), VNC (**B**), AcBNC (**C**), and AcVNC (**D**) acquired by atomic force microscopy.

The FTIR spectra of native and acetylated bacterial and vegetal nanocelluloses are shown in Figure 3. All samples exhibit absorbances typical of cellulose, i.e., those found in the 3600–3000 cm^{-1} interval ascribed to the stretching vibration of -OH groups present in carbohydrates; bands found in the 3000–2800 cm^{-1} region attributed to the stretching of C-H linkages of methyl and methylene groups of cellulose; and other signals within the 1450–900 cm^{-1} range, commonly associated with the symmetrical bending of CH_2 group, cellulose C-O-C bridges, C-O stretching and other absorptions typical of β-linked glucose polymers [27,32,33]. In addition, both acetylated nanocelluloses FTIR spectra present signals characteristic of ester groups (e.g., stretching of C=O at 1735 cm^{-1} and stretching of C-O at 1248 cm^{-1}), confirming the occurrence of the modification [34,35]. For both acetylated nanocelluloses, the attained degree of substitution (DS) determined by saponification was in the range of 0.15–0.20.

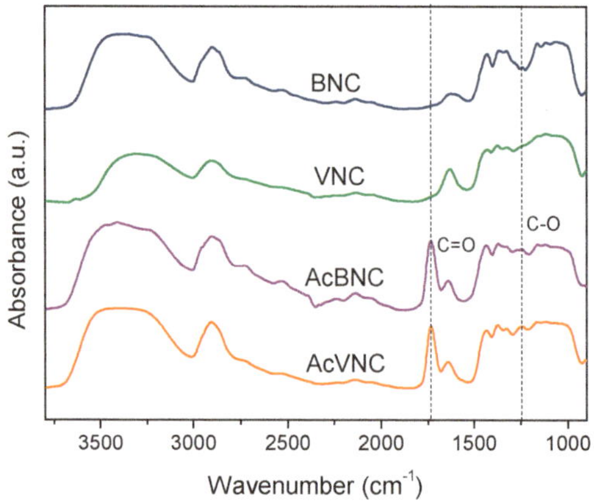

Figure 3. FTIR spectra of BNC, VNC, AcBNC, and AcVNC. Curves were shifted along the y-axis for clarity.

Figure 4 presents the X-ray diffractograms of the different samples. Four signals characteristic of the crystalline peaks of cellulose I can be observed in BNC and AcBNC samples centered at $2\theta = 14.9°$, 17.2°, 23.1°, and 34.9° in accordance with previous reports [35–37]. The fifth signal characteristic of cellulose I samples sometimes reported at 20.5° (usually a low-intensity signal) is not clearly defined in the bacterial nanocelluloses diffractograms. The crystallinity index (CrI) of BNC and AcBNC calculated by Segal's empirical method was (85 ± 1) %, in agreement with previous values calculated for microbial nanocelluloses derived from Kombucha tea production using the same method [38,39].

In reference to VNC and AcVNC samples, three characteristic cellulose I intense signals are observed centered at $2\theta = 15.8°$ (resulting from the overlapping of peaks centered at $2\theta = 14.9°$ and 16.3°), 22.7°, and 34.6° [36]. The calculated crystallinity index for VNC and AcVNC was (76 ± 2) %, in accordance with previous reports for vegetal nanocelluloses [31,40], which are typically less crystalline than nanocelluloses of bacterial origin [41,42]. As described by other authors, the higher crystallinity of bacterial nanocelluloses is also illustrated in their better-resolved diffractograms [40–42].

On the other hand, the fact that the CrIs of both acetylated nanocelluloses were similar to those of their native counterparts suggests that chemical modification was limited to the surface and more accessible amorphous domains of the cellulose, and that the crystalline arrangement of the inner layers of the fibrils was not significantly altered upon acetylation [35,40,41].

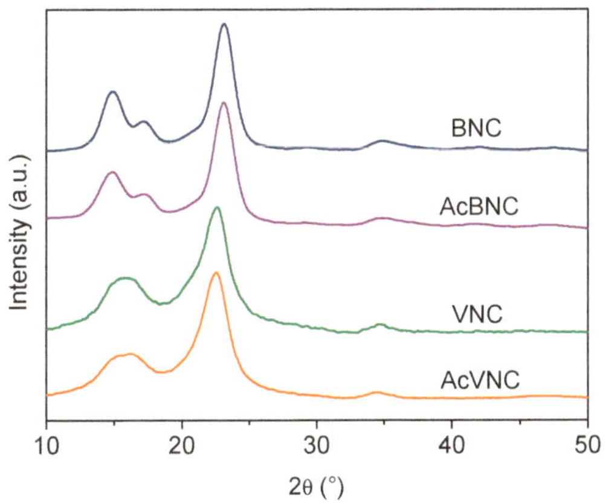

Figure 4. X-ray diffractograms of BNC, VNC, AcBNC, and AcVNC. Curves were shifted along the y-axis for clarity.

3.2. *PLA/Nanocelluloses Composites*

Figure 5 collects representative photographs of neat PLA and all the composite films obtained by intensive mixing and compression molding with the fillers incorporated through the three different strategies assayed. The direct introduction of dried milled bacterial or vegetal nanocelluloses into the PLA matrix resulted in a poor dispersion of the filler (PLA/BNC-D and PLA/VNC-D in Figures 5A and 5B, respectively), with the appearance of nanocellulose aggregates as dots of varying size all over the film.

Aggregates of nanocelluloses smaller than 0.5 mm were mainly observed, although larger ones can be also distinguished (with sizes up to 4 mm in the case of PLA/BNC-D, and up to 2 mm for PLA/VNC-D), affecting the optical properties of the films.

On the other hand, all composites obtained through masterbatching exhibit much smaller aggregates (especially in the case of BNC-based composites), highlighting an improved dispersion of the nanofillers. Results suggest that the strategy of preparing PLA/nanocellulose masterbatches by solvent casting successfully allowed feeding the nanocellulose into the processing equipment in a dry state with limited hornification by placing PLA chains between cellulose nanofibrils. Regarding the effects of acetylation, the images in Figure 5 show that chemical modification further improved the dispersion of both nanocelluloses within the PLA matrix compared to their native counterparts, according to a higher compatibility between the hydrophobic matrix and the hydrophobized nanocelluloses, as previously reported by other authors for similar systems [34,35,40]. Tingaut et al. (2010) evaluated the quality of filler dispersion within PLA using transparency as an indicator of good dispersibility. The authors highlighted the beneficial effect of microfibrillated cellulose acetylation on the reduction in the presence of aggregates [43]. The effect of filler loading is also evident from the images in Figure 5, with films showing reduced translucency as nanocellulose content was increased. Filler aggregation during films' manufacturing compromises their optical properties, negatively affecting the product appearance [44].

Representative cryo-fractured surfaces of PLA and PLA/nanocellulose composites are presented in Figure 6. Large irregular aggregates of nanocelluloses can be observed in the ESEM micrographs of composites obtained by the direct introduction of dried milled BNC and VNC into PLA (Figure 6A,B, upper rows). In accordance with the observations from Figure 5, BNC aggregates are larger than VNC ones and they have the typical multilayered arrangement of pure dried BNC films [45], with no evidence of PLA intercalation. Results are consistent with previous observations for composites obtained by mixing dried milled BNC and PLA [46,47]. The limited dispersion of the milled reinforcements can be attributed to the hornification of nanocelluloses taking place upon drying, as well as to the inherent incompatibility between components, resulting in a poor adhesion between phases as evidenced by the presence of gaps at the filler/matrix interface, some of them pointed to with arrows in Figure 6A,B, upper rows. Evident voids around cellulosic aggregates have previously been associated with poor adhesion between PLA and the filler [48,49].

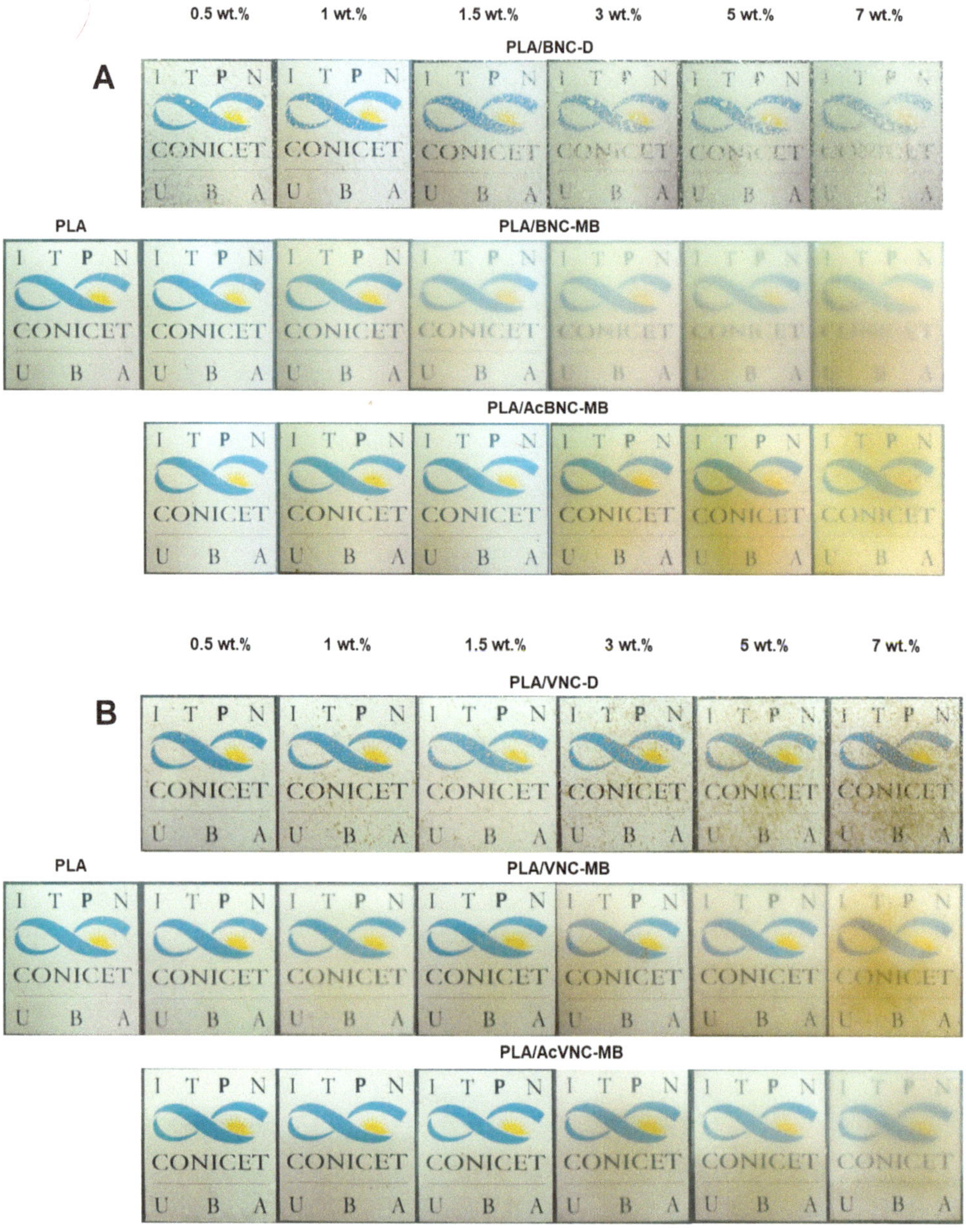

Figure 5. Photographs of neat PLA, PLA/BNC-D, PLA/BNC-MB, and PLA/AcBNC-MB (**A**), and PLA/VNC-D, PLA/VNC-MB, and PLA/AcVNC-MB (**B**) composite films obtained using different filler incorporation strategies and nanocelluloses contents.

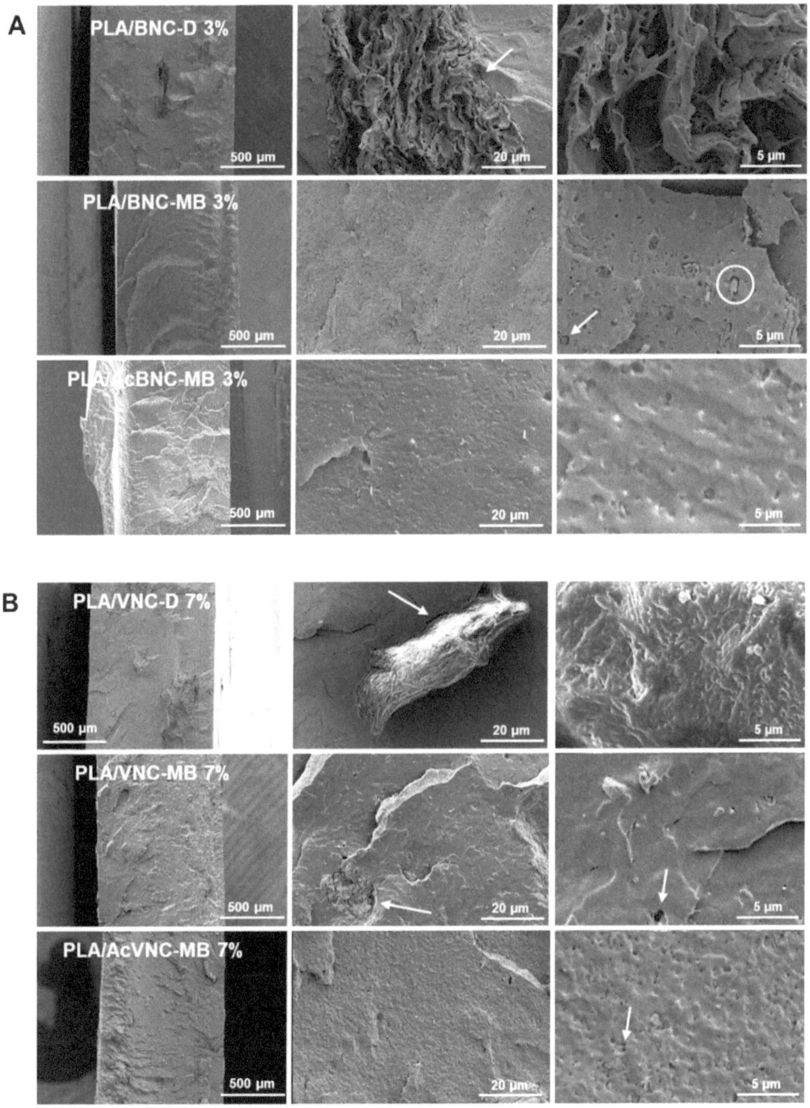

Figure 6. ESEM micrographs of cryo-fractured surfaces of PLA/BNC-D, PLA/BNC-MB, and PLA/AcBNC-MB composite films with 3 wt.% of bacterial nanocellulose (**A**), and PLA/VNC-D, PLA/VNC-MB, and PLA/AcVNC-MB composite films with 7 wt.% of vegetal nanocellulose (**B**). Circles and arrows were added to point gaps at the filler/matrix interface.

Regarding composites obtained through masterbatching, a significant reduction in the size of the aggregates is observed (Figure 6A,B, middle rows), and in the case of BNC, the typical morphology of pure multilayered BNC aggregates is no longer distinguishable. This further indicates that the elaboration of a masterbatch by solvent casting before melt compounding succeeded in decreasing the hornification of both nanocelluloses and promoted a better dispersion of the fillers within the PLA matrix. Improved dispersion of nanocelluloses upon masterbatching is consistent with the results reported by Bagheriasl and co-workers, who evaluated the relative dispersion of cellulose nanocrystals introduced through direct mixing and masterbatching strategies into PLA matrix [50]. Regarding the effect of acetylation, no evident improvement was observed, with similar surface roughness and dispersion levels (Figure 6A,B, bottom rows) [34]. No significant effect of filler content was evidenced from ESEM fractographs either.

Figure 7 exhibits the light transmittance (%) curves of composite films collected in the range of 200–800 nm. For both nanocelluloses, and regardless of the strategy employed for their incorporation into the PLA matrix, Figure 7A,B show that composites displayed lower light transmittance values than PLA throughout the measured wavelength range. Additionally, in agreement with visible changes illustrated in Figure 5, light transmittance values significantly decreased as higher filler contents were used. The light transmittance of composites containing 5 wt.% and 7 wt.% of nanocellulose was especially low.

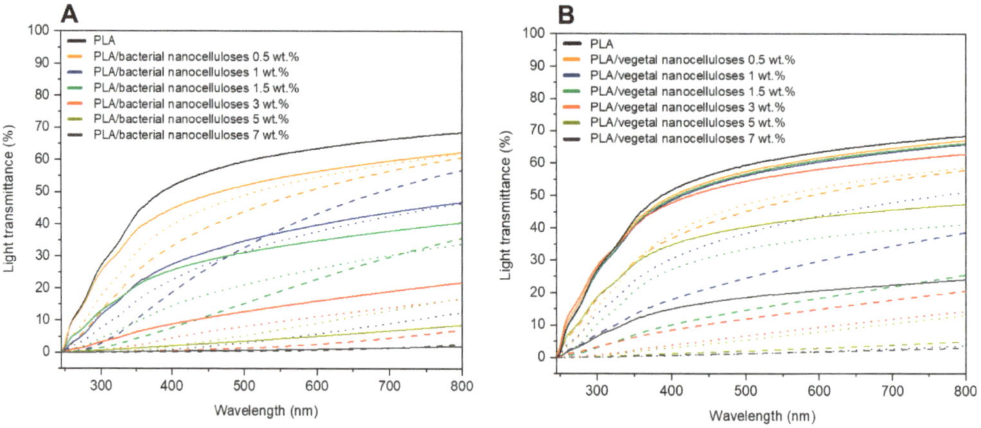

Figure 7. Light transmittance curves of neat PLA, PLA/BNC-D (solid lines), PLA/BNC-MB (dashed lines), and PLA/AcBNC-MB (dotted lines) (**A**); and PLA/VNC-D (solid lines), PLA/VNC-MB (dashed lines), and PLA/AcVNC-MB (dotted lines) (**B**) composite films obtained using different filler incorporation strategies and nanocellulose contents.

In terms of the effects of the strategies employed to incorporate the cellulosic nanofillers, most results collected in Figure 7 indicate that the direct introduction of both dried milled nanocelluloses resulted in composites with higher light transmittance values than composites produced by masterbatching, regardless of the filler content. Results may be attributed to the high hornification of dried milled nanocelluloses which, when directly introduced into the polymer melt, dispersed poorly, and composite materials with relatively extensive zones of the neat matrix with the high light transmittance characteristic of PLA were obtained. On the other hand, the preparation of composites by masterbatching through solvent casting promoted a better dispersion of less hornified nanocelluloses within the matrix, and the measured light transmittance values were the result of much more homogenous materials involving the PLA matrix and smaller nanocellulose aggregates. Regarding the effect of acetylation in composites prepared by masterbatching, in general, hydrophobized nanocelluloses led to composites with higher light transmittance curves than those involving native nanocelluloses, as previously reported for similar systems [34,35,40], and in accordance with observations from Figure 5. These results further suggest an improvement in the compatibility of nanocelluloses and PLA upon chemical modification superimposed to the reduced hornification of nanocelluloses derived from masterbatching.

Table 2 collects the color parameters of PLA and all composite films obtained. Compared to PLA, composite films show slightly reduced L*; similar a*; and higher b* values, resulting in ΔE values between 0.3 and 27.4. More pronounced changes in color parameters were observed as the filler content was increased, especially in more homogeneous composites prepared by masterbatching. Similar findings were reported by other authors who developed composites based on PLA reinforced with bacterial nanocellulose [51]. One of the most common applications of PLA-based materials is food packaging, where visual properties play a key role in influencing consumer preference. For this use, lighter and more transparent materials with less saturated colors are favored [52]. Accordingly, regarding color, the composites with lower filler contents developed herein (e.g., those with nanocellulose contents ≤ 1.5 wt.%) would be more suitable for those applications.

Table 2. Color parameters of neat PLA, PLA/BNC-D, PLA/BNC-MB, PLA/AcBNC-MB, PLA/VNC-D, PLA/VNC-MB, and PLA/AcVNC-MB composite films obtained using different filler incorporation strategies and nanocellulose contents. ΔE values were calculated with respect to neat PLA.

Composite	L*	a*	b*	ΔE	Composite	L*	a*	b*	ΔE
PLA	80.8 ± 0.9	−0.2 ± 0.1	−1.5 ± 0.6	-	PLA	80.8 ± 0.9	−0.2 ± 0.1	−1.5 ± 0.6	-
PLA/BNC-D 0.5%	78.6 ± 1.5	−0.6 ± 0.5	−0.5 ± 0.4	2.5	PLA/VNC-D 0.5%	80.6 ± 0.5	−0.2 ± 0.0	−1.9 ± 0.3	0.0
PLA/BNC-D 1%	79.3 ± 0.6	−0.5 ± 0.0	−0.8 ± 0.3	1.7	PLA/VNC-D 1%	79.7 ± 0.5	−0.2 ± 0.0	−1.2 ± 0.3	1.1
PLA/BNC-D 1.5%	78.7 ± 1.4	−0.5 ± 0.0	1.9 ± 1.0	4.0	PLA/VNC-D 1.5%	78.5 ± 0.9	−0.1 ± 0.1	−0.7 ± 0.4	2.4
PLA/BNC-D 3%	77.9 ± 1.1	−0.6 ± 0.0	1.9 ± 1.1	4.5	PLA/VNC-D 3%	76.4 ± 0.8	0.0 ± 0.0	0.8 ± 0.5	5.0
PLA/BNC-D 5%	79.9 ± 0.5	−0.5 ± 0.0	3.4 ± 0.4	5.0	PLA/VNC-D 5%	75.5 ± 1.2	0.2 ± 0.1	2.5 ± 0.8	6.7
PLA/BNC-D 7%	79.5 ± 0.5	−0.2 ± 0.2	6.2 ± 0.9	7.8	PLA/VNC-D 7%	73.9 ± 1.6	0.5 ± 0.2	4.4 ± 0.1	9.1
PLA/BNC-MB 0.5%	79.3 ± 1.1	−0.2 ± 0.0	−0.2 ± 0.5	2.0	PLA/VNC-MB 0.5%	80.7 ± 0.8	−0.2 ± 0.1	−1.8 ± 0.6	0.3
PLA/BNC-MB 1%	78.1 ± 0.5	−0.5 ± 0.0	1.4 ± 0.4	4.0	PLA/VNC-MB 1%	79.2 ± 0.8	−0.3 ± 0.0	0.4 ± 0.2	2.5
PLA/BNC-MB 1.5%	75.4 ± 1.5	−0.5 ± 0.0	3.4 ± 0.4	7.3	PLA/VNC-MB 1.5%	78.9 ± 1.2	−0.3 ± 0.0	0.4 ± 0.8	2.7
PLA/BNC-MB 3%	77.2 ± 0.9	−0.2 ± 0.2	7.3 ± 0.5	9.5	PLA/VNC-MB 3%	75.6 ± 1.3	0.1 ± 0.1	4.4 ± 0.8	7.9
PLA/BNC-MB 5%	78.1 ± 1.1	−0.2 ± 0.1	8.2 ± 0.9	10.1	PLA/VNC-MB 5%	69.3 ± 0.7	1.2 ± 0.0	10.2 ± 0.3	16.5
PLA/BNC-MB 7%	72.5 ± 1.8	2.0 ± 0.4	14.5 ± 0.8	18.2	PLA/VNC-MB 7%	68.8 ± 1.0	1.7 ± 0.3	11.6 ± 1.0	17.9
PLA/AcBNC-MB 0.5%	79.3 ± 0.5	−0.3 ± 0.4	−0.5 ± 0.5	1.8	PLA/AcVNC-MB 0.5%	81.7 ± 0.5	−0.3 ± 0.0	−1.4 ± 0.3	0.9
PLA/AcBNC-MB 1%	77.4 ± 1.2	0.2 ± 0.1	2.4 ± 1.0	5.2	PLA/AcVNC-MB 1%	81.1 ± 0.5	−0.4 ± 0.0	−1.4 ± 0.3	0.4
PLA/AcBNC-MB 1.5%	78.2 ± 0.9	−0.2 ± 0.0	0.6 ± 0.6	3.4	PLA/AcVNC-MB 1.5%	80.7 ± 0.7	−0.3 ± 0.0	−1.2 ± 0.3	0.3
PLA/AcBNC-MB 3%	73.6 ± 0.5	1.0 ± 0.1	7.7 ± 0.4	11.8	PLA/AcVNC-MB 3%	78.9 ± 0.5	−0.4 ± 0.0	0.5 ± 0.4	2.8
PLA/AcBNC-MB 5%	68.6 ± 1.6	1.8 ± 0.3	12 ± 0.8	18.3	PLA/AcVNC-MB 5%	77.2 ± 0.5	−0.3 ± 0.1	2.2 ± 0.4	5.2
PLA/AcBNC-MB 7%	66.0 ± 0.7	3.0 ± 0.1	15.3 ± 0.3	27.4	PLA/AcVNC-MB 7%	72.8 ± 0.3	0.0 ± 0.1	6.1 ± 0.3	11.0

Tensile parameters values of neat PLA and the composites developed as a function of filler content and for the three strategies assayed are presented in Figure 8. Regardless of the type of nanocellulose used, most composites presented higher stiffness values than neat PLA, and an increasing trend of these values with nanocellulose content was observed as a result of the introduction of stiff fillers into a more compliant matrix [14]. In addition, an effect of the incorporation strategy was also evident in Young's modulus values, with higher values for the composites obtained through masterbatching, and no significant effect derived from nanofiller acetylation. The previous preparation of masterbatches by solvent casting enhanced the dispersion of the nanofillers by hindering their hornification during drying from a diluted system in the presence of PLA chains. This effect was particularly pronounced in the case of the composites containing BNC and AcBNC. As Young´s modulus is a mechanical parameter obtained at low deformation levels, it is mostly determined by the dispersion of the filler within the matrix and should not be affected by filler–matrix interaction [53]. Enhanced dispersion of nanocelluloses within PLA derived from masterbatching justifies the stiffness increases observed.

On the other hand, the direct incorporation of dried milled nanocelluloses into PLA led to a reduction in the material's strength. Tensile strength depends on the interaction between the components, which determines stress transfer from the matrix to the filler, and also on the filler dispersion within the matrix. The results obtained herein indicate an insufficient matrix ability for stress transfer to the hydrophilic fillers poorly dispersed in the hydrophobic polymer matrix [37,41]. A decreasing trend of strength with filler content was also observed—especially for bacterial nanocellulose composites—attributed to an increasing content of poorly dispersed filler. Alternatively, for VNC composites prepared by masterbatching, the neat PLA film strength value was restored at the highest filler loadings used (i.e., 5 wt.% and 7 wt.%), whereas the same strategy led to BNC composites with tensile strength values similar to that of PLA for all filler contents assayed. Compared to the composites produced by direct mixing of milled dried nanocelluloses with PLA, masterbatching led to stronger composites as a result of better nanofiller dispersion due to reduced hornification effects. Finally, with the aim to increase the interaction between both phases and further improve fillers dispersion, nanocelluloses were surface-acetylated prior to masterbatching. Although the dispersion of the fillers had been found to improve as a result of chemical modification (Figures 5 and 7), no significant changes in tensile strength were observed, suggesting that the interfacial adhesion between PLA and acetylated nanocelluloses was not significantly improved at the degree of substitution conferred to the nanofillers (i.e., DS: 0.15–0.20).

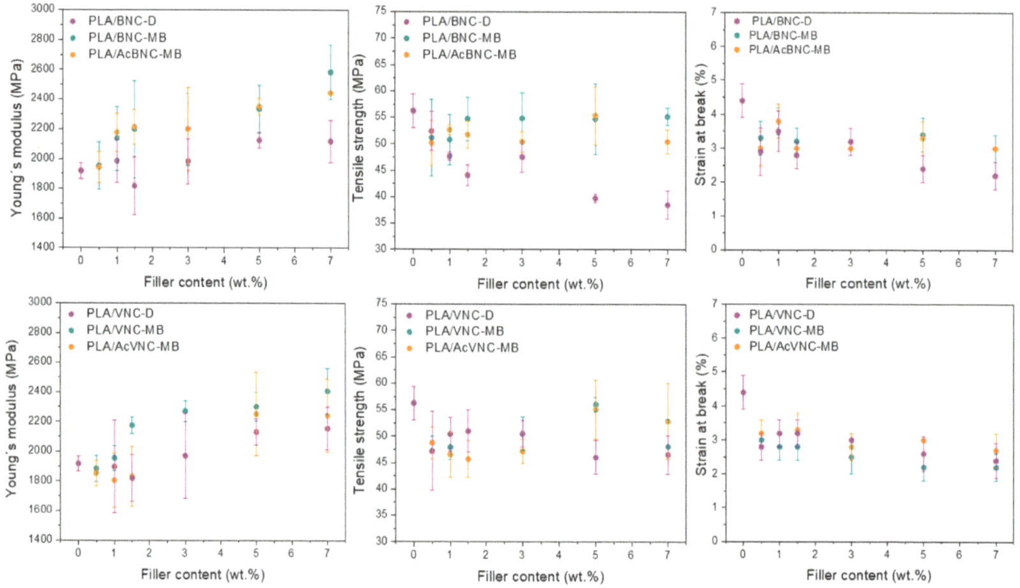

Figure 8. Mechanical properties of neat PLA, PLA/BNC-D, PLA/BNC-MB, PLA/AcBNC-MB, PLA/VNC-D, PLA/VNC-MB, and PLA/AcVNC-MB composite films obtained using different filler incorporation strategies and nanocellulose contents.

Finally, the ductility of all composites was significantly reduced compared to neat PLA, and a slight decreasing trend was found as nanocellulose contents were increased. These findings are consistent with results reported by others for similar systems [34,48,54–57]. The reduction in ductility exhibited by the composites in comparison to neat PLA is commonly observed in thermoplastic composites [58] and can be attributed to the restriction imposed by a stiffer filler to matrix plastic deformation, with the decrease in ductility more noticeable at higher loadings.

4. Conclusions

In the current contribution, bacterial nanocellulose obtained from a by-product of Kombucha tea production and vegetal nanocellulose isolated from milled rice husks, both in native and acetylated forms, were employed as fillers in PLA-based composite films prepared by intensive mixing followed by compression molding. Given the challenges associated with the proper dispersion of nanocelluloses—initially obtained as aqueous suspensions—in PLA by melt compounding processing, three different nanofibrils incorporation strategies at six different nanofiller loadings were investigated.

Morphological, optical, and mechanical characterization of the developed films revealed that the previous preparation of masterbatches by solvent casting successfully improved the dispersion of the nanocellulosic fillers within the polyester matrix compared to the direct mixing of dried milled nanocelluloses and PLA. These results were attributed to the limited hornification of the fillers achieved upon drying in the presence of PLA. The development of masterbatches by solvent casting from homogenized well-dispersed never-dried nanofibers suspensions involved drying nanocellulose suspensions in the presence of PLA chains which, according to the results obtained, must have interfered in the formation of intermolecular hydrogen bonding among cellulose fibrils, limiting hornification and promoting later easier dispersion of the nanofibrils in the final composites.

Regarding the effects of acetylation, chemical modification of nanocelluloses used in the masterbatches further improved the dispersion of both nanocelluloses within the PLA matrix, according to a higher compatibility between the hydrophobic matrix and the surface-hydrophobized nanocelluloses. Nonetheless, the enhanced dispersion of acetylated fillers did not lead to a clear improvement in the mechanical properties of the films. In particular, in reference to composite strength, the degree of substitution conferred to nanocelluloses may have not been enough to promote the interfacial adhesion required for effective stress transfer.

The influence of filler loading was also evident from the different characterization assays performed. Particularly important were its effects on increasing the materials' stiffness while reducing their translucency.

Overall, the preparation of masterbatches by solvent casting as a previous step to melt processing appears to be a promising strategy for developing PLA/nanocelluloses composites, combining the high dispersibility advantages of lab-scale solvent casting with an industrially viable processing technique. It is worth noting that the strategies used were assayed in composites containing nanofillers isolated from wastes of other processes (Kombucha and rice production), but they may be extended to other nanocellulose sources.

Author Contributions: Conceptualization, C.R.B. and M.L.F.; methodology, J.B., J.F.D., O.d.l.O. and M.A.P.; formal analysis, J.B., C.R.B. and M.L.F.; resources, M.A.P., C.R.B. and M.L.F.; writing—original draft preparation, J.B.; writing—review and editing, J.B., J.F.D., M.A.P., C.R.B. and M.L.F.; supervision, C.R.B. and M.L.F.; project administration, C.R.B. and M.L.F.; funding acquisition, C.R.B. and M.L.F. All authors have read and agreed to the published version of the manuscript.

Funding: This research was funded by Universidad de Buenos Aires (UBACyT 2018 No. 20020170100-696BA) and Agencia Nacional de Promoción Científica y Tecnológica of Argentina (PICT-2018-04217).

Institutional Review Board Statement: Not applicable.

Data Availability Statement: Data are contained within the article.

Acknowledgments: Authors acknowledge Silvana Arroyo for collecting X-ray diffractograms.

Conflicts of Interest: The authors declare no conflicts of interest.

References

1. Zhou, L.; Ke, K.; Yang, M.B.; Yang, W. Recent Progress on Chemical Modification of Cellulose for High Mechanical-Performance Poly(Lactic Acid)/Cellulose Composite: A Review. *Compos. Commun.* **2021**, *23*, 100548. [CrossRef]
2. Wu, Y.; Gao, X.; Wu, J.; Zhou, T.; Nguyen, T.T.; Wang, Y. Biodegradable Polylactic Acid and Its Composites: Characteristics, Processing, and Sustainable Applications in Sports. *Polymers* **2023**, *15*, 3096. [CrossRef] [PubMed]
3. Rajeshkumar, G.; Arvindh Seshadri, S.; Devnani, G.L.; Sanjay, M.R.; Siengchin, S.; Prakash Maran, J.; Al-Dhabi, N.A.; Karuppiah, P.; Mariadhas, V.A.; Sivarajasekar, N.; et al. Environment Friendly, Renewable and Sustainable Poly Lactic Acid (PLA) Based Natural Fiber Reinforced Composites—A Comprehensive Review. *J. Clean. Prod.* **2021**, *310*, 127483. [CrossRef]
4. Li, X.; Yu, J.; Meng, L.; Li, C.; Liu, M.; Meng, L. Study of PLA-Based Wood-Plastic Composites. *J. Phys. Conf. Ser.* **2022**, *2437*, 012029. [CrossRef]
5. Ashothaman, A.; Sudha, J.; Senthilkumar, N. A Comprehensive Review on Biodegradable Polylactic Acid Polymer Matrix Composite Material Reinforced with Synthetic and Natural Fibers. *Mater. Today Proc.* **2023**, *80*, 2829–2839. [CrossRef]
6. Vatansever, E.; Arslan, D.; Nofar, M. Polylactide Cellulose-Based Nanocomposites. *Int. J. Biol. Macromol.* **2019**, *137*, 912–938. [CrossRef] [PubMed]
7. Shen, R.; Xue, S.; Xu, Y.; Liu, Q.; Feng, Z.; Ren, H.; Zhai, H. Research Progress and Development Demand of Nanocellulose Reinforced Polymer Composites. *Polymers* **2020**, *12*, 2113. [CrossRef] [PubMed]
8. Bikiaris, N.D.; Koumentakou, I.; Samiotaki, C.; Meimaroglou, D.; Varytimidou, D.; Karatza, A.; Kalantzis, Z.; Roussou, M.; Bikiaris, R.D.; Papageorgiou, G.Z. Recent Advances in the Investigation of Poly(Lactic Acid) (PLA) Nanocomposites: Incorporation of Various Nanofillers and Their Properties and Applications. *Polymers* **2023**, *15*, 1196. [CrossRef]

9. Parvathy, S.U.; Hema, S.; Sajith, M.; Sulthan, R.; Sreelekshmi, C.; Sambhudevan, S.; Shankar, B. A Review on Barrier Properties of Nanocellulose and Polylactic Acid Composites. *IOP Conf. Ser. Mater. Sci. Eng.* **2022**, *1258*, 012017. [CrossRef]
10. Nicharat, A.; Sapkota, J.; Foster, E.J. Pre-mixing and masterbatch approaches for reinforcing poly(vinyl acetate) with cellulose based fillers. *Ind. Crop. Prod.* **2016**, *93*, 244–250. [CrossRef]
11. Fernandes Diniz, J.M.B.; Gil, M.H.; Castro, J.A.A.M. Hornification—Its Origin and Interpretation in Wood Pulps. *Wood Sci. Technol.* **2004**, *37*, 489. [CrossRef]
12. Silva, L.E.; dos Santos, A.d.A.; Torres, L.; McCaffrey, Z.; Klamczynski, A.; Glenn, G.; de Sena Neto, A.R.; Wood, D.; Williams, T.; Orts, W.; et al. Redispersion and structural change evaluation of dried microfibrillated cellulose. *Carbohydr. Polym.* **2021**, *252*, 117165. [CrossRef] [PubMed]
13. Wang, Q.; Yao, Q.; Liu, J.; Sun, J.; Zhu, Q.; Chen, H. Processing nanocellulose to bulk materials: A review. *Cellulose* **2019**, *26*, 13–14. [CrossRef]
14. Mokhena, T.C.; Sefadi, J.S.; Sadiku, E.R.; John, M.J.; Mochane, M.J.; Mtibe, A. Thermoplastic Processing of PLA/Cellulose Nanomaterials Composites. *Polymers* **2018**, *10*, 1363. [CrossRef] [PubMed]
15. Aigaje, E.; Riofrio, A.; Baykara, H. Processing, Properties, Modifications, and Environmental Impact of Nanocellulose/Biopolymer Composites: A Review. *Polymers* **2023**, *15*, 1219. [CrossRef] [PubMed]
16. Jonoobi, M.; Harun, J.; Mathew, A.P.; Oksman, K. Mechanical Properties of Cellulose Nanofiber (CNF) Reinforced Polylactic Acid (PLA) Prepared by Twin Screw Extrusion. *Compos. Sci. Technol.* **2010**, *70*, 1742–1747. [CrossRef]
17. Oksman, K.; Aitomäki, Y.; Mathew, A.P.; Siqueira, G.; Zhou, Q.; Butylina, S.; Tanpichai, S.; Zhou, X.; Hooshmand, S. Review of the Recent Developments in Cellulose Nanocomposite Processing. *Compos. Part A Appl. Sci. Manuf.* **2016**, *83*, 2–18. [CrossRef]
18. Olonisakin, K.; Fan, M.; Xin-Xiang, Z.; Ran, L.; Lin, W.S.; Zhang, W.; Wenbin, Y. Key Improvements in Interfacial Adhesion and Dispersion of Fibers/Fillers in Polymer Matrix Composites; Focus on PLA Matrix Composites. *Compos. Interfaces* **2022**, *29*, 1071–1120. [CrossRef]
19. Bovi, J.; Butto, M.; Arroyo, S.; Bernal, C.; Foresti, M.L. Bacterial Cellulose Nanofibrils from a By-Product of the Growing Kombucha Business: Obtention and Use in the Development of Poly (Lactic Acid) Composites. *Lat. Am. Appl. Res.* **2024**, *54*, 55–62. [CrossRef]
20. Delgado, J.F.; de la Osa, O.; Salvay, A.G.; Cavallo, E.; Cerrutti, P.; Foresti, M.L.; Peltzer, M.A. Reinforcement of Yeast Biomass Films with Bacterial Cellulose and Rice Husk Cellulose Nanofibres. *J. Polym. Environ.* **2021**, *29*, 3242–3251. [CrossRef]
21. Ávila Ramírez, J.A.; Juan Suriano, C.; Cerrutti, P.; Foresti, M.L. Surface Esterification of Cellulose Nanofibers by a Simple Organocatalytic Methodology. *Carbohydr. Polym.* **2014**, *114*, 416–423. [CrossRef] [PubMed]
22. Ilharco, L.M.; Garcia, A.R.; Lopes Da Silva, J.; Vieira Ferreira, L.F. Infrared Approach to the Study of Adsorption on Cellulose: Influence of Cellulose Crystallinity on the Adsorption of Benzophenone. *Langmuir* **1997**, *13*, 4126–4132. [CrossRef]
23. Segal, L.; Creely, J.; Martin, A.; Conrad, C. An empirical method for estimating the degree of crystallinity of native cellulose using the X-ray diffractometer. *Text. Res. J.* **1959**, *29*, 786–794. [CrossRef]
24. *ASTM D638-14*; Standard Test Method for Tensile Properties of Plastics. West Conshohocken, PA, USA: ASTM International, 2014.
25. Marin, D.C.; Vecchio, A.; Ludueña, L.N.; Fasce, D.; Alvarez, V.A.; Stefani, P.M. Revalorization of Rice Husk Waste as a Source of Cellulose and Silica. *Fibers Polym.* **2015**, *16*, 285–293. [CrossRef]
26. Ludueña, L.; Fasce, D.; Alvarez, V.A.; Stefani, P.M. Nanocellulose from Rice Husk Following Alkaline Treatment to Remove Silica. *BioResources* **2011**, *6*, 1440–1453. [CrossRef]
27. Paramasivan, M.; Kumar, T.S.S.; Chandra, T.S. Microbial synthesis of hydroxyapatite-nanocellulose nanocomposites from symbiotic culture of bacteria and yeast pellicle of fermented kombucha tea. *Sustainability* **2022**, *14*, 8144. [CrossRef]
28. Dima, S.O.; Panaitescu, D.M.; Orban, C.; Ghiurea, M.; Doncea, S.M.; Fierascu, R.C.; Nistor, C.L.; Alexandrescu, E.; Nicolae, C.A.; Trica, B.; et al. Bacterial Nanocellulose from Side-Streams of Kombucha Beverages Production: Preparation and Physical-Chemical Properties. *Polymers* **2017**, *9*, 374. [CrossRef]
29. Johar, N.; Ahmad, I.; Dufresne, A. Extraction, Preparation and Characterization of Cellulose Fibres and Nanocrystals from Rice Husk. *Ind. Crop. Prod.* **2012**, *37*, 93–99. [CrossRef]
30. Zuluaga, R.; Putaux, J.L.; Cruz, J.; Vélez, J.; Mondragon, I.; Gañán, P. Cellulose Microfibrils from Banana Rachis: Effect of Alkaline Treatments on Structural and Morphological Features. *Carbohydr. Polym.* **2009**, *76*, 51–59. [CrossRef]
31. Delgado, J.F. Desarrollo de Películas Multicomponentes a Base de Levadura. Evaluación de Sus Propiedades Físicas y Aplicaciones Potenciales. Ph.D. Thesis, Universidad Nacional de Quilmes, Buenos Aires, Argentina, 2021.
32. Ramírez Tapias, Y.A.; Peltzer, M.A.; Delgado, J.F.; Salvay, A.G. Kombucha Tea By-Product as Source of Novel Materials: Formulation and Characterization of Films. *Food Bioprocess. Technol.* **2020**, *13*, 1166. [CrossRef]
33. Castro-Guerrero, C.F.; Díaz-Guillén, M.R.; Delgado-Arroyo, F.; Rodas-Grapain, A.; Godavarthi, S. Purification of Cellulose from Rice Husk for the Synthesis of Nanocellulose. In Proceedings of the 16th International Conference of Nanotechnology, IEEE NANO 2016, Sendai, Japan, 22–25 August 2016.
34. Jonoobi, M.; Mathew, A.P.; Abdi, M.M.; Makinejad, M.D.; Oksman, K. A Comparison of Modified and Unmodified Cellulose Nanofiber Reinforced Polylactic Acid (PLA) Prepared by Twin Screw Extrusion. *J. Polym. Environ.* **2012**, *20*, 991–997. [CrossRef]
35. Ávila Ramírez, J.A.; Bovi, J.; Bernal, C.; Errea, M.I.; Foresti, M.L. Development of Poly(Lactic Acid) Nanocomposites Reinforced with Hydrophobized Bacterial Cellulose. *J. Polym. Environ.* **2020**, *28*, 61–73. [CrossRef]
36. Park, M.E.; Baker, S.; Himmel, J.O.; Parilla, P.A.; Johnson, D.K. Cellulose crystallinity index: Measurement techniques and their impact on interpreting cellulase performance. *Biotechnol. Biofuels* **2010**, *3*, 10. [CrossRef] [PubMed]

37. Arteaga-Ballesteros, B.E.; Guevara-Morales, A.; Martín-Martínez, E.S.; Figueroa-López, U.; Vieyra, H. Composite of Polylactic Acid and Microcellulose from Kombucha Membranes. *E-Polymers* **2021**, *20*, 15–26. [CrossRef]
38. Goh, W.N.; Rosma, A.; Kaur, B.; Fazilah, A.; Karim, A.A.; Bhat, R. Microstructure and Physical Properties of Microbial Cellulose Produced during Fermentation of Black Tea Broth (Kombucha). II. *Int. Food Res. J.* **2012**, *19*, 153–158.
39. Oliver-Ortega, H.; Geng, S.; Espinach, F.X.; Oksman, K.; Vilaseca, F. Bacterial Cellulose Network from Kombucha Fermentation Impregnated with Emulsion-Polymerized Poly(Methyl Methacrylate) to Form Nanocomposite. *Polymers* **2021**, *13*, 664. [CrossRef]
40. Tomé, L.C.; Pinto, R.J.B.; Trovatti, E.; Freire, C.S.R.; Silvestre, A.J.D.; Neto, C.P.; Gandini, A. Transparent Bionanocomposites with Improved Properties Prepared from Acetylated Bacterial Cellulose and Poly(Lactic Acid) through a Simple Approach. *Green Chem.* **2011**, *13*, 419–427. [CrossRef]
41. Cunha, A.G.; Zhou, Q.; Larsson, P.T.; Berglund, L.A. Topochemical Acetylation of Cellulose Nanopaper Structures for Biocomposites: Mechanisms for Reduced Water Vapour Sorption. *Cellulose* **2014**, *21*, 2773–2787. [CrossRef]
42. Klemm, D.; Kramer, F.; Moritz, S.; Lindström, T.; Ankerfors, M.; Gray, D.; Dorris, A. Nanocelluloses: A New Family of Nature-Based Materials. *Angew. Chem.-Int. Ed.* **2011**, *50*, 5438–5466. [CrossRef]
43. Tingaut, P.; Zimmermann, T.; Lopez-Suevos, F. Synthesis and characterization of bionanocomposites with tunable properties from poly(lactic acid) and acetylated microfibrillated cellulose. *Biomacromolecules* **2010**, *11*, 454–464. [CrossRef]
44. Yang, W.; Weng, Y.; Puglia, D.; Qi, G.; Dong, W.; Kenny, J.M.; Ma, P. Poly(lactic acid)/lignin films with enhanced toughness and anti-oxidation performance for active food packaging. *Int. J. Biol. Macromol.* **2020**, *144*, 102–110. [CrossRef] [PubMed]
45. Hervy, M.; Blaker, J.J.; Braz, A.L.; Lee, K.Y. Mechanical response of multi-layer bacterial cellulose nanopaper reinforced polylactide laminated composites. *Compos. Part A* **2018**, *107*, 155–163. [CrossRef]
46. Quero, F.; Nogi, M.; Yano, H.; Abdulsalami, K.; Holmes, S.M.; Sakakini, B.H.; Eichhorn, S.J. Optimization of the Mechanical Performance of Bacterial Cellulose/Poly(L-Lactic) Acid Composites. *ACS Appl. Mater. Interfaces* **2010**, *2*, 321–330. [CrossRef] [PubMed]
47. Zhang, X.; Li, W.; Ye, B.; Lin, Z.; Rong, J. Studies on Confined Crystallization Behavior of Nanobiocomposites Consisting of Acetylated Bacterial Cellulose and Poly (Lactic Acid). *J. Thermoplast. Compos. Mater.* **2013**, *26*, 346–361. [CrossRef]
48. Fortunati, E.; Armentano, I.; Iannoni, A.; Barbale, M.; Zaccheo, S.; Scavone, M.; Visai, L.; Kenny, J.M. New multifunctional poly (lactide acid) composites: Mechanical, antibacterial, and degradation properties. *J. Appl. Polym. Sci.* **2012**, *124*, 87–98. [CrossRef]
49. Mathew, A.P.; Oksman, K.; Sain, M. Mechanical properties of biodegradable composites from poly lactic acid (PLA) and microcrystalline cellulose (MCC). *J. Appl. Polym. Sci.* **2005**, *97*, 2014–2025. [CrossRef]
50. Bagheriasl, D.; Safdari, F.; Carreau, P.J.; Dubois, C.; Riedl, B. Development of Cellulose Nanocrystal-Reinforced Polylactide: A Comparative Study on Different Preparation Methods. *Polym. Compos.* **2017**, *40*, 342–350. [CrossRef]
51. Patwa, R.; Saha, N.; Sáha, P.; Katiyar, V. Biocomposites of Poly(Lactic Acid) and Lactic Acid Oligomer-Grafted Bacterial Cellulose: It's Preparation and Characterization. *J. Appl. Polym. Sci.* **2019**, *136*, 47903. [CrossRef]
52. Oliveira, A.C.S.D.; Ribeiro, M.N.; Ugucioni, J.C.; Rocha, R.A.D.; Borges, S.V. Consumer perception of biodegradable packaging for food. *Polímeros* **2024**, *33*, e20230045. [CrossRef]
53. Móczó, J.; Pukánszky, B. Polymer micro and nanocomposites: Structure, interactions, properties. *J. Ind. Eng. Chem.* **2008**, *14*, 535–563. [CrossRef]
54. Kale, R.D.; Gorade, V.G.; Madye, N.; Chaudhary, B.; Bangde, P.S.; Dandekar, P.P. Preparation and Characterization of Biocomposite Packaging Film from Poly(Lactic Acid) and Acylated Microcrystalline Cellulose Using Rice Bran Oil. *Int. J. Biol. Macromol.* **2018**, *118*, 1090–1102. [CrossRef] [PubMed]
55. Ren, H.; Li, S.; Gao, M.; Xing, X.; Tian, Y.; Ling, Z.; Yang, W.; Pan, L.; Fan, W.; Zheng, Y. Preparation and Characterization of Microcrystalline Cellulose/Polylactic Acid Biocomposite Films and Its Application in Lanzhou Lily (*Lilium davidii* var. unicolor) Bulbs Preservation. *Sustainability* **2023**, *15*, 13770. [CrossRef]
56. Johari, A.P.; Kurmvanshi, S.K.; Mohanty, S.; Nayak, S.K. Influence of Surface Modified Cellulose Microfibrils on the Improved Mechanical Properties of Poly (Lactic Acid). *Int. J. Biol. Macromol.* **2016**, *84*, 329–339. [CrossRef]
57. Sucinda, E.F.; Abdul Majid, M.S.; Ridzuan, M.J.M.; Cheng, E.M.; Alshahrani, H.A.; Mamat, N. Development and characterisation of packaging film from Napier cellulose nanowhisker reinforced polylactic acid (PLA) bionanocomposites. *Int. J. Biol. Macromol.* **2021**, *187*, 43–53. [CrossRef]
58. Fortunati, E.; Armentano, I.; Zhou, Q.; Iannoni, A.; Saino, E.; Visai, L.; Berglund, L.A.; Kenny, J.M. Multifunctional bionanocomposite films of poly(lactic acid), cellulose nanocrystals and silver nanoparticles. *Carbohydr. Polym.* **2012**, *87*, 1596–1605. [CrossRef]

polymers

MDPI

Article

Effect of Different Porous Size of Porous Inorganic Fillers on the Encapsulation of Rosemary Essential Oil for PLA-Based Active Packaging

Raúl Cerdá-Gandia [1,2], Ángel Agüero [1,3], Marina Patricia Arrieta [3,4,*] and Octavio Fenollar [1,*]

1 Instituto Universitario de Investigación de Tecnología de Materiales (IUITM), Universitat Politècnica de València (UPV), Plaza Ferrándiz y Carbonell 1, 03801 Alcoy, Spain; raucergl@epsa.upv.es (R.C.-G.); anagrod@upv.es (Á.A.)
2 FAPERIN S.L. Av. de los Trabajadores, 27, 03430 Onil, Spain
3 Departamento de Ingeniería Química Industrial y del Medio Ambiente, Escuela Técnica Superior de Ingenieros Industriales, Universidad Politécnica de Madrid (ETSII-UPM), C/José Gutiérrez Abascal 2, 28006 Madrid, Spain
4 Grupo de Investigación: Polímeros, Caracterización y Aplicaciones (POLCA), 28006 Madrid, Spain
* Correspondence: m.arrieta@upm.es (M.P.A.); ocfegi@upv.es (O.F.)

Abstract: Essential oils are interesting active additives for packaging manufacturing as they can provide the final material with active functionalities. However, they are frequently volatile compounds and can be degraded during plastic processing. In this work *Rosmarinus officinalis* (RO) essential oil was encapsulated into Diatomaceous earth (DE) microparticles and into Halloysite nanotubes (HNTs) and further used to produce eco-friendly active packaging based on polylactic acid (PLA). PLA-based composites and nanocoposites films based on PLA reinforced with DE + RO and HNTs + RO, respectively, were developed by melt extrusion followed by cast-film, simulating the industrial processing conditions. As these materials are intended as active food packaging films, the obtained materials were fully characterized in terms of their mechanical, thermal and structural properties, while migration of antioxidant RO was also assessed as well as the compostability at laboratory scale level. Both DE and HNTs were able to protect the *Rosmarinus officinalis* (RO) from thermal degradation during processing, allowing to obtain films with antioxidant properties as demonstrated by the antioxidant assays after the materials were exposed for 10 days to a fatty food simulant. The results showed that incorporating *Rosmarinus officinalis* encapsulated in either DE or HNTs and the good dispersion of such particles into the PLA matrix strengthened its mechanical performance and sped up the disintegration under composting conditions of PLA, while allowing to obtain films with antioxidant properties of interest as antioxidant active food packaging materials.

Keywords: polylactic acid; *Rosmarinus officinalis*; Diatomaceous earth (DE); Halloysite nanotubes (HNTs); active packaging; compostable

Citation: Cerdá-Gandia, R.; Agüero, Á.; Arrieta, M.P.; Fenollar, O. Effect of Different Porous Size of Porous Inorganic Fillers on the Encapsulation of Rosemary Essential Oil for PLA-Based Active Packaging. *Polymers* **2024**, *16*, 2632. https://doi.org/10.3390/polym16182632

Academic Editor: Miguel Ángel López Manchado

Received: 8 August 2024
Revised: 13 September 2024
Accepted: 16 September 2024
Published: 18 September 2024

1. Introduction

In the last decades, the continuing growth of an environmental concerns shown by society and its demands has propitiated that terms like circular economy, waste management, and sustainable consumption are the main research routes in the engineering and science fields. As a result of this, in the plastic sector, many works have focused on the study of bio-based and biodegradable polymers as alternatives to traditional plastics, particularly in short-term applications, to reduce the consumption of fossil-based resources and the plastics waste accumulation in the environment, avoiding micro-plastic generation. To produce fully sustainable biopolymeric formulations, natural additives are also required, such as vegetable oils as sustainable plasticizers or natural nanoclays as sustainable reinforcements, to produce high environmental performance composites or nanocomposites able to replace traditional plastics [1–3]. These naturally occurring materials also use to show some common characteristics, such as biodegradability, nontoxicity, biocompatibility and, in terms of processability, have been proven as a viable solution to substitute those obtained from nonrenewable sources in many industries [4]. As the food industry is one of the most involved sectors in the production of short-term plastics products, needing thousands of tons of plastics (traditionally obtained from fossil sources) to produce food packaging [5,6], the food packaging industry is one of the most involved sector focus on substituting the non-renewable and non-degradable polymers for sustainable biopolymeric based formulations [7].

Among all biopolymers, poly (lactic acid) (PLA) is the biobased and biodegradable polymer that has been positioned in the packaging industry as the most promising biopolymer to replace fossil-based polymers currently used in this field due to its ease of processing, availability in the market and economic competitiveness [6,8]. However, PLA possesses some disadvantages for food packaging, and this is why PLA-based composite and/or nanocomposites have gained industrial attention for food packaging purposes, since they are able to improve the mechanical and thermal properties of neat PLA. This industry has also been affected in the last years by a steadily increasing consumers demand for the use of more natural additives in processed food products [9].

In this sense, not only does the introduction of this eco-friendly material play a key role in the current global paradigm of the transition to a circular economy model, but also the possibility of providing these materials with active functionalities that allow the reduction of additives into the foodstuff. In this context, since lipid oxidation is one of the major reasons that compromise the food's sensory performance which could also be harmful to consumers' health, a new tendency in food packaging materials is to incorporate antioxidants into the packaging formulations instead of directly into the foodstuff [10,11]. In fact, active packaging is a technology thought to extend the food's shelf life by promoting positive interaction between the packaging material and the foodstuff [12].

Therefore, besides the desirable characteristics for foodstuff packaging that biopolymers can supply, such as good processability, enough transparency, and adequate mechanical resistance; some natural essential oils (EOs) in eco-efficient composites are interesting due to their known active antimicrobial and/or antioxidant properties [13]. A feasible route to design active packaging materials is by the incorporation of active agents embedded into the polymeric matrix that would be part of the packaging system [14]. Thus, in concordance with the consumers' increasing interest in non-processed and high nutritional food products, this has led to concepts such as active packaging totally linked with the sustainable concept in packed food [15]. In this context, essential oils (EOs) are naturally produced by plants and consist of complex lipids that are rich in biologically active compounds such as polyphenolic acids [16]. Some EOs obtained from thyme, oregano, sage, coriander, garlic, clove or rosemary have been well studied because of their effective antimicrobial and antioxidant properties [12,17,18]. In addition, most EOs are classified as Generally Recognized as Safe (GRAS) by the U.S. Food and Drug Administration [19].

As PLA is a biopolymer already integrated into the packaging industry, PLA, its blends, or biocomposites and bionanocomposites have been also successfully combined with different EOs. In this sense, Sharma et al. [20] concluded that the addition of Thyme and Clove essential oils on PLA/poly (butylen alipate-co-terephtalate) (PBAT) blend films processed by solvent casting has a great influence on the morphological and mechanical performance, while providing the material with antibacterial properties. Similar results were obtained in other work when Cinnamon and Eucalyptus essential oils were incorporated instead [21]. Meanwhile, Wang et al. [22] have blended PLA with Perilla essential oil by electrospinning, which resulted in a potential fresh-keeping packaging extending the shelf-life of chilled chicken by 12 days. Cheran et al. prepared nano film-coated paper based on PLA-nanocellulose incorporated with Thyme oil, Cinnamon oil, clove oil, and Rosemary oil and observed an increase in the hydrophobicity, while the materials showed active properties such as antimicrobial and insecticidal performance [23].

However, for the industrial production of PLA-based materials, they should be processed by melt-blending approaches, but as EOs are highly volatile, a high amount of them could be lost during the PLA-EO melt-processing, as it was already observed in PLA-EO formulations [24].

Among other essential oils, *Rosmarinus officinalis* L., an aromatic plant rich in phenolic compounds from the *Lamiaceae* family that originates from the Mediterranean region, is an interesting spice commercially available for use as an antioxidant. Rosemary extracts have been widely used in food preservation since they have shown their effectiveness in preventing oxidation and microbial contamination [25]. In this regard, there are some components present in rosemary extracts, i.e., carnosol, rosmanol, rosmarinic acid, and carnosic acid, that are effective for the antioxidative stabilization process for biopolymers [25,26].

Therefore, the introduction of EOs in the production process, due to their high volatility, usually involves a loss of some part of it, while the hydrophobic nature of the EOs could be adverse for a good dispersion in the PLA biopolymeric matrix, and consequently, the release of EOs is hardly homogenous and controlled. For this reason, encapsulation technology is a recurrent solution to protect active compounds from thermal degradation during PLA-based formulation processability and promote a dispersed phase in the polymeric matrix [12,27]. Regarding biopolymers reinforcement, nanoclays are the most prospective fillers due to their surface chemistry and morphology [28], which can increase PLA-based nanocomposite performance [12,18].

Halloysite nanotubes (NHTs) are naturally formed aluminosilicate ($Al_2Si_2O_5(OH)_4 \cdot nH_2O$) clay with a characterized nanotubular hollow structure. On the external surface of the HNTs are located Si–O–Si groups, whereas the inner surface consists of an Al–OH array [29,30]. Due to this tubular shape with a negatively charged outside surface and a higher concentration of –OH groups in the inner lumen, HNTs can be easily dispersed into polymeric matrices. Commercially it can be found with a length of 50–500 nm, an external diameter of 20–200 nm, and an internal diameter of 10–70 nm [31]. HNTs have received considerable attention as nanofiller for enhancing the mechanical, thermal, and crystallization properties of several polymers such as PLA [32], chitosan [33], poly (ε-caprolactone) (PCL) [34], or polyamides (PA) [30,35]. As HNTs belong to GRAS kaolin clays for food contact materials owing to their non-toxicity and biocompatibility, they have been also employed as an encapsulation system for active packaging applications [36–38].

Diatomaceous earth (DE) is also a naturally occurring clay that can be found in geological deposits. DE is basically fossilized skeletons of unicellular microorganisms living in either fresh or seawater all over the world. Although there exist around 30,000 diatom species, all of them present a silica shell with different structures and regularly organized slits or pores [39,40]. These microstructures are composed mainly of amorphous silica ($SiO_2 \cdot nH_2O$), have a large effective surface area, above 200 $m^2\ g^{-1}$, and a high absorption capacity due to their nanoporous with a range size from 10 to 1000 nm [41,42]. DE is a well-recognized encapsulating material with applications in filter, gas sensors, or drug delivery systems [30,43]. They have also been used as reinforcement/encapsulating particles in green composites based on PLA [44,45].

The purpose of this study is to reinforce PLA matrix with two natural clays, with different size ranges, as encapsulating systems for food active film packaging production by a melt-extrusion approach. For this, HNTs (nanoparticles) and DE (microparticles) were loaded with an active agent and then incorporated into the PLA matrix to produce active films. *Rosmarinus officinalis* essential oil (RO) was the active agent selected due to its local abundance and for its widely reported active (antioxidant and antibacterial) properties [46,47]. The HNTs nanoparticles were characterized by DLS, while both HNT nanoparticles and DE microparticles were observed by transmission electron microscopy (TEM). PLA is an elsewhere known biopolymer, and the contribution of the loaded HNTs and DE particles was studied in terms of structural, mechanical, and thermal properties by means of scanning electron microscopy, tensile test measurements, differential scanning calorimetry, and thermogravimetric analysis. The antioxidant activity of the RO and the effectiveness of the two encapsulating systems were analyzed by measuring the antioxidant activity of the films exposed to a fatty food simulant through the measurement of DPPH scavenging activity, while the disintegrability under composting conditions was conducted at a laboratory scale level to corroborate the biodegradable character of the developed materials.

2. Materials and Methods

2.1. Materials

The PLA used was Ingeo™ Biopolymer 2003D commercialized in pellets form by NatureWorks LLC (Minnetonka, MN, USA). This PLA grade has a melt flow rate (MFR) at 210 °C of 6 g 10 min^{-1} and a true density of 1.24 g cm^{-3}, which makes it adequate for suitable for melt blending process. *Rosmarinus officinalis* essential oil (RO), CAS 8000-25-7, was purchased from Esencias Martinez Lozano S.A. (Murica, Spain). According to the producer, this EO has a density of 0.892–0.910 g cm^{-3}, comes from plant leaves, and its principal chemical compounds are α-pinene + α-thuyene (18–26%), 1-8-cineole (17–25%), camphre (12.5–22%), and camphene (8–13%). Halloysite nanotubes (HNTs) in powder form, CAS 685445, were provided by Sigma-Aldrich S.A. (Madrid, Spain). According to specifications, HNTs density is 2.53 g cm^{-3}, while nanotubes diameters are between 30 and 70 nm, with a length of 1–3 μm, an effective surface area of 64 m^2 g^{-1} and a pore volume of 1.26–1.34 mL g^{-1}. Meanwhile, Diatomaceous earth (DE), CAS 68855-54-9, with a density of 2.36 g cm^{-3}, is also distributed in powder form by Sigma-Aldrich S.A. (Madrid, Spain). This product is a mixture of diatoms species shells with a wide variety of particle sizes and geometric shapes, but disc and tubular forms are predominant, with an average particle size of around 125 μm. The ethanol with 99.8% purity, and The radical 2,2-diphenyl-1-picrylhydrazyl (DPPH) were also obtained from Sigma-Aldrich S.A.

2.2. Essential Oil Encapsulation

Firstly, to eliminate residual moisture due to the hydrophilic nature of these silica particles, HNTs and DE were dried at 150 °C for 3 h. Then the RO was encapsulated into HNTs or DE particles using a vacuum procedure according to previous reports [28,37] with slight modifications. Briefly, a solution of rosemary essential oil (RO) at 25 wt.% in ethanol was prepared and fully mixed by vortex-stirring for 30 min. Subsequently, each encapsulation material (HNTs or DE) was separately added to the RO solution at a weight ratio 1:0.2 and stirred at room temperature at 500 rpm for 60 min followed by ultrasonic homogenization in a Thermo Fisher Scientific Q125 (Waltham, MA, USA) at 10% of amplitude equipped with a tapered microtip CL-18 for 20 min in an ice bath, while an on/off pulse mode was chosen to prevent the overheating of the system. The resultant suspensions were subjected to vacuum for 1 h at 30 °C, and then cycled back to atmospheric pressure to replace the entrapped air in the particles lumens with the oil solution. Afterward, both mixtures were again homogenized by ultrasonication for 15 min. Finally, the remaining liquid was separated by centrifugation, washed with ethanol to remove the unencapsulated RO from the external HNTs and DE, and finally filtered to obtain a powder. In Figure 1 there is a schematic representation of the encapsulation process. Before processing with PLA matrix, the loaded HNTs and DE particles were dried in an air-circulating oven at 80 °C for 12 h [31].

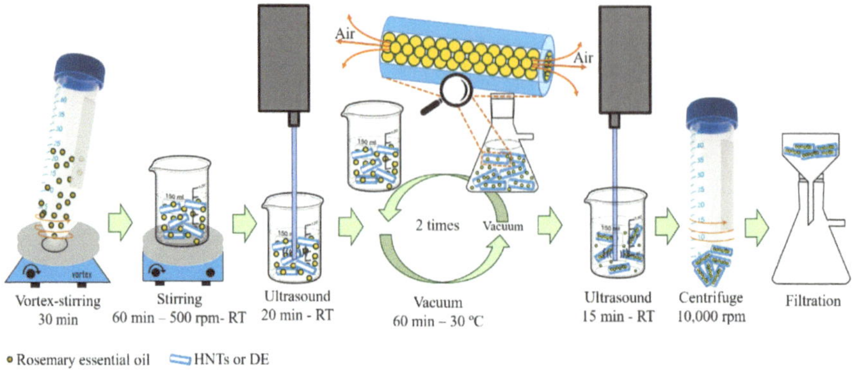

Figure 1. Schematic representation of the HNTs or DEs encapsulation process with RO.

2.3. Nanoparticles Sizes Characterization

The hydrodynamic particle sizes of the HNTs in water (1 mg mL^{-1}) were measured at 20 °C with a dynamic light scattering (DLS) analyzer (BeNano 90, Nane Vita, Madrid, Spain) equipped with a 50 mW solid-state laser beam at 671 nm and with a detection angle of 90°. The reported particle size was the average value of at least five consecutive runs.

Transmission electron microscopy (TEM) images of HNTs and DE containing the RO essential oil were obtained by transmission electron microscopy (TEM, JEOL JEM-1010) operating at 100 kV. One droplet of DE or HNT aqueous suspension was deposited on carbon-coated copper grids and allowed to dry at room temperature for 25 min before observations.

2.4. Films Processing

Once the HNTs and DE particles were loaded with the RO, namely HNTs + RO and DE + RO respectively, they were used to reinforce PLA. Thus, PLA was loaded with 1, 3 and 5 wt.% of HNTs + RO and DE + RO. The as-received PLA pellets were previously dried at 60 °C for 12 h in a dehumidifier MDEO from Industrial Marsé S.L. (Barcelona, Spain). The selected HNTs + RO and DE + RO contents were chosen since some authors have reported that over a 3–5% nanoclays content in biopolymer composites, some filler aggregates can occur, and general material performance is compromised [48,49]. In addition to PLA-HNTs + RO and PLA-DO formulations, pristine PLA film and formulations with only RO incorporated at three concentrations (0.5, 1, and 3 wt.%) were also produced for comparison and labeled as PLA-RO 0.5, PLA-RO 1, and PLA-RO 3 respectively.

The different amounts of materials were mechanically mixed and were then melt-compounded in a twin-screw extruder from Construcciones Mecánicas Dupra S.L. (Alicante, Spain). The rotating speed was set at 20 rpm and the temperature profile from the feeding hopper to the extrusion die was: 185 °C, 190 °C, 195 °C, and 200 °C. After the extrusion, materials were pelletized in an air-knife unit for further processing in a cast-film extruder from Eutotech S.A. (San Martino in Riu, Italy) with a 200–205 °C temperature profile from the feeding to the head. The extrusion speed was 20 rpm and the calendar and drag speed were adjusted to obtain an average thickness of 30–50 μm approximately [50]. The resultant film rolls were attained and stored at room temperature for 15 days before testing.

2.5. Film Characterization

2.5.1. Scanning Electron Microscopy

The surface morphology of the cross-section of cryo-fractured samples using liquid N_2 was evaluated by JEOL JSM-6300 scanning electron microscopy (SEM) (JeolUSA Inc., Peabody, MA, USA). The samples were previously sputtered with a gold-palladium alloy in an EMITECH sputter coating SC7620 model from Quorum Technologies, Ltd. (Lewes, UK).

2.5.2. Mechanical Test

Tensile properties of the films produced were determined in a universal test machine ELIB 30 from S.A.E. Ibertest (Madrid, Spain). The tests were carried out at room conditions following the guideline of ISO 527-2:2012 [51] for rectangular samples sizing 10 × 150 mm^2, the selected load cell was 5 kN and a crosshead speed of 2 mm min^{-1}. At least five samples of each formulation were tested.

2.5.3. Thermal Properties

2.5.3.1. Differential Scanning Calorimetry

To study the thermal transition of the different formulations, DSC analyses were carried out in a Differential Scanning Calorimeter DSC 821 from Mettler-Toledo Inc. (Schwerzenbach, Switzerland). Samples with an average weight of 6–8 mg of each formulation were placed in an aluminum-sealed 40 μL crucible and subjected to a single-step cycle from 25 °C to 300 °C and run at a range of 10 °C min^{-1} in an air atmosphere. Crystallinity degree (χ_c%) was calculated following Equation (1):

$$\chi_C = \left[\frac{|\Delta H_m| - |\Delta H_{CC}|}{\Delta H_m^0 (1 - w)} \right] \times 100 \quad (1)$$

where ΔH_m (J g^{-1}) is the melting peak enthalpy, ΔH_{CC} (J g^{-1}) is the cold crystallization peak enthalpy, ΔH_m^0 (J g^{-1}) is the theoretical value of the melting enthalpy of a fully crystalline PLA polymer, reported in the literature as 93.7 J g^{-1} [52], and w represents the weight fraction of PLA in the sample.

2.5.3.2. Thermogravimetric Analysis

The thermal stability of the formulations was determined by thermogravimetric analysis (TGA) using a thermobalance TGA PT1000 from Linseis Inc. (Selb, Germany). For each formulation, a sample with an average weight of 5–10 mg was placed in standard alumina 90 μL crucibles and subjected to a thermal program from 40 °C to 800 °C with a heating rate of 20 °C min^{-1}.

2.5.4. Antioxidant Activity

The antioxidant activity of the RO released from films into a food simulant in this work was determined by the DPPH inhibition method [10,14]. With this aim, samples of 15 mm × 15 mm of each formulation were submerged in 4.5 mL of fatty foodstuff simulant (Simulant D1, 50% v/v ethanol). The DPPH radical scavenging activity (RSA) of the simulant D1 was measured after 10 days at 40 °C in contact with the active films. For this purpose, 2.7 mL of the food simulant was mixed with 3 mL of 2 mM DPPH solution and maintained in darkness for 15 min at room temperature. The DPPH absorbance was recorded at the wavelength of 517 nm using a UV–VIS Varian Cary spectrophotometer (Varian, Palo Alto, CA, USA). The antioxidant capacity of the films was expressed as a percentage of inhibition of DPPH radical by the determination of the radical scavenging activity (RSA) using the Equation (2):

$$RSA\ (\%) = \frac{A_0 - A_t}{A_0} \cdot 100 \quad (2)$$

where A_0 is the absorbance of the control and A_t is the absorbance of the food simulant after the migration test and its reaction with DPPH. At least three different samples migrated from each film were measured.

2.5.5. Disintegration under Composting Conditions

Following ISO 20200 [53], the disintegrability of a polymer based material can be considered as the mass loss of a sample buried in a controlled compost soil. To this end, square samples of 25 × 25 mm^2 of each film formulation were placed in a bioreactor, prepared as indicated by the cited standard, and kept at 58 °C and water content in the soil of around 55%. The disintegration process under composting conditions was carried out during four weeks. Every week (7 days) samples were unburied, washed with distilled water, and dried at 40 °C for two days to be reweighed in an analytic thermobalance. The weight loss percent during the biodegradation process was calculated using Equation (3):

$$Weight\ loss\ (\%) = \frac{W_0 - W_t}{W_0} \cdot 100 \quad (3)$$

where W_0 is the weight of the sample before disintegration and W_t represents the weight after a disintegration time t.

2.5.6. Statistical Analysis

The significant differences in mechanical, thermal and antioxidant properties were statistically assessed at 95% confidence level using a one-way analysis of variance (ANOVA) according to Tukey's test employing OriginPro2018 software (OriginLab, Northampton, MA, USA).

3. Results

3.1. Particles Characterization

3.1.1. Dynamic Light Scattering

The hydrodynamic sizes of the pristine HNTs and HNTs encapsulating RO were determined by DLS, which showed a monomodal size distribution showing no significant differences between them with well-defined peaks between 155 nm to 410 nm (Figure 2a) centered at 254 nm in the case of HNTs and at 257 nm for RO loaded HNTs. However, the DLS technique is designed to calculate the hydrodynamic diameter of spherical nanoparticles, and thus, to corroborate the tubular shape as well as the actual dimensions of HNTs, TEM observations were also conducted. Additionally, as DE are microparticles, it was not possible to measure the particles through DLS and they were directly observed by TEM.

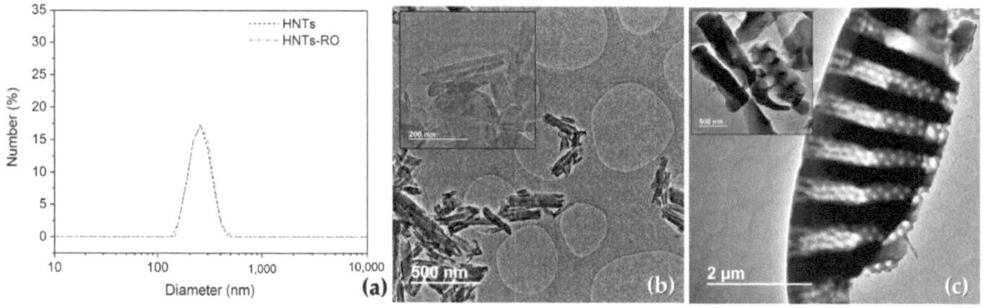

Figure 2. (**a**) DLS of HNTs and HNTs + RO as well as TEM images of (**b**) DE microparticles and (**c**) HNTs nanoparticles.

3.1.2. Transmission Electron Microscopy

The TEM images of the obtained HNTs and DE after the encapsulation treatment are shown in Figure 2b,c. As expected, the presence of RO could not be distinguished as no significant differences were observed between either HNTs or DE before the encapsulation process (without EO, not shown) and the HNTs and DE containing the RO. Although HNTs (Figure 2b) tended to aggregate in the TEM grid as a consequence of their high hydrophilicity [54], the TEM analysis confirmed the dimension of the nanotubes observed by DLS and from the zoom image (Figure 2b) of HNTs the tubular structure can be clearly seen. In the case of DE (Figure 2c), it showed microparticles with a high characteristic porosity that showed different shapes and a wide variety of sizes in good accordance with previous work [41].

3.2. Films Characterization

3.2.1. Visual Appearance

Figure 3 shows the visual appearance of the produced neat PLA film, Rosemary essential oil (RO) incorporated PLA films (PLA-RO), and PLA films loaded with Diatomaceous earth (DE) microparticles and Halloysite nanotubes (HNTs) encapsulating Rosemary essential oil (PLA-NHTs + RO and PLA-DE + RO films). The PLA film sample, at the top of the image (Figure 3a), is characterized by a high transparency. The three images below correspond to the film of PLA with the direct incorporation of RO (Figure 3b–d). The concentration of this essential oil added to the PLA matrix in this study seems low enough not to affect significantly the transparency of the material. Continuing, in the middle row of the figure are the images of the PLA films with RO encapsulated into HNTs at different contents (Figure 3e–g). The presence of the RO loaded HNTs nanoparticles is observable even at low concentrations, showing the PLA-HNTs + RO 1 certain yellowing compared with neat PLA film. This is more evident when the HNTs + RO is increased, acquiring besides a greater opacity at higher contents. Finally, when the RO is encapsulated in the DE the modification of the optical appearance of the PLA is more marked (Figure 3h–j) and related to the loss of transparency, getting more opaque aspect when more DE + RO amount is added.

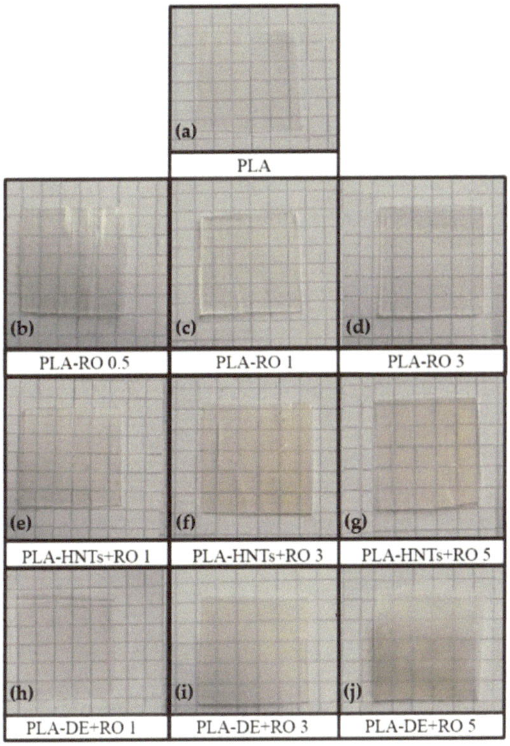

Figure 3. Visual appearance of the obtained films. (**a**) PLA; (**b**) PLA-RO 0.5; (**c**) PLA-RO 1; (**d**) PLA-RO 3; (**e**) PLA-HNTs + RO 1; (**f**) PLA-HNTs + RO 3; (**g**) PLA-HNTs + RO 5; (**h**) PLA-DE + RO 1; (**i**) PLA-DE + RO 3; and (**j**) PLA-DE + RO 5.

3.2.2. Scanning Electron Microscopy

Figure 4 shows the SEM micrographs of the cross cryo-fractured sections of the films developed here. From the cryo-fracture surface of the neat PLA film (Figure 4a), it can be seen that the typical rigid brittle and smooth fracture surface of PLA [10,54]. The incorporation of essential oil in 0.5 wt.% practically did not produce changes on the fracture surface (Figure 3b). Meanwhile, higher contents of rosemary essential oil of 3 wt.% (Figure 4c) and 5 wt.% (Figure 4d) produced homogeneous cryo-fracture surfaces with more plastic deformation, producing a somewhat plasticization effect as was already observed in PLA blended with other essential oils, such as limonene [24,55], carvacrol [56], thymol [18], or cinnamaldehyde [18]. The no apparent phase separation suggests good compatibility between RO and PLA matrix.

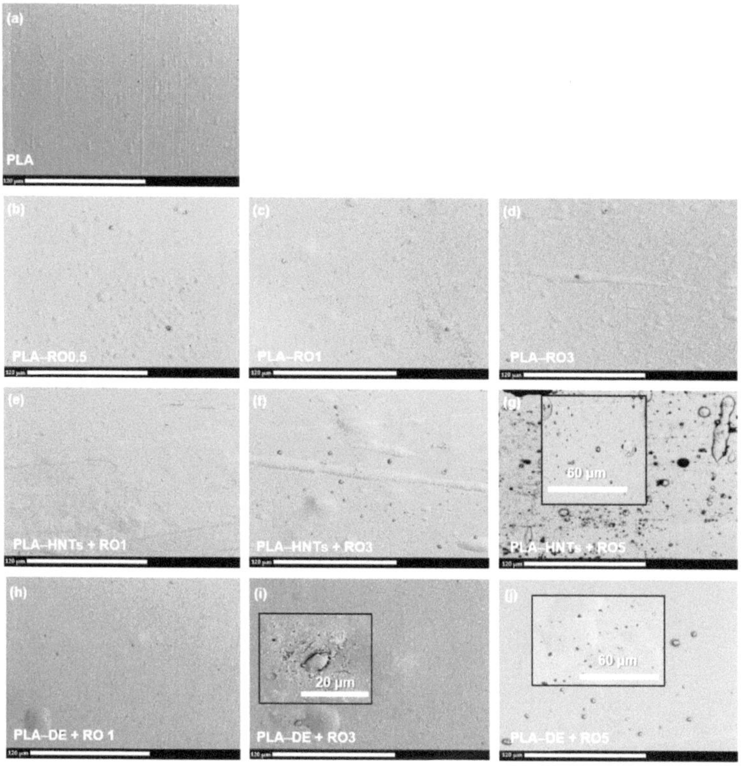

Figure 4. SEM images of the cryo-fracture section of the obtained films. (**a**) PLA; (**b**) PLA-RO 0.5; (**c**) PLA-RO 1; (**d**) PLA-RO 3; (**e**) PLA-HNTs + RO 1; (**f**) PLA-HNTs + RO 3; (**g**) PLA-HNTs + RO 5; (**h**) PLA-DE + RO 1; (**i**) PLA-DE + RO 3; and (**j**) PLA-DE + RO 5.

In the case of nanocomposites and composites, it is widely known that the particle dispersion is one of the major challenges to obtaining improved properties in the final material, particularly when using highly hydrophilic nanoparticles such as HNTs, which tend to form aggregates in the polymeric matrix when processed by conventional melt blending approaches [54,57]. From the SEM observations, it can be seen that both particles appeared well dispersed in the melt-blended polymeric matrix, which influenced that the films were still transparent after the addition of the HNTs and DE particles (see Figure 3).

The fracture surfaces of HNTs-based nanocomposites (Figure 4e–g) and DE-based composites (Figure 4h–j), showed that both particles reduced the smoothness and the materials showed more roughness as the particle amount increased in the film formulations.

In HNTs-based nanocomposites, some small white spots on the cross-section were observed which became denser with increasing HNTs concentration in the formulation from 1 wt.% (Figure 4e) to 3 wt.% (Figure 4f), while micro-voids are present in the nanocomposite with the higher concentration used here of 5 wt.% (Figure 4g). Similar findings were obtained by Rojas-Lema et al. who observed that when PLA was loaded with HNTs above 5 wt.%, the HNTs pull-out due to their poor interfacial adhesion with the PLA polymeric matrix which is intensified as a consequence of high HNTs loadings which can form agglomerates [57]. Similarly, De Silva et al. also observed an increase in the pores and microvoids formation with increasing HNTs content, particularly at PLA loading levels higher than 5 wt.%, and attributed these findings to an increasing detachment of the HNTs from the PLA matrix due to HNTs aggregation which ultimately leads to an increasing phase separation [58]. DE-based nanocomposites also showed the increasing density of the small white spots on the cross-section with an increasing amount of DE in the formulation. While the PLA loaded with 1 wt.% (Figure 4h) shows a homogeneous and regular matrix, for composites with DE higher loadings of 3 wt.% it is possible to identify individual DE particles and the surroundings of these microparticles show a small gap between them and the PLA polymeric matrix (see zoom image in Figure 4f). Similar findings have been already observed in a previous work [41]. Conversely, micro-voids are present in the nanocomposite with the higher concentration used here of 5 wt.% (Figure 4g), these unexpected results, in which better compatibility between the PLA polymeric matrix and DE at higher loading levels (5 wt.%) could be related to the higher amount of RO in this formulation that improves the compatibility between the microparticles and the polymeric matrix.

3.2.3. Mechanical Characterization

Tensile test results of the manufactured cast-extruded films are gathered in Table 1. Firstly, it can be observed that the neat PLA film showed the characteristic brittleness of this polymer, with a tensile elastic modulus (E) of 3700 MPa, tensile strength at yield (σ_y) of 38.5 MPa and a very low elongation at break (ε_b) of 2.3% [59]. It can be realized that the direct addition of a small percent of RO (0.5 and 1 wt.%) into the film formulation exerts similar and no significant ($p > 0.05$) variations in the strength at yield and elongation at break values, remaining around 38–39 MPa and 1.5%, respectively, through a small increase in the elastic modulus (E), reaching values of around 4000 MPa. But, when the RO content increases to 3 wt.% (PLA-RO 3), certain plasticization of the PLA matrix is produced with an increment in the flexibility of the films and a decrease in tensile strength. Specifically, σ_y values significantly ($p < 0.05$) decrease to 26 MPa, while elongation at break significantly ($p < 0.05$) enhances to 3.6%. As a consequence, the elastic modulus returns close to neat PLA values, reaching even significantly ($p < 0.05$) lower values. This reduction of the elastic modulus is due to the directly related decrease in strength and the increase in ductility, since the tensile modulus represents a coefficient of these parameters in the elastic behavior of the material [60]. The plasticization effect that EOs produce in PLA matrices has already been studied and reported in other works [18,24,61]. Some authors have concluded that the incorporation of EOs may interfere with polymer chains, reducing the intermolecular forces and, thus, acting as a plasticizer with a consequent improvement of the chain mobility and the flexibility of the material [12,24].

Table 1. Mechanical properties of the developed films obtained from tensile test measurements.

Film	*E* (MPa)	σ_y (MPa)	ε_b (%)
PLA	$3700 \pm 80^{a,b,c}$	38.5 ± 0.5^{a}	$2.3 \pm 0.1^{a,b}$
PLA-RO 0.5	$3880 \pm 80^{d,e}$	39.8 ± 0.4^{a}	1.7 ± 0.2^{c}
PLA-RO 1	3980 ± 40^{d}	38.4 ± 0.5^{a}	1.2 ± 0.2^{a}
PLA-RO 3	3120 ± 30^{f}	26.0 ± 0.9^{b}	3.6 ± 0.7^{d}
PLA-HNTs + RO 1	$3770 \pm 40^{a,e}$	28.8 ± 0.2^{b}	$3.1 \pm 0.3^{b,d}$
PLA-HNTs + RO 3	3590 ± 80^{b}	36.6 ± 0.4^{a}	2.7 ± 0.4^{b}
PLA-HNTs + RO 5	2940 ± 60^{g}	25.9 ± 0.2^{b}	$2.9 \pm 0.3^{b,d}$
PLA-DE + RO 1	4110 ± 70^{h}	46.9 ± 0.7^{c}	4.8 ± 0.4^{e}
PLA-DE + RO 3	4550 ± 60^{i}	39.2 ± 0.4^{a}	$1.6 \pm 0.2^{c,d}$
PLA-DE + RO 5	3670 ± 40^{c}	29.7 ± 0.6^{d}	4.3 ± 0.7^{e}

$^{a–i}$ Different letters within the same column show statistically significant differences between formulations ($p < 0.05$).

With regards to the PLA film with RO encapsulated into the two different particles (HNTs and DE), it was evidenced that each type of particle employed has a different effect. In this sense, the incorporation of 1 wt.% of HNTs induces a slight but significant ($p < 0.05$) increase in the elongation at break, and a significant ($p < 0.05$) reduction in the tensile strength value, reaching values close to that of PLA-RO 3, indicating that the RO essential oil role is more relevant than the HNTs particles, resulting in a plasticization of the polymeric matrix. However, HNTs seem effective in protecting the RO from thermal degradation since for a lower amount supplied, the results are very similar. The HNTs contribution is noticeable for the intermediate content formulation, PLA-HNTs + RO 3, which showed the most balanced mechanical performance with a tensile strength, and elastic modulus close to that of PLA while somewhat improving of the elongation at break. Generally, when a nanofiller is introduced and well dispersed in a polymeric matrix, due to its extensive surface area able to interact with the polymeric matrix, a greater difference in the material properties is caused [62]. Instead, at the highest HNTs content, which is in PLA-HNTs + RO 5, the enhanced parameters tended to decrease. The results obtained here are totally in concordance with the structural defects shown in SEM as well as with the literature. For instance, Rojas-Lema et al. [57] found similar results in plasticized PLA/OLA films with HNTs contents above 3 phr. Additionally, another study has reported that 3 wt.% of HNTs resulted in the optimal mechanical properties in PLA-based films, forming agglomerates if this content is exceeded [63].

On other hand, the reinforcement effect of the DE filler of the composite film is noticeable even with the lowest content. In terms of mechanical performance, PLA-DE + RO 1 resulted as the optimum formulation of composite with these microparticles added, with values of 47 MPa for tensile strength, 4.85% for elongation at break, and 4110 MPa for elastic modulus. This is indicative of a more homogenous dispersion of the DE in the PLA matrix compared with HNTs. Besides, the good compatibility between DE particles and PLA has been reported by other authors [44], highlighting that the high porosity of the diatoms allows polymer chains to enter and go out, providing a good particle/matrix adhesion. With a higher content of encapsulating filler (PLA-DE + RO 3) an interesting phenomenon seems to occur. At this amount DE tends to self-aggregate and form a dispersed phase [49,64]. Hence, the tensile properties of PLA-DE + RO 3 are significantly ($p < 0.05$) reduced with respect to PLA-DE + RO 1, with a strength value of 39.2 MPa and elongation at break of 1.6%, whereas rigidity still grow, leading to a modulus of around 4550 MPa. Finally, due to the low retention capacity of DEs when the addition of this filler with RO up to 5 wt.% (i.e., PLA-DE + RO 5), it seems that the already mentioned high porous microstructure of the DE particles is not able to retain as much RO essential oil [65], and this means a major amount of essential oil at the particles surrounding. This excess of essential oil is presumably released slowly from the DE particles and interacts with the

PLA chains, plasticizing the material and leading to a better compatibility between DE and the polymeric matrix.

3.2.4. Thermal Properties

3.2.4.1. Differential Scanning Calorimetric

Differential scanning calorimetric (DSC) analyses were carried out to study the different thermal transitions of the PLA-RO films and the influence of the encapsulation of the RO in HNTs and DE, and Table 2 lists the main thermal parameters from DSC of the PLA-RO, PLA-HNTs + RO, and PLA-DE + RO developed films as well as the neat PLA film. The glass transition temperature T_g, melting temperature T_m, and crystallinity grade (χ_c) of the PLA were at 60.3 °C, 153.5 °C, and 12.0%, respectively.

Table 2. DSC thermal parameters of the developed films.

Film	T_g (°C)	T_{cc} (°C)	ΔH_{cc} (J g^{-1})	T_m (°C)	ΔH_m (J g^{-1})	χ_c (%)
PLA	60.3 ± 0.8 ^{a}	123.5 ± 1.3 ^{a}	9.5 ± 0.8	153.5 ± 0.9 ^{a}	20.7± 0.7	11.9 ± 0.1
PLA-RO 0.5	58.1 ± 1.4 ^{a}	114.0 ± 0.9 ^{b,c}	18.7 ± 0.7	153.6 ± 1.1 ^{a}	27.3 ± 1.1	9.2 ± 0.2
PLA-RO 1	59.1 ± 1.2 ^{a}	114.0 ± 1.1 ^{b,c}	21.2 ± 0.9	147.8 ± 0.8 ^{b,c}	29.4 ± 0.9	8.9 ± 0.2
PLA-RO 3	58.4 ± 1.4 ^{a}	111.6 ± 1.4 ^{b,c}	20.7 ± 0.6	148.2 ± 1.3 ^{b,c}	26.0 ± 0.8	5.9 ± 0.7
PLA-HNTs + RO 1	58.1 ± 1.1 ^{a}	111.3 ± 0.8 ^{b,d}	16.6 ± 0.8	147.7 ± 1.1 ^{c}	27.8 ± 0.7	12.1 ± 0.3
PLA-HNTs + RO 3	59.2 ± 1.1 ^{a}	115.1 ± 1.2 ^{b,c}	21.5 ± 0.8	148.7 ± 1.2 ^{b,c}	22.8 ± 0.9	1.4 ± 0.4
PLA-HNT + RO 5	56.6 ± 1.3 ^{a}	112.0 ± 1.1 ^{b,e}	26.7 ± 0.9	147.7 ± 0.9 ^{b,c}	27.7 ± 1.1	1.2 ± 0.3
PLA-DE + RO 1	58.9 ± 0.9 ^{a}	118.1 ± 0.9 ^{c}	11.0 ± 1.2	152.1 ± 0.8 ^{a,b,d}	22.7 ± 0.8	12.6 ± 0.4
PLA-DE + RO 3	59.3 ± 1.2 ^{a}	115.3 ± 1.3 ^{b,c}	17.4± 1.1	148.2 ± 1.4 ^{c,d}	22.9 ± 0.9	6.1 ± 0.2
PLA-DE + RO 5	57.7 ± 1.1 ^{a}	108.6 ± 1.1 ^{d,e}	12.8 ± 1.2	153.4 ± 1.2 ^{a}	28.4 ± 0.7	17.6 ± 0.7

$^{a–e}$ Different letters within the same column show statistically significant differences between formulations ($p < 0.05$).

Concerning the direct addition of RO, essential oil molecules interact directly with the polymer chains and the thermal parameters modifications are directly related to the RO content. In Figure 5, the DSC first heating thermograms corresponding to neat PLA and PLA-RO at different percentages are plotted. Therefore, it is possible to observe a similar and no significant decrease in the T_g ($p > 0.05$) and a small but significant decrease in T_m ($p < 0.05$) for all the PLA-RO films but a higher reduction in crystallinity (χ_c) for a greater amount of essential oil, down to a 6.7% for PLA-RO 3. A significant ($p < 0.05$) reduction in the cold crystallization temperature (T_{cc}) was observed for all samples containing the RO either non-encapsulated or encapsulated in HNTs and DE. This behavior could highlight the compatibility between the RO and both particles, HNTs and RO, with the polymeric matrix, where as a result of the decrease in T_g, a subsequent lowering in the transport barrier for crystallization takes place, leading to a lower cold crystallization temperature, as it was already observed for PLA-Limonene blends [24]. These changes in thermal properties are in concordance with the tensile test results previously discussed. It is well studied that essential oils provide a plasticizing effect on PLA matrices, decreasing the glass transition temperature and crystallinity due to the increase in the free volume and chain mobility [61].

With the encapsulation of the RO in HNTs particles the modification in the DSC transition temperatures is close to those obtained with the direct addition of the essential oil to PLA (Figure 6). The T_g and T_m values decreased from 60.3 °C to 58–59 °C and from 153.5 °C to 147–148 °C, respectively, whether a 1–3 wt.% of RO or 1–3 wt.% of HNTs + RO are added, indicating that the essential oil is mainly what modifies the polymer matrix. In fact, the poor interactions between HNTs and PLA have been reported by Montava-Jorda et al. [54], while here it seems that the RO is increasing the interaction between PLA and HNTs. For the PLA-HNTs + RO 5, the inherent increment of encapsulated vegetable oil follows the tendency and produces a higher reduction but not significant ($p > 0.05$) of the T_g value (56.6 °C). Attending to the degradation of crystallinity (χ_c), a remarkable decrease was obtained, with values around 1%. This is a consequence of the increment of the crystallization enthalpy (ΔH_{cc}), raised to 26.7 J g^{-1}, which is almost triple that of the neat PLA (9.5 J g^{-1}). This could be attributed to the fact that HNTs avoid the crystallization of the PLA during the cooling of the material during the film-forming process and this is why it crystallizes during the DSC heating. Additionally, there is a reduction in the T_{cc}, which can be attributed to the presence of less perfect crystals as reported by Chen et al. [66] due to the presence of HNTs. On the other side, Lee et al. studied the encapsulation of thyme oil in HNTs and concluded that the thyme oil molecules were successfully loaded in the lumen of the nanoclays without producing any modification to the nanotubes [9]. Less perfect crystals, which start to form at a lower temperature, and a considerable amount of oil trapped in the HNTs agglomerates, induce a longer time for full crystallization, which causes the high reduction in the crystallinity already mentioned. Besides, this is totally in concordance with the double melting peak that appeared in the PLA-HNTs + RO 5 thermograph (Figure 6), in which less perfect crystals melt at significantly lower ($p < 0.05$) temperatures, while the most perfect crystals melt at higher values.

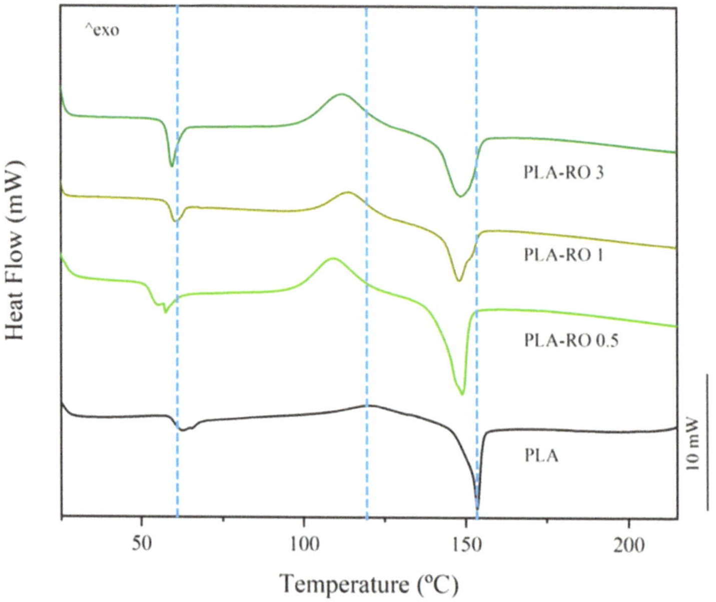

Figure 5. DSC first heating scan of the neat PLA and PLA with direct addition of RO films.

Using DE + RO as an encapsulating filler material did not suppose a significant difference in terms of the transition temperatures compared with HNTs. One can see (Figure 6) how, for PLA-DE + RO 1 and PLA-DE + RO 3, the advance of the glass transition and melting temperatures is practically the same as that of the HNTs encapsulating particle counterpart, although somewhat less pronounced. As occurs with PLA-HNTs + RO 5, the high content of essential oil in PLA-DE + RO 5 translates into an even greater advancement of temperatures, with no significant ($p > 0.05$) change in the T_g of 58.0 °C and a significant reduction of the T_m ($p < 0.05$). It can be concluded that neither micro- nor nanoclays modify the polymer structure but rather it is the RO that does it. Nevertheless, the release mechanism of RO from the particles seems to play a significant role in the final crystallinity of the material. The χ_c of the PLA-DE + RO 3 is around 7%, which is 42% lower than that of neat PLA. However, for the PLA-DE + RO 5 films, this value is increased to 22.4%, almost twice that of neat PLA. This tendency is parallel to the results obtained in the tensile test, with a reduction of ductility and strength for a 3 wt.% DE + RO and then an enhancement of these parameters for the 5 wt.% DE + RO. It can be hypothesized that the small phase rich in essential oil formed in PLA-DE + RO 3 due to a complete dispersion within the polymeric matrix interrupts the continuity and inhibits the crystal growth. Whereas with 5 wt.% of DE + RO, the presence of a higher amount of essential oil interacting with PLA chains allows the formation of a second population of crystals, which is also evidenced by the double peak in the melting enthalpy curve.

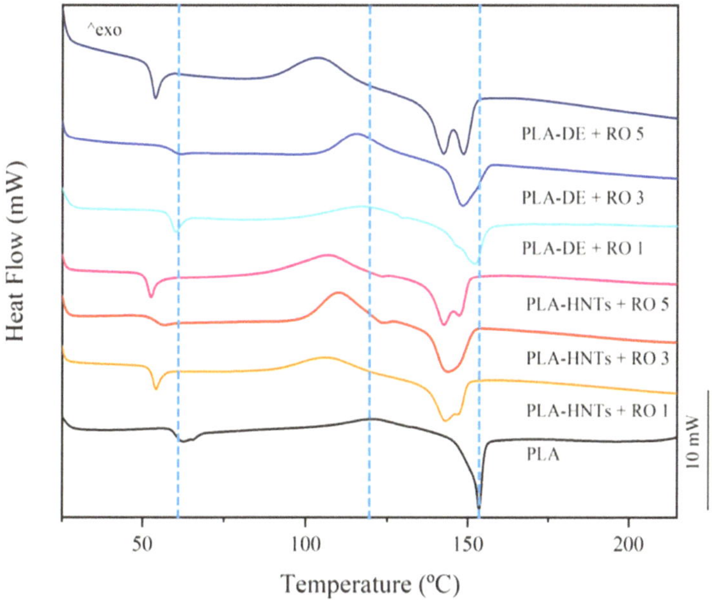

Figure 6. DSC first heating scan of the neat PLA, PLA-HNTs + RO, and PLA-DE + RO films.

3.2.4.2. Thermogravimetric Analysis

TGA assays were conducted to study the influence of the RO and the use of HNT_S and DE as encapsulation fillers on the thermal stability of the PLA films. The TGA curves for PLA, PLA-RO, PLA-HNTs + RO, and PLA-DE + RO at different loading levels are shown in Figure 7, whereas the main values obtained from these curves are summarized in Table 3.

Table 3. TGA results of the developed films.

Film	$T_{10\%}$ (°C)	T_{deg}	Residual Mass (%)
PLA	322 ± 1.2 [a]	366.94 ± 1.1 [a,b,c]	1.6 ± 0.2
PLA-RO 0.5	328. ± 1.2 [b]	354.09 ± 0.9 [a]	0.4 ± 0.3
PLA-RO 1	328.7 ± 0.9 [b]	361.27 ± 1.1 [a,b,c]	0.4 ± 0.2
PLA-RO 3	343.6 ± 1.1 [c]	359.64 ± 0.7 [a,b]	0.3 ± 0.2
PLA-HNTs + RO 1	322.1 ± 0.9 [a]	353.29 ± 1.1 [a]	0.6 ± 0.3
PLA-HNTs + RO 3	316.6 ± 1.5 [d]	362.55 ± 1.4 [a,b,c]	3.0 ± 0.1
PLA-HNTs + RO 5	313.8 ± 1.4 [d]	372.61 ± 1.2 [b,c]	5.0 ± 0.2
PLA-DE + RO 1	350.0 ± 1.1 [e]	378.49 ± 1.2 [c]	1.2 ± 0.3
PLA-DE + RO 3	306.2 ± 1.2 [f]	360.81 ± 0.9 [a,b,c]	2.3 ± 0.2
PLA-DE + RO 5	328.3 ± 0.9 [b]	358.76 ± 1.1 [a,b]	2.1 ± 0.1

[a–f] Different letters within the same column show statistically significant differences between formulations ($p < 0.05$).

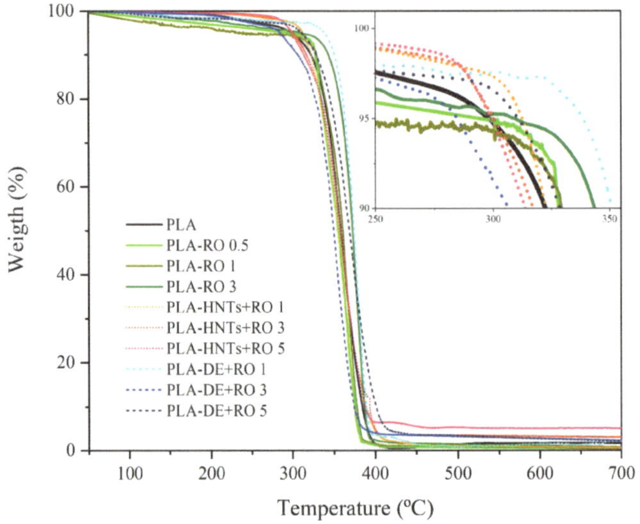

Figure 7. TGA curves and DTG (zoom image) of films.

It can be observed that all the films presented a single-step pattern of mass loss. This similarity in the thermal stability of the materials developed here and the PLA was as expected for PLA [22]. The onset degradation temperature of PLA, meaning when 10% of mass loss occurs ($T_{10\%}$), was at ~322 °C and the maximum thermal degradation temperature (T_{deg}) was above 378 °C.

The direct addition of a small percent of RO provided a thermal stability improvement, and it is more noticeable for the highest contest, resulting in a $T_{10\%}$ at ~343 °C, significantly higher ($p < 0.05$) for PLA-RO 3 as a consequence of the thermal protection of an antioxidant compound on the polymeric matrix provided by the RO. In this context, similar delays in the onset degradation temperature of PLA have been observed in PLA added with garlic acid [50] and PLA added with catechin [14], attributing this enhancement to the intrinsic antioxidant activity of the polyphenols.

Regarding the nanoparticles, the thermogravimetric curve of PLA-HNTs + RO 1 presented a shift towards a closer value to that of PLA, indicating that RO trapped within the lumen of the nanoclays is not able to modify the polymeric matrix. By checking the curves for higher contents of HNTs loaded with RO, it can be seen that there is some shift to lower temperatures. The RO encapsulated within HNTs tend to remain the RO in the lumen, and thus avoids the thermal protection of PLA provided by RO. Meanwhile, HNTs induce a degradation temperature advance. The residual mass obtained for these samples corresponds to the amount of filler. Higher loading levels of HNTs (3 and 5 wt.%) lead to a small but significant ($p < 0.05$) reduction in the onset thermal degradation, but being thermally stable at temperatures considerably higher than the processing temperature used.

For microparticles, in the SEM and mechanical results section, it has been described that the apparent good dispersion and optimum matrix/oil/particle proportion of the PLA-DE + RO 1 can also be observed in the TGA curve. The PLA-DE + RO 1 onset degradation temperature turned out to be the highest, reaching ~350 °C. Small amounts of inorganic materials such as clays in a polymer matrix used to provide an increment in thermal stability [67]. Then in the PLA-DE + RO 3, there is a reduction in the $T_{10\%}$ ($p < 0.05$) and T_{deg} values, probably due to the lesser interaction between PLA matrix and DE at this proportion, as already discussed in SEM results. As for the PLA-DE + RO 5, the modified polymer matrix by the excess of essential oil counteracts the positive effect of the DE particle interaction with the polymeric matrix, and the thermal stability resultant is close to that of the neat PLA.

3.2.5. Antioxidant Activity

Considering that lipids are one of the main targets of oxidative reactions and, thus, the lipid oxidation process is responsible for a major problem in both natural and processed foodstuffs [10,14], the antioxidant activity of the developed RO-containing films was studied in direct contact with a fatty food simulant (simulant D1).

The antioxidant activity of *Rosmary officinalis* L. essential oil has been atributed to a cooperative results of its compositions [68]. Thus, the antioxidant activity was indirectly measured through the DPPH method, and the radical scavenging activity (RSA%) results are provided in Table 4. The RSA of pristine RO was 43.2 ± 1.9% showing values in the range of those observed in the literature [68,69].

Table 4. RSA results of the developed films after 10 contact days in Simulant D1 at 40 °C.

Film	RSA (%)
PLA	-
PLA-RO 0.5	12.7 ± 2.5 ^{a}
PLA-RO 1	22.1 ± 3.1 ^{b}
PLA-RO 3	32.9 ± 1.9 ^{c,d}
PLA-HNTs + RO 1	37.3 ± 2.1 ^{e}
PLA-HNTs + RO 3	22.4 ± 3.2 ^{b,f}
PLA-HNTs + RO 5	43.8 ± 1.6 ^{e}
PLA-DE + RO 1	43.3 ± 1.4 ^{e}
PLA-DE + RO 3	28.8 ± 1.3 ^{d,f}
PLA-DE + RO 5	27.6 ± 0.9 ^{g}

$^{a–g}$ Different letters within the same column show statistically significant differences between formulations ($p < 0.05$).

All formulations containing RO, either directly added or encapsulated into HNTs or DE, showed antioxidant activity, highlighting that RO retains its antioxidant ability after thermal processing of the films by melt-extrusion followed by cast-film. The antioxidant activity of directly added RO to PLA films (PLA-RO) is comparable to other systems based on plasticized PLA incorporated with similar amounts of rosemary essential oil [70]. Therefore, the present results show that part of the RO was able to support the thermal processing and then also act as an effective antioxidant agent to protect packed food from oxidative processes. The antioxidant activity was higher in the films in which the RO was encapsulated, particularly in PLA-HNTs + RO 5 and PLA-DE + RO 1, since the EO directly added into the PLA matrix suffered thermal degradation due to the volatility of RO. While for nanoparticles, the higher loading levels of 5 wt.% lead to a higher antioxidant activity, as expected, in microparticles, the best results were observed for the lowest loading levels used of 1 wt.%. These findings are directly related to the already commented loss of RO from DE at higher loading levels, as it is a microparticle. Meanwhile, the RO in HNTs is better retained in the lumen of HNTs. Moreover, the results are in good agreement with the results of the onset degradation temperature tendency as the RO provided thermal protection due to the antioxidant activity.

It should be highlighted that the antioxidant activity of the encapsulated PLA-HNTs + RO 5 and PLA-DE + RO 1 is significantly higher ($p < 0.05$) than that of other systems based on PLA and rosemary essential oil as the fillers can protect the essential oil from thermal degradation.

3.3. Disintegration under Composting Conditions

The compostability of the films with RO and encapsulated RO in HNTs and DE was studied by a disintegration test in controlled composting conditions. The percentage weight loss of all the samples during the disintegration process is gathered in Figure 8, while the visual appearance of the recovered films at different disintegration days after exposure to the composting medium is shown in Figure 9. During the first week, normally related to the induction period, all the formulations lost the same mass of approximately 10%, which is slightly higher than neat PLA, showing that RO speeds up the disintegration phenomenon, as was already observed for other essential oils such as limonene [71] or carvacrol [12]. In the second week, significant differences started to be appreciated between the weight losses of the samples. In particular, formulations with the highest loading of both clays (PLA-HNTs + RO 5 and PLA-DE + RO 5) presented a disintegration rate close to 30%, whereas the rest did not exceed 20%. This increase in disintegration may be propinated by the increment of the moisture absorption resulting from the intrinsic hydrophilic character of the clays [72].

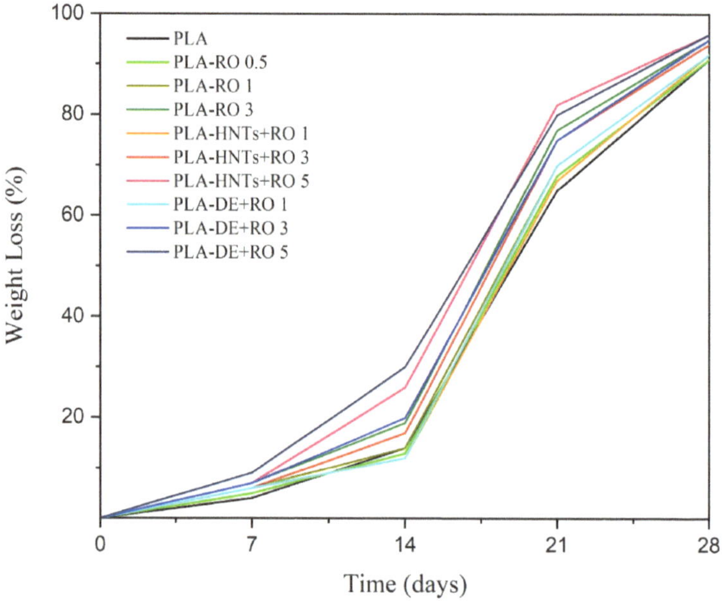

Figure 8. Degree of disintegration under composting conditions as a function of time of the developed films.

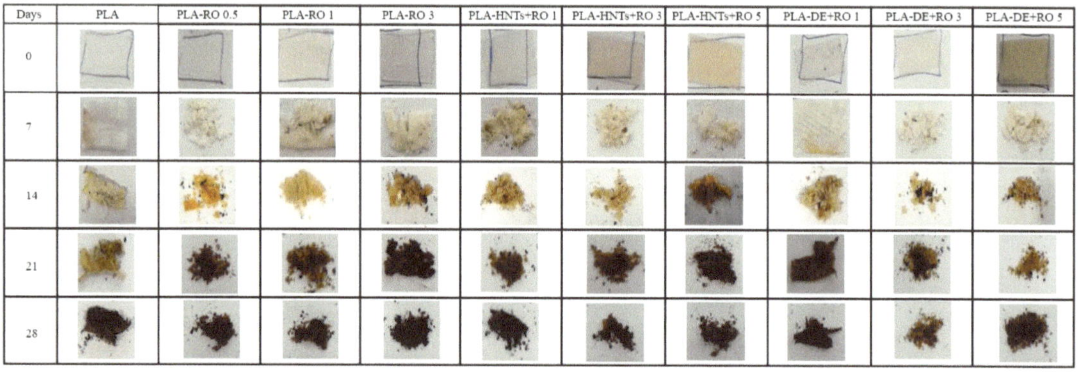

Figure 9. Visual appearance of films before and after different recovered days of disintegration under composting conditions.

After 21 days, the disintegration rate of all the films was over 50%. It is worthy to note that the film with the highest content of non-encapsulated essential oil reached a weight loss similar to those just mentioned with the highest DE or HNTs content, thus evidencing that the essential oil also favors faster disintegradability due to the increased polymer chain mobility as a consequence of the already mentioned plasticizing effect of RO for the PLA matrix. Although more days were necessary in this case, it was due to the lower presence of RO in the PLA-RO 3 than in PLA-HUNTS + RO 5 and PLA-DE + RO 5. The increment of free volume, with more space between polymer chains that plasticization supposes, leads to an easier diffusion of water molecules and hence facilitates the hydrolysis activity to shorten the polymer chain and make the enzymatic disintegration promotion of the PLA matrix.

At the end of the burial period of 28 days, all the samples were almost disintegrated, with remaining mass of less than 10% (more than 90% of the weight was lost), which is the frequent value used in the literature to consider the materials disintegrated [10,73]. Figure 9 shows the variation of the appearance of the samples as the weeks progressed, where it can be seen how this small amount of mass remained, turning to a sand-kind aspect.

4. Conclusions

Rosmarinus officinalis (RO) essential oil was successfully encapsulated into DE microparticles as well as into HNTs nanoparticles and further used to produce PLA-based composites and nanocomposites, respectively. The materials were developed by melt extrusion followed by cast film, simulating the industrial processing conditions. Both DE and HNTs were able to protect the *Rosmarinus officinalis* (RO) from thermal degradation during processing, allowing to obtain films with antioxidant properties, as demonstrated by the antioxidant assays after the materials were exposed for 10 days to a fatty food simulant. The RO encapsulated into either DE or HNTs showed a reinforcing effect, particularly at 3 wt.% loading levels. However, from a structural point of view DE microparticles showed less adhesion to the PLA matrix than the smaller HNTs. In fact, DE at higher loading levels of 1 wt.% showed structural defects, while HNTs showed agglomerations at higher loading levels of 3 wt.%. Meanwhile, RO produced to a certain extent a plasticization effect, improving the interaction between DE or HNTs and the PLA polymeric matrix.

The films developed here showed their potential for sustainable films for biodegradable antioxidant food packaging applications.

Author Contributions: Conceptualization, R.C.-G., M.P.A. and O.F.; methodology, O.F. and M.P.A.; software, R.C.-G. and Á.A.; validation, R.C.-G., Á.A., M.P.A. and O.F.; formal analysis, R.C.-G., Á.A., M.P.A. and O.F.; investigation, R.C.-G., Á.A., M.P.A. and O.F.; resources, M.P.A. and O.F.; data curation, R.C.-G., Á.A., M.P.A. and O.F.; writing—original draft preparation, R.C.-G., Á.A., M.P.A. and O.F.; writing—review and editing, M.P.A. and O.F.; visualization, R.C.-G., Á.A., M.P.A. and O.F.; supervision, M.P.A. and O.F.; project administration, M.P.A. and O.F.; funding acquisition, M.P.A. and O.F. All authors have read and agreed to the published version of the manuscript.

Funding: This research was funded by the Spanish Ministry of Science and Innovation (MICINN) through PID-AEI project (grants PID2020-116496RB-C22 and PID2021-123753NA-C32) funded by MCIN/AEI/10.13039/501100011033 and by ERDF "A way of making Europe" by the "European Union" and TED-AEI Project (grant TED2021-129920A-C43) and Research Consolidator Project (grant CNS2022-136064) funded by MCIN/AEI/10.13039/501100011033 and by the "European Union NextGenerationEU/PRTR". Ángel Agüero acknowledges the Margarita Salas postdoctoral grant from the Ministerio de Universidades, Spain, funded by the European Union-Next Generation EU.

Institutional Review Board Statement: Not applicable.

Data Availability Statement: Data are contained within the article.

Acknowledgments: Microscopy Services at UPV are acknowledged for their help in collecting and analyzing images.

Conflicts of Interest: Author Raúl Cerdá-Gandia was employed by the company FAPERIN S.L. Av. The remaining authors declare that the research was conducted in the absence of any commercial or financial relationships that could be construed as a potential conflict of interest.

References

1. Alam, M.; Akram, D.; Sharmin, E.; Zafar, F.; Ahmad, S. Vegetable Oil Based Eco-Friendly Coating Materials: A Review Article. *Arab. J. Chem.* **2014**, *7*, 469–479. [CrossRef]
2. Gbadeyan, O.J.; Sarp, A.; Glen, B.; Sithole, B. Nanofiller/Natural Fiber Filled Polymer Hybrid Composite: A Review. *J. Eng. Sci. Technol. Rev.* **2021**, *14*, 61–74. [CrossRef]
3. George, A.; Sanjay, M.R.; Srisuk, R.; Parameswaranpillai, J.; Siengchin, S. A Comprehensive Review on Chemical Properties and Applications of Biopolymers and Their Composites. *Int. J. Biol. Macromol.* **2020**, *154*, 329–338. [CrossRef] [PubMed]
4. Sadasivuni, K.K.; Saha, P.; Adhikari, J.; Deshmukh, K.; Ahamed, M.B.; Cabibihan, J.J. Recent Advances in Mechanical Properties of Biopolymer Composites: A review. *Polym. Compos.* **2019**, *41*, 32–59. [CrossRef]
5. Siracusa, V.; Rocculi, P.; Romani, S.; Rosa, M.D. Biodegradable Polymers For Food Packaging: A Review. *Trends Food Sci. Technol.* **2008**, *19*, 634–643. [CrossRef]
6. Luzi, F.; Torre, L.; Kenny, J.M.; Puglia, D. Bio- and Fossil-Based Polymeric Blends and Nanocomposites for Packaging: Structure-property Relationship. *Materials* **2019**, *12*, 471. [CrossRef]
7. Han, B.; Chen, P.; Guo, J.; Yu, H.; Zhong, S.; Li, D.; Liu, C.; Feng, Z.; Jiang, B. A Novel Intelligent Indicator Film: Preparation, Characterization, and Application. *Molecules* **2023**, *28*, 3384. [CrossRef]
8. Arrieta, M.P.; Samper, M.D.; Aldas, M.; López, J. On the use of Pla-PHB Blends for Sustainable Food Packaging Applications. *Materials* **2017**, *10*, 1008. [CrossRef]
9. Lee, M.H.; Park, H.J. Preparation of Halloysite Nanotubes Coated with Eudragit for a Controlled Release of Thyme Essential Oil. *J. Appl. Polym. Sci.* **2015**, *132*, 42771. [CrossRef]
10. Agüero, A.; Lascano, D.; Ivorra-Martinez, J.; Gómez-Caturla, J.; Arrieta, M.P.; Balart, R. Use of Bacterial Cellulose Obtained from Kombucha Fermentation in Spent Coffee Grounds for Active Composites Based on Pla and Maleinized Linseed Oil. *Ind. Crops Prod.* **2023**, *202*, 116971. [CrossRef]
11. García-Arroyo, P.; Arrieta, M.P.; Garcia-Garcia, D.; Cuervo-Rodríguez, R.; Fombuena, V.; Mancheño, M.J.; Segura, J.L. Plasticized Poly(Lactic Acid) Reinforced with Antioxidant Covalent Organic Frameworks (COFs) as Novel Nanofillers Designed for Non-Migrating Active Packaging Applications. *Polymer* **2020**, *196*, 122466. [CrossRef]
12. Faba, S.; Arrieta, M.P.; Romero, J.; Agüero, Á.; Torres, A.; Martínez, S.; Rayón, E.; Galotto, M.J. Biodegradable nanocomposite poly(lactic acid) Foams Containing Carvacrol-Based Cocrystal Prepared by Supercritical CO_2 Processing for Controlled Release in Active Food Packaging. *Int. J. Biol. Macromol.* **2024**, *254*, 127793. [CrossRef] [PubMed]
13. Sivakanthan, S.; Rajendran, S.; Gamage, A.; Madhujith, T.; Mani, S. Antioxidant and Antimicrobial Applications of Biopolymers: A Review. *Food Res. Int.* **2020**, *136*, 109327. [CrossRef] [PubMed]
14. Arrieta, M.P.; Castro-López, M.D.M.; Rayón, E.; Barral-Losada, L.F.; López-Vilariño, J.M.; López, J.; González-Rodríguez, M.V. Plasticized Poly(Lactic Acid)-Poly(Hydroxybutyrate) (Pla-PHB) Blends Incorporated with Catechin Intended for Active Food-Packaging Applications. *J. Agric. Food Chem.* **2014**, *62*, 10170–10180. [CrossRef] [PubMed]
15. Krepker, M.; Shemesh, R.; Danin Poleg, Y.; Kashi, Y.; Vaxman, A.; Segal, E. Active Food Packaging Films with Synergistic Antimicrobial Activity. *Food Control* **2017**, *76*, 117–126. [CrossRef]
16. Bhavaniramya, S.; Vishnupriya, S.; Al-Aboody, M.S.; Vijayakumar, R.; Baskaran, D. Role of Essential oils in Food Safety: Antimicrobial and Antioxidant Applications. *Grain Oil Sci. Technol.* **2019**, *2*, 49–55. [CrossRef]
17. Gachkar, L.; Yadegari, D.; Rezaei, M.; Taghizadeh, M.; Astaneh, S.; Rasooli, I. Chemical and Biological Characteristics of *Cuminum cyminum* and *Rosmarinus officinalis* essential oils. *Food Chem.* **2007**, *102*, 898–904. [CrossRef]
18. Villegas, C.; Arrieta, M.P.; Rojas, A.; Torres, A.; Faba, S.; Toledo, M.J.; Gutierrez, M.A.; Zavalla, E.; Romero, J.; Galotto, M.J.; et al. Pla/Organoclay Bionanocomposites Impregnated with Thymol and Cinnamaldehyde by Supercritical Impregnation for Active and Sustainable Food Packaging. *Compos. Part B Eng.* **2019**, *176*, 107336. [CrossRef]
19. Ruiz-Navajas, Y.; Viuda-Martos, M.; Sendra, E.; Perez-Alvarez, J.A.; Fernández-López, J. In Vitro Antibacterial and Antioxidant Properties of Chitosan Edible Films Incorporated with Thymus moroderi or Thymus piperella essential oils. *Food Control* **2013**, *30*, 386–392. [CrossRef]
20. Sharma, S.; Barkauskaite, S.; Duffy, B.; Jaiswal, A.K.; Jaiswal, S. Characterization and Antimicrobial Activity of Biodegradable Active Packaging Enriched with Clove and Thyme Essential Oil for Food Packaging Application. *Foods* **2020**, *9*, 1117. [CrossRef]
21. Sharma, S.; Barkauskaite, S.; Jaiswal, S.; Duffy, B.; Jaiswal, A.K. Development of Essential Oil Incorporated Active Film Based on Biodegradable Blends of Poly (Lactide)/Poly (Butylene Adipate-Co-Terephthalate) for Food Packaging Application. *J. Packag. Technol. Res.* **2020**, *4*, 235–245. [CrossRef]
22. Wang, D.; Sun, Z.; Sun, J.; Liu, F.; Du, L.; Wang, D. Preparation and Characterization of Polylactic Acid Nanofiber Films Loading Perilla Essential oil for Antibacterial Packaging of Chilled Chicken. *Int. J. Biol. Macromol.* **2021**, *192*, 379–388. [CrossRef] [PubMed]
23. Cheran, E.; Sharmila Rahale, C.; Divyabharathi, P.; Viswanathan, C.; Narayanan, L. Corn Cob Nanocellulose Packaging for Increasing the Shelf Life of Food Products. *Int. J. Biol. Macromol.* **2024**, *268*, 131403. [CrossRef] [PubMed]
24. Arrieta, M.P.; López, J.; Ferrándiz, S.; Peltzer, M.A. Characterization of Pla-Limonene Blends for Food Packaging Applications. *Polym. Test.* **2013**, *32*, 760–768. [CrossRef]
25. Nieto, G.; Ros, G.; Castillo, J. Antioxidant and Antimicrobial Properties of Rosemary (*Rosmarinus officinalis*, L.): A review. *Medicines* **2018**, *5*, 98. [CrossRef]

26. Vasile, C.; Stoleru, E.; Darie-Niţa, R.N.; Dumitriu, R.P.; Pamfil, D.; Tarţau, L. Biocompatible Materials Based on Plasticized Poly(Lactic Acid), Chitosan and Rosemary Ethanolic Extract I. Effect of Chitosan on the Properties of Plasticized Poly(Lactic Acid) Materials. *Polymers* **2019**, *11*, 941. [CrossRef]
27. Arrieta, M.P.; García, A.D.; López, D.; Fiori, S.; Peponi, L. Antioxidant bilayers based on PHBV and Plasticized Electrospun Pla-PHB Fibers Encapsulating Catechin. *Nanomaterials* **2019**, *9*, 346. [CrossRef]
28. Lisuzzo, L.; Cavallaro, G.; Pasbakhsh, P.; Milioto, S.; Lazzara, G. Why Does Vacuum Drive to the Loading of Halloysite Nanotubes? The Key Role of Water Confinement. *J. Colloid Interface Sci.* **2019**, *547*, 361–369. [CrossRef]
29. Vergaro, V.; Abdullayev, E.; Lvov, Y.M.; Zeitoun, A.; Cingolani, R.; Rinaldi, R.; Leporatti, S. Cytocompatibility and Uptake of Halloysite Clay Nanotubes. *Biomacromolecules* **2010**, *11*, 820–826. [CrossRef]
30. Cerdá-Gandia, R.; Ivorra Martínez, J.; Agüero, Á.; Quiles-Carrillo, L.; Gomez-Caturla, J.; Fenollar, O.; Arrieta, M.P. Processing of Biopolymers Loaded with Porous Inorganic Fillers Encapsulating Active Substance for Active Food Packaging Applications. In *Biopolymers: Synthesis, Properties, and Emerging Applications*; Elsevier: Amsterdam, The Netherlands, 2023; Volume 1.
31. Price, R.R.; Gaber, B.P.; Lvov, Y. In-Vitro Release Characteristics Of Tetracycline HCl, Khellin and Nicotinamide Adenine Dineculeotide from Halloysite; a Cylindrical Mineral. *J. Microencapsul.* **2001**, *18*, 713–722.
32. Gorrasi, G.; Pantani, R.; Murariu, M.; Dubois, P. Pla/Halloysite Nanocomposite Films: Water Vapor Barrier Properties and Specific Key Characteristics. *Macromol. Mater. Eng.* **2014**, *299*, 104–115. [CrossRef]
33. Liu, M.; Zhang, Y.; Wu, C.; Xiong, S.; Zhou, C. Chitosan/Halloysite Nanotubes Bionanocomposites: Structure, Mechanical Properties and Biocompatibility. *Int. J. Biol. Macromol.* **2012**, *51*, 566–575. [CrossRef] [PubMed]
34. Torres, E.; Fombuena, V.; Valles-Lluch, A.; Ellingham, T. Improvement of Mechanical and Biological Properties of Polycaprolactone Loaded with Hydroxyapatite and Halloysite Nanotubes. *Mater. Sci. Eng. C Mater. Biol. Appl.* **2017**, *75*, 418–424. [CrossRef] [PubMed]
35. Hao, A.; Wong, I.; Wu, H.; Lisco, B.; Ong, B.; Sallean, A.; Butler, S.; Londa, M.; Koo, J.H. Mechanical, Thermal, and Flame-Retardant Performance of Polyamide 11–Halloysite Nanotube Nanocomposites. *J. Mater. Sci.* **2014**, *50*, 157–167. [CrossRef]
36. Lee, M.H.; Kim, S.Y.; Park, H.J. Effect of Halloysite Nanoclay on the Physical, Mechanical, and Antioxidant Properties of Chitosan Films Incorporated with Clove Essential Oil. *Food Hydrocoll.* **2018**, *84*, 58–67. [CrossRef]
37. Lee, M.H.; Seo, H.S.; Park, H.J. Thyme Oil Encapsulated in Halloysite Nanotubes for Antimicrobial Packaging System. *J. Food Sci.* **2017**, *82*, 922–932. [CrossRef]
38. Gorrasi, G. Dispersion of Halloysite Loaded with Natural Antimicrobials into Pectins: Characterization and Controlled Release Analysis. *Carbohydr. Polym.* **2015**, *127*, 47–53. [CrossRef]
39. Lopez-Cebral, R.; Peng, G.; Reys, L.L.; Silva, S.S.; Oliveira, J.M.; Chen, J.; Silva, T.H.; Reis, R.L. Dual Delivery of Hydrophilic and Hydrophobic Drugs from Chitosan/Diatomaceous Earth Composite Membranes. *J. Mater. Sci. Mater. Med.* **2018**, *29*, 21. [CrossRef]
40. Delasoie, J.; Zobi, F. Natural Diatom Biosilica as Microshuttles in Drug Delivery Systems. *Pharmaceutics* **2019**, *11*, 537. [CrossRef]
41. Aguero, A.; Quiles-Carrillo, L.; Jorda-Vilaplana, A.; Fenollar, O.; Montanes, N. Effect of Different Compatibilizers on Environmentally Friendly Composites from Poly(Lactic Acid) and Diatomaceous earth. *Polym. Int.* **2019**, *68*, 893–903. [CrossRef]
42. Uthappa, U.T.; Sriram, G.; Brahmkhatri, V.; Kigga, M.; Jung, H.-Y.; Altalhi, T.; Neelgund, G.M.; Kurkuri, M.D. Xerogel Modified Diatomaceous Earth Microparticles for Controlled Drug Release Studies. *New J. Chem.* **2018**, *42*, 11964–11971. [CrossRef]
43. Hakovirta, M.; Aksoy, B.; Hakovirta, J. Self-Assembled Micro-Structured Sensors For Food Safety in Paper Based Food Packaging. *Mater. Sci. Eng. C Mater. Biol. Appl.* **2015**, *53*, 331–335. [CrossRef] [PubMed]
44. Cacciotti, I.; Mori, S.; Cherubini, V.; Nanni, F. Eco-Sustainable Systems Based on Poly(Lactic Acid), Diatomite and Coffee Grounds Extract for Food Packaging. *Int. J. Biol. Macromol.* **2018**, *112*, 567–575. [CrossRef] [PubMed]
45. Aggarwal, S.; Johnson, S.; Saloni, D.; Hakovirta, M. Novel 3D Printing Filament Composite Using Diatomaceous Earth and Polylactic Acid for Materials Properties And Cost Improvement. *Compos. Part B Eng.* **2019**, *177*, 107310. [CrossRef]
46. Kabouche, Z.; Boutaghane, N.; Laggoune, S.; Kabouche, A.; Aitkaki, Z.; Benlabed, K. Comparative Antibacterial Activity of Five Lamiaceae Essential Oils from Algeria. *Int. J. Aromather.* **2005**, *15*, 129–133. [CrossRef]
47. Yinrong Lu, L.; Foo, Y. Antioxidant Activities of Polyphenols from Sage (*Salvia officinalis*). *Food Chem.* **2001**, *75*, 197–202.
48. Othman, S.H.; Hassan, N.; Talib, R.A.; Kadir Basha, R.; Risyon, N.P. Mechanical and Thermal Properties of Pla/Halloysite Bio-Nanocomposite Films: Effect of Halloysite Nanoclay Concentration and Addition Of Glycerol. *J. Polym. Eng.* **2017**, *37*, 381–389. [CrossRef]
49. Gonzalez, L.; Aguero, A.; Quiles-Carrillo, L.; Lascano, D.; Montanes, N. Optimization of the Loading of an Environmentally Friendly Compatibilizer Derived from Linseed Oil in Poly(Lactic Acid)/Diatomaceous Earth Composites. *Mater.* **2019**, *12*, 1627. [CrossRef]
50. Quiles-Carrillo, L.; Montanes, N.; Lagaron, J.; Balart, R.; Torres-Giner, S. Bioactive Multilayer Polylactide Films with Controlled Release Capacity of Gallic Acid Accomplished by Incorporating Electrospun Nanostructured Coatings and Interlayers. *Appl. Sci.* **2019**, *9*, 533. [CrossRef]
51. *ISO 527-2:2012*; Plastics-Determination of Tensile Properties Part 2: Test Conditions for Moulding and Extrusion Plastics. International Organization for Standardization: Geneva, Switzerland, 2012.
52. Saeidlou, S.; Huneault, M.A.; Li, H.; Park, C.B. Poly(Lactic Acid) Crystallization. *Prog. Polym. Sci.* **2012**, *37*, 1657–1677. [CrossRef]

53. *ISO 20200:2023*; Plastics–Determination of the Degree of Disintegration of Plastic Materials under Composting Conditions in a Laboratory-Scale Test. International Organization for Standardization: Geneva, Switzerland, 2023.
54. Montava-Jorda, S.; Chacon, V.; Lascano, D.; Sanchez-Nacher, L.; Montanes, N. Manufacturing and Characterization of Functionalized Aliphatic Polyester from Poly(Lactic Acid) with Halloysite Nanotubes. *Polymers* **2019**, *11*, 1314. [CrossRef] [PubMed]
55. Fortunati, E.; Luzi, F.; Puglia, D.; Dominici, F.; Santulli, C.; Kenny, J.M.; Torre, L. Investigation of Thermo-Mechanical, Chemical and Degradative Properties of Pla-Limonene Films Reinforced with Cellulose Nanocrystals Extracted from Phormium Tenax leaves. *Eur. Polym. J.* **2014**, *56*, 77–91. [CrossRef]
56. Armentano, I.; Fortunati, E.; Burgos, N.; Dominici, F.; Luzi, F.; Fiori, S.; Jiménez, A.; Yoon, K.; Ahn, J.; Kang, S.; et al. Bio-Based Pla_PHB Plasticized Blend Films: Processing and Structural Characterization. *LWT Food Sci. Technol.* **2015**, *64*, 980–988. [CrossRef]
57. Rojas-Lema, S.; Quiles-Carrillo, L.; Garcia-Garcia, D.; Melendez-Rodriguez, B.; Balart, R.; Torres-Giner, S. Tailoring the Properties of Thermo-Compressed Polylactide Films for Food Packaging Applications by Individual and Combined Additions of Lactic Acid Oligomer and Halloysite Nanotubes. *Molecules* **2020**, *25*, 1976. [CrossRef] [PubMed]
58. De Silva, R.T.; Pasbakhsh, P.; Goh, K.L.; Chai, S.-P.; Chen, J. Synthesis and Characterisation of Poly (Lactic Acid)/Halloysite Bionanocomposite Films. *J. Compos. Mater.* **2014**, *48*, 3705–3717. [CrossRef]
59. Gomez-Caturla, J.; Montanes, N.; Quiles-Carrillo, L.; Balart, R.; Garcia-Garcia, D.; Dominici, F.; Puglia, D.; Torre, L. Development of Biodegradable Pla Composites And Tangerine Peel Flour with Improved Toughness Containing a Natural-Based Terpenoid. *Express Polym. Lett.* **2023**, *17*, 789–805. [CrossRef]
60. Armentano, I.; Fortunati, E.; Burgos, N.; Dominici, F.; Luzi, F.; Fiori, S.; Jimenez, A.; Yoon, K.; Ahn, J.; Kang, S.; et al. Processing and Characterization of plasticized Pla/PHB Blends for Biodegradable Multiphase Systems. *Express Polym. Lett.* **2015**, *9*, 583–596. [CrossRef]
61. Ahmed, J.; Hiremath, N.; Jacob, H. Antimicrobial, Rheological, and Thermal Properties of Plasticized Polylactide Films Incorporated with Essential Oils to Inhibit *Staphylococcus aureus* and *Campylobacter jejuni*. *J. Food Sci.* **2016**, *81*, E419–E429. [CrossRef]
62. Pirani, S.I.; Krishnamachari, P.; Hashaikeh, R. Optimum Loading Level of Nanoclay in Pla Nanocomposites. *J. Thermoplast. Compos. Mater.* **2013**, *27*, 1461–1478. [CrossRef]
63. Othman, S.H.; Ling, H.N.; Talib, R.A.; Naim, M.N.; Risyon, N.P.; Saifullah, M. Pla/MMT and Pla/Halloysite Bio-Nanocomposite Films: Mechanical, Barrier, and Transparency. *J. Nano Res.* **2019**, *59*, 77–93. [CrossRef]
64. Silverajah, V.S.; Ibrahim, N.A.; Zainuddin, N.; Yunus, W.M.; Hassan, H.A. Mechanical, Thermal and Morphological Properties of Poly(Lactic Acid)/Epoxidized Palm Olein Blend. *Molecules* **2012**, *17*, 11729–11747. [CrossRef] [PubMed]
65. Xiao, L.; Zhao, M.; Hu, H. Study on Graphene Oxide Modified Inorganic Phase Change Materials and Their Packaging Behavior. *J. Wuhan Univ. Technol.-Mater. Sci. Ed.* **2018**, *33*, 788–792. [CrossRef]
66. Chen, Y.; Geever, L.M.; Killion, J.A.; Lyons, J.G.; Higginbotham, C.L.; Devine, D.M. Halloysite Nanotube Reinforced Polylactic Acid Composite. *Polym. Compos.* **2017**, *38*, 2166–2173. [CrossRef]
67. Carrasco, F.; Gámez-Pérez, J.; Santana, O.O.; Maspoch, M.L. Processing of poly(Lactic Acid)/Organomontmorillonite Nanocomposites: Microstructure, Thermal Stability And Kinetics of the Thermal Decomposition. *Chem. Eng. J.* **2011**, *178*, 451–460. [CrossRef]
68. Wang, W.; Wu, N.; Zu, Y.; Fu, Y. Antioxidative Activity of *Rosmarinus officinalis* L. Essential Oil Compared to Its Main Components. *Food Chem.* **2008**, *108*, 1019–1022. [CrossRef]
69. Zeid, A.; Karabagias, I.K.; Nassif, M.; Kontominas, M.G. Preparation and Evaluation of Antioxidant Packaging Films Made of Polylactic Acid Containing Thyme, Rosemary, and Oregano Essential Oils. *J. Food Process. Preserv.* **2019**, *43*, e14102. [CrossRef]
70. Darie-Niţă, R.N.; Vasile, C.; Stoleru, E.; Pamfil, D.; Zaharescu, T.; Tarţău, L.; Tudorachi, N.; Brebu, M.A.; Pricope, G.M.; Dumitriu, R.P.; et al. Evaluation of the Rosemary Extract Effect on the Properties of Polylactic Acid-Based Materials. *Materials* **2018**, *11*, 1825. [CrossRef] [PubMed]
71. Arrieta, M.P.; López, J.; Hernández, A.; Rayón, E. Ternary Pla–PHB–Limonene Blends Intended for Biodegradable Food Packaging Applications. *Eur. Polym. J.* **2014**, *50*, 255–270. [CrossRef]
72. Alakrach, A.M.; Noriman, N.Z.; Dahham, O.S.; Hamzah, R.; Alsaadi, M.A.; Shayfull, Z.; Syed Idrus, S.Z. Chemical and Hydrophobic Properties of Pla/HNTs-ZrO_2Bionanocomposites. *J. Phys. Conf. Ser.* **2018**, *1019*, 012065. [CrossRef]
73. Luzi, F.; Fortunati, E.; Jiménez, A.; Puglia, D.; Pezzolla, D.; Gigliotti, G.; Kenny, J.M.; Chiralt, A.; Torre, L. Production and Characterization of Pla_PBS Biodegradable Blends Reinforced with Cellulose Nanocrystals Extracted From Hemp Fibres. *Ind. Crops Prod.* **2016**, *93*, 276–289. [CrossRef]

polymers

Review

A Review of Natural Fiber-Reinforced Composites for Lower-Limb Prosthetic Designs

Angel D. Castro-Franco [1], Miriam Siqueiros-Hernández [1,*], Virginia García-Angel [1], Ismael Mendoza-Muñoz [1], Lidia E. Vargas-Osuna [1] and Hernán D. Magaña-Almaguer [2]

1 Facultad de Ingeniería, Universidad Autónoma de Baja California, Mexicali 21280, Baja California, Mexico; dcastro88@uabc.edu.mx (A.D.C.-F.); virginia.garcia@uabc.edu.mx (V.G.-A.); ismael.mendoza@uabc.edu.mx (I.M.-M.); lidia.vargas@uabc.edu.mx (L.E.V.-O.)

2 Tecnológico Nacional de México, IT de Mexicali, Mexicali 21376, Baja California, Mexico; maganahernan@itmexicali.edu.mx

* Correspondence: miriam.siqueiros@uabc.edu.mx; Tel.: +52-(686)-123-61-94

Abstract: This paper presents a comprehensive review of natural fiber-reinforced composites (NFRCs) for lower-limb prosthetic designs. It covers the characteristics, types, and properties of natural fiber-reinforced composites as well as their advantages and drawbacks in prosthetic designs. This review also discusses successful prosthetic designs that incorporate NFRCs and the factors that make them effective. Additionally, this study explores the use of computational biomechanical models to evaluate the effectiveness of prosthetic devices and the key factors that are considered. Overall, this document provides a valuable resource for anyone interested in using NFRCs for lower-limb prosthetic designs.

Keywords: natural fiber-reinforced composites; lower-limb prosthesis; mechanical properties; finite-element analysis; biomechanics; polymer matrix composites; ISO standards; fatigue strength; impact resistance; computational modeling

Citation: Castro-Franco, A.D.; Siqueiros-Hernández, M.; García-Angel, V.; Mendoza-Muñoz, I.; Vargas-Osuna, L.E.; Magaña-Almaguer, H.D. A Review of Natural Fiber-Reinforced Composites for Lower-Limb Prosthetic Designs. *Polymers* **2024**, *16*, 1293. https://doi.org/10.3390/polym16091293

Academic Editors: Miguel Ángel López Manchado, Laura Peponi and Luigi Torre

Received: 4 April 2024
Revised: 23 April 2024
Accepted: 25 April 2024
Published: 5 May 2024

1. Introduction

Lower-limb amputation is a complex and devastating issue affecting millions of people worldwide [1–6]. Prosthetic devices are the most practical choice for restoring ambulatory motor function in individuals with lower-limb amputations [7–9]. However, in developing countries, most prosthetic users are compelled to use rudimentary and inefficient designs because of their low acquisition costs [10,11]. This lack of affordable and efficient prosthetic devices perpetuates the marginalization of individuals with lower-limb amputations, subsequently affecting their musculoskeletal system and mobility [12–21].

In developed countries, the sustainability of materials and manufacturing components used in prosthetic devices is usually not a primary concern in the prosthetic design process [22]. Consequently, the most functional prostheses are those with the most advanced components and naturally higher costs [7,8,10,11,23–28]. Therefore, proposing alternative, sustainable, and cost-effective materials for lower-limb prosthetic designs could provide affordable prosthetic solutions for the most vulnerable populations.

Recent advances in natural fiber-reinforced composites (NFRCs) have demonstrated good mechanical and ecological properties compared with synthetic fibers (such as carbon and glass fibers), while also being economically accessible and abundant [29–34]. To mention a few, fibers such as kenaf [8,35–37], jute [38], ramie [39], and hemp [40] have highlighted their mechanical performance and versatility, comparing favorably against synthetic fibers in terms of their mechanical properties and common applications.

The aim of this review is to methodically evaluate NFRCs (natural fiber-reinforced composites) as a potential substitute material for use in lower-limb prosthetics. This study establishes the groundwork for future research and development in this sector by offering a thorough examination of the characteristics, manufacturing methods, advantages, and drawbacks of NFRCs, along with evaluating the pertinent international standards and available computational biomechanical modeling techniques to validate the performance and usability of devices. This work supports the development and production of prosthetic devices that are affordable, eco-friendly, and high-performing, improving the lives of those who have lost limbs, especially in situations where resources are scarce.

2. Background

2.1. Lower-Limb Amputation and Prosthetic Options

Lower-limb amputations can be caused by trauma, illnesses, or congenital problems, having a serious negative impact on both physical and emotional health [41–48]. Prosthetic devices have been created to solve this problem, and can be customized to match the unique demands of each user [7–11]. Finding a suitable fit for the user and assuring the durability of the device remain challenges, despite advances in prosthetic technology [18–21].

One major challenge associated with prosthetic devices is their high acquisition costs, ranging from several thousands to tens of thousands of dollars, depending on the type and complexity of the device. This cost can limit access to prosthetic devices for many individuals, particularly those with inadequate health insurance coverage [10,11,49–52]. Programs such as Medicare (https://www.medicare.gov, accessed on 12 May 2023) and Medicaid (https://medicaid.gov, accessed on 12 May 2023) in the United States may offer some relief; however, the costs associated with prosthetic devices may not be fully covered. In addition, restrictions on the types of devices covered or the frequency with which they can be replaced may limit access. Nevertheless, it must be highlighted that these government medical programs may vary in their availability and quality from one country to another, or, in some cases, may not exist, especially in developing countries [53].

In low- and middle-income countries, the high acquisition costs of prosthetic devices present a significant barrier to their accessibility [10,49,53,54]. Although efforts have been made to develop low-cost prosthetic devices using alternative materials and manufacturing processes, these devices may not offer the same level of functionality or durability as more expensive options [8,55,56].

2.2. Transtibial Prosthetic Design Considerations

Transtibial prostheses are lower-limb prosthetic devices that replace the missing or amputated foot and ankle. The design of transtibial prostheses involves several key components, including the socket, pylon or shank, and the foot or ankle [7,8,27,28,57].

The socket component (Figure 1a) is crucial for the prosthesis to fit the residual limb securely and comfortably. The socket must be made to offer a stable attachment point for the rest of the prosthesis and fit snugly over the residual limb [8,11,27,28]. A good fit is essential to ensure that the prosthesis performs as intended and does not irritate or disturb the user.

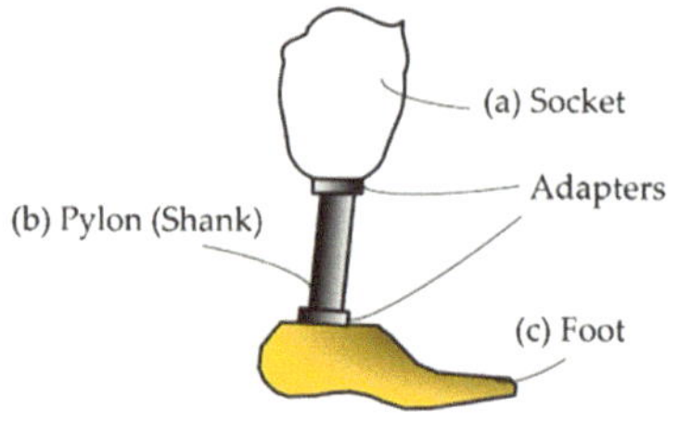

Figure 1. Main components of a passive transtibial prosthesis: (a) the socket, the part of the prosthesis that fits over the residual limb (the remaining part of the amputated leg); (b) the shank (or pylon), a rigid component that extends from the socket down to the foot; (c) the foot, the component of the prosthesis that mimics the function of a natural foot, absorbs shock during walking, and provides stability. Reprinted from ref. [58].

Connecting the socket to the foot or ankle component is performed using the pylon or shank component (Figure 1b) of the prosthesis. For weight-bearing activities, such as walking or running [11], this component must provide sufficient support and stability [7]. The materials utilized for this component must be sufficiently sturdy to withstand continuous loading cycles without failure [10].

The foot component (Figure 1c) of the prosthesis is designed to replicate the function of the biological foot as much as possible [59–64]. This might involve incorporating features such as shock absorption, flexibility, and articulation into the design [7]. The materials used for this element must also be carefully selected to ensure that they are sufficiently durable to withstand repeated use over time [50]. Table 1 describes some details of the commercial prosthetic feet currently available on the market, as well as the different materials used to manufacture them.

Table 1. Commercial prosthetic feet from different manufacturers.

Brand	Model	Material	Reference (Access Date)	Country
Blatchford	Elan	Carbon Fiber	www.blatchfordmobility.com (17 April 2024)	UK
Blatchford	Epirus	Carbon Fiber	www.blatchfordmobility.com (17 April 2024)	UK
Blatchford	Stellar	Nylon	www.blatchfordmobility.com (17 April 2024)	UK
Össur	Balance Foot S	Glass Fiber	www.ossur.com (17 April 2024)	US
Össur	Vari-Flex	Carbon Fiber	www.ossur.com (17 April 2024)	US
Ottobock	Kintrol	Glass Fiber	www.ottobock.com (17 April 2024)	DE
Ottobock	Restore	Glass Fiber	www.ottobock.com (17 April 2024)	DE

Hydraulic systems, advanced sensors, and control systems are just a few more elements that transtibial prostheses may contemplate if necessary, increasing the cost for the user [11]. These features can provide the prosthesis with more control and allow a more fluid gait pattern.

Transtibial prostheses must be carefully designed considering characteristics such as the proper fit, support, functionality, and material selection [11]. In recent years, advances in both materials science and engineering have achieved more sophisticated prosthetic devices, but to advance the field of prosthetics and improve outcomes for those with amputated lower limbs, continued work in this area is crucial.

2.3. Materials Used in Prosthetic Manufacturing

For many years, conventional materials, such as metals and plastics, have mainly been used in the manufacture of prosthetics [7,11,65]. Although they are strong and long-lasting, metals, such as aluminum and titanium, can also be heavy and unwieldy [10,26,66]. In contrast, while plastics such as polypropylene and polyethylene are lightweight, they cannot be as strong or long-lasting as metals and usually deteriorate with time [7,10,65,67,68].

In contrast to conventional materials, synthetic composites offer several benefits. They are created by fusing a matrix substance, such as epoxy resin, with synthetic fibers such as carbon or glass fibers [29,69–73]. Excellent strength-to-weight ratios and customizability for particular design requirements are notable features of synthetic composites [70,74–76]. They also exhibit good fatigue characteristics and are corrosion-resistant [77–80]. Although they can be expensive [71,81–87], synthetic composites may not be biodegradable [8,27,28].

NFRCs have many advantages over synthetic and conventional materials [30–32]. Natural fibers, such as kenaf, flax, or jute, can be used to create these types of composites [29,73,88–90]. Natural fiber-reinforced composites have exceptional mechanical qualities, including a low density, biodegradability, and high strength-to-weight and impact–resistance ratios [29,33,34,90]. Additionally, compared with synthetic composites, they are less expensive to manufacture [33].

Ensuring the consistent quality and performance of NFRCs is one of the key issues regarding these materials. Natural fibers can have different qualities depending on factors such as their moisture content and the harvesting conditions [89,91–93]. Nevertheless, recent studies have shown that high-quality natural fiber-reinforced composites may be fabricated to match the specifications for prosthetic designs with fiber treatment and a good composite composition [30,31,79,94,95].

Each material has its own unique advantages and disadvantages in prosthetic manufacturing. Conventional materials are durable but may be heavy or uncomfortable for users, and they offer excellent strength-to-weight ratios but may not be biodegradable and can be expensive, whereas natural fiber-reinforced composites offer a lightweight and biodegradable option with excellent mechanical properties but require careful quality control during production.

3. Natural Fiber-Reinforced Composites

3.1. Characteristics of Natural Fibers

The utilization of natural fibers as reinforcement materials in composite materials has garnered significant attention owing to their distinctive properties [73]. Derived from diverse sources (Figure 2), such as plants, animals, and minerals [69,90,96], natural fibers have become increasingly popular in recent years owing to their cost-effectiveness, biodegradability, and renewability [30–32].

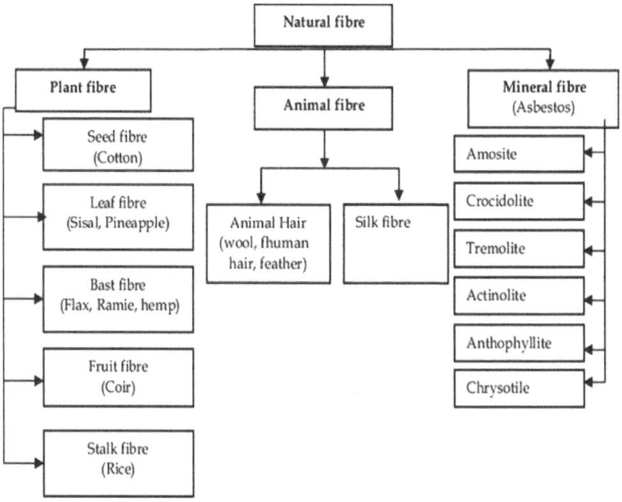

Figure 2. Classification of natural fibers based on their origin: plant, animal, or mineral [29]. Reprinted from Materials Today: Proceedings, Vol. 50, Sandeep Kumar, Alakesh Manna, Rakesh Dang, A review on applications of natural fiber-reinforced composites (NFRCs), Pages 1632–1636, Copyright (2022), with permission from Elsevier.

The remarkable characteristic of natural fibers lies in their exceptional strength-to-weight ratio, particularly in jute, flax, and hemp fibers [38,79,90,97]. Owing to their robust tensile strength, natural fibers are highly suitable for integration into composite materials, particularly in applications where weight reduction is paramount, such as the development of lower-limb prosthetics. Furthermore, the low density of natural fibers presents an additional advantage, as they significantly contribute to reducing the overall weight in weight-sensitive applications [29,33,34]. The mechanical and physical properties of the most popular natural fibers are shown in Table 2.

Table 2. Physical and mechanical properties of the most popular natural fibers [88,98–100].

Fiber	Diameter (μm)	Density (g/cm^3)	Tensile Strength (MPa)	Young's Modulus (GPa)	Elongation at Break (%)	Moisture Absorption (%)
Abaca	10–30	1.5	430–813	31.10–33.60	2.9–10	~
Bagasse	~	1.20	20–290	19–27	1.10	~
Bamboo	88–125	0.91–1.26	503	35.91	1.40	~
Banana	100–250	1.35	529–914	27–32	2.60–5.90	~
Basalt	17	2.8	4800	90	3.15	~
Coconut	150–250	1.15–1.25	131–220	4–6	15–40	10
Cotton	~	1.50–1.51	287–597	5.50–12.60	0.30–10	8–25
Flax	25	1.40–1.50	345–1500	27.60–80	1.20–3.20	7
Hemp	25–600	1.48–1.50	550–900	70	1.60–4	8
Henequen	~	1.20–1.40	430–570	10–16.30	3.70–5.90	~
Jute	25–250	1.30–1.48	393–800	0.13–27.60	1.16–1.80	12
Kenaf	~	1.25–1.40	284–930	0.13–26.50	1.16–1.80	~
Pineapple	50	1.44	413–1627	60–80	14.50	~
Ramie	20–280	1.30–1.50	400–938	61.40–128	3.60–3.80	12–17
Rice husk	~	0.50–0.70	~	~	~	~
Sisal	50–200	1.30–1.50	390–635	9.40–41	2–2.50	11
Softwood	~	1.50	1000	40	~	~
Viscose cord	~	~	593	11	11.4	~

Another noteworthy attribute of natural fibers is their biodegradability and renewability, rendering them an environmentally friendly alternative to synthetic fibers. By incorporating natural fibers into composite materials, the environmental impact of such materials can be reduced, thereby promoting sustainability [8,30–32].

Moisture absorption is a significant consideration in the utilization of natural fibers. Untreated natural fibers tend to exhibit higher moisture absorption than treated fibers [95]. This moisture absorption can cause dimensional changes and affect the mechanical properties of composites [33]. Techniques such as alkali treatment and the incorporation of hydrophobic additives or coatings can effectively mitigate moisture absorption [30,31,95].

Natural fibers possess distinctive characteristics that make them highly suitable for integration into composite materials, making them attractive substitutes for synthetic fibers. Incorporating natural fibers into composite materials not only offers the potential to reduce the environmental impact associated with these materials but also provides a sustainable solution for the design of lower-limb prosthetics.

3.2. *Types and Properties of Natural Fiber-Reinforced Composites*

Natural fiber-reinforced composites typically consist of natural fibers embedded within a polymer matrix [96]. The selection of the polymer matrix significantly influences the mechanical properties and overall performance of the composite [29,101–106]. The matrix materials commonly used in NFRCs include polypropylene (PP) [40], polyester (PET) [72,107,108], and epoxy [109–113], each offering distinct advantages in terms of their mechanical strength, durability, and compatibility with natural fibers [72,114,115].

While thermoplastic and thermosetting matrices are mostly used in NFRCs, bio-based epoxy resins derived from renewable sources such as vegetable oils offer a sustainable alternative for the matrix phase [116–119]. These bio-resins can potentially improve characteristics like the biodegradability and environmental impact of NFRCs when combined with natural-fiber reinforcements [118]. However, challenges related to the fiber–matrix compatibility, moisture sensitivity, and achieving the optimal mechanical performance still present difficulties that hinder their use in heavy-duty and long-lasting applications, such as in prosthetic devices.

The mechanical properties of NFRCs play a critical role in determining their suitability for lower-limb prosthetic designs. Parameters such as their tensile strength, flexural strength, impact resistance, and fatigue behavior are of utmost importance and must be carefully considered (Table 1).

Tensile strength represents the maximum stress that a material can withstand under tension before failure. NFRCs exhibit promising tensile strength owing to the reinforcing effect of their natural fibers. Several factors, including the fiber type, fiber content, fiber orientation, and fiber–matrix adhesion, influence the tensile properties of NFRCs [33,69,93,120,121]. Table 3 exhibits a summary of several NFRCs proposed and studied by different authors in detail.

Table 3. Tensile and flexural strengths of some NFRCs [68].

Authors	Matrix	Types of Fiber	Method of Fabrication
Aslan et al. [122]	Polypropylene matrix	Carbon/sisal, glass/sisal	Single-screw co-rotating extrusion method
Assarar et al. [123]	Epoxy matrix	Flax–carbon fiber	Platen press process
Campbell et al. [124]	Plant oil resin	Ramie/stockinet	Standard layup method
Chaudhary et al. [125]	Epoxy matrix	Flax hemp/jute/fiber	Hand-layup method
Essabir et al. [126]	Polypropylene matrix	Coir fiber	Twin-screw extrusion method
Gu et al. [127]	Epoxy matrix	Ramie fiber	Vacuum infusion process
Indra Reddy et al. [128]	Epoxy matrix	Pineapple, glass, and jute fiber	Hand-layup method
Jagannatha et al. [129]	Epoxy matrix	Glass/carbon	Vacuum bagging technique
Lee et al. [130]	Polypropylene matrix	Kenaf/jute	Hot-pressing method
Rahman et al. [131]	Vinyl-ester matrix	PALF	Hand-layup method
Saba et al. [132]	Epoxy matrix	Kenaf fiber	Hand-layup technique
Sekaran et al. [133]	Epoxy matrix	Sisal fiber and aloe vera	Hand-layup method
Shanmugam et al. [134]	Polyester matrix	Jute fiber palmyra and leaf stalk fiber	Compression molding
Shih et al. [135]	Poly-lactic acid	PALF/chopsticks	Counter-rotating internal mixing
Shrivastava et al. [136]	Epoxy resin	Coir–glass	Hand-layup method
Sreekumar et al. [137]	Polyester matrix	Sisal fiber	Resin transfer molding technique
Widhata et al. [138]	Methyl methacrylate	Water hyacinth	Compression molding
Yan et al. [139]	Epoxy matrix	Flax/linen/bamboo	Vacuum bagging process
Yang et al. [140]	Polypropylene matrix	Hemp fiber	Twin-screw extrusion

Flexural strength characterizes a material's ability to resist deformation when subjected to bending. NFRCs exhibit considerable flexural strength, which renders them suitable for load-bearing applications. The flexural properties of NFRCs are influenced by factors such as the fiber content, fiber length, fiber orientation, and matrix properties [33,74,93,141,142]. Attaining proper fiber–matrix interactions and optimal fiber dispersion within the matrix are crucial for achieving enhanced flexural strength.

The impact resistance of NFRCs is of utmost importance in prosthetic design because it determines the material's capability to absorb and dissipate energy during dynamic loading [7,24,25]. As well as the flexural and tensile strength, the impact properties of NFRCs are influenced by the fiber selection, chemical treatment, and matrix properties [30,31,93]. Comprehensive assessments and optimization of impact resistance are essential to ensure the reliability and functionality of lower-limb prosthetics.

3.3. Manufacturing Methods for Natural Fiber-Reinforced Composites

The manufacturing process for natural fiber-reinforced composites (NFRCs) is similar to that of conventional composites, differing primarily in the utilization of natural fibers instead of synthetic fibers. Various manufacturing methods exist for producing NFRCs, each with its own set of advantages and disadvantages [29,33,88,121]. The selection of a specific method depends on its intended application and production requirements [120].

The hand-layup approach (Figure 3) is the most widely used manufacturing method. It entails manually inserting fibers into a mold, followed by applying resin to help the fibers join. This approach is well suited for small-scale production because it is simple and economical [121,143,144].

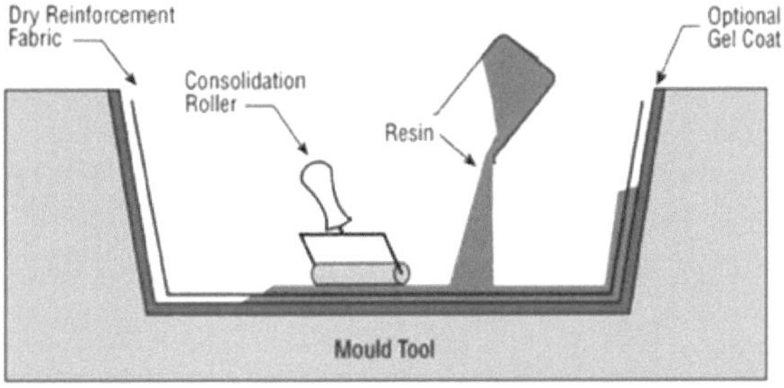

Figure 3. Hand-layup method for manufacturing composite materials [145]. Reprinted from Comprehensive Composite Materials, Author(s): D. Cripps, T.J. Searle, J. Summerscales, Title of chapter: Comprehensive Composite Materials, Pages 737–761, Copyright (2000), with permission from Elsevier.

Compression molding is another process used in the production of NFRCs. Using this technique, the fibers are placed in a mold and compressed under pressure to release trapped air. The mold is then filled with resin, which is then heated to start the resin's curing process. Compression molding permits the creation of intricately formed components, and is appropriate for high-volume production [33,89,121,146].

The resin transfer molding (RTM) method (Figure 4) is employed to produce composites with high strength and stiffness. In this approach, the fibers are positioned within a mold and resin is injected under pressure. The mold is then heated to promote the curing of the resin. RTM has widespread applications in the automotive and aerospace industries [71,147–150].

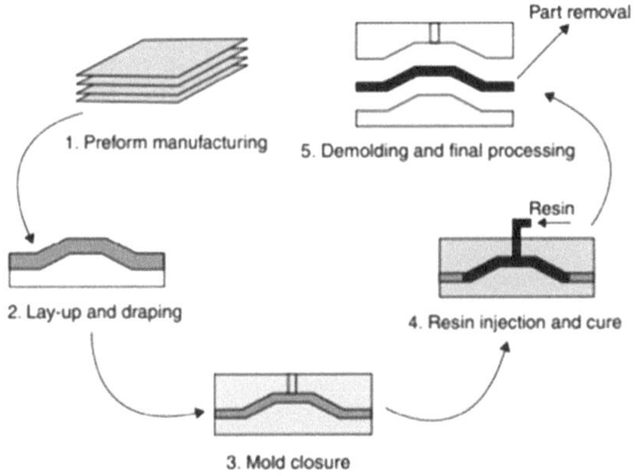

Figure 4. Resin transfer molding (RTM) method for manufacturing composite materials [151]. Reprinted from Design for Sustainability, Author(s): Nadlene Razali, Muhd Ridzuan Mansor, Ghazali Omar, Syed Ahmad Faiz Syed Kamarulzaman, Mohd Hanafee Zin, Nadia Razali, Title of chapter: Design for Sustainability, Pages 395–413, Copyright (2021), with permission from Elsevier.

Pultrusion is a continuous manufacturing method that is extensively used for the production of fiber-reinforced composites. In this technique, the fibers are pulled through a resin bath and subsequently passed through a heated die, resulting in resin curing. Pultrusion is well suited for creating components with a consistent cross-section and is commonly employed in the construction sector [89,120,121,146,147].

Finally, the filament winding method is employed to manufacture cylindrical NFRC parts, including pipes and tanks. This process involves winding fibers around a mandrel in a specific pattern, followed by the application of a resin to facilitate fiber bonding. After curing, the mandrel is removed, leaving the final product [71,120,146,147,152].

4. Lower-Limb Prosthetic Design Using Natural Fiber-Reinforced Composites

4.1. Advantages and Drawbacks of Natural Composites in Prosthetic Design

NFRCs have garnered attention in the field of lower-limb prosthetic design owing to their unique properties and potential benefits. However, like other existing materials, they have their own advantages and drawbacks, which must be carefully considered and addressed in the design and manufacturing processes to fully exploit their potential in prosthetic applications.

Advantages:

1. Low Weight and High Strength-to-Weight Ratio: As mentioned before, NFRCs have the remarkable ability to have both a minimal weight and a high strength-to-weight ratio [29,33,34,55]. This quality is very helpful for designing lower-limb prosthetics because it makes it possible to create lightweight, comfortable prosthetics without sacrificing strength and longevity [7–9,11,66].
2. Energy Return and Shock Absorption: The mechanical properties of natural composites, including their ability to store and release energy, contribute to enhanced energy return and shock absorption [30,31,55,56]. This feature is crucial in lower-limb prosthetics, as it mimics the natural gait cycle and improves the overall walking efficiency [7,10,24–26,58].
3. Sustainability and Environmental Friendliness: Natural fibers used as composite reinforcements, such as kenaf and flax, are renewable resources that provide a sustainable alternative to synthetic fibers [8,29–33,35–37,98,119]. The utilization of natural composites in prosthetic design aligns with the increasing demand for eco-friendly materials and reduces the reliance on non-renewable resources.

Drawbacks:

1. Moisture Absorption: Natural fibers have a propensity to absorb moisture [89,93], which can lead to dimensional changes and diminished mechanical properties of the composites [33]. This drawback presents a challenge for prosthetic designs, as exposure to moisture can affect the long-term performance and durability of prosthetic devices [50]. Detailed data on different natural fibers are displayed in Table 1.
2. Limited Durability: Compared to synthetic-fiber-reinforced composites, natural composites may exhibit lower durability and resistance to wear and tear [29,34,153]. The natural fibers used in reinforcement may degrade over time [94], affecting the overall lifespan of prosthetic devices.
3. Variability in Mechanical Properties: Natural fibers, which are organic materials, inherently exhibit variability in their mechanical properties [89,93]. This variability can pose challenges in achieving consistent and predictable performance with natural fiber-reinforced composites in prosthetic designs. This necessitates careful selection and quality control of the natural fibers to ensure consistent mechanical properties and high performance of the prosthetic devices.

4.2. Cases of Natural Fiber-Reinforced Composites Used in Prosthetic Designs

Nurhanisah et al. [8] proposed the use of a kenaf-fiber-reinforced composite material for the fabrication of transtibial prosthetic sockets (Figure 5). Their proposed design exhibited favorable results in terms of mechanical properties and comfort, making it suitable for providing good strength and added aesthetic value. Additionally, the proposed design demonstrated environmental friendliness compared to fiberglass-based sockets.

In a study by Irawan et al. [55], the manufacturing of lower-limb prosthetic sockets using ramie fibers and epoxy composites was suggested. Their results showed that such sockets had a significant impact on comfort because of their light weight, strength, and flexibility compared to fiberglass sockets. Furthermore, sockets fabricated with the ramie fibers exhibited a considerably lower weight than those made with fiberglass, with a difference of 186 g, representing a 46.26% reduction. As mentioned several times before, the strength-to-weight ratio exhibited by natural fibers is one of their key advantages over synthetic fibers [33,34], which is essential for maintaining the prosthetic weight within an appropriate range, and thus avoiding increased energy expenditure and excessive stress on the user's residual limb [11,66].

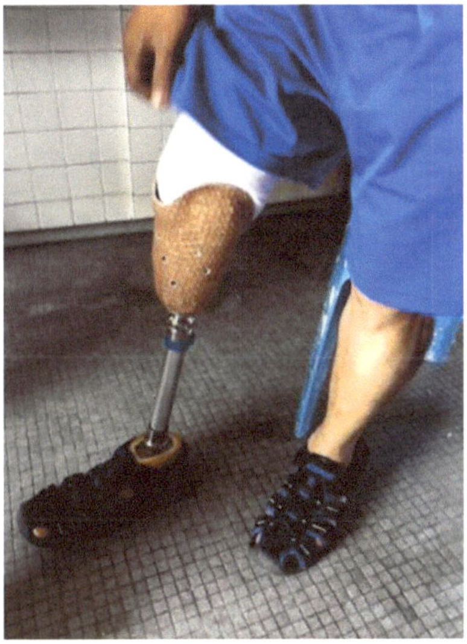

Figure 5. Prosthetic socket made from a kenaf-woven-fabric composite designed by Nurhanisah et al. Reprinted from ref. [8].

Moreover, a study by Mankai et al. [56] showed that esparto fibers (*Stipa tenacissima*) are a promising alternative material for manufacturing prosthetic sockets. Their fatigue testing results revealed the viscoelastic behavior of the material and estimated its lifespan to be 2,325,000 cycles, satisfying 77.5% of the ISO 10328 objective.

These studies provide evidence for the potential of NFRCs in lower-limb prosthetic designs. The utilization of these composites can result in lightweight, strong, and biocompatible prostheses. Furthermore, their incorporation into prosthetic designs offers environmental benefits, as natural fibers are renewable resources with a lower carbon footprint than synthetic fibers. This makes them a more sustainable option for prosthetic designs. In addition to their mechanical properties and sustainability advantages, natural fiber-reinforced composites can also provide aesthetic benefits by providing prosthetics with a more natural and organic appearance, thus mitigating the stigma associated with prosthetic use.

5. Evaluation of Natural Fiber-Reinforced Composite Prosthetics

5.1. Standards and Guidelines for Evaluating Prosthetic Devices

The evaluation of prosthetic devices is an important step in the design process because it ensures that the functional, safety, and performance requirements are met. Adhering to accepted norms and criteria is essential for prostheses constructed of NFRCs to guarantee their safe and efficient use.

One widely recognized standard for evaluating prosthetic devices is ISO 10328, which provides comprehensive guidelines for the mechanical testing of lower-limb prostheses [154,155]. This standard defines the testing procedures for various mechanical properties, including static strength, fatigue strength, and impact resistance. Furthermore, ISO 22675 offers guidelines specifically for testing ankle/foot prostheses, encompassing durability, strength, and range-of-motion requirements [156]. Both ISO 10328 and 22675 establish frameworks for evaluating prosthetic devices, ensuring the adherence to safety and performance requirements. Widely recognized and employed by manufacturers, clinicians, and researchers in the field of prosthetics, these standards guarantee the safety, efficacy, and reliability of prosthetic devices for individuals with lower-limb amputations.

ISO 10328 outlines procedures for both the static and dynamic testing of lower-limb prostheses, encompassing mechanical properties such as static strength, fatigue strength, and impact resistance (Figure 6). Static strength testing involves subjecting the prosthetic device to increasing loads until failure, whereas fatigue strength testing simulates cyclic loading to mimic normal usage stresses. Impact resistance testing involves dropping a weight onto the device to simulate the impact of a fall.

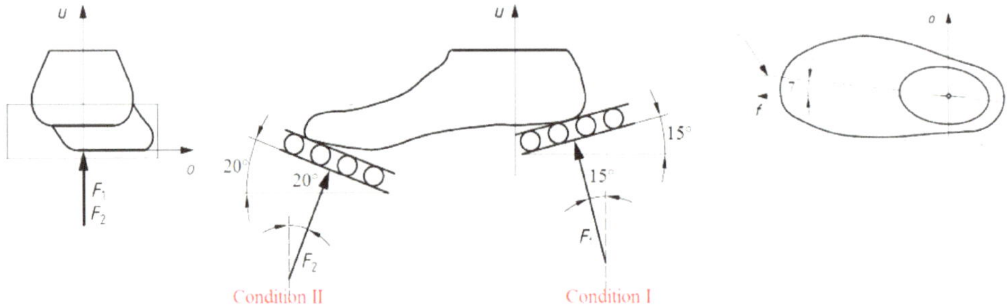

Figure 6. Alignment and angles of test loads for a prosthetic foot, established by ISO 10328 [157]. Reprinted from Procedia Engineering, Vol. 10, C. Colombo, E.G. Marchesin, L. Vergani, E. Boccafogli, G. Verni, Study of an ankle prosthesis for children: adaptation of ISO 10328 and experimental tests, Pages 3510–3517, Copyright (2011), with permission from Elsevier.

ISO 22675 provides guidelines for testing ankle–foot prostheses, including the requirements for durability, strength, and range of motion (Figure 7). This standard presents testing procedures for both cyclic and static loading, aiming to replicate the load conditions experienced during the stance phase of an individual's gait. Cyclic loading tests involve applying loads to the device cyclically to simulate typical usage stresses, whereas static loading tests apply loads at specific points in the gait cycle to simulate the maximum loads at those instances.

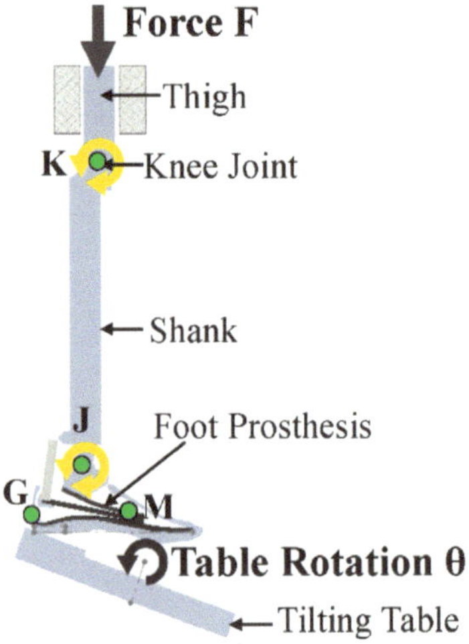

Figure 7. Alignment and angles of test loads for a prosthetic foot established in ISO 22675. Reprinted from ref. [158].

Numerous other studies have also employed the ISO 10328 [10,56,154,159,160] and 22675 [160–162] standards to assess the mechanical properties of prosthetic devices. These standards offer a valuable framework for evaluating the safety and performance of such devices, enjoying broad recognition and adoption by researchers, clinicians, and manufacturers in the prosthetics field. Additionally, organizations such as the American Orthotic and Prosthetic Association (AOPA) (https://www.aopanet.org/, accessed on 1 August 2023) [160,162] and KS P 8403 [163,164] have established guidelines for prosthetic device design, fabrication, fitting, clinical evaluation, and follow-up care.

Although these criteria and recommendations offer a framework for assessing prosthetic devices, it is crucial to keep in mind that not all device types or materials necessarily fall under their purview. Additional testing and assessments may be required for NFRC prostheses to guarantee their compliance with appropriate safety and performance standards.

5.2. Using Computational Biomechanical Models to Assess Prosthetic Devices

Computational biomechanical models have become a prevalent tool for assessing the performance of prosthetic devices, enabling simulations of the interaction between the device and the residual limb, as well as the forces and stresses during walking. These models offer the opportunity to optimize prosthetic designs and evaluate device performance under diverse conditions [165–167] (Figure 8).

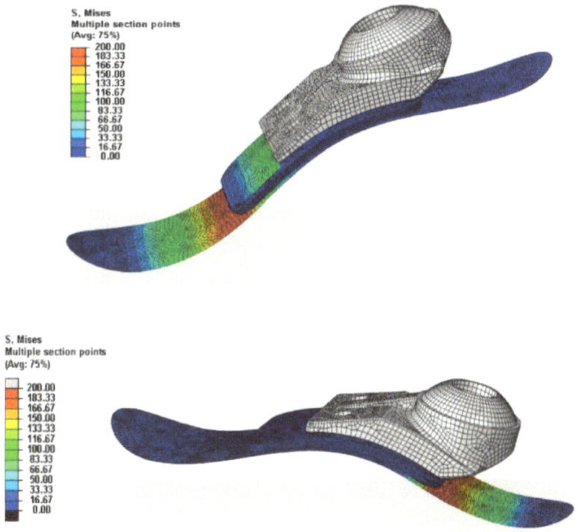

Figure 8. Illustrative finite-element analysis of a prosthetic foot designed and studied by Song et al. [164] (Note: The data shown in this figure is for illustrative purposes only and is not relevant to the present discussion). Reproduced with permission from Youngnam Song et al., International Journal of Precision Engineering and Manufacturing; published by Springer Nature, 2019.

In the context of NFRC prosthetics, computational biomechanical models allow the assessment of the impact of various design parameters on device performance (Figure 9). Parameters such as the fiber orientation, number of layers, and composite laminate thickness can be evaluated using these models [168,169]. Additionally, these models facilitate assessments of the effects of different loading conditions on device performance, such as walking on different terrains or at varying speeds [165–167].

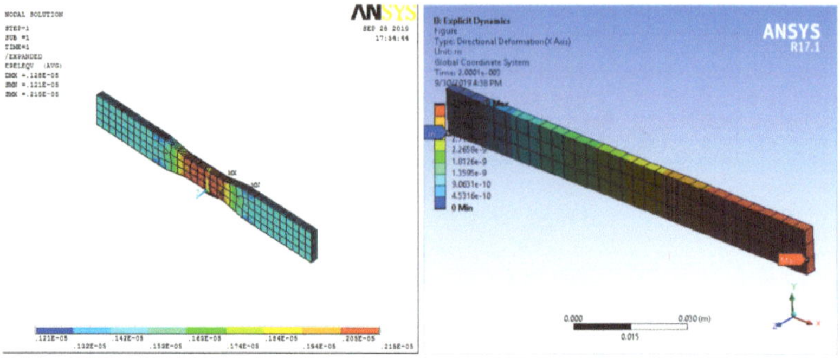

Figure 9. Illustrative tensile and flexural finite-element analysis results for an NFRC studied by Balasubramanian et al. [170] (Note: The data shown in this figure is for illustrative purposes only and is not relevant to the present discussion). Reprinted from Materials Today: Proceedings, Vol. 28, K. Balasubramanian, N. Rajeswari, K. Vaidheeswaran, Analysis of mechanical properties of natural fiber composites by experimental with FEA, Pages 1149–1153, Copyright (2020), with permission from Elsevier.

An inherent challenge in using computational biomechanical models for prosthetic designs is the requirement of accurate input data. This includes information about the geometry and material properties of both the prosthetic device and the residual limb. However, advances in imaging technology and material characterization techniques have led to increasingly accurate data acquisition [166,171,172].

The application of computational biomechanical models is a potent tool for designing and evaluating NFRC prosthetics. By utilizing these models, the design of a device can be optimized, and its performance can be evaluated under diverse conditions, ultimately leading to improved prosthetic devices for individuals with lower-limb amputations [67,167,173].

Finite-element analysis (FEA) is a widely used computational biomechanical tool for the design and evaluation of prosthetic devices. FEA is a numerical method that can simulate complex mechanical systems, including prosthetic devices and the human body, by dividing them into simpler elements. The mathematical equations applied to these elements allow FEA to predict the system behavior under different loading conditions [166,174,175]. FEA enables the assessment of mechanical performance in the context of prosthetic designs and evaluation under different specific loading scenarios, including walking, running, and leaping. Additionally, this computational biomechanical tool makes it easier to optimize designs of devices by analyzing the effects of many design factors, such as the material qualities, shape, and thickness.

An advantage of FEA is its ability to assess the stress and strain distribution within the prosthetic device and residual limb. This information helps in identifying areas of high stress or strain that could lead to failure or discomfort, prompting design optimization to mitigate these concerns. FEA has been extensively used in the design and evaluation of NFRCs [120,143,144,162,168–170]. For instance, it has been employed to evaluate the impact of different fiber orientations and laminate configurations on the mechanical performance of devices.

Even though computational biomechanics possesses multiple advantages, such as evaluating prosthetic device performance under different loading conditions, there are limitations. The accuracy of the results depends on the accuracy of the input data and assumptions made in the model, which can introduce errors and uncertainties. Acquiring accurate input data, such as the geometry and material properties, may be experimentally challenging. Moreover, the complexity of the models used in computational biomechanics can make interpreting and understanding the underlying mechanisms difficult. Computational biomechanics can also be computationally expensive and time-consuming, which may restrict their use in certain applications. Finally, it is essential to recognize that computational biomechanics cannot fully replace experimental testing, and that the results of computational models should be validated experimentally.

6. Future Directions and Conclusions

The integration of NFRCs in lower-limb prosthetic designs has demonstrated significant promise in recent times. These materials have several advantages over conventional prosthetic materials, including enhanced strength, durability, and biocompatibility. Nevertheless, there remains considerable scope for further exploration to fully unlock the potential of these materials.

A compelling avenue for future research is the development of novel NFRC materials with improved properties. This may entail exploring new natural fibers, such as bamboo or hemp, or devising innovative processing techniques to enhance the mechanical properties of existing fibers. Furthermore, it is imperative to evaluate the performance of NFRC prostheses through both computational modeling and experimental testing. This comprehensive assessment is pivotal to ensure the safety and efficacy of these materials across a diverse range of applications.

In conclusion, the integration of NFRCs in lower-limb prosthetic designs holds promise for transformative advancements in the field. These materials exhibit numerous advantages over conventional prosthetic materials, including increased strength, durability, and biocompatibility. Furthermore, they have the potential to offer enhanced cost-effectiveness and eco-friendliness compared to traditional alternatives.

However, to achieve widespread adoption in prosthetic designs, several challenges must be resolved. The foremost task involves the development of new materials with improved properties. Additionally, it is imperative to develop innovative prosthetic devices that effectively capitalize on the distinctive attributes of these materials. Concurrently, ongoing assessments of their safety and efficacy through the combined use of computational models and experimental testing remain crucial.

Notwithstanding these challenges, the potential benefits of NFRCs engender a compelling domain for future research. Through sustained exploration and refinement, these materials could significantly enhance the quality of life of millions of individuals worldwide who rely on prosthetic devices to preserve their mobility and independence.

Author Contributions: Conceptualization, A.D.C.-F. and M.S.-H.; methodology, A.D.C.-F. and M.S.-H.; investigation: A.D.C.-F.; writing—original draft, A.D.C.-F., M.S.-H. and V.G.-A.; writing—review and editing, A.D.C.-F., M.S.-H., V.G.-A., I.M.-M., L.E.V.-O. and H.D.M.-A. All authors have read and agreed to the published version of the manuscript.

Funding: This research received no external funding.

Acknowledgments: The authors would like to thank the National Council for Science and Technology of Mexico (CONACYT) for the Ph.D. Scholarship support given to Angel Daniel Castro Franco (CVU No. 983649). The authors are also greatly grateful to the Universidad Autónoma de Baja, California for facilitating the access and use of their facilities and equipment to carry out this research.

Conflicts of Interest: The authors declare no conflicts of interest.

References

1. Ahmad, N.; Thomas, G.N.; Gill, P.; Chan, C.; Torella, F. Lower limb amputation in England: Prevalence, regional variation and relationship with revascularisation, deprivation and risk factors. A retrospective review of hospital data. *J. R. Soc. Med.* **2014**, *107*, 483–489. [CrossRef] [PubMed]
2. Imam, B.; Miller, W.C.; Finlayson, H.C.; Eng, J.J.; Jarus, T. Incidence of lower limb amputation in Canada. *Can. J. Public Health* **2017**, *108*, 374–380. [CrossRef] [PubMed]
3. Behrendt, C.A.; Sigvant, B.; Szeberin, Z.; Beiles, B.; Eldrup, N.; Thomson, I.; Venermo, M.; Altreuther, M.; Menyhei, G.; Nordanstig, J.; et al. International Variations in Amputation Practice: A VASCUNET Report. *Eur. J. Vasc. Endovasc. Surg.* **2018**, *56*, 391–399. [CrossRef] [PubMed]
4. Moini, M.; Rasouli, M.R.; Khaji, A.; Farshidfar; Heidari, P. Patterns of extremity traumas leading to amputation in Iran: Results of Iranian National Trauma Project. *Chin. J. Traumatol. = Zhonghua Chuang Shang Za Zhi* **2009**, *12*, 77–80. Available online: http://www.ncbi.nlm.nih.gov/pubmed/19321050 (accessed on 13 January 2022). [PubMed]

5. Rouhani, A.; Mohajerzadeh, S. An epidemiological and etiological report on lower extremity amputation in northwest of Iran. *Arch. Bone Jt. Surg.* **2013**, *1*, 103–106. Available online: http://www.ncbi.nlm.nih.gov/pubmed/25207299 (accessed on 13 January 2022). [PubMed]
6. Shaw, J.; Challa, S.; Conway, D.; Liu, M.; Haonga, B.; Eliezer, E.; Morshed, S.; Shearer, D. Quality of life and complications in lower limb amputees in Tanzania: Results from a pilot study. *Lancet Glob. Health* **2018**, *6*, S18. [CrossRef]
7. Gard, S. Prosthetic Devices and Methods. In *Wiley Encyclopedia of Biomedical Engineering*; John Wiley & Sons, Inc.: Hoboken, NJ, USA, 2006; Volume 1960, no. 4; pp. 1–21. [CrossRef]
8. Nurhanisah, M.H.; Hashemi, F.; Paridah, M.T.; Jawaid, M.; Naveen, J. Mechanical properties of laminated kenaf woven fabric composites for below-knee prosthesis socket application. *IOP Conf. Ser. Mater. Sci. Eng.* **2018**, *368*, 012050. [CrossRef]
9. Strait, E.; McGimpsey, G.; Bradford, T. Limb Prosthetics Services and Devices. White Paper; January 2006; pp. 1–35. Available online: https://www.nist.gov/system/files/documents/2017/04/28/239_limb_prosthetics_services_devices.pdf (accessed on 13 January 2022).
10. Lee, W.C.C.; Zhang, M.; Chan, P.P.Y.; Boone, D.A. Gait analysis of low-cost flexible-shank transtibial prostheses. *IEEE Trans. Neural Syst. Rehabil. Eng.* **2006**, *14*, 370–377. [CrossRef] [PubMed]
11. DeWees, T. Transtibial prosthetics. In *Orthotics and Prosthetics in Rehabilitation*; Elsevier: Amsterdam, The Netherlands, 2020; pp. 605–634. [CrossRef]
12. Boulton, A.J.; Vileikyte, L.; Ragnarson-Tennvall, G.; Apelqvist, J. The global burden of diabetic foot disease. *Lancet* **2005**, *366*, 1719–1724. [CrossRef] [PubMed]
13. Vileikyte, L. Diabetic foot ulcers: A quality of life issue. *Diabetes. Metab. Res. Rev.* **2001**, *17*, 246–249. [CrossRef] [PubMed]
14. Renström, P.; Grimby, G.; Morelli, B.; Palmertz, B. Thigh muscle atrophy in below-knee amputees. *Scand. J. Rehabil. Med. Suppl.* **1983**, *9*, 150–162. [PubMed]
15. Jaegers, S.M.; Arendzen, J.H.; de Jongh, H.J. Changes in hip muscles after above-knee amputation. *Clin. Orthop. Relat. Res.* **1995**, *319*, 276–284. Available online: http://www.ncbi.nlm.nih.gov/pubmed/7554640 (accessed on 3 February 2022). [CrossRef]
16. Burger, H.; Valenčič, V.; Marinček, Č.; Kogovšek, N. Properties of musculus gluteus maximus in above-knee amputees. *Clin. Biomech.* **1996**, *11*, 35–38. [CrossRef]
17. Schmalz, T.; Blumentritt, S.; Reimers, C.D. Selective thigh muscle atrophy in trans-tibial amputees: An ultrasonographic study. *Arch. Orthop. Trauma Surg.* **2001**, *121*, 307–312. [CrossRef] [PubMed]
18. Henson, D.P.; Edgar, C.; Ding, Z.; Sivapuratharasu, B.; Le Feuvre, P.; Finnegan, M.E.; Quest, R.; McGregor, A.H.; Bull, A.M. Understanding lower limb muscle volume adaptations to amputation. *J. Biomech.* **2021**, *125*, 110599. [CrossRef] [PubMed]
19. Geurts, A.C.H.; Mulder, T.W. Reorganisation of Postural Control Following Lower Limb Amputation: Theoretical Considerations and Implications for Rehabilitation. *Physiother. Theory Pract.* **1992**, *8*, 145–157. [CrossRef]
20. Silverman, A.K.; Fey, N.P.; Portillo, A.; Walden, J.G.; Bosker, G.; Neptune, R.R. Compensatory mechanisms in below-knee amputee gait in response to increasing steady-state walking speeds. *Gait Posture* **2008**, *28*, 602–609. [CrossRef] [PubMed]
21. Clites, T.R.; Carty, M.J.; Ullauri, J.B.; Carney, M.E.; Mooney, L.M.; Duval, J.F.; Srinivasan, S.S.; Herr, H.M. Proprioception from a neurally controlled lower-extremity prosthesis. *Sci. Transl. Med.* **2018**, *10*, eaap8373. [CrossRef] [PubMed]
22. Cummings, D. Prosthetics in the developing world. *Prosthet. Orthot. Int.* **1996**, *20*, 51–60. [CrossRef]
23. McDonald, C.L.; Kramer, P.A.; Morgan, S.J.; Halsne, E.G.; Cheever, S.M.; Hafner, B.J. Energy expenditure in people with transtibial amputation walking with crossover and energy storing prosthetic feet: A randomized within-subject study. *Gait Posture* **2018**, *62*, 349–354. [CrossRef] [PubMed]
24. Macfarlane, P.A.; Nielsen, D.H.; Shurr, D. Mechanical gait analysis of transfemoral amputees: SACH foot versus the Flex-Foot. *J. Prosthet. Orthot.* **1997**, *9*, 144–155. [CrossRef]
25. Powers, C.M.; Torburn, L.; Perry, J.; Ayyappa, E. Influence of prosthetic foot design on sound limb loading in adults with unilateral below-knee amputations. *Arch. Phys. Med. Rehabil.* **1994**, *75*, 825–829. Available online: http://www.ncbi.nlm.nih.gov/pubmed/8024435 (accessed on 7 January 2022). [CrossRef] [PubMed]
26. Klodd, E.; Hansen, A.; Fatone, S.; Edwards, M. Effects of prosthetic foot forefoot flexibility on gait of unilateral transtibial prosthesis users. *J. Rehabil. Res. Dev.* **2010**, *47*, 899–910. [CrossRef] [PubMed]
27. Joshi, S.; Drzal, L.; Mohanty, A.; Arora, S. Are natural fiber composites environmentally superior to glass fiber reinforced composites? *Compos. Part A Appl. Sci. Manuf.* **2004**, *35*, 371–376. [CrossRef]
28. Etcheverry, M.; Barbosa, S.E. Glass Fiber Reinforced Polypropylene Mechanical Properties Enhancement by Adhesion Improvement. *Materials* **2012**, *5*, 1084–1113. [CrossRef] [PubMed]
29. Kumar, S.; Manna, A.; Dang, R. A review on applications of natural Fiber-Reinforced composites (NFRCs). *Mater. Today Proc.* **2022**, *50*, 1632–1636. [CrossRef]
30. Dixit, S.; Goel, R.; Dubey, A.; Shivhare, P.R.; Bhalavi, T. Natural Fibre Reinforced Polymer Composite Materials—A Review. *Polym. Renew. Resour.* **2017**, *8*, 71–78. [CrossRef]
31. ArunKumar, D.; Kaushik, V.; Raghavedra, R.P. Tensile and Impact properties of jute/glass and jute/carbon fiber reinforced polypropylene. *J. Polym. Compos.* **2016**, *4*, 35–39.
32. Mohammed, A.A.; Bachtiar, D.; Siregar, J.P.; Rejab, M.R.M.; Hasany, S.F. Physicochemical Study of Eco-Friendly Sugar Palm Fiber Thermoplastic Polyurethane Composites. *BioResources* **2016**, *11*, 9438–9454. [CrossRef]

33. Zhao, X.; Copenhaver, K.; Wang, L.; Korey, M.; Gardner, D.J.; Li, K.; Lamm, M.E.; Kishore, V.; Bhagia, S.; Tajvidi, M.; et al. Recycling of natural fiber composites: Challenges and opportunities. *Resour. Conserv. Recycl.* **2022**, *177*, 105962. [CrossRef]
34. Liu, C.; Luan, P.; Li, Q.; Cheng, Z.; Sun, X.; Cao, D.; Zhu, H. Biodegradable, Hygienic, and Compostable Tableware from Hybrid Sugarcane and Bamboo Fibers as Plastic Alternative. *Matter* **2020**, *3*, 2066–2079. [CrossRef]
35. Saba, N.; Paridah, M.T.; Jawaid, M. Mechanical properties of kenaf fibre reinforced polymer composite: A review. *Constr. Build. Mater.* **2015**, *76*, 87–96. [CrossRef]
36. Chen, Y.; Sun, L.; Chiparus, O.; Negulescu, I.; Yachmenev, V.; Warnock, M. Kenaf/Ramie Composite for Automotive Headliner. *J. Polym. Environ.* **2005**, *13*, 107–114. [CrossRef]
37. Yuhazri, M.Y.; Phongsakorn, P.T.; Sihombing, H.; Jeefferie, A.R.; Perumal, P.; Kamarul, A.M.; Rassiah, K. Mechanical properties of kenaf/polyester composites. *Int. J. Eng. Technol.* **2011**, *11*, 127–131.
38. Ku, H.; Wang, H.; Pattarachaiyakoop, N.; Trada, M. A review on the tensile properties of natural fiber reinforced polymer composites. *Compos. Part B Eng.* **2011**, *42*, 856–873. [CrossRef]
39. Batu, T.; Lemu, H.; Sirhabizuh, B. Study of the Performance of Natural Fiber Reinforced Composites for Wind Turbine Blade Applications. *Adv. Sci. Technol. Res. J.* **2020**, *14*, 67–75. [CrossRef] [PubMed]
40. Sullins, T.; Pillay, S.; Komus, A.; Ning, H. Hemp fiber reinforced polypropylene composites: The effects of material treatments. *Compos. Part B Eng.* **2017**, *114*, 15–22. [CrossRef]
41. Preece, R.A.; Dilaver, N.; Waldron, C.A.; Pallmann, P.; Thomas-Jones, E.; Gwilym, B.L.; Norvell, D.C.; Czerniecki, J.M.; Twine, C.P.; Bosanquet, D.C. A Systematic Review and Narrative Synthesis of Risk Prediction Tools Used to Estimate Mortality, Morbidity, and Other Outcomes Following Major Lower Limb Amputation. *Eur. J. Vasc. Endovasc. Surg.* **2021**, *62*, 127–135. [CrossRef] [PubMed]
42. Columbo, J.A.; Davies, L.; Kang, R.; Barnes, J.A.; Leinweber, K.A.; Suckow, B.D.; Goodney, P.P.; Stone, D.H. Patient Experience of Recovery After Major Leg Amputation for Arterial Disease. *Vasc. Endovasc. Surg.* **2018**, *52*, 262–268. [CrossRef] [PubMed]
43. Franklin, H.; Rajan, M.; Tseng, C.-L.; Pogach, L.; Sinha, A. Cost of lower-limb amputation in U.S. veterans with diabetes using health services data in fiscal years 2004 and 2010. *J. Rehabil. Res. Dev.* **2014**, *51*, 1325–1330. [CrossRef] [PubMed]
44. McDonald, C.L.; Westcott-McCoy, S.; Weaver, M.R.; Haagsma, J.; Kartin, D. Global prevalence of traumatic non-fatal limb amputation. *Prosthet. Orthot. Int.* **2020**, *45*, 105–114. [CrossRef] [PubMed]
45. Dillingham, T.R.; Pezzin, L.E.; MacKenzie, E.J. Limb Amputation and Limb Deficiency: Epidemiology and Recent Trends in the United States. *South. Med. J.* **2002**, *95*, 875–883. [CrossRef] [PubMed]
46. Calle-Pascual, A.L.; Redondo, M.J.; Ballesteros, M.; Martinez-Salinas, M.A.; Diaz, J.A.; De Matias, P.; Calle, J.R.; Gil, E.; Jimenez, M.; Serrano, F.J.; et al. Nontraumatic lower extremity amputations in diabetic and non-diabetic subjects in Madrid, Spain. *Diabetes Metab.* **1997**, *23*, 519–523. Available online: http://www.ncbi.nlm.nih.gov/pubmed/9496558 (accessed on 15 February 2022). [PubMed]
47. Ebskov, B.; Ebskov, L. Major lower limb amputation in diabetic patients: Development during 1982 to 1993. *Diabetologia* **1996**, *39*, 1607–1610. [CrossRef] [PubMed]
48. Tentolouris, N.; Al-Sabbagh, S.; Walker, M.G.; Boulton, A.J.M.; Jude, E.B. Mortality in Diabetic and Nondiabetic Patients After Amputations Performed From 1990 to 1995. *Diabetes Care* **2004**, *27*, 1598–1604. [CrossRef] [PubMed]
49. Vázquez, E.; Wiecher, E. *Los Amputados y su Rehabilitación*; Academia Nacional de Medicina: Mexico City, Mexico, 2016.
50. Carroll, K.; Rheinstein, J.; Pollard, E. *Understanding and Selecting Prosthetic Feet*, 4th ed.; Elsevier Inc.: Amsterdam, The Netherlands, 2019. [CrossRef]
51. Lawall Prosthetics & Orthotics. How Much Does a Prosthetic Leg Cost? The Cost of Prosthetic Legs. Available online: https://www.lawall.com/blog/how-much-does-a-prosthetic-leg-cost (accessed on 30 January 2023).
52. Luxmed Protez. Prosthetic Leg Cost. Available online: https://luxmedprotez.com/en/prosthetic-leg-prices/ (accessed on 30 January 2023).
53. Lavery, L.A.; van Houtum, W.H.; Armstrong, D.G.; Harkless, L.B.; Ashry, H.R.; Walker, S.C. Mortality following lower extremity amputation in minorities with diabetes mellitus. *Diabetes Res. Clin. Pract.* **1997**, *37*, 41–47. [CrossRef] [PubMed]
54. Instituto Nacional de Estadística y Geografía. *La Discapacidad en México, Datos al 2014. Versión 2017*; INEGI Aguascalientes: Aguascalientes, Mexico, 2017.
55. Irawan, A.P.; Soemardi, T.P.; Widjajalaksmi, K.; Reksoprodjo, A.H.S. Gait analysis of the prosthesis prototype made from the natural fiber reinforced composite. In Proceedings of the Apchi-Ergofuture 2010, Bali, Indonesia, 2–6 August 2010; pp. 37–43.
56. Mankai, W.; Brahim, S.B.; Smida, B.B.; Cheikh, R.B.; Chafra, M. Mechanical behavior of a lower limb prosthetic socket made of natural fiber reinforced composite. *J. Eng. Res.* **2021**, *9*, 269–277. [CrossRef]
57. Blatchford. Below Knee Prosthesis: Standard Features of a Below Knee Prosthesis. Available online: www.blatchford.co.uk/prosthetics/ (accessed on 3 February 2022).
58. Abbod, E.A.; Resan, K.K. Review on the Interface Pressure Measurement for Below Knee Prosthetic Socket. *IOP Conf. Ser. Mater. Sci. Eng.* **2021**, *1094*, 012036. [CrossRef]
59. Attwells, R.L.; Birrell, S.A.; Hooper, R.H.; Mansfield, N.J. Influence of carrying heavy loads on soldiers' posture, movements and gait. *Ergonomics* **2006**, *49*, 1527–1537. [CrossRef]
60. Birrell, S.A.; Haslam, R.A. The effect of military load carriage on 3-D lower limb kinematics and spatiotemporal parameters. *Ergonomics* **2009**, *52*, 1298–1304. [CrossRef]
61. Hansen, A.H.; Childress, D.S. Effects of adding weight to the torso on roll-over characteristics of walking. *J. Rehabil. Res. Dev.* **2005**, *42*, 381. [CrossRef] [PubMed]

62. Doyle, S.S.; Lemaire, E.D.; Besemann, M.; Dudek, N.L. Changes to level ground transtibial amputee gait with a weighted backpack. *Clin. Biomech.* **2014**, *29*, 149–154. [CrossRef] [PubMed]
63. Schnall, B.L.; Hendershot, B.D.; Bell, J.C.; Wolf, E.J. Kinematic analysis of males with transtibial amputation carrying military loads. *J. Rehabil. Res. Dev.* **2014**, *51*, 1505–1514. [CrossRef] [PubMed]
64. Schnall, B.L.; Dearth, C.L.; Elrod, J.M.; Golyski, P.R.; Koehler-McNicholas, S.R.; Ray, S.F.; Hansen, A.H.; Hendershot, B.D. A more compliant prosthetic foot better accommodates added load while walking among Servicemembers with transtibial limb loss. *J. Biomech.* **2020**, *98*, 109395. [CrossRef] [PubMed]
65. Tang, P.C.; Ravji, K.; Key, J.J.; Mahler, D.B.; Blume, P.A.; Sumpio, B. Let Them Walk! Current Prosthesis Options for Leg and Foot Amputees. *J. Am. Coll. Surg.* **2008**, *206*, 548–560. [CrossRef] [PubMed]
66. Gailey, R.S.; Nash, M.S.; Atchley, T.A.; Zilmer, R.M.; Moline-Little, G.R.; Morris-Cresswell, N.; Siebert, L.I. The effects of prosthesis mass on metabolic cost of ambulation in non-vascular trans-tibial amputees. *Prosthet. Orthot. Int.* **1997**, *21*, 9–16. [CrossRef] [PubMed]
67. Balaramakrishnan, T.M.; Natarajan, S.; Srinivasan, S. Roll-over shape of a prosthetic foot: A finite element evaluation and experimental validation. *Med. Biol. Eng. Comput.* **2020**, *58*, 2259–2270. [CrossRef] [PubMed]
68. Malalli, C.S.; Ramji, B. Mechanical characterization of natural fiber reinforced polymer composites and their application in Prosthesis: A review. *Mater. Today Proc.* **2022**, *62*, 3435–3443. [CrossRef]
69. Stokke, D.D.; Wu, Q.; Han, G. Wood and Natural Fiber Composites: An Overview. In *Introduction to Wood and Natural Fiber Composites*; Wiley: Hoboken, NJ, USA, 2013; pp. 1–17. [CrossRef]
70. Callister, W.J. *Fundamentals of Materials Science and Engineering: An Interactive eText*, 5th ed.; John Wiley & Sons, Inc.: New York, NY, USA, 2001.
71. Chung, D. Composite Materials. In *Kirk-Othmer Encyclopedia of Chemical Technology*; John Wiley & Sons, Inc.: Hoboken, NJ, USA, 2015; Volume 82, no. 35; pp. 1–39. [CrossRef]
72. Hsissou, R.; Seghiri, R.; Benzekri, Z.; Hilali, M.; Rafik, M.; Elharfi, A. Polymer composite materials: A comprehensive review. *Compos. Struct.* **2021**, *262*, 113640. [CrossRef]
73. Tan, Q.; Wu, C.; Li, L.; Shao, W.; Luo, M. Nanomaterial-Based Prosthetic Limbs for Disability Mobility Assistance: A Review of Recent Advances. *J. Nanomater.* **2022**, *2022*, 3425297. [CrossRef]
74. Yang, Y.; Boom, R.; Irion, B.; van Heerden, D.-J.; Kuiper, P.; de Wit, H. Recycling of composite materials. *Chem. Eng. Process. Process. Intensif.* **2012**, *51*, 53–68. [CrossRef]
75. Otani, L.B.; Pereira, A.H.A.; Melo, J.D.D.; Amico, S.C. Elastic Moduli characterization of composites using the Impulse Excitation Technique. *Mater. Sci. Eng. Phys.* **2014**, *36*. [CrossRef]
76. Ibrahim, I.D.; Jamiru, T.; E Sadiku, R.; Kupolati, W.K.; Agwuncha, S.C.; Ekundayo, G. The use of polypropylene in bamboo fibre composites and their mechanical properties—A review. *J. Reinf. Plast. Compos.* **2015**, *34*, 1347–1356. [CrossRef]
77. Hale, D.K. The physical properties of composite materials. *J. Mater. Sci.* **1976**, *11*, 2105–2141. [CrossRef]
78. Jones, R.M. *Mechanics of Composite Materials*; CRC Press: Boca Raton, FL, USA, 2018.
79. Yan, D.-X.; Ren, P.-G.; Pang, H.; Fu, Q.; Yang, M.-B.; Li, Z.-M. Efficient electromagnetic interference shielding of lightweight graphene/polystyrene composite. *J. Mater. Chem.* **2012**, *22*, 18772–18774. [CrossRef]
80. Sahmaran, M.; Li, V.C.; Andrade, C. Corrosion Resistance Performance of Steel-Reinforced Engineered Cementitious Composite Beams. *ACI Mater. J.* **2008**, *105*, 243. [CrossRef] [PubMed]
81. Budinski, K. *Engineering Materials: Properties and Selection*, 2nd ed.; Reston Publishing Company Inc.: Reston, VA, USA, 1983.
82. Morton, J.; Ho, H.; Tsai, M.Y.; Farley, G.L. An Evaluation of The Iosipescu Specimen for Composite Materials Shear Property Measurement. *Compos. Mater.* **1992**, *26*, 708–750. [CrossRef]
83. De Araújo, M. Natural and man-made fibres: Physical and mechanical properties. In *Fibrous and Composite Materials for Civil Engineering Applications*; Elsevier: Amsterdam, The Netherlands, 2011; pp. 3–28. [CrossRef]
84. Husić, S.; Javni, I.; Petrović, Z.S. Thermal and mechanical properties of glass reinforced soy-based polyurethane composites. *Compos. Sci. Technol.* **2005**, *65*, 19–25. [CrossRef]
85. Singh, T.J.; Samanta, S. Characterization of Kevlar Fiber and Its Composites: A Review. *Mater. Today Proc.* **2015**, *2*, 1381–1387. [CrossRef]
86. Rezaei, F.; Yunus, R.; Ibrahim, N.A. Effect of fiber length on thermomechanical properties of short carbon fiber reinforced polypropylene composites. *Mater. Des.* **2009**, *30*, 260–263. [CrossRef]
87. Davis, D.C.; Wilkerson, J.W.; Zhu, J.; Ayewah, D.O. Improvements in mechanical properties of a carbon fiber epoxy composite using nanotube science and technology. *Compos. Struct.* **2010**, *92*, 2653–2662. [CrossRef]
88. Kozlowski, R.; Wladyka-Przybylak, M. Uses of Natural Fiber Reinforced Plastics. In *Natural Fibers, Plastics and Composites*; Springer: Boston, MA, USA, 2004; pp. 249–274. [CrossRef]
89. Stokke, D.D.; Wu, Q.; Han, G. Natural Fiber and Plastic Composites. In *Introduction to Wood and Natural Fiber Composites*; Wiley: Hoboken, NJ, USA, 2013; pp. 237–285. [CrossRef]
90. Thapliyal, D.; Verma, S.; Sen, P.; Kumar, R.; Thakur, A.; Tiwari, A.K.; Singh, D.; Verros, G.D.; Arya, R.K. Natural Fibers Composites: Origin, Importance, Consumption Pattern, and Challenges. *J. Compos. Sci.* **2023**, *7*, 506. [CrossRef]
91. Tajvidi, M.; Takemura, A. Recycled Natural Fiber Polypropylene Composites: Water Absorption/Desorption Kinetics and Dimensional Stability. *J. Polym. Environ.* **2010**, *18*, 500–509. [CrossRef]

92. Beg, M.; Pickering, K. Reprocessing of wood fibre reinforced polypropylene composites. Part II: Hygrothermal ageing and its effects. *Compos. Part A Appl. Sci. Manuf.* **2008**, *39*, 1565–1571. [CrossRef]
93. Gogna, E.; Kumar, R.; Anurag; Sahoo, A.K.; Panda, A. *A Comprehensive Review on Jute Fiber Reinforced Composites*; Springer: Berlin/Heidelberg, Germany, 2019; pp. 459–467. [CrossRef]
94. Wei, L.; McDonald, A.G.; Freitag, C.; Morrell, J.J. Effects of wood fiber esterification on properties, weatherability and biodurability of wood plastic composites. *Polym. Degrad. Stab.* **2013**, *98*, 1348–1361. [CrossRef]
95. Kumar, A.P.; Singh, R.P.; Sarwade, B.D. Degradability of composites, prepared from ethylene–propylene copolymer and jute fiber under accelerated aging and biotic environments. *Mater. Chem. Phys.* **2005**, *92*, 458–469. [CrossRef]
96. Bongarde, U.S.; Shinde, V.D. Review on natural fiber reinforcement polymer composites. *Int. J. Eng. Sci. Innov. Technol.* **2014**, *3*, 431–436.
97. Chethan, N.; Nagesh, S.; Babu, L.S. Mechanical behaviour of Kenaf-Jute-E-glass reinforced hybrid polymer composites. *Mater. Today Proc.* **2020**, *46*, 4454–4459. [CrossRef]
98. Lopresto, V.; Leone, C.; De Iorio, I. Mechanical characterisation of basalt fibre reinforced plastic. *Compos. Part B Eng.* **2011**, *42*, 717–723. [CrossRef]
99. Bledzki, A.K.; Gassan, J. Composites reinforced with cellulose based fibres. *Prog. Polym. Sci.* **1999**, *24*, 221–274. [CrossRef]
100. Vigneshwaran, S.; Sundarakannan, R.; John, K.; Johnson, R.D.J.; Prasath, K.A.; Ajith, S.; Arumugaprabu, V.; Uthayakumar, M. Recent advancement in the natural fiber polymer composites: A comprehensive review. *J. Clean. Prod.* **2020**, *277*, 124109. [CrossRef]
101. Arabpour, A.; Shockravi, A.; Rezania, H.; Farahati, R. Investigation of anticorrosive properties of novel silane-functionalized polyamide/GO nanocomposite as steel coatings. *Surf. Interfaces* **2020**, *18*, 100453. [CrossRef]
102. Zheng, S.; Bellido-Aguilar, D.A.; Hu, J.; Huang, Y.; Zhao, X.; Wang, Z.; Zeng, X.; Zhang, Q.; Chen, Z. Waterborne bio-based epoxy coatings for the corrosion protection of metallic substrates. *Prog. Org. Coat.* **2019**, *136*, 105265. [CrossRef]
103. Ray, B. Temperature effect during humid ageing on interfaces of glass and carbon fibers reinforced epoxy composites. *J. Colloid Interface Sci.* **2006**, *298*, 111–117. [CrossRef] [PubMed]
104. Xu, Y.; Chung, D.; Mroz, C. Thermally conducting aluminum nitride polymer-matrix composites. *Compos. Part A Appl. Sci. Manuf.* **2001**, *32*, 1749–1757. [CrossRef]
105. Davim, J.; Reis, P. Study of delamination in drilling carbon fiber reinforced plastics (CFRP) using design experiments. *Compos. Struct.* **2003**, *59*, 481–487. [CrossRef]
106. Mukherjee, M.; Das, C.; Kharitonov, A. Fluorinated and oxyfluorinated short Kevlar fiber-reinforced ethylene propylene polymer. *Polym. Compos.* **2006**, *27*, 205–212. [CrossRef]
107. Armstrong, D.P.; Chatterjee, K.; Ghosh, T.K.; Spontak, R.J. Form-stable phase-change elastomer gels derived from thermoplastic elastomer copolyesters swollen with fatty acids. *Thermochim. Acta* **2020**, *686*, 178566. [CrossRef]
108. Senthilkumar, K.; Saba, N.; Chandrasekar, M.; Jawaid, M.; Rajini, N.; Alothman, O.Y.; Siengchin, S. Evaluation of mechanical and free vibration properties of the pineapple leaf fibre reinforced polyester composites. *Constr. Build. Mater.* **2018**, *195*, 423–431. [CrossRef]
109. Hsissou, R.; Abbout, S.; Seghiri, R.; Rehioui, M.; Berisha, A.; Erramli, H.; Assouag, M.; Elharfi, A. Evaluation of corrosion inhibition performance of phosphorus polymer for carbon steel in [1 M] HCl: Computational studies (DFT, MC and MD simulations). *J. Mater. Res. Technol.* **2020**, *9*, 2691–2703. [CrossRef]
110. Datsyuk, V.; Trotsenko, S.; Trakakis, G.; Boden, A.; Vyzas-Asimakopoulos, K.; Parthenios, J.; Galiotis, C.; Reich, S.; Papagelis, K. Thermal properties enhancement of epoxy resins by incorporating polybenzimidazole nanofibers filled with graphene and carbon nanotubes as reinforcing material. *Polym. Test.* **2019**, *82*, 106317. [CrossRef]
111. Hsissou, R.; Elharfi, A. Rheological behavior of three polymers and their hybrid composites (TGEEBA/MDA/PN), (HGEMDA/MDA/PN) and (NGHPBAE/MDA/PN). *J. King Saud Univ. Sci.* **2020**, *32*, 235–244. [CrossRef]
112. Parida, S.P.; Jena, P.C. Preparation of epoxy-glass composites with graphene and flyash filler. *Mater. Today Proc.* **2020**, *26*, 2328–2332. [CrossRef]
113. Jin, X.; Guo, N.; You, Z.; Wang, L.; Wen, Y.; Tan, Y. Rheological properties and micro-characteristics of polyurethane composite modified asphalt. *Constr. Build. Mater.* **2019**, *234*, 117395. [CrossRef]
114. Saba, N.; Jawaid, M. A review on thermomechanical properties of polymers and fibers reinforced polymer composites. *J. Ind. Eng. Chem.* **2018**, *67*, 1–11. [CrossRef]
115. Dang, Z.-M.; Yuan, J.-K.; Zha, J.-W.; Zhou, T.; Li, S.-T.; Hu, G.-H. Fundamentals, processes and applications of high-permittivity polymer–matrix composites. *Prog. Mater. Sci.* **2012**, *57*, 660–723. [CrossRef]
116. Pokharel, A.; Falua, K.J.; Babaei-Ghazvini, A.; Acharya, B. Biobased Polymer Composites: A Review. *J. Compos. Sci.* **2022**, *6*, 255. [CrossRef]
117. Ortiz, P.; Wiekamp, M.; Vendamme, R.; Eevers, W. Bio-based epoxy resins from biorefinery by-products. *BioResources* **2019**, *14*, 3200–3209. [CrossRef]
118. Thomas, J.; Patil, R. Enabling Green Manufacture of Polymer Products via Vegetable Oil Epoxides. *Ind. Eng. Chem. Res.* **2023**, *62*, 1725–1735. [CrossRef]
119. Lascano, D.; Valcárcel, J.; Balart, R.; Quiles-Carrillo, L.; Boronat, T. Manufacturing of composite materials with high environmental efficiency using epoxy resin of renewable origin and permeable light cores for vacuum-assisted infusion molding. *Ingenius* **2020**, *23*, 62–73. [CrossRef]

120. Sridhar, I.; Adie, P.; Ghista, D. Optimal design of customised hip prosthesis using fiber reinforced polymer composites. *Mater. Des.* **2010**, *31*, 2767–2775. [CrossRef]
121. Guild, F.J.; Taylor, A.C.; Downes, J. Composite Materials. In *Encyclopedia of Maritime and Offshore Engineering*; John Wiley & Sons, Ltd.: Chichester, UK, 2017; Volume 82, no. 35; pp. 1–14. [CrossRef]
122. Aslan, M.; Tufan, M.; Küçükömeroğlu, T. Tribological and mechanical performance of sisal-filled waste carbon and glass fibre hybrid composites. *Compos. Part B Eng.* **2018**, *140*, 241–249. [CrossRef]
123. Assarar, M.; Zouari, W.; Sabhi, H.; Ayad, R.; Berthelot, J.-M. Evaluation of the damping of hybrid carbon–flax reinforced composites. *Compos. Struct.* **2015**, *132*, 148–154. [CrossRef]
124. Campbell, A.I.; Sexton, S.; Schaschke, C.J.; Kinsman, H.; McLaughlin, B.; Boyle, M. Prosthetic limb sockets from plant-based composite materials. *Prosthet. Orthot. Int.* **2012**, *36*, 181–189. [CrossRef] [PubMed]
125. Chaudhary, V.; Bajpai, P.K.; Maheshwari, S. An Investigation on Wear and Dynamic Mechanical behavior of Jute/Hemp/Flax Reinforced Composites and Its Hybrids for Tribological Applications. *Fibers Polym.* **2018**, *19*, 403–415. [CrossRef]
126. Essabir, H.; Bensalah, M.; Rodrigue, D.; Bouhfid, R.; Qaiss, A. Structural, mechanical and thermal properties of bio-based hybrid composites from waste coir residues: Fibers and shell particles. *Mech. Mater.* **2016**, *93*, 134–144. [CrossRef]
127. Gu, Y.; Tan, X.; Yang, Z.; Li, M.; Zhang, Z. Hot compaction and mechanical properties of ramie fabric/epoxy composite fabricated using vacuum assisted resin infusion molding. *Mater. Des.* **2014**, *56*, 852–861. [CrossRef]
128. Reddy, M.I.; Kumar, M.A.; Raju, C.R.B. Tensile and Flexural properties of Jute, Pineapple leaf and Glass Fiber Reinforced Polymer Matrix Hybrid Composites. *Mater. Today Proc.* **2018**, *5*, 458–462. [CrossRef]
129. Jagannatha, T.D.; Harish, G. Mechanical properties of carbon/glass fiber reinforced epoxy hybrid polymer composites. *Int. J. Mech. Eng. Robot. Res.* **2015**, *4*, 131–137.
130. Lee, B.-H.; Kim, H.-J.; Yu, W.-R. Fabrication of long and discontinuous natural fiber reinforced polypropylene biocomposites and their mechanical properties. *Fibers Polym.* **2009**, *10*, 83–90. [CrossRef]
131. Rahman, A.; Rahman, M.Z.A.A.; Zaidi, A.M.A.; Rahman, I.A. Analysis of Comparison between Unconfined and Confined Condition of Foamed Concrete Under Uni-Axial Compressive Load. *Am. J. Eng. Appl. Sci.* **2010**, *3*, 68–72. [CrossRef]
132. Saba, N.; Paridah, M.; Abdan, K.; Ibrahim, N. Effect of oil palm nano filler on mechanical and morphological properties of kenaf reinforced epoxy composites. *Constr. Build. Mater.* **2016**, *123*, 15–26. [CrossRef]
133. Sekaran, A.S.J.; Kumar, K.P.; Pitchandi, K. Evaluation on mechanical properties of woven aloevera and sisal fibre hybrid reinforced epoxy composites. *Bull. Mater. Sci.* **2015**, *38*, 1183–1193. [CrossRef]
134. Shanmugam, D.; Thiruchitrambalam, M. Static and dynamic mechanical properties of alkali treated unidirectional continuous Palmyra Palm Leaf Stalk Fiber/jute fiber reinforced hybrid polyester composites. *Mater. Des.* **2013**, *50*, 533–542. [CrossRef]
135. Shih, Y.-F.; Chang, W.-C.; Liu, W.-C.; Lee, C.-C.; Kuan, C.-S.; Yu, Y.-H. Pineapple leaf/recycled disposable chopstick hybrid fiber-reinforced biodegradable composites. *J. Taiwan Inst. Chem. Eng.* **2014**, *45*, 2039–2046. [CrossRef]
136. Shrivastava, R.; Telang, A.; Rana, R.; Purohit, R. Mechanical Properties of Coir/G Lass Fiber Epoxy Resin Hybrid Composite. *Mater. Today Proc.* **2017**, *4*, 3477–3483. [CrossRef]
137. Sreekumar, P.; Joseph, K.; Unnikrishnan, G.; Thomas, S. A comparative study on mechanical properties of sisal-leaf fibre-reinforced polyester composites prepared by resin transfer and compression moulding techniques. *Compos. Sci. Technol.* **2007**, *67*, 453–461. [CrossRef]
138. Widhata, D.; Ismail, R.; Sulardjaka. Water Hyacinth (Eceng Gondok) As Fibre Reinforcement Composite for Prosthetics Socket. *IOP Conf. Ser. Mater. Sci. Eng.* **2019**, *598*, 012127. [CrossRef]
139. Yan, L.; Chouw, N.; Yuan, X. Improving the mechanical properties of natural fibre fabric reinforced epoxy composites by alkali treatment. *J. Reinf. Plast. Compos.* **2012**, *31*, 425–437. [CrossRef]
140. Yang, Z.; Peng, H.; Wang, W.; Liu, T. Crystallization behavior of poly(ε-caprolactone)/layered double hydroxide nanocomposites. *J. Appl. Polym. Sci.* **2010**, *116*, 2658–2667. [CrossRef]
141. Ahmadijokani, F.; Shojaei, A.; Dordanihaghighi, S.; Jafarpour, E.; Mohammadi, S.; Arjmand, M. Effects of hybrid carbon-aramid fiber on performance of non-asbestos organic brake friction composites. *Wear* **2020**, *452–453*, 203280. [CrossRef]
142. Yuan, H.; Zhang, S.; Lu, C.; He, S.; An, F. Improved interfacial adhesion in carbon fiber/polyether sulfone composites through an organic solvent-free polyamic acid sizing. *Appl. Surf. Sci.* **2013**, *279*, 279–284. [CrossRef]
143. Hatti, P.S.; Somanakatti, A.B.; Rakshith, M. Investigation on tensile behavior of glass-fiber reinforced polymer matrix composite with varying orientations of fibers. *Mater. Today Proc.* **2021**, *54*, 137–140. [CrossRef]
144. Puttaraju, D.; Hanumantharaju, H. Finite element analysis and validation of tensile properties of carbon fiber reinforced polymer matrix composites. *Mater. Today Proc.* **2022**, *62*, 2800–2807. [CrossRef]
145. Cripps, D.; Searle, T.J.; Summerscales, J. Open Mold Techniques for Thermoset Composites. In *Comprehensive Composite Materials Encyclopædia*; Talreja, R., Månson, J.-A., Eds.; Polymer Matrix Composites; Elsevier Science: Oxford, UK, 2000; Volume 2, Chapter 21; pp. 737–761. [CrossRef]
146. Barbero, E.J. *Introduction to Composite Materials Design*, 2nd ed.; CRC Press: Boca Raton, FL, USA, 2011.
147. Rajak, D.K.; Pagar, D.D.; Kumar, R.; Pruncu, C.I. Recent progress of reinforcement materials: A comprehensive overview of composite materials. *J. Mater. Res. Technol.* **2019**, *8*, 6354–6374. [CrossRef]
148. Oksman, K. High Quality Flax Fibre Composites Manufactured by the Resin Transfer Moulding Process. *J. Reinf. Plast. Compos.* **2001**, *20*, 621–627. [CrossRef]

149. Rouison, D.; Sain, M.; Couturier, M. Resin transfer molding of natural fiber reinforced composites: Cure simulation. *Compos. Sci. Technol.* **2004**, *64*, 629–644. [CrossRef]
150. Nawaz, A.; Islam, B.; Khattak, M.S.; Ali, L.; Saleem, U.; Ullah, A.; Ijaz, M.Z.; Mao, W.G. Polyester Usage in Manufacturing of Electrical and Mechanical Products and Assemblies. In *Polyester—Production, Characterization and Innovative Applications*; InTech Open: London, UK, 2018. [CrossRef]
151. Razali, N.; Mansor, M.R.; Omar, G.; Kamarulzaman, S.A.F.S.; Zin, M.H.; Razali, N. Out-of-autoclave as a sustainable composites manufacturing process for aerospace applications. In *Design for Sustainability*; Elsevier: Amsterdam, The Netherlands, 2021; pp. 395–413. [CrossRef]
152. Kassapoglou, C. *Design and Analysis of Composite Structures*; American Institute of Aeronautics and Astronautics (AIAA): Reston, VA, USA, 2010; ISBN 9781600867804.
153. Zhao, X.; Li, K.; Wang, Y.; Tekinalp, H.; Richard, A.; Webb, E.; Ozcan, S. Bio-treatment of poplar via amino acid for interface control in biocomposites. *Compos. Part B Eng.* **2020**, *199*, 108276. [CrossRef]
154. Taboga, P.; Grabowski, A.M. Axial and torsional stiffness of pediatric prosthetic feet. *Clin. Biomech.* **2017**, *42*, 47–54. [CrossRef] [PubMed]
155. *ISO 10328*; Prosthethics—Structural Testing of Lower-Limb Prostheses—Requirements and Test Methods. ISO: Geneva, Switzerland, 2006.
156. *ISO 22675*; Prosthetics—Testing of Ankle-Foot Devices and Foot Units—Requirements and Test Methods. ISO: Geneva, Switzerland, 2006.
157. Colombo, C.; Marchesin, E.; Vergani, L.; Boccafogli, E.; Verni, G. Study of an ankle prosthesis for children: Adaptation of ISO 10328 and experimental tests. *Procedia Eng.* **2011**, *10*, 3510–3517. [CrossRef]
158. Tabucol, J.; Brugo, T.M.; Povolo, M.; Leopaldi, M.; Oddsson, M.; Carloni, R.; Zucchelli, A. Structural FEA-Based Design and Functionality Verification Methodology of Energy-Storing-and-Releasing Prosthetic Feet. *Appl. Sci.* **2021**, *12*, 97. [CrossRef]
159. Colombo, C.; Marchesin, E.; Vergani, L.; Boccafogli, E.; Verni, G. Design of an ankle prosthesis for swimming and walking. *Procedia Eng.* **2011**, *10*, 3503–3509. [CrossRef]
160. Hamzah, M.; Gatta, A. Design of a Novel Carbon-Fiber Ankle-Foot Prosthetic using Finite Element Modeling. *IOP Conf. Ser. Mater. Sci. Eng.* **2018**, *433*, 012056. [CrossRef]
161. Starker, F.; Blab, F.; Dennerlein, F.; Schneider, U. A Method for Sports Shoe Machinery Endurance Testing: Modification of ISO 22675 Prosthetic Foot Test Machine for Heel-to-toe Running Movement. *Procedia Eng.* **2014**, *72*, 405–410. [CrossRef]
162. Santana, J.P.; Beltran, K.; Barocio, E.; Lopez-Avina, G.I.; Huegel, J.C. Development of a Low-Cost and Multi-Size Foot Prosthesis for Humanitarian Applications. In Proceedings of the 2018 IEEE Global Humanitarian Technology Conference (GHTC), Radnor, PA, USA, 12–15 October 2023; pp. 1–8.
163. *KS P 8403*; Prosthetic Feet and Ankle Joints. Korean Industrial Standards: Seoul, Republic of Korea, 2017.
164. Song, Y.; Choi, S.; Kim, S.; Roh, J.; Park, J.; Park, S.H.; Yoon, J. Performance Test for Laminated-Type Prosthetic Foot with Composite Plates. *Int. J. Precis. Eng. Manuf.* **2019**, *20*, 1777–1786. [CrossRef]
165. Miller, K. Computational biomechanics for medicine. *Int. J. Numer. Methods Biomed. Eng.* **2011**, *27*, 345–346. [CrossRef]
166. Simkins, D.C.C.; Alford, J.B.B. The Role of Computational Tools in Biomechanics. In *Biomechanics of the Female Pelvic Floor*; Elsevier: Amsterdam, The Netherlands, 2016; pp. 351–366. [CrossRef]
167. Balaramakrishnan, T.M.; Natarajan, S.; Sujatha, S. Biomechanical design framework for prosthetic feet: Experimentally validated non-linear finite element procedure. *Med. Eng. Phys.* **2021**, *92*, 64–70. [CrossRef] [PubMed]
168. Parashar, S.; Chawla, V.K. Evaluation of fiber volume fraction of kenaf-coir-epoxy based green composite by finite element analysis. *Mater. Today Proc.* **2022**, *50*, 1265–1274. [CrossRef]
169. Lian, W.; Yao, W. Fatigue life prediction of composite laminates by FEA simulation method. *Int. J. Fatigue* **2010**, *32*, 123–133. [CrossRef]
170. Balasubramanian, K.; Rajeswari, N.; Vaidheeswaran, K. Analysis of mechanical properties of natural fibre composites by experimental with FEA. *Mater. Today Proc.* **2019**, *28*, 1149–1153. [CrossRef]
171. Birrer, R.B.; Buzermanis, S.; DellaCorte, M.P.; Grisalfi, P.J. *Biomechanics of Running*; Medical Publishing Division: Oxford, UK, 2001.
172. Rietman, H.S.; Postema, K.; Geertzen, J.H.B. Gait analysis in prosthetics: Opinions, ideas and conclusions. *Prosthet. Orthot. Int.* **2003**, *27*, 76–77. [CrossRef] [PubMed]
173. Balaramakrishnan, T.M.; Natarajan, S.; Sujatha, S. Design of a biomimetic sach foot: An experimentally verified finite element approach. *J. Biomim. Biomater. Biomed. Eng.* **2020**, *45*, 22–30. [CrossRef]
174. Wittek, A.; Miller, K. Computational biomechanics for medical image analysis. In *Handbook of Medical Image Computing and Computer Assisted Intervention*; Elsevier: Amsterdam, The Netherlands, 2020; pp. 953–977. [CrossRef]
175. Paknys, R. Finite Element Method. In *Applied Frequency-Domain Electromagnetics*; John Wiley & Sons, Ltd.: Chichester, UK, 2016; pp. 233–267. [CrossRef]

polymers

MDPI

Article

Development of Ultraviolet-Shielding Bamboo/Silk Fibroin Hybrid Films with Good Mechanical Properties: A Proof Study on Human Keratinocyte Cells

Silvia Bittolo Bon [1], Valeria Libera [1], Maria Rachele Ceccarini [2,3], Rocco Malaspina [1], Michela Codini [2] and Luca Valentini [3,*]

[1] Dipartimento di Fisica e Geologia, Università degli Studi di Perugia, Via A. Pascoli, 06123 Perugia, Italy; silvia.bittolo@gmail.com (S.B.B.); valeria.libera@unipg.it (V.L.); rocco.malaspina@dottorandi.unipg.it (R.M.)
[2] Department of Pharmaceutical Science, University of Perugia, 06123 Perugia, Italy; mariarachele.ceccarini@unipg.it (M.R.C.); michela.codini@unipg.it (M.C.)
[3] Civil and Environmental Engineering Department and INSTM Research Unit, University of Perugia, Strada di Pentima 8, 05100 Terni, Italy
* Correspondence: luca.valentini@unipg.it

Abstract: In this study, we report the preparation and characterization of water-stable films with UV-shielding and good mechanical properties, exploiting the synergistic effect of regenerated silk fibroin and bamboo-derived cellulose. Silk fibroin (SF)/bamboo (B) hybrid films are achieved by solubilizing both silk and bamboo fibers in formic acid with added $CaCl_2$. Infrared spectroscopy indicates that SF, when combined with bamboo, undergoes a conformational transition, providing evidence of an increase in SF crystallinity. Exploiting the intrinsic absorption of SF in the ultraviolet region, UV–Vis spectroscopy was used to assess the glass transition temperature (T_g) of SF/B films, showing a decrease in T_g by increasing the SF content. The addition of 10 wt% SF to the B matrix improved the elastic modulus by about 10% while conserving the strain at break with respect to the neat B films, increasing the UV shielding properties, while water absorption suggested the material's hydrophilic and swelling capacity even after one month. The hybrid films showed, under solar irradiation, a photoprotective behavior on keratinocyte human cells by increasing cellular viability. These findings may find potential applications in functional fabrics.

Keywords: silk fibroin; bamboo; UV shielding; mechanical properties; cells

Citation: Bittolo Bon, S.; Libera, V.; Ceccarini, M.R.; Malaspina, R.; Codini, M.; Valentini, L. Development of Ultraviolet-Shielding Bamboo/Silk Fibroin Hybrid Films with Good Mechanical Properties: A Proof Study on Human Keratinocyte Cells. *Polymers* **2024**, *16*, 2244. https://doi.org/10.3390/polym16162244

Academic Editor: Andrzej Puszka

Received: 6 July 2024
Revised: 1 August 2024
Accepted: 5 August 2024
Published: 7 August 2024

1. Introduction

Silk fibers spun by silkworms are, among all natural polymers, those with the best biological and mechanical properties [1–3]. The intrinsic high mechanical strength and elongation at break of silk, along with its biocompatibility [4–7], can be potentially used for multifunctional high-performance textiles. The main drawback is the fact that spiders and silkworms produce silk for their biological needs, and, thus, upscaling production is extremely challenging as the variability in the climatic conditions where silkworms take their food to spin these fibers creates an intrinsic variation in the mechanical properties of the final product [8–11]. Currently, the use of regenerated silk fibroin (SF) obtained from natural fibers is a viable method for the large-scale production of natural polymers. The principal bottleneck of this approach is the variability in the properties depending, for example, on the humidity that plasticizes the silk, reducing the tensile strength, the degumming process, which affects conformational transition, making the fibers soluble in water [12–15], and the dissolving process, which, finally, destroys the original multiscale fibrillar structure, increasing the brittleness [16].

Polymer blending is a practical approach that is currently adopted to obtain SF-based composites with desired properties [17,18]. Among all biopolymers, SF films' properties can be improved by blending SF with cellulose—a biocompatible, low-cost, and abundant natural polymer [19–21]. SF with UV-absorbing properties can be obtained by tuning the conformational transition [22]. Bamboo is a lignocellulosic green raw material that, for its fast growth rate, abundance, and outstanding mechanical properties, is currently widely used, from building applications to the textile sector [23–26].

It is well known that solar UV radiation is the most important risk factor for skin diseases, with UVA (320–400 nm) and UVB (280–320 nm) components [26]. UV-blocking effects are based on UV absorption and UV-scattering materials [27]. Thus, the development of novel materials protecting one from ultraviolet (UV) rays represents the frontier for these applications. Since cellulose lacks UV absorption properties despite its intrinsic mechanical properties, combining bamboo-derived cellulose with the natural photoprotective properties of SF could be a viable approach to developing bio-based photoprotective films. In the past, we realized that sunscreens prevented the damage produced by UV radiation on epidermal cells. The sustainable nature of the proposed materials as well as the easy processing method may be considered valid alternatives to the metal oxide additives currently used in cosmetics to block UV rays, which have the drawback of inducing the production of reactive oxygen species, stimulating oxidative stress in tissues.

2. Materials and Methods

2.1. Bamboo Solution Preparation

Commercial bamboo (B) yarns (supplied by a local farm) were used to produce films using a solvent-casting method: 200 mg of bamboo yarn was cut in small pieces and solubilized in 10 mL of formic acid (FA) through magnetic stirring while keeping the solution at 80 °C for 2 h (bamboo content in the FA solution: 20 mg/mL).

2.2. Silk Fibroin and Silk Fibroin/Bamboo Hybrid Solution Preparation

SF was previously extracted from *Bombyx mori* cocoons (supplied from a local farm) in hot deionized water and sodium bicarbonate (supplied by Sigma Aldrich, St. Louis, MO, USA): 5 silk cocoons were soaked in 200 mL of boiling deionized water containing 5 g of sodium bicarbonate for 45 min; the silk fibers were then rinsed with deionized water; and the complete treatment was repeated two times to separate SF from all the sericin component of the silk cocoon. SF was left at room temperature until its complete drying. SF was solubilized in formic acid (purity 98%, supplied by Sigma Aldrich) and calcium chloride ($CaCl_2$, Sigma Aldrich) at room temperature for 5 min using magnetic stirring: 5 mL of FA with $CaCl_2$ was used to dissolve 700 mg of SF, and the weight ratio between SF and $CaCl_2$ was 70:30 (silk content in the FA solution: 140 mg/mL). Hybrid solutions were prepared mixing neat SF solution and B solution, with two different volume ratios considered: SF/B 10/90 and SF/B 50/50. The solutions were mixed at room temperature with magnetic stirring (mixing time: 5 min).

2.3. Silk Fibroin and Silk Fibroin/Bamboo Hybrid Film Preparation

The SF solution was poured in an uncovered Petri dish, and the FA was left to evaporate at room temperature for 12 h and annealed at 40 °C for 2 h to achieve complete solvent removal. Hybrid films were obtained by casting the solutions in Petri dishes with a diameter of 5 cm. The Petri dishes were left open at room temperature to allow complete solvent evaporation and then annealed at 40 °C for 2 h.

2.4. FTIR Characterization

Infrared spectra were recorded in transmission mode using a Fourier transform spectrometer from Jasco (Oklahoma City, OK, USA) (model FTIR 615). The spectra recorded in the 4000−400 cm^{-1} spectral range with a resolution of 2 cm^{-1} were averaged from 300 scans. To estimate the different components of amide I spectral profiles for the samples containing SF, a curve-fitting procedure was employed. Each component was assigned a

Gaussian line shape, a full width at half height (FWHH) fixed at 20 cm^{-1}, and the weight was determined without constraints.

2.5. Degradation and Mechanical Characterization

To investigate degradation, SF and SF/B films (circular shape with a diameter of 5 cm and thickness of 300 μm) were immersed at room temperature in 10 mL of phosphate-buffered saline (PBS) with a pH of 7.4. At a designated time, the samples were washed with distilled water, dried in a desiccator, and weighed to estimate the weight variation.

A tensile testing machine (Lloyd Instr. LR30 K, Fareham, UK) was used to measure the mechanical properties of the films. The films were cut to have rectangular-shaped samples with the following dimensions: 1 cm × 3 cm × 300 μm. The samples were stretched with a strain rate of 5 mm/min using a 500 N load cell. The mechanical characteristics (i.e., Young's modulus, tensile strength, and strain at break) were then calculated from the stress–strain curves.

2.6. Optical and Thermal Characterization

UV–Visible (UV–Vis) absorption measurements were performed using a Jasco V-570 spectrophotometer. A 0.01 mm path-length quartz cuvette was used in order to obtain the maximum signal-to-noise ratio. Each spectrum was collected in the range from 190 to 800 nm, with a scan speed of 100 nm/min. A Peltier temperature controller was used to change the temperature from 24 °C to 100 °C. To obtain information on the glass transition temperature (Tg), UV−Vis spectroscopy was employed to observe changes in the absorption spectrum due to the structural reorganization of polymers when heated above the Tg.

This property was quantified by the deviation metric of the temperature (DM_T), defined as follows [28–30]:

$$DM_T = \sum_{\lambda_{min}}^{\lambda_{max}} [I_{RT}(\lambda) - I_T(\lambda)]^2$$

where $I_{RT}(\lambda)$ and $I_T(\lambda)$ are, respectively, the absorption intensity of the film at room temperature and at temperature T. The sum on the wavelength is taken on the full optical sweep around the absorption peak. The sum of a range of wavelengths affected by this variation was, for each sample, as follows: 250–350 nm, 400–500 nm, 350–500 nm, and 300–400 nm for the B, SF/B 10/90, SF/B 50/50, and SF, respectively.

2.7. Differential Scanning Calorimetry

Conventional differential scanning calorimetry (DSC) measurements were also performed with Perkin-Elmer Pyris (Waltham, MA, USA) on all the samples (10 mg), with a scan rate of 10 °C/min.

Measurements were taken in a temperature range from 40 °C to 300 °C. The instrument was calibrated using indium and zinc. To measure the film, an aluminum sample/reference pan was used.

2.8. Cell Culture

The HaCaT cell line comprises spontaneously transformed aneuploid immortal keratinocyte cells derived from adult human skin. Cells were purchased at passage level 46 from the Istituto Zooprofilattico Sperimentale della Lombardia e dell'Emilia Romagna (I.Z.S.L.E.R.) "Bruno Ubertini" (Brescia, Italy) and were used for experiments until 59 passages, routinely checking for their morphology. HaCat were grown in standard conditions, as previously described [31].

2.9. Irradiation and MTT Assay

During the experimental procedure, human keratinocyte cell lines were firstly trypsinized upon reaching 80% confluence, then harvested and counted using a trypan blue exclusion assay in a Bürker chamber. Cells were seeded using 35 mm dishes at three different concentrations based on the incubation time: 4×10^5 cells/dish, 3×10^5 cells/dish, and 2×10^5 cells/dish, respectively, for 4, 24, and 48 h of incubation after constant and fixed irradiation (220 sec) under a 300 W Xenon light (ThermoOriel solar simulator model 69907 (Ljubljana, Slovenia)), as previously described in ref. [31]. All the cells were irradiated with the same condition: 1X pre-warmed PBS, which maintains cell viability during radiation exposure and does not absorb in the UV–Vis spectrum. The doses of radiation during 220 s exposure were 2.96 kJ/m^2 (UVC), 6.95 kJ/m^2 (UVB), 21.9 kJ/m^2 (UVA), and 110 kJ/m^2 (VIS-nIR), respectively. A negative control (CTR-) was kept under a polystyrene cover during solar simulation under the same conditions as the samples, which were covered, respectively, with different proportions of SF and B (SF, SF/B 50/50, SF/B 10/90, and B). A positive control (CTR +) was exposed to the solar simulator without any protection. The four patches with different proportions of SF and B were used, likewise, with a lid to cover the upper part of the dish, protecting the cell monolayer from the UV-nIR rays. Each dish was irradiated for 220 s, which corresponds approximately to an hour and a half of sun exposure to UV-B rays. Immediately after irradiation, the PBS was aspirated, and the DMEM complete medium was replaced. The HaCat cells were left for 4, 24, and 48 h, respectively, in an incubator with standard conditions. An MTT assay was performed to test cell viability. The stock solution (5 mg/mL) was freshly prepared and diluted 1:10 in a cell medium. The final concentration in each dish was 0.5 mg/mL. After 3 h in the incubator, the supernatant was completely eliminated. To lyse the cells, 1 mL of dimethyl sulfoxide (DMSO) was used for 30 min at 37 °C. The absorbance was measured in a 96-well plate at 540 nm by a spectrophotometer (Tecan Austria GmbH (Grödig, Austria), Model INFINITE 200 PRO). The experiments were performed in a biological triplicate. The % of viable cells was compared to CTR- using the following formula:

$$\text{Viable cells (\%)} = (\text{OD\_sample}/\text{OD\_CTR}-) \times 100$$

where OD is the optical density.

2.10. Statistical Analysis

In the statistical analysis for the comparison between multiple groups, a two-way ANOVA was conducted with significance thresholds of * $p < 0.05$, ** $p < 0.01$, *** $p < 0.001$, and **** $p < 0.0001$.

3. Results

In this study, SF/B films were obtained via the dissolution of SF and bamboo fibers in FA. Thus, we could obtain information on the structure of both the bamboo and the regenerated silk films after the treatment in formic acid by means of FTIR. The infrared spectra of the B, SF, and SF/B samples are reported in Figure 1a. The bamboo sample showed characteristic absorption peaks for C−O−C, hemicellulose acetyl, C-H, and O-H cellulose stretching at 1160 cm^{-1}, 1716 cm^{-1}, 2902 cm^{-1}, and 3334 cm^{-1} [32–34], respectively. The typical absorption peaks at 1606 and 1509 cm^{-1} of lignin [34] were not detected, indicating that lignin was removed from the bamboo after the FA treatment.

FTIR spectroscopy was also employed to analyze how the SF's secondary structure was modified by the interaction with bamboo (Figure 1a). In the amide I (1600−1700 cm^{-1}) and amide II (1500−1650 cm^{-1}) FTIR regions [22], the absorption bands at 1640−1654 cm^{-1} and 1535−1545 cm^{-1} were assigned to a random-coil secondary structure, whereas the β-sheets gave rise to the bands at 1610−1630 cm^{-1} and 1510−1520 cm^{-1} [35–37], respectively. By fitting the amide I region with these components, we estimated the relative percentages of different secondary structures (β-sheets, turns, and random coils). As shown in Figure 1b, the addition of B produced an increase in the fraction of the peak components in the ranges of 1620–1625 (amide I) and 1697–1703 (amide I) related to the β-sheet structure. Specifically, the β-sheet content in the SF/B 10/90 and 50/50 films was found to be 18% and 13%, respectively, while an 11% content was estimated for the neat SF sample.

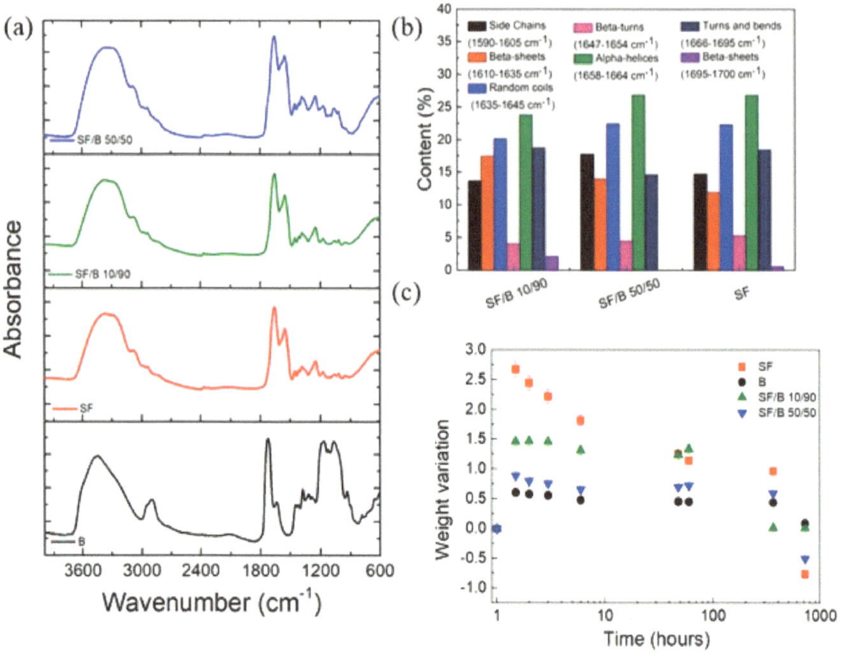

Figure 1. (**a**) FTIR spectra of the prepared films and (**b**) relative content of the secondary structures of SF obtained by curve-fitting the amide I region in the 1610–1700 cm^{-1} range according to that reported by Lu et al. [35]. (**c**) Degradation behaviors of B, SF, and SF/B films in PBS.

Then, we investigated the stability of the SF-based films in a PBS environment, since it was desirable to prove that they would be used in a biological environment (e.g., water) whilst avoiding their degradation [38,39]. All the samples (Figure 1c) showed an increase in their mass after a few hours of immersion, with this effect being more pronounced for the films containing SF. This was due to the water-trapping effect exerted by the Ca^{2+} ions dispersed into the fibroin. Ca^{2+} ions can trap water molecules due to the Ca ions' capability to coordinate 6–8 water molecules via the oxygen atoms [40,41]. Moreover, the SF/B 10/90 film was stable even after 1 month, while an appreciable weight loss between 44 and 50% for the SF and SF/B 50/50 films was observed.

Figure 2 shows the tensile tests for all the prepared samples. The mechanical characterization of the B (Figure 2a) and SF neat films (Figure 2b) shows very different behaviors: the bamboo films show high tensile strength values that are combined with a very low strain at break; the SF films, on the contrary, are characterized by a lower tensile strength and a high strain at break values. The influences of the amounts of added SF on the tensile strength, Young's modulus, and strain at break of the B films were analyzed (Figure 2c,d), and they are reported in the box charts in Figure 2e–g. The pure B films presented a Young modulus of 1.2 $\pm$ 0.1 GPa (Figure 2e) and a strain at break (Figure 2g) of 6.2 $\pm$ 1.5%. When the concentration of SF increased from 10 to 50%, the Young modulus of the hybrid films and the strain at break varied from 1.3 $\pm$ 0.1 GPa to 55 $\pm$ 7 MPa and from 6.8 $\pm$ 1.2% to 127 $\pm$ 40%, respectively. The nanocomposite films achieved a maximal tensile strength of 29.8 $\pm$ 1.3 MPa when the addition of SF was 10% (Figure 2f). The improvement in the elastic modulus without a significant reduction in the strain at break were indicative of the fact that SF was dispersed homogeneously in the B films. The increase in the strain at break along with the tensile strength decrease as the SF was increased were due to the predominant mechanical behavior of the SF.

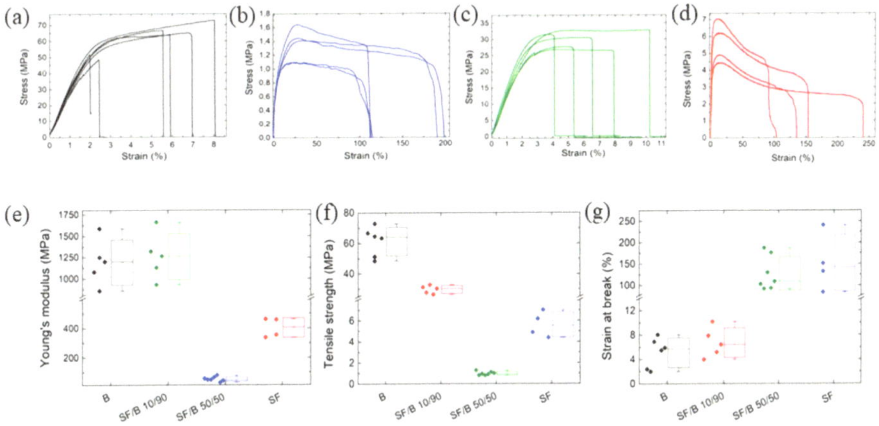

Figure 2. (**a**–**d**) Stress–strain curves of B, SF, SF/B 10/90, and SF/B 50/50 films, respectively. Influence of the SF content on the (**e**) Young modulus, (**f**) tensile strength, and (**g**) strain at break of all the analyzed samples.

In Figure 3a, the UV–Vis absorbance spectra of the prepared samples are reported. The samples show absorptions in the UVC (100–280 nm), UVB (280–315 nm), and UVA (315–400 nm) regions, with these two last components being the most important for UV-shielding applications [42]. The results confirm that the absorption of the B film is evident in these regions but lower compared to samples containing SF. Bamboo's UV-shielding ability is attributed to its lignin content [42]. The UV–Vis absorption spectra of the SF and SF/B films, on the contrary, indicate saturation around 280 nm, which is likely resulting from aromatic amino acids such as tyrosine, phenylalanine, and tryptophan, present along the SF chain. [43]. The spectra of the SF/B films are significantly influenced by the SF content, indicating that the SF is a major contributor to UV shielding in the UVB and UVA regions. This result indicates that the SF within the SF/B samples maintains its conformation and that B does not alter the SF spectrum.

Conventional methods, such as differential scanning calorimetry (DSC), used to measure glass transition in bulk systems, lack the sensitivity needed to detect this transition in films [44]. Absorption spectroscopy can detect transitions between various electronic states of samples in their absorption region [45]. Alterations in the absorption spectrum result from the molecular-scale reorganization of polymers as the latter are warmed above the T_g. To determine the T_g of the films, we performed UV absorption spectroscopy while increasing the temperature from 25 °C to 100 °C. The deviation metric as a function of temperature is reported in Figure 3b.

The B and SF/B 10/90 films (Figure 3b) exhibited bi-linear curve fitting, with T_g at around 80 °C. In accordance with the DSC thermograms reported below, the deviation metric revealed, for the SF and SF/B 50/50 films, the presence of a pre-transition stage at lower temperatures (Figure 3b). In particular, for the SF/B 50/50 film, we observed a transition in the deviation metric at ≈50 °C and the T_g at 72 °C. For the SF films, a smoother pre-transition at ≈60 °C was recorded, with the glass transition occurring at 80 °C. The occurrence of transitions recorded before the Tg can be ascribed to the strong water absorption capacity of the Ca^{2+} ions dispersed in the SF films. Water adsorbed on the Ca^{2+} ions forms a hydrated shell that interacts with the amide or hydroxyl groups in the silk proteins. The hydration of Ca^{2+} triggers the average distance of silk chains, thus governing their mobility. Regarding the T_g values measured for the SF, these data were consistent with the literature [46–48], which reports a T_g of 178 °C for dry fibroin films, which, in the presence of bound water, is reduced to 80 °C.

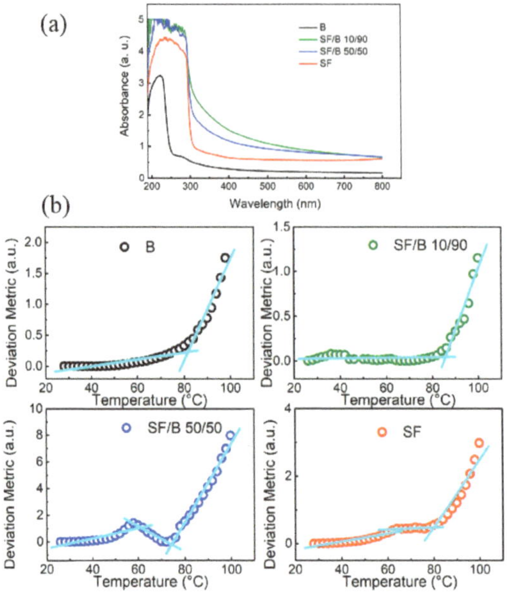

Figure 3. (**a**) UV–Vis spectra and (**b**) evolution of the deviation metric as a function of temperature of B, SF/B 10/90, SF/B 50/50, and SF films, respectively.

The DSC measurements revealed an endothermic peak between 100 °C and 250 °C for the samples containing SF (Figure 4). This peak was derived from water removal [47]. In particular, the endothermic peak at about 150 °C observed in the SF film was likely related to the evaporation of bound water [48]. This peak was shifted to a higher temperature (e.g., 200 °C) in the SF/B 10/90 sample, suggesting that the higher β-sheet content induced a stronger interaction between fibroin and the water molecules [49]. In the temperature range between 200 °C and 300 °C, another endothermic peak appeared to be caused by the degradation of silk I and silk II [49].

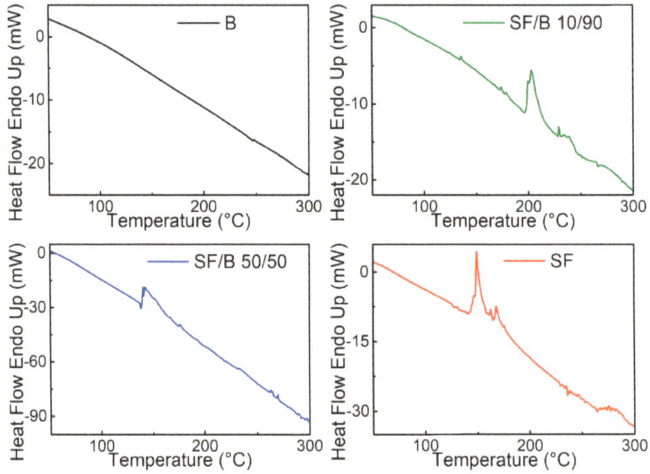

Figure 4. DSC thermograms of the B, SF/B 10/90, SF/B 50/50, and SF films, respectively.

As in Figure 3a, where the SF showed strong UV absorbance even at the lowest content, these results indicate that the SF/B films are the best candidates for shielding human skin cells from UV radiation. Using an MTT assay, we tested HaCat cell viability after 220 s of irradiation with a cover (CTR−), and we set this as 100% (Figure 5, in white). At the same time, we verified the viability without any type of protection (CTR+), and we observed a gradual decrease just after 4 h from irradiation ($p < 0.05$) and a strong worsening after 24 h and 48 h ($p < 0.0001$). This effect was probably due to nuclear DNA damage and reactive oxygen species production, such as H_2O_2, leading to cellular necrosis. HaCat cells covered with the SF (100%) showed a viability comparable to that of CTR−, meaning that no cytotoxic effect was registered. The same results were obtained with SF/B 50/50 and SF/B 10/90 used as a cover, with a cell viability always over 92%, even after 48 h from irradiation. On the contrary, when we used B as a lid, the cell viability decreased slightly after 4 h (92.1%) and considerably after 24 h and 48 h by, respectively, 21.5% and 13.2%, both comparable to that of CTR+. The cell viability remained similar to that of CTR− in the presence of SF, even in low concentrations. This means that only SF can suppress biochemical UV-induced cell damage due to its UV-blocking activity.

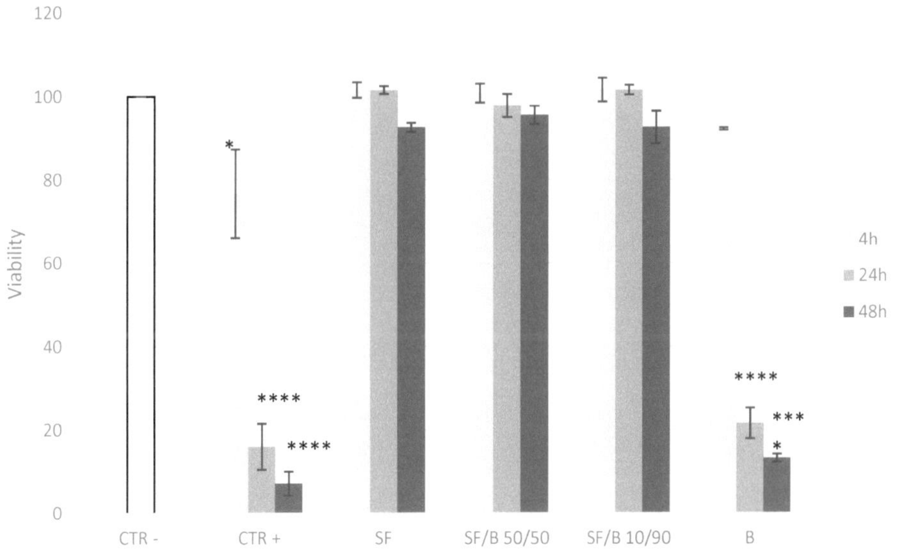

Figure 5. MTT assay on HaCaT cells after 4, 24, and 48 h of 220 s irradiation. * $p < 0.05$, *** $p < 0.001$, and **** $p < 0.0001$.

4. Conclusions

In this study, we obtain a hydrophilic, insoluble film with UV-shielding properties by mixing 10 wt% redissolved fibers of silk and bamboo in formic acid. The results suggest that the 10 wt% of SF embedded in the B matrix improved the elastic modulus and strain at break of bamboo by ≈10%, retaining its conformational structure. Moreover, the combination of SF and B acted as a UV-blocking filter, preserving the viability of human keratinocyte HaCaT cells under exposure to the UV component of the solar spectrum. Because of the easy processability of the solution, these findings pave the way for the formulation of environmentally sustainable sunscreens, which may be expanded to wound-healing management.

Author Contributions: All the authors have approved the final version of the manuscript. The authors equally contributed to the conceptualization, writing—original draft, writing—review and editing, and methodology of this study. L.V. acted as the lead in funding acquisition and supervision. S.B.B., M.R.C., and V.L. had leading roles in the data curation, formal analysis, and investigation steps. R.M. and M.C. supported the authors in the investigation and writing—review and editing—stages. All authors have read and agreed to the published version of the manuscript.

Funding: This work was funded by the European Union's NextGenerationEU, under the Italian Ministry of University and Research (MUR)'s National Innovation Ecosystem grants ECS00000041—VITALITY—CUP J97G22000170005 and CUP B43C22000470005. This study also received funding from the European Union's Next-GenerationEU National Recovery and Resilience Plan (NRRP), Mission 4 Component 2, investment n. 1.1, Prin 2022, Prometheus for "4D printing self-deploying bio-enabled polymer scaffolds for the non-invasive treatment of bleeding intestinal ulcers", with grants 2022BZLTTK and CUP I53D23002200006.

Data Availability Statement: Data are contained within the article.

Conflicts of Interest: The authors declare no conflicts of interest.

Ethical Statement: The HaCaT cell line was obtained from the Istituto Zooprofilattico Sperimentale della Lombardia e dell'Emilia Romagna (I.Z.S.L.E.R.) 'Bruno Ubertini' (Brescia, Italy).

References

1. Yang, Y.; Greco, G.; Maniglio, D.; Mazzolai, B.; Migliaresi, C.; Pugno, N.; Motta, A. Spider (*Linothele megatheloides*) and silkworm (*Bombyx mori*) silks: Comparative physical and biological evaluation. *Mater. Sci. Eng. C* **2020**, *107*, 110197. [CrossRef] [PubMed]
2. Wegst, U.G.K.; Bai, H.; Saiz, E.; Tomsia, A.P.; Ritchie, R.O. Bioinspired structural materials. *Nat. Mater.* **2015**, *14*, 23–36. [CrossRef] [PubMed]
3. Greco, G.; Pantano, M.F.; Mazzolai, B.; Pugno, N.M. Imaging and mechanical characterization of different junctions in spider orb webs. *Sci. Rep.* **2019**, *9*, 5776. [CrossRef] [PubMed]
4. Greco, G.; Mastellari, V.; Holland, C.; Pugno, N.M. Comparing modern and classical perspectives on spider silks and webs. *Perspect. Sci.* **2021**, *29*, 133–156. [CrossRef]
5. Holland, C.; Numata, K.; Rnjak-Kovacina, J.; Seib, F.P. The Biomedical Use of Silk: Past, Present, Future. *Adv. Healthc. Mater.* **2019**, *8*, 1800465. [CrossRef] [PubMed]
6. Dellaquila, A.; Greco, G.; Campodoni, E.; Mazzocchi, M.; Mazzolai, B.; Tampieri, A.; Pugno, N.M.; Sandri, M. Optimized production of a high-performance hybrid biomaterial: Biomineralized spider silk for bone tissue engineering. *J. Appl. Polym. Sci.* **2020**, *137*, 48739. [CrossRef]
7. Lewis, R.V. Spider silk: Ancient ideas for new biomaterials spider silk: Ancient ideas for new biomaterials. *Chem. Rev.* **2006**, *106*, 3762–3774. [CrossRef] [PubMed]
8. Greco, G.; Pugno, N.M. Mechanical properties and Weibull scaling laws of unknown spider silks. *Molecules* **2020**, *25*, 2938. [CrossRef] [PubMed]
9. Greco, G.; Wolff, J.; Pugno, N.M. Strong and tough silk for resilient attachment discs: The mechanical properties of piriform silk, in the spider Cupiennius salei. *Front. Mater.* **2020**, *7*, 138. [CrossRef]
10. Greco, G.; Mirbaha, H.; Schmuck, B.; Rising, A.; Pugno, N. Artificial and natural silk materials have high mechanical property variability regardless of sample size. *Sci. Rep.* **2022**, *12*, 3507. [CrossRef]
11. Vollrath, F.; Madsen, B.; Shao, Z. The effect of spinning conditions on the mechanics of a spider's dragline silk. *Proc. R. Soc. B Biol. Sci.* **2001**, *268*, 2339–2346. [CrossRef] [PubMed]
12. Cunniff, P.M.; Fossey, S.A.; Auerbach, M.A.; Song, J.W.; Kaplan, D.L.; Adams, W.W.; Eby, R.K.; Mahoney, D.; Vezie, D.L. Mechanical and thermal properties of dragline silk from the spider *Nephila clavipes*. *Polym. Adv. Technol.* **1994**, *5*, 401–410. [CrossRef]
13. Bucciarelli, A.; Greco, G.; Corridori, I.; Pugno, N.M.; Motta, A. A design of experiment rational optimization of the degumming process and its impact on the silk fibroin properties. *ACS Biomater. Sci. Eng.* **2021**, *57*, 1374–1393. [CrossRef] [PubMed]
14. Bucciarelli, A.; Greco, G.; Corridori, I.; Motta, A.; Pugno, N.M. Tidy dataset of the experimental design of the optimization of the alkali degumming process of *Bombyx mori* silk. *Data Brief* **2021**, *38*, 107294. [CrossRef]
15. Kundu, B.; Rajkhowa, R.; Kundu, S.C.; Wang, X. Silk fibroin biomaterials for tissue regenerations. *Adv. Drug. Deliver. Rev.* **2013**, *65*, 457–470. [CrossRef] [PubMed]
16. Huang, J.; Qin, J.; Zhang, P.; Chen, X.; You, X.; Zhang, F.; Zuo, B.; Yao, M. Facile preparation of a strong chitosan-silk biocomposite film. *Carbohydr. Polym.* **2020**, *229*, 115515. [CrossRef] [PubMed]
17. Lv, L.; Han, X.; Zong, L.; Li, M.; You, J.; Wu, X.; Li, C. Biomimetic hybridization of Kevlar into silk fibroin: Nanofibrous strategy for improved mechanic properties of flexible composites and filtration membranes. *ACS Nano* **2017**, *11*, 8178–8184. [CrossRef] [PubMed]
18. Kostag, M.; Jedvert, K.; El Seoud, O.A. Engineering of sustainable biomaterial composites from cellulose and silk fibroin: Fundamentals and applications. *Int. J. Biol. Macromol.* **2021**, *167*, 687–718. [CrossRef] [PubMed]

19. Feng, Y.; Li, X.; Li, M.; Ye, D.; Zhang, Q.; You, R.; Xu, W. Facile preparation of biocompatible silk fibroin/cellulose nanocomposite films with high mechanical performance. *ACS Sustain. Chem. Eng.* **2017**, *5*, 6227–6236. [CrossRef]
20. Zhou, L.; Wang, Q.; Wen, J.; Chen, X.; Shao, Z. Preparation and characterization of transparent silk fibroin/cellulose blend films. *Polymer* **2013**, *54*, 5035–5042. [CrossRef]
21. Ceccarini, M.R.; Ripanti, F.; Raggi, V.; Paciaroni, A.; Petrillo, C.; Comez, L.; Donato, K.; Bertelli, M.; Beccari, T.; Valentini, L. Development of Salmon Sperm DNA/Regenerated Silk Bio-Based Films for Biomedical Studies on Human Keratinocyte HaCaT Cells under Solar Spectrum. *J. Funct. Biomater.* **2023**, *14*, 280. [CrossRef] [PubMed]
22. Das, M.; Chakraborty, D. Evaluation of improvement of physical and mechanical properties of bamboo fibers due to alkali treatment. *J. Appl. Polym. Sci.* **2008**, *107*, 522–527. [CrossRef]
23. Jain, S.; Kumar, R.; Jindal, U.C. Mechanical behavior of bamboo and bamboo composites. *J. Mater. Sci.* **1992**, *27*, 4598–4604. [CrossRef]
24. Kerr, N.; Capjack, L.; Fedosejevs, R. Ability of textile covers to protect artifacts from ultraviolet radiation. *J. Am. Inst. Conserv.* **2000**, *39*, 345–353. [CrossRef]
25. Liese, W. Bamboo as carbon-sink—Fact or fiction? In Proceedings of the 8th World Bamboo Congress, Bangkok, Thailand, 16–19 September 2009.
26. D'Orazio, J.; Jarrett, S.; Amaro-Ortiz, A.; Scott, T. UV radiation and the skin. *Int. J. Mol. Sci.* **2013**, *14*, 12222–12248. [CrossRef] [PubMed]
27. Belzile, C.; Vincent, W.F.; Kumagai, M. Contribution of absorption and scattering to the attenuation of UV and photosynthetically available radiation in Lake Biwa. *Limnol. Oceanogr.* **2002**, *47*, 95–107. [CrossRef]
28. Turi, E. *Thermal Characterization of Polymeric Materials*; Elsevier: Amsterdam, The Netherlands, 2012.
29. Muller, C.; Andersson, L.M.; Garriga, M.; Campoy-quiles, M.; Peña-Rodríguez, O.; Inganas, O. Determination of Thermal Transition Depth Profiles in Polymer Semiconductor Films with Ellipsometry. *Macromolecules* **2013**, *46*, 7325–7331. [CrossRef]
30. Ceccarini, M.R.; Chiesa, I.; Ripanti, F.; Cardinali, M.A.; Micalizzi, S.; Scattini, G.; De Maria, C.; Paciaroni, A.; Petrillo, C.; Comez, L.; et al. Electrospun Nanofibrous UV Filters with Bidirectional Actuation Properties Based on Salmon Sperm DNA/Silk Fibroin for Biomedical Applications. *ACS Omega* **2024**, *9*, 6025. [CrossRef] [PubMed]
31. Pei, W.; Shang, W.; Liang, C.; Jiang, X.; Huang, C.; Yong, Q. Using lignin as the precursor to synthesize Fe3O4@lignin composite for preparing electromagnetic wave absorbing lignin-phenol-formaldehyde adhesive. *Ind. Crops Prod.* **2020**, *154*, 112638. [CrossRef]
32. Fang, L.; Zeng, J.; Liao, X.; Zou, Y.; Shen, J. Tensile shear strength and microscopic characterization of veneer bonding interface withpolyethylene film as wood adhesive. *Sci. Adv. Mater.* **2019**, *11*, 1223–1231. [CrossRef]
33. Zheng, L.; Lu, G.; Pei, W.; Yan, W.; Li, Y.; Zhang, L.; Huang, C.; Jiang, Q. Understanding the relationship between the structural properties of lignin and their biological activities. *Int. J. Biol. Macromol.* **2021**, *190*, 291–300. [CrossRef] [PubMed]
34. Lu, Q.; Hu, X.; Wang, X.; Kluge, J.A.; Lu, S.; Cebe, P.; Kaplan, D.L. Water-insoluble silk films with silk I structure. *Acta Biomater.* **2010**, *6*, 1380–1387. [CrossRef]
35. Hu, X.; Kaplan, D.; Cebe, P. Determining Beta-Sheet Crystallinity in Fibrous Proteins by Thermal Analysis and Infrared Spectroscopy. *Macromolecules* **2006**, *39*, 6161–6170. [CrossRef]
36. Partlow, B.P.; Hanna, C.W.; Rnjak-Kovacina, J.; Moreau, J.E.; Applegate, M.B.; Burke, K.A.; Marelli, B.; Mitropoulos, A.N.; Omenetto, F.G.; Kaplan, D.L. Highly tunable elastomeric silk biomaterials. *Adv. Funct. Mater.* **2014**, *24*, 4615–4624. [CrossRef] [PubMed]
37. Jin, H.J.; Park, J.; Valluzzi, R.; Cebe, P.; Kaplan, D.L. Biomaterial films of *Bombyx mori* silk fibroin with poly(ethylene oxide). *Biomacromolecules* **2004**, *5*, 711–717. [CrossRef] [PubMed]
38. Jin, H.J.; Park, J.; Karageorgiou, V.; Kim, U.-J.; Valluzzi, R.; Cebe, P.; Kaplan, D. Water-Stable Silk Films with Reduced β-Sheet Content. *Adv. Funct. Mater.* **2005**, *15*, 1241–1247. [CrossRef]
39. Yu, X.; Hu, Y.; Shi, H.; Sun, Z.; Li, J.; Liu, H.; Lyu, H.; Xia, J.; Meng, J.; Lu, X.; et al. Molecular Design and Preparation of Protein-Based Soft Ionic Conductors with Tunable Properties. *ACS Appl. Mater. Interfaces* **2022**, *14*, 48061–48071. [CrossRef]
40. Liu, J.; Chen, Q.; Liu, Q.; Zhao, B.; Ling, S.; Yao, J.; Shao, Z.Z.; Chen, X. Intelligent Silk Fibroin Ionotronic Skin for Temperature Sensing. *Adv. Mater. Technol.* **2020**, *5*, 2000430. [CrossRef]
41. Wang, J.; Wang, Y.; Wu, Y.; Zhao, W. A Multilayer Transparent Bamboo with Good Optical Properties and UV Shielding Prepared by Different Lamination Methods. *ACS Sustain. Chem. Eng.* **2022**, *10*, 6106–6116. [CrossRef]
42. Sionkowska, A.; Planecka, A. The influence of UV radiation on silk fibroin. *Polym. Degrad. Stab.* **2011**, *96*, 523–528. [CrossRef]
43. Danley, R.L.; Reader, J.R.; Schaefer, J.W. Differential Scanning Calorimeter. U.S. Patent US5842788 A, 1 December 1998.
44. Schroeder, B.C.; Chiu, Y.-C.; Gu, X.; Zhou, Y.; Xu, J.; Lopez, J.; Lu, C.; Toney, M.F.; Bao, Z. Non-Conjugated Flexible Linkers in Semiconducting Polymers: A Pathway to Improved Processability without Compromising Device Performance. *Adv. Electron. Mater.* **2016**, *2*, 1600104. [CrossRef]
45. Hu, X.; Kaplan, D.; Cebe, P. Effect of water on the thermal properties of silk fibroin. *Thermochim. Acta* **2007**, *461*, 137–144. [CrossRef]
46. Kenjiro, Y.; Kana, I.; Hiroyasu, M.; Takaaki, H.; Numata, K. Influence of Water Content on the β-Sheet Formation, Thermal Stability, Water Removal, and Mechanical Properties of Silk Materials. *Biomacromolecules* **2016**, *17*, 1057–1066.
47. Anuduang, A.; Loo, Y.Y.; Jomduang, S.; Lim, S.J.; Wan Mustapha, W.A. Effect of Thermal Processing on Physico-Chemical and Antioxidant Properties in Mulberry Silkworm (*Bombyx mori* L.) Powder. *Foods* **2020**, *9*, 871. [CrossRef] [PubMed]

48. Hu, X.; Shmelev, K.; Sun, L.; Gil, E.S.; Park, S.H.; Cebe, P.; Kaplan, D.L. Regulation of silk material structure by temperature-controlled water vapor annealing. *Biomacromolecules* **2011**, *12*, 1686–1696. [CrossRef]
49. Lu, Q.; Zhang, B.; Li, M.; Zuo, B.; Kaplan, D.L.; Huang, Y.; Zhu, H. Degradation mechanism and control of silk fibroin. *Biomacromolecules* **2011**, *12*, 1080–1086. [CrossRef]

polymers

Article

Composites Based on Eucalyptus Nitens Leaves and Natural Rubber as a Valuable Alternative for the Development of Elastomeric Materials with Low Microbiological Impact

Héctor Aguilar-Bolados [1,*], Natacha Rosales-Charlin [1], Claudia Pérez-Manríquez [2,*], Solange Torres-Galan [2], Mohamed Dahrouch [3], Raquel Verdejo [4], Marianella Hernández Santana [4] and Jose Becerra [2]

1 Departamento de Polímeros, Facultad de Ciencias Químicas, Universidad de Concepción, Concepcion 3349001, Chile; nrosales2019@udec.cl
2 Departamento de Fitoquímica, Facultad de Ciencias Naturales y Oceanográficas, Universidad de Concepción, Concepcion 3349001, Chile
3 Departamento de Química Orgánica, Facultad de Ciencias Químicas, Universidad de Concepción, Concepcion 3349001, Chile; mdahrouch@udec.cl
4 Instituto de Ciencia y Tecnología de Polímeros (ICTP), CSIC, Juan de la Cierva, 28006 Madrid, Spain; marherna@ictp.csic.es (M.H.S.)
* Correspondence: haguilar@udec.cl (H.A.-B.); claudiaperez@udec.cl (C.P.-M.); Tel.: +56-41-220-4171 (H.A.-B.)

Abstract: The forest industry produces several low-value by-products, such as bark, sawdust, limbs, and leaves, that are not ultimately disposed of and remain in the forests and sawmill facilities. Among these by-products are leaves, which contain not only cellulose fibers and lignin but also essential oils such as terpenes. These are biosynthesized in a similar way as cis-1,4-polyisoprene. In this context, this work evaluates the use of screened and unscreened dried Eucalyptus nitens leaves in natural rubber. Among the most relevant results of this work is a significant increase in mechanical properties, such as tensile strength and elongation at break, reaching values of 9.45 MPa and 649% of tensile strength and elongation at break, respectively, for a sample of natural rubber containing sieved dried leaves of Eucalyptus nitens. In addition, it is observed that the content of this vegetable filler allows for inhibiting the antibacterial effect of vulcanized rubber against several bacteria, such as Bacillus subtilis, Staphylococcus aureus, Escherichia coli K 12, Escherichia coli FT 17 and Pseudomonas fluorescens. These results are promising because they not only add value to a by-product of the forestry industry, improving the mechanical properties of natural rubber from a sustainable approach but also increase the affinity of rubber with bacterial microorganisms that may play a role in certain ecosystems.

Keywords: Natural Rubber; Eucalyptus nitens; mechanical properties

Citation: Aguilar-Bolados, H.; Rosales-Charlin, N.; Pérez-Manríquez, C.; Torres-Galan, S.; Dahrouch, M.; Verdejo, R.; Hernández Santana, M.; Becerra, J. Composites Based on Eucalyptus Nitens Leaves and Natural Rubber as a Valuable Alternative for the Development of Elastomeric Materials with Low Microbiological Impact. *Polymers* **2024**, *16*, 2215. https://doi.org/10.3390/polym16152215

Academic Editor: Wencai Wang

Received: 5 July 2024
Revised: 29 July 2024
Accepted: 31 July 2024
Published: 3 August 2024

1. Introduction

Natural rubber (NR) is a complex material with unique properties, such as elasticity and flexibility, which make it challenging to replace. In addition, it is also a sustainable material, as it is not derived from fossil sources. In fact, natural rubber could be considered a carbon-positive material since *cis*-1,4-polyisoprene is mainly derived from the *Hevea brasiliensis* tree and produced by a metabolic pathway called Rubber-transferase [1].

One opportunity for NR compounds to be included in the circular economy approach is to improve their sustainability, as the NR compounds must be vulcanized and fillers added to achieve higher mechanical performance. This vulcanization process imparts several properties to natural rubber-based compounds, including resistance to biodegradation. In addition, materials used as fillers, such as carbon black or silica, have a negative carbon footprint, which adds to the environmental impact of these rubber compounds [1–3].

Due to the new challenges of our society, new trends have emerged, such as the circular economy, eco-design concepts that promote a sustainable approach, prioritizing the development of technologies that have a lower environmental impact based on the 10R concepts, and a reduced carbon footprint [4]. In this context, there is growing interest in sustainable materials that can be used as fillers in NR. Among the materials used as fillers, forest by-products stand out, mainly leaves and branches from monocultures, such as Eucalyptus nitens [4]. In addition, this lignocellulosic material possesses essential oils that have a biological role based on the semiochemical properties of this species, which are compounds or mixtures released by an organism that affect the behavior of other individuals [5].

Eucalyptus species that are more abundant are Eucalyptus globulus and Eucalyptus nitens. Eucalyptus globulus is a widely utilized medicinal plant. It contains bioactive compounds, including antioxidants, antimicrobials, and phytoremediators [6–8]. In addition, it has a wide range of biological activities, including antimicrobial, fungicidal, insecticidal, herbicidal, and nematocidal properties [9–11]. Its essential oil can also be used as an herbicide and food preservative [11,12]. Its use is significant for reducing pest resistance and minimizing the environmental impact of synthetic pesticides. In particular, compounds such as 1,8-cineole (eucalyptol) are promising candidates for the development of new pharmaceuticals and agrochemicals [13].

On the other hand, various materials obtained from vegetable sources have been used as fillers in natural rubber to obtain composites. This approach includes cellulose [14–16], sawdust [17,18], and others. In general, the key challenge is to increase the affinity of the filler for the polymer matrix. It is also known that leaves from plants containing traces of essential oils have a greater affinity for natural rubber. This affinity is due to the fact that essential oils are mainly composed of terpenes. In this context, this work focused on obtaining materials based on NR and sifted or unsifted *Eucalyptus nitens* leaves to demonstrate the affinity between the filler and natural rubber. The aim was to obtain rubber materials and examine their mechanical and morphological properties, as well as their resistance to microorganisms such as bacteria. To the best of our knowledge, this type of material is tested for the first time to obtain composites, and interesting results have been obtained to conceptualize new insights with a sustainable approach.

2. Materials and Methods

2.1. Materials

The materials used to prepare the composite specimens were natural rubber (NR) (SGR-10) obtained from Química Miralles S.A. Industry (Quilicura, Chile), and zinc oxide (ZnO), stearic acid (SA), powdered sulfur (S), N-cyclohexyl-2-mercaptobenzotiazole (MBTS), tetramethylthiuram disulfide (TMTD), which were were purchased from Sigma Aldrich (St. Louis, MO, USA), and all reagents were of ACS grade ($\geq$95%).

2.2. Preparation of Lignocellulosic Fillers

Juvenile leaves of Eucalyptus nitens were collected from the "*La Cantera y el Guindo*" located in Concepción, Bio-Bío region, Chile. They were then dried in an oven at 45 °C for 14 days. The dried leaves were ground in an automatic laboratory mill PRO-02; half of the ground leaves were sifted through a #50 mesh sieve, and the other half was left unsifted.

2.3. Determination of Essential Oil Content in Filler

A yield analysis of the 3 types of samples, dry leaves (DL), sifted leaves (SF), and unsifted leaves (USF), was performed using the same conditions of water, temperature, time, and amount of sample by using the method based on hydrodistillation with a Clevenger apparatus. This procedure consisted of a hydrodistillation that allowed to obtain the essential oils present in the samples. The principle of this separation technique is that the steam generated entrains the essential oils present in the samples studied; this oil is then condensed and collected for subsequent analysis. For this purpose, Eucalyptus nitens leaf-based material was placed in a 3-L balloon connected to a cooling column and an essence extractor (Clevenger apparatus). This mixture was allowed to boil until the volume of extracted essential oils remained stable, which took about 3 h. At the end of the extraction period, the aqueous solution was immediately collected in a decanting funnel wrapped in aluminum and allowed to stand for a few minutes. The hydrolate, which corresponds to an aqueous solution of the hydro-soluble compounds of *Eucalyptus nitens*, was then separated from the essential oil and transferred to a clean vial. Sodium sulfate was added to the vial to remove excess water. The amount of essential oil was then measured with a 10 mL graduated cylinder and stored in a new vial with a clean cap, wrapped in aluminum, and stored at room temperature.

GC-MS analysis of the essential oils was performed using a gas chromatograph (Agilent 7890A, Santa Clara, CA, USA) with a splitless injector (250 °C) and a mass detector (Agilent 5975C). An HP-5MS capillary column (30 m × 0.25 mm × 0.25 μm) and helium gas (constant flow rate 1 mL/min) were used for separation. The temperature program was 5 min at 100 °C, 100–275 °C at 13 °C/min and 32 min at 275 °C. The detection range was m/z 50–550. The injector temperature was set to 250 °C. The components were identified by comparing the mass spectra with records from NIST 17 (NIST/EPA/NIH Mass Spectral Library 2017), and by comparing the spectra obtained with those reported in the literature [19].

2.4. Composites Preparation

The NR rubber composites were prepared using a ZL-3018 two-roll mill (Zhongli Instrument Technology Co., Ltd., Dongguan, China) according to the formulations described in Table 1 at room temperature for 20 min. The mixing consisted of dispersing first the activators (ZnO and stearic acid) into the NR rubber, then incorporating the filler (USF or SF), followed by the addition of accelerators (MBTS and TMTD) and sulfur (S).

Table 1. Formulation of composites based on NR.

Ingredient	NR	NR/USF10	NR/USF20	NR/USF40	NR/SF10	NR/SF20	NR/SF40
NR	100	100	100	100	100	100	100
ZnO	5	5	5	5	5	5	5
SA	2	2	2	2	2	2	2
MBTS	2	2	2	2	2	2	2
TMTD	1	1	1	1	1	1	1
USF	-	10	20	40	-	-	-
SF	-	-	-	-	10	20	40
S	2	2	2	2	2	2	2

2.5. Characterization

The characteristics of the vulcanization systems, minimum torque (M_L), maximum torque (M_H) scorch time (t_{s2}), and vulcanization time (t_{90}), were determined using a ZL-3001 moving matrix rheometer (Zhongli Instrument Technology Co. Ltd., China) for 30 min at 160 °C. Vulcanization of the composites was performed using a ZL-3022 laboratory hydraulic press (Zhongli Instrument Technology Co. Ltd., Dongguan, China) for 15 min at 120 Kg/cm^2 pressure and 180 °C [20]. The hardness was determined using a rectangular die of 11.5 × 13 × 0.2 cm according to ASTM 2240-15.

The stress-strain test was performed using a Shimadzu EZ-X L 200 V instrument with a load cell of 500 N at 100 mm/min, according to ASTM D 412. Resilience (ASTM D2632-15, West Conshohocken, PA, USA) and hardness Shore A (ASTM D2240-15) were also determined using tester equipment provided by Zhongli Instrument Technology Co., Ltd., (Dongguan, China). The morphology of the samples was examined by scanning electron microscopy (SEM) using a Zeiss (Oberkocken, Germany) scanning electron microscope, model Gemini SEM 360, equipped with an Oxford Instrument (Abingdon, UK) EDS detector.

The samples were then covered with an ultrathin gold (Au) film. The acceleration voltage was 10 kV.

A model FT/IR-4X spectrometer, Jasco (Tokyo, Japan), was used for infrared analysis in attenuated total reflection mode (FTIR-ATR). Spectra were recorded in the range of 500 to 4000 cm^{-1}. Dynamic mechanical analysis (DMA) was performed using a dynamic mechanical analyzer MCR 702e MultiDrive Anton Paar (Graz, Austria).

2.6. Biological Tests

A 6 mm punch was used to cut and autoclave 150 mm Grade 1 filter papers and replicates of the 7 samples, as well as uncured rubber samples. The 6 mm filter papers were impregnated with 8 µL of *Eucalyptus nitens* essential oil mixed 50% with ethanol, in addition to papers impregnated with ethanol only, and allowed to dry in a hood near a burner for 15 min. While the dilutions of bacteria were being prepared, the test tubes with the growth of bacterial colonies of *Pseudomonas fluorescens, Bacillus subtilis, Staphylococcus aureus, Escherichia coli* K-12, and *Escherichia coli* FT-17 in distilled water were taken and transferred with a Pasteur pipette to another test tube containing autoclaved distilled water until the appropriate turbidity pattern was obtained to be compared with the 0.5 McFarland's standard tube. Using seeding handles, the culture of each bacterium was spread over the whole of the plates containing Mueller-Hinton agar, and then the rubber samples with lignocellulosic filler and the already-dried filter papers were labeled and placed. *Pseudomona fluorescens, Bacillus subtilis,* and *Staphylococcus aureus* were incubated at 30 °C and the two *E. coli* strains at 37 °C. Measurements were made at the inhibition halos presented on the plates after 24 h

3. Results and Discussion

Table 2 shows the compounds identified by gas chromatography of the essential oils extracted from the unsifted (USF) and sifted (SF) Eucalyptus nitens dry-leaf-based powders. This table suggests a probable structure, approximate molecular weight, compound type, and area percentage of the major species detected. It is possible to observe that during the processing of the dry leaf powder, the content of the species changes; those with lower molar mass and lower boiling point tend to decrease their content, while those with higher boiling points tend to increase their content (Table 2). This could be explained by the fact that the grinding and sieving processes facilitate the evaporation of more volatile compounds. This is evident, for example, for 1R-α-pinene and eucalyptol (cineol), which is a volatile monoterpene, compared to argomadendrene and globulol, which are sesquiterpenes of higher molecular weight and less volatile [21,22] The identified compounds mostly have structures related to the cross-linking process, mainly because they have double bonds.

Table 2. Main compounds of the five species at major concentrations in DL, USF, and SF determined by gas chromatography using a qualification score, QUAL, ≥90%.

Compound Name	Boiling Point (°C)	Molecular Formula and Molecular Weight (Da)	Percentage Area (%)		
			DL	USF	SF
1R-α-Pinene	155	$C_{10}H_{16}$, 136	21.1	8.96	5.17
Eucalyptol (Cineol)	177	$C_{10}H_{18}O$, 154	56.93	38.75	37.58
α-Terpineol acetate	219	$C_{12}H_{20}O_2$, 196	5.39	5.56	6.21
Aroma-dendrene	258	$C_{15}H_{24}$, 204	1.6	14.84	15.77
Globulol	283	$C_{15}H_{26}$, 222	5.2	9.23	10.04

Figure 1 shows the FTIR spectra of the NR samples and those containing sifted (SF) and unsifted (USF) vegetable fiber-based fillers. As can be seen, all spectra exhibit dominant signals characteristic of NR, such as those observed at 2961 cm^{-1} and 1524 $cm^{-1,}$ corresponding to CH stretching and C-C vibration, respectively. In addition, absorption bands were observed at 1447 cm^{-1} and 1375 cm^{-1}, which were associated with CH_2 and CH_3 deformations, respectively. Meanwhile the signals at 974 cm^{-1}, 842 cm^{-1} and 560 cm^{-1} correspond to out-of-plane bending [23]. It is interesting to note that the main changes observed in the samples are subtle but associated with stretching of the C=S groups, and the characteristic absorption bands for this group are those recorded at 1391 cm^{-1}, 1032 $cm^{-1,}$ and 725 cm^{-1} [24,25]. The band observed at 1032 cm^{-1} decreased significantly in samples containing fillers, which could be associated with the interaction between the filler and the polymer matrix.

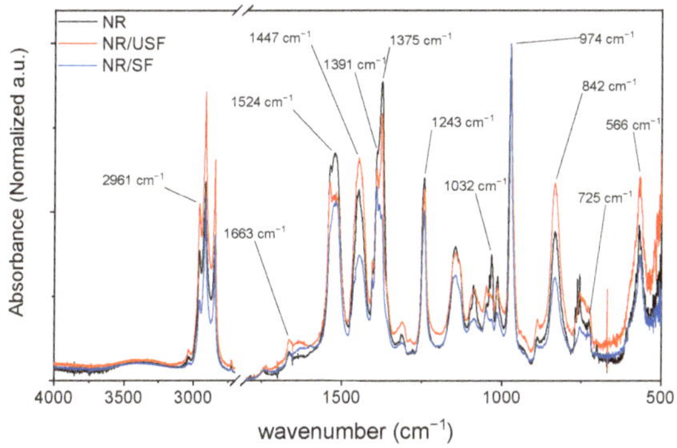

Figure 1. FTIR spectra of NR composites containing USF and SF.

3.1. Curing Curves and Vulcanization Parameters of NR-Based Composites

The values of minimum torque (M_L), maximum torque (M_H), scorch time 2 (t_{s2}), and optimum vulcanization time (t_{90}) of the NR composites are described in Table 3. It can be observed that the content of sifted and unsifted eucalyptus dry leaf powder has no significant effect on the M_H recorded. Similarly, it is pertinent to mention that the M_L slightly increased, which could be attributed to the effect of the vegetable filler content on the torque, which, by the hydrodynamic effect, gives more stiffness to the rubber system. However, the t90 also increases as the filler content increases, while the MH is not significant, being lower than the MH of unfilled NR. This suggests a slowing down of the vulcanization rate, probably due to the effect of the filler, which inhibits the diffusion of curing agents [26]. In addition, its chemical nature could probably play a role in such a process [1]. In fact, it is well known that vegetable fillers are made of wood, their pH is below 5.0, and they tend to be hygroscopic, properties that could interfere with the vulcanization process [27,28] In addition, since wood is mainly composed of cellulose and lignin, and the latter is used as a sorbent material, it is likely that the curing agent will be adsorbed by the filler, hindering the vulcanization process [29].

Table 3. Curing data of the NR composites.

Sample	t_{s2} (min:s)	t_{90} (min:s)	M_L (Nm)	M_H (Nm)
NR	0:55 ± 0:01	1:26 ± 0:08	0.043 ± 0.001	0.886 ± 0.068
NR/USF10	0:59 ± 0:01	2:18 ± 0:13	0.059 ± 0.001	0.800 ± 0.001
NR/USF20	1:09 ± 0:01	2:09 ± 0:08	0.067 ± 0.005	0.852 ± 0.036
NR/USF40	1:15 ± 0:04	2:07 ± 0:03	0.088 ± 0:003	0.821 ± 0.033
NR/SF10	1:13 ± 0:03	2:04 ± 0:07	0.070 ± 0.001	0.774 ± 0.063
NR/SF20	1:16 ± 0:06	2:14 ± 0:08	0.090 ± 0.007	0.780 ± 0.039
NR/SF40	1:17 ± 0:02	2:56 ± 0:20	0.101 ± 0.000	0.857 ± 0.001

3.2. Mechanical Properties, Hardness and Resilience of NR Composites

Table 4 shows the modulus values at different elongations (M100, M300, and M500), tensile strength (TS), and elongation at break (EB) of the rubber composites. The unfilled rubber sample has a maximum elongation of only 145%, and its M100 is 1.26 MPa. However, when the filler was added, the maximum elongation increased and reached 699% for the sample containing 40 phr of sifted leaf powder. Similarly, the unsifted samples showed an increase in elongation. This indicates that this lignocellulosic material, which contains traces of terpenes, tends to be compatible with NR and promotes the cohesion of the polymer chains through discrete cross-linking. It should be noted that monoterpenes and sesquiterpenes originate from a metabolic pathway related to that of NR and are structurally similar in polarity [30]. Moreover, the higher elongation recorded for the compounds containing sifted fibers, smaller than that of unsifted fibers, allows the system to present a homogeneous character in particle size distribution, which also contributes to achieving a homogeneous distribution and higher phase continuity than that of those composites containing unsifted filler [31]. Table 4 also shows the hardness and resilience values of the rubber compounds. Hardness tends to show a marginal increase with the content of sifted and unsifted EAF powder, a behavior attributed to the presence of filler in the composite composition of a harder material, such as the vegetable filler used in this research, than that of NR. Meanwhile, resilience tends to gradually decrease as the filler content increases. The filler appears to have a discrete effect on the viscoelastic properties of NR. The loss of resilience as a result of increasing the filler content is attributed to the fact that the filler occupies the free volume of the rubber, which hinders the movement of the rubber chains.

Table 4. Mechanical properties of NR composites consisting of moduli at 100% (M100), 200% (M200), and 300% (M300) of elongation, tensile strength (TS), elongation at break (EB), Shore A Hardness, and resilience.

Sample	M100 (MPa)	M200 (MPa)	M300 (MPa)	TS (MPa)	EB (%)	Shore A Hardness	Resilience (%)
NR	1.26 ± 0.01			1.71 ± 0.039	145 ± 3	41.6 ± 0.4	74.1 ± 0.9
NR/USF10	1.42 ± 0.08	2.40 ± 0.12		2.84 ± 0.045	240 ± 14	43.6 ± 1.0	69.7 ± 0.5
NR/USF20	1.56 ± 0.072	2.69 ± 0.21		2.69 ± 0.21	200 ± 6	46.2 ± 0.3	68.3 ± 0.3
NR/USF40	1.16 ± 0.04	1.81 ± 0.06	2.52 ± 0.083	6.42 ± 0.18	626 ± 21	43.2 ± 1.0	60.7 ± 0.5
NR/SF10	1.40 ± 0.029			1.98 ± 0.13	170 ± 8	42.7 ± 0.8	70.3 ± 0.9
NR/SF20	1.54 ± 0.03	2.48 ± 0.03	3.55 ± 0.03	11.98 ± 0.81	707 ± 41	44.0 ± 0.5	67.8 ± 0.9
NR/SF40	1.53 ± 0.06	2.44 ± 0.08	3.35 ± 0.10	9.45 ± 0.29	699 ± 25	44.1 ± 0.5	63.5 ± 4.0

Figure 2 shows the SEM images of the unsifted and sifted fillers. As can be seen, in the case of the unsifted fillers, the fibers have a wide size distribution, with some of them having a high aspect ratio. In contrast, the sifted filler had a more homogeneous size distribution, demonstrating the effectiveness of the milling and sieving processes.

Figure 3 shows the SEM micrographs of the fracture cross-section of NR and NR composite specimens containing 20 phr sifted and unsifted fillers, as well as composites that showed significant differences in mechanical properties. The phase homogeneity of rubber is altered by the presence of the filler material. As seen in Figure 3d–f, the samples containing USF show particles and fibers randomly distributed in the rubber matrix, while those containing sifted filler (Figure 3g–i) show particles of comparable diameters distributed in the polymer matrix. It is worth mentioning that although the phase continuity is modified by the continuity of the filler, it can be inferred that the filler has a certain affinity for the polymeric phase. This is supported by the shape of the analyzed zone, where it was observed that the filler maintained cohesion with the polymer matrix at the time of fracture (Figure 3f,i). This could explain the significant improvement in the mechanical properties of composites containing vegetable fillers compared to NR.

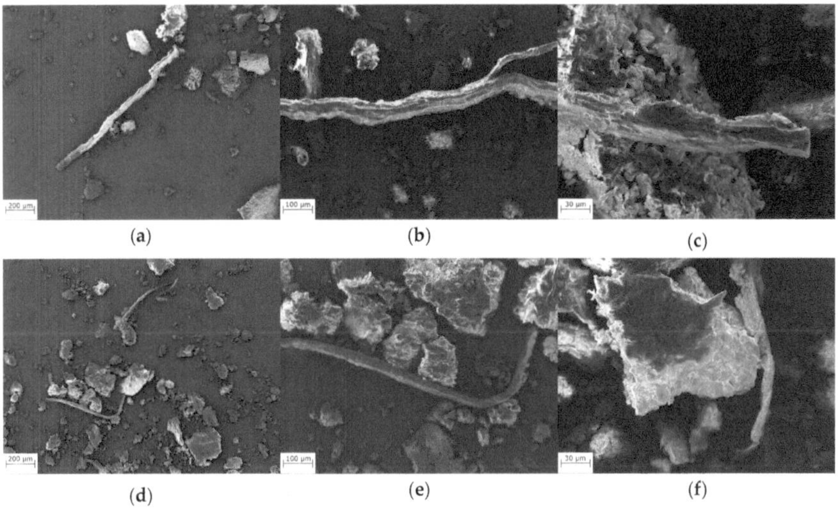

Figure 2. SEM images of USF (**a**–**c**) and SF (**d**–**f**).

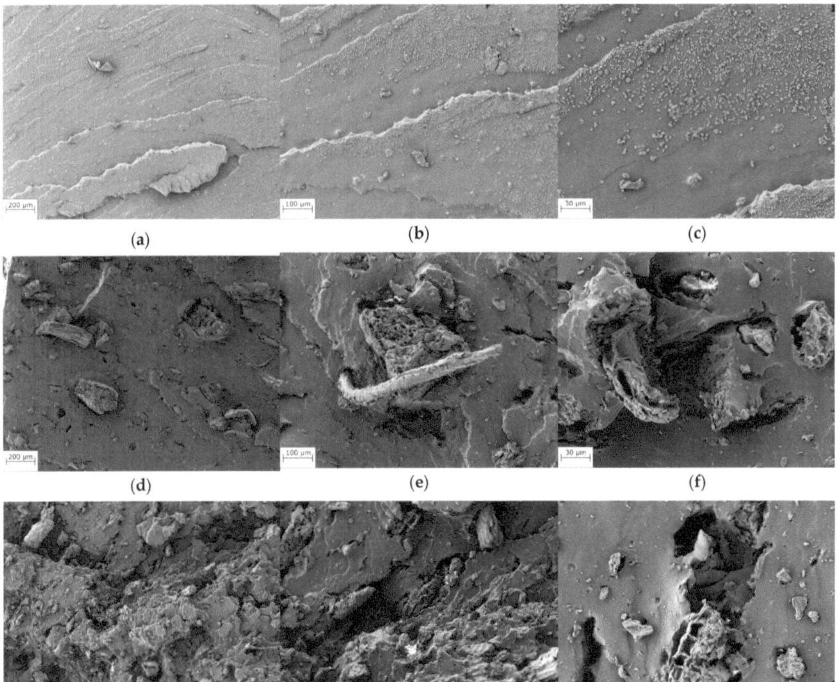

Figure 3. SEM images of NR (**a**–**c**), NR/USF (**d**–**f**), and NR/SF (**g**–**i**).

3.3. Dynamic Mechanical Analysis of Rubber Composites

Figure 4 shows the storage, loss modulus, and loss factor of the NR-based samples and filler-containing composites obtained by DMA.

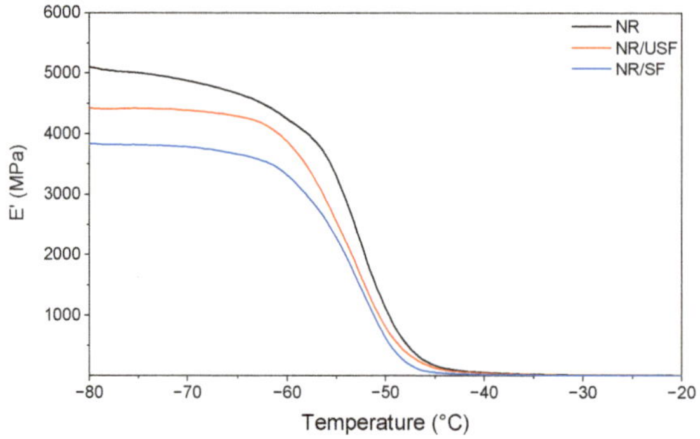

(**a**)

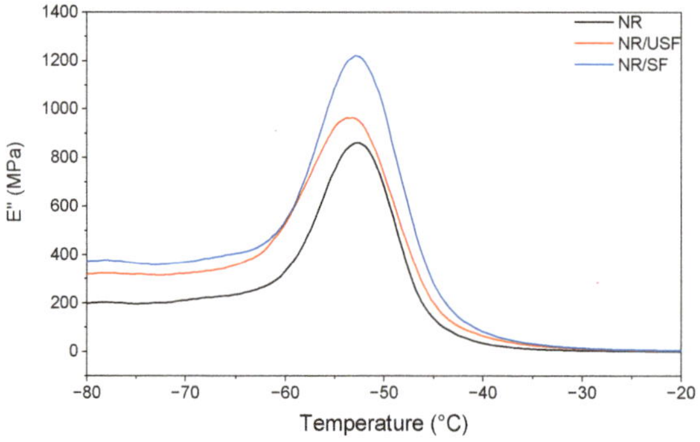

(**b**)

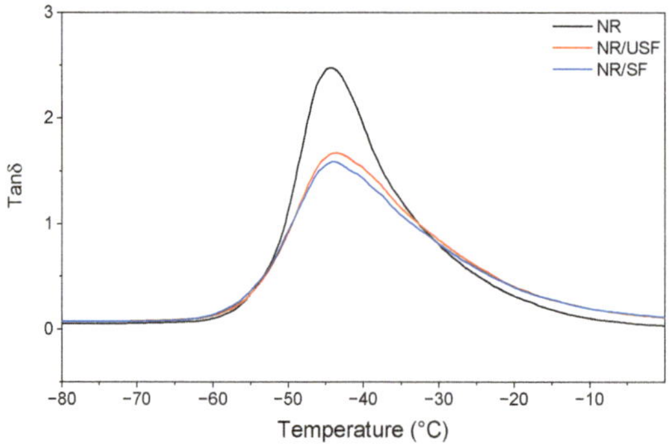

(**c**)

Figure 4. Dynamic mechanical analysis of NR, NR/USF, and NR/SF composites: (**a**) storage modulus (E′), (**b**) loss modulus (E″), and (**c**) loss factor (tanδ).

In Figure 4a, it is evident that the storage modulus of the samples with the vegetable filler is compared to that of NR. This increase can be attributed to the reinforcing role of the fibers and their specific compatibility with the matrix. Furthermore, the higher increase observed in the sample containing the sifted filler provides further evidence of this compatibility. This increase is likely due to the greater regularity in the size and smaller size distribution of the filler material, which facilitates the interaction between the polymer and the filler.

The loss moduli (Figure 4b) of the composites with fillers at low temperatures are comparable to and higher than those of the unfilled NR. However, there was a significant difference between the loss moduli of the composites with sifted and unsifted fillers. This suggests that the morphology and size distribution of the fillers influence stress dissipation. It is important to note that the curves reach their maxima at approximately −53 °C, which corresponds to the glass transition temperature (Tg) of these materials.

3.4. Study of Microorganism Affinity to Natural Rubber-Based Materials

Table 5 displays the average inhibition radius values of vulcanized NR for the bacteria *Bacillus subtilis*, *Staphylococcus aureus*, *Escherichia coli* K 12, *Escherichia coli* FT 17, and *Pseudomonas fluorescens*. It is evident that rubber additives, such as accelerator systems and activators, inhibit bacterial growth, indicating their toxicity. The inhibition is more pronounced for *Staphylococcus aureus*, while *Pseudomonas fluorescens* is less affected by these additives. This inhibitory effect contrasts with the behavior of the studied species against pristine unvulcanized NR, where proliferation occurs without disturbance and no inhibition halo is observed. It is worth noting that *Pseudomonas* has been used for the biological degradation of NR [32].

Table 5. Inhibition diameters of NR-based materials.

	Inhibition Diameter (cm)			
	Vulcanized NR	**Unvulcanized NR**	**NR/USF**	**NR/SF**
Bacillus subtilis	2.65	0	0	0
Staphylococcus aureus	3.10		0.67	0.62
Escherichia Coli K 12	1.22		0	0
Escherichia Coli FT 17	1.22		0	0
Pseudomonas fluorescens	0.623		0	0

In order to determine the level of inhibition caused by the filler, an equation has been developed to compare the inhibition diameters of vulcanized rubber with those of rubber containing the filler. This comparison is expressed as the material affinity index for microorganisms (*MA index*), which is defined by Equation (1).

$$MA\ index = \frac{(D_0 - D_c)}{D_0} \times 100 \quad (1)$$

where D_0 is the inhibition diameter of the unfilled material, and D_c is the inhibition diameter of the composite. This index was estimated for each of the microorganisms mentioned in Table 5. It can be observed in Figure 5 that the *MA* indexes of the NR-based composite containing unsifted and sifted fillers showed more affinity of the microorganism to this type of material. This indicates that the filler content produces a change in the behavior of the microorganisms toward the composite, reaching maximum proliferation. This indicates that the antimicrobial nature of vulcanized natural rubber decreases, which is probably promoted by the presence of biomolecules such as polysaccharides, which act as carbon sources, favoring bacterial growth [33].

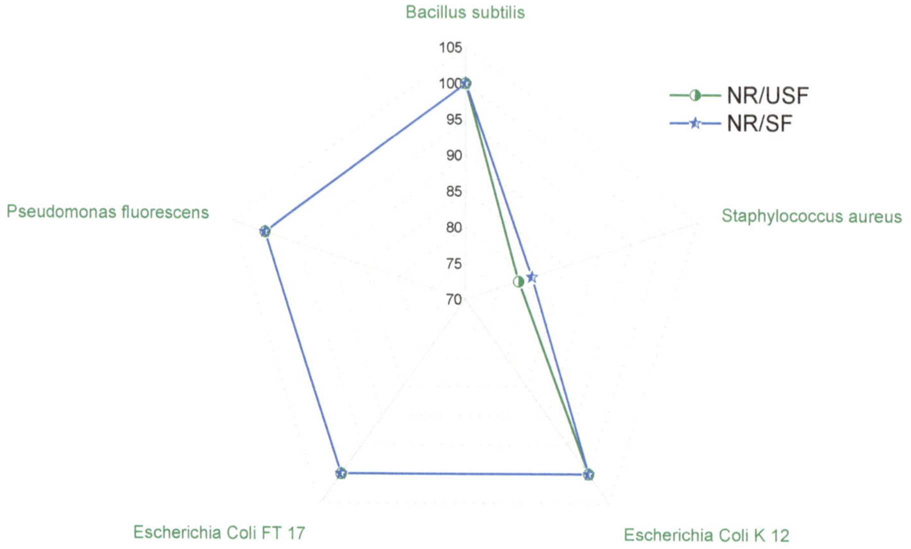

Figure 5. Radial plot showing the MA index for NR/USF and NR/SF, both containing 40 phr of filler.

Based on the results obtained, it was found that although the microorganisms studied were pathogenic, a variety of microorganisms grew in the presence of the leaf-based filler. This indicates that the filler provides a carbon source, such as a polysaccharide (cellulose), which promotes the proliferation of bacteria. Since Eucalyptus nitens leaves are a complex forestry by-product containing cellulose, lignin, and essential oil, more may not only provide a better affinity for microorganisms but also improve the performance of natural rubber composites. This aspect is interesting because this material can be used in applications where bacterial proliferation is required or minimal impact on bacterial colonies is required, such as a specific ecosystem.

4. Conclusions

Composites based on natural rubber and a filler derived from Eucalyptus nitens leaves were prepared. A discrete slowing down of the cross-linking process and a decrease in the maximum modulus were observed, which can be attributed to the chemical structure of the plant filler as well as to the adsorption of the cross-linking agents. The addition of this filler to rubber promoted an increase in the mechanical properties, significantly increasing the elongation at break and tensile strength. It was demonstrated that the presence of a filler prepared from dry shredded and sifted leaves promoted a greater increase in mechanical properties than the unsifted filler. On the other hand, the vegetable filler content facilitated bacterial proliferation, increasing the affinity index of the material for microorganisms such as Bacillus subtilis, Staphylococcus aureus, Escherichia coli K 12, Escherichia coli FT 17, and Pseudomonas fluorescens. This higher affinity was attributed to the chemical nature of the filler, which is composed of macromolecules that would act as a carbon source for bacteria. These results are interesting because materials that have a higher affinity for microorganisms, such as bacteria, may have a lower environmental impact on various ecosystems.

Author Contributions: Conceptualization, H.A.-B. and C.P.-M.; methodology, H.A.-B. and C.P.-M.; validation, M.H.S. and R.V.; formal analysis, S.T.-G. and N.R.-C.; investigation, H.A.-B. and N.R.-C.; resources: H.A.-B., C.P.-M., J.B., M.H.S., and R.V.; data curation: N.R.-C.; writing—original draft preparation: H.A.-B., N.R.-C., M.D., and C.P.-M.; writing—review and editing: H.A.-B., M.H.S., and R.V.; supervision: H.A.-B. All authors have read and agreed to the published version of the manuscript.

Funding: PID2022-143107OB-I00 funded by MCIN/AEI/10.13039/501100011033 and the European Union NextGeneration/PRTR. BIP40046955 research project granted by Gobierno Regional del Biobío (Chile) and the VRID Research project VRID N°2024001180INV (Universidad de Concepción). FONDEQUIP EQM220083 funded by Subdirectorate of Centers and Associative Research (SCIA), National Agency for Research and Development (ANID).

Institutional Review Board Statement: Not applicable.

Data Availability Statement: The data presented in this study are available on request from the corresponding author.

Acknowledgments: M. Hernández, Santana, and R. Verdejo would like to acknowledge the State Research Agency of Spain (AEI) and the European Union for a grant PID2022-143107OB-I00 funded by MCIN/AEI/10.13039/501100011033 and by the European Union NextGeneration/PRTR. H. Aguilar-Bolados, C. Perez-Manriquez, and J. Becerra thank the BIP40046955 research project granted by Gobierno Regional del Biobío (Chile) and H. Aguilar-Bolados also acknowledges to the VRID Research project VRID N°2024001180INV (Universidad de Concepción) and thanks to Center for Spectroscopy and Microscopy (CESMI), Vicerrectoría de Investigación y Desarrollo (VRID), Universidad de Concepción, Concepción 3349001 Chile. HA-B thanks to FONDEQUIP EQM220083 funded by (SCIA, ANID).

Conflicts of Interest: The authors declare no conflicts of interest.

References

1. Aguilar-Bolados, H.; Bascuñan-Heredia, A.; Alvarez, G. Sustainable Approach of the Natural Rubber. In *Green-Based Nanocomposite Materials and Applications*; Engineering Materials; Springer: Cham, Switzerland, 2023. [CrossRef]
2. Avalos-Belmontes, F.; González, F.J.; López-Manchado, M.Á. Introduction to Green Based Nanocomposites: From Polymer Discovery to Eco-friendly Materials. In *Green-Based Nanocomposite Materials and Applications*; Engineering Materials; Springer: Cham, Switzerland, 2023. [CrossRef]
3. Utrera-Barrios, S.; Bascuñan, A.; Verdejo, R.; López-Manchado, M.Á.; Aguilar-Bolados, H.; Hernández Santana, M. Sustainable Fillers for Elastomeric Compounds. In *Green-Based Nanocomposite Materials and Applications*; Engineering Materials; Springer: Cham, Switzerland, 2023. [CrossRef]
4. Morseletto, P. Targets for a circular economy. *Resour. Conserv. Recycl.* **2020**, *153*, 104553. [CrossRef]
5. Surbhi; Kumar, A.; Singh, S.; Kumari, P.; Rasane, P. Eucalyptus: Phytochemical composition, extraction methods and food and medicinal applications. *Adv. Tradit. Med.* **2023**, *23*, 369–380. [CrossRef]
6. Liu, S.; Zhao, C.; Cao, Y.; Li, Y.; Zhang, Z.; Nie, D.; Tang, W.; Li, Y. Comparison of Chemical Compositions and Antioxidant Activity of Essential Oils from Litsea Cubeba, Cinnamon, Anise, and Eucalyptus. *Molecules* **2023**, *28*, 5051. [CrossRef] [PubMed]
7. Elaissi, A.; Rouis, Z.; Mabrouk, S.; Bel Haj Salah, K.; Aouni, M.; Khouja, M.L.; Farhat, F.; Chemli, R.; Harzallah-Skhiri, F. Correlation between chemical composition and antibacterial activity of essential oils from fifteen Eucalyptus species growing in the Korbous and Jbel Abderrahman arboreta (North East Tunisia). *Molecules* **2012**, *17*, 3044–3057. [CrossRef] [PubMed]
8. Nouri-Ganbalani, G.; Ebadollahi, A.; Nouri, A. Chemical Composition of the Essential Oil of *Eucalyptus procera* Dehnh. and Its Insecticidal Effects Against Two Stored Product Insects. *J. Essent. Oil-Bear. Plants* **2016**, *19*, 1234–1242. [CrossRef]
9. Shala, A.Y.; Gururani, M.A. Phytochemical properties and diverse beneficial roles of eucalyptus globulus labill: A review. *Horticulturae* **2021**, *7*, 450. [CrossRef]
10. Chandorkar, N.; Tambe, S.; Amin, P.; Madankar, C. A systematic and comprehensive review on current understanding of the pharmacological actions, molecular mechanisms, and clinical implications of the genus Eucalyptus. *Phytomedicine Plus* **2021**, *1*, 100089. [CrossRef]
11. Santos, B.; Farias, J.H.A.; Simões, M.M.; Medeiros, M.A.A.; Alves, M.S.; Diniz, A.F.; Soares, A.P.O.; Cavalcante, A.P.T.M.; Silva, B.J.N.; Almeida, J.C.S.; et al. Evaluation of the antimicrobial activity of Eucalyptus radiata essential oil against *Escherichia coli* strains isolated from meat products. *Braz. J. Biol.* **2024**, *84*, e281361. [CrossRef] [PubMed]
12. Elangovan, S.; Mudgil, P. Antibacterial Properties of Eucalyptus globulus Essential Oil against MRSA: A Systematic Review. *Antibiotics* **2023**, *12*, 474. [CrossRef]
13. Batish, D.R.; Singh, H.P.; Kohli, R.K.; Kaur, S. Eucalyptus essential oil as a natural pesticide. *For. Ecol. Manag.* **2008**, *256*, 2166–2174. [CrossRef]
14. Kulshrestha, U.; Gupta, T.; Kumawat, P.; Jaiswal, H.; Ghosh, S.B.; Sharma, N.N. Cellulose nanofibre enabled natural rubber composites: Microstructure, curing behaviour and dynamic mechanical properties. *Polym. Test.* **2020**, *90*, 106676. [CrossRef]
15. Supanakorn, G.; Taokaew, S.; Phisalaphong, M. Ternary composite films of natural rubber, cellulose microfiber, and carboxymethyl cellulose for excellent mechanical properties, biodegradability and chemical resistance. *Cellulose* **2021**, *28*, 8553–8566. [CrossRef]
16. Lozada, E.R.; Gutiérrez Aguilar, C.M.; Jaramillo Carvalho, J.A.; Sánchez, J.C.; Barrera Torres, G. Vegetable Cellulose Fibers in Natural Rubber Composites. *Polymers* **2023**, *15*, 2914. [CrossRef]

17. Homkhiew, C.; Pianhanuruk, E.; Rawangwong, S.; Boonchouytan, W.; Numrat, R. Rubberwood sawdust filled natural rubber composites: Effects of filler loading and zinc oxide content. *IOP Conf. Ser. Mater. Sci. Eng.* **2020**, *773*, 012005. [CrossRef]
18. Kaewpruk, C.; Boopasiri, S.; Poonsawat, C.; Sae-Oui, P.; Siriwong, C. Utilization of Sawdust and Wood Ash as a Filler in Natural Rubber Composites. *ChemistrySelect* **2021**, *6*, 264–272. [CrossRef]
19. Troncoso, C.; Perez, C.; Hernandez, V.; Sanchez-Olate, M.; Rios, D.; Martin, A.S.; Becerra, J. Induction of defensive response in Eucalyptus globulus plants and its persistence in vegetative propagation. *Nat. Prod. Commun.* **2013**, *8*, 1934578X1300800330. [CrossRef]
20. Zhang, Z.; Guo, F.; Ke, Y.; Xiang, C.; Jia, X. Effect of vulcanization on deformation behavior of rubber seals: Thermal–mechanical–chemical coupling model, numerical studies, and experimental validation. *Mater. Des.* **2022**, *224*, 111314. [CrossRef]
21. Mulyaningsih, S.; Sporer, F.; Zimmermann, S.; Reichling, J.; Wink, M. Synergistic properties of the terpenoids aromadendrene and 1,8-cineole from the essential oil of eucalyptus globulus against antibiotic-susceptible and antibiotic-resistant pathogens. *Phytomedicine* **2010**, *17*, 1061–1066. [CrossRef]
22. Ololade, S.Z.; Olawore, O.N. Characterization of essential oil from the seed of *Eucalyptus cloeziana* and evaluation of its modes of medicinal potentials. *Edorium J. Infect. Dis.* **2017**, *3*, 1–8.
23. Rahmah, M.; Ahmad Khusyairi, A.R.; Nur Khairunnisya, H.; Siti Nurbalqis, M. Oven Ageing versus UV Ageing Properties of Natural Rubber Cup Lump/EPDM Rubber Blend with Mangosteen Powder (MPP) as Natural Antioxidant. *IOP Conf. Ser. Mater. Sci. Eng.* **2019**, *548*, 012014. [CrossRef]
24. Ariponnammal, S.; Chandrasekaran, S.; Sanjeeviraja, C. Low temperature photoluminescence studies on semiorganic tris thiourea copper (I) chloride single crystal. *Cryst. Res. Technol.* **2012**, *47*, 145–150. [CrossRef]
25. Kashte, S.; Arbade, G.; Sharma, R.K.; Kadam, S. Herbally Painted Biofunctional Scaffolds with Improved Osteoinductivity for Bone Tissue Engineering. *J. Biomim. Biomater. Biomed. Eng.* **2019**, *41*, 49–68. [CrossRef]
26. Aguila-Toledo, E.; Maldonado-Magnere, S.; Yazdani-Pedram, M.; Bascuñan-Heredia, A.; Dahrouch, M.R.; Molina, F.; Santana, M.H.; Verdejo, R.; Lopez-Manchado, M.A.; Aguilar-Bolados, H. Fluorosilicone Composites with Functionalized Graphene Oxide for Advanced Applications. *ACS Appl. Polym. Mater.* **2023**, *5*, 7755–7765. [CrossRef]
27. Geffert, A.; Geffertova, J.; Dudiak, M. Direct method of measuring the pH value of wood. *Forests* **2019**, *10*, 852. [CrossRef]
28. Akahori, Y.; Kawahara, S. Effect of water on the accelerated sulfur vulcanization of natural rubber. *Polym. Test.* **2023**, *123*, 108030. [CrossRef]
29. Fan, H.; Li, F.; Huang, H.; Yang, J.; Zeng, D.; Liu, J.; Mou, H. pH graded lignin obtained from the by-product of extraction xylan as an adsorbent. *Ind. Crops Prod.* **2022**, *184*, 114967. [CrossRef]
30. Mosquera, M.E.G.; Jiménez, G.; Tabernero, V.; Vinueza-Vaca, J.; García-Estrada, C.; Kosalková, K.; Sola-Landa, A.; Monje, B.; Acosta, C.; Alonso, R.; et al. Terpenes and Terpenoids: Building Blocks to Produce Biopolymers. *Sustain. Chem.* **2021**, *2*, 467–492. [CrossRef]
31. Samal, S. Effect of shape and size of filler particle on the aggregation and sedimentation behavior of the polymer composite. *Powder Technol.* **2020**, *366*, 43–51. [CrossRef]
32. Aboelkheir, M.G.; Bedor, P.B.; Leite, S.G.; Pal, K.; Toledo Filho, R.D.; Gomes de Souza, F. Biodegradation of Vulcanized SBR: A Comparison between Bacillus subtilis, Pseudomonas aeruginosa and Streptomyces sp. *Sci. Rep.* **2019**, *9*, 19304. [CrossRef]
33. Sichert, A.; Cordero, O.X. Polysaccharide-Bacteria Interactions from the Lens of Evolutionary Ecology. *Front. Microbiol.* **2021**, *12*, 705082. [CrossRef]

polymers

MDPI

Article

Sulphur Copolymers with Pyrrole Compounds as Crosslinking Agents of Elastomer Composites for High-Performance Tyres

Simone Naddeo, Vincenzina Barbera and Maurizio Galimberti *

Department of Chemistry, Materials and Chemical Engineering "G. Natta", Politecnico di Milano, Via Mancinelli 7, 20131 Milano, Italy; simone.naddeo@polimi.it (S.N.); vincenzina.barbera@polimi.it (V.B.)
* Correspondence: maurizio.galimberti@polimi.it

Abstract: Driving a car at extreme speeds, road holding, and sustainability do not go together well. Formula 1 racing is exciting but is not an example of sustainability. The aim of this work was to use materials, suitable for the treads of high-performance racing tyres, that can favour both high performance and sustainability. In particular, the objective was to achieve high dynamic rigidity at high temperatures (>100 °C) and a stable crosslinking network. A copolymer from an industrial waste such as sulphur and a comonomer from a circular biosourced material were used as the crosslinking agent of an elastomer composite based on poly(styrene-co-butadiene) from solution anionic polymerization and a carbon black with a high surface area. The biosourced circular material was 1,6-*bis*(2,5-dimethyl-1*H*-pyrrol-1-yl)hexane (HMDP), the di-pyrrole derivative of hexamethylenediamine. Two poly(S-co-HMDP) copolymers, with different S/HMDP ratios (6 and 8.9, Copolymer 1 and Copolymer 2) were carefully characterized by means of ^{1}H-, ^{13}C-, 2D^{1}H-^{1}H-COSY and 2D ^{1}H-^{13}C HSQC NMR. The comparison of the spectra highlighted the substitution with sulphur of the β-position of the pyrrole ring: mono-substitution largely prevailed in Copolymer 1 and also bi-substitution in Copolymer 2. The copolymers were used as additives in the vulcanization system. Compared with a reference composite, they allowed us to achieve more efficient vulcanization, a higher density of the crosslinking network, higher dynamic rigidity, better ultimate tensile properties, and better stability of the crosslinking network at high temperatures. Compared with a traditional oil-based crosslinking agent for elastomer composites with high rigidity and a stable structure at high temperatures, such as the perthiocarbamate 6-((dibenzylcarbamothioyl)disulfaneyl)hexyl 1,3-diphenylpropane-2-sulfinodithioate, the poly(S-co-HMDP) copolymers led to higher dynamic rigidity and better ultimate tensile properties. These improvements occurring simultaneously are definitely unusual. This work paves the way for the upcycling of circular materials in a large-scale application such as in tyres.

Keywords: inverse vulcanization; sulphur copolymers; rubber compounds

Citation: Naddeo, S.; Barbera, V.; Galimberti, M. Sulphur Copolymers with Pyrrole Compounds as Crosslinking Agents of Elastomer Composites for High-Performance Tyres. *Polymers* **2024**, *16*, 2802. https://doi.org/10.3390/polym16192802

Academic Editor: Bernhard V. K. J. Schmidt

Received: 30 July 2024
Revised: 26 September 2024
Accepted: 27 September 2024
Published: 3 October 2024

1. Introduction

The key feature of elastomers is the entropic elasticity. The elastomers chains, above their glass transition temperature, can be repeatedly stretched to at least twice their original length and recover immediately and forcibly their initial size and shape, upon removing any stress on them [1,2]. To achieve this behaviour, elastomer chains must be linked together. As a matter of fact, the most important unsaturated and saturated elastomers, poly(1,4-cis-isoprene) from natural sources (natural rubber, NR) [3] and poly(ethene-co-propene) (EPR) from insertion polymerization [4], respectively, can have entropic elasticity also in the uncrosslinked state. Both NR [5–7] and EPR [8–10] have the ability to rapidly crystallize under strain, and the crystallites act as physical crosslinks. The strain-induced crystallization of NR was attributed to the high stereoregularity of the polymer chains [3,11–13] and to the small melting entropy [14,15]. More recently, the NR chains have been shown to be linked together, through their chain ends, which are dimethylallyl groups modified with proteins and with phospholipids containing fatty acid esters [16–21]. The ionic and hydrogen bonds formed by these polar groups are responsible for the crosslinking and the saturated fatty acids that act as nucleating agents [22,23]. Natural rubber samples from *Hevea Brasiliensis*, dandelion, and guayule all exhibit entropic elasticity in their pristine state [24,25]. Commercially available EPR does not show entropic elasticity in the uncured state. The strain-induced crystallization of EPR was reported in the patent and scientific literature for samples having ethene sequences that are able to crystallize under strain but amorphous at rest and propene sequences whose isotacticity does not disturb the crystallization [8–10].

It is indeed highly desirable to have elastomers with entropic elasticity in the uncured state. However, to achieve the performance required by their demanding applications, elastomer chains must be chemically crosslinked. Crosslinking dates back to Mesoamericans [26]. Nowadays, many different systems are used to promote the crosslinking of the elastomer chains [27]: they are based on peroxides, phenolic resins, metal oxides, silanes, quinones, maleimides, and particularly, sulphur. The sulphur-based systems contain other chemicals [28]. Indeed, the active sulphurating species must be formed, and to do that, activators and primary and secondary accelerators must be used. The activators are oxides or carboxylates of divalent metals, such as ZnO, zinc stearates, and zinc octanoates. Primary accelerators are sulphenamides, thiazoles, and carbamates. Vulcanization leads to the formation of sulphur bridges between the polymer chains, made by one, two, or more sulphur atoms. The properties of the elastomer composites depend on the length of the sulphur bridges: higher elasticity is achieved with short bridges, at the expense of the fatigue resistance. Hence, a longer service life is achieved with longer bridges. The sulphur bridges experience a dynamic equilibrium: they evolve towards more stable but shorter sequences. Moreover, the chemical bond established by sulphur with the polymer chains is reversible [29–31]. It would be highly desirable to have a crosslinking network with a constant structure all over the whole service life of the rubber compound, particularly when it is applied to tyres. This need is particularly relevant in the case of high-performance sport tyres. The objective to have a crosslinking network with higher stability was achieved by adding further ingredients to the vulcanization system, able to establish permanent chemical bonds with the elastomer chains. A typical chemical substance is a perthiocarbamate, 1,6-*bis*(N,N′-dibenzylthiocarbamoyldithio)-hexane (DBTH) [32–34], whose chemical structure is shown in Figure 1.

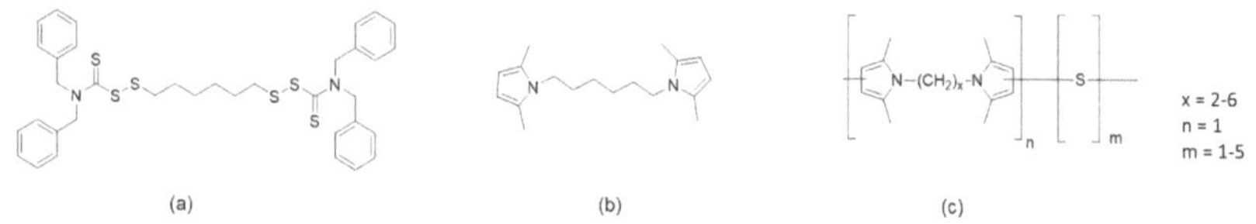

Figure 1. The chemical structure of (**a**) 6-((dibenzylcarbamothioyl)disulfaneyl)hexyl 1,3-diphenylpropane-2-sulfinodithioate (DBTH), (**b**) 1,6-*bis*(2,5-dimethyl-1*H*-pyrrol-1-yl)hexane (HMDP), and (**c**) poly(S-co-HMDP).

During the sulphur-based vulcanization of an unsaturated elastomer, the cleavage of the sulphur–sulphur bonds releases the amines, which act as accelerators, and the dithioalkane reacts with the elastomer chains, forming bridges more stable than the polysulphidic ones [32]. DBTH is not efficient when it is used as the only vulcanizer [33], but in the presence of sulphur and a sulphenamide [32–34], the same crosslink density as the reference compound (without DBTH) and excellent retention of the physical properties were obtained, with the dynamic behaviour remaining almost at the original level even after severe cure conditions [32]. DBTH leads to a higher modulus and lower hysteresis of the composites [34]. However, DBTH presents some technical drawbacks. Its behaviour in vulcanization depends on the formulation of the vulcanization system, and the length of the bridge cannot be tuned. Moreover, it is an oil-based chemical, which cannot be recycled.

This work had the objective of enhancing the dynamic rigidity and the stability of the crosslinking network of elastomer composites suitable for high-performance tyre treads. This objective was pursued by using an innovative crosslinking agent that could be completely circular and biobased. A sulphur copolymer with a di-pyrrole compound was used as an ingredient of a traditional vulcanization system, based on sulphur and a sulphenamide. Sulphur is largely available, in millions of tons, from oil desulphuration [35]. It is unexpensive and not toxic, but it is flammable, and there is a need to find large-scale applications to remove the accumulated stockpiles. The di-pyrrole compound was 1,6-*bis*(2,5-dimethyl-1H-pyrrol-1-yl)hexane (hexamethylenediaminopyrrole, HMDP), whose chemical structure is shown in Figure 1, the di-pyrrole derivative of hexamethylenediamine, which could become available from the hydrolysis of end-of-life poly(amide-6,6). The di-pyrrole compound was prepared through the reaction of the diamine with 2,5-hexanedione. The authors reported on the synthesis of 2,5-hexanedione from dimethylfuran [36]. The sulphur copolymer with HMDP, poly(S-co-HMDP), was prepared through the reaction of sulphur and HMDP in the typical experimental conditions of the so-called inverse vulcanization [35,37–46], which means by simply adding HMDP to molten sulphur. The chemical structure of the copolymer is shown in Figure 1. The synthesis and characterization of poly(S-co-HMDP) has been recently reported elsewhere [47]. Two copolymer samples were used in this work, with different amounts of sulphur. The comparative characterization of the two samples is presented. The elastomer composite had a typical formulation for a high-performance tyre tread: it was based on poly(styrene-co-butadiene) from solution anionic polymerization (SBR) with a high styrene content and a carbon black with a surface area much higher than a standard ASTM N100 carbon black such as CRX1391. The poly(S-co-HMDP) copolymer was used as an ingredient of the vulcanization system, and its behaviour was compared to that of DBTH. HMDP was used as well in its pristine state. In the following text, DBTH, HMDP, and poly(S-co-HMDP) are referred to as "crosslinking agents". The characterization of the composites was carried out by determining the curing kinetics, the dynamic mechanical properties in the shear and the axial modes, and the tensile properties through quasi-static measurements. The structure and the robustness of the crosslinking network were investigated by means of swelling measurements and by performing curing tests at a high temperature (200 °C) for a prolonged time. The one in tyre treads is really a large-scale application: the tyre market is expected to reach 2.7 billion units in the forthcoming year [48].

2. Experimental Section

2.1. Materials

Synthesis of HMDP as the pyrrole compound (PyC): 1,6-*bis*(2,5-dimethyl-1H-pyrrol-1-yl)hexane (HMDP) was synthesized according to the literature [47], and 2,5-hexanedione and hexamethylenediamine from Sigma Aldrich were used as the substrates.

Synthesis of poly(S-co-PyC)copolymers: HMDP was used as the novel comonomer for the preparation of sulphur-based copolymers. Orthorhombic sulphur, purchased from ZOLFINDUSTRIA Pavia, Italy, was used as the second comonomer. Poly(S-co-HMDP) was synthesized according to the literature [47].

Preparation of elastomer composites. The solution of Styrene–Butadiene Rubber (SBR) was HP755B and was provided by JSR. The chemical composition of the polymeric matrix was the following: styrene 39.5% wt, vinyl groups 38.2% wt, and extension oil (TDAE, Treated Distillate Aromatic Extract) 37.5% wt. The reinforcing filler used was CRX 1391, purchased from Cabot Corporation (Boston, MA, USA), with a surface area of 1391 m^2/g. Stearic acid and Zinc salt were provided from Oleo s.r.l. N-(1,3-dimethylbutyl)-N′-phenyl-p-phenylenediamine (6PPD) was used as the typical antioxidant and was purchased from Eastman, Milano, Italy. Zinc Oxide was provided from Lanxess, with a 20% wt of polymeric agent as the dispersing agent. Hansen and Rosenthal provided the oil (Vivatec, Iserlohn, Germany). Sulphur was the same as was used for the preparation of the copolymers. Extension oil (Treated Distillate Aromatic Extract) was used as the processing oil and was purchased from Klaus Dahleke KG, Hamburg, Germany. The Kristalex^{TM} 5140 Hydrocarbon Resin was from Eastman, Kingsport, TN, USA. Accelerating agents like thiuram and Dibenzothiazyl disulphide (MBTS), 80% in weight, were provided by Shandong Yanggu Huatai Chemical Co., Ltd, Yanggu, China. DBTH (Vulkuren, Lanxess, Vienna, Austria) was used as the commercial crosslinking agent and was from Lanxess Deutschland GmbH, Cologne, Germany.

2.2. *Synthesis of 1,6-Bis(2,5-Dimethyl-1H-Pyrrol-1-yl)Hexane (HMDP)*

The synthesis of HMDP was the same as was reported in reference [47]. In brief, hexamethylenediamine (36.6 mmol, 4.27 g) and water (10 mL) were sequentially added to a two-necked round-bottom flask. The mixture was stirred at 37 °C for 30 min, until a homogeneous solution was obtained. Then, droppwised 2,5-hexanedione (73.25 mmol, 8.62 mL) was added to the solution. The obtained reaction mixture was heated to 120 °C and stirred for three hours. Finally, the obtained HMDP was cooled down to room temperature and then washed with distilled water to remove impurity traces. Filtration on a filter paper of the obtained HMDP was performed and dried at 1 atm pressure. A pale-white solid was obtained. HMDP was obtained in 93% of yield, 9.1 g, with an Atom Efficiency of 73%. The calculated E factor was estimated to be 0.46, considering only the organic compounds and neglecting the water used during the washing. The characterization of HMDP was performed by means of NMR spectroscopy and GC-Mass and ESI-Mass analyses.

^{1}H-NMR ($CDCl_3$), δ (ppm): 5.75 (s, 4H), 3.73–3.70 (m, 4H), 2.21 (s, 12H), 1.67–1.59 (m, 4H), 1.43–1.34 (m, 4H).

^{13}C-NMR ($CDCl_3$), δ (ppm): 127.28, 105.12, 43.60, 31.06, 26.84, 12.57.

GC-Mass: retention time = 24.20 min; molecular peak = 272 m/z.

ESI-Mass: $[M + H]^+$: Calculated exact mass = 273.22 m/z; found exact mass = 273.3 m/z.

2.3. *Preparation of Poly(S-co-PyC) Copolymers*

The syntheses reported below were performed several times, to check the reproducibility of the process and to have samples suitable to reproduce the results of the characterization. The samples used as crosslinking agents were prepared in the work described in a previous publication [47].

Synthesis of Poly(S-co-HMDP) Polymers

Sulphur was introduced and heated at 185 °C in a round-bottom flask for ten minutes until a viscous red liquid was obtained. Then, HMDP was added at 185 °C, and the reaction mixture was stirred for an additional ten minutes at 185 °C. The obtained crude product was then cooled down to room temperature. Poly(S-co-HMDP) copolymers were obtained as dark brown solids.

Poly(Sulphur-co-HMDP) copolymer with the molar ratio S/HMDP = 6 was obtained by using 1.40 g of sulphur (43.67 mmol) and 2.0 g of HMDP (7.34 mmol).

Poly(Sulphur-co-HMDP) copolymer with the molar ratio S/HMDP = 8.9 was synthesized with 2.10 g of sulphur (65.50 mmol) and 2.0 g of HMDP (7.34 mmol).

As commented below in the text, residual sulphur and HMDP were not observed at the end of the reactions.

2.4. *Preparation of Elastomer Composites*

The recipes of the elastomer composites are reported in Table 1.

The same molar amount of sulphur and crosslinking agents were used in all the elastomer composites. The composites were prepared in two steps: (i) non-productive mixing: the preparation of a masterbatch; (ii) productive mixing: the addition of the vulcanization agents.

Table 1. Recipes of the elastomer composites [a].

Entry	1	2	3	4	5
Crosslinking agent [b]	=	DBTH	HMDP	Copolymer 1	Copolymer 2
SBR	100	100	100	100	100
CB CRX 1391	35	35	35	35	35
Stearic acid	2	2	2	2	2
CB CRX 1391	25	25	25	25	25
Oil	16	16	16	16	16
Resin	16	16	16	16	16
Zinc salt	1	1	1	1	1
ZnO	2	2	2	2	2
6PPD	2	2	2	2	2
CB CRX 1391	5	5	5	5	5
Sulphur	0.70	0.70	1.25	0.69	0.51
MBTS [c] 80%	2	2	2	2	2
Thiuram	0.6	0.6	0.6	0.6	0.6
DBTH	0.00	2.00	0.00	0.00	0.00
HMDP	0.00	0.00	0.79	0.00	0.00
Poly(HMDP-co-Sulphur)Ratio S/-$(CH_2)_6$ = 6	0.00	0.00	0.00	1.36	0.00
HMDP	*0.00*	*0.00*	*0.00*	*0.80*	*0.00*
S_8	*0.00*	*0.00*	*0.00*	*0.56*	*0.00*
Poly(HMDP-co-Sulphur) Ratio S/-$(CH_2)_6$ = 8.9	0.00	0.00	0.00	0.00	1.53
HMDP	*0*	*0*	*0*	*0*	*0.79*
S_8	*0*	*0*	*0*	*0*	*0.74*

[a] The amounts are reported in phr; [b] for enhancing the stability of the crosslinking network; [c] dibenzothiazyl disulphide.

2.4.1. Non-Productive Mixing: Preparation of Masterbatch

A Banbury-type tangential internal mixer, having a volume of 1200 cm^3, was used for the preparation of the masterbatch. Particularly, two different phases were used.

Phase 1: SBR, 642.96 g, was masticated at 40 °C for 30 s at 75 rpm. Then, 225.04 g of CRX 1391 was added, and the compound was mixed for 50 s at 40 °C and 75 rpm. Then, stearic acid, 12.86 g, CRX 1391, 160.74 g, oil, 102.87 g, and resins, 102.87 g, were added, and the compound was mixed for 1 min and 15 s at 40 °C and 75 rpm. The compound was then discharged.

Phase 2: Before starting the second phase, the compound was remixed at 40 °C for 30 s at 75 rpm. Then, Zinc salt, 6.2 g, Zinc Oxide, 12.39 g, 6PPD, 12.39 g, and CRX 1391, 30.98 g, were added and mixed for 1 min and 15 s at 40 °C and 75 rpm. Finally, the compound was discharged at 155 °C for 55 s at 75 rpm. A total of 1.309 Kg of the masterbatch was obtained.

The same masterbatch was used for both the composites.

2.4.2. Final Mixing: Addition of the Vulcanization Agents

The masterbatch was first mixed in a Brabender®-type internal mixer, for 1 min at 80 °C, 50 rpm. MBTS, thiuram, sulphur, and either DBTH or HMDP or poly(S-co-HMDP) copolymers as the crosslinking agents were added. Mixing was then carried out for 2 min at 80 °C, 50 rpm. The composites were discharged and cooled at room temperature and finally homogenized by being passed 5 times on the two-roll mill, operating at 50 °C, with the front and the back roll rotating at 30 rpm and 38 rpm, respectively, and 1 cm as the nip between the rolls.

2.5. Characterization Methods

2.5.1. Characterization of HMDP

The instrument used for GC-MS analyses was an Agilent 5973 network mass selective detector (Agilent, Santa Clara, CA, USA) with a 6890 Series GC system mass spectrometer. The column used for all analyses was a J&W GC Column HP-5MS [(5%-phenyl)-methylpolysiloxane] 30 m, with a 0.25 mm internal diameter and a 0.25 μm film thickness.

ESI-Mass analyses were recorded with Electrospray ionization (ESI) with a Bruker Esquire 3000 plus ion-trap mass spectrometer instrument (Bruker, Billerica, MA, USA) equipped with an ESI Ion Trap LC/MSn System.

NMR. A Bruker 400 MHz instrument (100 MHz) was used for recording the ^{1}H-NMR and ^{13}C-NMR spectra of HMDP and poly(S-co-HMDP) copolymers. The temperature at which the spectra were recorded was 298 K, and 5 mg of PyC was dissolved in 1.0 mL of $CDCl_3$ for ^{1}H-NMR analysis. $CDCl_3$: δ_H = 7.26 ppm was used as the reference chemical shift of the residual solvent. For ^{13}C-NMR analysis, 40 mg of PyC was dissolved in $CDCl_3$. Chemical shifts were reported in ppm with the solvent residual peak as the internal standard ($CDCl_3$: δ_H = 77.16 ppm).

2.5.2. Characterization of Poly(S-co-PyC) Copolymers

The characterization methods for performing the elemental analysis, the differential scanning calorimetry, the gel permeation chromatography, and the infrared spectroscopy have been reported in a previous publication [47].

NMR analysis. 1D ^{1}H-NMR and ^{13}C-NMR spectra were recorded as reported above for HMDP. 2D ^{1}H-^{1}H COSY and ^{1}H-^{13}C HSQC were acquired with the same parameters that were already used for 1D experiments. The number of scans was adapted to the sample concentration to ensure a proper signal-to-noise ratio.

2.5.3. Characterization of Elastomeric Composites

Crosslinking. The crosslinking reaction of the rubber composites was performed at a temperature of 170 °C for a duration of 20 min. A sample of 5 g of crude rubber compound was placed into a Rubber Process Analyser (RPA) rheometer, from Alpha Technologies, Hudson, OH, USA. Initially, a strain sweep was performed at low deformation levels (0.1–25% strain) prior to the crosslinking stage. The sample was then maintained at 50 °C for 10 min and subjected to an additional strain sweep, also conducted at 50 °C. Then, the crosslinking process was carried out at 170 °C for 20 min, with an oscillation angle of 6.98% and a frequency of 1.7 Hz. During this process, the torque versus time curve was recorded, along with key parameters such as the minimum achievable torque (ML), the maximum achievable torque (MH), the time required to reach a torque equal to ML + 1 (tS1), and the time necessary to attain 90% of the maximum torque (T90).

Analysis of the crosslink network. Vulcanized composite samples were swollen in n-heptane for two days in a nitrogen atmosphere to allow the diffusion of reagents. The n-heptane was then removed, and the samples were washed with light petroleum and dried overnight at room temperature under reduced pressure. The samples were then immersed in toluene (200 mL) and placed in a glass tube flushed with nitrogen. The tube was then sealed and left in the dark for 72 h, the time required for equilibrium swelling to occur. After this time, the samples were dried by blotting with filter paper. Finally, the samples were quickly sealed in a clean container and weighed. The samples were then dried under vacuum at 70 °C for 24 h to remove the solvent and weighed again in the dried state. This is completed to obtain the amount of sorbed solvent and the weight of the dry network. The Flory–Rehner equation was applied to calculate the crosslinking network:

$$v_e = \frac{-\left[\ln(1-V_r) + V_r + X_1 V_r^2\right]}{\frac{V_1\left(v_r^{1/3} - V_r\right)}{2}}$$

where ve is defined as the effective number of chains in a real network per unit volume; χ_1 is the polymer–solvent interaction parameter, and V_1 is the molecular volume of the solvent. V_r is defined as the volume fraction of polymer in a swollen network in equilibrium with pure solvent. The mathematical definition is reported in the following equation:

$$V_r = \frac{\frac{weight\ of\ dry\ rubber}{density\ of\ dry\ rubber}}{\frac{weight\ of\ dry\ rubber}{density\ of\ dry\ rubber} + \frac{weight\ of\ solvent\ absorbed\ by\ sample}{density\ of\ dry\ solvent}}$$

The procedure for measuring mono- and di-sulfidic crosslinks is very similar.

A total of 100 mg of crosslinked composite was placed in a 20 mL beaker with 100 mL of heptane. The sample was allowed to stand for 24 h. After this time, 3.8 mL of propanethiol and 4 mL of piperidine were added and left at room temperature for 2 h. The mixture was then washed 3 times in 50 mL of n-heptane and then filtered. The solid was left in n-heptane for another 24 h and then washed in light petroleum. The sample was filtered under vacuum and dried under reduced pressure for 24 h.

Dynamic mechanical analysis in the shear mode. In the RPA instrument reported above, curing was first performed. On the crosslinked samples, a first strain sweep was applied at 50 °C with a strain amplitude ranging from 0.1% to 25%. Then the instrument was set at the minimum strain amplitude (0.11), and the sample was subjected to these conditions to achieve a fully equilibrated system. Finally, the shear storage modulus G', the shear loss modulus G'', and the ratio G''/G' (tan δ) were detected and measured under strain-sweep oscillation, having a strain amplitude ranging from 0.1% to 25% and a frequency of 1 Hz.

Dynamic mechanical analysis in the axial mode. The rubber composite taken from the two-roll mill was rolled up to prepare a long cylinder. The cylinder was cutinton smaller cylinders and vulcanized to give cylindrical specimens with a length of about 25 mm and a diameter of about 12 mm. Dynamic mechanical measurements were performed with an Instron dynamic device in the traction-compression mode, at the following temperatures: 70 °C, 100 °C, and 120 °C, as reported in the literature [49]. An axial compressive pre-strain of 25% was applied on the specimen followed by a sinusoidal axial strain sweep having an amplitude of ±3.5% with respect to the applied pre-strain, with a 100 Hz frequency. The dynamic mechanical storage modulus (E'), the loss factor (E''), and the ratio of E''/E' (tan δ) were registered.

Tensile tests. The composite from the two-roll mill was crosslinked in a vulcanizing press to produce a 1 mm thick vulcanized plate. ISO 37, the determination of the tensile properties of Vulcanized Rubber (ISO: Geneva, Switzerland, 2017), was used. Specimens were used for the tests. They were dumbbell-shaped, punched from the vulcanized plate, with a length of 10 ± 0.5 mm and a width of 4 ± 0.2 mm (type 3). On these specimens, the tensile tests were carried out with a Zwick universal testing machine (Zwick Roell Z001, Test expert, Ulm, Germany) and an optical extensometer. Stresses at 100, 200, and 300% elongation (σ_{100}, σ_{200}, σ_{300}, respectively) as the stress at break (σ_B), the elongation at break (ε_B), and the energy at break were determined. Three measurements were performed for each sample.

Test for the stability of the crosslinking network. Curing was carried out in the RPA instrument reported above, at 170 °C for 10 min, with an oscillation angle of 6.98% and a frequency of 1.7 Hz. The temperature was then increased to 200 °C, in 6 min, without subjecting the sample to oscillation. When the system reached the temperature of 200 °C, the crosslinked sample was then held in the instrument at 200 °C for 20 min without oscillation.

3. Results and Discussion

3.1. Synthesis and Characterization of HMDP

HMDP was synthesized via a classical Paal–Knorr reaction. Hexamethylenediamine was first mixed in water and then heated up and mixed with 2,5-hexanedione. At the end of the reaction, the final crude product was only washed with distilled water due to the solubility of the two starting materials, in order to remove the unreacted substrates. The yield, 93%, was calculated by considering the final mmol of HMDP with respect to the starting mmols of hexamethylenediamine, the limiting reagent of the process. HMDP was obtained with a high Atom Economy and an almost null E factor. In Figure 2, the ^{1}H-NMR spectrum of HMDP is reported.

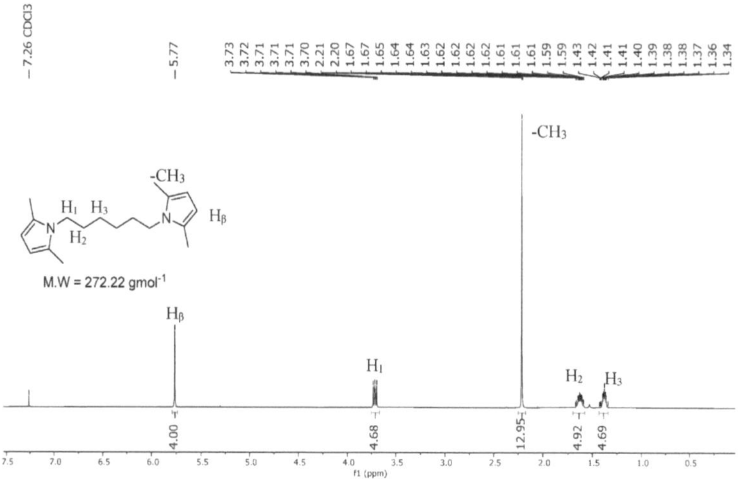

Figure 2. ^{1}H-NMR spectrum of HMDP.

3.2. Synthesis and Characterization of Poly(S-co-HMDP) Copolymers

The poly(S-co-HMDP) copolymers were obtained from the reaction of orthorhombic sulphur with the pyrrole derivative of hexamethylenediamine. HMDP was prepared as already reported [47], by dissolving the diamine in water, adding 2,5-hexanedione, and heating to 120 °C for 3 h, and it was obtained with a yield of up to 93%. The characterization was carried out by means of ^{1}H-NMR and DSC and reproduced the results already reported [47]. The ^{1}H-NMR spectrum is visible in Figure S1 in the Supporting Information.

The scheme for the synthesis of poly(S-co-HMDP) copolymers is shown in Figure 3.

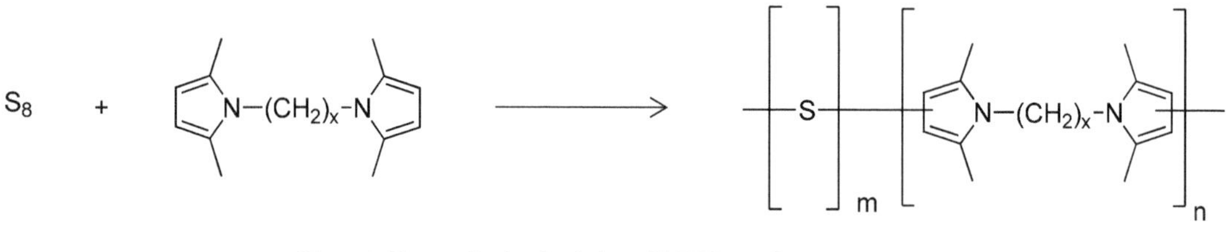

Figure 3. The synthesis of poly(s-co-HMDP) copolymer.

The conditions reported in the seminal work on inverse vulcanization [35] were used. Two copolymers were used in this work as crosslinking agents, with different sulphur/HMDP ratios: 6 and 8.9. In this text, they are referred to as Copolymer 1 and Copolymer 2, respectively. The preparation of these copolymers was performed several times, to check the reproducibility of the process, and the same results were obtained. Residual sulphur and HMDP were not observed at the end of the reaction. The whole process was thus characterized by a high yield and, by neglecting the water released as the co-product of the synthesis of the pyrrole compound, by almost null E Factor, down to 0.11 [50].

The characterization confirmed their reproducibility. The preparation and characterization of the samples used in this work have been already reported [47]. For the sake of clarity, some of the copolymers synthesized in our previous work [47] were used as novel crosslinking agents in the present work. Particularly, the elemental analysis results reported in Tables 2 and 3 and the gel permeation chromatography (GPC) results in Table 4 were the same as those already published in a scientific work [47]. Additionally, differential scanning calorimetry (DSC) and infrared (IR) analyses of the two copolymers are compared in the Supporting Information in Figures S1 and S2, respectively. In brief, evidence of pristine sulphur and HMDP was not found in the DSC traces. Additionally, the values of the glass transition temperatures (T_g) of the poly(S-co-HMDP) copolymers were observed. In particular, a T_g value of 30 °C was observed for the copolymer with a molar ratio sulphur/HMDP = 6, and a T_g value of 22 °C was observed for the copolymer with a molar ratio sulphur/HMDP = 8.9.

Table 2. Chemical composition of poly(S-co-HMDP) samples [a,b].

Sample	Carbon wt%	Hydrogen wt%	Nitrogen wt%	Sulphur wt%
Copolymer 1 [c]	40.0	6.1	5.0	20.3
Copolymer 2 [c]	49.8	7.9	6.0	23.2

[a] Data from ref. [47]; [b] from elemental analysis; [c] the variation coefficient was between 0.4 and 0.5 for C, 4.9 and 5 for H, 1.7 and 2 for N, 2.2 and 2.5 for S, with the greater values for the polymer with the lower contents.

Table 3. The elemental analysis of poly(S-co-HMDP) copolymers used in the present work.

Entry [a]	Carbon wt%	Standard Deviation	Hydrogen wt%	Standard Deviation	Nitrogen wt%	Standard Deviation	Sulphur wt%	Standard Deviation
1	40.0	±0.2	6.06	±0.3	4.96	±0.1	20.3	±0.5
3	49.8	±0.2	7.90	±0.3	6.05	±0.1	23.2	±0.5

[a] See Table 1 of reference [47] for the molar ratio of sulphur and HMDP used in the synthesis of the copolymers.

Table 4. The weight average molecular weight (M_w), the number average molecular weight (M_n), and the dispersity ($Đ = M_w/M_n$) of Copolymer 1 and Copolymer 2. The results of the two copolymers have already been published in a previous work [47].

Sample	Molar Ratio Sulphur/Pyrrole Compound	$M_w \times 10^{-3}$	$M_n \times 10^{-3}$	$Đ$ (M_w/M_n)
Copolymer 1	6.0	7.6	6.0	1.3
Copolymer 2	8.9	6.4	5.0	1.3

3.2.1. NMR Analysis

The NMR results, discussed in a previous publication [47], revealed that the sulphur essentially reacts with the β-position of the pyrrole ring. There were clear indications of a mono-substitution of the ring, and the di-substitution could not be excluded. In this work, the microstructures of Copolymer 1 and Copolymer 2 were carefully compared by means of ^{1}H-, ^{13}C-, 2D ^{1}H-^{1}H-COSY, and 2D ^{1}H-^{13}C HSQC NMR analyses. The different sulphur content in the copolymers is indeed a valuable tool to better elucidate the copolymer microstructure. For the sake of clarity, the attribution of the NMR peaks, also reported in ref. [47], is given in the text below.

3.2.2. ^{1}H-NMR Analysis

1D 1H-NMR. In Figure 4, the 1D ^{1}H-NMR spectra of Copolymer 1 (Figure 4a), Copolymer 2 (Figure 4b), and HMDP (Figure 4c) are shown.

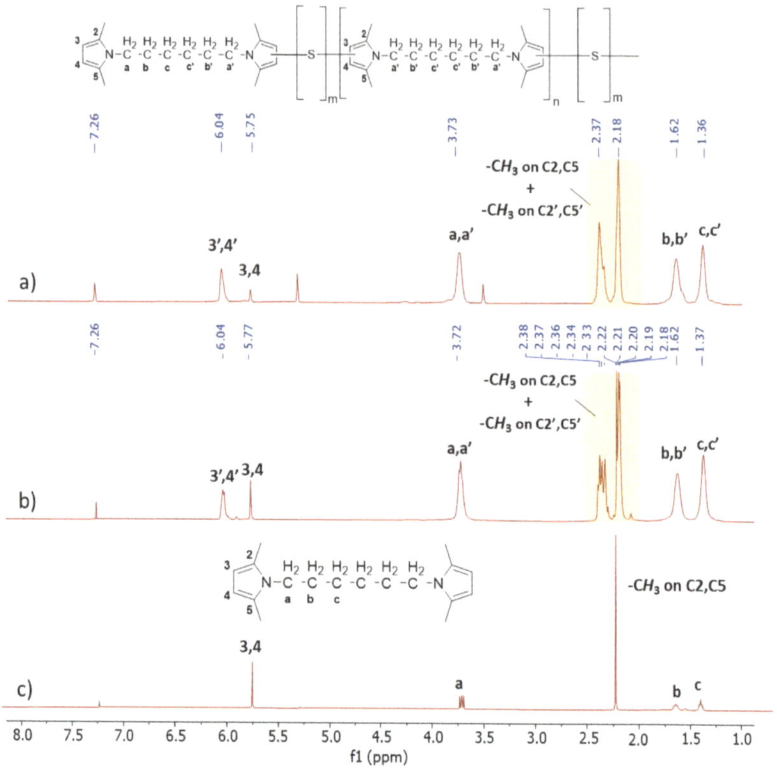

Figure 4. ^{1}H-NMR spectra ($CDCl_3$, 400 MHz) of (**a**) copolymer 2, (**b**) copolymer 1, and (**c**) of HMDP.

The reference for the attribution of the peaks was the ppm value of the hydrogen atom signal (7.26 ppm). The assignments of HMDP and Copolymer 1 confirm what has already been reported [47]. In particular, Figure 4c shows the sharp singlets and defined multiplets of the HMDP spectrum. The spectra of the copolymers appear different: multiplets are present along with a new broad peak.

By examining the aromatic part of the copolymers' spectra, the region between 5.5 and 6.7 ppm corresponding to heterocyclic aromatic protons, notable differences were observed between the spectra of Copolymer 1 and Copolymer 2. Specifically, variations were identified in the relative areas of the broad singlet at 5.8 ppm and the multiplet centred at 6.05 ppm. These signals were attributed [47] to the 3,4 protons in the β-positions of the pyrrole ring, without any covalent bonds with sulphur atoms, and to the 3′,4′ protons in the β-positions of the pyrrole ring containing sulphur atoms, respectively. On the basis of this assignment, it can be concluded that the amount of non-substituted pyrroles is lower in Copolymer 2, which indeed has a higher sulphur content.

In the spectrum of HMDP, the signals corresponding to the methyl groups on carbon atoms 2 and 5 are represented by a singlet at 2.21 ppm, and they are detected at a high field. In the spectra of the copolymers, this singlet transforms into a multiplet, with the respective signals detected in the spectre region between 2.17 ppm and 2.21 ppm. Additionally, a new set of peaks was found between 2.32 ppm and 2.37 ppm. In the spectrum of Copolymer 1, these signals can be attributed to the methyl groups on carbon atoms 2 and 5 of the pyrrole ring not substituted by sulphur atoms, the chain end, and to the methyl group on carbon atoms (2′ or 5′) next to a carbon atom in the β-position, which carries a sulphur atom (3′ or 4′). If the substitution with sulphur occurs at position 3′, the methyl group in position 5′ was detected with a higher field signal. On the other hand, the signal corresponding to the hydrogen atoms at positions 2 and 5 of the pyrrole ring is not detectable in the spectrum of Copolymer 2, probably due to the greater substitution with sulphur atoms. The signals corresponding to all the methylene units between the two pyrrole rings are at the same chemical shift as in the spectrum of HMDP but display greater complexity. This complexity may result from the overlap between the peaks arising from the methylene groups (a, b, c) adjacent to the pyrrole ring without sulphur atoms (the chain end) and the peaks associated with the methylene groups (a′, b′, c′) near the pyrrole ring containing sulphur atoms (in the repeating unit).

2D ^{1}H-^{1}H-COSY NMR. The assignments of the 1D ^{1}H-NMR were further investigated by performing the COSY experiment. The ^{1}H-^{1}H COSY spectra of the copolymers are shown in Figure 5, Figure 5a for Copolymer 1 and Figure 5b for Copolymer 2.

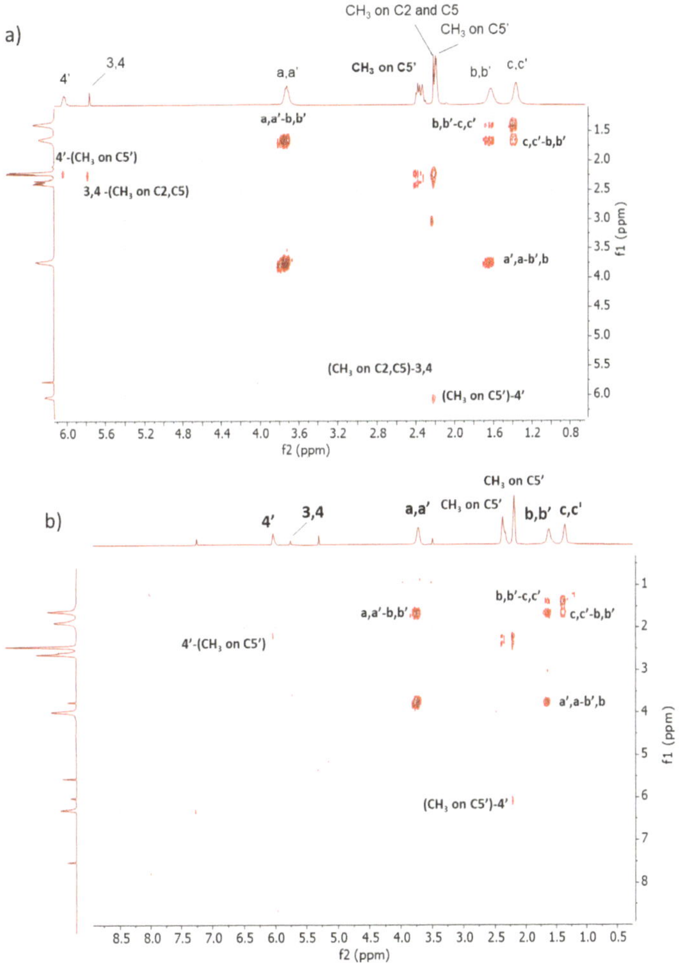

Figure 5. ^{1}H-^{1}H COSY spectrum in $CDCl_3$ of (**a**) Copolymer 1 and (**b**) Copolymer 2.

The comparison between the COSY spectra of the two copolymers reveals remarkable differences. The correlations between 4′-(CH3 on C5′) and 3,4-(CH3 on C2, C5) were already commented on for Copolymer 1 [47] and are confirmed in this spectrum. In the spectrum of Copolymer 2, only the 4′-(CH3 on C5) correlation is detectable. These findings confirm what was observed by examining the 1D ^{1}H-NMR spectra, i.e., the higher substitution with sulphur of the pyrrole ring in Copolymer 2.

3.2.3. ^{13}C-NMR Analysis

^{13}C-NMR spectra are in Figure 6, for Copolymer 1 (Figure 6a), Copolymer 2 (Figure 6b), and HMDP (Figure 6c).

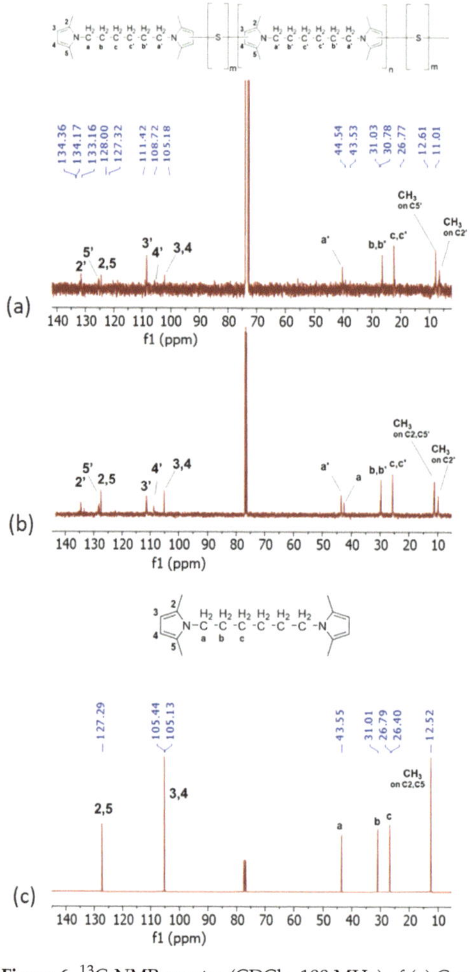

Figure 6. ^{13}C-NMR spectra ($CDCl_3$, 100 MHz) of (**a**) Copolymer 2, (**b**) Copolymer 1, and (**c**) HMDP.

The C2, C5, C3, and C4 carbon atoms of pristine HMDP, and so of the non-substituted pyrrole ring, are attributed to the peaks at 127 ppm and at 105 ppm, respectively. New signals are visible in the ^{13}C-NMR spectra of both copolymers. Particularly, C2′ at 134.17 ppm, C5′ at 127.13 ppm, C3′ at 111.23 ppm, and C4′ at 108.53 ppm are detected due to the interaction of the carbon atoms of the pyrrole ring with sulphur. The intensity of the peak due to C4′ is lower in the spectrum of Copolymer 2. The analysis of the spectrum of Copolymer 1 allowed us to hypothesize [47] that the carbon atoms of the pyrrole ring chemically react with sulphur atoms (3″ and 4″), which would decrease the intensity of the C4′ peak. Such intensity should be lower in the spectrum of a copolymer with a higher amount of sulphur, as observed in Figure 6a. The signals of 2″ and 5″ could fall in the region between 133 and 135 ppm.

In the spectrum of Copolymer 1, new signals were also observed in the typical NMR region for aliphatic hydrocarbons, between 25 and 45 and between 10 and 13 ppm. In particular, the signals of the methyls, either in the unsubstituted pyrrole (C2 and C5) or in the substituted pyrroles (C2′ and C5′), are at 12.43, 12.24, and 10.82 ppm. Interestingly, in the spectrum of Copolymer 2, the signal due to the aliphatic carbon in the unsubstituted pyrrole is lower in intensity if compared with the same signal in the spectrum of Copolymer 1. This suggests what was already seen in ^{1}H NMR for signals due to the end-chain carbons of the polymer.

HSQC experiments. HSQC experiments were performed in order to match the correlation between ^{1}H-NMR and ^{13}C-NMR. ^{1}H-^{13}C HSQC spectra are shown in Figure 7.

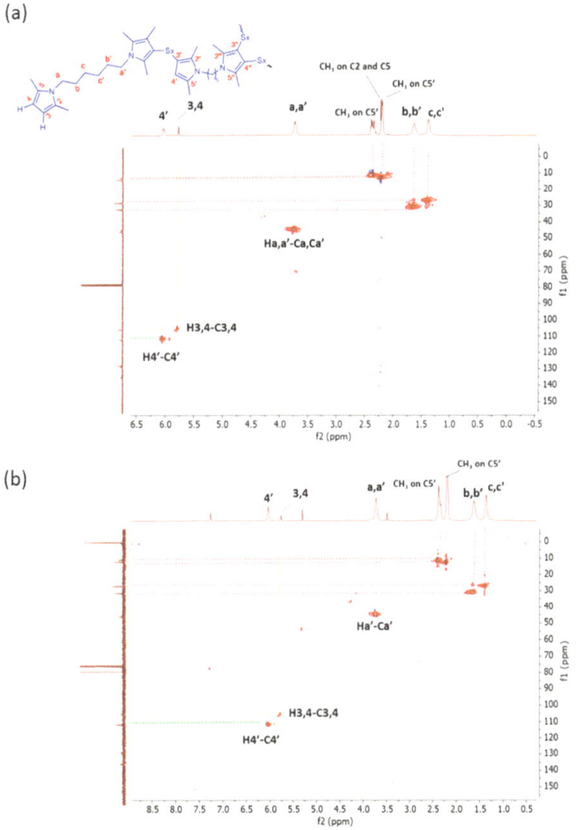

Figure 7. ^{1}H-^{13}C HSQC *spectra* (in $CDCl_3$) of (**a**) Copolymer 1 and (**b**) Copolymer 2.

Different correlations in the HSQC spectrum of Copolymer 1 were detected. Particularly, the H4-C4 correlation and the H4′-C4′ correlation were observed. In the spectrum of Copolymer 2, only the H4′-C4′ correlation is remarkable. In addition, a correlation between H-a′ and Ca′ is revealed, and it is possible to distinguish between the C-b′ and C-c′ methylenes due to their related correlations with H-b′ and H-c′ in both spectra. These results confirm the higher substitution with sulphur of the pyrrole ring in Copolymer 2, in the β-position.

The NMR findings indicate that the carbon atoms in the β-position of the pyrrole ring are the propagation site for the poly(S-co-HMDP) copolymer. The mono-substitution of the pyrrole ring seems to be largely favoured in Copolymer 1, whereas the substitution in both positions appears likely in Copolymer 2. The clear evidence that both the beta positions of the pyrrole ring can be attacked by sulphur is unprecedented and appears to be relevant for the development of pyrrole-based sulphur copolymers. It is worth adding that the reaction of sulphur and sulphur compounds with pyrrole rings has been documented, in the scientific literature, only by some of the authors [47,51] and indeed deserves further investigations.

3.3. Preparation and Characterization of Elastomer Composites

As mentioned in the Introduction, the objective of this work was to prepare an elastomer composite suitable for a high-performance tyre tread. The recipes are shown in Table 1. Typical ingredients were used: an SBR copolymer with a high styrene content, a high-surface-area carbon black, and a styrenic resin. These ingredients are suitable to enhance the hysteresis of the composite and hence the road grip of the tyre tread. A so-called efficient vulcanization system was used, with thiuram and a low sulphur/accelerator ratio, in order to reduce the length of the sulphur bridges between the polymer chains, thus enhancing the modulus of the composite and the stability of the crosslinking network. The elastomer composites were prepared via melt blending, by using an internal mixer and a traditional processing. The details can be found in Section 2.

Reference composites were prepared, with DBTH as the crosslinking agent to enhance the stability of the filler network as well as without any crosslinking agent. HMDP was used in the same molar amount as DBTH. The amount of poly(S-co-HMDP) copolymer was tuned to have the same HMDP content in the composite, and the amount of S_8 was tuned to have the same amount of sulphur in all the composites. Two samples of poly(S-co-HMDP) copolymers were used, with different S/HMDP ratios in the chain, 6.0 and 8.9 (Copolymer 1 and Copolymer 2).

3.3.1. Curing

The vulcanization reaction was carried out at 170 °C for 10 min. The values of the minimum modulus (M_L), maximum modulus (M_H), induction curing time (t_{s1}), optimum curing time (t_{90}), and curing rate are shown in Table 5. The vulcanization curves are shown in Figure S3 in the Supporting Information.

Table 5. Curing curve values derived from strain-sweep analysis.

Entry	1	2	3	4	5
Crosslinking agent	=	DBTH	HMDP	Copolymer 1	Copolymer 2
M_L	2.90	2.76	2.81	3.12	3.10
M_H	10.12	12.53	11.97	12.68	12.64
$M_H - M_L$	7.22	9.77	9.16	9.56	9.54
t_{s1}	2.69	2.76	2.56	2.44	2.45
t_{90}	8.30	7.80	8.23	8.21	7.83
$t_{90} - t_{s1}$	5.61	5.04	5.67	5.77	5.38
Curing rate [a]	1.29	1.94	1.61	1.66	1.77
M_H *Relative values*	100	124	118	125	125

Curing rate [a] = $M_H - M_L/t_{90} - t_{s1}$.

The addition of either DBTH or HMDP as the crosslinking agent did not appreciably affect the viscosity of the composites, as indicated by the M_L value. The slight increase in M_L observed for the composites from Entry 4 and Entry 5 could be due to the polymeric nature of the crosslinking agents.

Higher values of M_H and ($M_H - M_L$) were obtained in the presence of all the crosslinking agents, based either on DBTH or on HMDP. The value of M_H can be mainly attributed to the extent of the crosslinking network. However, the filler network does play a role, as the strain applied during the oscillation test is not sufficient to completely break it. As the filler content was the same in all the Entries, the comparison of the M_H values can give an indication of the extent of the crosslinking network, which can be thus assumed to be enhanced by the crosslinking agents. This induction was confirmed by the analysis of the crosslinking network, carried out with swelling measurements, discussed below in the text.

The use of either HMDP or poly(HMDP-co-sulphur) copolymers led to a slightly lower induction curing time (lower t_{s1}), whereas a lower optimum curing time (lower t_{90}) and a higher curing rate were obtained with DBTH. It is useful to remember that t_{s1} and t_{90} are determined from the torque values of the composites. The faster crosslinking (lower t_{s1}) observed in the presence of HMDP suggests that, with HMDP, crosslinking occurs via a radical-type mechanism that is known to occur in the first moments of a vulcanization [27]. DBTH seems to be more efficient when the vulcanization proceeds through the mechanism based on the active sulphurating species [27]. However, all the crosslinking agents lead to a more efficient vulcanization. It is worth commenting that the results obtained with pristine HMDP and with the copolymers, shown in Table 5, are in the same direction when compared with the reference compound. However, the differences are more pronounced with the copolymers. The poly(S-co-HMDP) copolymer is definitely more efficient as a crosslinking agent when its preparation is *ex-ante*. However, these results demonstrate that the HMDP can form the copolymer with sulphur in situ, during the preparation of the composite. The differences between the composites based on either Copolymer 1 or Copolymer 2 are slight. A higher sulphur content resulted in a lower optimum time of vulcanization and a higher curing rate.

3.3.2. Crosslinking Network

The analysis of the crosslinking network was performed by means of swelling measurements, by applying the Flory–Rehner theory [52]. The data of the total X-linking density and of the relative amount of nominal mono/di and poly-sulphides are shown in Table 6. The analysis of the bridges only nominally refers to sulphur sequences, as the crosslinking agents introduce methylene groups between the elastomer chains.

Table 6. The crosslink density of the elastomer composite of examples 1–5.

Entry	1	2	3	4	5
Crosslinking agent	=	DBTH	HMDP	Copolymer 1	Copolymer 2
Total X-link mol/g $\times 10^{-5}$	1.16	1.87	1.67	1.73	1.94
Mono and di-sulphide (wt %)	97.10	89.80	97.60	87.00	94.50
"Poly-sulphide" (wt %)	2.90	10.20	2.40	13.00	5.50

The crosslinking agents resulted in an increase in the total X-link concentration. These data appear to confirm the interpretations given above for the values of M_H. A higher relative amount of long crosslink bridges was found for the composites with DBTH and with Copolymer 1, i.e., the copolymer with shorter sulphur bridges. A low amount of long crosslink bridges was found for the composites with HMDP and with Copolymer 2, i.e., the copolymer with longer sulphur bridges. Only preliminary comments and hypotheses can be made, in order not to overstretch conclusions without sufficient support from experimental data. The difference between HMDP and Copolymer 1 could be due to the preparation ex ante of the copolymer. The difference between Copolymer 1 and Copolymer 2 could be due to the different lengths of the sulphur bridges. The longer bridges in Copolymer 2 could react with the thiol used in the procedure for the determination of the short bridges, as described in detail in Section 2. However, it is indeed worth repeating that these are only hypotheses to be verified in future experiments.

3.3.3. Dynamic Mechanical Characterization in the Shear Mode

A dynamic mechanical characterization in the shear mode was performed, by means of strain sweep tests, as described in detail in Section 2. The values of $G'_{\gamma min}$, $G'_{\gamma max}$, $\Delta G'$, $\Delta G'/G'$, G''_{max}, and Tan δ (G''/G') are shown in Table 7. The dependence of G′ on the strain amplitude is shown in Figure 8.

Table 7. The values of G′, ΔG′, G″, and tan δ from a strain-sweep analysis.

Entry	1	2	3	4	5
	Reference	DBTH	HMDP	Copolymer 1	Copolymer 2
$G'_{\gamma min\ (0.2\%)}$	4.58	4.60	4.63	5.36	5.37
$G'_{\gamma max\ (25\%)}$	0.63	0.75	0.73	0.80	0.79
$\Delta G'$	3.95	3.85	3.90	4.56	4.58
$\Delta G'/G'$	0.86	0.84	0.84	0.85	0.85
$G''_{(max)}$	0.77	0.71	0.76	0.84	0.83
Tan $\delta_{(max)}$	0.37	0.33	0.35	0.35	0.35

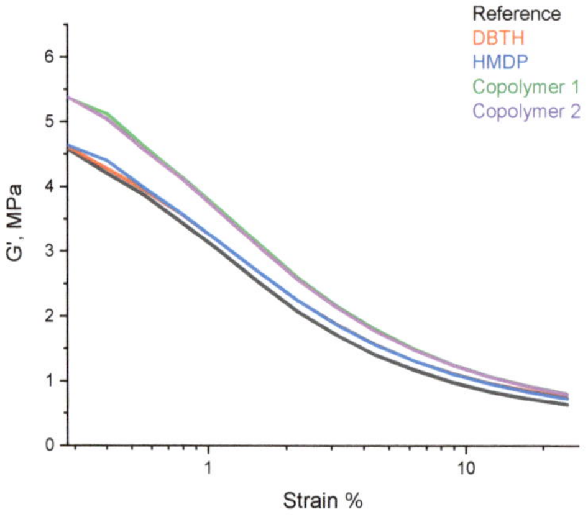

Figure 8. G′ vs. strain for the composites of Table 1.

Only preliminary comments can be made, as the experiments were not repeated a sufficient number of sufficient to allow a statistical analysis of the data. The values of G′ at the minimum strain were similar for the reference composite and for the composites with DBTH and HMDP. The poly(S-co-HMDP) copolymers prepared *ex-ante* gave rise to the highest values of G′ at the minimum strain. The values of G′ at the minimum strain give an indication of the extent of the filler network [53], which is unlikely to be remarkably affected by the crosslinking agents, used in a low amount. The values of G′ at the maximum strain are similar for all the composites, in particular for those containing a crosslinking agent. Hence, the composites with the crosslinking agents appear to have higher dynamic rigidity, and this is in line with the higher values of the crosslinking density, shown in Table 6. All the composites revealed, as expected, a reduction in the storage G′ modulus with the strain amplitude. The nonlinearity of the storage modulus is a phenomenon known as the Payne effect [54–56] and is due to the disruption of the filler network, which was above its percolation threshold in all the composites. The values of ΔG′/G′, a normalized index of the Payne effect, were similar for all the composites, and the values of $Tan\delta_{max}$ were lower for the composites with the crosslinking agent. These findings suggest that the higher value of G′ at the minimum strain for the composites with the poly(S-co-HMDP) copolymers could be due to the higher rigidity of these composites rather than to the poorer distribution of carbon black.

3.3.4. Dynamic Mechanical Characterization in the Axial Mode

Dynamic mechanical properties from axial compression tests were determined as described in Section 2. The values of storage modulus E′, loss modulus E″, and tan δ (E″/E′) were obtained at 70 °C, 100 °C, and 120 °C. These temperatures were chosen in view of the target final application of elastomer composites, in high-performance tyre treads. The dependence on the temperature of E′ and tan δ is shown in Figure 9. The values of the parameters at all the temperatures can be seen in Table S1 in the Supporting Information. The composites with a crosslinking agent reveal higher dynamic rigidity and a lower tan delta than the reference composite. The highest values of the dynamic rigidity are shown by the composites with the poly(S-co-HMDP) copolymers. To explain the higher rigidity due to the poly(S-co-HMDP) copolymers prepared *ex-ante*, one could hypothesize the grafting of the copolymers on the CB surface, as recently reported [57] for this type of copolymers. These findings seem to be in line with the results shown above on the density of the crosslinking network. The reduction in the dynamic rigidity with the temperature is higher for the copolymers, compared with DBTH. This suggests that the structure of the copolymers could be improved, for example, by tuning the amount of sulphur. The comparison of the results obtained with HMDP and with poly(S-co-HMDP) copolymers confirms what was found with the curing tests and also with the strain sweep experiments: HMDP can form the copolymer with sulphur in situ, during the preparation of the composite.

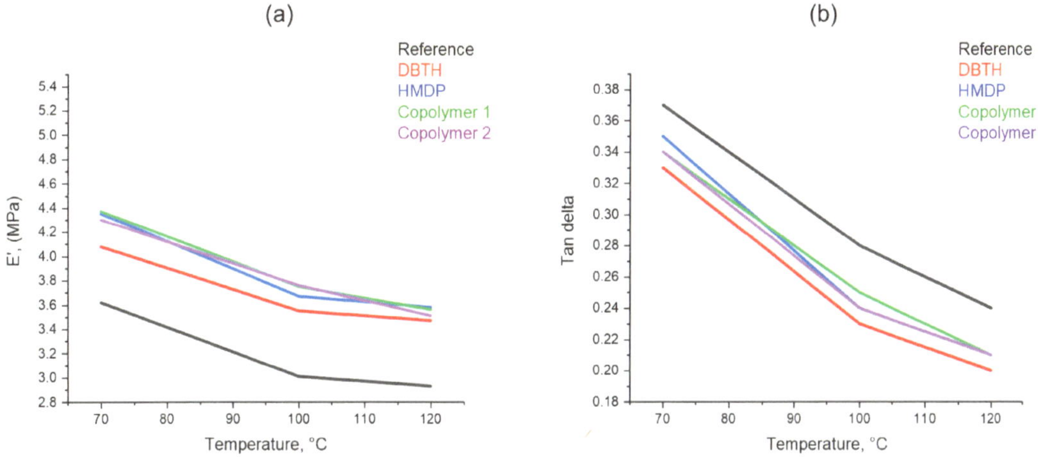

Figure 9. The dependence on temperature of (**a**) E′ and (**b**) Tan δ.

3.3.5. Tensile Tests

Tensile tests were performed on the composites of Table 1, with quasi-static measurements. The stress—strain curves are shown in Figure 10, and the values of σ_{100}, σ_{300}, σ_B, and ε_B are shown in Table S2 in the Supporting Information.

All the composites with a crosslinking agent revealed higher values of stresses, at all the elongations and at break, and the lower elongation at break, compared with the reference composite. The composites with the HMDP-based crosslinking agents showed better ultimate properties, i.e., higher stress, elongation, and energy at break, compared to the composite with HMDP. This result really deserves attention. Indeed, the better ultimate properties occur in the presence of the higher dynamic rigidity commented on in the paragraphs above. It is unusual to have the simultaneous improvement of these properties.

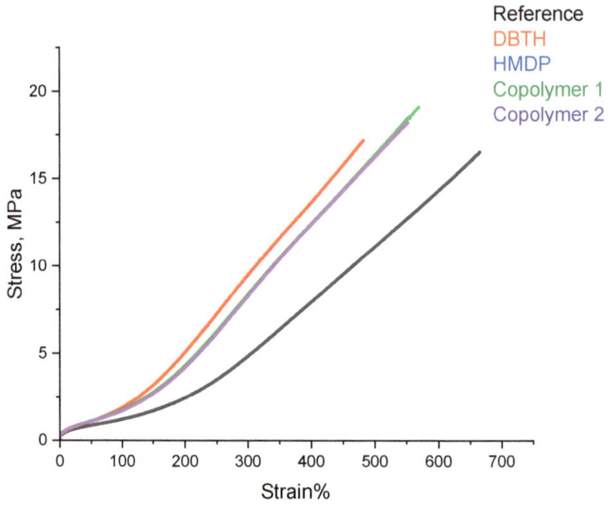

Figure 10. Stress vs. strain for the composites of Table 1.

3.3.6. Stability of the Crosslinking Network

To evaluate the stability of the crosslinking network, the cured composites were treated at high temperatures, as described in detail in Section 2. Curing was performed at 170 °C for 10 min, and then the composites were kept at 200 °C for 20 min, measuring the torque value. The values after the curing at 170 °C and after the treatment at 200 °C are shown in Table 8. The curves are shown in Figure S4 in the Supporting Information.

Table 8. Curing curve values derived from the strain-sweep analysis of elastomer composites of Table 1.

Entry	1	2	3	4	5
Crosslinking agent	=	DBTH	HMDP	Copolymer 1	Copolymer 2
S'_{max} [a]	10.5	13.1	12.5	12.8	12.9
S'_{end} [b]	8.9	11.2	10.6	11.3	11.2
$100\times \frac{S'max-S'end}{S'max}$	15.2	14.5	15.2	11.7	13.2

[a] Maximum value of torque, achieved after the curing at 170 °C; [b] value of torque measured after the treatment at 200 °C.

This test is preliminary and gives only an indication of the stability of the crosslinking network. However, it can be observed that the poly(S-co-HMDP) copolymers led to a lower reduction in the modulus. To explain this finding, it could be hypothesized that the pyrrole rings favour the stability of the dynamic rigidity. In particular, the lowest reduction in torque was obtained with the poly(S-co-HMDP) copolymer with the lower sulphur content, Copolymer 1. It could be assumed that shorter and hence more stable sulphur bridges are formed with Copolymer 1.

4. Conclusions

This work demonstrates that circular materials can be upcycled for preparing elastomer composites for high-performance tyre treads. Poly(S-co-HMDP) copolymers were used as the crosslinking agents of elastomer composites. The copolymers were made with a problematic waste such as sulphur and with the pyrrole derivative of a chemical that could be prepared from the hydrolysis of an end-of-life polyamide. The process, from the synthesis of HMDP to the isolation of the copolymers, was characterized by the absence of solvents and catalysts. The synthetic pathway was characterized by a high yield and an almost null E Factor, particularly down to 0.11 when calculated by considering only the organic part involved in the synthesis of HMDP. Two poly(S-co-HMDP) copolymers with different sulphur contents were used as novel crosslinking agents in the preparation of rubber compounds and composites suitable for high-performance tyres, achieving more efficient curing, a higher density of the crosslinking network, higher dynamic rigidity, better ultimate tensile properties, and higher stability of the crosslinking network, when compared with a reference composite with a traditional vulcanization system, without any further crosslinking agent. Similar results were obtained with the poly(S-co-HMDP) copolymers and DBTH, the crosslinking agent that has been used for decades on a commercial scale. The copolymers resulted in a better balance of dynamic rigidity and ultimate tensile properties, allowing for an unusual simultaneous enhancement. The results were obtained with pristine HMDP, and the HMDP copolymers were in the same direction, but the differences with the reference composite were more evident with the copolymers, prepared *ex-ante*. However, these findings allow us to hypothesize that HMDP, due to its high reactivity with sulphur, can form copolymers with sulphur in situ, during the preparation of the composite. The structure of the sulphur copolymers is tuneable, by modifying the amount of sulphur and the structure of the organic spacer. Hence, these results potentially pave the way for families of crosslinking agents.

Supplementary Materials: The following supporting information can be downloaded at: https://www.mdpi.com/article/10.3390/polym16192802/s1, Figure S1: DSC curves (second heating) of Copolymer 1 (black line) and Copolymer 2 (blue line); Figure S2: FT-IR spectra of (**a**) pristine HMDP, (**b**) poly(S-co-HMDP) material having molar ratio Sulphur/HMDP equal to 6 and (black line) and (**c**) poly(S-co-HMDP) material having molar ratio Sulphur/HMDP equal to 8.9 (blue line); Figure S3: Curing curves for the composites of Table 1. Curing at 170 °C for 20 min; Figure S4: Stability test curves for the composites of Table 1. Curing reaction was performed at 170 °C for 10 min and, subsequently, at 200 °C for 20 min. Table S1: Dynamic mechanical properties from axial measurements of composites in Table 1; Table S2: Tensile properties of composites in Table 1.

Author Contributions: Conceptualization, M.G. and V.B.; methodology, M.G., V.B. and S.N.; investigation, S.N.; resources, M.G.; writing—original draft preparation, M.G., V.B. and S.N.; writing—review and editing, M.G., V.B. and S.N.; supervision, M.G. and V.B. All authors have read and agreed to the published version of the manuscript.

Funding: This research received no external funding.

Institutional Review Board Statement: Not applicable.

Data Availability Statement: The data that support the findings of this study are available from the corresponding author upon reasonable request.

Acknowledgments: The authors wish to thank Pirelli Tyre for the financial support of the research and for financing the PhD of Simone Naddeo.

Conflicts of Interest: The authors declare no conflicts of interest.

References

1. Gent, A.N. Rubber Elasticity: Basic Concepts and Behavior. In *Science and Technology of Rubber*, 4th ed.; Mark, J.E., Erman, B., Roland, M., Eds.; Academic Press: San Diego, CA, USA, 2013; pp. 1–27. [CrossRef]
2. Erman, B.; Mark, J.E. The Molecular Basis of Rubberlike Elasticity. In *Science and Technology of Rubber*, 4th ed.; Mark, J.E., Erman, B., Roland, M., Eds.; Academic Press: San Diego, CA, USA, 2013; pp. 167–192. [CrossRef]
3. Eng, A.H.; Ong, E.L. Hevea Natural Rubber. In *Handbook of Elastomers*, 2nd ed.; Bhowmick, A.K., Stephens, H.L., Eds.; CRC Press: Boca Raton, FL, USA, 2000; Volume 61, pp. 29–59.
4. Karpeles, R.; Grossi, A.V. EPDM Rubber Technology. In *Handbook of Elastomers*, 2nd ed.; Bhowmick, A.K., Stephens, H.L., Eds.; CRC Press: Boca Raton, FL, USA, 2000; pp. 845–875.
5. Magill, J.H. Crystallization and morphology of rubber. *Rubber Chem. Technol.* **1995**, *68*, 507–539. [CrossRef]
6. Tosaka, M. Strain-induced crystallization of crosslinked natural rubber as revealed by X-ray diffraction using synchrotron radiation. *Polym. J.* **2007**, *39*, 1207–1220. [CrossRef]

7. Huneau, B. Strain-Induced Crystallization of natural rubber: A Review of X ray diffraction investigations. *Rubber Chem. Technol.* **2011**, *84*, 425–452. [CrossRef]
8. Galimberti, M.; Resconi, L.; Martini, E.; Guglielmi, F.; Albizzati, E. Elastomeric Copolymer of Ethylene with Alha Olefin. EP 0585658 B1, 28 January 1998.
9. Guerra, G.; Galimberti, M.; Piemontesi, F.; Ruiz de Ballesteros, O. Influence of regio- and stereoregularity of propene insertion on crystallization behaviour and elasticity of ethene-propene copolymers. *J. Am. Chem. Soc.* **2002**, *124*, 1566–1567. [CrossRef] [PubMed]
10. Guerra, G.; Ruiz de Ballesteros, O.; Venditto, V.; Galimberti, M.; Sartori, F. Pseudo-hexagonal crystallinity and thermal and tensile properties of ethene/propene copolymers. *J. Polym. Sci. Part B Polym. Phys.* **1999**, *37*, 1095–1103. [CrossRef]
11. Natta, G.; Allegra, G. Stereochemical aspects of crystalline synthetic macromolecules. *Tetrahedron* **1974**, *30*, 1987–2000. [CrossRef]
12. Brock, M.J.; Hackathorn, M.J. The synergistic effect of natural rubber on the crystallization of lithium poly(isoprene)S (coral rubber). *Rubber Chem. Technol.* **1972**, *45*, 1303–1314. [CrossRef]
13. Burfield, D.R.; Tanaka, Y. Cold crystallization of natural rubber and its synthetic analogues: The influence of chain microstructure. *Polymer* **1987**, *28*, 907–910. [CrossRef]
14. Corradini, P. Conformation of polymer molecules and entropy of melting. *J. Polym. Sci. Polym. Symp.* **1975**, *50*, 327–344. [CrossRef]
15. Allegra, G.; Bruzzone, M. Effect of entropy of melting on strain-induced crystallization. *Macromolecules* **1983**, *16*, 1167–1170. [CrossRef]
16. Chaikumpollert, O.; Yamamoto, Y.; Suchiva, K.; Kawahara, S. Protein-free natural rubber. *Colloid Polym. Sci.* **2012**, *290*, 331. [CrossRef]
17. Tanaka, Y. Structural characterization of natural polyisoprenes: Solve the mystery of natural rubber based on structural study. *Rubber Chem. Technol.* **2001**, *74*, 355. [CrossRef]
18. Kawahara, S.; Kakubo, T.; Nishiyama, N.; Tanaka, Y.; Isono, Y.; Sakdapipanich, J.T. Crystallization behavior and strength of natural rubber: Skim rubber, deproteinized natural rubber, and pale crepe. *J. Appl. Polym. Sci.* **2000**, *78*, 1510. [CrossRef]
19. Kawahara, S.; Kakubo, T.; Sakdapipanich, J.T.; Isono, Y.; Tanaka, Y. Characterization of fatty acids linked to natural rubber—Role of linked fatty acids on crystallization of the rubber. *Polymer* **2000**, *41*, 7483. [CrossRef]
20. Kawahara, S.; Isono, Y.; Kakubo, T.; Tanaka, Y.; Aik-Hwee, E. Crystallization Behavior and Strength of Natural Rubber Isolated from Different Hevea Clone. *Rubber Chem. Technol.* **2000**, *73*, 39. [CrossRef]
21. Gent, A.N.; Kawahara, S.; Zhao, J. Crystallization of cis- and trans-1,4-Polyisoprene Dispersed in SBR. *Rubber Chem. Technol.* **1998**, *71*, 668. [CrossRef]
22. Tanaka, Y.; Tarachiwin, L. Recent advances in structural characterization of natural rubber. *Rubber Chem. Technol.* **2009**, *82*, 283. [CrossRef]
23. Kakubo, T.; Matsuura, A.; Kawahara, S.; Tanaka, Y. Origin of Characteristic Properties of Natural Rubber—Effect of Fatty Acids on Crystallization of cis-1,4-Polyisoprene. *Rubber Chem. Technol.* **1998**, *71*, 70. [CrossRef]
24. Musto, S.; Barbera, V.; Maggio, M.; Mauro, M.; Guerra, G.; Galimberti, M. Crystallinity and Crystalline Phase Orientation of Poly(1,4-cis-isoprene) from *Hevea brasilliensis* and *Taraxacum kok-saghyz*. *Polym. Adv. Technol.* **2016**, *27*, 1082–1090. [CrossRef]
25. Musto, S.; Barbera, V.; Guerra, G.; Galimberti, M. Processing and strain induced crystallization and reinforcement under strain of poly(1,4-cis-isoprene) from Ziegler-Natta catalysis, *Hevea brasiliensis*, *Taraxacum kok-saghyz* and *Partenium argentatum*. *Adv. Ind. Eng. Polym. Res.* **2019**, *2*, 1–12. [CrossRef]
26. Hosler, D.; Burkett, S.L.; Tarkanian, M.J. Prehistoric polymers: Rubber processing in ancient Mesoamerica. *Science* **1999**, *284*, 1988–1991. [CrossRef]
27. Coran, A.Y. *Vulcanization in Science and Technology of Rubber*, 4th ed.; Mark, J.E., Erman, B., Roland, M., Eds.; Academic Press: San Diego, CA, USA, 2013; pp. 337–381.
28. Aprem, A.S.; Joseph, K.; Thomas, S. Recent developments in crosslinking of elastomers. *Rubber Chem. Technol.* **2005**, *78*, 458–488. [CrossRef]
29. Tanaka, Y. Recent advances in structural characterization of elastomers. *Rubber Chem. Technol.* **1991**, *64*, 325. [CrossRef]
30. Wolfman, G.; Hasenhidl, S.; Wolf, S. High resolution solid-state carbon-13 NMR studies of the crosslink structure in accelerated sulfur-vulcanized natural rubber. *Kautsch. Gummi Kunststoffe* **1991**, *44*, 118.
31. Milani, G.; Leroy, E.; Milani, F.; Deterre, R. Mechanistic modeling of reversion phenomenon in sulphur cured natural rubber vulcanization kinetics. *Polym. Test.* **2013**, *32*, 1052–1063. [CrossRef]
32. Buding, H.; Jeske, W.; Weidenhaupt, H.J. Vulcuren trial product KA 9188-A new bifunctional crosslinker for diene rubber. *Kautsch. Gummi Kunststoffe* **2001**, *54*, 8–13.
33. Jiang, Y.; Zhang, Y. Improving thermal oxidative aging resistance and anti-reversion property of natural rubber by adding a crosslinking agent. *J. Appl. Polym. Sci.* **2022**, *139*, 51882. [CrossRef]
34. Park, B.H.; Choi, S.S.; Kim, M.S.; Han, M.H. Effects of 1, 6-bis (N, N′-dibenzyl thiocarbamoyl dithio)-Hexane on Properties of Filled Natural Rubber Vulcanizates. *J. Ind. Eng. Chem.* **2002**, *8*, 197–202.
35. Chung, W.J.; Griebel, J.J.; Kim, E.T.; Yoon, H.; Simmonds, A.G.; Ji, H.J.; Pyun, J. The use of elemental sulphur as an alternative feedstock for polymeric materials. *Nat. Chem.* **2013**, *5*, 518–524. [CrossRef]

36. Naddeo, S.; Gentile, D.; Margani, F.; Prioglio, G.; Magaletti, F.; Galimberti, M.; Barbera, V. Pyrrole Compounds from the Two-Step One-Pot Conversion of 2,5-Dimethylfuran for Elastomer Composites with Low Dissipation of Energy. *Molecules* **2024**, *29*, 861. [CrossRef]
37. Sun, Z.; Xiao, M.; Wang, S.; Han, D.; Song, S.; Chen, G.; Meng, Y. Sulphur-rich polymeric materials with semi-interpenetrating network structure as a novel lithium–sulphur cathode. *J. Mater. Chem. A* **2014**, *2*, 9280–9286. [CrossRef]
38. Arslan, M.; Kiskan, B.; Cengiz, E.C.; Demir-Cakan, R.; Yagci, Y. Inverse vulcanization of bismaleimide and divinylbenzene by elemental sulphur for lithium sulphur batteries. *Eur. Polym. J.* **2016**, *80*, 70–77. [CrossRef]
39. Crockett, M.P.; Evans, A.M.; Worthington, M.J.; Albuquerque, I.S.; Slattery, A.D.; Gibson, C.T.; Chalker, J.M. Sulphur-Limonene Polysulfide: A material synthesized entirely from industrial by-products and its use in removing toxic metals from water and soil. *Angew. Chem. Int. Ed.* **2016**, *55*, 1714–1718. [CrossRef]
40. Shukla, S.; Ghosh, A.; Sen, U.K.; Roy, P.K.; Mitra, S.; Lochab, B. Cardanol benzoxazine-Sulphur Copolymers for Li-S batteries: Symbiosis of Sustainability and Performance. *ChemistrySelect* **2016**, *1*, 594–600. [CrossRef]
41. Zhang, Y.; Griebel, J.J.; Dirlam, P.T.; Nguyen, N.A.; Glass, R.S.; Mackay, M.E.; Pyun, J. Inverse vulcanization of elemental sulphur and styrene for polymeric cathodes in Li-S batteries. *J. Polym. Sci. Part A Polym. Chem.* **2017**, *55*, 107–116. [CrossRef]
42. Khawaja, S.Z.; Kumar, S.V.; Jena, K.K.; Alhassan, S.M. Flexible sulphur film from inverse vulcanization technique. *Mater. Lett.* **2017**, *203*, 58–61. [CrossRef]
43. Akay, S.; Kayan, B.; Kalderis, D.; Arslan, M.; Yagci, Y.; Kiskan, B. Poly (benzoxazine-co-sulphur): An efficient sorbent for mercury removal from aqueous solution. *J. Appl. Polym. Sci.* **2017**, *134*, 45306. [CrossRef]
44. Parker, D.J.; Jones, H.A.; Petcher, S.; Cervini, L.; Griffin, J.M.; Akhtar, R.; Hasell, T. Low cost and renewable sulphur-polymers by inverse vulcanisation, and their potential for mercury capture. *J. Mater. Chem. A* **2017**, *5*, 11682–11692. [CrossRef]
45. Thiounn, T.; Lauer, M.K.; Bedford, M.S.; Smith, R.C.; Tennyson, A.G. Thermally-healable network solids of sulphur-crosslinked poly (4-allyloxystyrene). *RSC Adv.* **2018**, *8*, 39074–39082. [CrossRef] [PubMed]
46. Smith, J.A.; Wu, X.; Berry, N.G.; Hasell, T. High sulphur content polymers: The effect of crosslinker structure on inverse vulcanization. *J. Polym. Sci. Part A Polym. Chem.* **2018**, *56*, 1777–1781. [CrossRef] [PubMed]
47. Naddeo, S.; Barbera, V.; Galimberti, M. A sulphur copolymer with a pyrrole compound for the crosslinking of unsaturated elastomer. *Polym. Chem.* **2024**, *15*, 3675–3690. [CrossRef]
48. Ramli, N.; Yusof, S. Economic and Market Trends of Specialty Rubber. In *Epoxidised Natural Rubber: Properties & Applications*; Springer Nature: Singapore, 2023; pp. 283–301. [CrossRef]
49. Farroni, F.; Russo, M.; Sakhnevych, A.; Timpone, F. TRT EVO. Advances in real-time thermodynamic tire modeling for vehicle dynamics simulations. *Proc. Inst. Mech. Eng. Part D J. Automob. Eng.* **2019**, *233*, 121–135. [CrossRef]
50. Sheldon, R.A. The E factor 25 years on: The rise of green chemistry and sustainability. *Green Chem.* **2017**, *19*, 18–43. [CrossRef]
51. Locatelli, D.; Bernardi, A.; Rubino, L.R.; Gallo, S.; Vitale, A.; Bongiovanni, R.; Galimberti, M. Biosourced Janus Molecules as Silica Coupling Agents in Elastomer Composites for Tires with Lower Environmental Impact. *ACS Sustain. Chem. Eng.* **2023**, *11*, 2713–2726. [CrossRef]
52. Flory, P.J.; Rehner, J., Jr. Statistical mechanics of crosslinked polymer networks I. Rubberlike elasticity. *J. Chem. Phys.* **1943**, *11*, 512–520. [CrossRef]
53. Tunnicliffe, L.B.; Busfield, J.J. Reinforcement of Rubber and Filler Network Dynamics at Small Strains. In *Designing of Elastomer Nanocomposites: From Theory to Applications*; Springer: Cham, Switzerland, 2017; pp. 71–102. [CrossRef]
54. Dillon, J.H.; Prettyman, I.B.; Hall, G.L. Hysteretic and elastic properties of rubberlike materials under dynamic shear stresses. *J. Appl. Phys.* **1944**, *15*, 309–323. [CrossRef]
55. Fletcher, W.P.; Gent, A.N. Nonlinearity in the dynamic properties of vulcanized rubber compounds. *Rubber Chem. Technol.* **1954**, *27*, 209–222. [CrossRef]
56. Payne, A.R. The dynamic properties of carbon black-loaded natural rubber vulcanizates. Part I. *J. Appl. Polym. Sci.* **1962**, *6*, 57–63. [CrossRef]
57. Wang, D.; Tang, Z.; Fang, S.; Wu, S.; Zeng, H.; Wang, A.; Guo, B. The use of inverse vulcanised polysulfide as an intelligent interfacial modifier in rubber/carbon black composites. *Carbon* **2021**, *184*, 409–417. [CrossRef]

polymers

MDPI

Article

Analysis on Isotropic and Anisotropic Samples of Polypropylene/Polyethyleneterephthalate Blend/Graphene Nanoplatelets Nanocomposites: Effects of a Rubbery Compatibilizer

Vincenzo Titone [1], Marilena Baiamonte [1], Manuela Ceraulo [1], Luigi Botta [1,2] and Francesco Paolo La Mantia [1,2,*]

[1] Department of Engineering, University of Palermo, V. le delle Scienze, 90128 Palermo, Italy; vincenzo.titone@unipa.it (V.T.); marilena.baiamonte@unipa.it (M.B.); manuela.ceraulo@unipa.it (M.C.); luigi.botta@unipa.it (L.B.)

[2] INSTM, Consortium on Materials Science and Technology, Via G. Giusti 9, 50125 Florence, Italy

* Correspondence: francescopaolo.lamantia@unipa.it

Abstract: Over the past few years, polymer nanocomposites have garnered a significant amount of interest from both the scientific community and industry due to their remarkable versatility and wide range of potential uses in various fields, including automotive, electronics, medicine, textiles and environmental applications. In this regard, this study focuses on the influence of a compatibilizer rubber on a nanocomposite incorporating graphene nanoparticles (GNPs), with a polymer matrix based on a blend of polypropylene (PP) and polyethylene terephthalate (PET). This effect has been investigated on both isotropic samples and on anisotropic/spun fiber samples. The influence of the compatibilizer rubber on morphological, rheological and mechanical properties was analysed and discussed. Mechanical and morphological properties were evaluated on both isotropic samples obtained by compression moulding and melt-spun fibers. The addition of the rubbery compatibilizer increased the viscosity, improving interfacial adhesion, and the same effect was observed for the melt strength and breaking stretching ratios. Mechanical properties, including the elastic modulus, tensile strength and elongation at break, improved in both types of samples but more significantly in the fibers. These improvements were attributed to the orientation of the matrix, the formation of PET microfibrils, and the reduction in the size of graphene nanoparticles due to the action of the elongational flow. This reduction, facilitated by the elongation flow and the action of the compatibilizer, improved matrix–nanofiller adhesion due to the increased contact area between the two polymeric phases and between the filler and matrix. Finally, a transition from brittle to ductile behaviour was observed, particularly in the system with the compatibilizer, attributed to defect reduction and improved stress transmission.

Keywords: compatibilization; graphene nanoplatelets (GNPs); polyethylene terephthalate (PET); polymer blends; polypropylene (PP)

Citation: Titone, V.; Baiamonte, M.; Ceraulo, M.; Botta, L.; Mantia, F.P.L. Analysis on Isotropic and Anisotropic Samples of Polypropylene/Polyethyleneterephthalate Blend/Graphene Nanoplatelets Nanocomposites: Effects of a Rubbery Compatibilizer. *Polymers* **2024**, *16*, 1092. https://doi.org/10.3390/polym16081092

Academic Editors: Luigi Torre, Miguel Ángel López Manchado and Laura Peponi

Received: 15 March 2024
Revised: 8 April 2024
Accepted: 12 April 2024
Published: 14 April 2024

1. Introduction

Polymer blends are a field of great interest for both scientific research and industry, as they offer a unique opportunity to combine the distinctive properties of different polymers [1,2]. However, this approach is not without its challenges, as a crucial problem emerges: the potential immiscibility between components [3]. In fact, this challenge requires innovative solutions, as the lack of miscibility between different polymers can lead to separate phases or regions of immiscibility, compromising the desired properties of the final material.

Among the many polymer blends investigated in the literature [4–8], a widely studied polymer blend concerns polypropylene (PP) and polyethylene terephthalate (PET) [9–12], two polymers with very different characteristics. Polypropylene offers remarkable impact and flexural strengths, while polyethylene terephthalate is distinguished by its excellent thermal properties and good chemical resistance [13]. The combination of these polymers promises materials with an optimal balance of strength, ductility and thermal stability [14]. However, the significant challenge of overcoming the inherent immiscibility between these two polymers stems from their different chemical structures and thermodynamic properties [3].

One strategy to address this problem is the introduction of compatibilizer agents. These agents play a crucial role in modifying intermolecular interactions between polymers, facilitating more homogenous mixing and ensuring a superior final performance.

In fact, many papers present in the literature show how the presence of compatibilizer agents in PP/PET blends promotes an effective interaction between the two surfaces [12,15–19]. Farikov et al. [15] observed a reduction in microfibril length attributed to the inhibitory effect of the compatibilizer on the coalescence process. Similarly, in our prior research [16], we noted a comparable trend: the microfibrils in ternary blends appeared to be shorter than those in binary blends. Mostovi et al. [17] demonstrated that microfibrils, when SEBS is present, exhibit lower aspect ratios, with a portion of SEBS adhered to the fracture surfaces, in contrast to the long, well-oriented microfibrils of PET without SEBS, as also corroborated by Park et al. [19] in PET/PP blended fibers utilizing a ternary ethylene ester acrylic–glycidyl methacrylate (EAG) copolymer. Of course, only a careful choice of compatibilizer agents can optimize the mechanical and thermal properties of these materials, making them more competitive.

Kraton™ polymers are styrene (ethylene-co-butylene)–styrene (SEBS) polymers functionalized with maleic anhydride (MA) grafted into a rubber intermediate block [20]. As previously reported in another paper [16], the presence of Kraton™ generates copolymers that act as a bridge between phases, enhancing the adhesion and properties of the polymer blend through a two-step reaction mechanism between the maleic ring and the terminal -OH group of PET.

In addition to the use of compatibilizer agents, another strategy to improve the properties of the polymer blends is the introduction of nanofillers that assume a key role in the improvement of these polymer blends. Indeed, as already reported in other works [21–23], graphene nanoplatelets (GNPs), with their unique properties, act as a structural reinforcement, bringing significant improvements to the mechanical and thermal properties of the composite material. In fact, in two distinct investigations, Inuwa et al. [21,22] examined the impact of incorporating styrene–ethylene–butylene–styrene grafted maleic anhydride (SEBS-g-MAH) and graphene nanoplatelets (GnPs) into a PP/PET blend. Their findings revealed that the nanoplatelets maintained their structural integrity and were uniformly dispersed within the polymer matrix, without experiencing an excessive exfoliation of the GnPs. Consequently, enhancements in mechanical properties were attributed to the inherent stiffness of the nanoplatelets and the efficient transfer of stress between the matrix and the filler. Furthermore, in a separate study, the researchers investigated the flammability characteristics of PP/PET/GnP nanocomposites, observing a notable enhancement in flammability parameters owing to the development of aligned, dense and well-organized carbon layers on the surface of the nanocomposites. Additionally, they noted an escalation in the effective thermal conductivity of the nanocomposites with increasing GnP loading.

For both polymer blends and nanopolymer blends, the final morphology is the key to obtain a material with the desired properties. To this aim, the elongational flow can play an important role. In fact, the flow involved in some processing operations, such as fiber spinning, film blowing and foaming, plays a crucial role in causing the preferential orientation of macromolecules along the flow direction. This flow-induced orientation significantly affects the deformation mechanism of the dispersed phase, thereby influencing the morphological evolution of blends and nanoblends, strongly affecting the final properties of the polymer system. Moreover, it has also evidenced that the elongational flow promotes the adhesion between the two, strongly affecting the final property phases of the polymer blends and between the polymer phase and the dispersed particles. This is due to the fact that the elongational flow is able to strongly decrease the dimension of the polymer phases and of nanofillers like GNP and nanoclay, increasing the area of the contact surface [24]. In our previous paper [16], it was demonstrated that the effect of the compatibilizer in a PP/PET blend that underwent elongational flow was negative as the increase in the mechanical properties was less than that observed for uncompatibilized blend. This behaviour was interpreted in terms of the reduced size of the initial PET droplet size in the ternary blend that gave rise to microfibrils of a shorter length than that of those formed during the spinning of the binary blend.

This study aims to evaluate the influence of a rubber compatibilizer and of the elongation flow on a GNP PP/PET nanocomposite. Our objective is to understand whether the presence of the compatibilizer affects the fibrillation efficiency of the dispersed PET phase. To this end, the mechanical and morphological properties of melt-spun fibers obtained with different hot drawing ratios were analysed and compared with isotropic samples. In addition, the rheological, morphological and mechanical properties of the isotropic samples obtained by compression moulding were evaluated.

2. Materials and Methods

2.1. Materials and Preparation

The main materials used in this work are as follows:

- Polypropylene (PP) with the trade name Moplen RP340H was purchased from LyondellBasell (LyondellBasell, Ferrara, Italy). It is a random polypropylene copolymer with a melt flow index (MFI) of 1.8 g/10 min (230 °C/2.16 kg) and a density of 0.90 g/cm) [25];
- Polyethylene terephthalate (PET) was obtained by grinding preforms for water bottles. Melt flow index (MFI) recorded at 275 °C under a load of 325 g (condition K) was 55 g/10 min [26];
- Kraton™ under the trade name Kraton™ FG1901 G was supplied from Kraton (Houston, TX, USA) as a dusted pellet. It is a clear, linear triblock copolymer based on styrene and ethylene/butylene with a polystyrene content of 30% [20];
- Graphene nanoplatelets (GNPs) under the trade name xGNP®, grade C, were supplied from XG Sciences Inc. (Lansing, MI, USA) with the following characteristics reported in a previous paper [23]: average diameter between 1 and 2 µm; an average thickness lower than 2 nm; and a specific surface area of about 750 m^2/g.

The nanocomposites were prepared by melt mixer process in a Brabender mixer mod. PLE330 (Brabender, Duisburg, Germany) operating at 270 °C at 60 rpm for 5 min. Prior to blending, both PET and GNPs were dried in a vacuum oven at 120 °C overnight. Table 1 shows the codes and compositions of the nanocomposites investigated.

Table 1. Nanocomposites' codes and compositions.

Sample Code	PP, wt%	PET, wt%	GNPs, wt%	Kraton™, wt%
PPG	75	25	2	-
PPGK	75	25	2	5

Mechanical properties were assessed on specimens obtained by compression moulding with a Carver laboratory hydraulic press (Carver, Wabash, IN, USA) at temperature of 270 °C and mould pressure of 300 psi for about 3 min. Fibers samples, ranging in diameter from 90 to 400 µm, were prepared at 270 °C utilizing the drawing module of a capillary viscometer (Rheologic 1000, CEAST, Turin, Italy).

Figure 1 illustrates the preparation and characterization of all the samples.

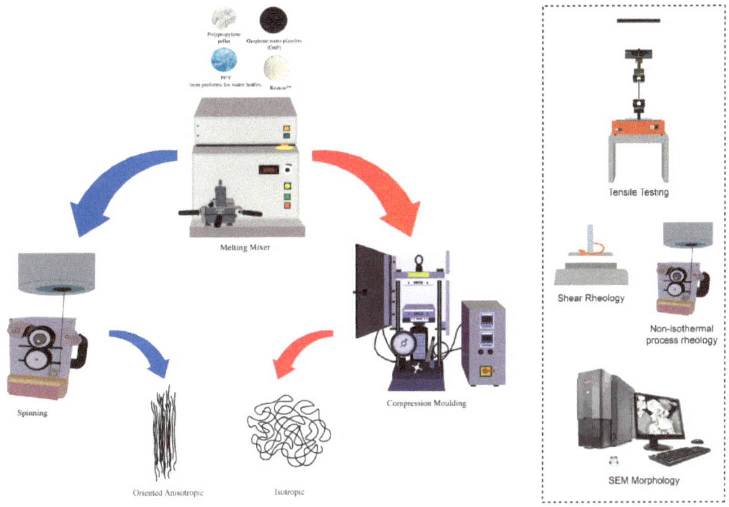

Figure 1. Image of the process of nanocomposites' preparation and characterization.

2.2. *Characterization*

Microscopic Raman examination was conducted at room temperature using a Renishaw InVia Raman microscope (Renishaw, Wotton-under-Edge, UK) equipped with a 532-nm Nd:YAG laser for excitation. Observations were carried out in the range of 3000–500 cm^{-1} with a spectral resolution of 0.5 cm^{-1}.

Characterization in shear flow was conducted using an ARES G2 (TA Instruments, New Castle, DE, USA) rotational rheometer. Shear viscosity was assessed within the angular frequency range of 0.1 to 100 rad/s, employing a parallel plate geometry with a spacing of 1.5 mm, a diameter of 25 mm and a strain of 5% at a temperature of 270 °C. Additionally, flow curves at elevated shear rates were obtained using a capillary viscometer (Rheologic 1000, CEAST) featuring a capillary diameter of 1 mm and a length-to-diameter ratio of 40. In this study, the Bagley correction, which has to be applied to compensate for the pressure drop at the entrance of the capillary when its length-to-diameter ratio value is low, was omitted due to the high value of the length-to-diameter ratio. The Rabinowitsch correction was applied in order to calculate the true shear rate considering the non-Newtonian behaviour of the melt.

To evaluate the spinnability of these systems, on-isothermal elongational flow tests were performed employing the same instrument. A drawing module composed of a sequence of pulleys was used to grab the hot filament, which gradually cooled in the air, and transport it to an end pulley rotating at a consistent velocity. The force at break, known as the melt strength (MS), of the molten filament was directly measured. The breaking stretching ratio (BSR), denoted as the ratio of drawing speed at break to extrusion speed at the die, was determined, as described in the literature previously [27], by the following equation:

$$\mathrm{BSR} = \frac{V_{roll}}{V_p \cdot \frac{D_p^2}{D_c^2}} \tag{1}$$

where V_{roll} represents the collecting speed, V_p signifies the capillary piston speed, D_p stands for the piston diameter and D_c denotes the capillary diameter.

Tensile tests were executed employing an Instron universal testing machine (Instron, mod. 3365, High Wycombe, UK) adhering to ASTM D638 [28]. The initial crosshead speed was set at 1 mm/min up to 3 min; afterwards, it was raised to 100 mm/min until failure of the specimen. The elastic modulus, E, tensile strength, TS, and elongation at break, EB, were determined with the mean value of eight measurements along with the corresponding standard deviation reported. Both isotropic sheets and oriented fibers were subject to characterization.

The drawing ratio, DR, of the fibers was computed, following the methodology detailed in a prior study [23], using the following formula:

$$\mathrm{DR} = \frac{D_0^2}{D_F^2} \tag{2}$$

where D_0 represents the capillary diameter and D_F signifies the diameter of the fibers.

The surface structure of the samples was analysed with a Quanta 200F scanning electron microscope (manufactured by a company in Hillsboro, OR, USA). Before SEM analysis, all samples were fractured in a cryogenic liquid and then coated with a gold layer to increase their conductivity. Image analysis was conducted using imaging software known as Image J, V. 1.53 which is freely available online.

3. Results and Discussion

Figure 2 shows the Raman spectra for the GNP, PPG and PPGK samples.

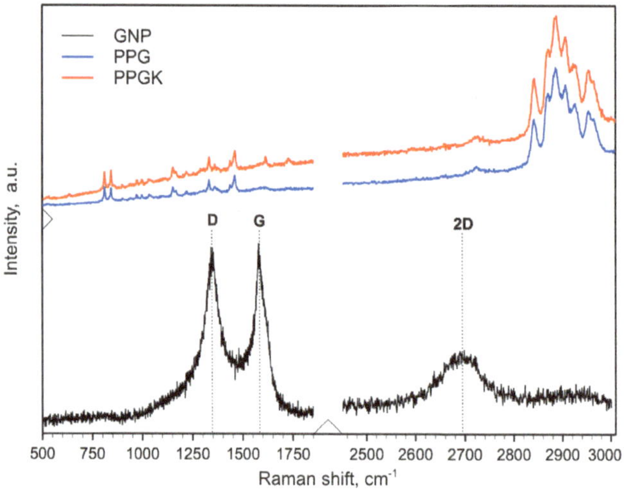

Figure 2. Raman spectra of GNP, PPG and PPGK samples.

The presence of graphene nanoparticles (GNPs) within these nanocomposites was confirmed by our analysis of the Raman spectra. The Raman spectra of the GNPs showed three distinctive peaks at about 1326, 1590 and 2673 cm^{-1}, associated with the D, G and 2D modes, respectively [29]. These peaks are also observed in the Raman spectra of the nanocomposites. The other peaks detected are related to the PP/PET blend.

Figure 3a–c show, respectively, the flow curves of the two nanocomposites measured with both the rotational rheometer and capillary viscometer (Figure 3a); the melt strength (MS) (Figure 3b); and the breaking stereo ratio (BSR) versus the shear rate (Figure 3c) for the two systems investigated.

Firstly, it can be seen that the curves obtained with the capillary viscometer do not coincide with the corresponding curves obtained with the rotational rheometer. This result is in agreement with those found in other similar studies [30–32] on heterogeneous, multiphase systems in which the Cox–Merz law is not respected. As regards to the significant increase in viscosity observed in the nanocomposite in the presence of the styrene (ethylene-co-butylene)–styrene copolymer grafted with maleic anhydride (SEBS-g-MA), this may be attributed to the better adhesion between the two phases. This improved adhesion, as previously reported in the literature [16], may be associated with the presence of the compatibilizer and the formation of copolymers which act as an adhesion promoter, reducing the interfacial tension and preventing coalescence. Consequently, the presence of Kraton™ provides more favourable support for optimizing the interaction between the components, contributing significantly to their flow behaviour.

The melt strength (MS) and breaking stretching ratio (BSR) are critical factors in some industrial processes involving elongational flow, particularly non-isothermal elongational flow, such as operations like spinning and film blowing.

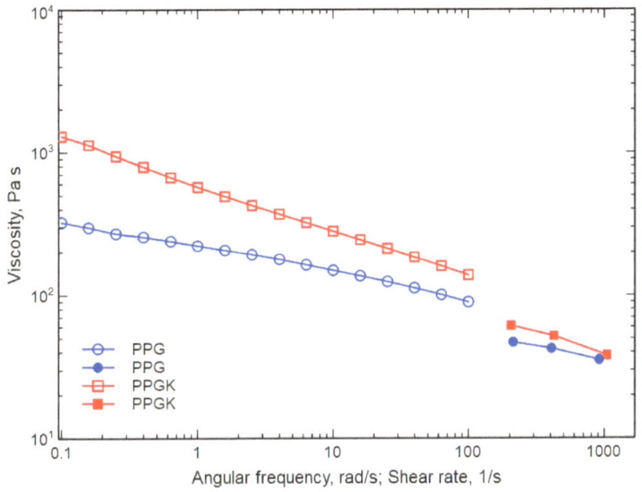

(a)

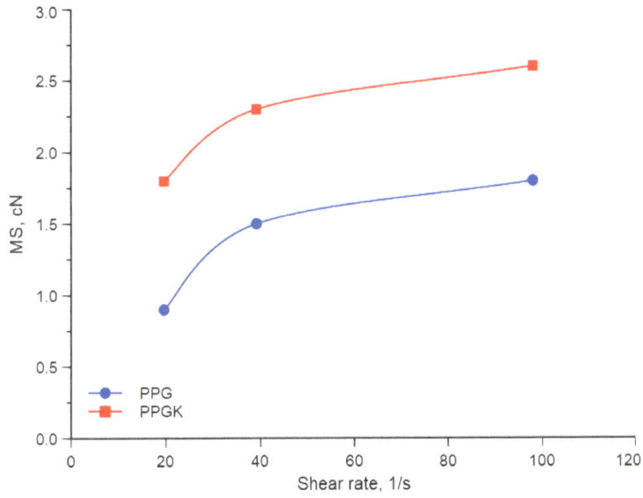

(b)

Figure 3. *Cont.*

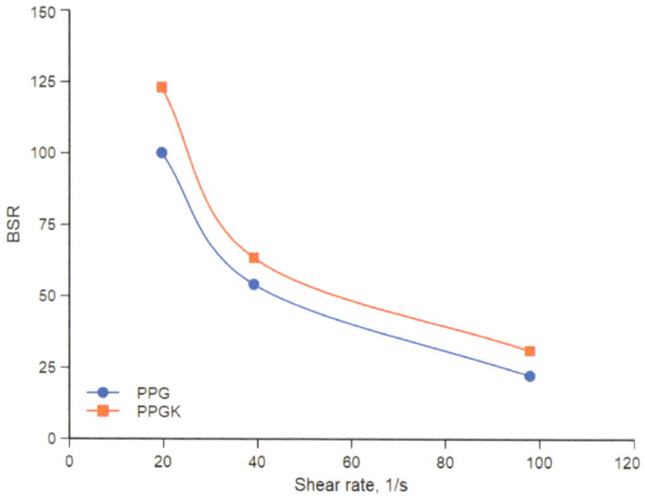

(c)

Figure 3. (**a**) Flow curves of nanocomposites: data represented with hollow shapes are acquired from the rotational rheometer and data represented by filled-in shapes are acquired from the capillary viscometer. (**b**) Melt strength (MS) versus shear rate of PPG and PPGK. (**c**) Breaking stretching ratio (BSR) versus shear rate of PPG and PPGK.

As expected, the melt strength (see Figure 3b) increases as the shear rate gradient increases, while the breaking stretching ratio decreases as it increases (see Figure 3c). Specifically, it is observed that the system with the compatibilizer exhibits higher values than the system without the compatibilizer, both in terms of the melt strength and breaking stretching ratios. Interestingly, this effect is mainly attributable to the increase in viscosity in the case of the melt strength, and, in the case of the breaking stretching ratio, the improvement in stretchability is due to the enhancement of adhesion at the interface of PP/PET.

Figure 4 shows typical stress–strain curves for the two nanocomposite systems obtained by compression moulding, while Table 2 summarizes the average values with their respective standard deviations.

Table 2. Tensile properties of isotropic sheets.

Sample Code	E, MPa	TS, MPa	EB, %
PPG	455 ± 14	22.7 ± 0.9	6.7 ± 1.4
PPGK	509 ± 21	27.1 ± 1.4	16.7 ± 2.1

It is evident from the stress–strain curve that material properties such as the elastic modulus, tensile strength and elongation at break significantly improve in the presence of the compatibilizer. More specifically, the most significant improvements are seen in the elongation at break, with a change from about 6.7% in the absence of the compatibilizer to about 16.7% in its presence. This result could also be attributed to the same reason as previously reported, i.e., the improved adhesion between the two phases. On the other hand, a relative increase of 11% in the elastic modulus and a 19% increase in the tensile strength are observed.

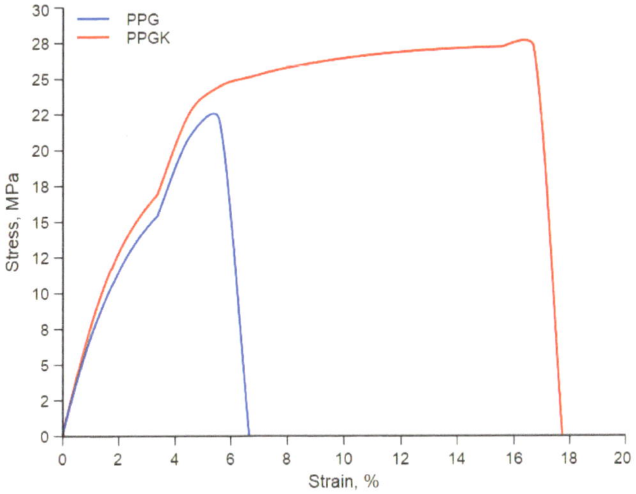

Figure 4. Typical stress–strain curves of isotropic sheets.

Figure 5 shows a morphological analysis (SEM) of and the normal distribution curves for the isotropic samples.

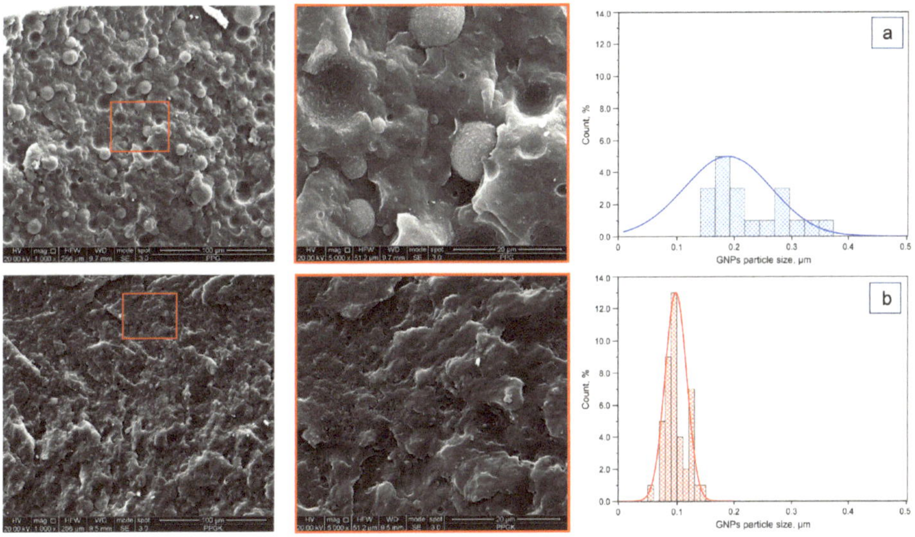

Figure 5. SEM micrographs and normal distribution curves of the isotropic sheets: (**a**) PPG; (**b**) PPGK.

The effect of the presence of the compatibilizer is well evident on both the PET phase and on the GNPs. Indeed, in the binary blend, the two phases are easily distinguished, the dispersed PET phase shows a large range of the dimensions of the particless and adhesion between the two phases is observed. In contrast, the two phases are almost indistinguable in the compatibilized blend. This is, of course, the result of the better adhesion between the PP and PET phases due to the formation of copolymers created by the reaction between the compatibilizer and the PET macromolecules. The GNPs are essentially concentrated in the PET phase and their dimension is strongly reduced in the compatibilized blend.

Moreover, their distribution is also much more tight in the ternary blend, as is evident from the diameter distribution curves reported in Figure 5. The average diameter is about 0.2 um in the non-compatibilized blend, while it is only 0.1 um for the blend in the presence of the Kraton™. It is possible to interpret these data by considering the ternary blend has a larger viscosity and is able to undergo a lot of stress to break the GNPs.

The values for the elastic modulus, E, tensile strength, TS, and elongation at break, EB, of the fibers are shown in Figure 6a–c versus the drawing ratio (DR).

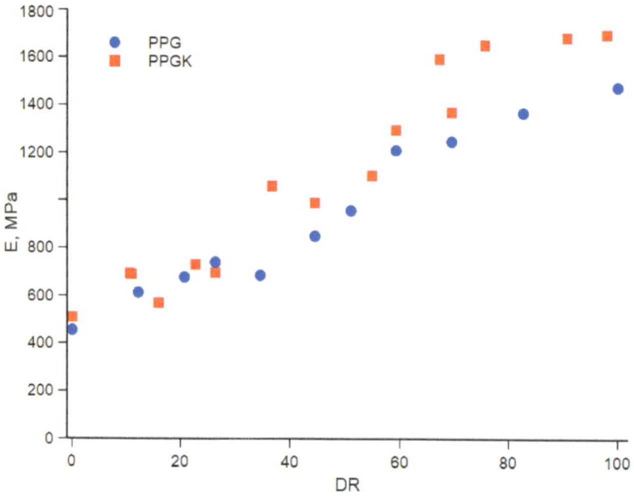

(**a**)

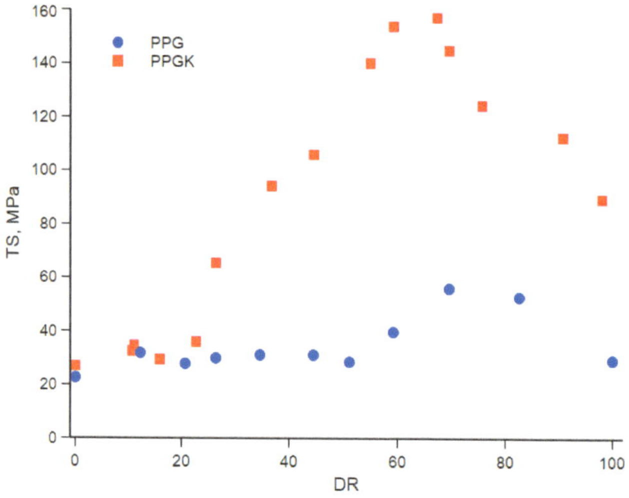

(**b**)

Figure 6. *Cont.*

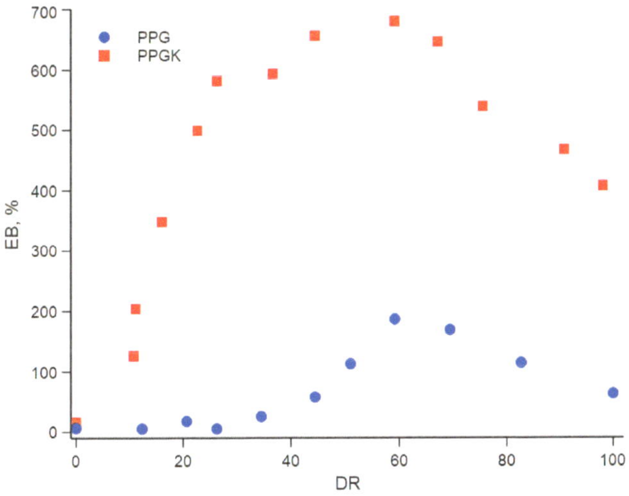

(c)

Figure 6. (**a**) Elastic modulus (E) versus drawing ratio (DR). (**b**) Tensile strength (TS) versus drawing ratio (DR). (**c**) Elongation at break (EB) versus drawing ratio (DR).

From Figure 6a, it can be seen that the elastic modulus increases with the hot drawing ratio. This is due to the orientation of the macromolecules, which is as a result of the applied elongation flow. However, the increase in the elastic modulus for both systems becomes negligible when the drawing ratio is more than about 60 and both curves tend to a plateau. On the other hand, the tensile strength and elongation at break (see Figure 6b,c) show a complex trend. Both the tensile strength and elongation at break first increase with the drawing ratio, reach a maximum and then start to decrease. This behaviour has already been noted in amorphous polymers, polymer blends and nanocomposites [23,30,31] and has been interpreted considering many factors. Firstly, the ordered morphology obtained during spinning, which aligns the macromolecules along the direction of flow, facilitates the sliding of the macromolecules during tensile testing. In addition, the microfibrils formed under the action of the non-isothermal elongational flow do not act as defects, unlike isotropic samples, during deformation [33]. Additionally, the large contact area between the matrix and PET fibrils reduces the impact of poor adhesion. However, once a certain drawing ratio (DR) is reached, in this case 60, the decrease in the elongation at break is probably due to the better orientation of the matrix, which makes the sample more brittle.

In Figure 7, the SEM micrographs of the fibers obtained at a DR = 60 and 100 for both samples are reported.

For both fibers at different drawing ratios, the PET phase exhibits a fibrillar morphology. However, in the non-compatibilized blend, these fibers are distinctly visible and separated from the matrix, whereas in the compatibilized blend, they are thoroughly embedded and firmly adherent. This improved adhesion undoubtedly contributes to the superior mechanical properties observed in the compatibilized blend compared to those of the binary blend. Furthermore, the graphene nanoplatelets remain smaller than those observed in isotropic samples. This reduction in size can be attributed to the elongational flow, which has the capability to further fragment the nanoplatelets into smaller dimensions. In fact, as can be seen in Figure 7a–d, as the drawing ratio increases, the particles undergo greater fragmentation, becoming smaller and thus less distinguishable. The box plot diagrams in Figure 8 clearly illustrate the different GNPs' particle size distributions.

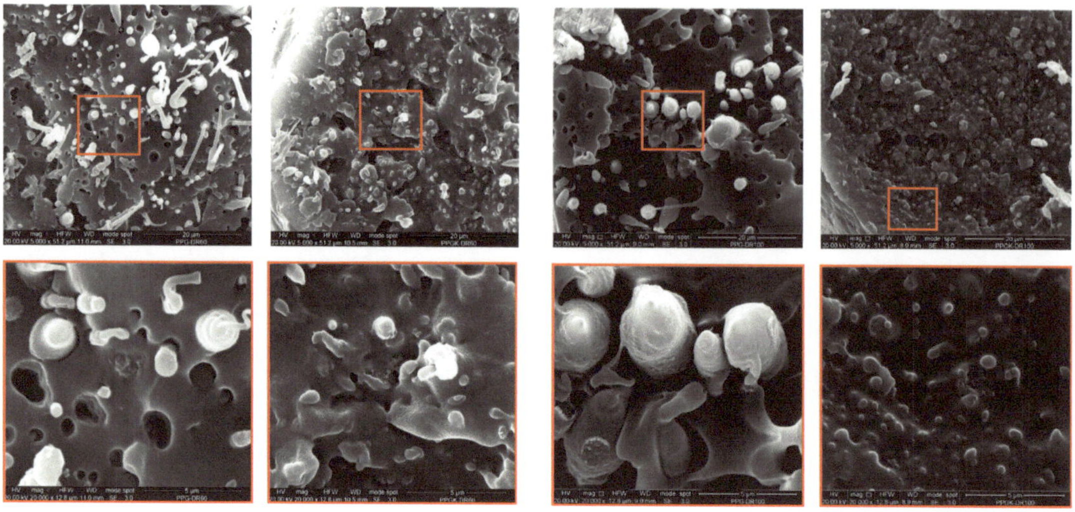

Figure 7. SEM micrographs of the anisotropic samples at DR = 60: (a) PPG and (b) PPGK; and DR = 100: (c) PPG and (d) PPGK.

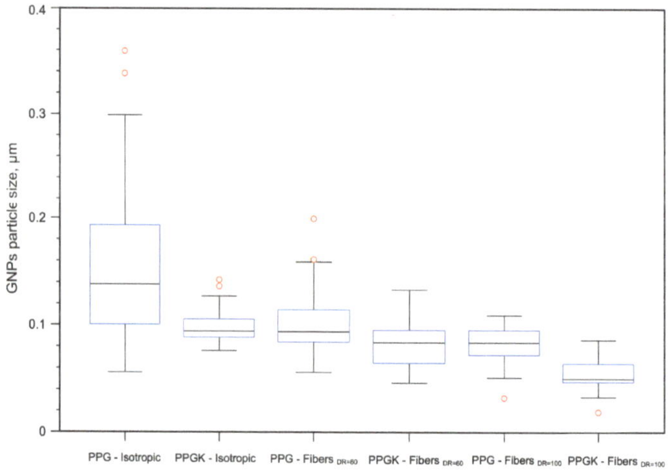

Figure 8. Boxplot of GNPs' particle size distributions.

The isotropic PPG initially shows a wider size distribution (from 0.05 to 0.30 um), and the particle size of the GNPs (with an average value of 0.188 um) is larger than that of the isotropic PPGK (with an average value of 0.098 um). However, for fibers with a drawing ratio (DR) of 60, a reduction in the width of the size distribution is observed compared to the non-compatibilized isotropic sample (from 0.05 to 0.15). This reduction remains almost constant for both samples at the same DR; however, it is noted that the average particle size of the GNPs decreases for the compatibilized sample (0.08 um against 0.11 um). Similar results are found for both the PPG and PPGK with a DR of 100, showing a further decrease in the width of the size distribution and a decrease in the average particle size. This behaviour can be attributed to the action of the elongation flow, which further fragments the nanoparticles into smaller sizes with the increasing drawing ratio.

4. Conclusions

In this study, GNP nanocomposites, with and without the addition of a compatibilizer, were prepared by melt mixing with the aim of investigating the influence of a rubbery compatibilizer on the elongation flow of such systems. The most relevant results of this study are summarized in the radar chart in Figure 9 as follows.

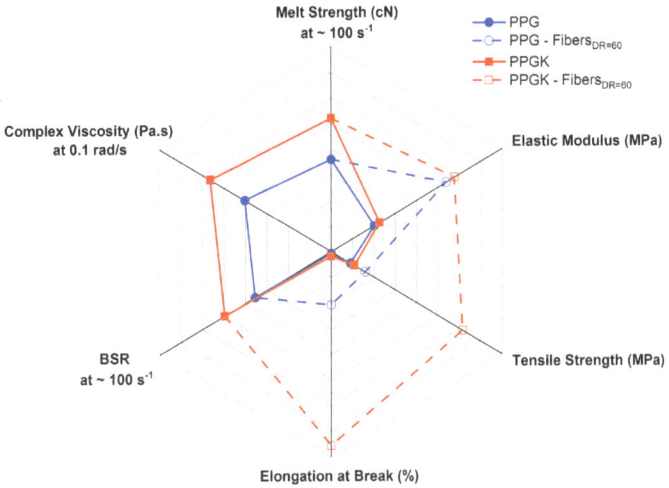

Figure 9. The properties of PPGK and PPG.

In detail, the introduction of the compatibilizer in the nanocomposite blend results in an increase in melt viscosity compared to that of the corresponding sample without it, thus ensuring high melt strength values, which are essential for processing in some industrial processes, like spinning and film blowing. In addition, slight improvements in the BSR are shown due to better interfacial adhesion between the two phases. The mechanical properties of the isotropic sample, as can be seen from Figure 8, show slight improvements in the elastic modulus, tension and elongation at break. However, the most significant improvements, in terms of the elastic modulus, but especially the tensile strength and elongation at break, were observed on the melt-spun fibers. Indeed, the elongational flow acts on two different levels as follows:

1. The chains of the PP matrix are oriented along the spinning line and the dispersed PET particles are elongated and oriented along the draw direction, reinforcing the polymer system and increasing the contact area between the two phases;
2. The nanoplates are fragmented into smaller particles, increasing the contact area between the polymer and nanofiller.

Both of these actions give rise to a more ordered and homogeneous morphology with very high values at the interfaces. This morphology results in brittle to ductile behaviour and better values for the elastic modulus, tensile strength and elongation at break.

The compatibilizer and non-isothermal elongational flow affect the adhesion between the two phases of the blend and between the matrix and the graphene nanoplatelets; the fibrillation of the minor component of the blend; and the size reduction of the nanofiller. These actions result in the change of a more rigid polymer system into a more deformable polymer system, which is a very surprising and interesting result.

Author Contributions: Conceptualization, F.P.L.M.; methodology, F.P.L.M.; investigation, V.T, M.B., M.C. and L.B.; data curation, V.T.; writing—original draft preparation, V.T. and F.P.L.M.; writing—review and editing, V.T., M.B., M.C., L.B. and F.P.L.M.; supervision, F.P.L.M. All authors have read and agreed to the published version of the manuscript.

Funding: This research received no external funding.

Institutional Review Board Statement: Not applicable.

Data Availability Statement: Data are contained within the article.

Acknowledgments: The authors would like to thank the INSTM for financial support.

Conflicts of Interest: The authors declare no conflicts of interest.

References

1. Utracki, L.A. (Ed.) *Polymer Blends Handbook*; Springer: Dordrecht, The Netherlands, 2003; ISBN 978-1-4020-1114-6.
2. Barlow, J.W.; Paul, D.R. Polymer Blends and Alloys—A Review of Selected Considerations. *Polym. Eng. Sci.* **1981**, *21*, 985–996. [CrossRef]
3. Olabis, O. *Polymer-Polymer Miscibility*; Elsevier: Amsterdam, The Netherlands, 2012; ISBN 0-323-15307-0.
4. Srivastava, S.; Tchoudakov, R.; Narkis, M. A Preliminary Investigation of Conductive Immiscible Polymer Blends as Sensor Materials. *Polym. Eng. Sci.* **2000**, *40*, 1522–1528. [CrossRef]
5. Taktak, S.; Ghorbel, N.; Hammami, H.; Fakhfakh, S.; Rondot, S.; Jbara, O. Characterization of Compatibility of Polypropylene/Poly(Butylene Succinate) Blends: Impact of Weight Ratios on Interfacial Polarization. *J. Mol. Liq.* **2022**, *368*, 120633. [CrossRef]
6. Carrot, C.; Mbarek, S.; Jaziri, M.; Chalamet, Y.; Raveyre, C.; Prochazka, F. Immiscible Blends of PC and PET, Current Knowledge and New Results: Rheological Properties. *Macromol. Mater. Eng.* **2007**, *292*, 693–706. [CrossRef]
7. Mistretta, M.; Morreale, M.; Botta, L.; Ceraulo, M.; Fontana, P.; La Mantia, F. "Compatibilization" through Elongational Flow Processing of LDPE/PA6 Blends. *Materials* **2018**, *11*, 2375. [CrossRef]
8. Rosales, C.; Aranburu, N.; Otaegi, I.; Pettarin, V.; Bernal, C.; Müller, A.J.; Guerrica-Echevarría, G. Improving the Mechanical Performance of LDPE/PP Blends through Microfibrillation. *ACS Appl. Polym. Mater.* **2022**, *4*, 3369–3379. [CrossRef]
9. Thakre, A.R.; Baxi, R.N.; Walke, P.V.; Gupta, P. Mechanical and Thermal Behavior of PP/PET Blends. *Int. J. Innov. Eng. Sci.* **2018**, *3*, 2456–3463.
10. Kharghanian, M.; Perchicot, R.; Irusta, S.; Argon, C.Y.; Leonardi, F.; Dagreou, S. Manufacture and Rheological Behavior of All Recycled PET/PP Microfibrillar Blends. *Polym. Eng. Sci.* **2023**, *63*, 1702–1715. [CrossRef]
11. Neve, H.; Thorat, S.; Radhakrishnan, S. Melt Processing Behaviour and Structure Development in Recycled PP Blends with Recycled PET. *Prog. Rubber Plast. Recycl. Technol.* **2023**, *39*, 141–155. [CrossRef]
12. Si, X.; Guo, L.; Wang, Y.; Lau, K. Preparation and Study of Polypropylene/Polyethylene Terephthalate Composite Fibres. *Compos. Sci. Technol.* **2008**, *68*, 2943–2947. [CrossRef]
13. Calhoun, A. Polypropylene. In *Multilayer Flexible Packaging*; Elsevier: Amsterdam, The Netherlands, 2010; pp. 31–36, ISBN 978-0-8155-2021-4.
14. Visakh, P.M. Polyethylene Terephthalate: Blends, Composites, and Nanocomposites—State of Art, New Challenges, and Opportunities. In *Poly (Ethylene Terephthalate) Based Blends, Composites and Nanocomposites*; Elsevier: Amsterdam, The Netherlands, 2015; pp. 1–14, ISBN 978-0-323-31306-3.
15. Fakirov, S.; Bhattacharyya, D.; Lin, R.J.T.; Fuchs, C.; Friedrich, K. Contribution of Coalescence to Microfibril Formation in Polymer Blends during Cold Drawing. *J. Macromol. Sci. Part B* **2007**, *46*, 183–194. [CrossRef]
16. Mantia, F.; Ceraulo, M.; Giacchi, G.; Mistretta, M.; Botta, L. Effect of a Compatibilizer on the Morphology and Properties of Polypropylene/Polyethylentherephthalate Spun Fibers. *Polymers* **2017**, *9*, 47. [CrossRef] [PubMed]
17. Mostofi, N.; Nazockdast, H.; Mohammadigoushki, H. Study on Morphology and Viscoelastic Properties of PP/PET/SEBS Ternary Blend and Their Fibers. *J. Appl. Polym. Sci.* **2009**, *114*, 3737–3743. [CrossRef]
18. Ekinci, A.; Öksüz, M.; Ates, M.; Aydin, I. Thermal and Mechanical Properties of Polypropylene/Post-Consumer Poly (Ethylene Terephthalate) Blends: Bottle-to-Bottle Recycling. *J. Polym. Res.* **2022**, *29*, 433. [CrossRef]
19. Park, Y.W.; Park, M.; Kim, H.Y.; Kim, H.C.; Lim, J.C.; Jin, F.-L.; Park, S.-J. Thermal and Curl Properties of PET/PP Blend Fibres Compatibilized with EAG Ternary Copolymer. *Bull. Mater. Sci.* **2018**, *41*, 104. [CrossRef]
20. KRATONTM FG. Available online: https://kraton.com/products/kraton-fg/ (accessed on 15 January 2024).
21. Inuwa, I.M.; Hassan, A.; Wang, D.-Y.; Samsudin, S.A.; Mohamad Haafiz, M.K.; Wong, S.L.; Jawaid, M. Influence of Exfoliated Graphite Nanoplatelets on the Flammability and Thermal Properties of Polyethylene Terephthalate/Polypropylene Nanocomposites. *Polym. Degrad. Stab.* **2014**, *110*, 137–148. [CrossRef]
22. Inuwa, I.M.; Hassan, A.; Samsudin, S.A.; Haafiz, M.K.M.; Jawaid, M.; Majeed, K.; Razak, N.C.A. Characterization and Mechanical Properties of Exfoliated Graphite Nanoplatelets Reinforced Polyethylene Terephthalate/Polypropylene Composites. *J. Appl. Polym. Sci.* **2014**, *131*, app.40582. [CrossRef]
23. Mantia, F.P.L.; Titone, V.; Milazzo, A.; Ceraulo, M.; Botta, L. Morphology, Rheological and Mechanical Properties of Isotropic and Anisotropic PP/rPET/GnP Nanocomposite Samples. *Nanomaterials* **2021**, *11*, 3058. [CrossRef] [PubMed]
24. Arrigo, R.; Malucelli, G.; Mantia, F.P.L. Effect of the Elongational Flow on the Morphology and Properties of Polymer Systems: A Brief Review. *Polymers* **2021**, *13*, 3529. [CrossRef]

25. Moplen RP340H. Available online: https://www.lyondellbasell.com/en/polymers/p/moplen-rp340h/0e002116-5935-44aa-Bee7-B2c007895821 (accessed on 15 January 2024).
26. *ASTM D1238-10*; Standard Test Method for Melt Flow Rates of Thermoplastics by Extrusion Plastometer. ASTM International: West Conshohocken, PA, USA, 2010; Volume 83, p. 20.
27. Titone, V.; Botta, L.; Mistretta, M.C.; La Mantia, F.P. Influence of a Biodegradable Contaminant on the Mechanical Recycling of a Low-Density Polyethylene Sample. *Polym. Eng. Sci.* **2024**, *64*, 845–851. [CrossRef]
28. *ASTM D638-14*; Standard Test Method for Tensile Properties of Plastics. ASTM International: West Conshohocken, PA, USA, 2015. Available online: http://www.dept.aoe.vt.edu/~aborgolt/aoe3054/manual/expt5/D638.38935.pdf (accessed on 15 January 2024).
29. Botta, L.; Scaffaro, R.; Sutera, F.; Mistretta, M. Reprocessing of PLA/Graphene Nanoplatelets Nanocomposites. *Polymers* **2017**, *10*, 18. [CrossRef] [PubMed]
30. La Mantia, F.P.; Arrigo, R.; Morreale, M. Effect of the Orientation and Rheological Behaviour of Biodegradable Polymer Nanocomposites. *Eur. Polym. J.* **2014**, *54*, 11–17. [CrossRef]
31. La Mantia, F.P.; Ceraulo, M.; Mistretta, M.C.; Botta, L. Effect of the Elongational Flow on Morphology and Properties of Polypropylene/Graphene Nanoplatelets Nanocomposites. *Polym. Test.* **2018**, *71*, 10–17. [CrossRef]
32. Ren, J.; Krishnamoorti, R. Nonlinear Viscoelastic Properties of Layered-Silicate-Based Intercalated Nanocomposites. *Macromolecules* **2003**, *36*, 4443–4451. [CrossRef]
33. La Mantia, F.P.; Fontana, P.; Morreale, M.; Mistretta, M.C. Orientation Induced Brittle—Ductile Transition in a Polyethylene/Polyamide 6 Blend. *Polym. Test.* **2014**, *36*, 20–23. [CrossRef]

polymers

MDPI

Article

Influence of Twin Screw Extrusion Conditions on MWCNT Length and Dispersion and Resulting Electrical and Mechanical Properties of Polycarbonate Composites

Petra Pötschke *, Tobias Villmow, Beate Krause and Bernd Kretzschmar

Leibniz-Institut für Polymerforschung Dresden e.V. (IPF), Hohe Straße 6, 01069 Dresden, Germany; krause-beate@ipfdd.de (B.K.); bkretzsch@online.de (B.K.)
* Correspondence: poe@ipfdd.de; Tel.: +49-351-4658-395

Abstract: The processing conditions were varied during the production of polycarbonate-based composites with the multiwalled carbon nanotubes (MWCNTs) Baytubes® C150 P (Bayer MaterialScience AG, Leverkusen, Germany), by melt mixing with an extruder on a laboratory scale. These included the screw design, rotation speed, throughput, feeding position and MWCNT content. Particular attention was paid to the shortening of the MWCNT length as a function of the conditions mentioned. It was found that there is a correlation between the applied specific mechanical energy (SME) during the melt mixing process and MWCNT dispersion, which was quantified by the agglomerate area ratio of the non-dispersed nanotubes based on optical microscopic analysis. The higher the SME value, the lower this ratio, which indicates better dispersion. Above an SME value of about 0.4 kWh/kg, no further improvement in dispersion was achieved. The MWCNT length, as measured by the quantitative analysis of TEM images of the MWCNTs dissolved from the composites, decreased with the SME value down to values of 44% of the original MWCNT length. At a constant loading of 3 wt.%, the tensile strength and tensile modulus were almost independent of the SME, while the elongation at break and notched impact strength showed an increasing trend. The variation in the feeding position showed that feeding the MWCNTs into a side feeder led to slightly better electrical and mechanical properties for both types of MWCNTs studied (Baytubes® C150 P and Nanocyl™ NC7000 (Nanocyl S.A., Sambreville, Belgium)). However, feeding into the hopper led to better CNT dispersion with Baytubes® C150 P, while this was the case with Nanocyl™ NC7000 when feeding into the side feeder. The screw profile had an influence on the dispersion, the MWCNT length and the electrical resistance, but only to a small extent. Distributive screws led to a greater shortening of the MWCNT length than dispersive screws. By varying the MWCNT content, it was shown that a greater MWCNT shortening occurred at higher loadings. Two-stage masterbatch dilution leads to stronger shortening than composite production with direct MWCNT incorporation.

Keywords: polycarbonate; multiwalled carbon nanotubes; melt mixing; dispersion; electrical conductivity; nanotube length; mechanical properties

Citation: Pötschke, P.; Villmow, T.; Krause, B.; Kretzschmar, B. Influence of Twin Screw Extrusion Conditions on MWCNT Length and Dispersion and Resulting Electrical and Mechanical Properties of Polycarbonate Composites. *Polymers* **2024**, *16*, 2694. https://doi.org/10.3390/polym16192694

Academic Editor: Xufeng Dong

Received: 31 July 2024
Revised: 12 September 2024
Accepted: 17 September 2024
Published: 24 September 2024

1. Introduction

In polymer nanocomposites, the full potential of the nanofillers can be only exploited if the particles are dispersed on a nanoscale and suitably distributed in the matrix. Then, a high interfacial area is created which at the same time means that the polymer chains around the nanoparticles are restricted by the contact with the filler surface and display different properties than in the matrix. When the property of electrical conductivity of carbon nanoparticles should be introduced into the composite, then a network of conductive fillers has to be established. This can be formed either by direct contact between neighboring nanoparticles or using hopping and tunneling effects between neighboring filler particles.

Carbon nanotubes (CNTs) provide a special challenge in dispersion into individualized nanoparticles. Due to their synthesis process and their cylindrical shape, they always exist as primary agglomerates, which can have quite different morphologies [1]. Their dispersibility mainly depends on their structural characteristics, like their length, diameter, waviness, branching, defect density and surface characteristics including functional groups. Their achievable state of dispersion in different media depends in addition on the matrix viscosity and surface energy properties, resulting in different interactions, but also very strongly on the processing conditions. Depending on all these influencing factors, not only can the state of dispersion be influenced significantly, but also the length of the CNTs can be reduced in different amounts.

Many investigations were performed on the influence of melt mixing conditions on dispersion and electrical properties on a small scale, such as using microcompounders. For polycarbonate (PC)-based composites with multiwalled CNTs (MWCNTs), Alig et al. [2] performed dielectric measurements on composites prepared with the masterbatch approach using different mixing times and rotation speeds. They concluded that near the percolation threshold (here, 1.0 wt.%), an increase in mixing time could improve dispersion in such a way that non-percolated systems got percolated. Decreased conductivity values when increasing the mixing speed or mixing time at a higher loading (1.5 wt.%) were explained with more pronounced nanotube shortening. However, no measurements on the nanotube length were performed. Lin et al. [3] compared three different types of miniature mixers when diluting an MWCNT masterbatch and measured the MWCNT length after processing by applying atomic force microscopy (AFM). Therefore, the composites were dissolved in methylene chloride, filtered, treated with sodium–hydroxyl–ethanol solution under sonification, washed, dried and dispersed in chloroform. Silicon wafers were dip coated in these dispersions and studied. Using this procedure, it was shown that after the melt mixing, the MWCNTs possessed length values between 66% and 74% of their initial lengths. This indicated that shortening occurred, however it was not very different when using the different small-scale mixers. The electrical percolation thresholds were lower for the composites produced with custom-built mixers (APAM and MBM) than with the commercial DACA microcompounder, even if the dispersion (investigated by TEM) was best for the DACA mixer. Thus, it was concluded that small agglomerates may help to form the electrical network.

In polyamide (PA)-based composites prepared by direct incorporation of MWCNT Nanocyl™ NC7000 nanotubes, Krause et al. [4] varied the mixing speed and mixing temperature and studied electrical percolation and dispersion. At constant mixing conditions, PA66 as a matrix resulted in a lower electrical percolation threshold than PA6, and increasing the screw speed in PA6 increased the threshold. The dispersion of 5 wt.% MWCNTs in PA6 improved when increasing the mixing speed. After calculating the mixing energy input for each mixing process, the increase in the state of dispersion with mixing energy was accompanied by an initial decrease in electrical volume resistivity followed by an increase when performing the mixing with a higher-energy input. Also, it was concluded that more pronounced CNT shortening at higher-mixing energy input may have been the reason for that resistivity increase.

In a study on high-impact polystyrene by McClory et al. [5], with MWCNTs from a Chinese producer, variation in screw speed (20 rpm to 150 rpm) at otherwise constant conditions resulted in the best dispersion at an intermediate screw speed (100 rpm). At higher speeds, the percolation concentration was significantly increased, which was assigned to a lower disentanglement of MWCNT agglomerates at such high speeds.

Kasaliwal et al. [6,7] studied polycarbonates with MWCNT Baytubes® C150 HP in detail to determine the dispersion mechanisms of primary nanotube agglomerates by varying the mixing speed, mixing time and melt temperature during small-scale compounding. The area ratio of the remaining agglomerates, as determined from thin sections using transmission light microscopy, was used was used to quantify the state of dispersion. From the relative change in the area ratio vs. mixing time at different mixing speeds and a proposed model, the time coefficients of the two mainly considered processes of rupture and erosion were calculated, from which the fractions of rupture and erosion mechanisms during agglomerate dispersion were estimated. At low mixing speeds, dispersion was found to be governed by both mechanisms, whereas the dominance of the rupture process increased with increasing mixing speed. Further, the relationship between electrical resistivity and dispersion was studied, indicating that a dependency on the amount of dispersed nanotubes was found only in a certain range of states of dispersion. The electrical resistivity of the composites decreased up to a specific mechanical energy (SME) of 1400 J/cm^3, whereas at higher mixing energies (up to about 55,000 J/cm^3), reached by either enhancing the mixing speed or mixing time, a plateau with low electrical resistivities was found.

In another study by Kasaliwal et al. [8,9], the molecular weight of the PC and thus the melt viscosity of the matrix material was changed, resulting in better dispersion at higher matrix viscosity when using constant mixing conditions but worse considering comparable shear stress values. To selectively study the influence of the molecular weight, the melt mixing temperatures of the three used PCs were adjusted to have similar viscosities, and melt mixing was performed at a constant mixing speed. As investigated on two viscosity levels, the composites based on the low-molecular weight matrix showed smaller undispersed primary agglomerates as compared to composites with higher molecular weight matrices. This result was discussed to indicate the role of matrix infiltration in primary nanotube agglomerates as the first step of dispersion, which is faster at lower molecular weights or lower matrix viscosities. The resistivity values of composites prepared using low-viscosity matrices were lower than those of composites from a high-viscosity matrix.

Such an influence of the matrix viscosity was also studied for other polymer matrices. Socher et al. presented a study comparing the matrix polymers polyamide 12, polybutylene terephthalate, PC, poly(ether ether ketone) and low-density polyethylene with three viscosity levels each in composites with MWCNT Baytubes® C150 P [10]. Huge differences in the electrical percolation thresholds were found using the same polymer matrix at different viscosity grades. The composites with the low-viscosity matrix always resulted in the lowest percolation threshold. The state of primary MWCNT agglomerate dispersion increased with matrix viscosity due to the higher input of mixing energy. This study also focused on the effect of the shortening of nanotubes due to breaking during processing, using two viscosity grades of PC. To evaluate such influences, a more sophisticated method was developed by Krause et al. [11] using the TEM analysis of single nanotubes dissolved from the composite sample that was applied in this study. The effect of nanotube shortening was quantified using two different viscosity grades of PC. Due to the higher mixing energy input, the nanotube shortening was more pronounced in the high-viscosity matrix, which partially explained the higher percolation threshold.

The relationships between the rotation speed during small-scale mixing (varied between 25 rpm and 400 rpm) and different properties of composites of polycaprolactone (PCL) and 0.5 wt.% Nanocyl™ NC7000 was studied in a comprehensive manner by Pötschke et al. [12]. The properties studied included the molecular weight of PCL after processing, PCL melting and crystallization behavior, MWCNT macrodispersion, MWCNT length distribution, electrical resistivity, and oscillatory melt rheological properties of the composites. The MWCNT dispersion was improved with increasing rotation speed but leveled off starting at about 100 rpm. The molecular weight of the PCL as well as crystallization and melting behavior did not change significantly when varying the rotation speed. The storage modulus G′ and complex viscosity η^* at 0.1 rad/s increased up to a rotation speed of about 75 rpm, illustrating improved dispersion. When the speed values were further increased, both values decreased. This was attributed to more pronounced MWCNT shortening, as quantified by TEM. Both effects, improved dispersion and nanotube shortening, were reflected in the electrical resistivity of compression-molded samples, which showed a minimum of resistivity at the rotation speed of 75 rpm, corresponding to a specific mechanical energy input of 0.47 kWh/kg. Mayoral et al. studied the influence of mixing speed (50 to 200 rpm), mixing time (1–30 min) and MWCNT loading (0.5 to 2.0 wt.%) in the batch mixing of polypropylene (PP)–MWCNT composites, with a focus on the agglomerate sizes and their distributions, melt rheology and electrical resistivity.

Concerning MWCNT dispersion, the highest rotor speed at the longest mixing time resulted in being the best, whereas the highest electrical conductivity was achieved at a medium speed and mixing time. The effects were explained with a suitable hierarchical arrangement of MWCNT agglomerates, smaller MWCNT bundles and individual MWCNTs, leading to the best conductive network; CNT shortening was not considered [13]. More examples of the influence of material properties and small-scale mixing conditions on dispersion and electrical and mechanical properties can be found in the literature [14], whereby the study of CNT shortening during processing is quite limited.

Even if the main results of such small-scale mixing experiments can be applied to laboratory twin screw extrusion, as commonly used in composite fabrication by melt mixing processes, much less studies have concerned the influence of twin screw processing conditions on the CNT dispersion, nanotube length and electrical resistivity or mechanical properties of the composites. The first examples were given by Villmow et al. [15] on poly(lactic acid) (PLA) with Nanocyl™ NC7000, who studied the influence of the screw profile, temperature profile and rotation speed (100 rpm and 500 rpm) using a Berstorff ZE25 extruder (KraussMaffei Berstorff, Laatzen, Germany) on the state of dispersion in masterbatches containing 7.5 wt.% and 15 wt.% MWCNT. These masterbatches were then diluted to composites with 0.75 wt.% MWCNT under varied process conditions. Interestingly, the state of MWCNT dispersion in the diluted composites was predominated by the state of filler dispersion in the masterbatches, independent of the mixing conditions during dilution. It was found that a high rotation speed that still ensured a certain residence time of the melt combined with a screw profile containing mainly mixing elements were convenient to disperse and distribute the MWCNTs in the PLA matrix as well during the masterbatch production as the dilution step. The temperature profile showed less influence; however, a profile with increasing temperatures along the screw resulted in slightly better nanotube dispersion. In their study, it was shown that differential scanning calorimetry could be used to judge the state of MWCNT dispersion in the masterbatches where microscopic methods had problems in distinguishing different states of dispersion. As more CNTs are dispersed, the higher the nucleating action of the CNT surfaces on the PLA.

In a later study by Villmow et al. [16], similar processing was adapted to PCL melt mixed with Nanocyl™ NC7000, with a special focus on the influence of screw configuration on the residence time, SME and the resulting macroscopic CNT dispersion in masterbatches containing 7.5 wt.%. In addition, the rotation speed and throughput were varied and samples along the screw were taken. The main output of their study was that the processing conditions had a strong influence on the residence time (t_R) of the extrudates and on the CNT dispersion. Both an increase in the rotation speed and throughput resulted in a decrease in t_R, whereas the use of back-conveying elements and the extension of the processing length showed the opposite effect. With increasing rotation speed, higher SME inputs and a significant increase in CNT dispersion were observed, whereas an increase in throughput resulted in worse dispersion. The use of distributive screw configurations containing mixing elements can further promote CNT dispersion. The best dispersion was found when using an extended distributive screw having a length-to-diameter ratio (L/D) of 48 instead of 36.

Finally, using the same equipment, the influence of the feeding position of the MWCNTs on the dispersion and properties of composites was also studied. Müller et al. [17] investigated polypropylene-based composites with both, Baytubes® C150 P and Nanocyl™ NC7000. Both MWCNT materials were fed at selected concentrations either in the hopper or using a side feeder under otherwise identical extrusion conditions. The results led to the conclusion that the more compact Baytubes® C150 P agglomerates should be added into the hopper, as the dispersion was better, the electrical resistivities measured on compression and injection-molded samples were lower, and the elastic modulus, yield strength and impact strength were higher as compared to side feeding. On the other hand, for the more loosely packed Nanocyl™ NC7000 agglomerates, addition using the side feeder led to better dispersion, lower electrical resistivity and higher mechanical properties. This agrees with the different dispersibility and agglomerate strengths of both MWCNT materials as shown in a dispersion study by Krause et al. [18] and illustrates that the structure of the primary agglomerates of MWCNTs has to be taken into account when optimizing processing conditions.

A study by Benedito et al. [19,20] used commercial or self-prepared thermoplastic polyurethane (TPU)–MWCNT masterbatches (15 wt.% NC7000) to dilute them in the co-rotating twin screw extruder Coperion ZSK25 under different compounding parameters. As shown in [19], using light microscopy images on 1 wt.% composites, the dispersion was significantly better in composites prepared at 600 rpm than in those made at 200 rpm. Dynamic rheology and dynamic mechanical analysis (DMA) were used to quantify differences at varied loadings. In [20], a masterbatch was prepared with a screw, comprising a combination of high-shear dispersive elements, distributive elements and mixing elements under a high screw speed of 600 rpm. The dilution was performed using the named screw and one with mainly transport elements, resulting in a lower residence time and mechanical energy at speeds of 200 rpm and 600 rpm. The screw design was shown to have a significant impact on the MWCNT dispersion, with significantly better dispersion when using the screw design with a higher energy input. At higher screw speeds, better dispersions were achieved, as quantified using a term called "agglomeration density". On the other hand, the samples with the worser dispersion, prepared at a low speed or using the low-mechanical energy screw design showed higher electrical conductivity. The shortening of MWCNTs due to breakup was hypothesized.

Another example is a later study by Mack et al. [21] on PC-based composites with 1, 3 and 5 wt.% MWCNT Nanocyl™ NC7000 processed in a twin screw extruder Leistritz ZSE 27HP-52D with two different screw configurations, screw speeds and throughputs, resulting in different SMEs at different conditions. Therefore, screw 1 was designed to have mainly kneading elements and transportation elements and screw 2 had some kneading elements replaced by other elements. The volume resistivity of the injection-molded PC-1 wt.% MWCNT composites varied only slightly with the conditions, whereas at 3 wt.% loading, the resistivity was independent of the process conditions. The modulus and strength of the composites were enhanced with filler addition, whereby at 5 wt.% loading, the strengths of the composites produced at higher SMEs (both screws, screw speed of 1100 rpm and throughput of 7.5 kg/h) were higher than those produced at lower SMEs (both screws, lower rotation speed of 500 rpm and higher throughput of 12.5 kg/h). The highest value was achieved for screw 2 at the lower throughput and the higher screw speed, which however, was not the sample that had experienced the highest SME. At the same filler loading, composites processed with higher SMEs showed, at all MWCNT concentrations, higher notched impact strength values which were assigned to the better filler dispersion. The same group also studied composites based on polystyrene (PS) and Baytubes® C150 P [22]. Here, 1, 2, 3, 5 and 7.5 wt.% MWCNTs were incorporated into PS in the twin screw extruder at varied speeds, throughputs and extruder barrel temperatures at two levels each. The higher SME at an enhanced processing speed was found to have the largest effect on enhancing nanotube dispersion. The electrical volume resistivity was significantly lower in the 2 wt.% MWCNT samples processed at 1100 rpm compared to those produced at 500 rpm. The influence of screw speed on the properties using the same equipment was also studied for composites based on polypropylene and 2 wt.% MWCNT Nanocyl™ NC7000 [23]. Better dispersion was found at the higher speed combined with a higher modulus, slightly higher tensile strength and slightly reduced notched impact strength in injection-molded composites. In 2023, Ezat et al. studied the effect of screw configuration using the intermeshing twin screw extruder PRISM-TSE-16-TC, using PP and acid-purified MWCNTs from Cheap Tubes, modified with a maleic anhydride-grafted PP as a compatibilizer [24]. Among the three designed screws, the high-intensity configuration with a higher number of kneading blocks achieved the best dispersion and lowest percolation concentration at 2 wt.%, compared to a standard screw and the screw designed to establish chaotic mixing conditions. The tensile properties were only slightly affected by the screw configuration. The tensile modulus was slightly higher when using the medium-intensity (standard) screw. Also, the tensile strength of the PP was most improved by the MWCNTs when incorporated using that standard screw.

In all these laboratory-scale studies, a low focus was put on the effect of the extrusion step on the CNT length, even if a possible shortening was partially named as the reason for unexpected results (e.g., [20]). However, to understand the relationships between the processing conditions, morphology and properties of nanocomposites completely, the change in the initial length of the reinforcing nanotubes and the effect of processing conditions on that phenomenon should also be considered. Studies are needed not only on small-scale mixing devices, but also on laboratory-scale extrusion (LSE). In addition to the already named studies made for small-scale mixing experiments, Krause et al. [25] studied the shortening of MWCNTs Nanocyl™ NC7000 ball-milled at different times (5 h, 10 h) after their melt extrusion with PC on a laboratory scale. They found that after melt compounding, the mean CNT lengths were shortened to 31%, 50% and 66% of the initial lengths of NC7000, NC7000-5 h and NC7000-10 h, respectively. In small-scale experiments, Liebscher et al. [26] compared the lengths of NC7000 MWCNTs in PC composites with 0.5 and 1.0 wt.% CNTs when using two different mixing conditions. It was found that the shortening was slightly higher when harsher mixing conditions were applied, such as a higher rotation speed and longer mixing time. For composites with 0.5 wt.% MWCNTs, after the harsher conditions, the initial CNT length before compounding was reduced to 70%, whereas after the milder conditions, it was reduced to 76%, based on the x_{50} values. From comparing samples with 0.5 wt.% and 1.0 wt.%, the CNT length shortening was more pronounced at the higher loading. In addition, the authors explored that even if the CNTs preferred the PC component, blending with a second polymer, namely styrene–acrylonitrile copolymer (SAN), to obtain co-continuous blend morphologies further reduced the CNT length [26]. Finally, in the blends, the lengths were only 27% and 30% of the initial length at harsher and milder conditions for blends containing 0.5 wt.% CNTs, respectively. In contrast, Guo et al. found, for small-scale mixed PS/poly (methyl methacrylate) (PMMA) blends with MWCNTs showing matrix droplet morphologies, that relatively long MWCNTs broke less in the polymer blends than in the neat polymer droplet material [27]. This resulted in a reduction in the percolation threshold. This breakage depended on the amount of the component which the CNTs preferred (here, PMMA) and was lower when filled PMMA formed the droplets [27].

Based on the lack of information about CNT length changes in laboratory-scale melt-compounded composites, this study is intended to show comprehensive results obtained on PC-based composites with industrial multiwalled carbon nanotubes. The composites were obtained by laboratory-scale extrusion (LSE) compounding with varied rotation speeds, throughputs, feeding positions and screw profiles. Therefore, a special focus was laid on the study of the CNT length after compounding, which so far has not been systematically studied. In addition, the state of dispersion was studied using light microscopy, and the electrical and selected mechanical properties are presented.

2. Materials and Methods

2.1. Materials and Processing

As the polycarbonate (PC), medium-viscosity general-purpose Makrolon® 2600-grade (Bayer MaterialScience AG, Leverkusen, Germany), intended for injection molding with a melt volume–flow rate of 12.0 $cm^3/10$ min (1.2 kg, 300 °C [28]), was applied. As for the molecular weight of this material, a previous study indicated an M_n value of 9500 g/mol, an M_w value of 26,200 g/mol g/mol and a polydispersity index of 2.76 [9]. Further properties of this material are given in its datasheet [28].

As for the multiwalled carbon nanotubes (MWCNTs), the widely used, commercially available Baytubes® C150 P (Bayer MaterialScience AG, Leverkusen, Germany), with a carbon purity of ~95%, bulk densities of 120–170 kg/m^3, agglomerate sizes of 30–550 μm [18], a mean diameter of 10.5 nm and a mean length x_{50} of 770 nm [29], were used. From these data, a mean aspect ratio of 73 was calculated. More information about this material can be found in its datasheet [30]. In addition, the commercially available MWCNT material Nanocyl™ NC7000 (Nanocyl S.A. Sambreville, Belgium), with a carbon purity of >90%, a bulk density of 66 kg/m³ [31], an agglomerate size of >675 µm [18], a mean diameter of 10.0 nm and a mean length x_{50} of 1341 nm [29], was also applied for comparison in part of the study. It had a mean aspect ratio of 134. For more information about the material, please refer to its datasheet [32]. The electrical conductivity of these MWCNT powdery materials was measured using a powder cell, and at a pressure of 30 MPa, values of 15 S/cm for the Baytubes® C150 P and of 24 S/cm for the Nanocyl™ NC7000 were obtained [33]. Scanning electron microscopic images of the initial agglomerates of Baytubes® C150 P and Nanocyl™ NC7000 are given in [34]. More information about the structural details of the Baytubes® and Nanocyl™ MWCNT products can be found in a detailed study by Tessonnier et al. in [35].

To study the influence of the processing conditions in laboratory-scale extrusion (LSE), in a first set polycarbonate composites filled with 3 wt.% Baytubes® C150 P were prepared using a co-rotating twin screw extruder ZE25 (KraussMaffei Berstorff, Laatzen, Germany), with an L/D ratio of 48 and at a mean barrel temperature of 260 °C. A number of variations were carried out. On the one hand, using a fixed screw configuration, the influence of rotation speed, varying between 100 and 1000 rpm, and the effect of the throughput, varying for all rotation speeds between 5 and 15 kg/h, were investigated. PC pellets and the powdery MWCNT material were fed simultaneously into the hopper by gravimetric dosing. To study the influence of the feeding position, at a rotation speed of 500 rpm and a throughput of 5 kg/h, the Baytubes® C150 P or Nanocyl™ NC7000 were added either in the hopper together with PC granules or into the side feeder attached at an L/D of 14. In a second set, different screw configurations (SC, see Table 1) adapted from Ref. [16] were used, whereby the throughput was set to 5 kg/h and the rotation speed to 100 rpm. Pure PC pellets were also processed using SC-5 at 260 °C, 500 rpm and a 5 kg/h throughput.

Table 1. Screw configurations (according to the screw designs used in Ref. [16]).

Screw No.	L/D Ratio	Screw Design
SC-1	36	Dispersive: 4 kneading blocks and 2 back-conveying elements
SC-2	36	Dispersive: 4 kneading blocks and 6 back-conveying elements
SC-3	36	Distributive: 3 mixing elements and 2 back-conveying elements
SC-4	36	Distributive: 3 mixing elements and 8 back-conveying elements
SC-5	48	Distributive: 5 mixing elements and 8 back-conveying elements

To relate the achieved physical properties to the energy that was introduced during the mixing process, the specific mechanical energy (SME) was calculated based on the torque values recorded during the compounding, using the following equation [18]:

$$\mathrm{SME} = \frac{2\pi \mathrm{N}\tau}{\dot{\mathrm{m}}} \tag{1}$$

where N is the mixing speed, τ is the mean torque value during compounding and $\dot{m}$ is the throughput during the extrusion.

The PC composites with 3 wt.% extruded in LSE by variation in throughput and rotation speed were further diluted to lower loadings between 0.5 und 2 wt.%, using the conical co-rotating twin screw microcompounder Xplore 15 (Xplore, MD, Geleen, The Netherlands), having an inner volume of 15 cm³. A mixing temperature of 280 °C, mixing speed of 50 rpm and mixing time of 5 min were applied.

To study the influence of the MWCNT loading on the nanotube breakage of NC7000 in LSE, in a third set extruded composites based on PC Lexan 141R (Sabic, Riyadh, Saudi Arabia, melt volume–flow rate of 12 cm³/10 min at 1.2 kg and 300 °C, according to the datasheet [36]), with MWCNT Nanocyl™ NC7000 (in short NC7000), were studied.

Therefore, the Berstorff ZE25 extruder with an L/D ratio of 48 (SC-5) was again used with a throughput of 5 kg/h, rotation speed of 500 rpm and mean melt temperature of 260 °C. A masterbatch containing 7.5 wt.% CNTs was prepared, which was in a second extrusion step diluted to 4.0 and 2.0 wt.%. For comparison, composites filled with 2 and 4 wt.% NC7000 were prepared by direct CNT incorporation.

To measure electrical properties, the granules obtained from the extruded strands were compression-molded into plates (60 mm diameter, 0.5 mm thickness) using a Weber hot press (Model PW 40 EH, Paul Otto Weber GmbH, Remshalden, Germany). Compression molding was performed with a pressing temperature of 280 °C. The pressing speed was 6 mm/min, the pressing time 1 min and the pressing force was increased in steps up to 100 kN.

For mechanical testing, injection molding was performed using a Demag Ergotech 100/420–310 machine (Demag Plastics Machinery GmbH, Roßleben-Wiehe, Germany) at 290 °C melt temperature, injection speed of 35 mm/s and rotation speed of 200 rpm. Standard test bars for tensile testing and impact testing were produced.

2.2. Characterization Techniques

In order to access the state of nanotube dispersion, light microscopy (LM) in transmission mode was performed to investigate the macrodispersion of CNT agglomerates within the PC matrix. Thin sections with a thickness of 5 µm were cut from extruded strands using a microtome Leica 2055 (Leica Mikrosysteme Vertrieb GmbH, Bensheim, Germany) and were fixed on glass slides. The LM images were recorded using a BH2 microscope combined with a DP71 camera (both Olympus, Deutschland GmbH, Hamburg, Germany). The agglomerate area ratio A_A was determined from the light micrographs by calculating the ratio of the area of remaining agglomerates A_{Agg} to the total area of the micrograph A_0 (~0.6 mm^2) using the software ImageJ Version 1.43 g. In addition, the numbers of agglomerates were detected. Only agglomerates larger than 5 µm were considered for this calculation. For the quantification, at least six cuts from different positions were investigated for each sample, and the standard deviation between these cuts is given.

For the determination of the nanotube length distribution after melt compounding in the polycarbonate matrix, a method described by Krause et al. [11] was used (see Figure 1). The polycarbonate composites containing 0.5 wt.% to 7.5 wt.% CNTs were dissolved in chloroform overnight followed by a treatment of 3 min in an ultrasonic bath. For the TEM investigations, a drop taken out from the freshly prepared dispersion (0.03 g CNT/l chloroform) was placed on a TEM grid with a carbon coating and dried in the air. In the TEM images collected with a Libra200 (Carl Zeiss Industrielle Messtechnik GmbH, Oberkochen, Germany), the nanotube lengths were measured on approximately 200–300 particles by applying the software SCANDIUM 5.1 (Olympus Soft Imaging Solutions GmbH, Hamburg, Germany). The full visible length of each separated nanotube, not touching the edge of the image, was thereby recorded by applying the polyline function. In order to measure the lengths of very long nanotubes, images were stitched together as needed. The results are given as number distributions with class sizes of 100 nm. To quantify the length distributions of the nanotubes, the typical distribution parameters (quantile) x_{10}, x_{50} and x_{90} were calculated based on the cumulative frequency distributions, indicating that 10%, 50% and 90% of the nanotube lengths were smaller than the given value. For interpretation, mainly x_{50} was used. The TEM images on which the CNT lengths were measured are available as a dataset in [37].

To perform electrical volume resistivity measurements, rectangular samples (ca. 30 × 4 × 0.5 mm^3) were cut from the compression-molded plates. For measurements, a 4-point test fixture (gold contact wires with a distance of 16 mm between the source electrodes and 10 mm between the measuring electrodes) combined with a Keithley electrometer 6517A (Keithley Instruments, Cleveland, OH, USA) was used (filled symbols in the plots) in accordance with ASTM D 4496 [38]. For electrical resistivity values higher than 1 × 10^7 Ohm cm, a Keithley 8009 Resistivity Test Fixture (Keithley Instruments, Cleveland,

OH, US) based on ring electrodes was used for measurement on pressed circular plates (open symbols), according to ASTM D 257 [39]. In the plots, the geometric mean values and the standard deviations of 4–8 measurements are shown.

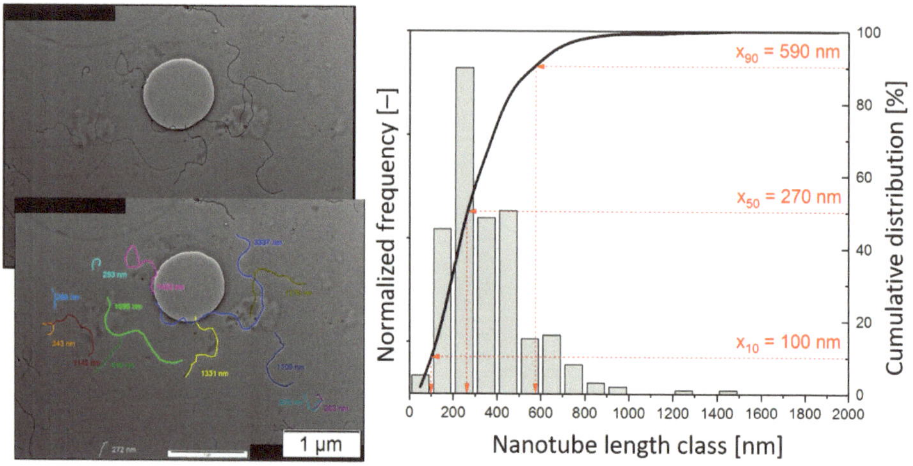

Figure 1. Schema of CNT length measurement: TEM micrograph containing the measured CNTs, a nanotube length histogram, the cumulative frequency distribution and the characteristic length values (here, with the example of processed Nanocyl^{TM} NC7000 MWCNTs).

Uniaxial tensile tests were performed on injection-molded dogbone type (1A) specimens according to EN ISO 527-2/1A/50 [40] using a Universalprüfmaschine (Zwick GmbH & Co. KG, Ulm, Germany), which was equipped with a Multisens extensometer and a load cell of 20 kN. The tests were carried out at a crosshead speed of 50 mm/min. The values are the mean values of 8–10 samples. The E-modulus was measured at a crosshead speed of 1 mm/min.

The Charpy notched impact strength (notched a_{cu}) was measured according to ISO 179/1eA [41] using a 4 J pendulum and a test velocity of 2.9 m/s, whereas the Charpy impact strength (a_{cu}) was measured according to ISO 179/1eU, using a 15 J pendulum and a test velocity of 3.8 m/s. Whereas pure PC showed a hinge break, all composites had a complete sample break. All mechanical tests were performed after conditioning the samples at 23 °C @ 60% r. H. for at least 2 days.

3. Results

3.1. Influence of Throughput and Rotation Speed

3.1.1. CNT Dispersion and Length

In the first set, during the preparation of PC composites filled with 3 wt.% MWCNTs, the rotation speed, throughput and feeding position were varied. The composites were investigated with regard to the MWCNT macrodispersion, as well as the electrical and mechanical properties, and for selected composites, the nanotube length distributions were also determined.

The dependencies of the MWCNT agglomerate area ratio on the screw rotation speed at different throughputs are illustrated in Figure 2. Thereby, no composite could be produced at 15 kg/h and 100 rpm and no composite was manufactured at 5 kg/h and 750 rpm. Feeding of the MWCNTs was performed together with the PC pellets in the hopper. The area ratio was generally lower at a lower throughput, which was combined with a longer residence time and higher SME values. The largest decrease was between 50 and 250 rpm. At 5 kg/h throughput, the agglomerate area ratio leveled off starting at 500 rpm, at 10 kg/h, at 750 rpm, and at 15 kg/h, the increase in rotation speed resulted up to 1000 rpm in slight decreases. In addition, the light microscopy images of thin sections shown in Figure 3 illustrate the very different states of CNT macrodispersion at different rotation speeds. Significantly more remaining undispersed CNT agglomerates were seen when extrusion was performed at 100 rpm as compared to 1000 rpm. However, even at 1000 rpm, not all agglomerates were destroyed and some, even of a relatively large size, were still seen.

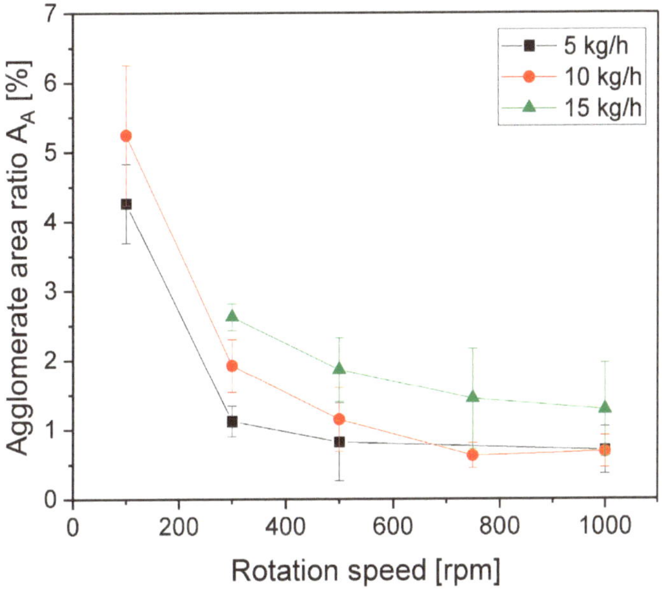

Figure 2. Agglomerate area ratio A_A vs. rotation speed for PC/3 wt.% MWCNT composites at different throughputs, with the feeding of Baytubes® C150 P at the hopper, for screw No. SC-5.

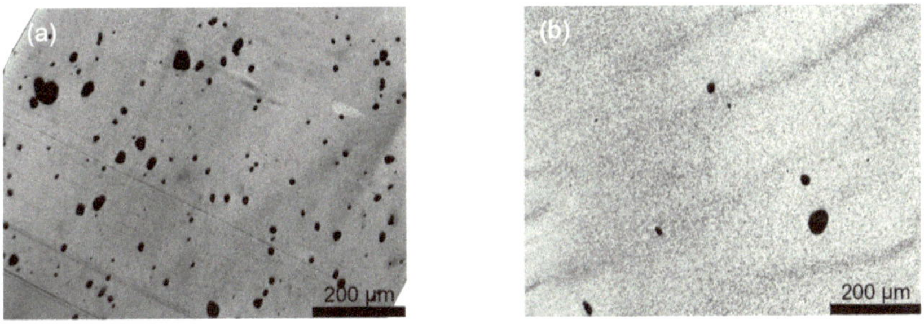

Figure 3. Light microscopy images of extruded PC/3 wt.% composites prepared at two processing conditions using screw SC-5, with the feeding of Baytubes® C150 P at the hopper: (**a**) 100 rpm, 5 kg/h, SME = 0.16 kWh/kg, A_A = 4.3%; (**b**) 1000 rpm, 5 kg/h, SME = 0.90 kWh/kg, A_A = 0.7%.

Furthermore, the nanotube length distributions in the composites were determined using TEM image analysis of the extracted CNTs and are shown for the example of different rotation speeds at a throughput of 5 kg/h in Figure 4. Compared to the initial length distribution of the pure MWCNT Baytubes® (as shown in [1]) with x_{10}, x_{50} and x_{90} values of 280 nm, 770 nm and 2410 nm, respectively, the length was significantly reduced after processing. Especially long tubes were shortened and the distributions shifted to lower length values with a higher proportion of CNTs with shorter lengths.

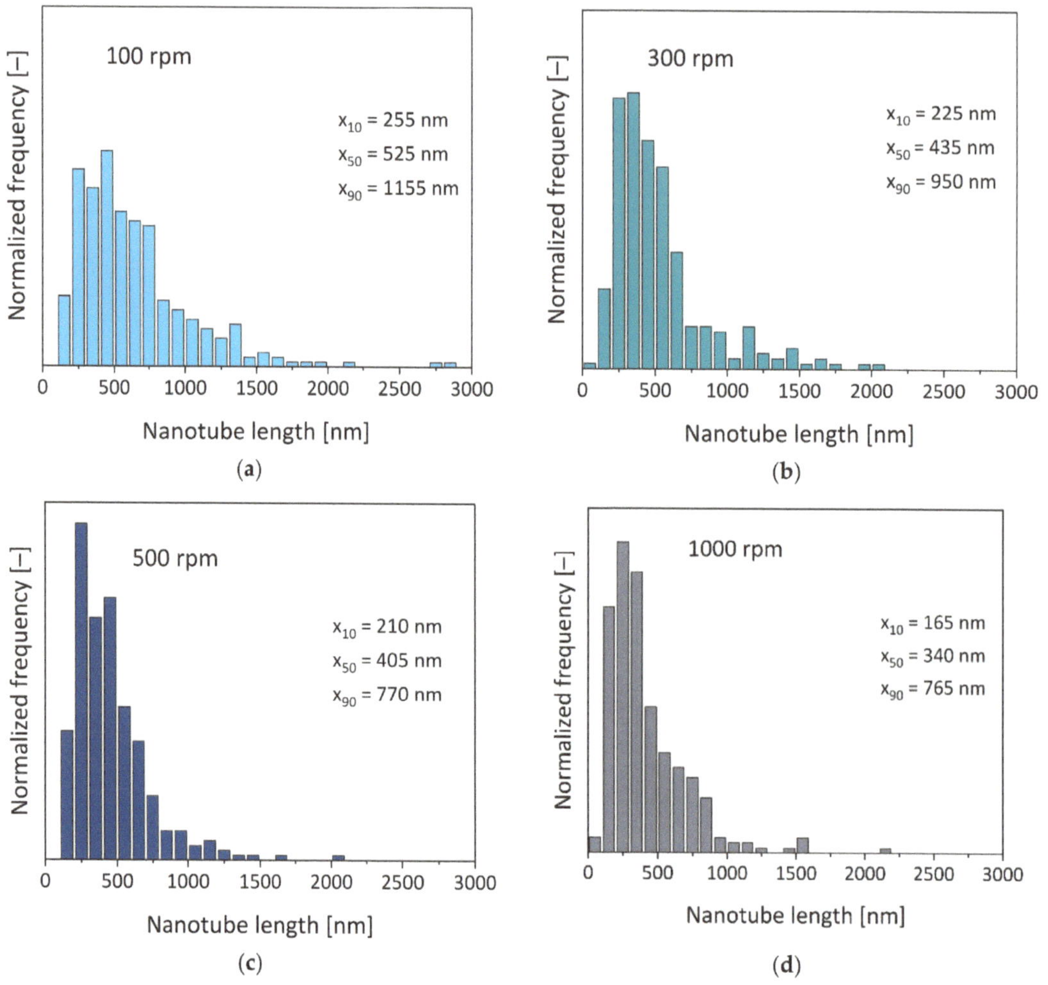

Figure 4. CNT nanotube length distributions at different rotation speeds for PC/3 wt.% Baytubes® C150 P composites at 5 kg/h throughput, with the feeding of MWCNTs at the hopper, for screw No. SC-5: (**a**) 100 rpm, SME = 0.16 kWh/kg; (**b**) 300 rpm, SME = 0.39 kWh/kg; (**c**) 500 rpm, SME = 0.55 kWh/kg; (**d**) 1000 rpm, SME = 0.90 kWh/kg.

The results in Figure 5 show the dependence of the x_{50} length values on rotation speed. As expected and described before by different authors [1,3,10–12,25,26,42], CNT shortening during the melt compounding took place. From comparing the different rotation speeds (at a throughput of 5 kg/h), a decrease in the nanotube lengths with the rotation speeds could be found. The mean x_{50} and x_{90} values decreased from 525 nm and 1155 nm (100 rpm) to 340 nm and 765 nm (1000 rpm), respectively. Therefore, a significant reduction in the aspect ratio was observed. At a constant rotation speed of 300 rpm, CNT shortening was more pronounced at a lower throughput, which correlated with the increasing SME values.

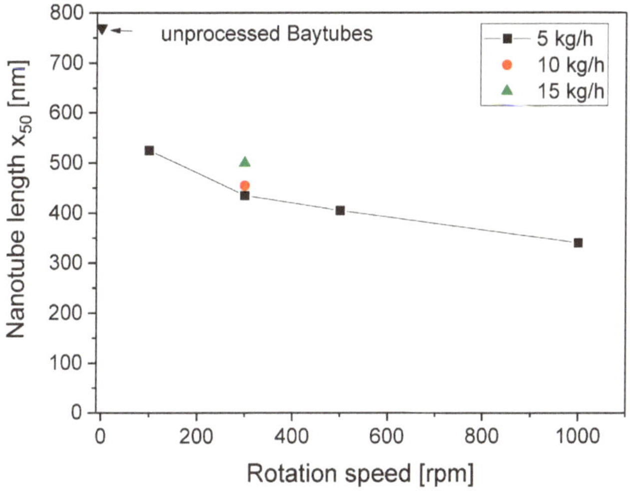

Figure 5. CNT nanotube length x_{50} vs. rotation speed for PC/3 wt.% MWCNT composites at different throughputs, with the feeding of Baytubes® C150 P at the hopper, for screw No. SC-5.

3.1.2. Electrical Resistivity

The influence of rotation speed during compounding on the electrical properties was studied for composites prepared at different throughputs using screw No. SC-5. The volume resistivity values of the PC/3 wt.% composites showed a tendency to decrease with rotation speed (Figure 6). At 15 kg/h, a slight increase was seen when enhancing the rotation speed from 750 rpm to 1000 rpm, and at 5 kg/h, the value at 300 rpm showed the lowest resistivity value. Under this extrusion condition, the nanotube length x_{50} of 435 nm was measured, corresponding to a mean aspect ratio of 41 (compared to that of 73 for unprocessed nanotubes). The agglomerate area ratio A_A for this sample was 1.1%, which indicates that a worse dispersion compared to the sample produced at 1000 rpm with a ratio of 0.7% was achieved.

From plotting the values vs. the SME (Figure 7), the same general decreasing tendency was seen up to an SME of ca. 0.3 kWh/kg. The lowest resistivities were measured for composites produced with SMEs in the range of 0.3–0.4 kWh/kg; at a higher SME, a slight increase in electrical resistivity was observed. Interestingly, this SME value range was quite similar to the observed minimum in resistivity of PCL–MWCNT composites, which were produced at a small scale by variation in the rotation speed. There, the minimum was found at 0.47 kWh/kg [12]. Also in the PA6/MWCNT composites, which were also processed at a small scale by varying the rotation speed and mixing time, a minimum in resistivity was observed at an SME of 0.5 kWh/kg [1]. These findings reflect the balance between better CNT dispersion with increasing SME, resulting in reduced resistivity and the parallel ongoing reduction in the CNT length negatively affecting the resistivity. As both parameters are important for the formation of the conductive CNT network in the composites, minima of the resistivity were found.

Since the electrical resistivity values for PC/3 wt.% MWCNT composites prepared at a throughput of 5 kg/h did not show a clear tendency, these composites were diluted to lower CNT contents using a microcompounder, and electrical percolation curves were determined, as shown in Figure 8. The lowest electrical percolation thresholds of 0.63 and 0.75 wt.% were determined for the composites prepared in LSE at 100 and 300 rpm, respectively. If the PC-3 wt.% composite was melt mixed in LSE at 500 rpm, the electrical percolation of the dilutions occurred between 1 and 1.25 wt.%. For the composites fabricated at the highest rotation speed of 1000 rpm, the highest electrical percolation threshold at 1 wt.% was found. For 300 rpm, the resistivity values at loadings up to 1.5 wt.% CNTs were the lowest, illustrating that this composite had the best balance between dispersion and CNT length to reach low electrical resistivities at low loadings. Thus, the lowest resistivity value observed at 3 wt.% loading at this rotation speed (see Figures 6 and 7) was also reflected in the percolation behavior of the dilutions.

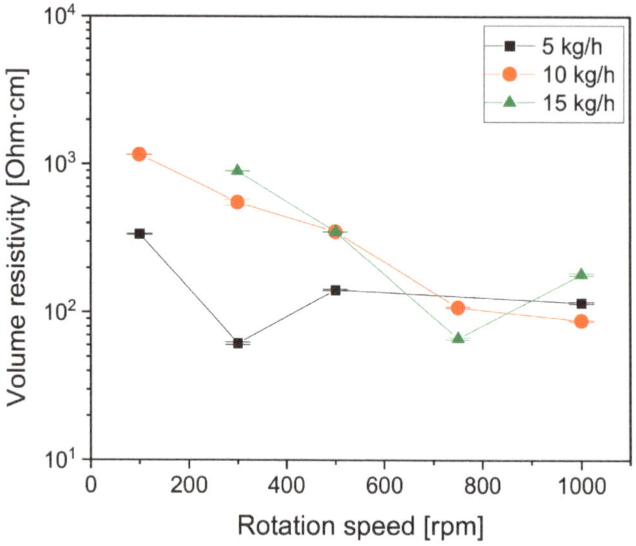

Figure 6. Electrical volume resistivity vs. rotation speed for PC/3 wt.% MWCNT composites at different throughputs, with the feeding of Baytubes® C150 P at the hopper, for screw No. SC-5 (compression-molded plates).

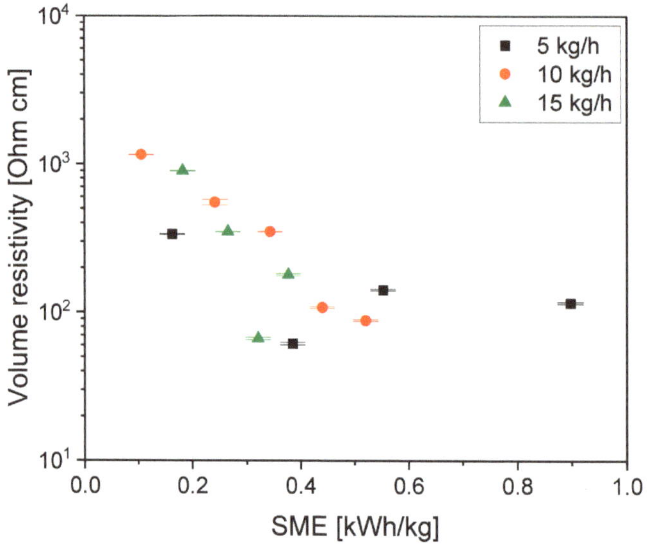

Figure 7. Electrical volume resistivity vs. SME for PC/3 wt.% MWCNT composites at different throughputs, with the feeding of Baytubes® C150 P at the hopper, for screw No. SC-5 (compression-molded plates).

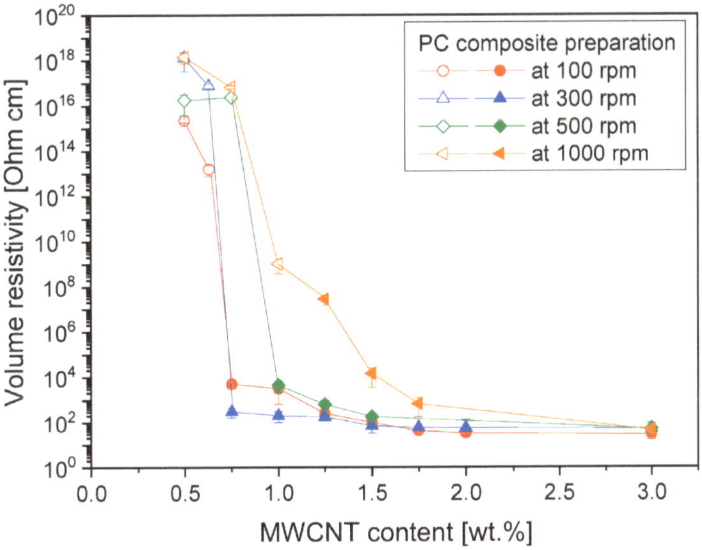

Figure 8. Electrical volume resistivity vs. CNT content for composites diluted using a microcompounder from extruded PC/3 wt.% Baytubes® C150 P MWCNT composites prepared at different rotation speeds, with a throughput of 5 kg/h and using screw No. SC-5. Open symbols indicate measurements using Keithley 8009 and closed symbols with Keithley 6517A test configurations.

Additional melt rheological studies at 280 °C on the PC/3 wt.% MWCNT composites are shown in the Supplementary Materials in Figures S1–S5.

3.1.3. Mechanical Properties

On the samples of this series, tensile tests were performed, and from the corresponding stress–strain curves, the elastic modulus, tensile strength and elongation at break were extracted (see Figure 9). Pure extruded PC showed a distinct yield point above which a cold drawing with nearly constant stress vs. strain was developed, which changed at about 90% strain into strain-hardening behavior. Most samples filled with 3 wt.% CNTs showed only cold-drawing behavior. Thus, the tensile strength as the maximum stress in the stress–strain curve in most samples represents the stress values at the yield point, whereas for pure PC, it is the stress at break. The samples prepared at 15 kg/h with 750 and 1000 rpm showed a break partially or completely before reaching the yield point and are therefore marked with an asterix *. When plotting the tensile strength of the composites vs. the SME applied during compounding, except for these two samples, slightly higher values than those measured for pure PC (63.7 ± 2.3 MPa) were found. Thereby the standard deviations of the tensile strength were between 0.1 and 3.6 MPa, those of the elastic modulus between 13 and 36 MPa. There was no significant dependence on the SME, a finding which was also reported by Mack et al. [21] for PC composites with 3 wt.% NC7000. The modulus was enhanced, on average by about 20%, as compared to pure extruded PC (2300 ± 12.4 MPa), again showing no significant differences when increasing the SME, similar to the report by Mack et al. [21].

Figure 10a shows the elongation at break as obtained from the tensile tests and Figure 10b, the notched impact strength at different rotation speeds and different throughputs. For processed unfilled PC, the values of the elongation at break at 119 ± 12% and the notched impact strength at 80 ± 7 kJ/m^2 were measured. With the incorporation of CNTs, both values decreased significantly. Both properties were strongly dependent on the state of CNT dispersion and network formation. Typically, a strong decrease in elongation at break is found in composites above the concentration of (electrical) percolation, as illustrated in [43] for PC/MWCNTs. The sharp decrease in elongation at break starting in the electrical percolation range seems to be connected to the restricted typical deformation behavior of the polymer chains as soon as a CNT network is interacting. At the CNT content of 3 wt.% selected in this study, all composites were conductive; however, they had different conductivities (as seen in Figure 6), illustrating different network structures.

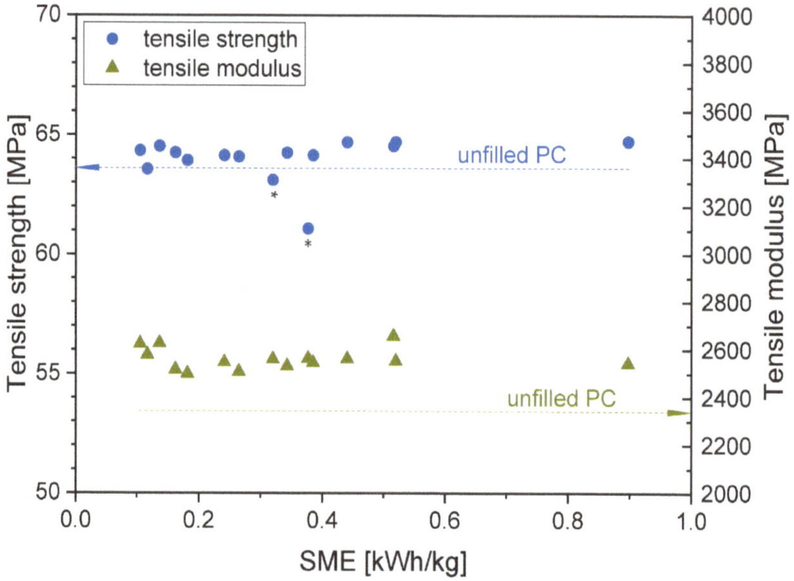

Figure 9. Tensile strength and tensile modulus vs. SME for PC/3 wt.% MWCNT composites at different rotation speeds and throughputs, with feeding at the hopper, for screw No. SC-5. The asterisk * marks the two samples with break partially or completely before reaching the yield point.

When the rotation speed was increased, different developments were seen at the three studied throughputs. At the lowest throughput of 5 kg/h, an increase in elongation at break was found, even if the values had a relatively large standard deviation. At 10 kg/h, all values were lower than at 5 kg/h; however, up to 750 rpm, an increase with the rotation speed was noted. At 15 kg/h, an increase in elongation at break between 300 and 500 rpm was found at the level between the two other throughputs, followed by a strong decrease. For the values at higher rotation speeds, it has to be noted that the samples showed a break before reaching the yield point, such that the resulting type of stress–strain curve was different. This behavior may be attributed to a mechanical degradation of the PC by molecular weight reduction, which was not studied.

For the notched impact strength (Figure 10b), an increase with the rotation speed was again observed at 5 kg/h and 10 kg/h, whereas at 15 kg/h, a decreasing tendency was found, possibly again connected to PC degradation. Mack et al. [21] found a similar drastic decrease in the notched impact strength compared to the initial PC, and also an increase with the rotation speed, but no significant change with the throughput.

From plotting both properties vs. the SME, no clear picture was seen (Figure 10c). However, there was a generally increasing tendency for both values with the SME when disregarding the two values of elongation at break for the samples with changed stress–strain behavior (marked by asterix *).

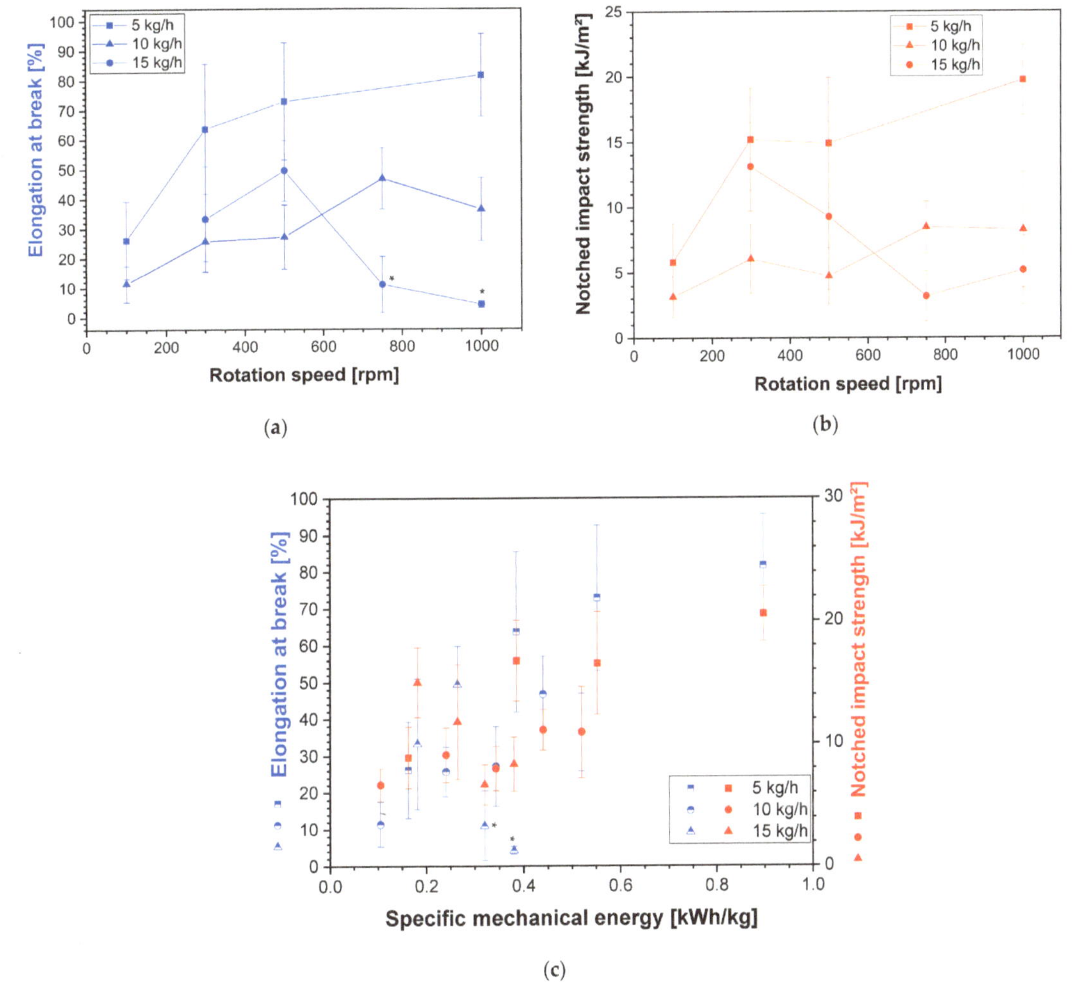

Figure 10. (**a**) Elongation at break and (**b**) notched impact strength vs. rotation speed for PC/3 wt.% MWCNT composites at different throughputs, and (**c**) elongation at break and notched impact strength vs. SME; with feeding at the hopper, for screw No. SC-5. The asterisk * marks the two samples with break partially or completely before reaching the yield point.

3.2. Variation in Feeding Position

Furthermore, the influence of the feeding position was investigated at a rotation speed of 500 rpm, throughput of 5 kg/h and for screw No. SC-5. The feeding of CNT powder and PC granules into the hopper occurred as two separate streams. As CNTs, Baytubes® C150 P and Nanocyl™ NC7000 with a CNT content of 3 wt.% in a PC composite were used. The side feeder was located at an L/D of 14. The SME was nearly the same for all four preparation processes and was calculated to be around 0.5 kWh/kg. The results are summarized in Table 2.

It was found that feeding in the hopper for Baytubes® C150 P resulted in better CNT dispersion, as indicated by a lower value of the agglomerate area ratio A_A. In contrast, with MWCNT NC7000, feeding into the side feeder led to significantly better CNT dispersion. Interestingly, Müller et al. already described the same tendencies when incorporating both kinds of MWCNTs into PP [17]. The differences can be explained by the different agglomerate structures in both MWCNT materials. Compared with side feeding, the incorporation of CNTs at the hopper at the beginning of the extrusion process led to a significantly longer residence time of the CNTs in the polymer melt and induced, especially in the melting zone of the extruder, higher shear stresses. The more compact Baytubes® C150 P material required higher shear stresses for the dispersion of its agglomerates, whereas MWCNT NC7000 has a less dense and combed yarn structure [18] and needed a comparatively lower energy input for dispersion. Thus, feeding into the already molten polymer is favorable in the context of dispersion.

Table 2. Properties of composites (PC/3 wt.% MWCNT) produced under variation in the feeding position.

Feeding	A_A [%]	ρ [Ohm·cm]	CNT Length x_{50} [nm]	E_t [GPa]	σ_{max} [MPa]	ε_{Break} [%]	a_{cu} [kJ/m²]	Notched a_{cu} [kJ/m²]
Baytubes® C150 P								
hopper	0.8 ± 0.3	42.4 ± 1.6	405	2.7	64.5 ± 0.2	26.1 ± 6.9	170.7 ± 90.7	6.9 ± 1.4
side feeder	1.3 ± 0.8	11.6 ± 1.1	440	2.7	64.9 ± 0.2	73.6 ± 14.3	329.2 ± 12.1	8.7 ± 1.9
Nanocyl™ NC7000								
hopper	1.1 ± 0.3	17.6 ± 1.2	337	2.7	65.7 ± 1.4	11.2 ± 6.1	123.0 ± 84.9	7.3 ± 1.9
side feeder	0.2 ± 0.2	24.2 ± 1.4	494	2.7	66.7 ± 0.1	30.6 ± 11.6	240.2 ± 112.1	10.1 ± 2.9

For both kinds of MWCNTs, the value of the electrical volume resistivity ρ was lower for the composites with poorer CNT dispersion, which can be interpreted in the context of the shortening of the CNT length during melt processing. The CNT shortening was more pronounced by the feeding of CNTs in the hopper as also clearly seen in the CNT length distribution of the extracted CNTs in Figure 11. CNT incorporation using the side feeder led both MWCNT types to have slightly higher values of tensile strength (σ_{max}), and significantly higher values of elongation at break ε_{Break} and impact as well as notched impact strength a_{CU} of their composites. The values of the tensile modulus E_t and strength σ_{max} were not influenced by the feeding position. It can be summarized that the side feeding led to slightly better electrical and mechanical properties for both kinds of MWCNTs.

3.3. *Influence of Screw Configuration*

In the second set of experiments, the selected screw configuration (under otherwise constant set extrusion parameters) had only a very low impact on the different studied properties of PC/3 wt.% Baytubes® C150 P composites, as illustrated in Table 3. This was also due to the fact that the SME did not differ much (0.11 to 0.16 kWh/kg, see Figure 12) under the selected conditions of a rotation speed of 100 rpm and throughput of 5 kg/h. The agglomerate area ratio A_A varied between 3.1 and 3.9%. The mechanical properties did not differ much, but had relatively large standard deviations. Similar results of a low influence of the screw configuration on different composite properties were also found in a study on PLA/MWCNT NC7000 composites by Villmow et al. [15].

Table 3. Properties of composites (PC/3 wt.% Baytubes® C150 P MWCNTs) produced under variations in the screw configuration and feeding at the hopper, with a rotation speed of 100 rpm and throughput of 5 kg/h.

Screw	A_A [%]	ρ [Ohm·cm] #	CNT Length x_{50} [nm]	E_t [GPa]	σ_{max} [MPa]	ε_{Break} [%]	a_{cu} [kJ/m²]	Notched a_{cu} [kJ/m²]
SC-1	3.4 ± 0.6	62.3 ± 1.2		2.6	51.4 ± 1.1	20.1 ± 12.6	312.3	9.1 ± 2.1
SC-2	3.1 ± 0.7	105.5 ± 1.6	604	2.6	50.9 ± 0.7	25.4 ± 9.6	293.9 ± 53.3	10.6 ± 3.3
SC-3	3.9 ± 0.5	259.8 ± 1.6		2.6	50.8 ± 2.2	38.8 ± 16.8	293.7 ± 48.1	8.1 ± 2.5
SC-4	3.8 ± 0.9	82.5 ± 1.5	450	2.6	50.9 ± 0.7	23.5 ± 7.6	289.2 ± 104.9	9.5 ± 3.1
SC-5	4.2 ± 0.5	297.2 ± 3.4	525	2.6	50.8 ± 0.8	31.6 ± 13.1	318.2 ± 2.4	10.1 ± 2.3

on compression-molded plates.

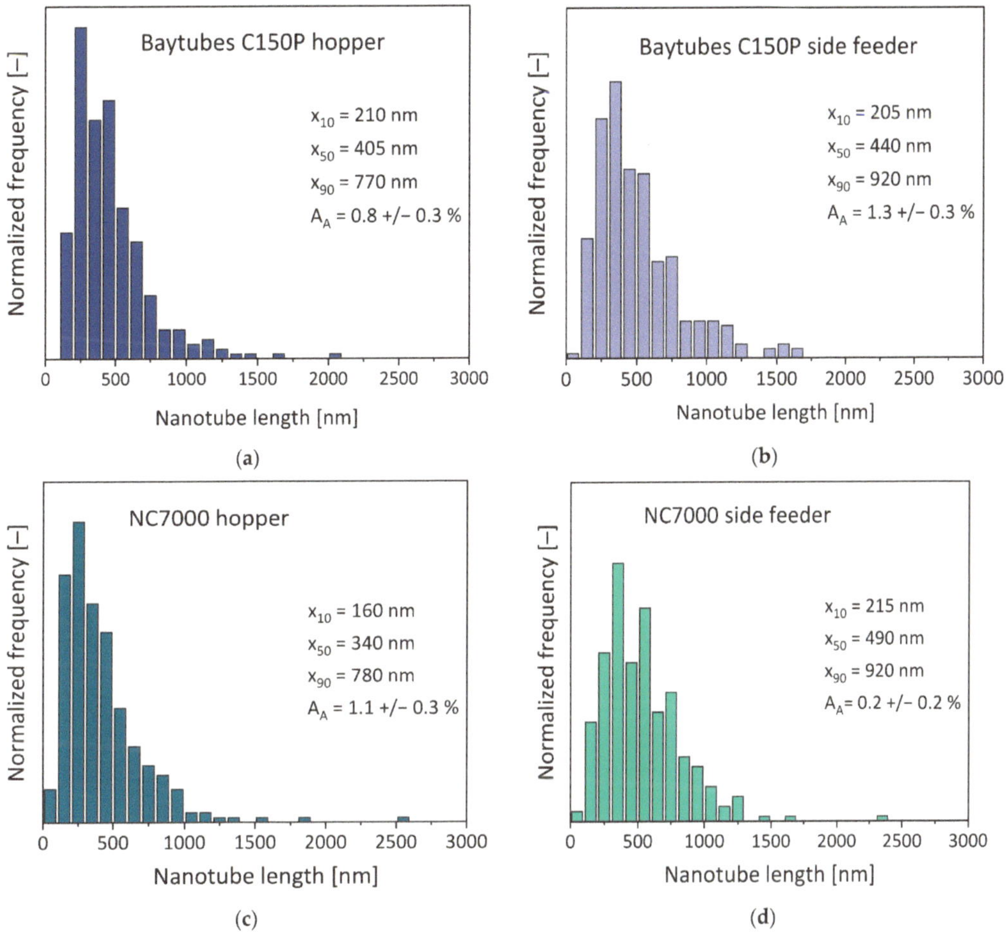

Figure 11. MWCNT length distributions for PC/3 wt.% MWCNT composites prepared using different feeding positions, at 500 rpm and 5 kg/h, for screw No. SC-5: (**a**) Baytubes® C150 P at the hopper; (**b**) Baytubes® C150 at the side feeder; (**c**) Nanocyl™ NC7000 at the hopper; (**d**) Nanocyl™ NC7000 at the side feeder.

CNT shortening was slightly less pronounced when using the dispersive screw SC-2 compared to the distributive screw and that with the longer screw length. A comparison of the CNT length distribution of MWCNTs extracted from the composites prepared using SC-2 and SC-4 is given in Figure 13, and that for SC-5 is represented by Figure 4a. The MWCNTs extracted after applying SC-4 showed the highest population of nanotubes in the lower length classes, even if some CNTs with higher lengths were also measured. The distributions were more shifted to higher CNT length classes for SC-5 and even more for SC-2.

3.4. Relation between SME, Dispersion and Nanotube Length

In order to summarize all the results using the variation in rotation speed, throughput, feeding position and screw configuration concerning their influence on dispersion and MWCNT shortening, Figure 14 shows the dependence of the agglomerate area ratio A_A and the CNT length x_{50} on the applied SME for all the composites with 3 wt.% MWCNTs Baytubes® C150 P. Interestingly, the data fit together in a general tendency. The plot also contains the mixing conditions. Looking at the effect of processing conditions on the CNT macrodispersion, it was found that the agglomerate area ratio A_A decreased with increasing SME and leveled off at an SME input of around 0.4 kWh/kg. A reduction in A_A from 5.2% at an SME of 0.11 kWh/kg to an A_A of 0.7% at 0.90 kWh/kg was found. The length reduction was also significant, and at the highest used SME of 0.90 kWh/kg, the length was only about 44% of the original length value of the pure Baytubes® material (corresponding to a mean aspect ratio of 32), whereas at an SME of 0.16 kWh/kg, it was about 64% (aspect ratio of 47). There was a significant length decrease up to about 0.24 kWh/kg and a much lower decrease when further enhancing the SME. In summary, increasing the SME above about 0.4 kWh/kg did not improve dispersion significantly and the final length of the CNTs of about 350 nm seemed to be reached.

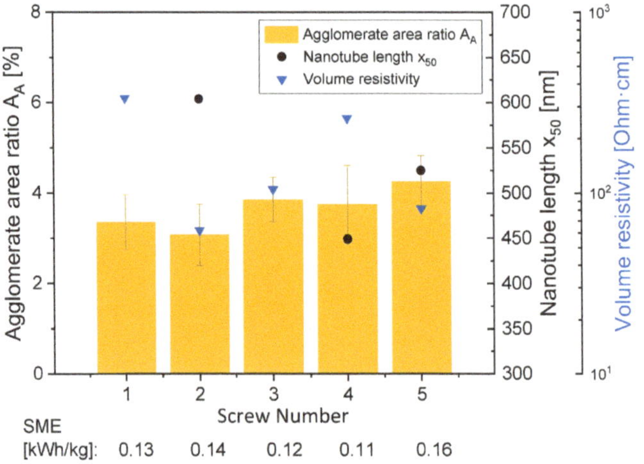

Figure 12. Electrical volume resistivity, agglomerate area ratio and CNT length x_{50} at different screw configurations for PC/3 wt.% Baytubes® C150 P composites (hopper feeding, 100 rpm, 5 kg/h). Standard deviations for volume resistivity values are given in Table 3.

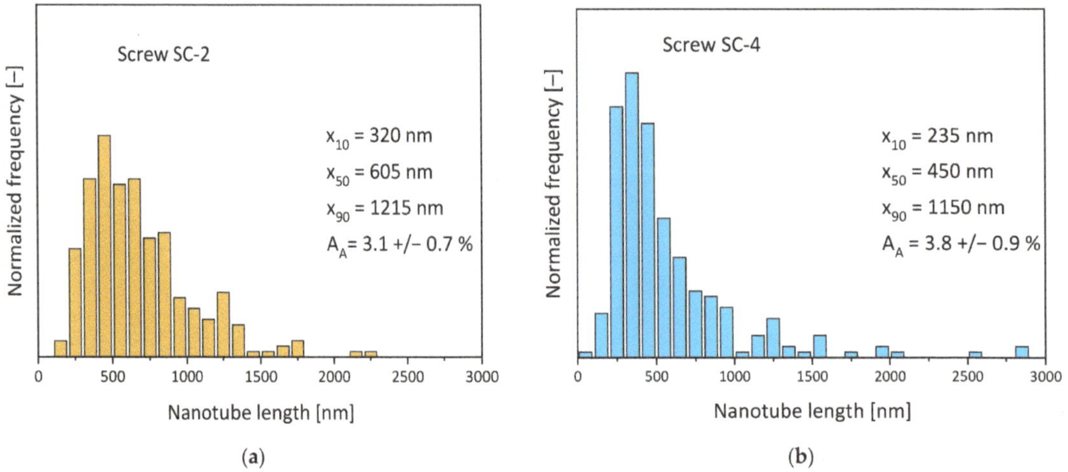

Figure 13. CNT length distributions for PC/3 wt.% Baytubes® C150 P composites prepared using different screw configurations, at 100 rpm and 5 kg/h, with feeding at the hopper: (**a**) screw SC-2, SME = 0.14 kWh/kg; (**b**) screw SC-4, SME = 0.11 kWh/kg (for SC-5, see Figure 4a).

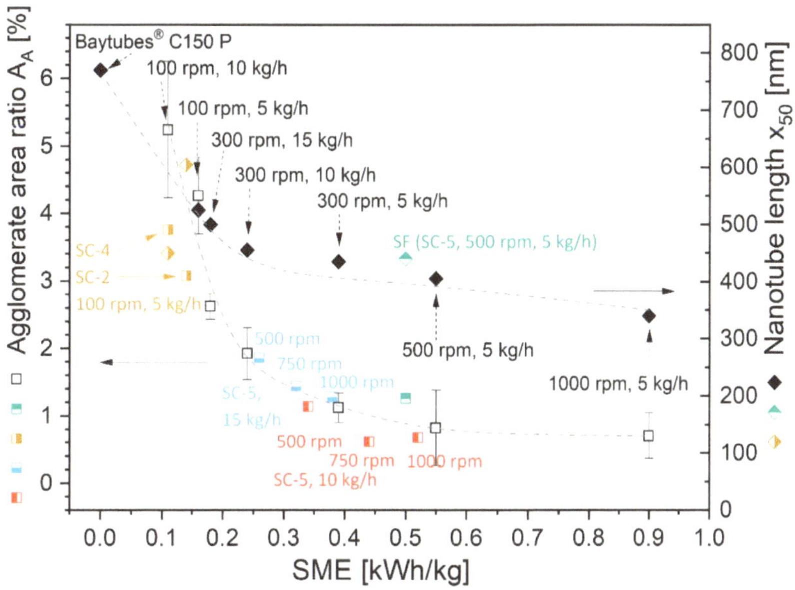

Figure 14. Agglomerate area ratio A_A and nanotube length x_{50} vs. SME for PC/3 wt.% MWCNT composites at different rotation speeds, throughputs, MWCNT feedings and screw designs. Values (in black) for screw SC-5 with directly feeding Baytubes® C150 P replotted with permission from Ref. [1] RSC 2013. Newly added values (direct feeding) in different colors (standard deviations of A_A shown before): brown symbols screws SC-2 and SC-4 at 100 rpm, 5 kg/h; light blue symbols, screw SC-5 at 500–1000 rpm, 15 kg/h; red symbols, screw SC-5 at 500–1000 rpm, 10 kg/h; and green symbols, side feeding (SF) screw SC-5 at 500 rpm, 5 kg/h. The lines are meant to guide the eyes.

3.5. CNT Shortening at Different Loadings

Whereas in the small-scale mixing of PC (grade Makrolon® 2205, low melt viscosity) with Nanocyl™ NC7000 MWCNTs, results comparing 0.5 and 1.0 wt.% were already reported [26] (Figure 15), no results could be found for LSE. As expected, under constant mixing conditions (here shown for the parameter set 2, with mixing at 5 min, at 250 rpm, using a 15 cm^3 DSM Xplore microcompounder), the length shortening was significantly higher at the higher loading. At 1.0 wt.% CNTs, the x_{50} value decreased from 1341 nm of the initial MWCNTs to 749 nm, which equals a reduction to 56% of the initial length. At 0.5 wt.% of MWCNT loading, the reduction was only to 70%.

To see that effect in LSE, in which higher shear stresses and more pronounced shortening may have been expected, composites with the same type of MWCNTs in another type of PC were prepared in a third set. A masterbatch with 7.5 wt.% Nanocyl™ NC7000 MWCNTs was diluted to 2 wt.% and 4 wt.% composites, and the CNT lengths of those composites were compared to those of composites prepared by direct incorporation of the MWCNTs under the same conditions, as used for masterbatch preparation and dilution. From comparing the masterbatch and the two composites with lower loadings, all prepared under the same set conditions by direct MWCNT incorporation (Figures 16 and 17), it was seen that the CNT length decreased to about the same value when using 7.5 wt.% or 4.0 wt.% (ca. 24%, mean aspect ratio of 32). By incorporating 2 wt.% directly, the shortening was lower and resulting in x_{50} values of 33% of the initial length (aspect ratio of 44). It was also seen that the shortening was much higher in the LSE compared to the small-scale mixing studied before and discussed above. The composites prepared by dilution of the masterbatch, which had experienced shear stresses of extrusion twice, showed even higher shortening, especially seen at 2 wt.% loading, where the x_{50} value was only 27% (aspect ratio of 36) of the value of the initial MWCNTs.

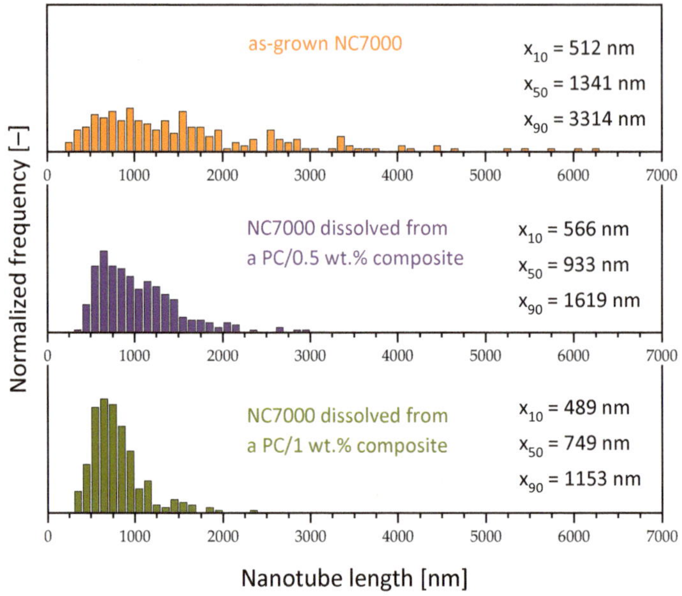

Figure 15. CNT length distributions of as-grown Nanocyl™ NC7000 MCWNTs (adapted from [11] ELSEVIER 2011) and PC (Makrolon®2205) composites filled with 0.5 and 1 wt.% NC7000 prepared at Xplore 15 (260 °C, 250 rpm, 15 min, PM2) (adapted from [26] JOHN WILEY AND SONS 2017).

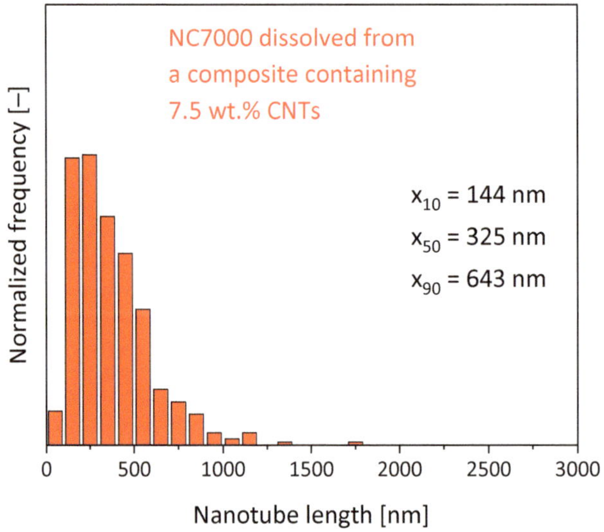

Figure 16. CNT length distributions of NC7000 MWCNTs dissolved from a PC/7.5 wt.% NC7000 masterbatch (prepared using Berstorff ZE25, SC-5, 260 °C, 5 kg/h, 500 rpm).

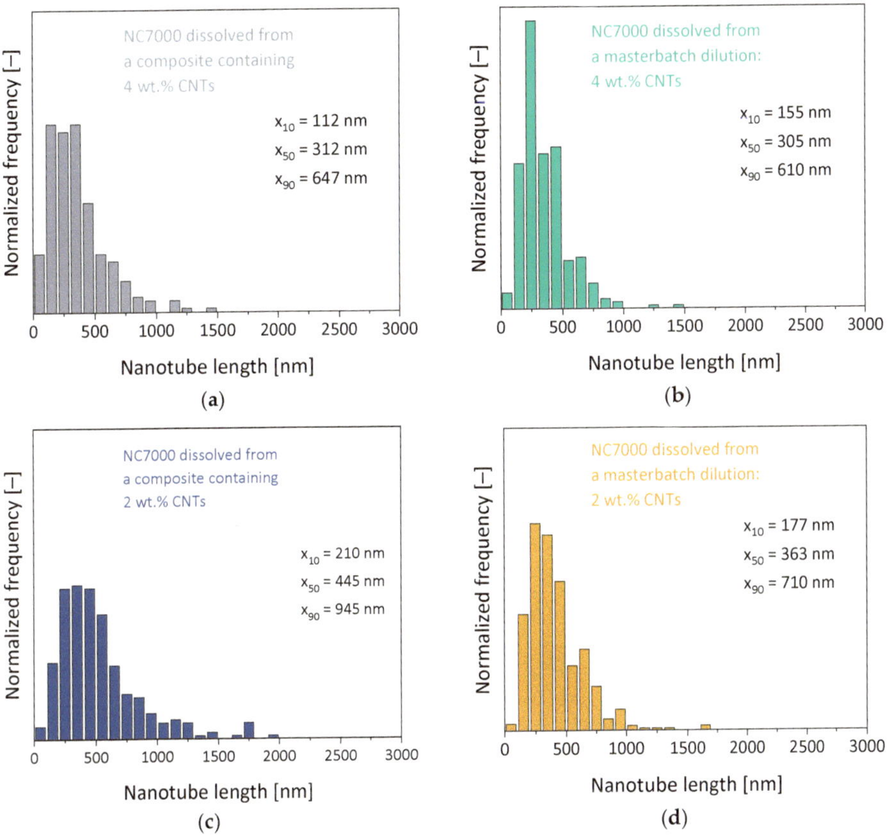

Figure 17. CNT length distributions of (**a**) directly compounded PC/4 wt.% NC7000; (**b**) a composite prepared by dilution of a masterbatch (7.5 wt.%) to 4 wt.%; (**c**) directly compounded PC/2 wt.% NC7000; (**d**) a composite prepared by masterbatch dilution to 2 wt.% (Berstorff ZE25, SC-5, 260 °C, 5 kg/h, 500 rpm).

This illustrates a significant and much higher relative shortening compared to the MWCNTs of the Baytubes® type which were studied before and had a lower initial length. Interestingly the CNT length shortening seems to have leveled off to length values of about 300 to 400 nm (related to x_{50}), a value which was also found for other CNT materials, matrix polymers and mixing conditions [1,10,12,25]. It may be concluded that longer CNTs show more pronounced shortening than shorter ones, a finding which was already reported by Rhue et al. [44] for ball-milled CNTs and by Krause et al. [25] for differently ball-milled CNTs incorporated in PC.

The comparison between the lengths at different loadings also showed that there seemed to be a problem with the selected dissolving method in extracting the CNTs properly especially at higher MWCNT loadings, where the CNTs were much more agglomerated. Such agglomerates may contain long nanotubes to a particular extent. As such agglomerates are not dispersed during the mild ultrasound treatment they could have been neglected due to their sedimentation in the dispersion by the applied preparation method for the TEM grids. Thus, some of the longer nanotubes may not have been available for the quantitative TEM analysis. This is indicated by the x_{50} and the x_{90} length values measured in the diluted composite with 2 wt.%, where a higher length was determined than in the starting masterbatch with 7.5 wt.% MWCNTs.

4. Summary and Conclusions

For composites of PC with MWCNTs, the presented results show that there is a strong influence of the twin screw extrusion conditions at the laboratory scale during their preparation on the achieved state of filler dispersion and the nanotube length. Both properties influence the electrical resistivity of the composites and the mechanical properties. The specific mechanical energy (SME) introduced during compounding is a good measure to quantify the effect of variations in screw speed, throughput, feeding position and screw profile on the mentioned composite properties.

On composites with 3 wt.% MWCNTs of the Baytubes® C150 P type, all being in the electrically conductive range above the electrical percolation threshold, it was shown that dispersion was enhanced with rotation speed and throughput, feeding the material in the hopper, and using a dispersive screw configuration. As measured using transmission light microscopy, the agglomerate area ratio varied between 5.2 and 0.6%, indicating that some undispersed MWCNT material was still existent in the extrudates. The dispersion increased with the SME being up to about 0.4 kwh/kg and then leveled off. The initial MWCNT length of 770 nm decreased with the rotation speed, up to values of 340 nm, whereby the decrease was more pronounced at lower throughputs. This decrease was stronger when the MWCNT material was fed to the hopper together with the PC granules. The latter result was also found for MWCNTs of the Nanocyl™ NC7000 type, whereby the absolute length decrease was much higher (up 25% of the initial length) for Nanocyl™ NC7000 as compared to that of the Baytubes® C150 P material (up to 53%). Thereby, the final lengths for both MWCNT types were in the range of 340 to 500 nm (based on the x_{50} values of the length distributions obtained from the TEM images of the dissolved MWCNTs), which seems to be a value at which the shortening leveled off. The electrical volume resistivity of compression-molded samples tended to decrease with rotation speed and the SME with a slight minimum in the range of 0.3 to 0.5 kWh/kg. This variation was between 60 and 1200 Ohm·cm, meaning more than one decade. It was nearly independent of screw configuration. The named SME range seems to be an optimum between good dispersion and not too harsh CNT shortening. Concerning the mechanical properties, the tensile modulus and strength were not significantly influenced by the processing conditions, whereas the elongation at break and notched impact strength tended to increase with the SME.

In addition, it was shown in the example of composites with MWCNTs of the Nanocyl™ NC7000 type that the MWCNT length reduction also depended on the concentration of the nanotubes in the mixing recipe. The shortening tended to be higher at a higher CNT loading and more pronounced if a masterbatch dilution process was used as compared to the direct CNT incorporation to obtain the same MWCNT concentration.

This study indicates that the effect of carbon nanotube shortening is significant in the twin screw extrusion of composites and needs to be considered when optimizing processing conditions to achieve materials with high electrical performance and/or high mechanical properties. The shown effects may be even more pronounced when composites with (lower) concentrations nearer to the electrical percolation threshold are considered, as the nanotube networks are much more sensitive to small structural differences when they are looser. These results can provide a basis for giving recommendations for the selection of proper processing conditions during composite production. However, the shaping conditions into final parts, e.g., by injection molding, also need to be considered. Here, it was interesting to study to what extent the different properties obtained in the granules after composite production could be modified by the injection molding conditions. In this context, attention should also be paid to the changes in the molecular weight of the matrix polymers involved in this processing.

Supplementary Materials: The following supporting information can be downloaded at: https://www.mdpi.com/article/10.3390/polym16192694/s1, Figure S1: Complex melt viscosity |Eta*| vs. frequency for processed PC and PC/3 wt.% MWCNT composites prepared in LSE at different rotation speeds; Figure S2: Storage modulus G′ vs. frequency for PC/3 wt.% MWCNT composites prepared in LSE at different rotation speeds; Figure S3: Loss modulus G″ vs. frequency for PC/3 wt.% MWCNT composites prepared in LSE at different rotation speeds; Figure S4: Tan delta vs. frequency for PC/3 wt.% MWCNT composites prepared in LSE at different rotation speeds; Figure S5: Storage modulus G′ vs. loss modulus G″ for processed PC and PC/3 wt.% MWCNT composites prepared in LSE at different rotation speeds.

Author Contributions: Conceptualization, T.V. and P.P.; methodology, T.V., P.P. and B.K. (Bernd Kretzschmar); software, B.K. (Beate Krause); validation, P.P. and B.K. (Beate Krause); formal analysis, T.V. and P.P.; investigation, T.V., B.K. (Beate Krause) and B.K. (Bernd Kretzschmar); resources, P.P.; data curation, T.V., B.K. (Beate Krause) and P.P.; writing—original draft preparation, P.P.; writing—review and editing, B.K. (Beate Krause), T.V. and B.K. (Bernd Kretzschmar); visualization, T.V., B.K. (Beate Krause) and P.P.; supervision, P.P.; project administration, P.P.; funding acquisition, P.P. All authors have read and agreed to the published version of the manuscript.

Funding: This research received funding from the German "Bundesministerium für Wirtschaft und Technology" under the grant number BMWi/AiF IGF 336 ZBG.

Institutional Review Board Statement: Not applicable.

Data Availability Statement: The TEM images of MWCNTs detached from PC composites to determine the MWCNT length distributions are openly available in Zenodo at https://doi.org/10.5281/zenodo.11400466, Ref. [37]. All other raw data are available in Zenodo at https://doi.org/10.5281/zenodo.13309199, Ref. [45]. This reference also contains the data sheets for the materials used.

Acknowledgments: We thank Manuela Heber (IPF Dresden) for generating the TEM images used for the CNT length investigations. The mechanical measurements were performed in the IPF testing laboratory by Marion Kretzschmar. We also thank Frieder Pursche for performing the extrusion experiments and Andreas Scholze for the injection molding of the test specimens for the mechanical tests (all IPF Dresden).

Conflicts of Interest: The authors declare no conflicts of interest. The funders had no role in the design of the study; in the collection, analyses, or interpretation of data; in the writing of the manuscript, or in the decision to publish the results.

References

1. Krause, B.; Mende, M.; Petzold, G.; Boldt, R.; Pötschke, P. Characterization of dispersability of industrial nanotube materials and their length distribution before and after melt processing. In *Carbon Nanotube-Polymer Composites*; Tasis, D., Ed.; Royal Society of Chemistry: Cambridge, UK, 2013; pp. 212–233. [CrossRef]
2. Pötschke, P.; Dudkin, S.M.; Alig, I. Dielectric spectroscopy on melt processed polycarbonate-multiwalled carbon nanotube composites. *Polymer* **2003**, *44*, 5023–5030. [CrossRef]
3. Lin, B.; Sundararaj, U.; Pötschke, P. Melt mixing of polycarbonate with multi-walled carbon nanotubes in miniature mixers. *Macromol. Mater. Eng.* **2006**, *291*, 227–238. [CrossRef]
4. Krause, B.; Pötschke, P.; Häußler, L. Influence of small scale melt mixing conditions on electrical resistivity of carbon nanotube-polyamide composites. *Compos. Sci. Technol.* **2009**, *69*, 1505–1515. [CrossRef]
5. McClory, C.; Pötschke, P.; McNally, T. Influence of Screw Speed on Electrical and Rheological Percolation of Melt-Mixed High-Impact Polystyrene/MWCNT Nanocomposites. *Macromol. Mater. Eng.* **2011**, *296*, 59–69. [CrossRef]
6. Kasaliwal, G.; Göldel, A.; Pötschke, P. Influence of processing conditions in small-scale melt mixing and compression molding on the resistivity and morphology of polycarbonate-MWNT composites. *J. Appl. Polym. Sci.* **2009**, *112*, 3494–3509. [CrossRef]
7. Kasaliwal, G.R.; Pegel, S.; Göldel, A.; Pötschke, P.; Heinrich, G. Analysis of agglomerate dispersion mechanisms of multiwalled carbon nanotubes during melt mixing in polycarbonate. *Polymer* **2010**, *51*, 2708–2720. [CrossRef]
8. Kasaliwal, G.R.; Villmow, T.; Pegel, S.; Pötschke, P. Influence of Material and Processing Parameters on Carbon Nanotube Dispersion in Polymer Melts. In *Polymer-Carbon Nanotube Composites: Preparation, Properties and Applications*; McNally, T., Pötschke, P., Eds.; Woodhead Publishing: Philadelphia, PA, USA, 2011; pp. 92–132.
9. Kasaliwal, G.R.; Göldel, A.; Pötschke, P.; Heinrich, G. Influences of polymer matrix melt viscosity and molecular weight on MWCNT agglomerate dispersion. *Polymer* **2011**, *52*, 1027–1036. [CrossRef]
10. Socher, R.; Krause, B.; Müller, M.T.; Boldt, R.; Pötschke, P. The influence of matrix viscosity on MWCNT dispersion and electrical properties in different thermoplastic nanocomposites. *Polymer* **2012**, *53*, 495–504. [CrossRef]
11. Krause, B.; Boldt, R.; Pötschke, P. A method for determination of length distributions of multiwalled carbon nanotubes before and after melt processing. *Carbon* **2011**, *49*, 1243–1247. [CrossRef]

12. Pötschke, P.; Villmow, T.; Krause, B. Melt mixed PCL/MWCNT composites prepared at different rotation speeds: Characterization of rheological, thermal, and electrical properties, molecular weight, MWCNT macrodispersion, and MWCNT length distribution. *Polymer* **2013**, *54*, 3071–3078. [CrossRef]
13. Mayoral, B.; Lopes, J.; McNally, T. Influence of Processing Parameters During Small-Scale Batch Melt Mixing on the Dispersion of MWCNTs in a Poly(propylene) Matrix. *Macromol. Mater. Eng.* **2014**, *299*, 609–621. [CrossRef]
14. Sathyanarayana, S.; Hübner, C. Thermoplastic Nanocomposites with Carbon Nanotubes. In *Structural Nanocomposites;* Njuguna, J., Ed.; Springer: Berlin/Heidelberg, Germany, 2013; pp. 19–60. [CrossRef]
15. Villmow, T.; Pötschke, P.; Pegel, S.; Häussler, L.; Kretzschmar, B. Influence of twin-screw extrusion conditions on the dispersion of multi-walled carbon nanotubes in a poly(lactic acid) matrix. *Polymer* **2008**, *49*, 3500–3509. [CrossRef]
16. Villmow, T.; Kretzschmar, B.; Pötschke, P. Influence of screw configuration, residence time, and specific mechanical energy in twin-screw extrusion of polycaprolactone/multi-walled carbon nanotube composites. *Compos. Sci. Technol.* **2010**, *70*, 2045–2055. [CrossRef]
17. Müller, M.T.; Krause, B.; Kretzschmar, B.; Pötschke, P. Influence of feeding conditions in twin-screw extrusion of PP/MWCNT composites on electrical and mechanical properties. *Compos. Sci. Technol.* **2011**, *71*, 1535–1542. [CrossRef]
18. Krause, B.; Mende, M.; Pötschke, P.; Petzold, G. Dispersability and particle size distribution of CNTs in an aqueous surfactant dispersion as a function of ultrasonic treatment time. *Carbon* **2010**, *48*, 2746–2754. [CrossRef]
19. Benedito, A.; Buezas, I.; Giménez, E.; Galindo, B. Dispersion and characterization of Thermoplastic Polyurethane/Multiwalled Carbon Nanotubes in co-rotative twin screw extruder. *AIP Conf. Proc.* **2010**, *1255*, 227–230. [CrossRef]
20. Benedito, A.; Buezas, I.; Giménez, E.; Galindo, B.; Ortega, A. Dispersion and characterization of thermoplastic polyurethane/multiwalled carbon nanotubes by melt mixing. *J. Appl. Polym. Sci.* **2011**, *122*, 3744–3750. [CrossRef]
21. Mack, C.; Sathyanarayana, S.; Weiss, P.; Mikonsaari, I.; Hübner, C.; Henning, F.; Elsner, P. Twin-screw extrusion of multi walled carbon nanotubes reinforced polycarbonate composites: Investigation of electrical and mechanical properties. *IOP Conf. Ser. Mater. Sci. Eng.* **2012**, *40*, 012020. [CrossRef]
22. Sathyanarayana, S.; Olowojoba, G.; Weiss, P.; Caglar, B.; Pataki, B.; Mikonsaari, I.; Hübner, C.; Henning, F. Compounding of MWCNTs with PS in a Twin-Screw Extruder with Varying Process Parameters: Morphology, Interfacial Behavior, Thermal Stability, Rheology, and Volume Resistivity. *Macromol. Mater. Eng.* **2012**, *298*, 89–105. [CrossRef]
23. Sathyanarayana, S.; Hübner, C.; Diemert, J.; Pötschke, P.; Henning, F. Influence of peroxide addition on the morphology and properties of polypropylene-multiwalled carbon nanotube nanocomposites. *Compos. Sci. Technol.* **2013**, *84*, 78–85. [CrossRef]
24. Ezat, G.S.; Kelly, A.L.; Youseffi, M.; Coates, P.D. Effect of extrusion cycles on the dispersion and properties of polypropylene/multi-walled carbon nanotube composite. *AIP Conf. Proc.* **2023**, *2607*, 070004. [CrossRef]
25. Krause, B.; Villmow, T.; Boldt, R.; Mende, M.; Petzold, G.; Pötschke, P. Influence of dry grinding in a ball mill on the length of multiwalled carbon nanotubes and their dispersion and percolation behaviour in melt mixed polycarbonate composites. *Compos. Sci. Technol.* **2011**, *71*, 1145–1153. [CrossRef]
26. Liebscher, M.; Domurath, J.; Krause, B.; Saphiannikova, M.; Heinrich, G.; Pötschke, P. Electrical and melt rheological characterization of PC and co-continuous PC/SAN blends filled with CNTs: Relationship between melt-mixing parameters, filler dispersion, and filler aspect ratio. *J. Polym. Sci. Part B Polym. Phys.* **2018**, *56*, 79–88. [CrossRef]
27. Guo, J.; Briggs, N.; Crossley, S.; Grady, B.P. A new finding for carbon nanotubes in polymer blends. *J. Thermoplast. Compos. Mater.* **2018**, *31*, 110–118. [CrossRef]
28. Datasheet Bayer MaterialScience AG, Polycarbonate Makrolon 2600, Edition 28 September 2011.
29. Castillo, F.Y.; Socher, R.; Krause, B.; Headrick, R.; Grady, B.P.; Prada-Silvy, R.; Pötschke, P. Electrical, mechanical, and glass transition behavior of polycarbonate-based nanocomposites with different multi-walled carbon nanotubes. *Polymer* **2011**, *52*, 3835–3845. [CrossRef]
30. Bayer MaterialScience AG, Data Sheet Baytubes®C150P, Edition 24 February 2009.
31. Pegel, S. *Komposite aus Polycarbonat und Kohlenstoff-Nanoröhren: Morphologie und elektrische Leitfähigkeit bei thermoplastischer Verarbeitung*; Dr. Hut: München, Germany, 2011.
32. Nanocyl, S.A. Technical Data Sheet: NC7000™, V08, 12 July 2016. Available online: https://www.nanocyl.com/wp-content/uploads/2016/07/DM-TI-02-TDS-NC7000-V08.pdf (accessed on 28 January 2021).
33. Krause, B.; Boldt, R.; Häußler, L.; Pötschke, P. Ultralow percolation threshold in polyamide 6.6/MWCNT composites. *Compos. Sci. Technol.* **2015**, *114*, 119–125. [CrossRef]
34. Socher, R.; Krause, B.; Boldt, R.; Hermasch, S.; Wursche, R.; Pötschke, P. Melt mixed nanocomposites of PA12 with MWNT: Influence of MWNT and matrix properties on macrodispersion and electrical properties. *Compos. Sci. Technol.* **2011**, *71*, 306–314. [CrossRef]
35. Tessonnier, J.-P.; Rosenthal, D.; Hansen, T.W.; Hess, C.; Schuster, M.E.; Blume, R.; Girgsdies, F.; Pfänder, N.; Timpe, O.; Su, D.S.; et al. Analysis of the structure and chemical properties of some commercial carbon nanostructures. *Carbon* **2009**, *47*, 1779–1798. [CrossRef]
36. Sabic, Europe Technical Data Sheet, Polycarbonate LexanTM Resin 141R, Revision 11 March 2024. Available online: https://www.sabic.com/en/products/documents/lexan-resin_141r_europe_technical_data_sheet/en (accessed on 10 September 2024).

37. Krause, B.; Heber, M. Transmission electron microscopy (TEM) images of multiwalled carbon nanotubes (MWCNT) detached from polycarbonate (PC) composites [Data set]. Zenodo. 2024. Available online: https://doi.org/10.5281/zenodo.11400466 (accessed on 31 May 2024).
38. *ASTM D4496*; Standard Test Method for D-C Resistance or Conductance of Moderately Conductive Materials. ASTM International: West Conshohocken, PA, USA, 2022.
39. *ASTM D257*; Standard Test Methods for DC Resistance or Conductance of Insulating Materials. ASTM International: West Conshohocken, PA, USA, 2021.
40. *EN ISO 527-2*; Plastics—Determination of Tensile Properties—Part 2: Test Conditions for Molding and Extrusion Plastics. ISO: Geneva, Switzerland, 2012.
41. *ISO 179*; Plastics—Determination of Charpy Impact Properties—Part 1: Non-Instrumented Impact Test. International Organization for Standardization—ISO Central Secretariat: Vernier, Switzerland, 2010.
42. Talò, M.; Krause, B.; Pionteck, J.; Lanzara, G.; Lacarbonara, W. An updated micromechanical model based on morphological characterization of carbon nanotube nanocomposites. *Compos. Part B Eng.* **2017**, *115*, 70–78. [CrossRef]
43. Pegel, S.; Villmow, T.; Kasaliwal, G.R.; Pötschke, P. Polymer-Carbon Nanotubes Composites: Melt Processing, Properties and Applications. In *Synthetic Polymer-Polymer Composites*; Bhattacharyya, D., Fakirov, S., Eds.; Carl Hanser Verlag: München, Germany, 2012; pp. 145–192.
44. Rhue, M.A.; Zaqout, M.; Bavlnka, C.; Crossley, S.; Grady, B.P. Length reduction kinetics of multiwalled carbon nanotubes correlated to planetary ball mill impact energy. *Fuller. Nanotub. Carbon Nanostructures* **2024**, *32*, 8–21. [CrossRef]
45. Krause, B.; Pötschke, P. Raw data collection for the publication P. Pötschke, T. Villmow, B. Krause and B. Kretzschmar, Influence of Twin-screw Extrusion Conditions on MWCNT Length and Dispersion and Resulting Electrical and Mechanical Properties of Polycarbonate Composites. Zenodo. 2024. Available online: https://doi.org/10.5281/zenodo.13309199 (accessed on 10 September 2024).

polymers

MDPI

Article

Synergistic Effect of CNT and N-Doped Graphene Foam on Improving the Corrosion Resistance of Zn Reinforced Epoxy Composite Coatings

Yana Mao [1], Shufu Liu [2,*], Shizhong Liu [1], Guodong Wu [3], Qi Liu [3] and Xusheng Du [4,*]

1 School of Civil Engineering, Lanzhou Jiaotong University, Lanzhou 730070, China; 0116016@stu.lzjtu.edu.cn (Y.M.); liusz@mail.lzjtu.cn (S.L.)
2 Gansu Province Highway Aviation Tourism Investment Group Co., Ltd., Lanzhou 730030, China
3 China Communications Construction Road & Bridge Southern Engineering Co., Ltd., Beijing 101100, China; 447094717@163.com (G.W.); 15561429472@163.com (Q.L.)
4 Institute of Advanced Wear & Corrosion Resistant and Functional Materials, Jinan University, Guangzhou 510632, China
* Correspondence: liusf_1@163.com (S.L.); xdusydjn@email.jnu.edu.cn (X.D.)

Abstract: The synergistic effect of CNT and three-dimensional N-doped graphene foam (3DNG) on improving corrosion resistance of zinc-reinforced epoxy (ZRE) composite coatings was studied in this work. Although CNT itself was demonstrated to be effective to promote the anti-corrosion performance of the ZRE coating, the incorporation of additional 3DNG leads to further enhancement of its corrosion resistance under the synergistic effect of the hybrid carbon nanofillers with different dimensions. Both the content of the carbonaceous fillers and the ratio between them affected the performance of the coating. The optimal content of hybrid filler in the coating was determined to be only 0.1% with 3DNG:CNT = 1:3. With the modification of hybrid fillers, the corrosion current of the coating could be reduced by more than six orders of magnitude. Additionally, the immersion test of the pre-scratched coating directly demonstrated the evident contribution of the hybrid fillers to the sacrificial anode-based surface protection mechanism of the coating. These results confirmed the synergistic effect of the hybrid 1D and 3D carbonaceous fillers on promoting the corrosion inhibition of their coating, which could be promising for application in other functional composites.

Keywords: epoxy composite coating; corrosion; three-dimensional graphene; carbon nanotube; hybrid fillers; self-repairing

Citation: Mao, Y.; Liu, S.; Liu, S.; Wu, G.; Liu, Q.; Du, X. Synergistic Effect of CNT and N-Doped Graphene Foam on Improving the Corrosion Resistance of Zn Reinforced Epoxy Composite Coatings. *Polymers* **2024**, *16*, 3513. https://doi.org/10.3390/polym16243513

Academic Editors: Miguel Ángel López Manchado, Laura Peponi and Luigi Torre

Received: 27 October 2024
Revised: 10 December 2024
Accepted: 14 December 2024
Published: 17 December 2024

1. Introduction

Zinc-rich epoxy (ZRE) composites have been widely used as the anti-corrosive coating for metallic substrates [1]. Besides the common barrier protection mechanism of the composite primers, the cathodic protection effect of zinc powders in the composite coating also contributes significantly to the corrosion resistance of the coating. Due to its excellent chemical stability, high conductivity, environmental friendliness and high surface area, carbon nanomaterials have been extensively used as functional fillers in the polymer composite coatings [2], including those filled with and without zinc powders.

There are many types of carbon nanomaterials, including both 1D, 2D and 3D ones. The 1D CNTs have been added into ZRE coatings to improve their anti-corrosion properties [3,4]. It was reported that, with the incorporation of 2 wt% CNTs, the hybrid epoxy coating has the highest anti-corrosion performance [4]. Different from 1D CNT, 2D graphene (G) has more advantages, as it is capable of providing an additional barrier to the coating due to its own excellent impermeability to corrosives [2,5–7]. Therefore, graphene and its derivatives, including graphene oxide and reduced graphene oxide (rGO), have been utilized to incorporate into ZRE coatings [8–18]. In a recent work, GO was in situ reduced and modified with the curing agent of epoxy during the fabrication of ZRE paint, and at an optimal content of 0.3 wt%, the corrosion current of the resulting coating was reduced to 1.69×10^{-12} A·cm^{-2}, which is five orders of magnitude lower than that of neat ZRE, demonstrating the great positive effect of graphene on the cathodic protection performance of the coating [9]. Moreover, in order to utilize the 3D structure of the unique carbon fillers, ZRE coatings modified with 3D graphene and N-doped reduced graphene oxide (3DNG) were developed very recently [19]. Thanks to their spacial effect on reduction in graphene agglomeration, the construction of a conduction network in the composite [20,21], their ZRE coating was promoted evidently at a very low filler content, which is superior to that modified with the conventional 2D graphene sheets.

The hybrid fillers with different dimensions have been demonstrated to have synergistic effects on various properties of their composites, including the mechanical [22], electrical [23] and thermal conduction and wear resistance [24]. The synergistic effect of 0D carbon dots and 2D BN nanosheets on improving the corrosion resistance of their epoxy coating was demonstrated [25]. Both 0D silicone and nanodiamond have been used to modify 2D graphene oxide for fabrication of epoxy vinyl ester composite coatings as well, and their excellent long-term anti-corrosive performance was confirmed [26]; the impedance value $|Z|_{0.01Hz}$ of the coating was ~2–3 orders larger than that of the pure polymer coating after soaking for 120 days in a salt solution. On the basis of previous reports, it can be speculated that one can utilize the hybrid filler of 3D G and 1D CNT to promote the cathodic protection of the ZRE coating. Unfortunately, little research on this aspect has been reported so far.

In this work, the effect of the hybrid filler of 3DNG and CNT on improving the corrosion resistance of zinc-rich epoxy coatings will be studied. Based on the previous work on the 3D graphene filler-modified ZRE coating, CNT will be additionally incorporated to promote the cathodic protection function of the ZRE coating to the metallic substrate, and the synergistic effect of these two nanocarbon fillers with different dimensions will be investigated. The content of the hybrid filler will be optimized to maximize this effect. Moreover, the anti-corrosion behavior of the coating will be studied with various methods, including the long-term immersion test and the element mapping of the scratch-defected ZRE coating samples after immersion.

2. Materials and Methods

2.1. Preparation of the Carbon Nanomaterial-Modified ZRE Coatings

The 3DNG was prepared by the flame combustion of GO coating polyurethane foam according to the method reported previously [20]. A certain amount of 3DNG or CNT (length of 0.5–2 µm) was added into the paint diluent, followed by the addition of epoxy resin E44 and zinc micro-flakes (lateral size: 5~10 µm). The mixture was stirred vigorously for 2 h. Polyamide 651 was then added into the mixture (its ratio to epoxy resin = 1:2) and continuously stirred for 10 min to obtain the carbon nanomaterial-modified ZRE paints. Zinc flakes and epoxy diluent accounted for 35% and 20% of the total paint, respectively. The fabrication process of the ZRE paints modified by different contents of hybrid fillers (3DNG:CNT = 1:3) was consistent with that for ZRE modified with a single filler. According to the type and content of carbon nanofillers, i.e., 0.2 wt% CNT, 0.2 wt% 3DNG, 0.2 wt% (3DNG + CNT), 0.1 wt% (3DNG + CNT) and 0.05 wt% (3DNG + CNT), the resulting ZRE paints were labeled as '02_CNT ZRE', "02_3DNG ZRE', '02_3DNG-CNT ZRE', '01_3DNG-CNT ZRE' and '005_3DNG-CNT ZRE', respectively. Neat ZRE without adding any carbon nanofiller was also prepared as the reference sample. Rectangular Q235 steel plates were used as metal substrates, which were ultrasonically cleaned with acetone after sandblasting. The paints were applied onto the metal surface with an applicator and cured at 60 °C for 24 h.

2.2. Characterization Methods

A typical three-electrode system was assembled with a saturated calomel electrode (SCE) as the reference electrode, platinum sheet as the counter electrode and the coated Q235 plates as the working electrode, whose exposed area in the corrosive solution was 1 cm^2. A 3.5 wt% NaCl solution was used as the corrosive solution for the corrosion tests. The electrochemical impedance spectroscopy (EIS) and potentiodynamic polarization (or so-called Tafel) curves of the ZRE coatings were measured on a CHI760E electrochemical analyzer (Chenhua CHI Co., Shanghai, China). The EIS plots of the ZRE coatings under different immersion times were carried out in the frequency range of 10^5~10^{-2} Hz, with a sinusoidal voltage disturbance amplitude of 5 mV. ZView software was used to fit the EIS data. The Tafel tests were also performed before and after 300 h soaking in the salt solution. For the immersion test of the scratch-defected coating sample, the scratch was made by a blade to expose the Q235 steel substrate. The morphology of the sample was observed on a scanning electron microscope (SEM; ULTRA55, Zeiss, Germany).

3. Results and Discussion

3.1. Anti-Corrosion Performance of Hybrid Carbon-Modified ZRE Coatings

In consideration of the practical application of ZRE paints, it is highly demanding for reducing the content of expensive graphene and its derivatives in the ZRE premiers. According to the recent work on 3DNG-modified ZRE coatings, its optimal content was determined to be 0.05 wt% [19]. On the base of this work, a small amount of commercial CNTs (0.15 wt%) was further added into the paint to observe if there was any hybrid effect between these two carbon nanomaterials with different dimensions. Obviously, the content of the total hybrid carbon nanofiller in the ZRE paint would be 0.2 wt%. The Tafel curve of the resulting 02_3DNG-CNT ZRE is shown in Figure 1a, and its corrosion current density (Icorr) and polarization resistance (R_L) were calculated to be 3.07×10^{-12} $A{\cdot}cm^{-2}$ and 1.39×10^{11} $\Omega{\cdot}cm^2$, respectively (Table 1). Compared to the ZRE modified with 0.05 wt% 3DNG, i.e., 9.56×10^{-12} $A{\cdot}cm^{-2}$ for Icorr and 4.55×10^{11} $\Omega{\cdot}cm^2$ for R_L [19], the decreased Icorr and increased R_L of 02_3DNG-CNT ZRE confirmed that it promoted corrosion resistance. Besides the corrosion current intensity and polarization resistance, the low frequency impedance modulus ($|Z|_{0.01Hz}$) obtained from the EIS data was another important parameter, which is widely used to evaluate the surface protection of a coating. Evidently, the larger $|Z|_{0.01Hz}$ shown in the Bode plot of the 02_3DNG-CNT ZRE coating in Figure 1b was also consistent with the Tafel analysis results. In fact, CNTs had been demonstrated to be effective nanofillers in reducing the porosity of the coating, hindering the diffusion of corrosion species, thus improving the anti-corrosion property of their coatings [4]. Moreover, 1D CNTs could work well with 3D graphene to form the conductive network to activate more zinc particles in the epoxy matrix to exert their cathodic protection function.

Table 1. Icorr, Ecorr and RL of neat ZRE and nanocarbon-modified ZRE coating samples.

Samples	I_{corr} ($A{\cdot}cm^{-2}$)	E_{corr} (V)	R_L ($\Omega{\cdot}cm^2$)
neat ZRE	6.11×10^{-6}	−0.646	$5.67{\times}10^3$
02_3DNG-CNT ZRE	3.07×10^{-12}	−0.290	$1.39{\times}10^{11}$
02_3DNG ZRE	9.63×10^{-10}	−0.885	$5.09{\times}10^7$
02_CNT ZRE	6.11×10^{-6}	−1.196	$6.47{\times}10^3$

As the synergistic effect of fillers with different dimensions on various performance of the polymer composites had been demonstrated previously [22–26], one was expected to see whether such an effect contributes to the improvement of the corrosion resistance of ZRE coatings. ZRE coatings incorporated with the same content of a single carbon filler were fabricated, and their corrosion resistances were characterized. As shown in Figure 1 and Table 1, compared with that of 02_3DNG-CNT ZRE, both Tafel curves of 02_3DNG ZRE and 02_CNT ZRE moved upwards, and their Icorr values were evidently larger than that of 02_3DNG-CNT ZRE, especially for that of 02_CNT ZRE. The Icorr value of 02_CNT ZRE was even the same as that of neat ZRE, indicating the neglect contribution of CNTs to the cathodic protection of the ZRE coating. Such an effect of CNTs on ZRE was similar to the report by [27], where single-walled CNTs were utilized. Remarkably, it could be found that a reduction of more than six orders of magnitude in Icorr was realized, thanks to the utilization of hybrid fillers in the ZRE coating, as well as an increase of approximately eight orders of magnitude in RL, as shown in Table 1. The Bode plot shown in Figure 1b also indicates that the 02_3DNG-CNT ZRE had a much larger $|Z|_{0.01Hz}$ than both 02_3DNG ZRE and 02_CNT ZRE. All these results demonstrate the synergistic effect of 3DNG and CNTs on enhancing the corrosion protection of ZRE coatings.

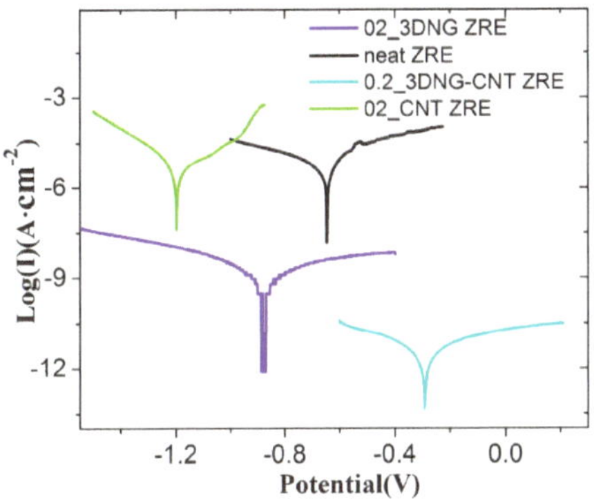

(**a**)

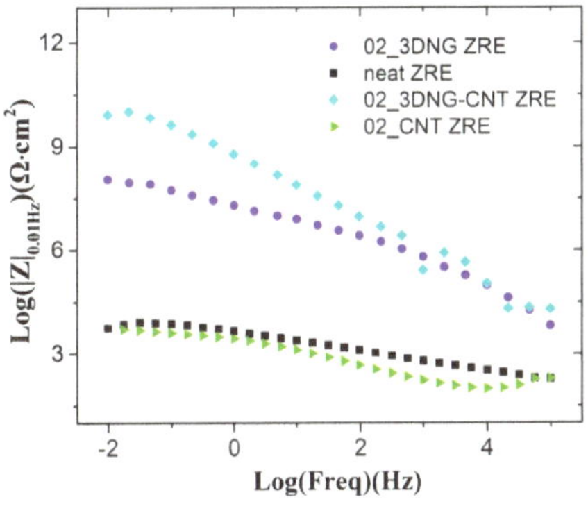

(**b**)

Figure 1. (**a**) Tafel curves and (**b**) Bode plots of neat ZRE and hybrid filler-modified ZRE coatings.

Based on the aforementioned results, the effects of different hybrid filler contents on the corrosion inhibition performance of ZRE coatings were studied. Interestingly, as shown in Figure 2, the Icorr of the coating was further reduced to 2.84×10^{-12} A·cm^{-2} when the content of the hybrid filler was reduced by half to 0.1 wt%, and its R_L increased accordingly as well (Table 2), indicating its improved anti-corrosion performance at an even lower content of hybrid filler. However, with a further decrease of the hybrid filler content to 0.05 wt%, its Icorr was greatly increased to 1.08×10^{-10} A·cm^{-2}, which was 38-fold that of 01_3DNG-CNT ZRE (Table 2). This meant that the effect of the hybrid filler on improving the anti-corrosion performance of 3DNG-CNT ZRE depended on its content, and the optimal content of the hybrid filler in this work was 0.1 wt%, which was comprised of only 0.025 wt% 3DNG and 0.075 wt% CNT.

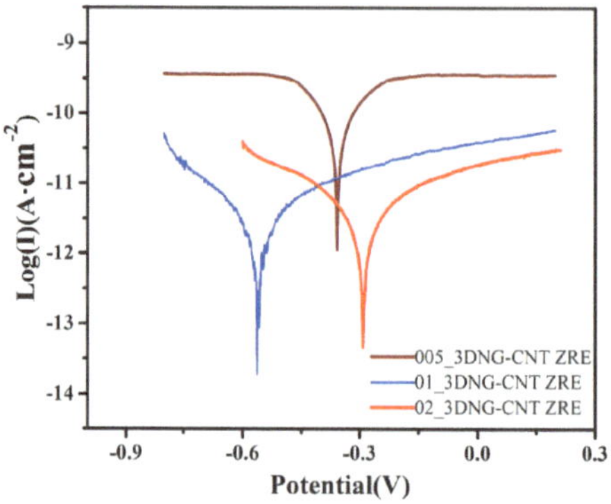

Figure 2. Tafel curves of nanocarbon-modified ZRE-coated Q235 samples.

Table 2. Icorr, Ecorr and RL of the nanocarbon hybrid filler-modified ZRE coatings.

Samples	I_{corr} (A·cm^{-2})	E_{corr} (V)	R_L (Ω·cm^2)
005_3DNG-CNT ZRE	1.08×10^{-10}	−0.357	4.03×10^{9}
01_3DNG-CNT ZRE	2.84×10^{-12}	−0.561	1.47×10^{11}
02_3DNG-CNT ZRE	3.07×10^{-12}	−0.290	1.39×10^{11}

3.2. Long-Term Immersion Test of Hybrid Filler-Modified Coatings

The $|Z|_{0.01Hz}$ values of the 01_3DNG-CNT ZRE coating were monitored for up to 300 h to study its long-term anti-corrosion behavior. As shown in Figure 3, the $|Z|_{0.01Hz}$ ~ immersion time curves of the hybrid filler-modified coating dropped with the immersion time initially. It reached a relatively stable state later during the entire 300 h immersion test. This might be caused by the corrosion products in situ generated after the rapid corrosion of zinc particles, which hindered the further corrosion of the steel substrate. Moreover, it could be observed that its $|Z|_{0.01Hz}$ values were always larger than those of neat ZRE, indicating it promoted long-term corrosion inhibition performance.

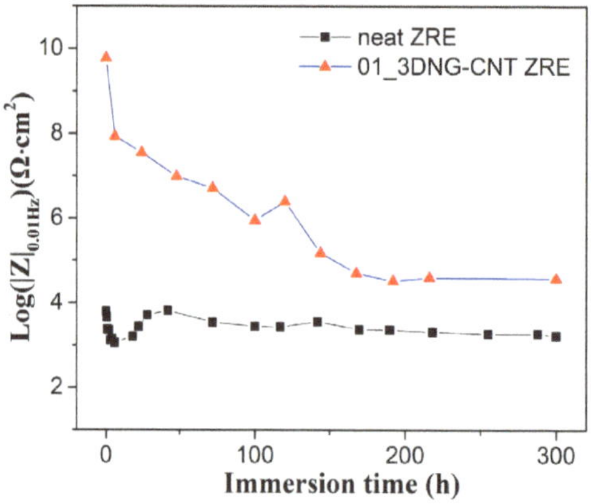

Figure 3. The dependence of $|Z|_{0.01Hz}$ of neat ZRE and 01_3DNG-CNT ZRE coatings on the immersion time.

The potentiodynamic polarization test of the hybrid filler-modified ZRE-coated Q235 samples was also performed to characterize its long-term corrosion resistance behavior. Figure 4 shows the Tafel curves of the 01_3DNG-CNT ZRE coating after 300 h immersion in a 3.5 wt% NaCl solution, and the corresponding corrosion parameters are presented in Table 3. Evidently, 01_3DNG-CNT ZRE displayed a much smaller Icorr value (6.94×10^{-7} A·cm^{-2}) compared to that of neat ZRE (4.87×10^{-5} A·cm^{-2}). Meanwhile, its R_L was still much larger than that of neat ZRE. This confirmed its superior long-term anti-corrosion performance as well.

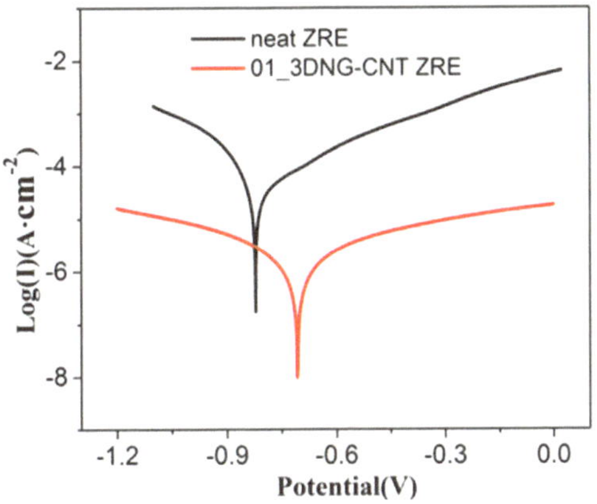

Figure 4. Tafel curves of neat ZRE and 01_3DNG-CNT ZRE coatings after 300 h immersion in 3.5 wt% NaCl solution.

Table 3. Icorr, Ecorr and RL of neat ZRE and 01_3DNG-CNT ZRE-coated Q235 samples after 300 h immersion in 3.5 wt% NaCl solution.

Samples After 300 h Immersion	I_{corr} (A·cm^{-2})	E_{corr} (V)	R_L (Ω·cm^2)
neat ZRE	4.87×10^{-5}	−0.821	7.76×10^2
01_3DNG-CNT ZRE	1.11×10^{-6}	−0.707	4.01×10^4

The Nyquist plots of the hybrid filler-modified ZRE coatings soaked in the corrosive solution at three time stages (25 h, 100 h and 300 h) are shown in Figure 5a. The schematic diagram of the equivalent analog circuit for fitting the EIS data is shown in Figure 5b. In the circuits, Rt relates to the electrochemical corrosion of zinc particles during the cathodic protection process, while the pore resistance (Rp) of the coating and the CPE-2 capacitance relate to the entire coating's impermeability [16]. The detailed as-fitted data for the samples are shown in Table 4. It can be observed that the Rt values of the 01_3DNG-CNT ZRE decreased with the extension of immersion, implying the deterioration of the corrosion resistance of the coating. Meanwhile, the CPE-2 values of the ZRE coating increased with the immersion time. It had been suggested that the water absorption behavior of the polymer coating could be evaluated with the CPE-2 value [9,28]. The increased CPE-2 values meant deteriorated impermeability of the 01_3DNG-CNT ZRE coating to water, which coincided with the increased Rp values as well (Table 4).

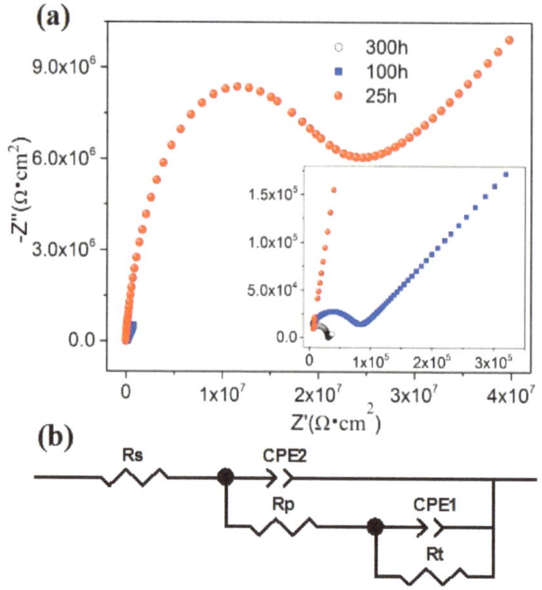

Figure 5. (**a**) Nyquist plots of 01_3DNG-CNT ZRE coating on Q235 sample immersed in NaCl solution at 25 h, 100 h and 300 h and (**b**) schematic diagrams of equivalent analog circuits for EIS analysis of the hybrid nanocarbon-modified ZRE.

Table 4. Equivalent circuit fitting of 01_3DNG-CNT ZRE-coated Q235 sample immersed in 3.5 wt% NaCl solution.

Time (h)	CPE-1 ($S{\cdot}cm^{-2}{\cdot}s^n$)	R_p ($\Omega{\cdot}cm^2$)	CPE-2 ($S{\cdot}cm^{-2}{\cdot}s^n$)	R_t ($\Omega{\cdot}cm^2$)
25	1.01×10^{-9}	1.39×10^{7}	5.95×10^{-8}	1.45×10^{15}
100	3.28×10^{-9}	8.11×10^{4}	3.46×10^{-6}	3.70×10^{14}
300	5.30×10^{-10}	2.94×10^{4}	3.33×10^{-4}	7.56×10^{8}

3.3. The Cathodic Protection Behavior of the Hybrid Filler-Modified Coating

In order to explore the cathodic protection behavior of the ZRE coating, the Q235 samples coated with either neat ZRE or 01_3DNG-CNT ZRE were deliberately scratched deeply enough to expose the steel substrate and immersed in a 3.5 wt% NaCl solution to detect their corrosion resistance mechanism. The inset images in Figures 6a and 7a show the digital photos of the exposed steel surface of neat ZRE and 01_3DNG-CNT ZRE-coated Q235 plates, respectively, after 12 h immersion in the corrosive solution. Obviously, after soaking for 12 h, the scratch area in the neat ZRE became totally brown, indicating the occurrence of severe corrosion in the steel substrate. In contrast, such a severe corrosion phenomenon was absent from the scratched 01_3DNG-CNTZRE sample (inset image in Figure 7a) soaked for the same time, and only a small area displayed a very light brown color. This directly demonstrated the better cathodic protection function of the 01_3DNG-CNT ZRE coating for the steel substrate.

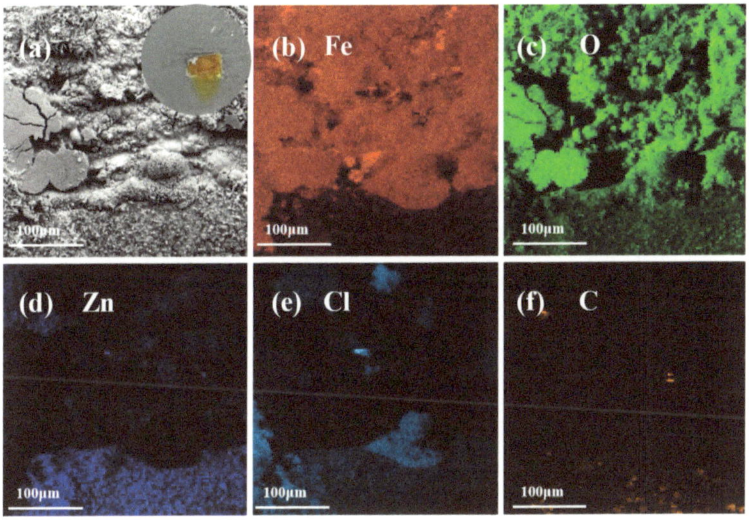

Figure 6. Images of scratched area of neat ZRE soaked in 3.5 wt% NaCl solution for 12 h: (**a**) SEM image and inset digital photo; the element mapping of (**b**) Fe, (**c**) O, (**d**) Zn, (**e**) Cl and (**f**) C.

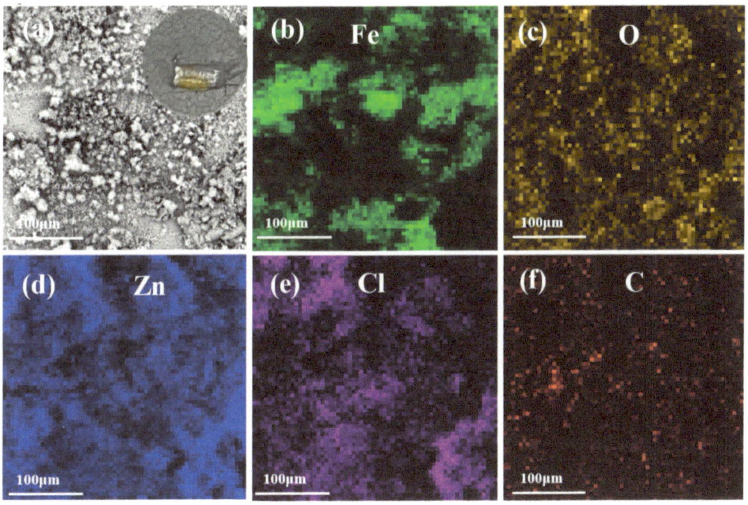

Figure 7. Images of scratched area of the hybrid filler-modified ZRE-coated steel sample soaked in 3.5 wt% NaCl solution for 12 h: (**a**) SEM image and inset digital photo and the element mapping of (**b**) Fe, (**c**) O, (**d**) Zn, (**e**) Cl and (**f**) C.

The SEM and elemental mapping analyses of the corroded scratch regions of both neat ZRE and the hybrid filler-modified ZRE coating were further carried out. As shown in Figure 6a, the major area of the corroded scratch of the sample was covered by the corrosion products of the steel substrate, for instance, products α-$Fe_2(OH)_3Cl$, Fe_3O_4 and α-Fe_2O_3 [26], except the one-third area at the bottom part, which was the unscratched part of the neat ZRE coating. The EDS mapping images of Figure 6b–d confirm this and demonstrate the severely corroded substrate of the neat ZRE-coated sample at the corrosive conditions. Different from the Fe-based major corrosion product of the steel substrate in the scratched neat ZRE-coated sample, the Zn element dominated in the element mapping image of the hybrid filler-modified ZRE-coated sample and was distributed relatively uniformly in the studied area, as shown in Figure 7d. This implied that the corrosion product generated from the Zn powders in the coating could form a continuous layer, which isolated the steel substrate from the corrosive solution and thus improved the corrosion resistance of the sample. The contents of the Cl element (Figure 7e) and C element (Figure 7f) obtained in the scratch area of the coating should majorly originate from the corrosion products of the Zn powders, such as Hydrozincite $Zn_5(CO_3)_2(OH)_6$ and Willemite ($Zn_5(OH)_8Cl_2 \cdot H_2O$) [19,29,30]. These test results demonstrate the self-repairing characteristic of the 01_3DNG-CNTZRE coating, whose mechanism is similar to those coatings based on the release of organic corrosion inhibitor molecules from the composite coating and isolation of steel substrate from further corrosion by the in situ formed solid Fe^{3+}-based complex compounds [31–33]. However, to the best of our knowledge, most of the research available so far on the epoxy coating incorporated with the corrosion inhibitors is not a zinc-rich epoxy composite coating. Due to the lack of the essential function of cathodic protection with the sacrificial anode provided by the large amount of Zn particles, they found little utilization as heavy-duty coatings.

Figure 8 illustrates the anti-corrosion mechanism of ZRE coatings modified with hybrid nanocarbon fillers. The incorporation of 3DNG into the ZRE coating helps to form electro-conductive paths between Zn particles and the steel substrate, as more conduction paths formed between the Zn particles and the steel substrate are conducive to activating more zinc particles to exert their cathodic protection function through the sacrificial-anode electrochemical corrosion reaction. In addition to this, the excellent impermeability of 3DNG to the corrosives also benefits the barrier property of its ZRE coating. However, taking into account the large size of 3DNG (micron scale), there are still some minor spaces to be filled to build a more effective conductive network. With further incorporation of CNTs, micropores in the coating can be eliminated more or less with the nanofillers, therefore hindering the corrosive invasion through the coating. Moreover, due to the electrical connection provided by the 1D CNTs, more Zn particles will be linked into the conductive network made of 3DNG, CNT, Zn particles and the steel substrate for exerting their cathodic protection function (see the red line in Figure 8), thereby maximizing the sacrificial-anode roles of zinc in the entire coating system.

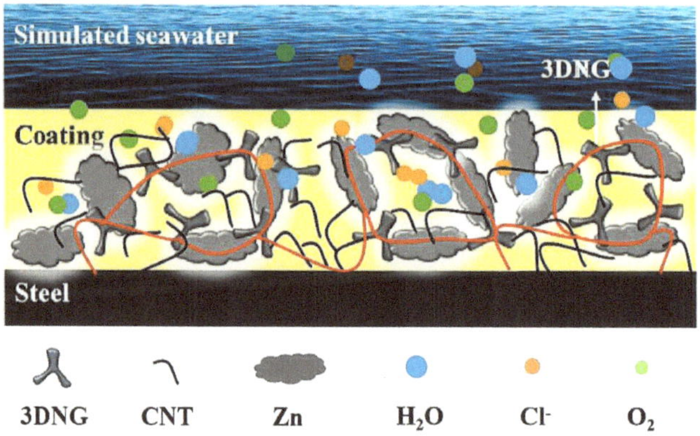

Figure 8. Schematic diagram of the anti-corrosive mechanism of 3DNG-CNT ZRE coating on steel in simulated seawater.

4. Conclusions

The effect of the hybrid filler of 3DNG and CNT on improving the cathodic protection function of a ZRE coating was exploited for the first time. Compared with neat ZRE and 0.2 wt% CNT-incorporated ZRE, the Icorr of ZRE modified with 0.2 wt% hybrid filler (0.05 wt% 3DNG and 0.15 wt% CNT) was reduced by more than six orders of magnitude.

Moreover, it was ~1/32 that of ZRE modified with the same content of 3DNG. Therefore, the synergistic effects of these two nanocarbon fillers with 1D and 3D dimensions on the cathodic protection of the ZRE coating were confirmed. The dependence of the hybrid filler-modified coating on its content was also revealed, and the optimal content of the hybrid filler was 0.1 wt%, which was comprised of only 0.025 wt% 3DNG and 0.075 wt% CNT. The long-term cathodic protection behavior of the 3DNG–CNT-modified ZRE coating was studied, and the corresponding corrosion mechanism was proposed. Additionally, the immersion test of the scratch-defected coating demonstrated the self-repairing capability of the hybrid filler-modified ZRE coating, which resulted from its promoted sacrificial-anode-based steel protection function. This work provides a versatile methodology to improve the cathodic protection function of a zinc-containing composite coating by utilizing the hybrid nanocarbon fillers with different dimensions, which might be applied to other composite coatings.

Author Contributions: Conceptualization, S.L. (Shufu Liu) and X.D.; methodology, Y.M., S.L. (Shizhong Liu) and G.W.; validation, G.W. and Q.L.; investigation, Y.M., S.L. (Shizhong Liu) and Q.L.; supervision, S.L. (Shufu Liu) and X.D.; writing, Y.M., S.L. (Shufu Liu) and X.D. All authors have read and agreed to the published version of the manuscript.

Funding: This research received no funding.

Institutional Review Board Statement: Not applicable.

Data Availability Statement: The original contributions presented in this study are included in the article; further inquiries can be directed to the corresponding authors.

Conflicts of Interest: Author Shufu Liu was employed by the company Gansu Province Highway Aviation Tourism Investment Group Co., Ltd. Author Guodong Wu and Qi Liu were employed by the company China Communications Construction Road & Bridge Southern Engineering Co., Ltd. The remaining authors declare that the research was conducted in the absence of any commercial or financial relationships that could be construed as a potential conflict of interest.

References

1. Golovin, V.A.; Tyurina, S.A. Current trends in the modification of zinc-filled polymer coatings. *Int. J. Corros. Scale Inhib.* **2020**, *9*, 372–393. [CrossRef]
2. Wu, Y.; Wu, Y.; Sun, Y.; Zhao, W.; Wang, L. 2D Nanomaterials Reinforced Organic Coatings for Marine Corrosion Protection: State of the Art, Challenges, and Future Prospectives. *Adv. Mater.* **2024**, *36*, 2470293. [CrossRef]
3. Cho, S.; Chiu, T.M.; Castaneda, H. Electrical and electrochemical behavior of a zinc-rich epoxy coating system with carbon nanotubes as a diode-like material. *Electrochim. Acta* **2019**, *316*, 189–201. [CrossRef]
4. Shen, W.; Feng, L.; Liu, X.; Luo, H.; Liu, Z.; Tong, P.; Zhang, W. Multiwall carbon nanotubes-reinforced epoxy hybrid coatings with high electrical conductivity and corrosion resistance prepared via electrostatic spraying. *Prog. Org. Coat.* **2016**, *90*, 139–146. [CrossRef]
5. Zhang, F.; Yang, K.; Liu, G.; Chen, Y.; Wang, M.; Li, S.; Li, R. Recent advances on graphene: Synthesis, properties and applications. *Compos. Part A Appl. Sci. Manuf.* **2022**, *160*, 107051. [CrossRef]
6. Zou, J.; Guan, J.; Wang, X.; Du, X. Corrosion and wear resistance improvements in NiCu alloys through flame-grown honeycomb carbon and CVD of graphene coatings. *Surf. Coat. Technol.* **2023**, *473*, 130040. [CrossRef]
7. Song, S.; Fan, X.; Yan, H.; Cai, M.; Huang, Y.; He, C.; Zhu, M. Facile fabrication of continuous graphene nanolayer in epoxy coating towards efficient corrosion/wear protection. *Compos. Commun.* **2023**, *37*, 101437. [CrossRef]
8. Cui, G.; Bi, Z.; Zhang, R.; Liu, J.; Yu, X.; Li, Z. A comprehensive review on graphene-based anti-corrosive coatings. *Chem. Eng. J.* **2019**, *373*, 104–121. [CrossRef]
9. Guan, J.; Wang, X.; Zou, J.; Yang, J.; Du, X. Significant promotion of the anti-corrosive performance of epoxy zinc-containing coating via in situ reduction of graphene oxide and simultaneous graft it with polyamide curing agent. *Diam. Relat. Mater.* **2023**, *138*, 110221. [CrossRef]
10. Navarchian, A.H.; Joulazadeh, M.; Karimi, F. Investigation of corrosion protection performance of epoxy coatings modified by polyaniline/clay nanocomposites on steel surfaces. *Prog. Org. Coat.* **2014**, *77*, 347–353. [CrossRef]
11. Shen, L.; Li, Y.; Zhao, W.; Miao, L.; Xie, W.; Lu, H.; Wang, K. Corrosion Protection of Graphene-Modified Zinc-Rich Epoxy Coatings in Dilute NaCl Solution. *ACS Appl. Nano Mater.* **2019**, *2*, 180–190. [CrossRef]
12. Zhu, X.; Yan, Q.; Cheng, L.; Wu, H.; Zhao, H.; Wang, L. Self-alignment of cationic graphene oxide nanosheets for anticorrosive reinforcement of epoxy coatings. *Chem. Eng. J.* **2020**, *389*, 124435. [CrossRef]

13. Huang, S.; Kong, G.; Yang, B.; Zhang, S.; Che, C. Effects of graphene on the corrosion evolution of zinc particles in waterborne epoxy zinc-containing coatings. *Prog. Org. Coat.* **2020**, *140*, 105531. [CrossRef]
14. Teng, S.; Gao, Y.; Cao, F.; Kong, D.; Zheng, X.; Ma, X.; Zhi, L. Zinc-reduced graphene oxide for enhanced corrosion protection of zinc-rich epoxy coatings. *Prog. Org. Coat.* **2018**, *123*, 185–189. [CrossRef]
15. Ramezanzadeh, B.; Moghadam, M.H.M.; Shohani, N.; Mandavian, M. Effects of highly crystalline and conductive polyaniline/graphene oxide composites on the corrosion protection performance of a zinc-rich epoxy coating. *Chem. Eng. J.* **2017**, *320*, 363–375. [CrossRef]
16. Zhou, S.; Wu, Y.; Zhao, W.; Yu, J.; Jiang, F.; Wu, Y.; Ma, L. Designing reduced graphene oxide/zinc rich epoxy composite coatings for improving the anticorrosion performance of carbon steel substrate. *Mater. Des.* **2019**, *169*, 107694. [CrossRef]
17. Lei, Y.; Liu, Y.; Liu, Q.; Xu, Z.; You, B.; Mo, X.; Wu, L.; Man, H.; Ma, W.; Wang, H. Nacre-mimetic Strategy to Fabricate Waterborne FrGO/ZREoxy Coatings with Dual Corrosion Protection. *ACS Appl. Mater. Interfaces* **2023**, *15*, 28570–28580. [CrossRef]
18. Zhan, Y.; Chen, Y.; Dong, H.; Li, Y.; Sun, A.; Chen, X.; Yang, X.; Zhu, F.; Jia, H. In situ growth of flower-like ZnO onto graphene oxide for the synergistically enhanced anti-corrosion ability of epoxy coating. *Ceram. Int.* **2024**, *50*, 5914–5926. [CrossRef]
19. Guan, J.; Du, X. Improving the anti-corrosive performance of epoxy/zinc composite coatings with N-doped 3D reduced graphene oxide. *Prog. Org. Coat.* **2024**, *186*, 107959. [CrossRef]
20. Du, X.; Liu, H.Y.; Mai, Y.-W. Ultrafast Synthesis of Multifunctional N-Doped Graphene Foam in an Ethanol Flame. *ACS Nano* **2016**, *10*, 453–462. [CrossRef]
21. Zhou, H.; Wang, H.; Du, X.; Zhang, Y.; Zhou, H.; Yuan, H.; Liu, H.-Y.; Mai, Y.-W. Facile fabrication of large 3D graphene filler modified epoxy composites with improved thermal conduction and tribological performance. *Carbon* **2018**, *139*, 1168–1177. [CrossRef]
22. Wei, L.; Liu, X.; Gao, Y.; Lv, X.; Hu, N.; Chen, M. Synergistic strengthening effect of titanium matrix composites reinforced by graphene oxide and carbon nanotubes. *Mater. Des.* **2021**, *197*, 109261. [CrossRef]
23. Tang, Z.-H.; Wang, D.-Y.; Li, Y.-Q.; Huang, P.; Fu, S.-Y. Modeling the synergistic electrical percolation effect of carbon nanotube/graphene/polymer composites. *Compos. Sci. Technol.* **2022**, *225*, 109496. [CrossRef]
24. Zhou, H.; Wang, H.; Du, X.; Mo, Y.; Yuan, H.; Liu, H.-Y. Hybrid three-dimensional graphene fillers and graphite platelets to improve the thermal conductivity and wear performance of epoxy composites. *Compos. Part A* **2019**, *123*, 270–277. [CrossRef]
25. Wan, S.; Chen, H.; Cai, G.; Liao, B.; Guo, X. Functionalization of h-BN by the exfoliation and modification of carbon dots for enhancing corrosion resistance of waterborne epoxy coating. *Prog. Org. Coat.* **2022**, *165*, 106757. [CrossRef]
26. Xu, C.-A.; Liang, W.; Ni, C.; Li, X.; Shi, X.; Chu, Z.; Hu, Y.; Yang, Z. Silicone and nano-diamond modified graphene oxide anticorrosive coating. *Surf. Coat. Technol.* **2024**, *479*, 130584. [CrossRef]
27. Tian, Y.; Bi, Z.; Cui, G. Study on the Corrosion Resistance of Graphene Oxide-Based Epoxy Zinc-Rich Coatings. *Polymers* **2021**, *13*, 1657. [CrossRef]
28. Yun, T.H.; Park, J.H.; Kim, J.-S.; Park, J.M. Effect of the surface modification of zinc powders with organosilanes on the corrosion resistance of a zinc pigmented organic coating. *Prog. Org. Coat.* **2014**, *77*, 1780–1788. [CrossRef]
29. Mouanga, M.; Berçot, P.; Rauch, J.Y. Comparison of corrosion behaviour of zinc in NaCl and in NaOH solutions. Part I: Corrosion layer characterization. *Corros. Sci.* **2010**, *52*, 3984–3992. [CrossRef]
30. Qin, Z.; Su, Y.; Bai, Y.; Lu, H.; Peng, T.; Zhong, H.; Chen, T.; Du, X. Improving the Corrosion Resistance of Zn-Rich Epoxy Coating with Three-Dimensional Porous Graphene. *Polymers* **2023**, *15*, 4302. [CrossRef]
31. Liu, S.; Jing, Y.; Zhang, T.; Zhang, J.; Xu, F.; Song, Q.; Ye, Q.; Liu, S.; Liu, W. Excellent tribological and anti-corrosion performances enabled by novel hollow graphite carbon nanosphere with controlled release of corrosion inhibitor. *Chem. Eng. J.* **2021**, *412*, 128648. [CrossRef]
32. Kamaruzzaman, W.M.I.W.M.; Fekeri, M.F.M.; Nasir, N.A.M.; Hamidi, N.A.S.M.; Baharom, M.Z.; Adnan, A.; Shaifudin, M.S.; Abdullah, W.R.W.; Wan Nik, W.M.N.; Suhailin, F.H.; et al. Anticorrosive and microbial inhibition performance of a coating loaded with *Andrographis paniculata* on stainless steel in seawater. *Molecules* **2021**, *26*, 3379. [CrossRef] [PubMed]
33. Udoh, I.I.; Shi, H.; Daniel, E.F.; Li, J.; Gu, S.; Liu, F.; Han, E.-H. Active anticorrosion and self-healing coatings: A review with focus on multi-action smart coating strategies. *J. Mater. Sci. Technol.* **2022**, *116*, 224–237. [CrossRef]

polymers

MDPI

Review

Carbon Nanodots-Based Polymer Nanocomposite: A Potential Drug Delivery Armament of Phytopharmaceuticals

Rabin Debnath [1,†], Abu Md Ashif Ikbal [1,†], Neeraj Kr. Ravi [1], Hanieh Kargarzadeh [2], Partha Palit [1,*] and Sabu Thomas [3,4,5,*]

1 Department of Pharmaceutical Sciences, Drug Discovery Research Laboratory, Assam University, Silchar 788011, India; rabindnath26@gmail.com (R.D.); abu.md.ashif.ikbal@aus.ac.in (A.M.A.I.); neerajrav2095@gmail.com (N.K.R.)
2 Center of Molecular and Macromolecular Studies, Polish Academy of Sciences, Sienkiewicza 112, 90-363 Lodz, Poland; hanieh.kargar@gmail.com
3 School of Energy Materials, School of Nanoscience and Nanotechnology, School of Polymer Science and Technology, School of Chemical Science and International, Inter University Centre for Nanoscience and Nantechnology (IIUCNN), Mahatma Gandhi University, Kottayam 686560, India
4 Department of Chemical Sciences, University of Johannesburg, Doornfontein, Johannesburg P.O. Box 17011, South Africa
5 TrEST Research Park, TC-4/2322, GEM Building, Opposite College of Engineering Trivandrum, Kulathoor Rd., Sreekariyam, Trivandrum 695016, India
* Correspondence: itspartha_p@yahoo.com (P.P.); sabuthomas@mgu.ac.in (S.T.)
† These authors contributed equally to this work.

Abstract: Carbon nanodots (CNDs) have garnered significant attention as viable drug delivery vehicles in recent years, especially in the field of phytomedicine. Although there is much promise for therapeutic applications with phytomedicine, its effectiveness is frequently restricted by its low solubility, stability, and bioavailability. This paper offers a thorough synopsis of the developing field of phytomedicine drug delivery based on CND. It explores CND synthesis processes, surface functionalization strategies, and structural and optical characteristics. Additionally, the advantages and difficulties of phytomedicine are examined, with a focus on the contribution of drug delivery methods to the increased effectiveness of phytomedicine. The applications of CNDs in drug delivery are also included in the review, along with the mechanisms that underlie their improved drug delivery capabilities. Additionally, it looks at controlled-release methods, stability augmentation, and phytomedicine-loading tactics onto CNDs. The potential of polymeric carbon nanodots in drug delivery is also covered, along with difficulties and prospective directions going forward, such as resolving toxicity and biocompatibility issues. In summary, the present review highlights the encouraging contribution of CNDs to the field of drug delivery, specifically in enhancing the potential of phytomedicine for therapeutic purposes.

Keywords: carbon nanodots; phytomedicine; polymeric carbon nanodots; drug delivery; CNDs

Academic Editor: Iolanda De Marco

Received: 11 November 2024
Revised: 31 December 2024
Accepted: 15 January 2025
Published: 29 January 2025

Citation: Debnath, R.; Ikbal, A.M.A.; Ravi, N.K.; Kargarzadeh, H.; Palit, P.; Thomas, S. Carbon Nanodots-Based Polymer Nanocomposite: A Potential Drug Delivery Armament of Phytopharmaceuticals. *Polymers* **2025**, *17*, 365. https://doi.org/10.3390/polym17030365

1. Introduction

1.1. Background

A class of nanomaterials known as carbon nanodots (CNDs) has caught the attention of scientists because of its special qualities and broad range of uses. The amazing features of CNDs, including their programmable surface chemistry, biocompatibility, and bio-adsorption, have led to increased interest in CNDs' application in phytopharmaceutical drug delivery [1]. However, there are still flaws and shortcomings in the recent way that drugs are currently delivered for different dosage forms, such as problems with therapeutic efficacy, sustained release, and specificity. Carbon nanodots have attracted considerable attention owing to their distinctive advantages compared to other formulations. These nanodots possess unique properties, such as adjustable size, surface chemistry, and fluorescence characteristics, rendering them highly versatile for diverse applications. Additionally, their biocompatibility is noteworthy, with certain carbon nanodots demonstrating minimal toxicity, a vital feature for their potential biomedical use. Importantly, their synthesis is relatively straightforward, often achievable from readily available starting materials through simple methods. These attributes contribute to the growing interest in carbon nanodots across various fields, promising novel solutions for a wide range of applications. These tiny carbon-based particles usually exhibit a spherical or quasi-spherical morphology and range in size from 1 to 10 nanometers [2,3]. Carbon nanodots, which are made up of carbon atoms arranged in a sp^2-hybridized structure similar to graphene, have attracted interest from a variety of scientific and technological fields [3]. Typical precursors comprise organic compounds such as glucose and citric acid, which offer a carbon-rich base for the creation of nanodots [4]. The size, surface chemistry, and the optical properties of the resultant carbon nanodots are all significantly influenced by the production process. The extraordinary optical qualities of carbon nanodots are one of their most notable characteristics. They are especially well-suited for use in bioimaging and noninvasive medical diagnostics because of their high fluorescence [5]. Their biocompatibility and imaging abilities suggest that they may be useful for targeted medication delivery and non-invasive diagnostics. Carbon nanodots have a high surface area and a distinct electrical structure that make them attractive candidates for a variety of uses [1]. Their surface is easily functionalized with various chemical groups, increasing their adaptability to many applications. Carbon nanodots have proven effective in sensing applications for a variety of analytes, such as metal ions, pH variations, and biomolecules. Surface functionalization is a simple process that enables customized interactions with particular target molecules, making them a useful platform tool in the production of sophisticated sensors for healthcare, environmental monitoring, and other industries [6]. Another application where carbon nanodots hold promise is catalysis. CNDs are effective catalysts for a variety of chemical processes due to their large surface area and distinctive electrical characteristics. Their catalytic potential is being investigated by researchers in a variety of processes, including environmental remediation and conventional chemical synthesis [7]. Carbon nanodots' versatility in catalysis highlights the contribution they provide to the advancement of effective and sustainable chemical transformations. Carbon nanodots are used in energy-related sectors in addition to sensing and catalysis. They are intriguing possibilities for usage in solar cells and supercapacitors due to their distinct electrical characteristics [8]. Researchers are examining strategies that exploit the capabilities of CNDs to optimize energy storage devices' efficiency and boost solar cells' performance. Carbon nanoparticles' biocompatibility and adjustable characteristics make them attractive for use in a variety of cutting-edge technologies [9]. A crucial feature of carbon nanodots that increases their potential for biological and medicinal applications is their biocompatibility. Because of their low toxicity, CNDs can be used in drug delivery systems to deliver therapeutic molecules to precise sites in a regulated way [10]. Their capacity to functionalize their surface with targeted ligands improves their precision in drug administration and presents a viable way to raise therapeutic treatment efficacy while lowering negative effects. The goals of carbon nanodot research continue to be improving synthesis techniques, finding new uses, and maximizing the qualities of the material for particular purposes. Current studies seek to learn more about the basic properties of CNDs, which will open the door to advancements in a variety of industries, including energy storage, sensing technologies, medicine, and catalysis. Carbon nanodot research is interdisciplinary, which highlights how revolutionary they can be for a variety of industries and how they can help create new kinds of innovative materials with never-before-seen capabilities [11]. Below, Figure 1 offers the different types of nanomaterials used for drug delivery.

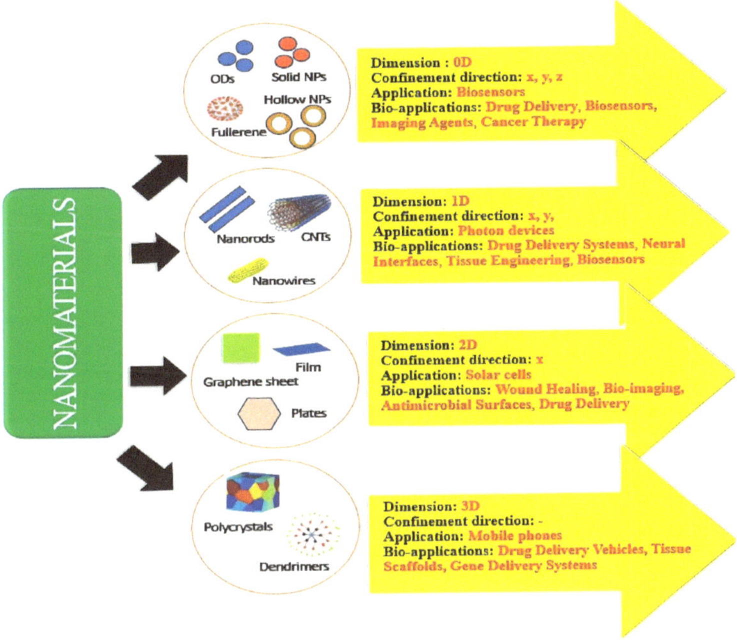

Figure 1. Different types of nanomaterials used in biomedical applications.

1.2. Significance of CNDs in Drug Delivery Systems

In contemporary medicine, drug delivery systems are essential because they provide creative ways to improve the efficacy and security of therapeutic interventions. The importance of drug delivery systems resides in their capacity to resolve several issues related to traditional drug administration, which eventually enhances treatment outcomes and patient experiences [12]. These technologies seek to improve drug bioavailability, target distribution to particular body locations, and optimize drug pharmacokinetics. Achieving the best possible drug concentration at the intended location with the least amount of systemic exposure is a major difficulty in traditional drug delivery. Drug delivery systems make the controlled release possible, guaranteeing a consistent and therapeutically effective drug concentration across the intended time [13]. By doing so, drug level variations are avoided, adverse effects are decreased, and patient compliance is increased. Drugs that are poorly soluble, for instance, can be encapsulated by liposomes and nanoparticles to prevent degradation and enhance absorption. By ensuring that a higher percentage of the supplied medicine reaches the target, this increase in bioavailability maximizes therapeutic effects [14]. This is especially important for treating diseases like cancer when localized therapy is preferred. For example, passive or active targeting methods can be used to create nanoparticles and liposomes so that they preferentially aggregate in tumor tissues, enhancing drug delivery to cancer cells while sparing healthy tissues. For instance, the blood–brain barrier prevents many medications from entering the central nervous system. Bypassing or penetrating these barriers with drug delivery devices creates new opportunities for treating diseases affecting the brain and other difficult anatomical regions [15]. Drug delivery systems are important for getting beyond the body's physiological barriers. Drug delivery systems are essential to the medical revolution because they improve the accuracy, effectiveness, and safety of therapeutic interventions. Because of its special qualities and biocompatibility, carbon nanodots (CNDs) have become one of the many nanomaterials being investigated for drug delivery [16]. Carbon nanodots' small size promotes interactions at the cellular and molecular levels, so medicines can be delivered to particular tissues or cells with accuracy and efficiency. One important aspect of CNDs' importance for drug administration is their biocompatibility [17]. Developing drug carriers that can successfully navigate the intricate biological environment and deliver therapeutic payloads without causing detrimental side effects requires biocompatibility [18]. The fluorescence in drug delivery systems primarily stems from their optical properties and creates opportunities for theragnostic applications, which combine therapy and diagnostics on a single platform, in addition to helping with bioimaging for real-time drug carrier tracking. Approaches to personalized medicine are made possible by the real-time visualization and monitoring of drug delivery mechanisms [19]. Despite limitations in emission of CNDs, which are usually absorbed by the tissues, Lee et al. demonstrated tissue imaging using nitrogen-doped carbon nanodots using photo acoustic imaging in the near IR region with high resolution. Authors even compared the response of nitrogen-doped CNDs with Gold Nanorods and Microbubbles which are used as contrast agents for photoacoustic imaging. CNDs showed higher heat conversion efficiency and stable photoacoustic signals than the other two contrast agents [20]. Carbon nanodots must be surface functionalized to be customized for a specific drug delivery application. Targeting ligands, medicinal compounds, or other biomolecules can be attached to CNDs by adding different functional groups on their surface; for example, carboxyl groups can be used to link anticancer drugs, and hydroxyl groups can facilitate the binding of DNA or RNA molecules for gene therapy [21]. Moreover, functionalized carbon nanodots can be designed to react to external factors like pH variations or certain enzymes, causing regulated drug release at the intended location. Carbon nanodots are used in drug delivery systems because of their ability to overcome biological barriers [22]. The central nervous system is protected from potentially dangerous substances by the blood–brain barrier (BBB). It is the highly discriminatory semipermeable border of endothelial cells that controls the transfer of solutes and chemicals between the central nervous system and the circulatory system. This protective function simultaneously poses a challenge to the delivery of therapeutic substances to the brain. Due of their small size and adaptable surface characteristics, carbon nanodots (CNDs) can be engineered to pass through the blood–brain barrier (BBB), opening up new therapeutic options for neurological disorders that are otherwise challenging to treat with traditional drug delivery techniques [23]. Additionally, there is plenty of room for loading and transporting therapeutic payloads on carbon nanodots due to their large surface area. CNDs offer a variety of drug-loading techniques, including encapsulation, adsorbing, and covalent attachment. In order to ensure that a sufficient amount of the therapeutic substance reaches the target site to exercise its intended effects, this loading capacity is crucial for optimizing drug delivery systems [24]. Beyond conventional small-molecule medications, carbon nanodots are incredibly versatile. A vast array of medicinal substances, including proteins, peptides, and nucleic acids (such as DNA and RNA), have been demonstrated to be delivered by them via their promise [25]. A major development in the realm of nucleic acid therapies and biological products is the capacity of CNDs to precisely deliver and protect sensitive biomolecules. To sum up, the importance of carbon nanodot-based drug delivery systems originates from their capacity to overcome the obstacles related to traditional drug administration [26]. CNDs are potential vehicles for targeted and regulated drug administration because of their special mix of characteristics: small size, biocompatibility, optical characteristics, surface functionalization, and barrier-crossing abilities. This field's continued study and development has the potential to completely transform medical care by providing more individualized and effective therapeutic interventions with fewer adverse effects and better patient outcomes. The use of carbon nanodots in drug delivery systems offers optimism for the advancement of medicine towards safer, more accurate, and effective treatment modalities as our understanding of them improves [27].

1.3. Emergence of Carbon Nanodots in Drug Delivery

The scientific and medical communities are particularly interested in the use of carbon nanodots (CNDs) in drug administration because they present a fresh and exciting opportunity to improve drug delivery methods [28]. These nanodots are excellent prospects for a variety of biomedical applications, especially in the field of drug administration, due to their unique qualities, which include biocompatibility, biodegradability, eco-friendly, green-synthesis techniques, fluorescence, and a high surface area [29]. Due to their low toxicity and ability to be produced from organic precursors like glucose or citric acid, these nanodots rarely cause harm to biological systems. Their incorporation into medication delivery systems, where materials must effortlessly interact with the complex biological environment without causing harm, depends on their biocompatibility [30]. However, the CND to be used in the clinical settings must undergo rigorous toxicity studies as the toxicity of a CND depends not only on its surface modifications; it also depends on the surface charge and the role of functional group restricted by the said charge. For a CND to be biocompatible and non-toxic, its core must not ionize into a toxic species and lead to in vivo toxicity despite no evidence of toxicity in vitro. Neutral or negatively charged CNDs are found to be less toxic, whereas positively charged CNDs are most cytotoxic by initializing oxidative stress. However, its natural compatibility strongly depends on the precursor and on the production method used. Various kinds of surface functional groups can be obtained by inducing different responses and effects from the biological environment [31]. The fluorescence creates opportunities for theragnostic applications, in addition to allowing for real-time drug carrier monitoring and imaging. Theragnostic is the process of combining therapy and diagnostics into one platform to provide individualized and focused successful medical care. The fluorescence of carbon nanodots makes drug transport processes easier to see, improving the accuracy and effectiveness of therapeutic actions [32]. Charges and BBB-based delivery refers to drug delivery strategies that utilize either electrostatic charges or interactions with the blood–brain barrier (BBB) for targeted delivery of therapeutic agents to specific sites, particularly the brain. Utilizing charges and the blood–brain barrier (BBB) for drug delivery offers significant advantages in targeting neurological disorders. Drugs can be precisely delivered to the brain by taking advantage of charge-based interactions, and doing so maximizes therapeutic efficacy and minimizes systemic side effects. Furthermore, BBB-based delivery techniques allow medicines to pass through the barrier, removing a significant obstacle in the treatment of disorders of the central nervous system [33]. Surface functionalization is an important factor in the adaptability of carbon nanodots in drug delivery. Different functional groups can be added to the surface of CNDs with ease, enabling the attachment of medicinal compounds, targeted ligands, or other biomolecules. Drug carriers with particular affinities to target cells or tissues can be designed with this customization, which enhances drug delivery selectivity and reduces off-target effects [28]. Moreover, functionalized carbon nanodots can be designed to react to outside stimuli like pH variations or certain enzymes, causing regulated drug release at the intended location. The ability to interact at the nanoscale provided by the small size of carbon nanodots makes medication administration more accurate and effective by enabling the distribution of therapeutic substances to target tissues. Due to their small size and specific surface characteristics, nanoparticles can enter anatomical regions that would otherwise be inaccessible for targeted medication delivery. In difficult situations such as neurodegenerative disorders, nanoparticles provide a viable avenue for therapeutic treatments by utilizing both active surface changes and passive mechanisms, such as size and the Enhanced Permeability and Retention (EPR) effect [30]. This capacity is especially important for treating neurological illnesses, as traditional drug delivery techniques might not be able to get past these obstacles. Therapeutic payloads can be loaded and transported with plenty of room due to the high surface area of carbon nanodots. Different approaches for drug loading are available depending on whether the drug is covalently bonded, adsorbed, or encapsulated on the surface of CNDs. To maximize drug delivery systems and make sure that a sufficient amount of the therapeutic agent reaches the target site to have the desired effects, this loading capacity is crucial. Moreover, carbon nanodots' growing importance in drug delivery is attributed to their ability to transport a diverse array of therapeutic molecules [34]. CNDs have demonstrated potential in delivering nucleic acids (such as DNA and RNA), proteins, and peptides in addition to conventional small-molecule medicines. This potential creates novel therapy options for cancer, genetic abnormalities, and other illnesses where the effectiveness of traditional medication delivery techniques may be constrained [25]. A major development in the realm of nucleic acid therapies and biologics is the capacity of CNDs to precisely deliver and protect sensitive biomolecules. To sum up, the application of carbon nanodots in drug delivery is a revolutionary step forward for the field of nanomedicine. Their nanoscale size, surface functionalization abilities, optical characteristics, and biocompatibility make them useful and efficient carriers of medicinal drugs. Carbon nanodots have the potential to completely transform drug delivery systems as long as an investigation into this field is conducted. They can provide novel approaches for the precise, targeted, and controlled delivery of medicines of various medical applications [35] in both life-threatening and chronic illnesses, including cancer, viral diseases, neurodegenerative disorders, etc.

2. Phytomedicine in Drug Delivery

A timeless field with roots in ancient wisdom, phytomedicine is embracing the cutting edge of modern drug delivery technologies while navigating the prevailing modern currents [36]. The combination of current novel drug delivery technology and phytomedicine may explore the Carbon Nanodots-Based Drug Delivery of Phytomedicine, lies at the crossroads of history and innovation, opening the door to therapeutic techniques that go beyond conventional practices [37]. The range of options available to us in phytomedicine becomes clear as we explore further. Beneath the abundance of bioactive substances is a dynamic interplay of holistic healing, an idea engrained in conventional medical practice. With a focus on overall health, phytomedicine aims to treat illnesses' underlying causes, as well as their symptoms [38]. The symbiotic link between sophisticated medication delivery technologies and phytomedicine is a prominent feature of the modern period [39]. Although the ancient healers had an innate understanding of the power of plant treatments, current technology allows us to fully realize their potential through targeted and deliberate delivery methods [40]. This combines the best aspects of conventional medicine with the cutting-edge techniques of contemporary medication delivery.

2.1. Carbon Nanodots (CNDs): Catalysts of Transformation

Carbon nanodots (CNDs), heralded as revolutionary agents in drug delivery, are fundamental to this metamorphosis. CNDs have their roots in nanotechnology and offer of several unique features. They are ideal phytomedicinal chemical carriers due to their large surface area, nanoscale particle size, and configurable surface functionalities. This increases the bar for phytomedicine since CNDs' biocompatibility and flexibility allow for more focused, accurate, and efficient drug administration [41].

2.1.1. Precision Delivery for Optimal Impact

The use of CNDs in medication delivery systems is going to revolutionize precision medicine. Targeting the delivery of phytomedicine to specific tissues or cells becomes possible. The utilization of ancient plant knowledge in conjunction with nanoscale precision maximizes therapeutic effects while mitigating adverse consequences [42]. Theoretical capabilities of traditional healers become tangible with the tailored delivery mechanisms made possible by CNDs [43,44]. The sophistication that may be gained when innovation and tradition come together is demonstrated by this exact delivery method.

2.1.2. Addressing Bioavailability Challenges

Though widely recognized for its comprehensive healing properties, phytomedicine encounters absorption challenges coupled with pharmacokinetic failure that could potentially diminish its therapeutic impact [45]. In this situation, CNDs are crucial for overcoming these barriers. Due to its ability to conjugate or encapsulate phytomedicinal compounds in nanodot flatform, it ensures that these bioactive substances are more soluble and stable and reach their target in a form that body can easily absorb [46]. The synergistic combination of phytomedicine and CNDs provides a modern answer to the historical problem of realizing the full potential of plant-derived therapies. CNDs are sentinels that protect stability and prolong the duration of therapeutic efficacy. CNDs provide protection against degradation, which has the potential to diminish the effectiveness of phytomedicines over time. These nanoscale entities stabilize the plant and guarantee that its medicinal properties hold true over time [47]. The resulting synergy bears witness to phytomedicine's adaptable character, which welcomes innovation while maintaining its core principles. The voyage transcends the traditional and modern dichotomies as we navigate this complex confluence, creating a story where the best practices from each era join together to provide the best possible therapeutic results. The investigation of "Carbon Nanodots-Based Drug Delivery of Phytomedicine" turns into a journey not just through the pages of history and the halls of contemporary science, but also towards a future in which the harmonious fusion of traditional knowledge and cutting-edge technology shapes the face of healthcare.

2.1.3. Guardians of Therapeutic Efficacy

Stability is essential for any drug delivery system to work, and in this regard, CNDs act as guardians. By stopping the degradation of phytomedicinal compounds, CNDs extend the duration of therapeutic activity. Although this property is not observed for every phytomedicine-derived CNDs, CND formulation of phytomedicine. This was observed in case of *Asparagus racemosus* [48]. This property is not observed in case of every CND derived from phytomedicine. But still CNDs act as a guardian of therapeutic efficacy. Despite absence of all anti-cancer active components in ginger–CDs except curcumin, they executed anti-cancer activity when prepared via hydrothermal method. Furthermore, Artemisiae Argyi Folium (AAF) was not indicated traditionally for anti-frostbite effects, still the CDs showed anti-frostbite activity by improving local inflammation and reducing blood glucose levels caused by frost bite despite the absence of isochlorogenic acid which is the major composition of AAF [49]. This stabilization ensures that phytomedicine's therapeutic potential persists throughout time, offering a dependable and durable solution for a variety of medical needs. In general, a high temperature-induced hydrothermal or calcination process degrades thermos-degradable flavonoids, terpenoids, and polyphenols to synthesize macromolecular carbon polymeric composites. The functional group present in the alkaloids, terpenoids, and flavonoids can functionalize the surface of polymeric carbon nanodots to offer significant and improved biological response.

Therefore, it is clear from this combination of phytomedicine and enhanced drug delivery, especially when seen through the lens of carbon nanodots-based drug delivery, that this is both a scientific study and a reflection of how medicine is evolving. This symbiotic partnership, with its traditional foundation and new drive, has the power to fundamentally alter the way we view healing. As we traverse this complicated intersection of conventional treatments and cutting-edge technologies, we get a glimpse of a future in which the best features of both worlds come together to give the best possible therapeutic outcomes.

2.2. *Overview of Phytomedicine*

In order to properly contextualize the intricate tale of "Carbon Nanodots-Based Drug Delivery of Phytomedicine", it is imperative that we first discuss the historical background and progression of phytomedicine. This overview serves as a bridge between the State-of-the-Art findings in the ground-breaking field of carbon nanodots (CNDs) and the conventional wisdom of plant-based therapy.

2.2.1. Historical Roots and Evolution

Throughout the ages of human civilization, phytomedicine has left enduring cultural imprints. Because they possessed an instinctive understanding of nature's pharmacy, healers from ancient times have exploited the medicinal properties of plants for millennia [50]. From Native American herbalists to Indian Ayurvedic sages, the lengthy history of phytomedicine attests to its ongoing efficacy. But as we learn more about "Carbon Nanodots-Based Drug Delivery of Phytomedicine", the historical continuum becomes more meaningful. It becomes proof of how flexible phytomedicine is, fitting in seamlessly with the evolving landscape of medical practices.

2.2.2. Bioactive Compounds in Phytomedicine

Examining the enormous pool of bioactive compounds found in phytomedicine reveals an unparalleled diversity in plant pharmacopoeia. Alkaloids, flavonoids, terpenoids, and polyphenols come together to provide a complex mosaic of therapeutic effects [51]. This symphony of bioactive molecules, with origins in conventional medicine, paves the way for the inclusion of carbon nanodots. Every class of chemicals engages in a cooperative interplay within the plant matrix, in addition to possessing distinct therapeutic characteristics. The originality of CNDs is based on the synergy that is the hallmark of phytomedicine.

As we explore the complex environment, it becomes clear that the bioactive substances in phytomedicine are not merely stand-alone treatments but rather integral parts of a whole healing process. Moreover, the interdependence within the plant matrix becomes a driving force behind the integration of contemporary medication delivery systems.

2.3. *Merits and Demerits of Phytomedicines*

2.3.1. Advantages of Phytomedicine

Phytomedicine is a medicinal method with several benefits that is based on the use of chemicals produced from plants [52]. These therapeutic substances have a strong cultural and traditional legitimacy due to their natural origin and long history of use. Plants include a wide range of bioactive chemicals, such as alkaloids, flavonoids, terpenoids, and polyphenols, which together contribute to their diverse and abundant medicinal capabilities [53]. This diversity highlights the holistic character of phytomedicine by enabling the treatment of various illnesses with a single plant or extract. Furthermore, in keeping with the tenets of personalized medicine, the holistic healing approach goes beyond treating symptoms to address the root causes of illnesses. When it comes to side effects and safety, phytomedicine frequently has an advantage over synthetic medications. The complex interplay between the many chemicals found in plants frequently reduces the negative effects of their isolated synthetic counterparts. Moreover, phytomedicine is an affordable and easily accessible alternative to traditional medicine in areas where healthcare practices are strongly rooted in traditional medicine [54]. Phytoextracts are a rich source of natural substances that can be used as carbon sources for nanoparticle synthesis, which is why they are used as precursors for the development of Carbon Nanodots (CNDs). Many organic components found in phytoextracts, including proteins, phenolic compounds, and carbohydrates, can be carbonized to produce polymeric carbon-based nanoparticles like CNDs, using a hydrothermal calcination process. This strategy provides an environmentally benign and sustainable way to synthesize

2.3.2. Challenges of Phytomedicine

Despite all of its benefits, phytomedicine faces a number of obstacles that require careful thought [55]. Among these difficulties, standardization and quality control rank first. The natural fluctuation in plant growth environments, harvesting schedules, and extraction techniques makes it more difficult to create constant potency and effectiveness. Strict protocols are therefore required to guarantee consistency throughout various phytomedicine batches. One of the main obstacles to the widespread application of phytomedicines in conventional medicine is the lack of knowledge regarding their complex mechanisms of action [56]. To properly understand these systems and incorporate them into traditional healthcare methods, more research is necessary. Another obstacle to the efficient application of phytomedicines is their bioavailability. When taken orally, many bioactive chemicals have low bioavailability, which restricts their potential applications as therapeutic agents. Developing novel strategies to improve these chemicals' availability and absorption within the body is necessary to meet this problem. Furthermore, much research is required to clarify and appropriately manage the possible interactions between phytomedicines and conventional pharmaceuticals [57]. Cultural prejudices and the various legal systems around the world could make it difficult for phytomedicine to become widely accepted, necessitating coordinated efforts to close these gaps.

2.4. Advantages of Carbon Nanodots-Based Drug Delivery for Phytomedicine

The incorporation of carbon nanodots into phytomedicine drug delivery systems represents a paradigm change, opening up new avenues for overcoming long-standing obstacles. There are numerous benefits, the most notable of which is the notable increase in bioavailability. Carbon nanodots have the capacity to enhance the solubility and stability of phytomedicines, thereby guaranteeing the best possible absorption and utilization within the body [58]. This discovery could greatly increase the medicinal benefits of substances obtained from plants. One of the most revolutionary benefits is that carbon nanodots can deliver drugs specifically to the right places. By means of careful engineering, these nanodots can be precisely engineered to overcome physiological barriers and target individual tissues or cells. By limiting off-target effects, this focused delivery maximizes therapeutic benefit and minimizes negative outcomes. By extending the release of phytomedicines over prolonged periods of time, carbon nanodots' controlled-release properties facilitate therapeutic optimization even more. In addition to increasing efficacy, this regulated distribution encourages patient compliance, which is essential for positive therapeutic results. In addition, the combination of phytomedicines and carbon nanodots offers the exciting possibility of synergistic effects. The utilization of these nanoscale carriers in conjunction with the complex blends of bioactive compounds present in plants has the potential to yield enhanced therapeutic effects, perhaps outperforming traditional drug delivery methods [40]. CNDs' surface functional groups, such as carboxyl, hydroxyl, and amino groups, render suitable anchoring with the bioreceptors in the target zone to elicit impressive pharmacological action. Positively charged CNDs can effectively electrostatically bind with negatively charged bacterial membranes to attenuate the microbial infection very rapidly [59]. This positively charged surface of CNDs also assists them to penetrate the BBB to interact with negatively charged beta-amyloid precipitated protein to attenuate the neurodegenerative disorders to a greater extent [60].

2.5. *Challenges of Carbon Nanodots-Based Drug Delivery for Phytomedicine*

Although carbon nanodots have several benefits in phytomedicine, integrating them is a difficult task. The most important of these is the absolute necessity of a thorough safety assessment. It is crucial to evaluate the possible toxicity and enduring impacts of carbon nanodots on human anatomy prior to their extensive clinical usage. Logistically difficult, the complicated synthesis of carbon nanodots demands careful attention to quality control and reproducibility when scaling up production for commercial application [61]. Clearance and biocompatibility concerns are important aspects that need careful examination. For safe and efficient use, it is essential to make sure that carbon nanodots blend in with the biological environment and to comprehend what happens to them inside the body. The complex regulatory environment around nanodot-based drug delivery devices calls for the creation of particular approval criteria [62]. Strict adherence to these recommendations is necessary to guarantee the moral and secure incorporation of this cutting-edge technology into standard medical procedures. Therefore, integrating carbon nanodots into phytomedicine drug delivery systems is a revolutionary step towards addressing long-standing obstacles. Although the benefits are great, resolving safety issues, producing nanodots on a large scale, and negotiating regulatory hurdles are essential stages in bringing this novel method to its full potential in clinical settings [63]. In order to guarantee the smooth integration of carbon nanodots with phytomedicine and launch a new era of precision medicine, a multidisciplinary and thorough approach is essential.

2.6. *Role of Drug Delivery Systems in Enhancing Phytomedicine Efficacy*

The medicinal potential of phytomedicine, which comes from plant sources, has been highly esteemed [64]. On the other hand, intrinsic difficulties such as low bioavailability, restricted targeting, and regulated release may compromise its effectiveness. One revolutionary way to address these issues and greatly improve the overall effectiveness of phytomedicine is the use of cutting-edge drug delivery systems, with a focus on those based on carbon nanodots [65].

2.6.1. Enhanced Bioavailability

Low bioavailability is a common problem for phytomedicines, mostly because of things like poor solubility. Because of their distinct surface characteristics and nanoscale size, carbon nanodots are effective carriers to overcome this constraint. By increasing the solubility of hydrophobic phytomedicine components, they can optimize bioavailability by facilitating improved absorption and distribution within the body [66].

2.6.2. Targeted Drug Delivery

The remarkable ability of carbon nanodots for targeted drug delivery is an important factor in optimizing treatment efficacy while reducing side effects. These nanodots' functionalization and surface alterations enable exact customization to target particular tissues or cells [22,67]. By ensuring that phytomedicines reach their intended site of action, this focused method maximizes therapeutic benefits and reduces off-target effects. By linking targeting molecules, such as antibodies and receptor ligands, to carbon nanodots help them locate specific cells in order to achieve targeted drugs delivery. Their surface is hydrophilic, which helps them attach to biological tissues in the body and avoid immune system clearance. By ensuring that the drug is released precisely where it is needed, innovative compounds that respond to pH changes can improve treatment and minimize adverse effects. Phytomolecule-derived pegylated CNDs can stabilize much better than empty CNDs without phytocompounds. Moreover, very low-sized carbon nanodots ranging from 10 to 30 nm, as evident from the literature, easily can uptake by the reticular endothelial cellular system without breakage [19] or damage for better efficacy including predicted delivery in low doses and frequency [68].

2.6.3. Controlled Release

Attaining long-term therapeutic concentrations is essential for maximum effectiveness. One unique benefit of carbon nanodots is their ability to release phytomedicines under controlled conditions. Modulating the release kinetics guarantees a steady and progressive delivery, which not only improves efficacy but also encourages patient compliance by lowering the frequency of doses [58].

2.6.4. Synergistic Effects

There is a chance that the combination of phytomedicines with carbon nanodots will have synergistic effects. Increased stability and activity may result from interactions between nanodots and the various bioactive substances found in phytomedicines [48]. When combined, these components have the potential to produce more potent therapeutic effects than those of single or conventional drug delivery methods.

2.6.5. Improving Cellular Uptake

Carbon nanodots' distinct surface characteristics and nanoscale size allow for improved cellular absorption of phytomedicines. This is especially important when it comes to intracellular targets. As effective carriers, the nanodots can help phytomedicines get through cell barriers and increase their total cellular bioavailability, which boosts their therapeutic effect. CNDs enhance cellular absorption of phytomedicines, potentially aiding uptake by macrophages and overcoming cellular barriers and could potentially aid in the uptake of phytomedicines by macrophages and other reticular endothelial system constituents [69].

Carbon nanodots (CNDs) have the potential to modulate and refine the immune system's responses. They can provide therapeutic benefits in diseases involving immunological dysregulation, such as infections or autoimmune diseases, by regulating immune activity and stimulating it when needed to avoid overreactions. Notwithstanding these positive results, more investigation is necessary to completely understand the immunomodulatory effects of CNDs and maximize their use in clinical contexts. Gaining insight into the ways in which CNDs engage with the immune system may open up new options for treatment approaches and illness control techniques [70].

2.6.6. Biocompatibility and Reduced Toxicity

Biocompatibility is essential for any medication delivery device to work. When thoughtfully developed, carbon nanodots exhibit low toxicity and great biocompatibility [71]. Their compatibility with biological systems reduces the possibility of negative consequences, and their biodegradability guarantees low long-term influence, all of which add to the overall safety of the phytomedicine delivery process. Specific solutions, such as surface modifications of carbon nanodots to reduce toxicity and enhance interaction with biological tissues helps in solving biocompatibility issues. Polyethylene glycol (PEG) is frequently applied to the surface of CNDs as a typical method of PEGylation, which reduces toxicity and increases stability in biological system. Moreover, CNDs can be coated with biopolymers, including hyaluronic acid or chitosan, to improve their uptake and decrease immunological responses by increasing their contact with cells. By functionalizing CNDs with targeting ligands, such as peptides or antibodies, they can bind to particular tissues or cells more specifically, minimizing off-target effects and enhancing biocompatibility [69]. The efficacy and safety of CNDs in drug delivery applications are greatly enhanced by these strategies.

2.6.7. Personalized Medicine Approach

The adaptability of carbon nanodots makes personalized medicine possible, enabling the customization of drug delivery systems according to the unique characteristics of each patient [72]. By tailoring the phytomedicine to each patient's unique needs and response, it might potentially minimize side effects and maximize therapeutic success.

To summarize, the application of drug delivery systems, especially those that make use of carbon nanodots goes beyond the traditional constraints connected to phytomedicine. Carbon nanodots present a novel platform for improving the effectiveness of phytomedicines by resolving issues with bioavailability, targeted distribution, controlled release, and promoting synergistic effects. In the field of natural therapies, this not only provides answers to current problems but also opens doors to a new era of precision and personalized therapy. It is highly promising that the integration of these cutting-edge drug delivery methods will transform the therapeutic landscape and enable phytomedicine to reach its full potential in clinical applications.

3. Carbon Nanodots: Properties and Synthesis

Carbon nanomaterials have been attracting a great deal of research interest in the past decades due to their unique properties such as biocompatibility, nontoxicity, high mechanical and thermal properties, easy functionalization, and possessing of photoluminescent nanomaterials. It also has strong absorption, high quantum yield, shows promising optical, energy, and biomedicine applications. Among all carbonaceous nanomaterials, carbon dots (CNDs), were first obtained during the purification of single-walled carbon nanotubes through preparative electrophoresis by Scrivens and co-workers in 2004 [73].

Over time, CNDs have been proven to have many excellent properties. Additionally, increasing interest has been observed in the utilization of CNDs in the fields of energy and catalysis [74,75], biological labelling, bioimaging, and gene/drug delivery [76,77]. This increased interest is attributed to their unique properties, such as water-solubility, low toxicity, high chemical stability, and easy functioning.

3.1. Sources, Structures, and Properties of Carbon Nanodots

Carbon-based substances like fruit, botanical extracts and citric acid are known as organic precursors in the synthesis of carbon nanodots (CNDs). On the other hand, inorganic precursors provide other elements that influence the properties of CNDs. The flexibility and environmental friendliness of CND synthesis techniques are demonstrated by the use of a variety of feedstock, including green carbon precursors like fruit peels and kitchen trash. Various chemical precursors have been identified for the synthesis of CNDs, including organic and inorganic precursors, such as ammonium citrate [78], ethylene glycol [79], citric acid [80], Ethylene diamine tetra acetic acid (EDTA) [81], phytic acid [82], phenylenediamine [83], thiourea [84], carbon nanotube [74,75], graphite [85], etc. Meanwhile, a large number of green carbon precursors have been used to produce CNDs, including fruits, fruit juices, and fruit peels [86–88]; animal and animal-derived materials, such as chicken eggs [89] and silkworm [90]; vegetables and spices [91], waste kitchen materials, like frying oil [92] or waste paper [93]; plant leaves and derivatives [94]; etc.

It is worth noting that, while CNDs typically showcase a dot-like structure in TEM, introducing unconventional precursors like rods, ribbons, and triangles can yield unexpected outcomes. Although this diversifies the appearance of CNDs, it concurrently heightens the challenge of uncontrollable preparation. Successful management of this process necessitates accurate prediction of CND formation and strategic selection of precursors [95].

CNDs are classified into three main groups, graphene quantum dots (GQDs), carbon nanodots (CNDs), and polymer dots (PDs) [96]. GQDs possess a crystalline structure and consist of a single or a few graphene layers with π conjugation. Chemically, it has more functional groups on the edges and shows an anisotropic shape, with lateral dimensions larger than the height. Mostly, these are synthesized in a circular or elliptical shape, whereas, CNDs are always spherical. CNDs are divided into two subgroups including carbon nanoparticles (CNPs) and carbon quantum dots (CQDs).

CNPs are amorphous without a clear crystal lattice and do not have any quantum confinement effect (QCE) while CQDs contain an obvious crystal lattice with sp^2/sp^3 carbons and have a quantum confinement effect [96,97]. It should be kept in mind that QCE is not dependent on the presence or absence of crystalline structure. When system dimensions are smaller than Bohr's radius of excitation, conduction and valence bands change from continuous to discrete energy levels restricting the movement of electrons in all directions. This leads to the quantum confinement effect (QCE). When size is higher than 5–7 nm, the particle behaves as a bulk and quantum confinement effects are lost [96]. CNDs have various functional groups on the surface, which play a leading role in the photoluminescence (PL) behavior of these CNDs. These functional groups could be oxygen-based, amino-based groups, polymer chains, etc. PDs are formed by the aggregation or crosslinking of monomers or linear polymers [98,99]. Figure 2 illustrates a typical structure of CNDs, GQDs, and PDs.

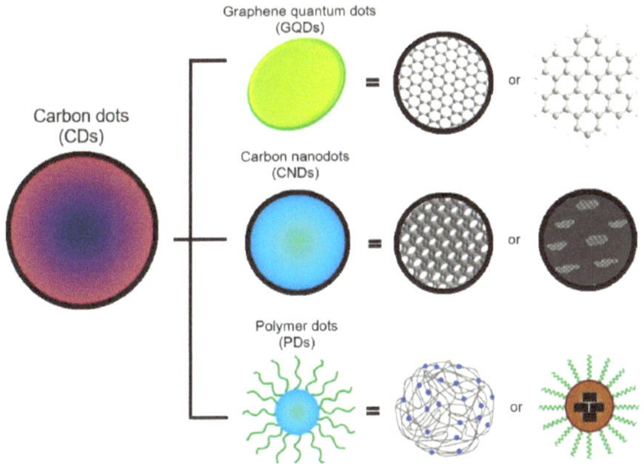

Figure 2. Schematic image of CNDs, GQDs, and PDs structure. Reproduced from ref. [99]. Copyright 2014. With permission from Springer [license no. 5945400714846].

One of the most interesting properties that CNDs have is their PL. It has been reported that CNDs are effective in photon-harvesting in the short-wavelength region (260 to 390 nm) because of $\pi-\pi^*$ transition of C=C bonds and n-π transition of C=O bonds. It is also a known fact that CNDs can emit wavelengths around the whole spectrum of visible light.

However, the PL of CNDs, particularly in GQDs and CQDs, is generally influenced by the QCE and surface states. These states encompass functionalities, defects, heteroatom doping, edge configurations and synthesis methods. The photophysical characteristics of CNDs, stemming from the QCE, are also significantly influenced by particle size. An increase in the size of CNDs leads to a proportional growth in both absorption and emission wavelengths. This size-dependent PL phenomenon is attributed to a reduction in the bandgap, a consequence of π-electron delocalization. Thus, when a particular emission wavelength is desired for CNDs, it necessitates specific shapes and sizes. Achieving this requires a meticulous selection of synthesis methods and precursors that offer control over the synthesis process for CNDs [100,101]. Additionally, the quantum yield (QY) which is related to the emission wavelength is low for CNDs and they are more efficient in absorption of long wavelengths. Whereas, GQDs have higher QYs due to the layered structure and better crystallinity. The QYs of CNDs depend on the synthesis method and carbon source. For instance, CNDs prepared with aliphatic compounds as the carbon source exhibit blue-green luminescence with the QYs of 100% [102], while those prepared with aromatic compounds exhibit yellow emission [103]. CNDs with red emission can be obtained by the addition of acid into the reaction mixture or using a different solvent, such as DMF, while the QYs are very low [104]. However, surface passivation is an efficient technique to improve the QYs and brightness of CNDs [97,105]. The optical properties and PL mechanism of the CNDs are extensively discussed in [95,99,105–109].

3.2. *Synthesis and Production of Carbon Nanodots*

CNDs can be prepared by two main synthetic methods: "top-down" and "bottom-up" methods or in another type of classification, chemical, and physical methods. Each of these methods has some pros and cons which can be chosen regarding the application.

The bottom-up method refers to a synthesis approach where nanomaterials are built by assembling smaller molecular or atomic precursors into larger structures. In this method, small organic molecules or atoms undergo chemical reactions to form complex nanostructures, such as carbon dots (CDs), gradually growing from the "bottom" (small units) to the "top" (complete nanostructure). The bottom-up techniques lie in chemical synthesis and employ pyrolysis and some chemical reactions in order to carbonize small organic molecules. Hydrothermal treatment, microwave, thermal decomposition, template routes, and plasma treatment are examples of bottom-up techniques to produce CNDs in which the precursor shows lower requirements of carbon sources. Bottom-up methods have the advantages of simple surface modification in one-pot and synthesis of variety of structures and functionality of CNDs. However, it is harder to modulate since side products can form that require further purification compared to top-down methods.

The top-down methods, by which CNDs are generally formed through the electrochemical, chemical or physical cutting processes of relatively microscopic carbon structures, such as graphene oxide, carbon nanotubes, graphite powders, etc., and the obtained CNDs resemble the structure of their precursors. The advantages of the top-down methods are the possibility of scaling up the production of CNDs and well-defined CND structures. Top-down methods include laser ablation, electrochemical oxidation, chemical oxidation, and ultrasonic synthesis. In the subsequent sections, a concise introduction to the selected top-down and bottom-up methods is provided.

Laser ablation is a straightforward method for producing CNDs in which carbon precursor is irradiated by a laser. It is environmentally friendly and scalable production method however the energy consumption is high, QY is low, and cannot assure control over the size of the nanoparticle [110–113]. The nucleation and growth of CNDs consequently its size and morphology depend on the laser pulse width. The CNDs' size and crystalline grains increased with increasing the laser pulse width [114].

Electrochemical synthesis is another top-down method which can be applied to different bulk carbon materials, such as graphite rod, carbon fiber, and carbon paste. In this method, two carbon-based electrodes are immersed in a water-based electrolyte. By applying redox potential, water electrolysis and H and OH radicals initiate at the edges, and the defects of electrodes act as electrochemical scissors to form CNDs. Similar to laser ablation, this method is also low-cost, scalable, and environmentally friendly, but it has poor size control. However, by tuning the parameters such as applied potential or starting materials CNDs with narrow size distribution can be obtained [115,116]. The major drawback of this method is that it is a tedious purification process, as the obtained CNDs via electrochemical oxidation are highly hydrophilic.

In chemical oxidation, a carbon precursor is oxidized by a strong oxidant to carbon dots. Various oxidants have been used to produce CNDs, such as acids (HNO_3, $H2SO_4$, $NaNO_3$, etc.), a mixture of $H_2SO_4/KMnO_4/H_2O_2$ [117], H_2O_2 [118] $NaOH/H_2O_2$ [119], and ozone [120]. The oxidation reaction usually requires purification in the case of acidic oxidation in order to remove the acid. Although chemical oxidation is an easy and low-cost method to produce CNDs at a large scale with high QYs, the lack of homogeneity in the size distribution of the resulting CNDs and the risk of burning or explosion are its major disadvantages [121].

Using ultrasonic technology, carbon dots (CNDs) are made by subjecting the reaction mixture to high-intensity ultrasonic waves. This causes cavitation and high-pressure vapor to develop, which facilitates the synthesis of CNDs [122]. This approach presents several benefits, including its affordability, ease of use, and the ability to adjust emission wavelengths to suit different synthesis conditions [123]. However, the efficacy of this method relies on factors such as ultrasonic duration, intensity, and temperature. Variables such as ultrasonic duration, intensity, and temperature can influence various properties of CNDs, including their size, emission wavelength, and surface characteristics. Longer ultrasonic exposure may result in larger CND sizes due to prolonged nucleation and growth, whereas higher intensities can augment cavitation intensity, impacting CND formation [124]. The method's affordability and simplicity make it a viable option for CND synthesis. Additionally, the ability to adjust emission wavelengths enhances its versatility for different applications [123]. Optimizing ultrasonic parameters can pose challenges, as it may require intricate adjustments and consume time. Furthermore, this technique may lack the precise control over reaction conditions offered by more sophisticated methods [124]. Despite these difficulties, the ultrasonic approach is favored due to its simplicity, cost-effectiveness, and capacity to alter CND characteristics. Although the approach has significant boundaries that may require careful parameter optimization to achieve the desired results, it is nevertheless a popular alternative for creating CNDs. The most common carbon sources for these techniques are carbon nanotubes, activated carbon and graphite.

Hydro/solvothermal methods are widely employed for the synthesis of carbon dots (CNDs), In hydrothermal methods, a water solution of the organic precursor is heated in a sealed reactor, while solvothermal methods use high-boiling organic solvents instead of water, offering better control and improved result [125]. Parameters such as temperature, pressure, precursor concentration, and reaction time significantly influence CND characteristics. Higher temperatures and longer reaction times generally result in smaller CND sizes due to enhanced nucleation and growth kinetics. The advantages of hydrothermal synthesis include its low cost, eco-friendliness, and ability to produce CNDs with tailored properties. However, disadvantages include the requirement for specialized equipment and potential challenges in parameter optimization [126]. Recently, spherical water-soluble CQDs (about 1–3 nm) have been prepared from lemon peel waste using a cost-effective hydrothermal approach. The produced CQDs were observed to have a quantum yield of about 14%

and excellent photoluminescence properties with high aqueous stability and oxygen-rich surface functionalities [127]. In another study, fluorescent CQDs (about 260–400 nm) were produced using *Tamarindus indica* leaves via a one-step hydrothermal treatment. These biocompatible CQDs were potentially applied in bio-imaging, disease diagnostics, sensing, and other analytical applications [128]. Citrus lemon juice was also used as a precursor of green synthesis of fluorescent CQDs (2–10 nm), through a one-pot hydrothermal approach. The prepared CQDs exhibited a 10.20% quantum yield. It was reported that the photoluminescence intensity depended on the pH of the solution and maximum intensity was obtained at pH of 6. Synthesized CQDs were also used for cell imaging and compared with the MTT assay to demonstrate its applicability as a fluorescent probe [88]. In another work, amorphous fluorescent CNDs from orange waste peels were produced using the hydrothermal treatment at 180 °C. Consequently, a composite of CNDs with zinc oxide was used as a photocatalyst for the degradation of naphthol blue-black azo dye under UV irradiation, and superior photocatalytic activity was reported [129].

Microwave (MW)-assisted method considered as a green bottom-up technique for CNDs production, providing several benefits including simplicity, speed, and precise control. MW reactors operate on the principles of dipolar polarization and ionic conduction [130,131]. It encompasses a broad spectrum of electromagnetic waves ranging from 1 mm to 1 m, providing accelerated energies ideal for breaking down chemical bonds within precursor molecules. The manipulation of MW parameters significantly impacts the size and attributes of the synthesized CNDs. Adjustments in MW power, synthesis duration, and precursor concentration influence the nucleation and growth kinetics of CNDs, thereby influencing their size distribution [132]. Notably, higher MW power and shorter synthesis durations often yield smaller CND sizes owing to accelerated heating rates and rapid reaction kinetics. Moreover, variations in precursor composition and solvent selection can further adjust CND characteristics, such as fluorescence emission, quantum yield, and surface properties [133]. For instance, it has been observed that the QY of CNDs increased with an increase in MW time, while the QY decreased with the decrease in the MW reaction temperature. MW is a rapid and effective and highly tunable technique for CNDs production. However, the QY is low, and it requires post-modification Moreover, it is high cost and energy consuming. Recently, highly stable and luminescent multicolor CQDs (4.85 ± 0.07 nm) were produced through microwave irradiation of *Cydonia oblonga* powder as a carbon source. Maximum emission intensity at 450 nm was observed for CQDs when it was excited at 350 nm and showed a quantum yield of 8.55% [134].

Pyrolysis/carbonization is considered a bottom-up method for CND production. It involves the decomposition of carbon precursors at high temperatures in the absence of oxygen. Various parameters, such as precursor type, temperature, heating rate, and duration, significantly influence CND size and properties. Higher pyrolysis temperatures typically lead to smaller CND sizes due to increased carbonization efficiency. The advantages of this method include its simplicity, scalability, and the ability to produce CNDs with tailored properties. However, challenges may arise in achieving precise control over CND size and surface functionalities [135].

Other preparation techniques have been also proposed, such as plasma treatment [136], supported synthesis [137–139], solution chemistry methods [140–143], cage-opening of fullerene [144], chemical vapor deposition [145], and template method [146]. The synthesized CNDs typically gain photoluminescence properties via oxidation with nitric acid and surface-passivation via diamine-terminated organic molecules [75]. Interestingly, most of the resulting CNDs exhibit blue or green emissions. The synthesis route and source of carbon are important factors, which affect the size, shape, and properties of the final product. The size of CNDs is very important for understanding quantum phenomena and also for biomedical applications and optoelectronics [107]. Generally, prepared CNDs, regardless of their preparation technique, have a mixture of sizes, which requires complex separation methods to obtain monodisperse CNDs. Some of the post-synthesis separation techniques include dialysis [147], chromatography [148,149] gel electrophoresis [75] and ultra-filtration [150]. Recently, CQDs were prepared by green ozone oxidation of lignite coal, which is abundant and cheap. In this case, also, synthesized CQDs (about 2–4 nm) were observed to be well dispersed in water, contain rich oxygen functional groups and excellent optical properties with a yield of 35%. It was observed that the fluorescence intensity of CQDs has a linear response to the Fe^{3+} concentration varying from 10 to 150 μmol/L. The detection limit was 0.26 μmol/L [119].

3.3. Surface Modification and Functionalization of Carbon Nanodots

Surface modification of CNDs is a popular strategy to tailor their properties for specific applications and to improve their performance. Surface functionalization/passivation and doping are two major techniques to chemically modify the surface of CNDs. CNDs' large numbers of surface functional groups, such as carboxylic groups, hydroxyl groups, amines, etc., make them hydrophilic and facilitate various surface functionalization and passivation. Surface functionalization/passivation of carbon dots refers to the process of treating the surface of CNDs to enhance their stability, biocompatibility, and optical properties. The introduction of various functional groups imposes different defects on the CND surface affecting its excitation energy and leading to large variations in fluorescence emissions. Passivation typically involves the functionalization or coating of the surface of CNDs with various organic, polymeric, inorganic or biological materials. It has several advantages: it prevents the surface defects from which the fluorescence originates; provides reactive sites for surface modification reactions and introduces new properties; enhances the potential application of CNDs for specific sensing, bioimaging, drug delivery, optoelectronics and other specific tasks [151].

The functionalization/passivation strategies can be classified as covalent and non-covalent modification. The covalent modification includes amidation, sulfonylation, esterification, sulfonylation, and copolymerization, while the non-covalent modification includes electrostatic interactions, and π interaction [152]. An effective surface passivation can lead to high fluorescence intensities and high quantum yield of CNDs [112]. Recently, Wang et al. have prepared CQDs with quantum yields of 60% through passivation with PEG_{1500N} [153]. It has been reported that electrochemiluminescence activities depended on the surface passivation of CQDs. Surface passivation reduces the electrochemiluminescence activities and enhances the fluorescence properties [154]. Meanwhile, ultra-small CNDs (0.9 nm) with QY of 47% have been produced through pyrolysis of anhydrous citric acid in organosilane at 240 °C for 1 min. The CNDs functionalized with organosilane can be directly fabricated into a hybrid fluorescent film or monolith via a simple heating process, without using any additional components. It can be further converted into silica-encapsulated NPs by hydrolyzing and co-condensing the CNDs with silica precursors for the biolabeling and imaging applications as well [155]. In another work, a methyl parathion sensor was established from CNDs functionalized with tyrosine methyl ester through hydrothermal reaction using citric acid as a carbon precursor. The functionalized CNDs show strong and stable photo-luminescence with a quantum yield of 3.8%. The obtained biosensors were used for the determination of different kinds of organophosphorus compounds [156]. Hydrothermal treatment of glucosamine produces amino-functionalized fluorescent CNDs. The obtained CNDs displayed stabilized green emission fluorescence at various excitation wavelengths and pH environments which are used to produce biosensors and selective detection of hyaluronidase [157]. In addition to the aforementioned modifying agents, polyethyleneimine [84], boronic acid [158],

NH_2-polyethylene-glycol (PEG_{200}) and N-acetyl-L-cysteine (NAC) [159] were also used to functionalize CNDs for the application in gene delivery and bioimaging, and blood sugar sensing and Hg (II) sensing, respectively.

Doping is another modification technique that involves the introduction of impurities or foreign atoms into the carbon matrix. It offers a powerful means to tailor the properties of CNDs and unlock new functionalities. Atomic doping enhances the florescence performance of CNDs and classified into metallic atoms (such as Cu, Fe, Zn, etc.) or nonmetallic atoms (nitrogen, phosphorus, selenium, silicon, etc.) [160]. Previous studies have demonstrated the effectiveness of doping in enhancing the optical, electronic, and catalytic properties of CNDs. For instance, nitrogen doping has been shown to improve the photoluminescence quantum yield of CNDs, making them more suitable for bioimaging and sensing applications. Similarly, phosphorus doping can enhance the photocatalytic activity of CNDs for pollutant degradation [161]. The advantages of doping include improved stability, enhanced functionality, and tunable properties, which broaden the scope of CND applications in sensing, imaging, drug delivery, and energy conversion. However, doping may introduce defects or alter the surface chemistry of CNDs, potentially affecting their biocompatibility and toxicity. Additionally, factors such as dopant type, concentration, synthesis method, and reaction conditions significantly influence the doping efficiency and resulting properties of CNDs [162]. The schematic representation of the chemistry of CNDs is given below in Figure 3.

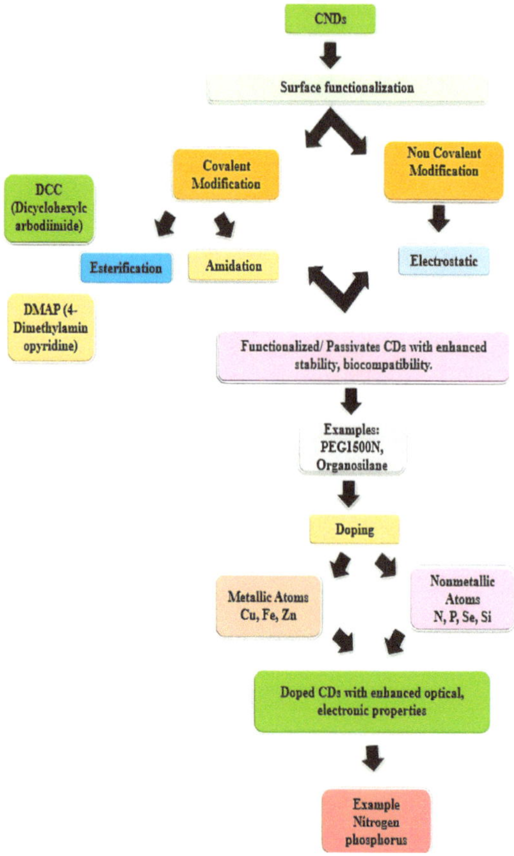

Figure 3. Schematic representation of the chemistry of CNDs.

4. Phytomedicine-Loaded Carbon Nanodots

Carbon nanodots loaded with phytomedicine are a State-of-the-Art example of how nanotechnology and natural medicine can work together to overcome many obstacles in traditional phytomedicine delivery. By utilizing the special qualities of carbon nanodots, this novel synergy improves the bioavailability, targeted administration, and general efficacy of chemicals originating from plants [163].

- Synthesis and Characterization

The process of creating carbon nanodots loaded with phytomedicine is a laborious one that usually uses hydrothermal or other specialized techniques. These nanodots are perfect carriers for phytomedicines because of their unique surface functions and nanoscale size. Carbon nanodots' surface characteristics can be adjusted to best encapsulate and distribute different kinds of bioactive substances [164].

- Improved Bioavailability

When it comes to overcoming the bioavailability constraints of phytomedicines, carbon nanodots are essential [165]. These nanodots improve the solubility of hydrophobic components and stop them from aggregating and degrading by conjugating or encasing plant-derived chemicals. The enhancement of solubility has a noteworthy impact on the bioavailability of phytomedicines loaded, guaranteeing effective absorption and dispersion throughout the biological system [166].

- Targeted Drug Delivery

One of the main benefits of personalized medicine is the ability to precisely target particular tissues or cells through the functionalization of carbon nanodots [167]. By interacting with particular biomolecules or receptors, surface changes let the loaded phytomedicines reach their targeted sites of action. By reducing adverse effects and off-target consequences, this focused administration maximizes therapeutic efficacy.

- Controlled-Release Kinetics

Carbon nanodots' controlled-release properties elevate the delivery of phytomedicine to a higher level. These nanodots allow for a controlled and prolonged release of the loaded phytomedicines by varying the surface characteristics or adding responsive components. By preserving ideal medication concentrations for prolonged periods of time, this controlled-release profile improves therapeutic effects and may lessen the frequency of administration [168].

- Biocompatibility and Safety

Because of their low toxicity and great biocompatibility, phytomedicine-loaded carbon nanodots are safe to use in biological systems. The effects of these nanodots on cellular architecture, metabolic pathways, and general physiological functions are the subject of much research [169,170]. Carbon nanodots' biodegradable nature adds to their safety profile and allays worries about long-term exposure.

- Intracellular Delivery and Cellular Uptake

Enhancing cellular absorption is a critical component for intracellular targets, and carbon nanodots' nanoscale size makes this possible. It is possible to effectively transport phytomedicines over cellular barriers, guaranteeing their delivery to targeted cell compartments. The greater potency and overall bioavailability of the loaded phytomedicines are a result of this increased cellular absorption [171].

- Encapsulation of Diverse Phytochemicals

Carbon nanodots loaded with phytomedicine are adaptable vehicles that can hold a wide variety of phytochemicals, such as polyphenols, terpenoids, alkaloids, and flavonoids.

This adaptability broadens the application of this delivery technology across different herbal medicines by enabling the encapsulation of complex mixes present in medicinal plants [172].

- Application in Multicomponent Phytomedicine Formulations

Phytomedicine-loaded carbon nanodots present an opportunity to formulate complicated, multicomponent delivery systems in the context of traditional herbal treatments that contain multiple active components. By delivering synergistic plant-derived components simultaneously, this method can enhance therapeutic effects and replicate the holistic aspect of conventional herbal preparations.

- Future Perspectives and Challenges

Natural therapies could undergo a revolution with the introduction of carbon nanodots loaded with phytomedicine into the medication delivery system. Research is still ongoing in several areas, including scalability, standardized synthesis techniques, and long-term safety assessments. Subsequent efforts will concentrate on optimizing these delivery systems based on nanodots for broad clinical use, with a focus on regulatory compliance and reproducibility [173]. To sum up, the creation of carbon nanodots loaded with phytomedicines is a cutting-edge in the field of sophisticated medication administration. With its capacity to target delivery, ensure regulated release, and overcome bioavailability limitations, this novel technique has enormous promise to further the medicinal applications of chemicals produced from plants. The current research in this area marks the beginning of a new era in which the combination of natural medicines and nanotechnology creates opportunities for improved treatment results.

4.1. Loading Strategies for Phytomedicine

Recent studies have thrown light on novel strategies and developments in this emerging field, offering insightful information about phytomedicine loading tactics onto carbon nanodots. Modern technologies and a greater comprehension of the interactions between nanomaterials have led to the adaptation and improvement of these tactics, which are essential for maximizing the effectiveness of medicine administration.

- Physical adsorption: recent developments

Current research has looked into creative approaches to improve phytomedicine's physical adsorption onto carbon nanodots [11]. Through specialized chemical interactions, surface modification of nanodots with particular functional groups, including amino or hydroxyl groups, has been studied to increase adsorption affinity [174]. Furthermore, sophisticated characterization methods have been used to accurately comprehend and control the non-covalent forces regulating adsorption, such as spectroscopic studies and computational modeling. The goal of these improvements is to get over issues with adsorption strength variability and environmental sensitivity.

- Covalent bonding: progress in precision

The goal of covalent bonding strategy advancements has been to increase loading-process precision. Researchers have looked into site-specific functionalization of carbon nanodots to facilitate selective covalent interaction with certain areas of phytomedicine molecules [175]. This focused strategy reduces inadvertent changes and maintains the integrity of the loaded chemicals. Furthermore, current studies have focused on the creation of green chemistry strategies that maximize covalent bonding without sacrificing the phytomedicine's therapeutic qualities [176]. These strategies make use of environmentally friendly solvents and reaction conditions.

- Encapsulation during nanodot synthesis: tailoring synthesis conditions

Recent developments in the field of encapsulation during nanodot synthesis have focused on optimizing synthesis conditions for enhanced control. Mathew et al. prepared chitosan/carbon dots to be used as a drug carrier for neurodegenerative diseases. Encapsulation of a phytomedicine in a carbon nanodot can be performed with the help of entrapment of drug in a polymer matrix and further encapsulating the drug–polymer matrix into a carbon nanodot [177]. Novel techniques like template-assisted procedures and microfluidic-assisted synthesis have been investigated to obtain a more accurate phytomedicine compound encapsulation. These methods address issues related to consistent distribution and possible changes in the therapeutic qualities of loaded phytomedicine during synthesis, providing scalability and reproducibility. Hollow CNDs can encapsulate the phytodrugs through hydrophobic and electrostatic interactions, pi-pi stacking, or hydrogen bonding, which needs more research for validation [35].

4.2. Stability and Bioavailability Enhancement

Recent work on phytomedicine-loaded carbon nanodots has yielded important advances to improve the stability of encapsulated bioactive chemicals, as well as novel approaches to increase bioavailability. These developments, which are based on cutting-edge techniques and technologies, provide an early look at more reliable and effective medication delivery methods in the future.

- Enhanced stability through nanoengineering

Utilizing cutting-edge nanoengineering methods to increase the stability of carbon nanodots loaded with phytomedicine has been the focus of recent research. Researchers have investigated how to produce protective shells around the nanodots by nanoencapsulation techniques, such as lipid-based formulations and polymeric coatings, to shelter the encapsulated phytomedicine from environmental stressors [178]. Furthermore, surface alterations like the application of silica or graphene oxide coatings have shown improved stability, delayed premature deterioration and maintained the bioactivity of the loaded chemicals.

- Smart nanomaterials for controlled release

Recent studies have investigated the incorporation of smart nanomaterials into carbon nanodots as a means of achieving controlled and sustained release mechanisms. The creation of intelligent drug delivery systems has sparked research into stimuli-responsive nanocomposites, such as pH-sensitive polymers and temperature-responsive hydrogels. These technologies provide customized and on-demand delivery of phytomedicine components by precisely controlling release kinetics in response to particular physiological situations.

- Advances in nanotoxicology and biocompatibility

Recent study has focused on ensuring the safety and biocompatibility of carbon nanodots loaded with phytomedicines. Utilizing cutting-edge imaging methods, including live-cell imaging and super-resolution microscopy, advanced nanotoxicology investigations have thoroughly evaluated the interactions between nanodots and biological systems at the cellular and subcellular levels [179]. This thorough knowledge has facilitated the creation of safer nanomaterials, which may reduce the possibility of cytotoxic effects and open the door to their use in biomedicine.

- Precision bioavailability enhancement

Improvements in lipid nanoparticle and micelle technologies have served as an inspiration for the development of nanoscale drug delivery vehicles, which have shown to better solubilize hydrophobic phytomedicine components and increase their bioavailability [180]. To further maximize bioavailability, studies have also looked at the use of ligand-targeted nanocarriers, which take use of ligand–receptor interactions to enable tailored distribution and uptake.

- Emerging trends and future directions

Trends in stability and improving bioavailability are moving in the direction of a multidisciplinary strategy. The amalgamation of computational modeling, artificial intelligence, and systems biology is increasingly common in order to forecast and enhance the behavior of nanomaterials in intricate biological contexts. Biosensors and imaging modalities are among the real-time monitoring approaches that are being used to offer dynamic insights into the in vivo fate of carbon nanodots loaded with phytomedicine. Another well-known trend in nanotechnology is the hunt for "green" materials. To address sustainability issues, researchers are investigating biodegradable nanomaterials and ecologically friendly synthesis techniques [181]. The goal of surface modification innovations like bioinspired coatings and biomimetic methods is to produce nanocarriers that are less immunogenic and more biocompatible. As the field advances, it is anticipated that these recent research insights and technological innovations will propel phytomedicine-loaded carbon nanodots towards broader clinical applications, offering more effective, safe, and targeted therapeutic interventions.

4.3. Controlled-Release Mechanisms

With a focus on accuracy, flexibility, and responsiveness, recent research has made significant advancements in the field of controlled-release mechanisms for carbon nanodots loaded with phytomedicine. The field of controlled drug delivery is being shaped by cutting-edge technologies and creative approaches, which provide insights into how nanotechnology might be used to maximize therapeutic results.

- Advanced surface engineering for tunable release

In order to give carbon nanodots variable releasing capabilities, recent investigations have investigated sophisticated surface engineering techniques [182]. Dynamic changes in release kinetics are possible by functionalizing the nanodot surface with stimuli-responsive polymers or biomimetic coatings. The focus has been on pH-responsive systems in particular, as they allow for controlled release in response to variations in pH that occur in various physiological compartments. The accuracy attained with these surface alterations guarantees a customized release profile in line with therapeutic requirements.

- Microfluidic-assisted synthesis for controlled encapsulation

One notable development in the field of encapsulation techniques is microfluidic-assisted synthesis. By allowing for exact encapsulation of phytomedicine components within carbon nanodots, this approach offers tight control over the production process. The continuous and uniform mixing of reactants made possible by microfluidic devices produces nanodots with consistent characteristics [183]. This degree of control makes drug-loaded nanodots more reproducible and helps to ensure a more regulated and predictable release of phytomedicine.

- Temperature-responsive systems for on-demand release

Temperature-responsive components incorporated into carbon nanodots have been the subject of recent studies, allowing for on-demand medication delivery [184]. These devices use the natural temperature fluctuations that occur in physiological settings to initiate regulated release. It is now possible to adjust the rate of drug release in nanodot formulations in response to particular thermal cues by using thermoresponsive polymers or materials that are susceptible to external heat sources. This strategy has potential for use in situations where exact control over release dynamics is essential.

- Advancements in biodegradable nanocarriers

A crucial factor in the development of controlled-release systems is biodegradability. The creation of biodegradable nanocarriers that can deliver phytomedicine in a regulated manner as they decompose over time has been the focus of recent developments. Researchers have looked into using biocompatible materials, including certain polymers and lipid-based nanoparticles, to make nanocarriers that break down gradually. This allows the release kinetics to be in line with therapeutic requirements and reduces the possibility of long-term accumulation issues [185].

- Emerging trends and future prospects

A convergence of disciplines characterizes emerging developments in controlled-release mechanisms for carbon nanodots loaded with phytomedicine. Predicting and optimizing release patterns based on a variety of criteria, such as pharmacological qualities, patient-specific data, and nanomaterial features, is becoming more popular by integrating nanotechnology with AI and machine learning [186].

In addition, the latest developments in real-time monitoring methods, such biosensors and in vivo imaging, provide dynamic insights into the spatiotemporal behavior of nanodots inside the body [187]. This real-time feedback loop makes adaptive release strategy modifications possible, improving the efficacy and adaptability of controlled drug delivery systems.

Emerging trends in the field suggest that regulated release mechanisms for carbon nanodots loaded with phytomedicines may change as a result of these recent discoveries and technological advancements. This will present prospects for personalized and precision medicine with improved therapeutic precision. An overview of the synthesis of CNDs from plants and raw materials as precursors is shown in Figure 4.

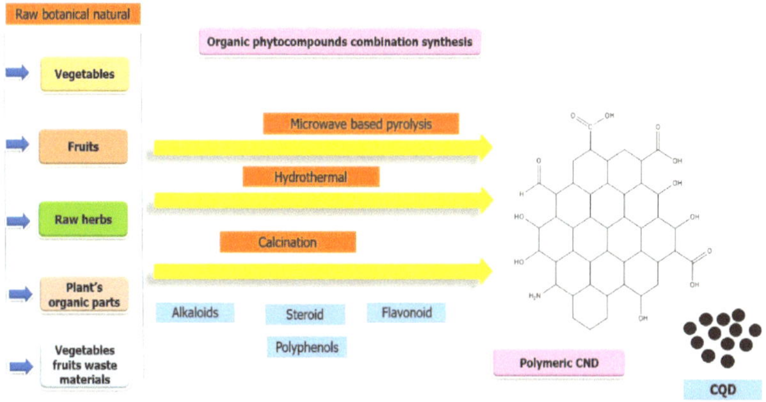

Figure 4. Overview of synthesis of CND from plants, with raw materials as precursors.

5. Polymeric Carbon Nanodots

5.1. Introduction to Polymer Nanocomposites

Over the past two decades, considerable scientific and technological interest has centered on polymer nanocomposites (PNCs). Initially, the focus was primarily on understanding synthetic and physical chemistry, colloidal particle physics, and various properties of these materials. Recently, there has been a shift towards comprehending and harnessing the unique physics of polymers within PNCs. This shift is driven by the realization that achieving truly engineered and functional nanocomposites requires a deeper understanding of their structure–property–processing relationship. Modern developments in the realm of nanomaterials include polymeric carbon nanodots (CNDs) and polymer nanocomposites, each of which offers a distinct set of qualities and uses. Since they are so adaptable, polymeric CNDs which are essentially nanoscale carbon-based particles have attracted a lot of attention [188]. They also have unique characteristics like high photoluminescence. However, by adding nanoscale fillers, such as nanoparticles or nanotubes, to polymer matrices, material characteristics can be improved, as in the case of polymer nanocomposites. Superior mechanical, thermal, and electrical properties are added to the resultant composite materials by the interaction of polymers and nanofillers [189]. Polymeric CNDs are distinguished by their distinct composition and structure. A majority of the carbon atoms in the core are grouped either crystalline or amorphously, resembling graphene. Functional groups including hydroxyl, carboxyl, and amino groups are present on the surface of these nanodots, which enhances their solubility, stability, and reactivity. The size and characteristics of the nanodots are affected by various techniques used in the synthesis of polymeric CNDs, including hydrothermal synthesis, microwave-assisted synthesis, and template-assisted synthesis [190]. Conversely, to produce materials with improved performance, polymers are blended with nanoscale fillers to make polymer nanocomposites. When nanofillers are added, the material behaves differently from a pure polymer, providing increases in stiffness, strength, and other important characteristics. Polymeric CNDs are distinguished by their distinct composition and structure. A majority of the carbon atoms in the core are grouped either crystalline or amorphously, resembling graphene [191]. On the other hand, polymer nanocomposites comprise a wider range of materials in which nanofillers enhance overall performance and polymers function as matrices. The addition of nanofillers improves the composite's mechanical, thermal, and electrical qualities, making it appropriate for use in the automotive, electronics, and aerospace industries. Melt blending, solution mixing, and in situ polymerization are the synthesis techniques used to create polymer nanocomposites; each has an impact on the properties of the finished product [192].

5.2. Structural Composition

Polymer nanocomposites' structural elements are distinguished from traditional composites by the accurate dispersion of nanofillers inside the polymer matrix. In this nanoscale configuration, the polymer matrix is finely interwoven with nanofillers, which can be nanoparticles, nanotubes, or other nanomaterials. Since it has a significant impact on the composite material's overall performance, this careful dispersion is essential. It is a laborious process with major consequences to distribute nanofillers uniformly throughout the polymer matrix. Given that they are nanoscale, the nanofillers are smaller than those that are usually employed in traditional composite materials. Its reduced size makes it possible for the polymer to be more thoroughly and uniformly integrated, which enhances its desirable qualities. A uniform dispersion of the nanofillers highlights the enhanced attributes, which include greater stiffness, strength, and thermal stability. This structural configuration at the nanoscale contrasts sharply with traditional composites, where the fillers are frequently bigger and less uniformly distributed. Uniform dispersion is more difficult to achieve in typical composites, and the unequal distribution of the filler elements can lead to variances in the properties of the final material. By optimizing the advantages of the nanofillers and enhancing the material's overall performance, polymer nanocomposites' nanoscale accuracy guarantees that the improved properties are consistently manifested throughout [193].

5.3. Various Synthetic Methods of Polymeric Carbon Nanodots

The process for generating polymeric carbon nanodots (CNDs) involves a range of techniques designed to produce carbon particles at the nanoscale with particular characteristics. Many well-established methods are used to synthesize these polymeric CNDs, which are preferred for their improved stability, biocompatibility, and flexible functionalization potential. By using microwave irradiation to speed the reaction kinetics, however, microwave-assisted synthesis produces polymeric CNDs with regulated characteristics quickly and effectively. Using templates like polymers or proteins, template-assisted synthesis allows for fine control over the morphology of the CNDs during production by dictating their size and shape [194]. Pyrolysis is a versatile process that uses a variety of carbon-rich sources and offers sustainability by carbonizing organic precursors which are often sourced from biomass under regulated circumstances. Using ultrasonic vibrations, bigger carbon structures can be gently and energy-efficiently broken down into nanoscale CNDs through the process of ultrasonic-assisted synthesis. With electrochemical synthesis, one can precisely regulate the synthesis conditions by inducing the creation of polymeric CNDs with the use of electrical energy [195]. Lastly, under benign and eco-friendly circumstances, photochemical synthesis uses light as an energy source to start reactions that result in the synthesis of polymeric CNDs. The resultant polymeric CNDs are adaptable for a variety of applications, including biological imaging, sensors, and optoelectronics, and can be further characterized and functionalized. The desired characteristics and planned uses of the polymeric CNDs inform the synthesis technique selection, which reflects the dynamic and varied nature of the nanomaterial sector. Various methods of synthesis of polymeric carbon nanodots are shown in Figure 5.

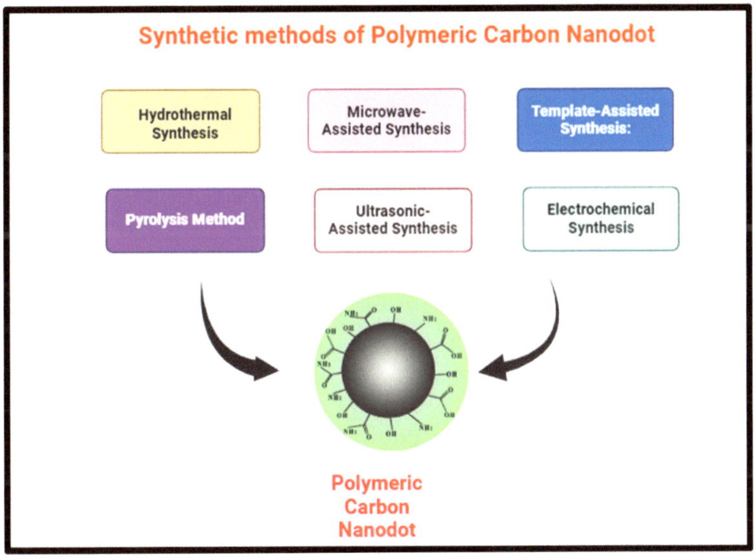

Figure 5. Various methods of synthesis of polymeric carbon nanodots.

5.3.1. Role of Polymer Nanocomposites in Drug Delivery

By addressing issues with traditional drug delivery systems, these nanocomposites, which are made of polymers and nanoscale materials, have opened the door for more accurate and regulated drug release. The potential of polymer nanocomposites to encapsulate and shield medications, preventing their untimely breakdown or excretion from the body, is one of its main benefits. This is especially important for medications that are difficult to degrade in an adverse biological environment or have low bioavailability. The medication is protected from the environment by nanocomposites, which also guarantee its stability until it reaches the intended location [196]. It is possible to modify the drug delivery system's characteristics by adding elements at the nanoscale, like nanoparticles, to polymer matrices. Some features, including targeting, can be built into these nanoparticles to allow the nanocomposites to selectively aggregate at the sick tissue while sparing healthy cells. By minimizing systemic side effects, this focused drug administration improves the overall therapeutic result [197]. In addition, these composites' nanoscale size makes it easier for them to pass across biological barriers like blood–brain barriers and cell membranes. This makes it possible to administer medications to regions that were previously unreachable or difficult to treat, so broadening the range of illnesses that can be treated, including neurological conditions. In order to achieve controlled and prolonged medication release, polymer nanocomposites are also essential [198]. The release kinetics of the medication enclosed can be precisely tailored by researchers by carefully crafting the polymer matrix and modifying the composition of the nanoparticles. This helps maintain therapeutic concentrations over an extended length of time, lowering the frequency of dose and increasing patient compliance. It is especially helpful for chronic illnesses that call for continuous and controlled drug administration. The capacity of polymer nanocomposites to encapsulate a broad variety of medicinal substances, such as proteins, nucleic acids, and tiny molecules, is another example of their versatility [19]. The creation of multifunctional drug delivery devices that can administer combination medicines is made possible by this flexibility. For example, a nanocomposite may carry a gene therapy payload and a chemotherapeutic medication at the same time, resulting in a synergistic effect that improves treatment outcomes. Another crucial component of polymer nanocomposites in drug administration is their biodegradability [199]. The fact that a lot of these polymers break down gradually into inert byproducts helps to prevent the delivery system from building up within the body. This feature is especially helpful since it reduces the possibility of long-term side effects and permits the carrier to be cleared after its medication-delivery purpose is completed. Drug delivery has advanced even further with the creation of stimuli-responsive polymer nanocomposites. These intelligent materials are able to react to particular bodily stimuli, such as variations in pH, temperature, or the presence of enzymes. This responsiveness improves accuracy and lessens off-target effects by enabling triggered medication release at the target spot [200]. For example, a medicine encapsulated in a nanocomposite that is made to withstand the acidic environment of tumors can release the drug only in the malignant tissue. Additionally, polymer nanocomposites have shown great promise in addressing drug resistance, a critical obstacle in the treatment of several illnesses, most notably cancer. These composites' multifunctionality enables the co-delivery of medications with various modes of action, addressing several drug resistance routes and raising the possibility of successful therapy. To sum up, polymer nanocomposites have a revolutionary function in drug delivery and provide a host of benefits over conventional drug delivery methods. These nanocomposites offer a platform for accurate, focused, and managed medication delivery. Future customized and extremely successful therapeutic interventions appear to be in the cards as this field of study progresses and more complex polymer nanocomposites are created [201].

5.3.2. Properties of Polymeric Carbon Nanodots

Polymeric carbon nanodots have outstanding photoluminescence, which is one of their prominent features. In the presence of visible or ultraviolet light, these nanodots can release visible light. Because CNDs emit light and can be adjusted in intensity, this feature is especially useful for bioimaging and fluorescence imaging applications, which require sensitive and accurate identification of biological processes and structures. Additionally, the capacity to regulate the emission wavelength. When these nanodots are subjected to ultraviolet or visible light, they can release visible light. The strong and adjustable emission of CNDs makes it possible to precisely and sensitively identify biological structures and processes, making this trait very beneficial for applications in fluorescence imaging and bioimaging [202]. Their further usefulness in multicolor imaging and diagnostics is increased by the controllable emission wavelength. Biocompatibility is another well-known characteristic of polymeric CNDs, which is essential for biomedical applications. Their composition guarantees low cytotoxicity and immunogenicity; they are frequently made from biocompatible polymers or carbon precursors [203]. Because of this, they can be used in a variety of biological settings, such as medication delivery, imaging in vivo and in vitro, and other biomedical applications. Polymeric CNDs have the potential to be safe and useful medical instruments because of their biocompatibility. The adaptable surface chemistry of polymeric carbon nanodots is another significant characteristic. It is simple to alter or functionalize the surface of CNDs with different groups, such as carboxyl, amino, or hydroxyl groups. The CNDs can be given extra functions by attaching particular ligands, biomolecules, or medications due to their adjustable surface chemistry. Functionalized polymeric CNDs exhibit great adaptability to individual requirements, as they can be customized for specialized applications like sensing, targeted medication delivery, or other uses [204]. Excellent water solubility is another feature of polymeric CNDs that makes them useful for biological and environmental applications. Because of their hydrophilic character, these nanodots are easily incorporated into biological systems by dispersing in aqueous solutions. This solubility is especially useful in drug administration applications, where stable physiological conditions and effective dispersion are essential for effective therapeutic results. Moreover, polymeric CNDs show stability under a range of environmental circumstances. Because of their resilience, they can tolerate variations in pH, temperature, and light exposure, thus guaranteeing their performance and functioning over time. The practical applications of CNDs in a variety of sectors, from electronics to biology, depend on this stability [205]. One of the notable characteristics that helps explain polymeric carbon nanodots' consistent and predictable performance is their homogeneity in size and shape. The manufacturing of nanodots with precise sizes and shapes is made possible by the controlled synthesis techniques used in their creation [206]. For uses like catalysis, sensing, or electrical devices where exact control over the characteristics of nanomaterials is critical this uniformity is vital [207]. To sum up, polymeric carbon nanodots have a number of amazing qualities that make them very appealing for a variety of uses. Their promise in several sectors like materials science, biomedical research, and sensing technology is enhanced by their photoluminescence, stability, water solubility, biocompatibility, and size uniformity. The special qualities of polymeric CNDs are anticipated to open up even more creative uses as research in this field develops, solidifying their status as useful nanomaterials across a wide range of scientific and technical fields [208].

5.4. *Applications of Polymeric Carbon Nanodots in Drug Delivery*

As a result of their distinct qualities and range of functions, polymeric carbon nanodots (CNDs) have become a viable nanomaterial for drug delivery applications. The use of polymeric CNDs in drug delivery has been discussed, along with their applications and underlying mechanics.

Applications for Drug Delivery Using Polymeric Carbon Nanodots:

- Delivery of Drugs with Specificity:

To distribute drugs in a targeted manner, polymeric CNDs can be functionalized with certain ligands, including peptides or antibodies. Target cells or tissues' surface receptors are the only receptors that the functionalized nanodots can bind to with selectivity. The tailored strategy improves overall treatment efficacy, minimizes off-target effects, and concentrates the therapeutic payload at the intended site to increase drug delivery efficiency [209].

- Imaging and Diagnosis:

Polymeric CNDs are perfect for imaging and diagnostic applications due to their exceptional photoluminescent characteristics. CNDs can allow for real-time medication delivery process monitoring when they are utilized as imaging agents. Furthermore, the vivid and adjustable fluorescence of CNDs can help see biological structures, which can help with illness monitoring and diagnostics [23,208].

- Theragnostic:

The potential for polymeric CNDs to integrate therapeutic and diagnostic functions onto a single platform is known as their theragnostic potential. Researchers can create systems that not only carry medications to the target region but also provide real-time feedback on the effectiveness of the treatment by combining therapeutic chemicals and imaging probes into the CNDs. Precision therapies and personalized treatment will be greatly impacted by this integrated approach [199].

- pH-Responsive Drug Release:

It is possible to incorporate pH-responsive components into polymeric CNDs due to their changeable surface chemistry. CNDs can undergo controlled disintegration or structural changes that release medications that are contained in acidic environments, like those seen in tumor tissues. By maximizing the delivery of therapeutic payloads to sick tissues and reducing exposure to healthy cells, this pH-responsive behavior improves drug release precision [210].

- Combination Therapy:

Combination therapy is made possible by the platform that polymeric CNDs provide, which enables the simultaneous loading of several therapeutic drugs with various modes of action. When treating complex diseases like cancer, where a combination of medicines may be more effective in reducing drug resistance and improving treatment results, this is very helpful.

- Intracellular Delivery:

Drug distribution into cells is made easier by the nanoscale size of CNDs, which allows them to pass through cell membranes. This characteristic makes medications that target intracellular processes more effective in treating patients by enabling the delivery of medications to particular cellular compartments or organelles.

In order to transfer stem cells to diabetic wounds, a novel pH and thermosensitive hydrogel made of chitosan (CTS) and sADM modified with carbon nanodots (CND) from onion peels is presented in the work by Bankoti et al. where enhanced diabetic wound healing can be achieved with the help of this hybrid hydrogel's several advantages, which include enhanced hydrophilicity, sustained release of ND, antibacterial and antioxidant qualities, promotion of stem cell delivery, and excellent biocompatibility and biodegradability. In addition to stimulating angiogenesis and scavenging reactive oxygen species (ROS), the hydrogel containing ND also makes it easier for stem cells to be encapsulated and delivered to the wound site, hastening wound closure and encouraging tissue regeneration without the formation of scars. In addition, the hydrogel has a moderate level of cytocompatibility and antimicrobial activity, which guarantees low cytotoxicity and supports cell viability. The study highlights the potential of capped or biopolymer-doped carbon nanodots in biomedical applications, especially advanced wound healing, where their multifunctional properties are crucial for tackling the intricate issues related to diabetic wounds, ultimately resulting in improved clinical outcomes [211].

5.5. Polymeric Carbon Nanodots' Mechanisms of Drug Delivery

Targeting passively: The Enhanced Permeability and Retention (EPR) effect, which is seen in tumors where leaky vasculature permits the preferential accumulation of nanoparticles, can be exploited by polymeric CNDs. CNDs can accumulate at the tumor site through passive targeting, which increases the amount of medication delivered to cancer cells.

Active targeting: Targeting ligand-functionalized polymeric CNDs allows active targeting. Ligands, which are recognized and bind to specific receptors on the surface of target cells, can be peptides or antibodies. This process facilitates the internalization of CNDs and the subsequent release of therapeutic payloads. By improving drug delivery's specificity through active targeting, systemic adverse effects are decreased.

Drug release that is pH-responsive: Polymeric CNDs have the ability to react to pH variations. It is possible for CNDs to experience structural alterations that result in the release of encapsulated medications in the acidic environment of tumors or certain cellular compartments. The accuracy and effectiveness of treatment are enhanced by this pH-responsive behavior, which guarantees that drug release is initiated at the targeted spot [212].

Cellular uptake and endocytosis: Polymeric CNDs' interactions with cells are influenced by their surface characteristics and nanoscale size. Different endocytic mechanisms, including caveolae- or clathrin-mediated endocytosis, can be used to internalize CNDs. CNDs' efficiency in intracellular drug delivery is attributed to their capacity to man oeuvre these cellular processes.

Sustained release: The medicine enclosed in CNDs can be released gradually due to the polymer matrix's architecture. Researchers can attain a regulated and extended-release profile by modifying the drug's interactions with the polymer or the polymer's rate of breakdown. When treating long-term illnesses that call for constant medication administration, this sustained release method is especially helpful [213].

Intracellular trafficking: Polymeric CNDs go through intracellular trafficking after being absorbed by cells. In order to maximize the effects of drug delivery, it is essential to comprehend the routes and destiny of CNDs within cells. Drug delivery can provide a greater therapeutic effect by optimizing drug release at certain subcellular sites through manipulation of CND intracellular trafficking.

To sum up, polymeric carbon nanodots have proven to have great promise for drug delivery applications. They provide a flexible platform for the precise and regulated release of medicinal molecules. Polymeric CNDs have distinct qualities like photoluminescence, biocompatibility, and programmable surface chemistry along with well-defined drug delivery systems that make them attractive options for furthering the science of nanomedicine. More advancements in the creation and use of polymeric CNDs are anticipated as this field of study develops, perhaps leading to the creation of more specialized and efficient drug delivery systems [214].

5.6. Polymeric Carbon Nanodots' Advanced Properties and Functionalities

Polymeric carbon nanodots (CNDs) display a variety of complex properties that go beyond simple photoluminescence, making them very desired in domains that need multifunctional and flexible materials. Polymeric CNDs' outstanding characteristics result from their nanoscale size, unique carbon-based architecture, and surface functionalization, which increase their applicability across disciplines [201].

5.6.1. Improved Photostability and Tunable Fluorescence

Polymeric CNDs differ from typical fluorescent dyes in terms of photostability, namely photobleaching resistance. Unlike organic fluorophores, which decay after prolonged exposure to light, polymeric CNDs retain their fluorescence via several excitation cycles. This makes them perfect for long-term imaging in biological and environmental settings [215]. Furthermore, scientists may control the fluorescence emitted by polymeric CNDs by altering synthesis parameters, surface functional groups, and dopants, allowing them to generate a wide range of colors. This tunability offers significant advantages for developing multicolor probes for sophisticated bioimaging and optical sensing applications [201].

5.6.2. Electronic Properties and Conductivity

Polymeric CNDs have unique electrical characteristics due to their conjugated π-electron systems, including semi-conductivity and high electron mobility. These properties, particularly in CNDs synthesized with graphitic structures, make them ideal for optoelectronic applications such as solar cells, LEDs, and photodetectors. Researchers can further alter the electronic characteristics of carbon frameworks by including heteroatoms like nitrogen, sulfur, or boron, leading in CNDs customized for certain electrical or catalytic functionalities [216]. Conductivity of CNDs helps in better interaction of drug-loaded CNDs with biological membranes, leading to improvement in release profile, as well as bioavailability [30].

5.6.3. Catalytic Activity and Environmental Applications

Research demonstrates that polymeric CNDs can work as catalysts or catalytic supports due to their large surface area, functionalizable surfaces, and capacity to effectively transport electrons. For example, nitrogen-doped CNDs have a high catalytic potential in oxygen reduction processes, which are important in fuel cell technology. In a similar way polymeric CNDs may help in the photocatalytic destruction of contaminants, providing a sustainable method to wastewater treatment [217,218]. Furthermore, the catalytic properties of CNDs enable it to scavenge reactive oxygen species with ease. This compliments the organ-protective abilities of phytomedicine-derived CNDs, which are advantageous for developing therapies especially for liver conditions and neurodegenerative disorders [2,219].

5.7. Enhanced Synthetic Approaches for Polymeric Carbon Nanodots

Beyond standard hydrothermal and microwave-assisted synthesis, novel and hybrid synthesis techniques are being developed to improve the characteristics and functions of polymeric CNDs.

5.7.1. Ultrathin Layered Synthesis

Ultrathin layered synthesis is a new approach for producing CNDs with precisely controlled thickness at the nanoscale. Under regulated circumstances, scientists may layer carbon precursors on substrates like silica or alumina to create ultra-thin CND layers with unique optical characteristics due to quantum confinement. This approach is especially beneficial in optoelectronics and biosensing, where precise control of CND thickness is required to tune electrical and photonic behavior [220].

5.7.2. Dopant-Driven Synthesis Techniques

Dopant-driven synthesis approaches have evolved that insert heteroatoms or functional groups into CND structures during the synthesis process rather than afterward. CNDs can be designed to have enhanced fluorescence, conductivity, or catalytic activity by adding nitrogen, boron, or sulfur during the initial synthesis stages. Dopant-driven techniques are especially useful in electrochemistry and catalysis, where heteroatom-rich CNDs outperform pure carbon equivalents [221].

5.7.3. Green Synthesis Using Biomass Precursors

Agricultural waste, plant extracts, and even food scraps are plentiful and sustainable sources of carbon. These green synthesis processes are not only cost-effective, but also waste-reducing, making them appealing for large-scale manufacturing. Furthermore, biomass-derived CNDs frequently demonstrate built-in biocompatibility, which increases their suitability for biomedical applications [222].

5.8. Functionalizing Polymeric Carbon Nanodots for Specific Applications

5.8.1. Surface Functionalization for Targeted Drug Delivery

Surface-functionalized CNDs can be modified with peptides, antibodies, or other biomolecules to improve medication targeting sick cells or tissues. By conjugating targeting ligands, CNDs may recognize and bind to specific cell receptors, enabling selective accumulation at certain areas, such as tumor tissues. This ligand-mediated targeting lowers off-target effects, systemic toxicity, and increases the therapeutic effectiveness of the drug payload [34].

5.8.2. Polymer Coatings for Improved Biocompatibility and Stability

Coating CNDs with biocompatible polymers such as polyethylene glycol (PEG) or polylactic acid (PLA) increases their stability and biocompatibility. Because of the "stealth" effect of the polymer layer, polymer-coated CNDs have a longer circulation lifetime in the bloodstream, are less vulnerable to immune detection, and aggregate more in target locations. Moreover, polymer coatings may help encapsulate a wide range of drugs, enabling CNDs to operate as dual or multi-drug delivery systems [223].

5.9. Polymer Nanocomposites: Innovation and Future Directions

The nanocomposites are still being widely researched, particularly in fields that benefit from their excellent mechanical, thermal, and electrical properties. Polymer nanocomposites are advancing, with a focus on improved material design, process optimization, and broadening their applications beyond traditional uses.

5.9.1. Self-Healing and Responsive Polymer Nanocomposites

Recent advances in polymer nanocomposites have led in the development of self-healing materials that can repair themselves. Composites like these can restore structural integrity after damage by integrating nanoparticles that trigger polymerization when a hole occurs or inserting microcapsules with healing agents. Self-healing nanocomposites are ideal for use in high-stress environments such as aerospace, where durability is critical [224]. Responsive polymer nanocomposites are designed to change their properties in reaction to environmental factors such as temperature, pH, and mechanical stress. These materials are helpful for applications that need adaptive properties, such as drug delivery systems that release therapeutic substances based on disease-specific conditions [225].

5.9.2. Bio-Based and Environmentally Friendly Nanocomposites

Bio-based nanocomposites made from renewable polymers and biodegradable materials are gaining popularity as a result of environmental measurements. Polymers derived from maize starch, cellulose, and polylactic acid are used to create eco-friendly nanocomposites that disintegrate at the end of their lifetimes, reducing environmental impact [226]. The use of naturally occurring nanofillers, such as cellulose nanocrystals and lignin nanoparticles, enhances the mechanical and thermal properties of bio-based composites. These materials have promise in packaging, pharmaceutical, and consumer product applications, where environmentally friendly materials are becoming increasingly popular [227].

5.9.3. Advanced Nanofiller Integration Techniques

Traditional techniques of incorporating nanofillers into polymer matrices frequently have problems in attaining uniform dispersion, which is critical for good composite characteristics. To overcome this issue, innovative technologies like electrospinning, shear mixing, and in situ polymerization have been developed to improve nanofiller distribution and integration. These approaches enable the production of nanosized reinforcing networks, resulting in composites with higher qualities when compared to those produced using traditional methods. This improved dispersion is critical in high-performance applications like flexible electronics and electromagnetic interference shielding [228].

5.10. Polymeric Carbon Nanodots for Sensor Technologies and Environmental

Polymeric CNDs have established themselves as useful components in sensor technology due to their fluorescence, conductivity, and ability to be selectively functionalized. CND-based sensors are versatile enough to detect a wide spectrum of chemical and biological analytes with good sensitivity and selectivity [229].

5.10.1. Fluorescence-Based Sensors

The photoluminescent characteristics of CNDs are critical to their use in fluorescence-based sensors. By altering their surface with analyte-specific recognition elements, CNDs can detect changes in fluorescence intensity, color, or wavelength in the presence of target molecules. For example, CNDs functionalized to detect heavy metals or poisons exhibit an apparent shift in fluorescence upon binding to the target, allowing for real-time monitoring of environmental contaminants [230].

5.10.2. Electrochemical Sensing Applications

CND electrochemical sensors are good for detecting biomolecules such as glucose or neurotransmitters due to their high conductivity and huge surface area. Surface functionalization with specialized enzyme immobilization, such as glucose oxidase for glucose detection, permits CND-based sensors to attain high sensitivity while maintaining low detection limits, which is critical for point-of-care diagnostics and environmental monitoring [231].

5.11. Polymeric Carbon Nanodots in Nanomedicine

Polymeric CNDs are increasingly used in nanomedicine due to their low toxicity, excellent biocompatibility, and adjustable characteristics. Photothermal treatment (PTT) is one area of active research in which CNDs absorb near-infrared (NIR) light and convert it to heat, therefore targeting cancer cells while minimizing damage to surrounding healthy tissues. Surface-functionalized CNDs may be adjusted to specific wavelengths, increasing their penetration and therapeutic efficacy [232]. When paired with chemotherapy (chemothermal treatment), CNDs enhance the cancer-killing effects, a method that is showing promise in early tests. Another promising area is gene and nucleic acid delivery. CNDs can be functionalized to bind siRNA or DNA, boosting cellular intake and minimizing genetic material destruction by circulatory enzymes. This feature solves a critical difficulty in gene therapy: delivering genetic material accurately to target cells while avoiding an immune reaction [233]. Furthermore, dual imaging and drug delivery applications are expanding, with CNDs acting as both a contrast agent for imaging and a carrier for drug administration. These multifunctional CNDs provide real-time tracking of medication distribution in the body, thus enhancing therapy efficacy and providing personalized treatment monitoring [234].

5.12. Structural Innovations in Polymer Nanocomposites

Developments in polymer nanocomposites include not just the incorporation of nanofillers, but also the engineering of internal structures to improve strength, flexibility, and functional capabilities. For example, layer-by-layer assembly processes allow for fine control over composite thickness and nanofiller arrangement, resulting in materials with unique mechanical characteristics, such as strong impact resistance or controlled elasticity. Core–shell nanocomposites, in which a core material is wrapped in a protective or functional shell, are being developed for application in harsh settings, such as aerospace. These composites can withstand high temperatures and prevent radiation deterioration, which is critical for materials used in satellite and spacecraft components. In addition, 3D printing technologies enable the development of complicated, customized nanocomposite structures. Researchers may change the qualities of 3D-printed materials for smart watches, flexible sensors, and sophisticated prostheses by changing the polymer matrix and filler content [235]. This versatility marks a huge step forward in material design, enabling for the exact customization of nanocomposites on demand.

Both polymeric CNDs and polymer nanocomposites have potential in sustainable agriculture and environmental applications. CNDs as Nano fertilizers provide a unique approach to precise nutrient delivery. Plants might take in vital nutrients more efficiently after loading them onto CNDs, decreasing fertilizer waste and lowering environmental impact. Some research suggests that CNDs can boost plant growth and stress tolerance, allowing crops to tolerate drought, pests, and other environmental stresses. In water treatment, polymeric CNDs are being developed as photocatalysts to break down organic contaminants and remove heavy metals [236]. CNDs' large surface area and adaptable reactivity make them perfect for breaking down pollutants when exposed to sunlight, which is both efficient and ecologically benign. In addition, membrane-based water purification systems that use polymer nanocomposites show tremendous potential for filtering out tiny pollutants, since the embedded nanofillers increase both filtration performance and mechanical durability [237]. Various applications of polymeric carbon nanocomposites are demonstrated in Figure 6.

Figure 6. Various applications of polymeric carbon nanocomposites.

6. Applications for Carbon Nanodots in Drug Delivery

6.1. Overview of Drug Delivery Systems

Drug delivery systems, or DDS, are essential to the progress of medicine because they aim to improve the convenience, safety, and effectiveness of therapeutic interventions. These systems entail the planning and execution of methods for releasing and transferring medications in a regulated way inside the body [238]. Their main objectives are to enhance medication pharmacokinetics, boost bioavailability, and guarantee targeted distribution to particular tissues or cells. Drug delivery systems comprise an array of methodologies, ranging from conventional oral tablets to cutting-edge nanotechnology-based carriers. This thorough analysis looks at the different kinds, intrinsic difficulties, and possible advancements in medication delivery systems in the future. In the past, oral pills, capsules, injections, and topical creams were the main traditional forms used for drug administration. These techniques had drawbacks despite their efficacy, such as unstable release profiles, low bioavailability, and poor drug solubility [239]. The development of sophisticated medication delivery systems has been prompted by the demand for more accurate and patient-friendly methods. There are several kinds of drug delivery systems. In order to increase patient compliance, oral drug administration options include both traditional tablets and capsules and controlled-release formulations. Rapid onset of action is ensured by injectable techniques like intramuscular and intravenous injections. Topical medication delivery offers localized therapies through the use of lotions, gels, ointments, and transdermal patches. Metered-dose inhalers and dry powder inhalers are two methods of inhalation medication administration that specifically target the respiratory system [240]. Liposomes, polymeric nanoparticles, and dendrimers are examples of nanoparticle-based drug delivery agents that provide targeted and controlled release. While implantable drug delivery systems, including drug-eluting implants, progressively release medications for localized treatment, microneedle-based systems enable transdermal drug delivery [241]. Approaches like antibody–drug conjugates (ADCs) and nanoparticle targeting, which improve specificity, are included in targeted medication delivery. Drug delivery systems have difficulties even with their importance. For drug carriers to be safely integrated into biological systems, biocompatibility and toxicity reduction are essential. For a treatment to be effective, it must maintain drug stability, achieve appropriate drug loading, and regulate release kinetics [242]. There are obstacles in overcoming biological barriers, like the blood–brain barrier, and immunogenicity is still an issue. Reproducibility and cost-effectiveness concerns must be addressed in order to scale up production for commercial use. Drug delivery has a bright future ahead of it with many developments and opportunities. Personalized medicine could lead to the creation of medication delivery methods specific to a person's genetic composition. Sensitive materials and sensor integration enable smart drug delivery systems to adjust to changes in the body's physiology. It is hoped that advanced nanocarriers would enable advances in gene and RNA therapy [243]. Emerging horizons include bioelectronics and digital medication delivery systems that may monitor patient responses and modify drug release accordingly. Furthermore, 3D-printing technology has the potential to completely transform the way drugs are delivered by enabling the development of personalized systems with exact control over composition and structure. Drug delivery systems have experienced tremendous development and now provide a variety of methods to improve therapeutic interventions. Drug delivery technologies are changing the face of medicine, from traditional techniques to cutting-edge nanotechnology. The future will be characterized by overcoming obstacles and adopting cutting-edge technologies, with an emphasis on smart systems, personalized medicine, and cutting-edge treatments [244]. Below, Figure 7 demonstrates the various applications of carbon nanodots.

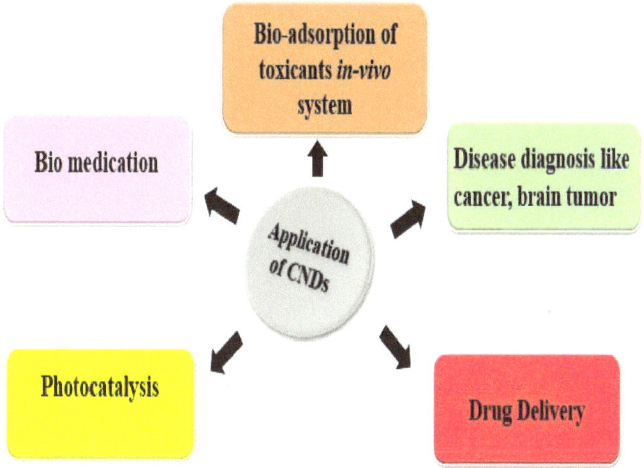

Figure 7. Different applications of carbon nanodots in biomedical domain.

6.2. Role of Carbon Nanodots in Enhancing Drug Delivery

Utilizing the special qualities of these nanoscale carbon-based materials, carbon nanodots (CNDs) can be used to improve the delivery of drugs, opening up new possibilities in nanomedicine and addressing issues related to traditional drug administration. CNDs are interesting candidates for drug delivery applications because of their unique properties, which include biocompatibility, fluorescence, and a high surface area [245]. The capacity of CNDs to increase the solubility and bioavailability of medications is one of their main advantages. Drug research frequently encounters the challenge of poor solubility, which restricts the potency of medicines [246]. Because of their large surface area, CNDs can be used as carriers for hydrophobic medications, improving their solubility and allowing enhanced body absorption. Furthermore, one essential quality that guarantees low toxicity and compatibility with biological systems is the biocompatibility of CNDs. Drug carriers must possess this attribute since any negative effects might compromise the treatment's effectiveness and safety. Because of their biocompatibility, CNDs can be used in a variety of drug delivery methods, offering an effective way to transfer therapeutic substances with a low chance of side effects [247]. Drug delivery systems are being revolutionized by the optical features of CNDs, especially their intense fluorescence, which helps with real-time imaging and tracking. Because CNDs glow, drug carriers can be seen, making it possible for researchers to track their movements throughout the body and evaluate how well drugs are delivered. This has great potential for therapeutic applications, as well as research and development, providing a way to optimize and customize drug administration based on patient reactions [248]. The function of CNDs in drug distribution is further enhanced by surface functionalization. Targeting ligands, medicinal compounds, or other biomolecules can be attached to CNDs by simply adding different functional groups on their surface. Because of their adaptability, medication carriers can be made to match the way they interact with particular cells or tissues [249]. Drug release can be regulated and targeted by engineering functionalized CNDs to react to external stimuli, such as pH changes or the presence of particular enzymes. This degree of accuracy in medication administration has a key role in reducing side effects and raising the therapeutic index of medications. Because CNDs are tiny, interactions at the cellular and molecular levels are facilitated, which is beneficial for drug delivery. They can cross physiological barriers, such as the blood–brain barrier, which is a difficult therapy issue for neurological illnesses; their nanoscale size makes it possible. The delivery of medications to the central nervous system by CNDs shows promise in treating diseases like brain tumors and neurodegenerative disorders [250]. Beyond conventional small-molecule medications, CNDs show promise in providing a broad spectrum of therapeutic substances. They are appropriate transporters for proteins, peptides, and nucleic acids (such as DNA and RNA) due to their flexibility. This skill is especially important in the emerging field of gene and RNA therapies, because successful interventions depend on the targeted delivery of genetic material. The potential of CNDs in developing nucleic acid therapies and biologics is highlighted by their capacity to distribute and protect sensitive biomolecules. Another domain in which CNDs excel is the recently developing subject of targeted medication delivery. The surface of CNDs can be functionalized with certain ligands to enable the construction of drug carriers that bind to target cells or tissues with preference [223]. By using a targeted strategy, medication distribution becomes more particular, and therapeutic agents are delivered to their intended location more precisely. For example, CNDs can be designed to precisely target cancer cells during cancer treatment, reducing harm to healthy tissues and enhancing the overall therapeutic result. The special combination of characteristics that carbon nanodots possess, i.e., biocompatibility, fluorescence, surface functionalization abilities, and nanoscale dimensions, defines their role in improving drug delivery. Their ability to tackle issues related to traditional drug administration is facilitated by these characteristics. Carbon nanodots have the potential to completely transform drug delivery systems as this field of study develops, providing cutting-edge approaches for the precise, targeted, and controlled distribution of therapeutic substances in a variety of medical applications [251]. Figure 8 demonstrates the role of CNDs in drug delivery enhancement.

Figure 8. Role of CNDs in drug delivery enhancement.

6.3. Mechanisms of Drug Delivery Using Carbon Nanodots

Since the introduction of nanotechnology, the field of drug delivery has advanced significantly. Of the various kinds of nanomaterials, carbon nanodots (CNDs) have attracted the most attention. On the nanoscale level, these carbon-based materials have special properties that make them attractive choices for enhancing drug delivery systems. In this extensive work, we explore the intricate mechanisms underlying CND-assisted drug delivery, examining how they can improve solubility, biocompatibility, surface functionalization, real-time imaging, and targeted drug administration [252].

6.3.1. Increasing Bioavailability and Solubility

CNDs solve the issue of many medicinal chemicals' poor solubility. Their ability to function as carriers for hydrophobic medicines stems from their enormous surface area and distinct surface chemistry. CNDs' hydrophilic surface interacts well with water, making it possible for stable bonds to form with hydrophobic medicines and increases their solubility. Because of its increased solubility, the medication has a higher bioavailability, which increases the amount of the medicine that enters the bloodstream and enhances its therapeutic effect [253].

6.3.2. Decreased Toxicity and Biocompatibility

Any successful drug delivery method must be biocompatible. As CNDs have low immunogenicity and cytotoxicity and are intrinsically biocompatible, they can be used in biological systems. They are a safe and well-tolerated substrate for drug administration because of their compatibility with biological surroundings, thus lowering the possibility of side effects. For CND-based drug delivery systems to be successfully integrated into clinical applications, this quality is necessary [254].

6.3.3. Imaging and Tracking in Real Time

CNDs' bright fluorescence is used to track and image drug distribution in real time. This optical characteristic makes it possible to monitor drug carriers inside the body non-invasively, giving important information on how they are distributed, accumulate, and are eliminated. In order to optimize drug delivery systems, real-time imaging is essential since it allows researchers to correctly monitor the movement of drug carriers and make sure they reach their intended targets. This capability holds promise for precise medication administration depending on the needs of each individual patient and tailored treatments [255].

6.3.4. Functionalization of Surfaces

Surface functionalization expands CNDs' potential for drug delivery. Through surface modification, CNDs can be designed to bind particular drug molecules, targeted ligands, or other biomolecules. This adaptation minimizes undesirable outcomes and improves drug-delivery precision by enabling selective targeting of specific cells or regions. Furthermore, it is possible to modify functionalized CNDs such that they react to outside stimuli like pH variations or the presence of specific enzymes, enabling precise drug release at the intended site. This flexibility is a valuable asset for accomplishing precise and efficient medication administration [256].

6.3.5. Nanoscale Aspects for Improved Cellular Communication

CNDs can interact with cellular and molecular targets in an efficient manner due to their nanoscale size. Due to their tiny size, CNDs are able to pass across physiological barriers such as cell membranes and the blood–brain barrier (BBB). Drug transport to the central nervous system is significantly hampered by the BBB in particular. CNDs are useful for treating neurological illnesses because of their capacity to pass through these barriers and their adjustable surfaces, which open up novel possibilities for drug delivery to the brain. Their nanoscale characteristics improve the effectiveness of medication delivery and cellular communication [58].

6.3.6. Transport of Different Therapeutic Agents

Delivering sensitive biomolecules while maintaining their functioning and integrity is made possible by CNDs, which offer a robust and biocompatible framework. Because of their versatility, they are useful instruments for developing biologics and nucleic acid therapeutics [257].

6.3.7. Mechanisms for Targeted Drug Delivery

CNDs' advanced surface functionalization methods allow for targeted medication delivery. CNDs can selectively bind to target cells or tissues by binding particular ligands to their surfaces, improving the accuracy of drug delivery. With CNDs specifically tailored to target cancer cells, this targeted strategy enhances therapeutic efficacy and decreases off-target effects, especially in cancer treatment. By raising the therapeutic index of drugs and reducing overall exposure, targeted drug delivery enhances patient outcomes [258].

7. Biomedical Applications for CNDS

CNDs have been widely used for various biomedical applications due to their excellent PL properties; wavelength-tunable emission properties; and unique electronic, mechanical, thermal, and chemical properties. In contrast to the traditional quantum dots (QDs), which essentially contain heavy metals, are known to show toxicity, and are environmentally hazardous, CNDs are safer and more non-toxic on both the cell and animal level, making them more environmentally friendly and biologically compatible for biomedical applications such as biosensing, bioimaging, and gene and drug delivery [259,260]. In this context, in the recent past, various kinds of sensors have been designed using CNDs to detect different targets, such as DNA [214], heavy metals [261], glucose [262], proteins [263], H_2O_2 [264], nitrite [265], phosphate [266], etc. For instance, CNDs prepared from flour and electrochemical carbonization of sodium citrate and urea show the PL emission which could be selectively quenched by Hg^{2+} with a detection limit of 0.5 nM and 3.3 nM, respectively [127,257]. Other heavy metals, such as Sn^{2+}, Fe^{3+}, Pb^{2+}, Cr^{6+}, Mn^{2+}, and Cu^{2+} [261], were also detected using CNDs. The water-soluble CQDs prepared from lemon peel were used to design a highly sensitive fluorescent probe for the detection of Cr^{6+} ions with a detection limit of about 73 nM. Additionally, these CQDs were immobilized over electrospun TiO_2 nanofibers. The photocatalytic activity for CQDs/TiO_2 composite was seen to be about 2.5 times higher than pure TiO_2 nanofibers when methylene blue dye was used as a model pollutant [126]. The CNDs modified with boronic acid have been utilized for nonenzymatic blood glucose sensing. This novel sensor could detect the glucose level in the range of 9-900 μM with a detection limit of 1.5 μM. The plasma glucose concentration determined by this method was observed to be in good agreement with the values measured by a commercial blood glucose monitor, thus proving the high efficiency of the produced CND-based sensor [267]. Recently, a CND-based fluorescence turn-on sensor was fabricated for hydrogen peroxide (H_2O_2) detection in aqueous solutions through a photo-induced electron transfer mechanism. The developed sensor exhibited good selectivity and sensitivity with a detection limit of 84 nM [268]. Table 1 presents examples of the application of CNDs in biosensors.

The application of CNDs as fluorescent labels for cellular imaging was first reported by Sun et al. CNDs are a promising candidate for bioimaging purposes because of their low side effects and toxicity, excellent water solubility, and visible-to-near-infrared (NIR) emission properties [141,263,269–271]. Jiang et al. [272] have reported the presence of photoluminescent (PL) CNDs in commercial Nescafe instant coffee with a size of 4.4 nm and QY of about 5.5%. Coffee-derived CNDs are directly applied in the imaging of carcinoma cells and small guppy fish without functionalization. CNDs are also passivated with PPEI-EI for two-photon luminescence microscopy, which is used to image human breast cancer MCF-7 cells [76]. CNDs exhibited bright PL both on the cytoplasm and in the cell membrane after a 2 h incubation at 37 °C. The cellular uptake of CNDs was temperature-dependent, with no internationalization observed at 4 °C. Similarly, Zhu et al. [273] prepared a two-photon "turn-on" fluorescent probe, where CNDs were used for imaging hydrogen sulfide in live cells and tissues.

Carbon dots are also proven to possess an excellent gene/drug-loading capability [274]. Besides their biocompatibility, nontoxicity, and photoluminescence properties, their small size and large surface area allow for rapid cellular uptake [275–277]. Recently, Lee et al. [274] demonstrated DOX delivery in vitro and in vivo using CNDs. The DOX was loaded on CNDs via electrostatic interactions with 95% loading efficiency and induced death of HepG2 and MCF-7 cancer cells, as well as tumors, in mice. Interestingly, pure CNDs preferably labeled the nucleus, whereas CNDs loaded with DOX were mainly distributed in the cytoplasm. Meanwhile, a CND coated alginate-based smart stimuli-responsive drug delivery system was proposed by Majumdar et al. [275]. Here, carbon dots were coated on the surface of alginate beads, and garlic extract (GE), which contains allicin, was taken as a model drug system. Interestingly, the amount of GE loaded on alginate beads coated with CNDs was 60% higher than uncoated alginate beads. The loaded system shows pH-dependent controlled drug release, which results in increased therapeutic efficiency and controlled drug release in relation to the amount of pathogen (MRSA) present in the target. Jiao et al. [276] developed a smart carrier for a redox-responsive controlled drug delivery system by grafting carboxyl-abundant CNDs to the surface of silica NPs, where drug release was seen to be highly pH-dependent. Here, DOX loaded onto the grafted silica have high drug loading, up to 13.1%, and exhibited a high cellular uptake and an excellent therapeutic effect on cancer cells by MTT (3-(4,5-Dimethylthiazol-2-yl)-2,5-diphenyl tetrazolium bromide) assay. Other than drug delivery, gene delivery has been also successfully demonstrated with positively charged CNDs by Yang et al. [277]. The CNDs with a positive charge link to plasmid DNA easily and efficiently transfect the therapeutic plasmid into cells with low cytotoxicity. Recently, Zhang et al. [278] demonstrated that hyaluronate (HA) and PEI-functionalized CNDs were internalized readily into the cytoplasm of cancer cells via HA-receptor-mediated endocytosis. These functionalized CNDs were found to have excellent gene condensation compatibility via electrostatic attraction and protective capacity by preventing nuclease degradation. Table 1 presents various examples of biomedical applications of CNDs.

Table 1. Examples of sources, preparation techniques, and biomedical applications of CNDs.

Sources of Carbon	Preparation Techniques	Size (nm)	Quantum Yield (%)	Color	Excitation Wavelength (nm)	Application	Ref.
Rose-heart radish	Hydrothermal	1.2–6.0	13.6	Blue	330	Sensing Fe^{+3}	[279]
Prunus persica (peach)	Hydrothermal	8	15	Blue	325	Cellular imaging and oxygen reduction reaction	[280]
Trapa bispinosa peel	Hydrothermal	5–10	0.1	Green	365	Cellular imaging	[281]
Saccharum officinarum juice	Hydrothermal	3	5.67	Blue	365	Cellular imaging of bacteria and yeast	[282]
Unripe fruit extract of Prunus mume	Hydrothermal	9	16	Blue	355	Cellular imaging	[283]
Apple juice	Hydrothermal	4.5	4.27	Blue	360	Imaging of mycobacterium and fungal cells	[86]
Chionanthus retusus fruit extract	Hydrothermal	5	9	Blue	365	Metal ion sensing and imaging of fungal cells	[284]
Pseudo-stem of banana	Hydrothermal	1–3	48	Green	-	Sensing Fe^{+3}, Imaging of Hela and MCF-7 cells *	[285]
Honey	Solvothermal	2	19.18	Blue	365	Sensing Fe^{+3} and imaging of Hep-2 and Hela cells *	[286]
Garlic	Hydrothermal	11	17.5	Blue	365	Cellular imaging and free radical scavenging	[287]
Sweet potato	Hydrothermal	3.39	8.64	Blue	365	Fe^{+3} sensing and cellular imaging	[288]
Walnut shell	Hydrothermal	3.4	-	Green	360–460	Cellular imaging	[289]
Glycerin and PEG	Microwave	3–4		Blue	365	Nitrite sensing	[290]
Bloomed algae	Microwave	8	13	Blue	365	In Vitro imaging	[291]
Tissue paper	Microwave	4.2	93	Blue	-	Determination of Glutathione	[292]
Kidney beans	Hydrothermal	20–30	8	green	340	Cellular imaging	[293]

Table 1. *Cont.*

Sources of Carbon	Preparation Techniques	Size (nm)	Quantum Yield (%)	Color	Excitation Wavelength (nm)	Application	Ref.
Water Chestnut and onion	Hydrothermal	3.5	12	Green-Blue	400–600	Sensing of Cu (II) and Imaging of Coenzyme A	[294]
Food waste-derived	Ultrasonic	4.6	2.85	Blue-Red	400–470	In vitro bioimaging	[295]
Beer	Gel filtration chromatography	2.5	7.39	Blue	360	Breast cancer cell imaging and drug delivery	[296]
Lignin biomass	Ultrasonic and hydrothermal	2–6	21	Blue Green Red	310–420–540	Cellular imaging	[297]
Onion waste	Hydrothermal	15	28	Blue Green Red	408–488–561	Sensoring of Fe^{3+} and cellular imaging	[298]
Bee pollens	Hydrothermal	1–2	6–12.8	Blue-green	365	Cellular imaging and catalysis	[299]
Coriander leaves	Hydrothermal	2.4	6.48	green	320	Sensoring of Fe^{3+} and cellular imaging	[300]
Grape seed	Microwave	1–8	31.79	multicolor	250–550	Nucleus imaging and pH sensing	[301]
Carrot	Hydrothermal	2.3	7.60	Blue	365	Drug delivery	[287]
Sugarcane molasses	Hydrothermal	1.9	5.8	Blue	365	Sensoring of Fe^{3+} and cellular imaging	[302]
Mango leaves	Microwave	2–8	-	Red	325	Cellular imaging and temperature sensors	[303]
Papaya juice	Hydrothermal	3	7	Blue Green Red	365 488 561	Cellular imaging	[304]
Latex	Microwave	2–8	-	green	360–520	Metal sensing and cellular imaging	[305]

Table 1. *Cont.*

Sources of Carbon	Preparation Techniques	Size (nm)	Quantum Yield (%)	Color	Excitation Wavelength (nm)	Application	Ref.
Mangosteen pulp	Hydrothermal	5	-	Blue	330	Sensoring of Fe^{3+} and cellular imaging	[306]
Lotus root	Microwave	9.41	19	Blue	360	Heavy metal ion detection and cellular imaging	[307]
Date kernel	Hydrothermal	2.5	12.5	Blue	365	Sensing of drugs and cellular imaging	[308]
Winter melon	Hydrothermal	4.5–5.2	7.51	Blue	360	Cellular imaging	[309]

* Hela, human epithelial carcinoma; HepG2, human hepatocarcinoma cells; MCF-7, breast cancer cell.

8. Challenges and Future Perspectives

8.1. Current Challenges in Carbon Nanodot-Based Drug Delivery

Carbon nanodots (CNDs) have emerged as promising candidates for drug delivery applications due to their unique physicochemical properties, biocompatibility, and potential for targeted therapy. However, several challenges persist in harnessing the full potential of CNDs for effective drug delivery systems. Understanding and addressing these challenges are critical for advancing this field towards practical biomedical applications.

One of the primary challenges in utilizing CNDs for drug delivery is the lack of standardized synthesis methods and characterization techniques. CNDs can be synthesized using various approaches, such as laser ablation, chemical oxidation, or microwave-assisted methods, leading to variations in size, surface chemistry, and optical properties. This diversity makes it difficult to establish universal protocols for drug loading, stability assessment, and performance evaluation [310]. Stability of CNDs under physiological conditions is crucial for their successful application in drug delivery. Many CNDs are prone to aggregation, which can compromise their dispersibility and effectiveness. Moreover, ensuring long-term biocompatibility and minimizing potential toxicity are ongoing challenges. The effects of CNDs on cellular function, metabolism, and immunogenicity require thorough investigation to ensure their safe use in vivo [311]. To improve CNDs' biocompatibility, researchers have concentrated on improving their surface functionalization and production techniques. Potential harmful effects are lessened by the selection of precursor materials and the use of biocompatible polymers. By adding hydrophilic functional groups to the surface, including hydroxyl or amino groups, CNDs' water solubility and biocompatibility are increased and negative interactions in biological systems are reduced.

Achieving high drug-loading capacity while maintaining controlled-release kinetics is another hurdle. As aforementioned, the surface chemistry of CNDs plays a pivotal role in drug loading efficiency and release behavior. Engineering CNDs with specific functional groups can enhance drug binding affinity and regulate release profiles. However, optimizing these parameters to achieve therapeutic concentrations at target sites remains a complex task. Enhancing targeted delivery and tissue-specific accumulation of drug-loaded CNDs is essential for maximizing therapeutic efficacy and minimizing off-target effects.

Scaling up the synthesis of CNDs for industrial production without compromising quality and cost-effectiveness is a practical challenge. Large-scale production methods must be developed to meet the demand for clinical translation. Additionally, cost-effective strategies for functionalizing CNDs and loading drugs efficiently are essential for widespread adoption in healthcare settings. Lastly, navigating regulatory pathways and demonstrating the safety and efficacy of CND-based drug delivery systems are critical steps towards clinical translation. Establishing robust preclinical models that accurately mimic human physiology and disease states is imperative. Addressing these regulatory and translational hurdles requires interdisciplinary collaboration between researchers, clinicians, and regulatory authorities [312,313].

8.2. Future Directions and Emerging Trends

Thanks to new developments in nanotechnology and developing trends, there is exciting potential for the use of carbon nanodots (CNDs) in medication delivery in the future. Looking ahead, several future directions and emerging trends are shaping the trajectory of CND-based drug delivery. Future efforts will concentrate on designing multifunctional CND-based nanoplatforms capable of simultaneous drug delivery, imaging, and therapy. Integrating functionalities such as targeting ligands, stimuli-responsive elements, and imaging agents into CNDs will enable precise control over drug release kinetics and enhance therapeutic efficacy [311]. Theragnostic CND platforms combining therapeutic and diagnostic capabilities will gain prominence. These systems can enable real-time monitoring of drug delivery, therapeutic response, and disease progression, thereby guiding personalized treatment strategies.

Smart CNDs provide focused and customized drug delivery, reducing adverse effects and maximizing therapeutic benefits. They do this by reacting to particular cues within the body, such as pH changes or the presence of biomolecules. Further advancements in surface engineering techniques will facilitate the development of CNDs with tailored physicochemical properties. Fine-tuning surface chemistry and structure can optimize biocompatibility, cellular uptake, and biodistribution, leading to improved pharmacokinetics and reduced off-target effects [314]. It is anticipated that the investigation of new biocompatible precursors and environmentally friendly synthesis techniques would solve environmental issues and improve the sustainability of CND-based drugs. Furthermore, research endeavors can concentrate on refining large-scale production techniques to expedite the transition of CNDs from laboratory-based to real-world clinical applications. Personalized drug delivery tactics based on unique patient characteristics may become possible as our understanding of the interactions between CNDs and biological systems expands. Customizing CNDs to a patient's unique traits could transform treatment modalities by increasing effectiveness and reducing side effects. Fundamentally, the convergence of innovation, sustainability, and personalized medicine holds the key to the future of CND-based medication delivery, offering revolutionary approaches to a variety of health issues. Collaborative endeavors between academia, industry, and regulatory agencies will be essential for bringing CND-based drug delivery technologies to the bedside [315].

8.3. Patents and Publications on CNDs

Scientific literatures on CNDs based studies (Table 2).

Table 2. Publications of CNDs.

No.	Title	Findings	Results	Discussions	Ref.
1	"Effect of carbon nano-dots (CNDs) on structural and optical properties of PMMA polymer composite"	Strong intermolecular contacts, an improved amorphous phase, well-dispersed CNDs, increased photoluminescence, a shifted refractive index, and well-defined electron transitions were all displayed by the PMMA/CNDs nanocomposite films that were created using the solution cast process.	Amorphous PMMA/CNDs nanocomposite films with enhanced complexation and optical characteristics were created in this work. Improved photoluminescence and UV-VIS absorption point to the material's applicability for photonic devices, LEDs, and other optoelectronic applications.	The stability, optical qualities, and amorphous phase of PMMA were all enhanced by the addition of CNDs. Their promise in nanotechnology devices is highlighted by their enhanced photoluminescence and UV-VIS absorption, which imply suitable for optoelectronic applications including LEDs and photodetectors.	[316]
2	"One-step synthesis and characterization of N-doped carbon nanodots for sensing in organic media"	The N-doped CNDs have useful applications in organic media without further functionalization due to their high quantum yield, excitation wavelength-dependent emission, and upconversion characteristics.	N-doped CNDs with 78% QY were created from PMA and showed excellent selectivity when it came to detecting nitroaromatic explosives by the quenching of fluorescence in organic environments.	In spite of their remarkable selectivity and possible commercial uses, such as self-cleaning surfaces, bioimaging of hydrophobic structures, and antiwetting, these N-doped CNDs hold great promise for the detection of nitroaromatic explosives.	[317]

Table 2. *Cont.*

No.	Title	Findings	Results	Discussions	Ref.
3	"pH-dependent synthesis of novel structure-controllable polymer carbon NanoDots with high acidophilic luminescence and super carbon dots assembly for white-light-emitting diodes:	Different nanodot structures are produced by pH-dependent synthesis. White light from SCNDs is appropriate for LEDs. Investigation by PL reveals distinct emission paths. For carbon nanodots, theoretical computations reveal information on their electrical properties.	Changing the pH led to different topologies of carbon nanodots. Super-small carbon nanodots (SCNDs) showed promise for LED technology at pH < 1. They radiated white light. Unique emission channels were found via PL research, which improved knowledge of carbon-based fluorescence.	pH regulation provides customized nanodot morphology, which is essential for a variety of uses. The white emission of SCNDs offers opportunities for affordable LEDs. Understanding of carbon-based fluorescence is advanced by insights into PL processes, which direct future study.	[318]
4	"Freestanding luminescent films of nitrogen-rich carbon nanodots toward large-scale phosphor-based white-light-emitting devices"	Under UV light, CNDs produced vivid visible light that was for phosphor applications. Solid-state quenching was avoided via large-scale freestanding luminous films distributed across a polymer matrix, allowing for flexible, scalable, and thermally robust solid-state lighting systems.	Nitrogen-rich carbon nanodots (CNDs) with a restricted size range and a well-developed graphitic structure were produced by carbonizing polyacrylamide using an emulsion template. A high quantum yield of 40% was achieved in the fabrication of large-scale luminous films	CNDs with desired features were produced via synthesis using an emulsion-templated carbonization process. Because polymer matrix dispersion avoided quenching, flexible lighting systems on a vast scale could be made possible. Under realistic circumstances, white LEDs showed steady emission spectra, demonstrating the promise of CND-based solid-state lighting.	[319]
5	"Photoluminescence of argan-waste-derived carbon nanodots embedded in polymer matrices"	Excitation-dependent emission was demonstrated by blue-emitting CND-polymer nanocomposites, with the blue spectrum exhibiting the highest emission. By placing CNDs within optically transparent matrices, PLQY was increased by two to three times, reaching a 29.6% improvement.	For use in photonic conversion layers on solar PV cells, luminescent carbon nanodots (CNDs) derived from argan waste were distributed in poly(styrene-co-acrylonitrile) and cyclo-olefin copolymer matrices to generate thin films with a 30% PL conversion efficiency.	After being distributed in transparent polymers, CNDs made from argan waste maintained their long-term luminescence characteristics and increased radiative efficiency by two to three times. Because thin films are easily processed, they may be used as photonic down-conversion layers to improve the efficiency of solar cells, especially when UV light is used.	[320]

Table 2. *Cont.*

No.	Title	Findings	Results	Discussions	Ref.
6	"Oxidative synthesis of highly fluorescent boron/nitrogen-co-doped carbon nanodots enabling detection of photosensitizer and carcinogenic dye"	P-CNDs with PEI passivation showed increased fluorescence after being synthesized in a simple manner. Protoporphyrin (PPD) introduction enabled fluorescence switch-off, allowing dye-doped nanoprobes with a limit of detection (LOD) of 9.9 pM−0.37 nM for Sudan red III (SRIII) and 15 pM for PPD.	The synthesis of boric acid and N-(4-hydroxyphenyl) glycine by hydrothermal oxidative method resulted in the straightforward production of carbon nanodots (CNDs) co-doped with silicon and nitrogen. polyethyleneimine (PEI) surface passivation improved fluorescence and monodispersity, resulting in polymerized CNDs (P-CNDs) with a 23.71% quantum yield.	Highly fluorescent B/N co-doped CNDs are easily synthesized and have potential uses in a number of fields. The surface passivation of PEI enhances monodispersity and fluorescence. PPD and SRIII may be detected with high sensitivity using dye-doped nanoprobes, indicating the possibility of useful sensing and detection applications.	[321]
7	"Fluorescent nitrogen-doped carbon nanodots synthesized through a hydrothermal method with different isomers"	N-oxide group production was aided by the o-PD precursor, whereas "lattice N" functionalities were inserted by hydrothermal synthesis using the m-PD precursor. While strongly quenched in propylene glycol methyl ether acetate (PGMEA), fluorescence was bright in polar solvents. Radiative emission from N-atom substitutions and N- and O-rich edge groups produced an ultrahigh quantum yield.	Using o-, m-, and p-phenylenediamine (PD) isomers, N-functionalized carbon nanodots (CNDs) were created hydrothermally, allowing for exact control of the N/C atomic ratio (20.2–25.7 at.%). In polar solvents, CNDs showed up to 99% ultrahigh quantum yield of strong fluorescence.	N-functionalized CNDs can have their characteristics precisely tuned by the hydrothermal process, which has applications in optical, sensing, energy storage/conversion, and biological devices. Strong fluorescence in polar solvents suggests that high-performing nanomaterials may find use in a range of fields.	[322]
8	"Synthesis of carbon nanodots from sugarcane syrup, and their incorporation into a hydrogel-based composite to fabricate innovative fluorescent microstructure polymer optical fibres"	The synthesis of CNDs is scalable, economical, and sustainable. Because of their N and O-rich edge groups, functionalized CNDs showed excellent quantum yields (85–99%), which increased their potential for use in optical and sensing applications.	Using a home microwave oven, CNDs with a 3 nm diameter and low polydispersity were created from sugarcane syrup. They were added to an optical fiber and fluorescent hydrogel composite and displayed fluorescence.	This work offers a green synthesis approach for CNDs that might potentially replace costly and harmful compounds used in optical fibers. The novel hydrothermal method improves the fluorescence and application of CNDs by precisely controlling N-functionalization.	[323]

Table 2. *Cont.*

No.	Title	Findings	Results	Discussions	Ref.
9	"Synthesis of highly stable red-emissive carbon polymer dots by modulated polymerization: from the mechanism to application in intracellular pH imaging"†	This study offers a green technique for the synthesis of CNDs that might replace costly, hazardous compounds. R-CPDs' surface state and crosslink enhanced emission (CEE) effect are the sources of their red emission. In optical fibers, they exhibit excellent stability, biocompatibility, and appropriateness for intracellular pH monitoring in HeLa cells. The novel hydrothermal method improves the fluorescence and application of CNDs by precisely controlling N-functionalization.	Red-emissive carbon polymer dots (R-CPDs) were synthesized at 80 °C and shown resistance to photobleaching, stability in high salinity, high pH sensitivity (pH 4–6), and adjustable solvent-color effect (λem 528–600 nm).	This study presents a straightforward, controlled technique for producing highly stable, biocompatible long-wavelength emitting R-CPDs. On the processes of photoluminescence, cellular uptake, and multifunctional uses, more study is required.	[324]
10	"Synthesis of surface molecularly imprinted poly-o-phenylene diamine/TiO_2/ carbon nanodots with a highly enhanced selective photocatalytic degradation of pendimethalin herbicide under visible light"	Adsorption and selectivity were enhanced by the imprinted cavities and particular recognition sites on MIP. Photocatalytic activity was increased by the redshifted absorption to visible areas and the lowered band gap energy. The primary species responsible for PM photodegradation were O_2% radicals.	Using PM herbicide as a template, a TiO_2/CNDs/MIP nanocomposite was created. Under visible light, it demonstrated a high adsorption capacity (86.1 mg/g), good selectivity, and increased photodegradation efficiency (95%).	The TiO_2/CNDs/MIP nanocomposite, with high stability and reusability, effectively adsorbs and degrades PM due to its unique structure. It offers a promising photocatalyst for environmental pollutant removal, leveraging lower energy to produce reactive species.	[325]
11	"Direct solvent-derived polymer-coated nitrogen-doped carbon nanodots with high water solubility for targeted fluorescence imaging of glioma"	N-CNDs demonstrated superior dispersibility, showed minimal cytotoxicity, and improved passive targeting to facilitate glioma fluorescence imaging. NMP was used in the synthesis as a source of carbon and nitrogen, as well as a solvent.	A direct solvothermal process was used to create pN-CNDs, which produced 5–15 nm particles with a quantum yield of 8.4%, sustained fluorescence, and great water solubility. They facilitated in vivo fluorescence imaging by penetrating glioma cells.	An effective method for producing functional carbon nanomaterials with promise for glioma-targeted imaging is the straightforward solvothermal synthesis of pN-CNDs. Their chemical makeup, growth, and targeting methods require more investigation.	[326]

Table 2. *Cont.*

No.	Title	Findings	Results	Discussions	Ref.
12	"Evolution and synthesis of carbon dots: from carbon dots to carbonized polymer dots"	Different from conventional CNDs, CPDs are characterized by partial carbonization of polymer clusters. The lack of control over structure and performance in current synthesis methods limits their applicability in biolabeling, sensing, LEDs, and other areas.	The unique polymer/carbon hybrid structure of CPDs, a novel type of carbon dots, is apparent. Different bottom-up synthesis techniques show how synthesis circumstances affect the structures and characteristics of CPDs.	Future studies should concentrate on comprehending the principles of CPD synthesis, reaction mechanisms, and formation processes in order to achieve regulated synthesis and maximize their potential for use in a variety of disciplines.	[188]
13	"Design, synthesis, and functionalization strategies of tailored carbon nanodots"	Tuning the emissive, electrochemical, and chiroptical characteristics of the CNDs was made possible by regulating the reaction conditions. Their surface chemistry was further modified by post-functionalization, which increased their potential for use in a variety of applications, including energy conversion, sensing, and imaging.	They synthesized nitrogen-doped carbon nanodots (CNDs) that generate blue light by a bottom-up, microwave-assisted hydrothermal process. These CNDs were effectively utilized in hybrid and composite systems and showed tunable optoelectronic characteristics.	CNDs are appropriate for biomedical and energy applications due to their low cost, low toxicity, and strong photostability. Controlling synthesis for improved structural and performance regulation should be the main goal of future study, since this will increase their usefulness in many other domains.	[327]
14	"Facile synthesis of multicolour photoluminescent polymer carbon dots with surface-state energy gap-controlled emission"	The multicolor photoluminescence of PCNDs was found to be mostly controlled by the surface state, namely the C=N functional groups. As the C=N concentration rose, the band gap shrunk, and the emission peak moved.	Hydroquinone and ethylenediamine were used to develop multicolor emissive PCNDs that emitted green, blue, and yellow fluorescence. These PCNDs demonstrated outstanding solubility, high stability, and wavelength-independent photoluminescence upon stimulation.	A brand-new, gentle, and simple process was created to create PCNDs with outstanding water solubility and brilliant, steady emissions. A thorough characterization revealed how important surface states are in dictating the photoluminescence characteristics of PCNDs.	[328]
15	"Synthesis, separation, and characterization of small and highly fluorescent nitrogen-doped carbon nanodots"	NCNDs demonstrated strong luminescence, excellent fluorescence quantum yields (up to 0.46), and ease of functionalization. The surface states, which are impacted by various emission centers and traps, have a significant impact on the fluorescence.	Using a microwave-assisted process, nitrogen-doped CNDs (NCNDs) were created, producing particles with a narrow size distribution, adjustable fluorescence emission, and superior water solubility. Their size and surface characteristics were further improved using size-exclusion chromatography.	NCNDs with exceptional optical characteristics were manufactured via a straightforward, programmable microwave-assisted approach that controlled both surface and size. These adaptable NCNDs may find use in biomedicine, bioimaging, and optoelectronics	[329]

Table 2. *Cont.*

No.	Title	Findings	Results	Discussions	Ref.
16	"Ultrahigh-yield synthesis of N-doped carbon nanodots with down-regulating ROS in zebrafish"	With a larger C=C percentage, the synthesis yield of CNDs rose by 3.3 times. By adding nitrogen in the forms of pyridinic-like N (74%) and NH2 (26%), the antioxidative qualities against ROS were strengthened.	A novel approach using carbon-carbon double bonds achieved a record-breaking 85.9% yield in synthesizing nitrogen-doped carbon nanodots (CNDs). These CNDs significantly reduced reactive oxygen species (ROS) by 68% in zebrafish.	A viable method for creating antioxidative nitrogen-doped CNDs is the idea of increasing synthesis yield via carbon–carbon double bonds. These CNDs may be used as nanodrugs to treat illnesses associated with ageing.	[330]
17	"A tumor microenvironment responsive mesoporous polydopamine theragnostic probe embedded with Gd/I-doped carbon nanodots for CT/MR/FL imaging and chemo/photothermal synergistic therapy"	A probe using mesoporous polydopamine (PDA) and Gd/I-doped carbon nanodots was developed for targeted tumor imaging and combined chemo/photothermal therapy. This probe responded to tumor microenvironment.	This probe increased the imaging contrast, as well as the therapeutic efficacy both in vitro and in vivo. Significant inhibition in tumor was observed when both chemotherapy and photothermal modalities were combined.	This theragnostic probe is a promising strategy for tumor imaging and personalized therapy. This can easily overcome limitations of single-modality cancer therapy whilst minimizing side effects.	[331]
18	"Printable Thermo- and Photo-stable Poly(D,L-lactide)/Carbon Nanodots Nanocomposites via Heterophase Melt-Extrusion Transesterification"	They developed Poly (D, L-lactide) (PDLA)/ carbon nanodots which had enhanced thermal and photostability. This was achieved through heterophase melt-extrusion transesterification. This also enhanced the mechanical properties of the CNDs significantly.	The resulting CNDs had showed excellent stability under induced photo and thermal stress with high resistance to degradation. They also retained their optical properties.	Retention of optical properties with high mechanical strength indicates utility in sensors and packaging materials. Melt-extrusion transesterification also provides a simple, scalable method to produce high performance CNDs.	[332]
19	"Decagram-Scale Synthesis of Multicolor Carbon Nanodots: Self-Tracking Nanoheaters with Inherent and Selective Anticancer Properties"	This study was able to develop a multicolor CND in decagram quantities. These CNDs had selective anticancer effects and self-tracking properties.	CNDs functioned as both fluorescent agent with strong photothermal conversion efficiency and nanoheaters for photothermal therapy against cancer whilst minimal toxicity to healthy cells. In vivo tests also confirmed these.	This can be an economical and scalable method to produce multicolor CNDs with dual properties that can aid in diagnostics, as well as therapy.	[333]
20	"Hyaluronic acid dressing of hydrophobic carbon nanodots: A self-assembling strategy of hybrid nanocomposites with theragnostic potential"	Authors introduced a self-assembly strategy to create hybrid nanocomposites by dressing CNDs with Hyaluronic Acid. This strategy enhanced the biocompatibility and stability of CNDs along with their use as a theranostic agent.	HA-CNDs had excellent water dispersibility and biocompatibility. This also had excellent tumor targeting abilities due to the affinity of hyaluronic acid towards CD44 receptors of cancer cells. In vitro and in vivo tests demonstrated their capabilities of diagnostics, as well as drug delivery therapy.	This hybrid system offered a simple multifunctional CND with theranostic applications especially in cancer diagnosis and treatment. This is a promising candidate for less invasive and personalized cancer diagnostics and therapy.	[334]

Table 2. *Cont.*

No.	Title	Findings	Results	Discussions	Ref.
21	"Charge-Convertible Carbon Dots for Imaging-Guided Drug Delivery with Enhanced in Vivo Cancer Therapeutic Efficiency"	This study developed charge-convertible CDs which enabled controlled drug release leading to improved tumor targeting efficiency. These also had fluorescence and MRI capabilities.	These CDs showed excellent in vivo performance with increased accumulation in tumor tissues. pH responsive behavior of these CDs allowed efficient drug release at acidic tumor microenvironment.	This presents a novel strategy for image guided drug delivery in cancer therapy with improved therapeutic outcomes and minimal side effects. Ability to release drug in acidic tumor microenvironment further makes this approach, a chance at personalized cancer therapy with higher efficacy.	[335]
22	"NIR Emissive Carbon Nanodots as a Tool to Mark Ribosomal RNA and Components using Super-Resolution Microscopy"	This study developed near infrared (NIR) emissive CNDs for labelling components of a live cell especially rRNA. These CNDs showed stability, biocompatibility and NIR fluorescence for high-resolution imaging of cellular structures.	CNDs were able to label ribosomal components, ribosome distribution within the cell, as well as visualization of rRNA. This provided insights into ribosomes that were not detectable with traditional imaging.	This highlights the importance of NIR emission CNDs in studying live cellular components with high resolution, especially ribosomal function and assembly. This marks a new approach for investigating molecular mechanisms of various synthesis and processes within the cell.	[336]
23	"Controlled delivery of sildenafil by β-Cyclodextrin-decorated sulfur-doped carbon nanodots: a synergistic activation of ROS signaling in tumors"	This study developed a drug delivery system for controlled release of sildenafil to enhance reactive oxygen species production in tumor cells with the help of β-cyclodextrin decorated sulfur-doped CNDs. This is an approach to improve synergistic ROS signaling activation by sildenafil.	The sulfur-doped CNDs delivered sildenafil in a controlled manner which enhanced the cytotoxic effects of sildenafil in tumor cells due to increased oxidative damage. These demonstrated improved therapeutic outcomes in comparison to free sildenafil, in vivo.	This is a promising approach for controlled drug delivery and tumor specific therapy to enhance anticancer activity by using ROS pathway.	[337]

Scientific literature presented in tabular form on Patents published in CNDs study (Table 3).

Table 3. Patents on CNDs.

No.	Title	Patent Number	Findings	Date	Ref.
1	"Metal-enhanced photoluminescence from carbon nanodots"	US10837904B2	The invention enhances the detectable emissions of carbon nanodots through Metal-Enhanced Fluorescence (MEF). By positioning carbon nanodots at an optimal distance from plasmon-supporting materials like silver island films, this technique significantly improves brightness, photostability, and detectability, making it highly effective for biological imaging applications.	17 November 2020	[338]

Table 3. *Cont.*

No.	Title	Patent Number	Findings	Date	Ref.
2	"Nanomaterials with enhanced drug delivery efficiency"	US11478433B2	Based on medicinal natural products, supramolecular particles improve bioavailability, increase stability in acidic conditions, and distribute therapeutic agents efficiently, leading to better treatment results for diabetes or tumors.	25 October 2022	[339]
3	"Carbon nanodot, and preparation method and application thereof"	CN102849722B	The technique solves quenching problems in the production of highly fluorescent carbon nanodots, which may be used for cryptography, photovoltaics, and biological imaging, among other things. It is straightforward and inexpensive.	29 August 2012	[340]
4	"Nanocarbon composite structure having ruthenium oxide trapped therein"	US7572542B2	The nanocarbon composite, incorporating ruthenium oxide within graphene via Ketjen black and ultracentrifugal reaction, exhibits enhanced electrochemical activity, making it suitable for high-capacity capacitor applications in electrical energy storage.	10 June 2005	[341]
5	"Traditional Chinese medicine bio-based carbon nanodots, preparation method thereof, fluorescent probe, traditional Chinese medicine pharmaceutical preparation and application"	CN111778018A	The innovation proposes a carbon nanodot based on ginsenoside that has a large number of surface functional groups that allow for flexible alterations and good stability. It is biocompatible with other substances and functions as a potential biological fluorescence probe. It also shows selective inhibition on PC12 cells.	8 June 2020	[342]
6	"Preparation and regulation method of high-colour quality fluorescent carbon nanodots"	CN109504375B	Using this technique, standard white-light emission and near-spectrum matching with sunlight are achieved, yielding high-quality fluorescent carbon nanodots with good color attributes.	12 December 2018	[343]
7	"Carbon nanodot compound and preparation method thereof, fluorescent powder and light source material"	CN106833631B	A stable carbon nanodot complex with a silicon dioxide covering is presented in this invention, providing improved fluorescence qualities and resilience to environmental influences.	4 February 2017	[344]
8	"Fused carbon dot, preparation method and application thereof"	CN113403068B	This innovation includes fused carbon dots manufactured by a straightforward, economical technique that exhibits improved near-infrared emission and good light–heat conversion.	16 June 2021	[345]

9. Conclusions

The utilization of carbon nanodots (CNDs) as carriers for phytomedicine delivery holds significant promise in overcoming the limitations associated with traditional drug delivery systems. By leveraging the unique properties of CNDs, such as their tunable surface chemistry; high loading capacity on the surface and using hollow CNDs and nanosized dots, and biocompatibility, researchers can enhance the solubility, stability, bioavailability, and biocompatibility of phytomedicinal compounds. When considered as a drug delivery armament for phytopharmaceuticals, CNDs offer a wide range of applications which can be coupled with its high precision owing to its surface modifications. Its size being in the nanometer range allows for ease of penetration of various physiological barriers, something that is traditionally a challenge for phytopharmaceuticals. This property of CNDs is particularly advantageous in the treatment of cancers and neurological disorders crossing BBB and interacting via potential charge against target tissues with a polymeric nanocomposite combination. The review has highlighted various strategies for loading phytomedicine onto CNDs, including encapsulation and surface modification, as well as mechanisms for controlled release. This helps in targeting ligands for delivering the phytopharmaceuticals, leading to an increase in efficacy, as well as imaging with enhanced resolutions, even imaging of processes going on inside a living cell. This review emphasizes the various methods of synthesis of CNDs from different starting materials, their advantages, and few limitations, and it moves on to further discuss the use of such CNDs in theragnostic applications, especially as a drug delivery system for phytopharmaceuticals. CNDs could solve some major traditional problems associated with phytopharmaceuticals, i.e., poor stability, low solubility, and most importantly, low bioavailability coupled with poor pharmacokinetics parameters. However, challenges such as biocompatibility and toxicity concerns need to be addressed through rigorous research and development efforts. Looking ahead, continued advancements in CND-based drug delivery systems are expected to revolutionize the field of phytomedicine, paving the way for novel therapeutic interventions with enhanced efficacy and reduced side effects. CNDs possessing large surface areas may be able to interact with target marker biomolecules for impressive remediation against the desirable health issues. Moreover, large surface area-based phytopharmaceutical-mediated CNDs can be used as potential therapeutics for treating inflammation, gut microbial disorders, diarrhea, and neurodegenerations owing to their bio-adsorbent, cleansing, and toxin-diffusing bioremediation properties. However, the toxicity, functionalization, yields, and scalability of most CNDs need to be optimized and evaluated for significant biomedical application with impressive remedy against chronic health disorders in clinical models.

Author Contributions: Conceptualization, S.T. and P.P.; validation, N.R.; data curation, modification, H.K.; data curation, R.D. and A.M.A.I.; writing—original draft preparation, R.D. and A.M.A.I.; writing—review and editing, P.P. and H.K.; visualization, H.K. and P.P.; supervision, P.P. and S.T. All authors have read and agreed to the published version of the manuscript.

Funding: This research received no external funding.

Acknowledgments: The authors are thankful to Mahatma Gandhi University, Kerala, India; Center of Molecular and Macromolecular Studies, Polish Academy of Sciences, Poland; and Assam University, Silchar, India for enormous support in designing this noteworthy review study. During the preparation of this work, the author(s) utilized OpenAI's QuillBot to improve English language, quality, and readability. Following the use of this tool/service, the author(s) reviewed, edited and validated the content of writeup as necessary and take(s) full responsibility for the publication's content.

Conflicts of Interest: The authors declare no conflict of interest.

List of Abbreviation

CND	Carbon nanodot
CNDs	Carbon nanodots
DNA	Deoxyribonucleic acid
RNA	Ribonucleic acid
BBB	Blood–brain barrier
EPR	Enhanced Permeability and Retention

EDTA	Ethylene Diamine Tetra Acetic acid
TEM	Transmission Electron Microscopy
GQDs	Graphene quantum dots
CNDs	Carbon dots
PDs	Polymer dots
CNPs	Carbon nanoparticles
QCE	Quantum confinement effect
PL	Photoluminescence
CD	Carbon dots
QY	Quantum yield
H_2SO_4	Sulfuric acid
HNO_3	Nitric acid
$NaNO_3$	Sodium nitrate
$KmNO_4$	Potassium permanganate
NaOH	Sodium hydroxide
H_2O_2	Hydrogen peroxide
CQDs	Carbon Quantum Dots
TiO_2	Titanium dioxide
MW	Microwave
NAC	N-acetyl-L-cysteine
PEG200	NH2-polyethylene-glycol
NPs	Nanoparticles
DDS	Drug delivery systems
DOX	Doxorubicin
ADCs	Antibody–drug conjugates
PNCs	Polymer nanocomposites
CTS	Chitosan
sADM	S-Adenosylmethionine
NDs	Nanodrugs
EPR	Enhanced Permeability and Retention
LEDs	Light-Emitting Diodes
PEG	Polyethylene glycol
PLA	Polylactic acid
PTT	Photothermal treatment
NIR	Near-infrared
EACs	Ehrlich ascites carcinoma cells
MTT	(3-(4,5-Dimethylthiazol-2-yl)-2,5-diphenyl tetrazolium bromide)
HA	Hyaluronate
PMMA	Polymethyl Methacrylate
NMP	N-Methyl-2-Pyrrolidone
N-CNDs	Nitrogen-Doped Carbon Nanodots
pN-CNDs	Phosphorus and Nitrogen-Doped Carbon Nanodots
CPD	Carbon polymer dots
SCNDs	Super-small carbon nanodots
UV	Ultraviolet
LOD	Limit of Detection
PEI	Polyethyleneimine
PPD	Protoporphyrin
PLQY	Photoluminescence Quantum Yield
P-CNDs	polymerized CNDs
SR III	Sudan red III
PGMEA	Propylene glycol methyl ether acetate
R-CPDs	Red-emissive carbon polymer dots
CEE	Crosslink enhanced emission

PCND	Photoluminescence CND
NCNDs	Nitrogen-doped CNDs
ROS	Reactive oxygen species
MEF	Metal-Enhanced Fluorescence

References

1. Manisha, H.; Swetha, P.D.P.; Shim, Y.-B.; Prasad, K.S. Revisiting fluorescent carbon nanodots for environmental, biomedical applications and puzzle about fluorophore impurities. *Nano-Struct. Nano-Objects* **2019**, *20*, 100391. [CrossRef]
2. Khan, S.; Dunphy, A.; Anike, M.S.; Belperain, S.; Patel, K.; Chiu, N.H.L.; Jia, Z. Recent Advances in Carbon Nanodots: A Promising Nanomaterial for Biomedical Applications. *Int. J. Mol. Sci.* **2021**, *22*, 6786. [CrossRef] [PubMed]
3. Su, M.; Hui, Y. Introduction to Carbon Structures. Undefined 2018. Available online: https://books.google.com/books?hl=en&lr=&id=0cN2DwAAQBAJ&oi=fnd&pg=RA1-PA3&dq=3.%09Meng%E2%80%90Chih,+S.;+Yung,+Hui,+Y.+Introduction+to+Carbon+Structures.+Carbon+Nanomaterials+for+Bioimaging,+Bioanal.,+The&ots=iRH5JD2AJ3&sig=niy68P9hCduZxe0-rlTVobQsfZM (accessed on 25 December 2024).
4. Đorđević, L.; Arcudi, F.; Prato, M. Preparation, functionalization and characterization of engineered carbon nanodots. *Nat. Protoc.* **2019**, *14*, 2931–2953. [CrossRef] [PubMed]
5. Sawalha, S.; Assali, M.; Nasasrah, A.; Salman, M.; Nasasrah, M.; Jitan, M.; Hilal, H.S.; Zyuod, A. Optical properties and photoactivity of carbon nanodots synthesized from olive solid wastes at different carbonization temperatures. *RSC Adv.* **2022**, *12*, 4490–4500. [CrossRef] [PubMed]
6. Vibhute, A.; Patil, T.; Gambhir, R.; Tiwari, A.P. Fluorescent carbon quantum dots: Synthesis methods, functionalization and biomedical applications. *Appl. Surf. Sci. Adv.* **2022**, *11*, 100311. [CrossRef]
7. Singh, A.; Kafle, S.R.; Sharma, M.; Kim, B.S. Comprehensive Review on Multifaceted Carbon Dot Nanocatalysts: Sources and Energy Applications. *Catalysts* **2023**, *13*, 1446. [CrossRef]
8. Daoudi, W.; Tiwari, A.; Tyagi, M.; Singh, P.; Saxena, A.; Verma, D.K.; Dagdag, O.; Sharma, H.K.; Fuertes, P.O.; Aatiaoui, A.E. Carbon dots nanoparticles: A promising breakthrough in biosensing, catalysis, biomedical and authers applications. *Nano-Struct. Nano-Objects* **2024**, *37*, 101074. [CrossRef]
9. Kumar, Y.R.; Deshmukh, K.; Sadasivuni, K.K.; Pasha, S.K.K. Graphene quantum dot based materials for sensing, bio-imaging and energy storage applications: A review. *RSC Adv.* **2020**, *10*, 23861–23898. [CrossRef] [PubMed]
10. Cadranel, A.; Margraf, J.T.; Strauss, V.; Clark, T.; Guldi, D.M. Carbon Nanodots for Charge-Transfer Processes. *Acc. Chem. Res.* **2019**, *52*, 955–963. [CrossRef]
11. Patra, J.K.; Das, G.; Fraceto, L.F.; Campos, E.V.R.; Rodriguez-Torres, M.D.P.; Acosta-Torres, L.S.; Diaz-Torres, L.A.; Grillo, R.; Swamy, M.K.; Sharma, S.; et al. Nano based drug delivery systems: Recent developments and future prospects. *J. Nanobiotechnol.* **2018**, *16*, 71. [CrossRef]
12. Sahu, T.; Ratre, Y.K.; Chauhan, S.; Bhaskar, L.V.K.S.; Nair, M.P.; Verma, H.K. Nanotechnology based drug delivery system: Current strategies and emerging therapeutic potential for medical science. *J. Drug Deliv. Sci. Technol.* **2021**, *63*, 102487. [CrossRef]
13. Stielow, M.; Witczyńska, A.; Kubryń, N.; Fijałkowski, Ł.; Nowaczyk, J.; Nowaczyk, A. The Bioavailability of Drugs-The Current State of Knowledge. *Molecules* **2023**, *28*, 8038. [CrossRef]
14. Mendoza-Muñoz, N.; Urbán-Morlán, Z.; Leyva-Gómez, G.; Zambrano-Zaragoza, M.D.L.L.; ¨Piñón-Segundo, E.; Quintanar-Guerrero, D. Solid Lipid Nanoparticles: An Approach to Improve Oral Drug Delivery. *J. Pharm. Pharm. Sci.* **2021**, *24*, 509–532. [CrossRef] [PubMed]
15. Manikkath, J.; Manikkath, A.; Lad, H.; Vora, L.K.; Mudgal, J.; Shenoy, R.R.; Ashili, S.; Radhakrishnan, R. Nanoparticle-Mediated Active and Passive Drug Targeting in Oral Squamous Cell Carcinoma: Current Trends and Advances. *Nanomedicine* **2023**, *18*, 2061–2080. [CrossRef]
16. Mohammed, J.; Desu, P.K.; Namratha, J.R.; Rao, G.S.N.K. Applications of Carbon Dots (CDs) in Drug Delivery. *Adv. Pharmacol. Pharm.* **2023**, *11*, 36–45. [CrossRef]
17. Bayda, S.; Amadio, E.; Cailotto, S.; Frión-Herrera, Y.; Perosa, A.; Rizzolio, F. Carbon dots for cancer nanomedicine: A bright future. *Nanoscale Adv.* **2021**, *3*, 5183–5221. [CrossRef] [PubMed]
18. Huang, W.-L.; Hsu, C.-J.; Wang, J.H.; Hsu, Z.-H.; Lu, K.-H.; Wang, W.-D.; Huang, C.-C. Low-Toxicity Antibacterial Carbon Nanodots for Wound Dressings and Food Packaging. *ACS Appl. Nano Mater.* **2023**, *6*, 19200–19209. [CrossRef]
19. Nair, A.; Haponiuk, J.T.; Thomas, S.; Gopi, S. Natural carbon-based quantum dots and their applications in drug delivery: A review. *Biomed. Pharmacother* **2020**, *132*, 110834. [CrossRef]
20. Lee, C.; Kwon, W.; Beack, S.; Lee, D.; Park, Y.; Kim, H.; Hahn, S.K.; Rhee, S.-W.; Kim, C. Biodegradable Nitrogen-Doped Carbon Nanodots for Non-invasive Photoacoustic Imaging and Photothermal Therapy. Undefined 2016. Available online: https://www.ncbi.nlm.nih.gov/pmc/articles/PMC5135443/ (accessed on 24 December 2024).

21. Ratre, P.; Nazeer, N.; Kumari, R.; Thareja, S.; Jain, B.; Tiwari, R.; Kamthan, A.; Srivastava, R.K.; Mishra, P.K. Carbon-Based Fluorescent Nano-Biosensors for the Detection of Cell-Free Circulating MicroRNAs. *Biosensors* **2023**, *13*, 226. [CrossRef] [PubMed]
22. Hosnedlova, B.; Kepinska, M.; Fernandez, C.; Peng, Q.; Ruttkay-Nedecky, B.; Milnerowicz, H.; Kizek, R. Carbon Nanomaterials for Targeted Cancer Therapy Drugs: A Critical Review. *Chem. Rec.* **2018**, *19*, 502–522. [CrossRef] [PubMed]
23. Guido, C.; Baldari, C.; Maiorano, G.; Mastronuzzi, A.; Carai, A.; Quintarelli, C.; De Angelis, B.; Cortese, B.; Gigli, G.; Palamà, I.E. Nanoparticles for Diagnosis and Target Therapy in Pediatric Brain Cancers. *Diagnostics* **2022**, *12*, 173. [CrossRef] [PubMed]
24. Hong, G.; Diao, S.; Antaris, A.L.; Dai, H. Carbon Nanomaterials for Biological Imaging and Nanomedicinal Therapy. *Chem. Rev.* **2015**, *115*, 10816–10906. [CrossRef] [PubMed]
25. Panwar, N.; Soehartono, A.M.; Chan, K.K.; Zeng, S.; Xu, G.; Qu, J.; Coquet, P.; Yong, K.-T.; Chen, X. Nanocarbons for Biology and Medicine: Sensing, Imaging, and Drug Delivery. *Chem. Rev.* **2019**, *119*, 9559–9656. [CrossRef] [PubMed]
26. Karami, M.H.; Abdouss, M.; Rahdar, A.; Pandey, S. Graphene quantum dots: Background; synthesis methods, and applications as nanocarrier in drug delivery and cancer treatment: An updated review. *Inorg. Chem. Commun.* **2024**, *161*, 112032. [CrossRef]
27. Dhamodharan, D.; Byun, H.-S.; Shree, M.V.; Veeman, D.; Natrayan, L.; Stalin, B. Carbon nanodots: Synthesis; mechanisms for bio-electrical applications. *J. Ind. Eng. Chem.* **2022**, *110*, 68–83. [CrossRef]
28. Sivasankarapillai, V.S.; Kirthi, A.V.; Akksadha, M.; Indu, S.; Dharshini, U.D.; Pushpamalar, J.; Karthik, L. Recent advancements in the applications of carbon nanodots: Exploring the rising star of nanotechnology. *Nanoscale Adv.* **2020**, *2*, 1760–1773. [CrossRef]
29. Ostadhossein, F.; Pan, D. Functional carbon nanodots for multiscale imaging and therapy. *WIREs Nanomed. Nanobiotechnol.* **2016**, *9*, e1436. [CrossRef] [PubMed]
30. Cohen, E.N.; Kondiah, P.P.D.; Choonara, Y.E.; Toit, L.C.D.; Pillay, V. Carbon Dots as Nanotherapeutics for Biomedical Application. *Curr. Pharm. Des.* **2020**, *26*, 2207–2221. [CrossRef] [PubMed]
31. Kuznietsova, H.; Géloën, A.; Dziubenko, N.; Zaderko, A.; Alekseev, S.; Lysenko, V.; Skryshevsky, V. In vitro and in vivo toxicity of carbon dots with different chemical compositions. *Discov. Nano* **2023**, *18*, 111. [CrossRef] [PubMed]
32. Chiari-Andréo, B.G.; Abuçafy, M.P.; Manaia, E.B.; da Silva, B.L.; Rissi, N.C.; Oshiro-Júnior, J.A.; Chiavacci, L.A. Drug Delivery Using Theranostics: An Overview of its Use, Advantages and Safety Assessment. *Curr. Nanosci.* **2020**, *16*, 3–14. [CrossRef]
33. Ding, S.; Khan, A.I.; Cai, X.; Song, Y.; Lyu, Z.; Du, D.; Dutta, P.; Lin, Y. Overcoming blood-brain barrier transport: Advances in nanoparticle-based drug delivery strategies. *Mater. Today* **2020**, *37*, 112–125. [CrossRef] [PubMed]
34. Gaur, M.; Misra, C.; Yadav, A.B.; Swaroop, S.; Maolmhuaidh, F.Ó.; Bechelany, M.; Barhoum, A. Biomedical Applications of Carbon Nanomaterials: Fullerenes, Quantum Dots, Nanotubes, Nanofibers, and Graphene. *Materials* **2021**, *14*, 5978. [CrossRef]
35. Yazdani, S.; Mozaffarian, M.; Pazuki, G.; Hadidi, N.; Villate-Beitia, I.; Zárate, J.; Puras, G.; Pedraz, J.L. Carbon-Based Nanostructures as Emerging Materials for Gene Delivery Applications. *Pharmaceutics* **2024**, *16*, 288. [CrossRef]
36. Wynn, S.G.; Fougere, B. The Roots of Veterinary Botanical Medicine. Undefined 2007. Available online: https://books.google.com/books?hl=en&lr=&id=iLbZDzumqt0C&oi=fnd&pg=PA33&dq=42.%09Wynn,+S.G.;+Foug%C3%A8re,+B.J.+The+roots+of+veterinary+botanical+medicine.+Veterinary&ots=qUPqPM3_bV&sig=Y-q9-yhOtEARWKMSgoIKEUrIRwQ (accessed on 26 December 2024).
37. Sun, L.; Zhang, R.; Zhang, T.; Liu, X.; Zhao, Y.; Yang, M.; Cheng, H.; Zhang, Q.; Zhang, Y.; Wu, X. Synthesis, applications and biosafety evaluation of carbon dots derived from herbal medicine. Undefined 2023. Available online: https://iopscience.iop.org/article/10.1088/1748-605X/acdeb8/meta (accessed on 26 December 2024).
38. Jamal, A. Embracing Nature's Therapeutic Potential: Herbal Medicine. *Int. J. Multidiscip. Sci. Arts* **2023**, *2*, 117–126. [CrossRef]
39. Chen, L.-H.; Hu, J.-N. Development of nano-delivery systems for loaded bioactive compounds: Using molecular dynamics simulations. *Crit. Rev. Food Sci. Nutr.* **2024**, 1–22. [CrossRef] [PubMed]
40. Gaikwad, D.; Sutar, R.; Patil, D. Polysaccharide mediated nanodrug delivery: A review. *Int. J. Biol. Macromol.* **2024**, *261*, 129547. [CrossRef]
41. Pandey, V.K.; Tripathi, A.; Taufeeq, A.; Dar, A.H.; Samrot, A.V.; Rustagi, S.; Malik, S.; Bhattacharya, T.; Kovacs, B.; Shaikh, A.M. Significance and applications of carbon dots in anti cancerous nanodrug conjugate development: A review. *Appl. Surf. Sci. Adv.* **2024**, *19*, 100550. [CrossRef]
42. Zeng, M.; Guo, D.; Fernández-Varo, G.; Zhang, X.; Fu, S.; Ju, S.; Yang, H.; Liu, X.; Wang, Y.-C.; Zeng, Y.; et al. The integration of nanomedicine with traditional chinese medicine: Drug delivery of natural products and other opportunities. *Mol. Pharm.* **2022**, *20*, 886–904. [CrossRef]
43. Hung, P.; Chang, W. Cardiovascular Medicine. Undefined 2004. Available online: https://journals.lww.com/co-pediatrics/fulltext/2004/10000/bibliography_current_world_literature.17.aspx (accessed on 26 December 2024).
44. Liu, M.; Anderson, R.; Lan, X.; Conti, P.S.; Chen, K. Recent advances in the development of nanoparticles for multimodality imaging and therapy of cancer. *Med. Res. Rev.* **2019**, *40*, 909–930. [CrossRef] [PubMed]
45. Jiang, H.; Zhang, W.; Li, X.; Xu, Y.; Cao, J.; Jiang, W. The anti-obesogenic effects of dietary berry fruits: A review. *Food Res. Int.* **2021**, *147*, 110539. [CrossRef] [PubMed]

46. Alexander, A.; Ajazuddin; Patel, R.J.; Saraf, S.; Saraf, S. Recent expansion of pharmaceutical nanotechnologies and targeting strategies in the field of phytopharmaceuticals for the delivery of herbal extracts and bioactives. *J. Control. Release* **2016**, *241*, 110–124. [CrossRef] [PubMed]
47. Singh, A.R.; Desu, P.K.; Nakkala, R.K.; Kondi, V.; Devi, S.; Alam, M.S.; Hamid, H.; Athawale, R.B.; Kesharwani, P. Nanotechnology-based approaches applied to nutraceuticals. *Drug Deliv. Transl. Res.* **2021**, *12*, 485–499. [CrossRef] [PubMed]
48. Naik, G.G.; Minocha, T.; Verma, A.; Yadav, S.K.; Saha, S.; Agrawal, A.K.; Singh, S.; Sahu, A.N. Asparagus racemosus root-derived carbon nanodots as a nano-probe for biomedical applications. *J. Mater. Sci.* **2022**, *57*, 20380–20401. [CrossRef]
49. Luo, W.K.; Zhang, L.L.; Yang, Z.Y.; Guo, X.H.; Wu, Y.; Zhang, W.; Luo, J.K.; Tang, T.; Wang, Y. Herbal medicine derived carbon dots: Synthesis and applications in therapeutics, bioimaging and sensing. *J. Nanobiotechnol.* **2021**, *19*, 320. [CrossRef] [PubMed]
50. Atran, S. The trouble with memes. *Hum. Nat.* **2001**, *12*, 351–381. [CrossRef]
51. Larit, F.; León, F. Therapeutics to Treat Psychiatric and Neurological Disorders: A Promising Perspective from Algerian Traditional Medicine. *Plants* **2023**, *12*, 3860. [CrossRef]
52. Lin, J.; Yin, X.; Zeng, Y.; Hong, X.; Zhang, S.; Cui, B.; Zhu, Q.; Liang, Z.; Xue, Z.; Yang, D. Progress and prospect: Biosynthesis of plant natural products based on plant chassis. *Biotechnol. Adv.* **2023**, *69*, 108266. [CrossRef]
53. Dar, R.A.; Shahnawaz, M.; Ahanger, M.A.; Majid, I.U. Exploring the Diverse Bioactive Compounds from Medicinal Plants: A Review. *J. Phytopharm.* **2023**, *12*, 189–195. [CrossRef]
54. Abdel-Aziz, S.M.; Aeron, A.; Kahil, T.A. Health Benefits and Possible Risks of Herbal Medicine. In *Microbes in Food and Health*; Springer: Cham, Switzerland, 2016; pp. 97–116. [CrossRef]
55. Cai, R.; Shan, Y.; Du, F.; Miao, Z.; Zhu, L.; Hang, L.; Xiao, L.; Wang, Z. Injectable hydrogels as promising in situ therapeutic platform for cartilage tissue engineering. *Int. J. Biol. Macromol.* **2024**, *261*, 129537. [CrossRef] [PubMed]
56. Parveen, B.; Parveen, A.; Parveen, R.; Ahmad, S.; Ahmad, M.; Iqbal, M. Challenges and opportunities for traditional herbal medicine today with special reference to its status in India. *Ann. Phytomed. Int. J.* **2020**, *9*, 97–112. [CrossRef]
57. Pandey, M.; Debnath, M.; Gupta, S.; Chikara, S.K. Phytomedicine: An ancient approach turning into future potential source of therapeutics. *J. Pharmacogn. Phytother.* **2011**, *3*, 27–37.
58. Naik, G.G.; Pratap, R.; Mohapatra, D.; Shreya, S.; Sharma, D.K.; Parmar, A.S.; Patra, A.; Sahu, A.N. From phytomedicine to photomedicine: Quercetin-derived carbon nanodots—Synthesis, characterization and healthcare applications. *J. Mater. Sci.* **2023**, *58*, 13744–13761. [CrossRef]
59. Lin, L.; Fang, M.; Liu, W.; Zheng, M.; Lin, R. Recent advances and perspectives of functionalized carbon dots in bacteria sensing. *Microchim. Acta* **2023**, *190*, 363. [CrossRef] [PubMed]
60. Yang, Q.; Li, R.; Hong, Y.; Liu, H.; Jian, C.; Zhao, S. Curcumin-Loaded Gelatin Nanoparticles Cross the Blood-Brain Barrier to Treat Stroke by Attenuating Oxidative Stress and Neuroinflammation. Available online: https://papers.ssrn.com/sol3/papers.cfm?abstract_id=4894156 (accessed on 26 December 2024).
61. Qiang, R.; Huang, H.; Chen, J.; Shi, X.; Fan, Z.; Xu, G.; Qiu, H. Carbon Quantum Dots Derived from Herbal Medicine as Therapeutic Nanoagents for Rheumatoid Arthritis with Ultrahigh Lubrication and Anti-inflammation. *ACS Appl. Mater. Interfaces* **2023**, *15*, 38653–38664. [CrossRef]
62. Viswanath, B.; Kim, S. Influence of nanotoxicity on human health and environment: The alternative strategies. *Rev. Environ. Contam. Toxicol.* **2017**, *242*, 61–104. [CrossRef]
63. Osazee, F.O.; Mokobia, K.E.; Ifijen, I.H. The Urgent Need for Tungsten-Based Nanoparticles as Antibacterial Agents. *Biomed. Mater. Devices* **2023**, *2*, 614–629. [CrossRef]
64. Khan, J.; Ikbal, A.B.U.M.D.A.I.; Debnath, B.; Rajkhowa, A.; Choudhury, P.D.; Sen, S.; Paul, K.; Choudhury, D.; Debsarkar, S.; Jamatia, K.; et al. Management of Diabetes Mellitus by Nano Based Drug Delivery with Special Reference to Phytosomes. *Pharm. Biosci. J.* **2021**, *9*, 11–28. [CrossRef]
65. Kaurav, H.; Verma, D.; Bansal, A.; Kapoor, D.N.; Sheth, S. Progress in drug delivery and diagnostic applications of carbon dots: A systematic review. *Front. Chem.* **2023**, *11*, 1227843. [CrossRef] [PubMed]
66. Gunasekaran, T.; Haile, T.; Nigusse, T.; Dhanaraju, M.D. Nanotechnology: An effective tool for enhancing bioavailability and bioactivity of phytomedicine. *Asian Pac. J. Trop. Biomed.* **2014**, *4*, S1–S7. [CrossRef] [PubMed]
67. Huang, H.; Li, S.; Chen, B.; Wang, Y.; Shen, Z.; Qiu, M.; Pan, H.; Wang, W.; Wang, Y.; Li, X. Endoplasmic reticulum-targeted polymer dots encapsulated with ultrasonic synthesized near-infrared carbon nanodots and their application for in vivo monitoring of Cu^{2+}. *J. Colloid Interface Sci.* **2022**, *627*, 705–715. [CrossRef]
68. Debnath, R.; Jamatia, K.; Choudhury, P.D.; Sen, S.; Saha, S.; Ikbal, A.M.A. Niosome: A Prominent Carrier in Advanced Drug Delivery System. *Pharm. Biosci. J.* **2023**. Available online: http://pharmabiosciencejournal.com/index.php/pbj/article/download/2314/1659 (accessed on 26 December 2024).
69. Brindhadevi, K.; Garalleh, H.A.L.; Alalawi, A.; Al-Sarayreh, E.; Pugazhendhi, A. Carbon nanomaterials: Types, synthesis strategies and their application as drug delivery system for cancer therapy. *Biochem. Eng. J.* **2023**, *192*, 108828. [CrossRef]

70. Annisa, W.D.; Permatasari, F.A.; Iskandar, F.; Rachmawati, H. Functionalized Phytochemicals-Embedded Carbon Dots Derived from Medicinal Plant for Bioimaging Application. *ACS Appl. Bio. Mater.* **2023**, *7*, 114–123. [CrossRef] [PubMed]
71. Busi, K.B.; Palanivel, M.; Ghosh, K.K.; Ball, W.B.; Gulyás, B.; Padmanabhan, P.; Chakrabortty, S. The Multifarious Applications of Copper Nanoclusters in Biosensing and Bioimaging and Their Translational Role in Early Disease Detection. *Nanomaterials* **2022**, *12*, 301. [CrossRef]
72. Cutshaw, G.; Uthaman, S.; Hassan, N.; Kothadiya, S.; Wen, X.; Bardhan, R. The Emerging Role of Raman Spectroscopy as an Omics Approach for Metabolic Profiling and Biomarker Detection toward Precision Medicine. *Chem. Rev.* **2023**, *123*, 8297–8346. [CrossRef] [PubMed]
73. Xu, X.; Ray, R.; Gu, Y.; Ploehn, H.J.; Gearheart, L.; Raker, K.; Scrivens, W.A. Electrophoretic Analysis and Purification of Fluorescent Single-Walled Carbon Nanotube Fragments. *J. Am. Chem. Soc.* **2004**, *126*, 12736–12737. [CrossRef] [PubMed]
74. Shinde, D.B.; Pillai, V.K. Electrochemical Preparation of Luminescent Graphene Quantum Dots from Multiwalled Carbon Nanotubes. *Chem. A Eur. J.* **2012**, *18*, 12522–12528. [CrossRef]
75. Tao, H.; Yang, K.; Ma, Z.; Wan, J.; Zhang, Y.; Kang, Z.; Liu, Z. In Vivo NIR Fluorescence Imaging, Biodistribution, and Toxicology of Photoluminescent Carbon Dots Produced from Carbon Nanotubes and Graphite. *Small* **2011**, *8*, 281–290. [CrossRef] [PubMed]
76. Cao, L.; Wang, X.; Meziani, M.J.; Lu, F.; Wang, H.; Luo, P.G.; Lin, Y.; Harruff, B.A.; Veca, L.M.; Murray, D.; et al. Carbon dots for multiphoton bioimaging. *J. Am. Chem. Soc.* **2007**, *129*, 11318–11319. [CrossRef] [PubMed]
77. Yang, S.-T.; Cao, L.; Luo, P.G.; Lu, F.; Wang, X.; Wang, H.; Meziani, M.J.; Liu, Y.; Qi, G.; Sun, Y.-P. Carbon dots for optical imaging in vivo. *J. Am. Chem. Soc.* **2009**, *131*, 11308–11309. [CrossRef]
78. Fang, L.; Zhang, L.; Chen, Z.; Zhu, C.; Liu, J.; Zheng, J. Ammonium citrate derived carbon quantum dot as on-off-on fluorescent sensor for detection of chromium(VI) and sulfites. *Mater. Lett.* **2017**, *191*, 1–4. [CrossRef]
79. Jaiswal, A.; Ghosh, S.S.; Chattopadhyay, A. One step synthesis of C-dots by microwave mediated caramelization of poly(ethylene glycol). *Chem. Commun.* **2012**, *48*, 407–409. [CrossRef]
80. Schneider, J.; Reckmeier, C.J.; Xiong, Y.; von Seckendorff, M.; Susha, A.S.; Kasák, P.; Rogach, A.L. Molecular Fluorescence in Citric Acid-Based Carbon Dots. *J. Phys. Chem. C* **2017**, *121*, 2014–2022. [CrossRef]
81. Liu, F.; Zhang, W.; Chen, W.; Wang, J.; Yang, Q.; Zhu, W.; Wang, J. One-pot synthesis of NiFe2O4 integrated with EDTA-derived carbon dots for enhanced removal of tetracycline. *Chem. Eng. J.* **2017**, *310*, 187–196. [CrossRef]
82. Wang, W.; Li, Y.; Cheng, L.; Cao, Z.; Liu, W. Water-soluble and phosphorus-containing carbon dots with strong green fluorescence for cell labeling. *J. Mater. Chem. B* **2014**, *2*, 46–48. [CrossRef]
83. Vedamalai, M.; Periasamy, A.P.; Wang, C.-W.; Tseng, Y.-T.; Ho, L.-C.; Shih, C.-C.; Chang, H.-T. Carbon nanodots prepared from o-phenylenediamine for sensing of Cu^{2+}ions in cells. *Nanoscale* **2014**, *6*, 13119–13125. [CrossRef]
84. Wang, L.; Bi, Y.; Gao, J.; Li, Y.; Ding, H.; Ding, L. Carbon dots based turn-on fluorescent probes for the sensitive determination of glyphosate in environmental water samples. *RSC Adv.* **2016**, *6*, 85820–85828. [CrossRef]
85. Joseph, J.; Anappara, A.A. White-Light-Emitting Carbon Dots Prepared by the Electrochemical Exfoliation of Graphite. *ChemPhysChem* **2016**, *18*, 292–298. [CrossRef] [PubMed]
86. Mehta, V.N.; Jha, S.; Basu, H.; Singhal, R.K.; Kailasa, S.K. One-step hydrothermal approach to fabricate carbon dots from apple juice for imaging of mycobacterium and fungal cells. *Sens. Actuators B Chem.* **2015**, *213*, 434–443. [CrossRef]
87. Humaera, N.A.; Fahri, A.N.; Armynah, B.; Tahir, D. Natural source of carbon dots from part of a plant and its applications: A review. *Luminescence* **2021**, *36*, 1354–1364. [CrossRef]
88. Prasannan, A.; Imae, T. One-Pot Synthesis of Fluorescent Carbon Dots from Orange Waste Peels. *Ind. Eng. Chem. Res.* **2013**, *52*, 15673–15678. [CrossRef]
89. Zhang, Z.; Sun, W.; Wu, P. Highly Photoluminescent Carbon Dots Derived from Egg White: Facile and Green Synthesis, Photoluminescence Properties, and Multiple Applications. *ACS Sustain. Chem. Eng.* **2015**, *3*, 1412–1418. [CrossRef]
90. Feng, J.; Wang, W.-J.; Hai, X.; Yu, Y.-L.; Wang, J.-H. Green preparation of nitrogen-doped carbon dots derived from silkworm chrysalis for cell imaging. *J. Mater. Chem. B* **2016**, *4*, 387–393. [CrossRef]
91. Yin, B.; Deng, J.; Peng, X.; Long, Q.; Zhao, J.; Lu, Q.; Chen, Q.; Li, H.; Tang, H.; Zhang, Y.; et al. Green synthesis of carbon dots with down- and up-conversion fluorescent properties for sensitive detection of hypochlorite with a dual-readout assay. *Analyst* **2013**, *138*, 6551. [CrossRef] [PubMed]
92. Xu, J.; Lai, T.; Feng, Z.; Weng, X.; Huang, C. Formation of fluorescent carbon nanodots from kitchen wastes and their application for detection of Fe^{3+}. *Luminescence* **2014**, *30*, 420–424. [CrossRef]
93. Wei, J.; Zhang, X.; Sheng, Y.; Shen, J.; Huang, P.; Guo, S.; Pan, J.; Liu, B.; Feng, B. Simple one-step synthesis of water-soluble fluorescent carbon dots from waste paper. *N. J. Chem.* **2014**, *38*, 906. [CrossRef]
94. Feng, Y.; Zhong, D.; Miao, H.; Yang, X. Carbon dots derived from rose flowers for tetracycline sensing. *Talanta* **2015**, *140*, 128–133. [CrossRef]
95. Wang, B.; Lu, S. The light of carbon dots: From mechanism to applications. *Matter* **2022**, *5*, 110–149. [CrossRef]

96. Ozyurt, D.; Al Kobaisi, M.; Hocking, R.K.; Fox, B. Properties; synthesis, and applications of carbon dots: A review. *Carbon Trends* **2023**, *12*, 100276. [CrossRef]
97. Lin, L.; Zhang, S. Creating high yield water soluble luminescent graphene quantum dots via exfoliating and disintegrating carbon nanotubes and graphite flakes. *Chem. Commun.* **2012**, *48*, 10177. [CrossRef]
98. Nie, H.; Li, M.; Li, Q.; Liang, S.; Tan, Y.; Sheng, L.; Shi, W.; Zhang, S.X.-A. Carbon Dots with Continuously Tunable Full-Color Emission and Their Application in Ratiometric pH Sensing. *Chem. Mater.* **2014**, *26*, 3104–3112. [CrossRef]
99. Zhu, S.; Song, Y.; Zhao, X.; Shao, J.; Zhang, J.; Yang, B. The photoluminescence mechanism in carbon dots (graphene quantum dots, carbon nanodots, and polymer dots): Current state and future perspective. *Nano Res.* **2015**, *8*, 355–381. [CrossRef]
100. Bailey, R.E.; Nie, S. Alloyed Semiconductor Quantum Dots: Tuning the Optical Properties without Changing the Particle Size. *J. Am. Chem. Soc.* **2003**, *125*, 7100–7106. [CrossRef]
101. Ma, P.; Sun, X.; Pan, W.; Yu, G.; Wang, J. Green and Orange Emissive Carbon Dots with High Quantum Yields Dispersed in Matrices for Phosphor-Based White LEDs. *ACS Sustain. Chem. Eng.* **2020**, *8*, 3151–3161. [CrossRef]
102. Han, Z.; Ni, Y.; Ren, J.; Zhang, W.; Wang, Y.; Xie, Z.; Zhou, S.; Yu, S.F. Highly efficient and ultra-narrow bandwidth orange emissive carbon dots for microcavity lasers. *Nanoscale* **2019**, *11*, 11577–11583. [CrossRef]
103. Yang, X.; Sui, L.; Wang, B.; Zhang, Y.; Tang, Z.; Yang, B.; Lu, S. Red-emitting, self-oxidizing carbon dots for the preparation of white LEDs with super-high color rendering index. *Sci. China Chem.* **2021**, *64*, 1547–1553. [CrossRef]
104. Zheng, X.T.; Ananthanarayanan, A.; Luo, K.Q.; Chen, P. Glowing Graphene Quantum Dots and Carbon Dots: Properties, Syntheses, and Biological Applications. *Small* **2014**, *11*, 1620–1636. [CrossRef] [PubMed]
105. Han, M.; Zhu, S.; Lu, S.; Song, Y.; Feng, T.; Tao, S.; Liu, J.; Yang, B. Recent progress on the photocatalysis of carbon dots: Classification, mechanism and applications. *Nano Today* **2018**, *19*, 201–218. [CrossRef]
106. Ai, L.; Yang, Y.; Wang, B.; Chang, J.; Tang, Z.; Yang, B.; Lu, S. Insights into photoluminescence mechanisms of carbon dots: Advances and perspectives. *Sci. Bull.* **2021**, *66*, 839–856. [CrossRef]
107. Li, H.; Kang, Z.; Liu, Y.; Lee, S.-T. Carbon nanodots: Synthesis; properties; applications. *J. Mater. Chem.* **2012**, *22*, 24230. [CrossRef]
108. Lim, S.Y.; Shen, W.; Gao, Z. Carbon quantum dots and their applications. *Chem. Soc. Rev.* **2015**, *44*, 362–381. [CrossRef]
109. Mintz, K.; Zhou, Y.; Nanoscale, R.L. Recent development of carbon quantum dots regarding their optical properties, photoluminescence mechanism, and core structure. Undefined 2019. Available online: https://pubs.rsc.org/en/content/articlehtml/2019/nr/c8nr10059d?casa_token=C4u7Aq_wlDoAAAAA:YydgJYE4Sjy4l_Jo5vTRD-HytoYlXLz7O-HMZi5tgx_U5bpTHPqMMSGYxhc5DdCuynRRW8LAUGoJ6Bs (accessed on 26 December 2024).
110. Sheng, H.; Pei, B.; Shi, C.; Jing, S. Preparation of fluorescent carbon nanoparticles by pulsed laser. *Chem. J. Chin. Univ. Chin.* **2009**, *30*, 1497–1500.
111. Hu, S.-L.; Niu, K.-Y.; Sun, J.; Yang, J.; Zhao, N.-Q.; Du, X.-W. One-step synthesis of fluorescent carbon nanoparticles by laser irradiation. *J. Mater. Chem.* **2009**, *19*, 484–488. [CrossRef]
112. Li, X.; Wang, H.; Shimizu, Y.; Pyatenko, A.; Kawaguchi, K.; Koshizaki, N. Preparation of carbon quantum dots with tunable photoluminescence by rapid laser passivation in ordinary organic solvents. *Chem. Commun.* **2011**, *47*, 932–934. [CrossRef] [PubMed]
113. Hu, S.; Liu, J.; Yang, J.; Wang, Y.; Cao, S. Laser synthesis and size tailor of carbon quantum dots. *J. Nanoparticle Res.* **2011**, *13*, 7247–7252. [CrossRef]
114. Ming, H.; Ma, Z.; Liu, Y.; Pan, K.; Yu, H.; Wang, F.; Kang, Z. Large scale electrochemical synthesis of high quality carbon nanodots and their photocatalytic property. *Dalton Trans.* **2012**, *41*, 9526. [CrossRef]
115. Bao, L.; Zhang, Z.; Tian, Z.; Zhang, L.; Liu, C.; Lin, Y.; Qi, B.; Pang, D. Electrochemical Tuning of Luminescent Carbon Nanodots: From Preparation to Luminescence Mechanism. *Adv. Mater.* **2011**, *23*, 5801–5806. [CrossRef]
116. Alaghmandfard, A.; Sedighi, O.; Rezaei, N.T.; Abedini, A.A.; Khachatourian, A.M.; Toprak, M.S.; Seifalian, A. Recent advances in the modification of carbon-based quantum dots for biomedical applications. *Mater. Sci. Eng. C* **2021**, *120*, 111756. [CrossRef]
117. Gu, S.; Hsieh, C.-T.; Chiang, Y.-M.; Tzou, D.-Y.; Chen, Y.-F.; Gandomi, Y.A. Optimization of graphene quantum dots by chemical exfoliation from graphite powders and carbon nanotubes. *Mater. Chem. Phys.* **2018**, *215*, 104–111. [CrossRef]
118. Hu, S.; Wei, Z.; Chang, Q.; Trinchi, A.; Yang, J. A facile and green method towards coal-based fluorescent carbon dots with photocatalytic activity. *Appl. Surf. Sci.* **2016**, *378*, 402–407. [CrossRef]
119. Liu, X.; Hao, J.; Liu, J.; Tao, H. Green synthesis of carbon quantum dots from lignite coal and the application in Fe^{3+}detection. *IOP Conf. Ser. Earth. Env. Sci.* **2018**, *113*, 012063. [CrossRef]
120. Ventrella, A.; Camisasca, A.; Fontana, A.; Giordani, S. Synthesis of green fluorescent carbon dots from carbon nano-onions and graphene oxide. *RSC Adv.* **2020**, *10*, 36404–36412. [CrossRef] [PubMed]
121. Sousa, H.B.A.; Martins, C.S.M.; Prior, J.A.V. You Don't Learn That in School: An Updated Practical Guide to Carbon Quantum Dots. *Nanomaterials* **2021**, *11*, 611. [CrossRef]
122. Wang, F.; Wang, S.; Sun, Z.; Zhu, H. Study on Ultrasonic Single-Step Synthesis and Optical Properties of Nitrogen-Doped Carbon Fluorescent Quantum Dots. *Fuller. Nanotub. Carbon Nanostruct.* **2014**, *23*, 769–776. [CrossRef]

123. Kumar, R.; Kumar, V.B.; Gedanken, A. Sonochemical synthesis of carbon dots, mechanism, effect of parameters, and catalytic, energy, biomedical and tissue engineering applications. *Ultrason. Sonochem.* **2020**, *64*, 105009. [CrossRef]
124. Xu, M.H.; Xu, M.; Li, Z.; Zhu, X.; Hu, N.; Wei, H.; Yang, Z.; Zhang, Y. Hydrothermal/Solvothermal Synthesis of Graphene Quantum Dots and Their Biological Applications. *Nano Biomed. Eng.* **2013**, *5*, 65–71. [CrossRef]
125. Mohammed, S.J.; Omer, K.M.; Hawaiz, F.E. Deep insights to explain the mechanism of carbon dot formation at various reaction times using the hydrothermal technique: FT-IR, (13)C-NMR, (1)H-NMR, and UV-visible spectroscopic approaches. *RSC Adv.* **2023**, *13*, 14340–14349. [CrossRef]
126. Tyagi, A.; Tripathi, K.M.; Singh, N.; Choudhary, S.; Gupta, R.K. Green synthesis of carbon quantum dots from lemon peel waste: Applications in sensing and photocatalysis. *RSC Adv.* **2016**, *6*, 72423–72432. [CrossRef]
127. Bano, D.; Kumar, V.; Singh, V.K.; Hasan, S.H. Green synthesis of fluorescent carbon quantum dots for the detection of mercury(ii) and glutathione. *N. J. Chem.* **2018**, *42*, 5814–5821. [CrossRef]
128. Tadesse, A.; RamaDevi, D.; Hagos, M.; Battu, G.R.; Basavaiah, K. Facile Green Synthesis of Fluorescent Carbon Quantum Dots from Citrus Lemon Juice for Live Cell Imaging. Undefined 2018. Available online: https://www.researchgate.net/profile/Mebrahtu-Hagos-Kahsay-2/publication/325651047_Facile_Green_Synthesis_of_Fluorescent_Carbon_Quantum_Dots_from_Citrus_Lemon_Juice_for_Live_Cell_Imaging/links/5b1a80e3a6fdcca67b67024c/Facile-Green-Synthesis-of-Fluorescent-Carbon-Quantum-Dots-from-Citrus-Lemon-Juice-for-Live-Cell-Imaging.pdf (accessed on 26 December 2024).
129. Polshettiwar, V.; Varma, R.S. Microwave-Assisted Organic Synthesis and Transformations using Benign Reaction Media. *Acc. Chem. Res.* **2008**, *41*, 629–639. [CrossRef]
130. de Medeiros, T.V.; Manioudakis, J.; Noun, F.; Macairan, J.-R.; Victoria, F.; Naccache, R. Microwave-assisted synthesis of carbon dots and their applications. *J. Mater. Chem. C Mater.* **2019**, *7*, 7175–7195. [CrossRef]
131. Kung, J.-C.; Tseng, I.-T.; Chien, C.-S.; Lin, S.-H.; Wang, C.-C.; Shih, C.-J. Microwave assisted synthesis of negative-charge carbon dots with potential antibacterial activity against multi-drug resistant bacteria. *RSC Adv.* **2020**, *10*, 41202–41208. [CrossRef]
132. Adeola, A.O.; Duarte, M.P.; Naccache, R. Microwave-assisted synthesis of carbon-based nanomaterials from biobased resources for water treatment applications: Emerging trends and prospects. *Front. Carbon* **2023**, *2*, 1220021. [CrossRef]
133. He, G.; Shu, M.; Yang, Z.; Ma, Y.; Huang, D.; Xu, S.; Wang, Y.; Hu, N.; Zhang, Y.; Xu, L. Microwave formation and photoluminescence mechanisms of multi-states nitrogen doped carbon dots. *Appl. Surf. Sci.* **2017**, *422*, 257–265. [CrossRef]
134. Ramezani, Z.; Qorbanpour, M.; Rahbar, N. Green synthesis of carbon quantum dots using quince fruit (Cydonia oblonga) powder as carbon precursor: Application in cell imaging and As^{3+} determination. *Colloids Surf. A Physicochem. Eng. Asp.* **2018**, *549*, 58–66. [CrossRef]
135. Qin, X.; Fu, C.; Zhang, J.; Shao, W.; Qin, X.; Gui, Y.; Wang, L.; Guo, H.; Chen, F.; Jiang, L.; et al. Direct preparation of solid carbon dots by pyrolysis of collagen waste and their applications in fluorescent sensing and imaging. *Front. Chem.* **2022**, *10*, 1006389. [CrossRef]
136. Jiang, H.; Chen, F.; Lagally, M.G.; Denes, F.S. New Strategy for Synthesis and Functionalization of Carbon Nanoparticles. *Langmuir* **2009**, *26*, 1991–1995. [CrossRef] [PubMed]
137. Bourlinos, A.B.; Stassinopoulos, A.; Anglos, D.; Zboril, R.; Georgakilas, V.; Giannelis, E.P. Photoluminescent Carbogenic Dots. *Chem. Mater.* **2008**, *20*, 4539–4541. [CrossRef]
138. Liu, R.; Wu, D.; Liu, S.; Koynov, K.; Knoll, W.; Li, Q. An Aqueous Route to Multicolor Photoluminescent Carbon Dots Using Silica Spheres as Carriers. *Angew. Chem. Int. Ed.* **2009**, *48*, 4598–4601. [CrossRef]
139. Zong, J.; Zhu, Y.; Yang, X.; Shen, J.; Li, C. Synthesis of photoluminescent carbogenic dots using mesoporous silica spheres as nanoreactors. *Chem. Commun.* **2011**, *47*, 764–766. [CrossRef] [PubMed]
140. Wang, J.; Xin, X.; Lin, Z. Cu2ZnSnS4 nanocrystals and graphene quantum dots for photovoltaics. *Nanoscale* **2011**, *3*, 3040. [CrossRef]
141. Hamilton, I.P.; Li, B.; Yan, X.; Li, L. Alignment of Colloidal Graphene Quantum Dots on Polar Surfaces. *Nano Lett.* **2011**, *11*, 1524–1529. [CrossRef]
142. Mueller, M.L.; Yan, X.; Dragnea, B.; Li, L. Slow Hot-Carrier Relaxation in Colloidal Graphene Quantum Dots. *Nano Lett.* **2010**, *11*, 56–60. [CrossRef] [PubMed]
143. Liu, R.; Wu, D.; Feng, X.; Müllen, K. Bottom-Up Fabrication of Photoluminescent Graphene Quantum Dots with Uniform Morphology. *J. Am. Chem. Soc.* **2011**, *133*, 15221–15223. [CrossRef] [PubMed]
144. Lu, J.; Yeo, P.S.E.; Gan, C.K.; Wu, P.; Loh, K.P. Transforming C60 molecules into graphene quantum dots. *Nat. Nanotechnol.* **2011**, *6*, 247–252. [CrossRef]
145. Fan, L.; Zhu, M.; Lee, X.; Zhang, R.; Wang, K.; Wei, J.; Zhong, M.; Wu, D.; Zhu, H. Direct Synthesis of Graphene Quantum Dots by Chemical Vapor Deposition. *Part. Part. Syst. Charact.* **2013**, *30*, 764–769. [CrossRef]
146. Kurdyukov, D.A.; Eurov, D.A.; Stovpiaga, E.Y.; Kirilenko, D.A.; Konyakhin, S.V.; Shvidchenko, A.V.; Golubev, V.G. Template synthesis of monodisperse carbon nanodots. *Phys. Solid State* **2016**, *58*, 2545–2549. [CrossRef]

147. Liu, H.; Ye, T.; Mao, C. Fluorescent Carbon Nanoparticles Derived from Candle Soot. *Angew. Chem.* **2007**, *119*, 6593–6595. [CrossRef]
148. Li, H.; He, X.; Kang, Z.; Huang, H.; Liu, Y.; Liu, J.; Lian, S.; Tsang, C.H.A.; Yang, X.; Lee, S. Water-Soluble Fluorescent Carbon Quantum Dots and Photocatalyst Design. *Angew. Chem.* **2010**, *122*, 4532–4536. [CrossRef]
149. Vinci, J.C.; Ferrer, I.M.; Seedhouse, S.J.; Bourdon, A.K.; Reynard, J.M.; Foster, B.A.; Bright, F.V.; Colón, L.A. Hidden Properties of Carbon Dots Revealed After HPLC Fractionation. *J Phys. Chem. Lett.* **2012**, *4*, 239–243. [CrossRef]
150. Zheng, X.T.; Than, A.; Ananthanaraya, A.; Kim, D.-H.; Chen, P. Graphene Quantum Dots as Universal Fluorophores and Their Use in Revealing Regulated Trafficking of Insulin Receptors in Adipocytes. *ACS Nano* **2013**, *7*, 6278–6286. [CrossRef]
151. Liu, W.; Li, C.; Ren, Y.; Sun, X.; Pan, W.; Li, Y.; Wang, J.; Wang, W. Carbon dots: Surface engineering and applications. *J. Mater. Chem. B* **2016**, *4*, 5772–5788. [CrossRef] [PubMed]
152. Wang, H.; Ai, L.; Song, Z.; Nie, M.; Xiao, J.; Li, G.; Lu, S. Surface Modification Functionalized Carbon Dots. *Chem. A Eur. J.* **2023**, *29*, e202302383. [CrossRef]
153. Wang, X.; Cao, L.; Yang, S.-T.; Lu, F.; Meziani, M.J.; Tian, L.; Sun, K.W.; Bloodgood, M.A.; Sun, Y.-P. Bandgap-like strong fluorescence in functionalized carbon nanoparticles. *Angew. Chem. Int. Ed. Engl.* **2010**, *49*, 5310–5314. [CrossRef]
154. Dong, Y.; Zhou, N.; Lin, X.; Lin, J.; Chi, Y.; Chen, G. Extraction of Electrochemiluminescent Oxidized Carbon Quantum Dots from Activated Carbon. *Chem. Mater.* **2010**, *22*, 5895–5899. [CrossRef]
155. Zou, Y.; Yan, F.; Zheng, T.; Shi, D.; Sun, F.; Yang, N.; Chen, L. Highly luminescent organosilane-functionalized carbon dots as a nanosensor for sensitive and selective detection of quercetin in aqueous solution. *Talanta* **2015**, *135*, 145–148. [CrossRef] [PubMed]
156. Hou, J.; Dong, J.; Zhu, H.; Teng, X.; Ai, S.; Mang, M. A simple and sensitive fluorescent sensor for methyl parathion based on l -tyrosine methyl ester functionalized carbon dots. *Biosens Bioelectron* **2015**, *68*, 20–26. [CrossRef]
157. Liu, S.; Zhao, N.; Cheng, Z.; Liu, H. Amino-functionalized green fluorescent carbon dots as surface energy transfer biosensors for hyaluronidase. *Nanoscale* **2015**, *7*, 6836–6842. [CrossRef] [PubMed]
158. Shen, P.; Xia, Y. Synthesis-Modification Integration: One-Step Fabrication of Boronic Acid Functionalized Carbon Dots for Fluorescent Blood Sugar Sensing. *Anal. Chem.* **2014**, *86*, 5323–5329. [CrossRef]
159. Gonçalves, H.; Jorge, P.A.S.; Fernandes, J.R.A.; da Silva, J.C.G.E. Hg(II) sensing based on functionalized carbon dots obtained by direct laser ablation. *Sens. Actuators B Chem.* **2010**, *145*, 702–707. [CrossRef]
160. Liang, S.; Wang, M.; Gao, W.; Zhao, X. Effects of elemental doping, acid treatment, and passivation on the fluorescence intensity and emission behavior of yellow fluorescence carbon dots. *Opt. Mater.* **2022**, *128*, 112471. [CrossRef]
161. Barman, M.K.; Jana, B.; Bhattacharyya, S.; Patra, A. Photophysical Properties of Doped Carbon Dots (N, P, and B) and Their Influence on Electron/Hole Transfer in Carbon Dots–Nickel (II) Phthalocyanine Conjugates. *J. Phys. Chem. C* **2014**, *118*, 20034–20041. [CrossRef]
162. Redondo-Fernandez, G.; Canga, J.C.; Soldado, A.; Encinar, J.R.; Costa-Fernandez, J.M. Functionalized heteroatom-doped carbon dots for biomedical applications: A review. *Anal. Chim. Acta* **2023**, *1284*, 341874. [CrossRef] [PubMed]
163. Das, P.; Ganguly, S.; Saravanan, A.; Margel, S.; Gedanken, A.; Srinivasan, S.; Rajabzadeh, A.R. Naturally Derived Carbon Dots In Situ Confined Self-Healing and Breathable Hydrogel Monolith for Anomalous Diffusion-Driven Phytomedicine Release. *ACS Appl. Bio. Mater.* **2022**, *5*, 5617–5633. [CrossRef]
164. Zahin, N.; Anwar, R.; Tewari, D.; Kabir, M.T.; Sajid, A.; Mathew, B.; Uddin, M.S.; Aleya, L.; Abdel-Daim, M.M. Nanoparticles and its biomedical applications in health and diseases: Special focus on drug delivery. *Environ. Sci. Pollut. Res.* **2019**, *27*, 19151–19168. [CrossRef] [PubMed]
165. Jain, K.K. Current status and future prospects of drug delivery systems. *Methods Mol. Biol.* **2014**, *1141*, 1–56. [CrossRef]
166. Chatterjee, P.; Dhibar, S. Nanomaterial marvels: Pioneering applications and cutting-edge advancements in drug delivery. *Nano Med. Mater.* **2023**, *3*, 220. [CrossRef]
167. Ranganathan, R.; Madanmohan, S.; Kesavan, A.; Baskar, G.; Krishnamoorthy, Y.R.; Santosham, R.; Ponraju, D.; Rayala, S.K.; Venkatraman, G. Nanomedicine: Towards development of patient-friendly drug-delivery systems for oncological applications. *Int. J. Nanomed.* **2012**, *7*, 1043–1060. [CrossRef]
168. Thakuria, A.; Kataria, B.; Gupta, D. Nanoparticle-based methodologies for targeted drug delivery—An insight. *J. Nanoparticle Res.* **2021**, *23*, 1–30. [CrossRef]
169. Ho, R.J.Y.; Chien, J. Trends in translational medicine and drug targeting and delivery: New insights on an old concept-targeted drug delivery with antibody-drug conjugates for cancers. *J. Pharm. Sci.* **2014**, *103*, 71–77. [CrossRef]
170. Manzari, M.T.; Shamay, Y.; Kiguchi, H.; Rosen, N.; Scaltriti, M.; Heller, D.A. Targeted drug delivery strategies for precision medicines. *Nat. Rev. Mater.* **2021**, *6*, 351–370. [CrossRef] [PubMed]
171. Khan, F.A.; Narasimhan, K.; Swathi, C.S.V.; Mustak, S.; Mustafa, G.; Ahmad, M.Z.; Akhter, S. 3D Printing Technology in Customized Drug Delivery System: Current State of the Art, Prospective and the Challenges. *Curr. Pharm. Des.* **2019**, *24*, 5049–5061. [CrossRef]

172. Zaidi, Z.; Maiti, N.; Ali, M.I.; Sharma, G.; Moin, S.; Padhy, H.; Balaji, G.L.; Sundramurthy, V.P. Fabrication, Characteristics, and Therapeutic Applications of Carbon-Based Nanodots. *J. Nanomater.* **2022**, *2022*, 8031495. [CrossRef]
173. Luong, A.D.; Buzid, A.; Luong, J.H.T. Important Roles and Potential Uses of Natural and Synthetic Antimicrobial Peptides (AMPs) in Oral Diseases: Cavity, Periodontal Disease, and Thrush. *J. Funct. Biomater.* **2022**, *13*, 175. [CrossRef]
174. Kashyap, B.K.; Singh, V.V.; Solanki, M.K.; Kumar, A.; Ruokolainen, J.; Kesari, K.K. Smart Nanomaterials in Cancer Theranostics: Challenges and Opportunities. *ACS Omega* **2023**, *8*, 14290–14320. [CrossRef] [PubMed]
175. Sameer, M.; Arif, Y.; Aqil, A.; Nadaf, A.; Rafiya, K.; Hasan, N.; Kesharwani, P.; Ahmad, F.J. Carbon nanodots as a remedial nanovesicles for drug delivery. *Eur. Polym. J.* **2023**, *200*, 112515. [CrossRef]
176. Merino, S.; Martín, C.; Kostarelos, K.; Prato, M.; Vázquez, E. Nanocomposite Hydrogels: 3D Polymer–Nanoparticle Synergies for On-Demand Drug Delivery. *ACS Nano* **2015**, *9*, 4686–4697. [CrossRef] [PubMed]
177. Mathew, S.A.; Praveena, P.; Dhanavel, S.; Manikandan, R.; Senthilkumar, S.; Stephen, A. Luminescent chitosan/carbon dots as an effective nano-drug carrier for neurodegenerative diseases. *RSC Adv.* **2020**, *10*, 24386. [CrossRef] [PubMed]
178. Fernandes, N.B.; Nayak, Y.; Garg, S.; Nayak, U.Y. Multifunctional engineered mesoporous silica/inorganic material hybrid nanoparticles: Theranostic perspectives. *Coord. Chem. Rev.* **2023**, *478*, 214977. [CrossRef]
179. Iqbal, B.; Ali, J.; Baboota, S. Recent advances and development in epidermal and dermal drug deposition enhancement technology. *Int. J. Dermatol.* **2018**, *57*, 646–660. [CrossRef] [PubMed]
180. Li, D.; Han, D.; Qu, S.-N.; Liu, L.; Jing, P.-T.; Zhou, D.; Ji, W.-Y.; Wang, X.-Y.; Zhang, T.-F.; Shen, D.-Z. Supra-(carbon nanodots) with a strong visible to near-infrared absorption band and efficient photothermal conversion. *Light. Sci. Appl.* **2016**, *5*, e16120. [CrossRef] [PubMed]
181. Mao, Q.-X.; Han, L.; Shu, Y.; Chen, X.-W.; Wang, J.-H. Improving the biocompatibility of carbon nanodots for cell imaging. *Talanta* **2016**, *161*, 54–61. [CrossRef]
182. Loukanov, A.; Mladenova, P.; Toshev, S.; Karailiev, A.; Ustinovich, E.; Nakabayashi, S. Real time monitoring and quantification of uptake carbon nanodots in eukaryotic cells. *Microsc. Res. Tech.* **2018**, *81*, 1541–1547. [CrossRef] [PubMed]
183. Park, Y.; Yoo, J.; Lim, B.; Kwon, W.; Rhee, S.-W. Improving the functionality of carbon nanodots: Doping and surface functionalization. *J. Mater. Chem. A Mater.* **2016**, *4*, 11582–11603. [CrossRef]
184. Henna, T.K.; Raphey, V.R.; Sankar, R.; Shirin, V.K.A.; Gangadharappa, H.V.; Pramod, K. Carbon nanostructures: The drug and the delivery system for brain disorders. *Int. J. Pharm.* **2020**, *587*, 119701. [CrossRef] [PubMed]
185. Saleemi, M.A.; Kong, Y.L.; Yong, P.V.C.; Wong, E.H. An overview of recent development in therapeutic drug carrier system using carbon nanotubes. *J. Drug Deliv. Sci. Technol.* **2020**, *59*, 101855. [CrossRef]
186. Akharame, M.O.; Fatoki, O.S.; Opeolu, B.O.; Olorunfemi, D.I.; Oputu, O.U. Oputu, Polymeric Nanocomposites (PNCs) for Wastewater Remediation: An Overview. *Polym. Plast. Technol. Eng.* **2018**, *57*, 1801–1827. [CrossRef]
187. Fu, S.; Sun, Z.; Huang, P.; Li, Y.; Hu, N. Some basic aspects of polymer nanocomposites: A critical review. *Nano Mater. Sci.* **2019**, *1*, 2–30. [CrossRef]
188. Xia, C.; Zhu, S.; Feng, T.; Yang, M.; Yang, B. Evolution and Synthesis of Carbon Dots: From Carbon Dots to Carbonized Polymer Dots. *Adv. Sci.* **2019**, *6*, 1901316. [CrossRef]
189. de Oliveira, A.D.; Beatrice, C.A.G. Polymer nanocomposites with different types of nanofiller. Undefined 2018. Available online: https://books.google.com/books?hl=en&lr=&id=wC-RDwAAQBAJ&oi=fnd&pg=PA103&dq=206.%09De,+Oliveira,+A.D.;+Beatrice,+C.A.+Polymer+nanocomposites+with+different+types+of+nanofiller.+Nanocomposites-recent+evolutions.+2018,+103-4.&ots=hny1nyScTT&sig=QWn_4NS5QDjV2g0wP3r4QnIDT_E (accessed on 26 December 2024).
190. Keledi, G.; Hári, J.; Pukánszky, B. Polymer nanocomposites: Structure; interaction; functionality. *Nanoscale* **2012**, *4*, 1919. [CrossRef] [PubMed]
191. Singh, R.K.; Kumar, R.; Singh, D.P.; Savu, R.; Moshkalev, S.A. Progress in microwave-assisted synthesis of quantum dots (graphene/carbon/semiconducting) for bioapplications: A review. *Mater. Today Chem.* **2019**, *12*, 282–314. [CrossRef]
192. Lan, G.; Yang, J.; Ye, R.; Boyjoo, Y.; Liang, J.; Liu, X.; Li, Y.; Liu, J.; Qian, K. Sustainable Carbon Materials toward Emerging Applications. *Small Methods* **2021**, *5*, 2001250. [CrossRef] [PubMed]
193. Rawat, P.; Nain, P.; Sharma, S.; Sharma, P.K.; Malik, V.; Majumder, S.; Verma, V.P.; Rawat, V.; Rhyee, J.S. An overview of synthetic methods and applications of photoluminescence properties of carbon quantum dots. *Luminescence* **2022**, *38*, 845–866. [CrossRef]
194. Wilczewska, A.Z.; Niemirowicz, K.; Markiewicz, K.H.; Car, H. Nanoparticles as drug delivery systems. *Pharmacol. Rep.* **2012**, *64*, 1020–1037. [CrossRef]
195. Zhou, Y.; Peng, Z.; Seven, E.S.; Leblanc, R.M. Crossing the blood-brain barrier with nanoparticles. *J. Control. Release* **2018**, *270*, 290–303. [CrossRef] [PubMed]
196. Shariati, L.; Esmaeili, Y.; Rahimmanesh, I.; Babolmorad, S.; Ziaei, G.; Hasan, A.; Boshtam, M.; Makvandi, P. Advances in nanobased platforms for cardiovascular diseases: Early diagnosis, imaging, treatment, and tissue engineering. *Environ. Res.* **2023**, *238*, 116933. [CrossRef]

197. Torchilin, V.P. Multifunctional, stimuli-sensitive nanoparticulate systems for drug delivery. *Nat. Rev. Drug. Discov.* **2014**, *13*, 813–827. [CrossRef] [PubMed]
198. Coelho, J.F.; Ferreira, P.C.; Alves, P.; Cordeiro, R.; Fonseca, A.C.; Góis, J.R.; Gil, M.H. Drug delivery systems: Advanced technologies potentially applicable in personalized treatments. *EPMA J.* **2010**, *1*, 164–209. [CrossRef] [PubMed]
199. Das, S.S.; Bharadwaj, P.; Bilal, M.; Barani, M.; Rahdar, A.; Taboada, P.; Bungau, S.; Kyzas, G.Z. Stimuli-Responsive Polymeric Nanocarriers for Drug Delivery, Imaging, and Theragnosis. *Polymers* **2020**, *12*, 1397. [CrossRef]
200. Benko, A.; Medina-Cruz, D.; Vernet-Crua, A.; O'Connell, C.P.; Świętek, M.; Barabadi, H.; Saravanan, M.; Webster, T.J. Nanocarrier drug resistant tumor interactions: Novel approaches to fight drug resistance in cancer. *Cancer Drug Resist.* **2021**, *4*, 264–297. [CrossRef]
201. Vithalani, R.; Patel, D.; Modi, C.K.; Suthar, D.H. Glowing photoluminescene in carbon-based nanodots: Current state and future perspectives. *J. Mater. Sci.* **2020**, *55*, 8769–8792. [CrossRef]
202. Ding, H.; Zhou, X.-X.; Wei, J.-S.; Li, X.-B.; Qin, B.-T.; Chen, X.-B.; Xiong, H.-M. Carbon dots with red/near-infrared emissions and their intrinsic merits for biomedical applications. *Carbon* **2020**, *167*, 322–344. [CrossRef]
203. Bagyalakshmi, S.; Sivakami, A.; Pal, K.; Sarankumar, R.; Mahendran, C. Manufacturing of electrochemical sensors via carbon nanomaterials novel applications: A systematic review. *J. Nanoparticle Res.* **2022**, *24*, 1–28. [CrossRef]
204. P., A.M.; Pardhiya, S.; Rajamani, P. Carbon Dots: An Excellent Fluorescent Probe for Contaminant Sensing and Remediation. *Small* **2022**, *18*, 2105579. [CrossRef]
205. Xiao, L.; Sun, H. Novel properties and applications of carbon nanodots. *Nanoscale Horiz.* **2018**, *3*, 565–597. [CrossRef]
206. Barhoum, A.; Meftahi, A.; Sabery, M.S.K.; Heravi, M.E.M.; Alem, F. A review on carbon dots as innovative materials for advancing biomedical applications: Synthesis, opportunities, and challenges. *J. Mater. Sci.* **2023**, *58*, 13531–13579. [CrossRef]
207. Gawrońska, M.; Kowalik, M.; Makowski, M. Recent advances in medicinal chemistry of ampicillin: Derivatives, metal complexes, and sensing approaches. *TrAC Trends Anal. Chem.* **2022**, *155*, 116691. [CrossRef]
208. Chen, B.B.; Liu, M.L.; Huang, C.Z. Recent advances of carbon dots in imaging-guided theranostics. *TrAC Trends Anal. Chem.* **2021**, *134*, 116116. [CrossRef]
209. Deb, A.; Konwar, A.; Chowdhury, D. pH-Responsive Hybrid Jute Carbon Dot-Cotton Patch. *ACS Sustain. Chem. Eng.* **2020**, *8*, 7394–7402. [CrossRef]
210. Pardo, J.; Peng, Z.; Leblanc, R.M. Cancer Targeting and Drug Delivery Using Carbon-Based Quantum Dots and Nanotubes. *Molecules* **2018**, *23*, 378. [CrossRef]
211. Chen, Y.; Li, X. The utilization of carbon-based nanomaterials in bone tissue regeneration and engineering: Respective featured applications and future prospects. *Med. Nov. Technol. Devices* **2022**, *16*, 100168. [CrossRef]
212. Sajjadi, M.; Nasrollahzadeh, M.; Jaleh, B.; Soufi, G.J.; Iravani, S. Carbon-based nanomaterials for targeted cancer nanotherapy: Recent trends and future prospects. *J. Drug Target.* **2021**, *29*, 716–741. [CrossRef]
213. Wang, K.; Gao, Z.; Gao, G.; Wo, Y.; Wang, Y.; Shen, G.; Cui, D. Systematic safety evaluation on photoluminescent carbon dots. *Nanoscale Res. Lett.* **2013**, *8*, 122. [CrossRef]
214. Wang, Y.; Wang, S.; Ge, S.; Wang, S.; Yan, M.; Zang, D.; Yu, J. Facile and sensitive paper-based chemiluminescence DNA biosensor using carbon dots dotted nanoporous gold signal amplification label. *Anal. Methods* **2013**, *5*, 1328. [CrossRef]
215. Koutsogiannis, P.; Thomou, E.; Stamatis, H.; Gournis, D.; Rudolf, P. Advances in fluorescent carbon dots for biomedical applications. *Adv. Phys. X* **2020**, *5*, 1758592. [CrossRef]
216. Tao, S.; Zhu, S.; Feng, T.; Xia, C.; Song, Y.; Yang, B. The polymeric characteristics and photoluminescence mechanism in polymer carbon dots: A review. *Mater. Today Chem.* **2017**, *6*, 13–25. [CrossRef]
217. Yin, Y.; Wang, J.; Li, T.; Hill, J.P.; Rowan, A.; Sugahara, Y.; Yamauchi, Y. Nanoarchitecturing Carbon Nanodot Arrays on Zeolitic Imidazolate Framework-Derived Cobalt—Nitrogen-Doped Carbon Nanoflakes toward Oxygen Reduction Electrocatalysts. *ACS Nano* **2021**, *15*, 13240–13248. [CrossRef] [PubMed]
218. Innocenzi, P.; Malfatti, L.; Carboni, D. Graphene and carbon nanodots in mesoporous materials: An interactive platform for functional applications. *Nanoscale* **2015**, *7*, 12759–12772. [CrossRef] [PubMed]
219. Yadav, P.K.; Chandra, S.; Kumar, V.; Kumar, D.; Hasan, S.H. Carbon Quantum Dots: Synthesis, Structure, Properties, and Catalytic Applications for Organic Synthesis. *Catalysts* **2023**, *13*, 422. [CrossRef]
220. Yeoh, G.H.; De Cachinho Cordeiro, I.M.; Wang, W.; Wang, C.; Yuen, A.C.Y.; Chen, T.B.Y.; Vargas, J.B.; Mao, G.; Garbe, U.; Chua, H.T. Carbon-based Flame Retardants for Polymers: A Bottom-up Review. *Adv. Mater.* **2024**, *36*, 2403835. [CrossRef] [PubMed]
221. Ma, C.; Zhu, Y.; Huang, S.; Di, W.; Jessie, Y.; Lu, C.; Yuyo, G.; Gemma, F.; Shi, L.; Bingjie, L. Exploration of the Synthesis, Properties, and Biological Applications of B-CDs Materials. *Authorea, preprint*. 2024. [CrossRef]
222. Qasim, M.; Clarkson, A.N.; Hinkley, S.F.R. Green Synthesis of Carbon Nanoparticles (CNPs) from Biomass for Biomedical Applications. *Int. J. Mol. Sci.* **2023**, *24*, 1023. [CrossRef]
223. Bhattacharya, T.; Shin, G.H.; Kim, J.T. Carbon Dots: Opportunities and Challenges in Cancer Therapy. *Pharmaceutics* **2023**, *15*, 1019. [CrossRef]

224. Thakur, V.K.; Kessler, M.R. Self-healing polymer nanocomposite materials: A review. *Polymer* **2015**, *69*, 369–383. [CrossRef]
225. Hsu, L.; Weder, C.; Rowan, S.J. Stimuli-responsive, mechanically-adaptive polymer nanocomposites. *J. Mater. Chem.* **2011**, *21*, 2812–2822. [CrossRef]
226. Gamage, A.; Thiviya, P.; Mani, S.; Ponnusamy, P.G.; Manamperi, A.; Evon, P.; Merah, O.; Madhujith, T. Environmental Properties and Applications of Biodegradable Starch-Based Nanocomposites. *Polymers* **2022**, *14*, 4578. [CrossRef] [PubMed]
227. M, P.K.C.; B, A.R.; Kumar, M.; Roopa, C.P. Natural nano-fillers materials for the Bio-composites: A review. *J. Indian Chem. Soc.* **2022**, *99*, 100715. [CrossRef]
228. Balasubramanian, K.B.N.; Ramesh, T. Role, effect, and influences of micro and nano-fillers on various properties of polymer matrix composites for microelectronics: A review. *Polym. Adv. Technol.* **2018**, *29*, 1568–1585. [CrossRef]
229. Gómez, I.J.; Sulleiro, M.V.; Mantione, D.; Alegret, N. Carbon Nanomaterials Embedded in Conductive Polymers: A State of the Art. *Polymers* **2021**, *13*, 745. [CrossRef] [PubMed]
230. Ali, Y.A.E.H.; Hejji, L.; Lahcen, A.A.; Perez-Villarejo, L.; Azzouz, A.; Kim, K.-H. Progress and prospects in the green synthesis of molecularly imprinted polymers for sorptive extraction and sensing applications toward emerging contaminants in various sample matrices. *TrAC Trends Anal. Chem.* **2024**, *170*, 117466. [CrossRef]
231. Rozhin, P.; Gamal, J.A.M.; Giordani, S.; Marchesan, S. Carbon Nanomaterials (CNMs) and Enzymes: From Nanozymes to CNM-Enzyme Conjugates and Biodegradation. *Materials* **2022**, *15*, 1037. [CrossRef] [PubMed]
232. Chen, H.; Luo, K.; Xie, C.; Zhou, L. Nanotechnology of carbon dots with their hybrids for biomedical applications: A review. *Chem. Eng. J.* **2024**, *496*, 153915. [CrossRef]
233. Bagheri, A.R.; Aramesh, N.; Bilal, M.; Xiao, J.; Kim, H.-W.; Yan, B. Carbon nanomaterials as emerging nanotherapeutic platforms to tackle the rising tide of cancer—A review. *Bioorganic Med. Chem.* **2021**, *51*, 116493. [CrossRef]
234. Malhan, A.; Guleria, M.; Das, U.; Singh, S.; Prajapati, B.G.; Mohite, P.; Bhattacharya, S.; Chidrawar, V.R.; Puri, A.; Datta, D. Navigating the future of cancer management through carbon nanodots: A review. *Nano-Struct. Nano-Objects* **2024**, *39*, 101217. [CrossRef]
235. Ponnamma, D.; Yin, Y.; Salim, N.; Parameswaranpillai, J.; Thomas, S.; Hameed, N. Recent progress and multifunctional applications of 3D printed graphene nanocomposites. *Compos. B Eng.* **2021**, *204*, 108493. [CrossRef]
236. Bazazi, S.; Hashemi, E.; Mohammadjavadi, M.; Saeb, M.R.; Liu, Y.; Huang, Y.; Xiao, H.; Seidi, F. Metal-organic framework (MOF)/C-dots and covalent organic framework (COF)/C-dots hybrid nanocomposites: Fabrications and applications in sensing, medical, environmental, and energy sectors. *Adv. Colloid Interface Sci.* **2024**, *328*, 103178. [CrossRef]
237. Ahmed, H.E.H.; Soylak, M. Exploring the potential of carbon quantum dots (CQDs) as an advanced nanomaterial for effective sensing and extraction of toxic pollutants. *TrAC Trends Anal. Chem.* **2024**, *180*, 117939. [CrossRef]
238. Yadav, T.C.; Bachhuka, A. Tuning foreign body response with tailor-engineered nanoscale surface modifications: Fundamentals to clinical applications. *J. Mater. Chem. B* **2023**, *11*, 7834–7854. [CrossRef] [PubMed]
239. Leppert, W.; Malec-Milewska, M.; Zajaczkowska, R.; Wordliczek, J. Transdermal and Topical Drug Administration in the Treatment of Pain. *Molecules* **2018**, *23*, 681. [CrossRef] [PubMed]
240. Sahana, S.; Gautam, A.; Singh, R.; Chandel, S. A recent update on development, synthesis methods, properties and application of natural products derived carbon dots. *Nat. Prod. Bioprospect.* **2023**, *13*, 51. [CrossRef]
241. Adepu, S.; Ramakrishna, S. Controlled Drug Delivery Systems: Current Status and Future Directions. *Molecules* **2021**, *26*, 5905. [CrossRef]
242. Zhang, H. Molecularly Imprinted Nanoparticles for Biomedical Applications. *Adv. Mater.* **2019**, *32*, e1806328. [CrossRef]
243. Das, A.; Adhikari, S.; Deka, D.; Baildya, N.; Sahare, P.; Banerjee, A.; Paul, S.; Bisgin, A.; Pathak, S. An Updated Review on the Role of Nanoformulated Phytochemicals in Colorectal Cancer. *Medicina* **2023**, *59*, 685. [CrossRef]
244. Walia, S.; Sharma, C.; Acharya, A. Biocompatible Fluorescent Nanomaterials for Molecular Imaging Applications, Nanomaterial-Based Biomedical Applications in Molecular Imaging. *Diagn. Ther.* **2020**, 27–53. [CrossRef]
245. Javed, M.N.; Dahiya, E.S.; Ibrahim, A.M.; Alam, M.S.; Khan, F.A.; Pottoo, F.H. Recent Advancement in Clinical Application of Nanotechnological Approached Targeted Delivery of Herbal Drugs. *Nanophytomedicine* **2020**, 151–172. [CrossRef]
246. Soni, V.; Raizada, P.; Singh, P.; Cuong, H.N.; Rangabhashiyam, S.; Saini, A.; Saini, R.V.; Van Le, Q.; Nadda, A.K.; Le, T.-T.; et al. Sustainable and green trends in using plant extracts for the synthesis of biogenic metal nanoparticles toward environmental and pharmaceutical advances: A review. *Environ. Res.* **2021**, *202*, 111622. [CrossRef] [PubMed]
247. John, V.L.; Nair, Y.; Vinod, T.P. Doping and Surface Modification of Carbon Quantum Dots for Enhanced Functionalities and Related Applications. *Part. Part. Syst. Charact.* **2021**, *38*, 2100170. [CrossRef]
248. Arsene, M.-L.; Răut, I.; Călin, M.; Jecu, M.-L.; Doni, M.; Gurban, A.-M. Versatility of Reverse Micelles: From Biomimetic Models to Nano (Bio)Sensor Design. *Processes* **2021**, *9*, 345. [CrossRef]
249. Valizadeh, A.; Asghari, S.; Abbaspoor, S.; Jafari, A.; Raeisi, M.; Pilehvar, Y. Implantable smart hyperthermia nanofibers for cancer therapy: Challenges and opportunities. *WIREs Nanomed. Nanobiotechnol.* **2023**, *15*, e1909. [CrossRef]

250. Vallejo, F.A.; Sigdel, G.; Veliz, E.A.; Leblanc, R.M.; Vanni, S.; Graham, R.M. Carbon Dots in Treatment of Pediatric Brain Tumors: Past, Present, and Future Directions. *Int. J. Mol. Sci.* **2023**, *24*, 9562. [CrossRef]
251. Yun, W.S.; Cho, H.; Jeon, S.I.; Lim, D.-K.; Kim, K. Fluorescence-Based Mono- and Multimodal Imaging for In Vivo Tracking of Mesenchymal Stem Cells. *Biomolecules* **2023**, *13*, 1787. [CrossRef] [PubMed]
252. Peng, G.; Fadeel, B. Understanding the bidirectional interactions between two-dimensional materials, microorganisms, and the immune system. *Adv. Drug Deliv. Rev.* **2022**, *188*, 114422. [CrossRef]
253. Lee, H.P.; Gaharwar, A.K. Light-Responsive Inorganic Biomaterials for Biomedical Applications. *Adv. Sci.* **2020**, *7*, 2000863. [CrossRef] [PubMed]
254. Bakhle, S.S.; Avari, J.G. Development and Characterization of Solid Self-emulsifying Drug Delivery System of Cilnidipine. *Chem. Pharm. Bull.* **2015**, *63*, 408–417. [CrossRef] [PubMed]
255. Yang, B.; Dong, Y.; Wang, F.; Zhang, Y. Nanoformulations to Enhance the Bioavailability and Physiological Functions of Polyphenols. *Molecules* **2020**, *25*, 4613. [CrossRef] [PubMed]
256. Mauro, N.; Utzeri, M.A.; Varvarà, P.; Cavallaro, G. Functionalization of Metal and Carbon Nanoparticles with Potential in Cancer Theranostics. *Molecules* **2021**, *26*, 3085. [CrossRef]
257. Ji, Z.; Yin, Z.; Jia, Z.; Wei, J. Carbon Nanodots Derived from Urea and Citric Acid in Living Cells: Cellular Uptake and Antioxidation Effect. *Langmuir* **2020**, *36*, 8632–8640. [CrossRef]
258. Ji, D.-K.; Dali, H.; Guo, S.; Malaganahally, S.; Vollaire, J.; Josserand, V.; Dumortier, H.; Ménard-Moyon, C.; Bianco, A. Multifunctional Carbon Nanodots: Enhanced Near-Infrared Photosensitizing, Photothermal Activity, and Body Clearance. *Small Sci.* **2021**, *2*, 2100082. [CrossRef]
259. Baker, S.N.; Baker, G.A. Luminescent Carbon Nanodots: Emergent Nanolights. *Angew. Chem. Int. Ed.* **2010**, *49*, 6726–6744. [CrossRef]
260. Kumar, A.; Chowdhuri, A.R.; Laha, D.; Mahto, T.K.; Karmakar, P.; Sahu, S.K. Green synthesis of carbon dots from Ocimum sanctum for effective fluorescent sensing of Pb^{2+} ions and live cell imaging. *Sens. Actuators B Chem.* **2017**, *242*, 679–686. [CrossRef]
261. Gogoi, N.; Barooah, M.; Majumdar, G.; Chowdhury, D. Carbon Dots Rooted Agarose Hydrogel Hybrid Platform for Optical Detection and Separation of Heavy Metal Ions. *ACS Appl. Mater. Interfaces* **2015**, *7*, 3058–3067. [CrossRef]
262. Shi, W.; Wang, Q.; Long, Y.; Cheng, Z.; Chen, S.; Zheng, H.; Huang, Y. Carbon nanodots as peroxidase mimetics and their applications to glucose detection. *Chem. Commun.* **2011**, *47*, 6695. [CrossRef]
263. Wu, Z.L.; Zhang, P.; Gao, M.X.; Liu, C.F.; Wang, W.; Leng, F.; Huang, C.Z. One-pot hydrothermal synthesis of highly luminescent nitrogen-doped amphoteric carbon dots for bioimaging from Bombyx mori silk—Natural proteins. *J. Mater. Chem. B* **2013**, *1*, 2868. [CrossRef]
264. Zhao, J.; Yan, Y.; Zhu, L.; Li, X.; Li, G. An amperometric biosensor for the detection of hydrogen peroxide released from human breast cancer cells. *Biosens Bioelectron* **2013**, *41*, 815–819. [CrossRef]
265. Lin, Z.; Xue, W.; Chen, H.; Lin, J.-M. Peroxynitrous-Acid-Induced Chemiluminescence of Fluorescent Carbon Dots for Nitrite Sensing. *Anal. Chem.* **2011**, *83*, 8245–8251. [CrossRef] [PubMed]
266. Zhao, H.X.; Liu, L.Q.; De Liu, Z.; Wang, Y.; Zhao, X.J.; Huang, C.Z. Highly selective detection of phosphate in very complicated matrixes with an off–on fluorescent probe of europium-adjusted carbon dots. *Chem. Commun.* **2011**, *47*, 2604. [CrossRef]
267. Miao, P.; Han, K.; Tang, Y.; Wang, B.; Lin, T.; Cheng, W. Recent advances in carbon nanodots: Synthesis, properties and biomedical applications. *Nanoscale* **2015**, *7*, 1586–1595. [CrossRef] [PubMed]
268. Ray, S.C.; Saha, A.; Jana, N.R.; Sarkar, R. Fluorescent Carbon Nanoparticles: Synthesis, Characterization, and Bioimaging Application. *J. Phys. Chem. C* **2009**, *113*, 18546–18551. [CrossRef]
269. Zhang, X.; Wang, S.; Zhu, C.; Liu, M.; Ji, Y.; Feng, L.; Tao, L.; Wei, Y. Carbon-dots derived from nanodiamond: Photoluminescence tunable nanoparticles for cell imaging. *J. Colloid Interface Sci.* **2013**, *397*, 39–44. [CrossRef]
270. Dong, Y.; Pang, H.; Yang, H.B.; Guo, C.; Shao, J.; Chi, Y.; Li, C.M.; Yu, T. Carbon-Based Dots Co-doped with Nitrogen and Sulfur for High Quantum Yield and Excitation-Independent Emission. *Angew. Chem. Int. Ed.* **2013**, *52*, 7800–7804. [CrossRef]
271. Xu, Y.; Wu, M.; Liu, Y.; Feng, X.Z.; Yin, X.B.; He, X.W.; Zhang, Y.K. Nitrogen-doped carbon dots: A facile and general preparation method, photoluminescence investigation, and imaging applications. *Chem. A Eur. J.* **2013**, *19*, 2276–2283. [CrossRef] [PubMed]
272. Jiang, C.; Wu, H.; Song, X.; Ma, X.; Wang, J.; Tan, M. Presence of photoluminescent carbon dots in Nescafe® original instant coffee: Applications to bioimaging. *Talanta* **2014**, *127*, 68–74. [CrossRef]
273. Zhu, A.; Luo, Z.; Ding, C.; Li, B.; Zhou, S.; Wang, R.; Tian, Y. A two-photon "turn-on" fluorescent probe based on carbon nanodots for imaging and selective biosensing of hydrogen sulfide in live cells and tissues. *Analyst* **2014**, *139*, 1945–1952. [CrossRef]
274. Lee, H.U.; Park, S.Y.; Park, E.S.; Son, B.; Lee, S.C.; Lee, J.W.; Lee, Y.-C.; Kang, K.S.; Kim, M.I.; Park, H.G.; et al. Photoluminescent carbon nanotags from harmful cyanobacteria for drug delivery and imaging in cancer cells. *Sci. Rep.* **2014**, *4*, 4665. [CrossRef]
275. Majumdar, S.; Krishnatreya, G.; Gogoi, N.; Thakur, D.; Chowdhury, D. Carbon-Dot-Coated Alginate Beads as a Smart Stimuli-Responsive Drug Delivery System. *ACS Appl. Mater. Interfaces* **2016**, *8*, 34179–34184. [CrossRef] [PubMed]

276. Jiao, J.; Liu, C.; Li, X.; Liu, J.; Di, D.; Zhang, Y.; Zhao, Q.; Wang, S. Fluorescent carbon dot modified mesoporous silica nanocarriers for redox-responsive controlled drug delivery and bioimaging. *J. Colloid Interface Sci.* **2016**, *483*, 343–352. [CrossRef] [PubMed]
277. Yang, X.; Wang, Y.; Shen, X.; Su, C.; Yang, J.; Piao, M.; Jia, F.; Gao, G.; Zhang, L.; Lin, Q. One-step synthesis of photoluminescent carbon dots with excitation-independent emission for selective bioimaging and gene delivery. *J. Colloid Interface Sci.* **2017**, *492*, 1–7. [CrossRef]
278. Zhang, M.; Zhao, X.; Fang, Z.; Niu, Y.; Lou, J.; Wu, Y.; Zou, S.; Xia, S.; Sun, M.; Du, F. Fabrication of HA/PEI-functionalized carbon dots for tumor targeting, intracellular imaging and gene delivery. *RSC Adv.* **2017**, *7*, 3369–3375. [CrossRef]
279. Liu, W.; Diao, H.; Chang, H.; Wang, H.; Li, T.; Wei, W. Green synthesis of carbon dots from rose-heart radish and application for Fe3+ detection and cell imaging. *Sensors Actuators B Chem.* **2017**, *241*, 190–198. [CrossRef]
280. Atchudan, R.; Edison, T.N.J.I.; Lee, Y.R. Nitrogen-doped carbon dots originating from unripe peach for fluorescent bioimaging and electrocatalytic oxygen reduction reaction. *J. Colloid Interface Sci.* **2016**, *482*, 8–18. [CrossRef]
281. Mewada, A.; Pandey, S.; Shinde, S.; Mishra, N.; Oza, G.; Thakur, M.; Sharon, M.; Sharon, M. Green synthesis of biocompatible carbon dots using aqueous extract of Trapa bispinosa peel. *Mater. Sci. Eng. C* **2013**, *33*, 2914–2917. [CrossRef] [PubMed]
282. Mehta, V.N.; Jha, S.; Kailasa, S.K. One-pot green synthesis of carbon dots by using Saccharum officinarum juice for fluorescent imaging of bacteria (*Escherichia coli*) and yeast (*Saccharomyces cerevisiae*) cells. *Mater. Sci. Eng. C* **2014**, *38*, 20–27. [CrossRef] [PubMed]
283. Atchudan, R.; Edison, T.N.J.I.; Sethuraman, M.G.; Lee, Y.R. Efficient synthesis of highly fluorescent nitrogen-doped carbon dots for cell imaging using unripe fruit extract of Prunus mume. *Appl. Surf. Sci.* **2016**, *384*, 432–441. [CrossRef]
284. Atchudan, R.; Edison, T.N.J.I.; Chakradhar, D.; Perumal, S.; Shim, J.-J.; Lee, Y.R. Facile green synthesis of nitrogen-doped carbon dots using Chionanthus retusus fruit extract and investigation of their suitability for metal ion sensing and biological applications. *Sens. Actuators B Chem.* **2017**, *246*, 497–509. [CrossRef]
285. Vandarkuzhali, S.A.A.; Jeyalakshmi, V.; Sivaraman, G.; Singaravadivel, S.; Krishnamurthy, K.R.; Viswanathan, B. Highly fluorescent carbon dots from Pseudo-stem of banana plant: Applications as nanosensor and bio-imaging agents. *Sensors Actuators B Chem.* **2017**, *252*, 894–900. [CrossRef]
286. Mohammadi, S.; Pashaee, A.; Amini, N.; Marzban, N.; Puttaiah, S.H.; Tang, V.T.; Maleki, A. Green carbon dots derived from honey, garlic, and carrot: Synthesis, characterization, and antibacterial properties. *Biomass Convers. Biorefinery* **2024**, 1–15. [CrossRef]
287. Zhao, S.; Lan, M.; Zhu, X.; Xue, H.; Ng, T.-W.; Meng, X.; Lee, C.-S.; Wang, P.; Zhang, W. Green Synthesis of Bifunctional Fluorescent Carbon Dots from Garlic for Cellular Imaging and Free Radical Scavenging. *ACS Appl. Mater. Interfaces* **2015**, *7*, 17054–17060. [CrossRef] [PubMed]
288. Shen, J.; Shang, S.; Chen, X.; Wang, D.; Cai, Y. Facile synthesis of fluorescence carbon dots from sweet potato for Fe3+ sensing and cell imaging. *Mater. Sci. Eng. C* **2017**, *76*, 856–864. [CrossRef]
289. Cheng, C.; Shi, Y.; Li, M.; Xing, M.; Wu, Q. Carbon quantum dots from carbonized walnut shells: Structural evolution, fluorescence characteristics, and intracellular bioimaging. *Mater. Sci. Eng. C* **2017**, *79*, 473–480. [CrossRef]
290. Chen, M.; Wang, W.; Wu, X. One-pot green synthesis of water-soluble carbon nanodots with multicolor photoluminescence from polyethylene glycol. *J. Mater. Chem. B* **2014**, *2*, 3937–3945. [CrossRef]
291. Ramanan, V.; Thiyagarajan, S.K.; Raji, K.; Suresh, R.; Sekar, R.; Ramamurthy, P. Outright Green Synthesis of Fluorescent Carbon Dots from Eutrophic Algal Blooms for In Vitro Imaging. *ACS Sustain. Chem. Eng.* **2016**, *4*, 4724–4731. [CrossRef]
292. Sivasankaran, U.; Jesny, S.; Jose, A.R.; Kumar, K.G. Fluorescence Determination of Glutathione Using Tissue Paper-derived Carbon Dots as Fluorophores. *Anal. Sci.* **2017**, *33*, 281–285. [CrossRef] [PubMed]
293. Tripathi, K.M.; Tran, T.S.; Tung, T.T.; Losic, D.; Kim, T. Water Soluble Fluorescent Carbon Nanodots from Biosource for Cells Imaging. *J. Nanomater.* **2017**, *2017*, 1–10. [CrossRef]
294. Hu, Y.; Zhang, L.; Li, X.; Liu, R.; Lin, L.; Zhao, S. Green Preparation of S and N Co-Doped Carbon Dots from Water Chestnut and Onion as Well as Their Use as an Off–On Fluorescent Probe for the Quantification and Imaging of Coenzyme A. *ACS Sustain. Chem. Eng.* **2017**, *5*, 4992–5000. [CrossRef]
295. Park, S.Y.; Lee, H.U.; Park, E.S.; Lee, S.C.; Lee, J.-W.; Jeong, S.W.; Kim, C.H.; Lee, Y.-C.; Huh, Y.S.; Lee, J. Photoluminescent Green Carbon Nanodots from Food-Waste-Derived Sources: Large-Scale Synthesis, Properties, and Biomedical Applications. *ACS Appl. Mater. Interfaces* **2014**, *6*, 3365–3370. [CrossRef]
296. Wang, Z.; Liao, H.; Wu, H.; Wang, B.; Zhao, H.; Tan, M. Fluorescent carbon dots from beer for breast cancer cell imaging and drug delivery. *Anal. Methods* **2015**, *7*, 8911–8917. [CrossRef]
297. Wang, R.; Zhang, S.; Zhang, J.; Wang, J.; Bian, H.; Jin, L.; Zhang, Y. State-of-the-art of lignin-derived carbon nanodots: Preparation, properties, and applications. *Int. J. Biol. Macromol.* **2024**, *273*, 132897. [CrossRef] [PubMed]
298. Bandi, R.; Gangapuram, B.R.; Dadigala, R.; Eslavath, R.; Singh, S.S.; Guttena, V. Facile and green synthesis of fluorescent carbon dots from onion waste and their potential applications as sensor and multicolour imaging agents. *RSC Adv.* **2016**, *6*, 28633–28639. [CrossRef]

299. Zhang, J.; Yuan, Y.; Liang, G.; Yu, S. Scale-Up Synthesis of Fragrant Nitrogen-Doped Carbon Dots from Bee Pollens for Bioimaging and Catalysis. *Adv. Sci.* **2015**, *2*, 1500002. [CrossRef] [PubMed]
300. Sachdev, A.; Gopinath, P. Green synthesis of multifunctional carbon dots from coriander leaves and their potential application as antioxidants, sensors and bioimaging agents. *Analyst* **2015**, *140*, 4260–4269. [CrossRef]
301. Li, J.; Xu, O.; Zhu, X. A facile green and one-pot synthesis of grape seed-derived carbon quantum dots as a fluorescence probe for Cu(ii) and ascorbic acid. *RSC Adv.* **2021**, *11*, 34107–34116. [CrossRef] [PubMed]
302. Huang, G.; Chen, X.; Wang, C.; Zheng, H.; Huang, Z.; Chen, D.; Xie, H. Photoluminescent carbon dots derived from sugarcane molasses: Synthesis, properties, and applications. *RSC Adv.* **2017**, *7*, 47840–47847. [CrossRef]
303. Gong, Y.; Han, Y.; Zhang, F.; Zhai, M.; Chen, X.; Long, Q.; Wang, Z.; Chen, F.; Yang, J. Color Tuning of Biomass-Derived Carbon Nanodots by Reaction Temperature Toward White Light-Emitting Diodes. *Nano* **2020**, *15*. [CrossRef]
304. Kasibabu, B.S.B.; D'souza, S.L.; Jha, S.; Kailasa, S.K. Imaging of Bacterial and Fungal Cells Using Fluorescent Carbon Dots Prepared from Carica papaya Juice. *J. Fluoresc.* **2015**, *25*, 803–810. [CrossRef] [PubMed]
305. Murugesan, B.; Sonamuthu, J.; Pandiyan, N.; Pandi, B.; Samayanan, S.; Mahalingam, S. Photoluminescent reduced graphene oxide quantum dots from latex of Calotropis gigantea for metal sensing, radical scavenging, cytotoxicity, and bioimaging in Artemia salina: A greener route. *J. Photochem. Photobiol. B Biol.* **2018**, *178*, 371–379. [CrossRef]
306. Yang, R.; Guo, X.; Jia, L.; Zhang, Y.; Zhao, Z.; Lonshakov, F. Green preparation of carbon dots with mangosteen pulp for the selective detection of Fe3+ ions and cell imaging. *Appl. Surf. Sci.* **2017**, *423*, 426–432. [CrossRef]
307. Gu, D.; Shang, S.; Yu, Q.; Shen, J. Green synthesis of nitrogen-doped carbon dots from lotus root for Hg(II) ions detection and cell imaging. *Appl. Surf. Sci.* **2016**, *390*, 38–42. [CrossRef]
308. Amin, N.; Afkhami, A.; Hosseinzadeh, L.; Madrakian, T. Green and cost-effective synthesis of carbon dots from date kernel and their application as a novel switchable fluorescence probe for sensitive assay of Zoledronic acid drug in human serum and cellular imaging. *Anal. Chim. Acta* **2018**, *1030*, 183–193. [CrossRef]
309. Feng, X.; Jiang, Y.; Zhao, J.; Miao, M.; Cao, S.; Fang, J.; Shi, L. Easy synthesis of photoluminescent N-doped carbon dots from winter melon for bio-imaging. *RSC Adv.* **2015**, *5*, 31250–31254. [CrossRef]
310. Shakeri, A.; Jarad, N.A.; Khan, S.; Didar, T.F. Bio-functionalization of microfluidic platforms made of thermoplastic materials: A review. *Anal. Chim. Acta* **2022**, *1209*, 339283. [CrossRef]
311. Chan, M.-H.; Chen, B.-G.; Ngo, L.T.; Huang, W.-T.; Li, C.-H.; Liu, R.-S.; Hsiao, M. Natural Carbon Nanodots: Toxicity Assessment and Theranostic Biological Application. *Pharmaceutics* **2021**, *13*, 1874. [CrossRef]
312. Reina, G.; González-Domínguez, J.M.; Criado, A.; Vázquez, E.; Bianco, A.; Prato, M. Promises, facts and challenges for graphene in biomedical applications. *Chem. Soc. Rev.* **2017**, *46*, 4400–4416. [CrossRef]
313. Kumar, A.; Sharipov, M.; Turaev, A.; Azizov, S.; Azizov, I.; Makhado, E.; Rahdar, A.; Kumar, D.; Pandey, S. Polymer-Based Hybrid Nanoarchitectures for Cancer Therapy Applications. *Polymers* **2022**, *14*, 3027. [CrossRef] [PubMed]
314. Kasouni, A.; Chatzimitakos, T.; Stalikas, C. Bioimaging Applications of Carbon Nanodots: A Review. *J. Carbon Res.* **2019**, *5*, 19. [CrossRef]
315. Kwee, Y.; Kristanti, A.N.; Siimon, K.; Aminah, N.S.; Fahmi, M.Z. Carbon Nanodots Derived from Natural Products. *S. Afr. J. Chem.* **2021**, *75*, 40–63.
316. Aziz, S.B.; Abdullah, O.G.; Brza, M.A.; Azawy, A.K.; Tahir, D.A. Effect of carbon nano-dots (CNDs) on structural and optical properties of PMMA polymer composite. *Results Phys.* **2019**, *15*, 102776. [CrossRef]
317. Cayuela, A.; Carrillo-Carrión, C.; Soriano, M.L.; Parak, W.J.; Valcárcel, M. One-Step Synthesis and Characterization of N-Doped Carbon Nanodots for Sensing in Organic Media. *Anal. Chem.* **2016**, *88*, 3178–3185. [CrossRef]
318. Lu, S.; Cong, R.; Zhu, S.; Zhao, X.; Liu, J.; Tse, J.S.; Meng, S.; Yang, B. pH-Dependent Synthesis of Novel Structure-Controllable Polymer-Carbon NanoDots with High Acidophilic Luminescence and Super Carbon Dots Assembly for White Light-Emitting Diodes. *ACS Appl. Mater. Interfaces* **2016**, *8*, 4062–4068. [CrossRef]
319. Kwon, W.; Do, S.; Lee, J.; Hwang, S.; Kim, J.K.; Rhee, S.-W. Freestanding Luminescent Films of Nitrogen-Rich Carbon Nanodots toward Large-Scale Phosphor-Based White-Light-Emitting Devices. *Chem. Mater.* **2013**, *25*, 1893–1899. [CrossRef]
320. Stan, C.S.; Elouakassi, N.; Albu, C.; Conchi, A.O.; Coroaba, A.; Ursu, L.E.; Popa, M.; Kaddami, H.; Almaggoussi, A. Photoluminescence of Argan-Waste-Derived Carbon Nanodots Embedded in Polymer Matrices. *Nanomaterials* **2023**, *14*, 83. [CrossRef]
321. Jahan, S.; Mansoor, F.; Naz, S.; Lei, J.; Kanwal, S. Oxidative Synthesis of Highly Fluorescent Boron/Nitrogen Co-Doped Carbon Nanodots Enabling Detection of Photosensitizer and Carcinogenic Dye. *Anal. Chem.* **2013**, *85*, 10232–10239. [CrossRef] [PubMed]
322. Yang, P.-C.; Ting, Y.-X.; Gu, S.; Gandomi, Y.A.; Hsieh, C.-T. Fluorescent nitrogen-doped carbon nanodots synthesized through a hydrothermal method with different isomers. *J. Taiwan Inst. Chem. Eng.* **2021**, *123*, 302–309. [CrossRef]
323. Perli, G.; Soares, M.C.P.; Cabral, T.D.; Bertuzzi, D.L.; Bartoli, J.R.; Livi, S.; Duchet-Rumeau, J.; Cordeiro, C.M.B.; Fujiwara, E.; Ornelas, C. Synthesis of Carbon Nanodots from Sugarcane Syrup, and Their Incorporation into a Hydrogel-Based Composite to Fabricate Innovative Fluorescent Microstructured Polymer Optical Fibers. *Gels* **2022**, *8*, 553. [CrossRef] [PubMed]

324. Xia, J.; Chen, S.; Zou, G.-Y.; Yu, Y.-L.; Wang, J.-H. Synthesis of highly stable red-emissive carbon polymer dots by modulated polymerization: From the mechanism to application in intracellular pH imaging. *Nanoscale* **2018**, *10*, 22484–22492. [CrossRef]
325. Shahnazi, A.; Nabid, M.R.; Sedghi, R. Synthesis of surface molecularly imprinted poly-o-phenylenediamine/TiO_2/carbon nanodots with a highly enhanced selective photocatalytic degradation of pendimethalin herbicide under visible light. *React. Funct. Polym.* **2020**, *151*, 104580. [CrossRef]
326. Wang, Y.; Meng, Y.; Wang, S.; Li, C.; Shi, W.; Chen, J.; Wang, J.; Huang, R. Direct Solvent-Derived Polymer-Coated Nitrogen-Doped Carbon Nanodots with High Water Solubility for Targeted Fluorescence Imaging of Glioma. *Small* **2015**, *11*, 3575–3581. [CrossRef] [PubMed]
327. Arcudi, F.; Đorđević, L.; Prato, M. Design, Synthesis, and Functionalization Strategies of Tailored Carbon Nanodots. *Acc. Chem. Res.* **2019**, *52*, 2070–2079. [CrossRef]
328. Han, L.; Liu, S.G.; Dong, J.X.; Liang, J.Y.; Li, L.J.; Li, N.B.; Luo, H.Q. Facile synthesis of multicolor photoluminescent polymer carbon dots with surface-state energy gap-controlled emission. *J. Mater. Chem. C* **2017**, *5*, 10785–10793. [CrossRef]
329. Arcudi, F.; Đorđević, L.; Prato, M. Synthesis, Separation, and Characterization of Small and Highly Fluorescent Nitrogen-Doped Carbon NanoDots. *Angew. Chem. Int. Ed.* **2016**, *55*, 2107–2112. [CrossRef]
330. Wang, L.; Li, B.; Li, L.; Xu, F.; Xu, Z.; Wei, D.; Feng, Y.; Wang, Y.; Jia, D.; Zhou, Y. Ultrahigh-yield synthesis of N-doped carbon nanodots that down-regulate ROS in zebrafish. *J. Mater. Chem. B* **2017**, *5*, 7848–7860. [CrossRef] [PubMed]
331. Zhao, J.; Dai, D.; Zhou, L.; Yu, Z.; Ma, J.; Yang, M.; Yi, C. A tumor microenvironment responsive mesoporous polydopamine theranostic probe embedded with Gd/I-doped carbon nanodots for CT/MR/FL imaging and chemo/photothermal synergistic therapy. *Carbon* **2024**, *224*, 119065. [CrossRef]
332. Mauro, N.; Utzeri, M.A.; Sciortino, A.; Cannas, M.; Messina, F.; Cavallaro, G.; Giammona, G. Printable Thermo-and Photo-stable Poly (D, L-lactide)/Carbon Nanodots Nanocomposites via Heterophase Melt-Extrusion Transesterification. Undefined 2022. Available online: https://www.sciencedirect.com/science/article/pii/S1385894722020204 (accessed on 25 December 2024).
333. Mauro, N.; Utzeri, M.A.; Sciortino, A.; Messina, F.; Cannas, M.; Popescu, R.; Gerthsen, D.; Buscarino, G.; Cavallaro, G.; Giammona, G. Decagram-Scale Synthesis of Multicolor Carbon Nanodots: Self-Tracking Nanoheaters with Inherent and Selective Anticancer Properties. *ACS Appl. Mater. Interfaces* **2022**, *14*, 2551–2563. [CrossRef]
334. Mauro, N.; Utzeri, M.A.; Drago, S.E.; Nicosia, A.; Costa, S.; Cavallaro, G.; Giammona, G. Hyaluronic acid dressing of hydrophobic carbon nanodots: A self-assembling strategy of hybrid nanocomposites with theranostic potential. Undefined 2021. Available online: https://www.sciencedirect.com/science/article/pii/S0144861721006007 (accessed on 25 December 2024).
335. Feng, T.; Ai, X.; An, G.; Yang, P.; Zhao, Y. Charge-Convertible Carbon Dots for Imaging-Guided Drug Delivery with Enhanced in Vivo Cancer Therapeutic Efficiency. *ACS Nano* **2016**, *10*, 4410–4420. [CrossRef] [PubMed]
336. Lamba, R.; Salam, A.; Anjum, F.; Yadav, A.; Garg, R.; Kaushik, K.; Sharma, S.; Nandi, C.K. NIR Emissive Carbon Nanodots as a Tool to Mark Ribosomal RNA and Components using Super-Resolution Microscopy. *Nanoscale* **2024**, *16*, 11739. [CrossRef]
337. Mauro, N.; Cillari, R.; Utzeri, M.A.; Costa, S.; Giammona, G.; Nicosia, A.; Cavallaro, G. Controlled delivery of sildenafil by β-Cyclodextrin-decorated sulfur-doped carbon nanodots: A synergistic activation of ROS signaling in tumors overexpressing PDE-5. *Int. J. Pharm.* **2023**, *645*, 123409. [CrossRef]
338. Available online: https://patentimages.storage.googleapis.com/62/e4/c2/0addaa60e6dfef/US10837904.pdf (accessed on 1 January 2023).
339. Available online: https://patentimages.storage.googleapis.com/03/96/a9/28ce82d2c74544/US11478433.pdf (accessed on 1 January 2023).
340. Available online: https://patentimages.storage.googleapis.com/82/4d/1a/d6b3bc5dd09ffb/CN102849722B.pdf (accessed on 1 January 2023).
341. Available online: https://patentimagezs.storage.googleapis.com/b6/28/0b/4149a74fffe4a8/US7572542.pdf (accessed on 1 January 2023).
342. Available online: https://patentimages.storage.googleapis.com/a9/f9/8a/f57f618a5f3384/CN111778018A.pdf (accessed on 1 January 2023).
343. Available online: https://patentimages.storage.googleapis.com/85/e0/b8/e07c9f2159c602/CN109504375B.pdf (accessed on 1 January 2023).
344. Available online: https://patents.google.com/patent/CN106833631B/en (accessed on 1 January 2023).
345. Available online: https://patents.google.com/patent/CN113403068B/en (accessed on 1 January 2023).

polymers

MDPI

Article

Hybrid Solid Polymer Electrolytes Based on Epoxy Resins, Ionic Liquid, and Ceramic Nanoparticles for Structural Applications

Bianca K. Muñoz [1,*], Jorge Lozano [1], María Sánchez [1,2] and Alejandro Ureña [1,2]

[1] Material Science and Engineering Area, ESCET, Universidad Rey Juan Carlos, C/Tulipán s/n, 28933 Móstoles, Spain; j.lozanoma@alumnos.urjc.es (J.L.); maria.sanchez@urjc.es (M.S.); alejandro.urena@urjc.es (A.U.)
[2] Instituto de Tecnologías Para la Sostenibilidad, Universidad Rey Juan Carlos, C/Tulipán s/n, 28933 Móstoles, Spain
* Correspondence: bianca.munoz@urjc.es; Tel.: +34-91-488-71-45

Abstract: Solid polymer electrolytes (SPE) and composite polymer electrolytes (CPE) serve as crucial components in all-solid-state energy storage devices. Structural batteries and supercapacitors present a promising alternative for electric vehicles, integrating structural functionality with energy storage capability. However, despite their potential, these applications are hampered by various challenges, particularly in the realm of developing new solid polymer electrolytes that require more investigation. In this study, novel solid polymer electrolytes and composite polymer electrolytes were synthesized using epoxy resin blends, ionic liquid, lithium salt, and alumina nanoparticles and subsequently characterized. Among the formulations tested, the optimal system, designated as L70P30ILE40Li1MAl2 and containing 40 wt.% of ionic liquid and 5.7 wt.% of lithium salt, exhibited exceptional mechanical properties. It displayed a remarkable storage modulus of 1.2 GPa and reached ionic conductivities of 0.085 mS/cm at 60 °C. Furthermore, a proof-of-concept supercapacitor was fabricated, demonstrating the practical application of the developed electrolyte system.

Keywords: solid polymer electrolyte; composite polymer electrolyte; structural supercapacitor; energy storage; structural devices

Citation: Muñoz, B.K.; Lozano, J.; Sánchez, M.; Ureña, A. Hybrid Solid Polymer Electrolytes Based on Epoxy Resins, Ionic Liquid, and Ceramic Nanoparticles for Structural Applications. *Polymers* **2024**, *16*, 2048. https://doi.org/10.3390/polym16142048

Academic Editor: Asterios (Stergios) Pispas

Received: 25 June 2024
Revised: 12 July 2024
Accepted: 16 July 2024
Published: 18 July 2024

1. Introduction

Multifunctional structural supercapacitors and batteries are considered an outstanding approach for energy storage applications in electrical vehicles (EVs) [1,2]. It is well-known that reducing weight in a vehicle can help reduce fuel consumption. This statement has also been estimated in terms of energy consumption (around 7%) when weight is reduced by 10% [3,4]. Based on this fact, electrical vehicles have different options for reducing weight, the most intuitive and well-established consist of using lightweight materials (fiber composites or aluminum matrix composites) [5]. Another potential way includes structural materials as energy storage devices, the most attractive route but still far from practical implementation and commercialization [4,5]. The synergic combination of lightweight-resistant materials as carbon fiber-reinforced polymers (CFRP), with their ability for energy storage (supercapacitors, batteries, and fuel cells), and detaching the inert battery mass (needed but useless) appears to be an ideal scenario for achieving a practical fuel independency [6]. Concerning the components of a multifunctional device, recent studies have been addressed regarding the modification of carbon fiber as electrodes, the development of solid polymer electrolytes (SPE), the optimization of glass or Kevlar fiber separators, the coupling of current collectors, etc.

The SPE are key components of the structural supercapacitors and batteries, and they are at their early stage of development since their ionic conductivities are still not comparable to liquid electrolytes. In this context, different approaches have been carried out so far with various degrees of success, revealing some technical issues, as the stability window and ionic conductivity, that still need to be solved in this technology before being considered as a competitive option. Hybrid electrolytes (HE) or composite polymer electrolytes (CPE) are electrolytes with more than two components that can be an alternative to traditional electrolytes to simultaneously improve ionic conductivity and mechanical properties. They have been classified as active and passive electrolytes, depending on the nature of their reinforcements and whether they participate or not in the conduction mechanism. Electrolytes with ceramic reinforcements that do not provide ionic movement by themselves, such as SiO_2 or TiO_2, are considered passive. Passive HEs have recently attracted the attention of researchers because they exhibit higher mechanical resistance and thermal stability than active HEs, and being less expensive and more accessible [7–11]. For instance, alumina has been reported to modify the ionic conductivity in polymer electrolytes by preventing the formation of crystals and improving mechanical properties due to their rigid nature [12–15]. Some examples of its use have been reported in poly(vinyl alcohol) (PVA) matrices, where it has been observed that using nano-alumina in 6 wt.% showed a remarkable increase in conductivity due to the increase of amorphous areas in the PVA. Above 6 wt.%, aggregation of alumina NPs that creates unsuitable paths for ion transfer has been detected [12]. Also, many studies have used alumina as reinforcement in a poly(ethylene oxide) (PEO) matrix, including lithium salt [16,17].

Epoxy resins are appropriate matrices for structural applications [18]. Those derived from Polyethylene Glycol Diglycidyl Ether (PEGDGE) or diglycidyl ether of bisphenol A (DGEBA) have been investigated as SPEs [19–23]. Kwon et al. reported the introduction of alumina nanowires (2–6 nm) in a epoxy resin/Li salt system. They have observed that increasing the amount of alumina used to increase the conductivity (0.29 $mS{\cdot}cm^{-1}$ at 25 °C), but above certain content can result in reducing the ionic dissociation due the NPs aggregation. Thus, a very high alumina content acts in detriment to the T_g of the resulting polymer [14]. Another successful example has been reported by Choi and coworkers, where SPEs based on DGEBA, succinonitrile and lithium salt afford ionic conductivities from 1×10^{-3} to 0.1 mS/cm, reaching a high Young's modulus of 1 GPa in some cases [24].

Recently, in our group, we have developed HEs based on a mixture of epoxy resins with PEGDGE, and a commercial epoxy resin (Araldite LY556) based on DGEBA, using a commercial hardener Araldite XB3473 and 4,4′-diaminodiphenyl sulfone. The additives used were the ionic liquid 1-Ethyl-3-methylimidazolium bis(trifluoromethylsulfonyl)imide (EMIMTFSI) and titania nanoparticles (21 nm). These electrolytes exhibited good mechanical properties ($T_g > 70$ °C and storage modulus at 30 °C > 1 GPa), as well as promising ionic conductivities [20]. Using the same epoxy resin system, a systematic study was carried out including alumina (particle size < 13 nm, 2–8 wt.% of nanoparticles content) [25]. In this study the best system found, containing 2 wt. % of alumina, was able to afford outstanding thermomechanical properties (T_g = 83 °C and E′ at 30 °C = 1.2 GPa) and modest ionic conductivities at room temperature ($\sigma_1 = 1.6 \times 10^{-3}$ mS/cm). This family of electrolytes was found to be very appropriate for structural applications, as it has been demonstrated in the fabrication of a structural supercapacitor using graphene-modified woven carbon fiber and woven lass fiber as separator, which exhibit a Young modulus of 24.0 ± 1.6 GPa and a tensile strength of 294 ± 9 MPa [26]. From a mechanical point of view, the CFRP obtained with this electrolyte formulation achieves a promising behavior for energy storage devices since, in some practical examples previously reported for batteries, the Young modulus and tensile strength are in the range of 1.8–25 GPa and 90–300 MPa, respectively [27].

Despite the good results obtained using the SPEs mentioned above [25], these formulations cannot be used for structural Li-ion batteries as they do not contain lithium ions.

Based on this limitation, herein we report a new family of SPEs and CPEs for structural supercapacitors and lithium-ion batteries. The new formulation of SPEs and CPEs, are also based on epoxy resin blends, ionic liquid, lithium salt, and alumina nanoparticles. This new electrolyte family represents a promising alternative in the development of more efficient and safer structural energy storage devices.

2. Materials and Methods

2.1. Materials

Solvents and reagents were purchased and used without further purifications. Epoxy resins Araldite LY556 based on DGEBA and its hardener XB3473, were purchased in Huntsman. The epoxy resin has an epoxy equivalent mass range of 183.48–188.67 epoxy equiv^{-1}. The amines equivalent mass range of the crossing agent was 82.6–89.3 g amine equiv^{-1}. The ratio Epoxy resin/hardener used was 100:23 wt.%, according to the supplier recommendations, and this system was identified as L. The reagents Poly(ethylenegycol) Diglycidyl Ether (PEGDGE, Mn 500), 4,4′-diaminodiphenyl sulfone (DDS), the lithium salt bis(trifluoromethylsulfonyl)imide lithium (LiTFSI, Li), propylene carbonate (PC), alumina nanoparticles (Al, 13 nm), copper chloride, (L)-ascorbic acid (L-AA), and PVP K90 were purchased from Sigma-Aldrich (St. Louis, MO, USA). Ionic liquid 1-ethyl-3-methyl-imidazolium bis(trifluoromethylsulfonyl) imide (ILE, 99%) was purchased from IOLITEC -Ionic Liquids Technologies GmbH (Heilbronn, Germany). The resin system PEGDGE/DDS used an epoxy/amine ratio of 100:35 wt.%, and it was identified as P. Woven carbon fabric HexForce® 48,193 plain 12 K, manufactured by Hexcel (Stamford, CT, USA), was used in this work. The glass fiber separator used (E-Fiberglass Woven Roving) was purchased from Castro Composites® (Pontevedra, Spain).

The nomenclature used for the reagents and epoxy resins are denoted in bold capital letters, using the acronyms shown above.

2.2. Methods

All the samples were characterized by dynamic mechanical thermal analysis (DMTA) Q800 V7.1 (TA Instruments, New Castle, DE, USA), cyclic voltammetry (CV, AUTOLAB PGSTAT302N, Herisau, Switzerland), linear sweep voltammetry (LSV, AUTOLAB PGSTAT302N, Herisau, Switzerland), electrochemical impedance spectroscopy (EIS, AUTOLAB PGSTAT302N, Herisau, Switzerland) and field-emission gun scanning electron microscopy (FEGSEM) Nova NanoSEM 230 apparatus from Philips (Amsterdam, The Netherlands).

2.2.1. Dynamic Mechanical Thermal Analysis (DMTA)

The thermo-mechanical properties were studied following ASTM 5418 [27] in a single cantilever mode over a temperature range from 25–275 °C, using a ramp of 2 °C/min and a frequency of 1 Hz. The thermal scanning was performed from −50 to 200 °C, to determine the glass transition temperature (T_g), which was measured as the maximum of the loss tangent curve (tan δ). In addition, the average modulus at room temperature allows evaluating the stiffness of the samples.

2.2.2. Cyclic Voltammetry (CV) and Linear Sweep Voltammetry (LSV)

The CV and LSV analyses were performed to determine the stability window and electrochemical behavior of the electrolytes. The measurements were performed in a potentiostat AUTOLAB PGSTAT302N with a software Nova 2.1. The CV parameters were from −1 to 1 V at 10 mVs^{-1} and for LSV the potential varied from 0 to 4 at the same scan rate.

2.2.3. Electrochemical Impedance Spectroscopy (EIS)

The samples for EIS were dried overnight in an oven at 80 °C after being painted with silver ink to favor the electrode–electrolyte contact. The EIS analysis was performed in a potentiostat AUTOLAB PGSTAT302N. The samples were sandwiched between two symmetric clean and polished stainless-steel electrodes and introduced in a homemade press hold sampler [25], to improve the electrolyte–electrode contact. The impedance of the samples was measured at room temperature using a frequency range between 1.0 MHz and 0.1 Hz and an amplitude of 30 mV. The Nyquist plots (imaginary contribution of impedance Z″ vs. real contribution Z′) showed a semicircle at high frequencies and an additional contribution at lower frequencies corresponding to the R_0 (CPE_1 R_1) (CPE_2 R_2)

equivalent circuit. The ionic conductivities were obtained from R_1 for the solid polymer electrolytes following Equation (1). The composite polymer electrolytes (those containing alumina nanoparticles) showed two conduction paths due to the reinforcement and two ionic conductivities were obtained from R_0 and R_1, respectively.

$$\sigma = \frac{d(cm)}{R(\Omega) \times A(cm^2)} \quad (1)$$

2.2.4. Chronoamperometry for Lithium Transference Number Calculations ($t_{Li}{}^{+}$)

The samples were sanded to obtain thin layers. The layers were dried in an oven and placed under vacuum overnight before preparing the coin cell battery CR2032 in Li/CPE/Li configuration. The coin cells were assembled in the glove box. The lithium transference number was measured following Bruce–Vincent–Evans equation at room temperature (2), where ΔV is 10 mV, I_0 and I_{SS} the initial and steady state currents, and R_0 and R_{SS} are the resistances obtained from EIS analysis before and after polarization. The coin cells were tested by dc polarization, applying 10 mV during 7200 s. EIS analysis to determine the ionic conductivity were acquired before and after the chronoamperometry.

$$t_{Li}{}^{+} \frac{I_{SS}(\Delta V - R_0 I_0)}{I_0(\Delta V - R_{SS} I_{SS})} \quad (2)$$

2.2.5. Field-Emission Gun Scanning Electron Microscopy (FEGSEM)

The electrolyte-free samples were analyzed morphologically using FEGSEM Nova NanoSEM 230 working at 5 kV and at 5 mm of distance. The samples were treated previously with ethanol to remove the ionic liquid following the procedure described in the literature [22]. This procedure included changing the solvent twice a day, for one week and drying the samples in a vacuum oven at 70 °C overnight. Cryofractured samples were analyzed after coating with gold in a gold sputtering (2 nm Au).

2.3. *Solid Polymer Electrolyte Preparation*

The solid polymer electrolytes were prepared following two different procedures described below, depending on whether the samples contain nanoparticles or not. The weight of reagents was calculated for 30 g of the total sample according to the wt.% content listed in Table 1. Once the samples were cured, they were prepared to adapt the form for each technique samplers.

Table 1. Composition of hybrid solid polymer electrolytes.

	wt.%					
Sample	**L**	**P**	**ILE**	**Li (1M)**	**PC**	**Al**
L65P35(ILE30)Li	41.80	22.50	30	5.7	-	-
L70P30(ILE40)Li	36.70	15.70	40	7.6	-	-
L65P35(ILE40)Li	34.20	18.20	40	7.6	-	-
L60P40(ILE40)Li	31.47	20.93	40	7.6	-	-
L70P30(ILE45)Li	33.55	12.95	45	8.5	-	-
L65P35(ILE45)Li	30.23	16.27	45	8.5	-	-
L60P40(ILE45)Li	27.90	18.60	45	8.5	-	-
L70P30(ILE50)Li	28.41	12.19	50	9.4	-	-
L65P35(ILE50)Li	26.40	14.20	50	9.4	-	-
L60P40(ILE50)Li	24.37	16.23	50	9.4	-	-
L70P30(ILE40)(PC5)Li	33.20	14.20	40	7.6	5	-
L65P35(ILE40)(PC5)Li	30.80	16.60	40	7.6	5	-
L65P35(ILE30)Li(Al2)	40.50	21.8	30	5.7	-	2
L70P30(ILE40)Li(Al2)	35.30	15.10	40	7.6	-	2
L70P30(ILE45)Li(Al2)	31.15	13.35	45	8.5	-	2
L70P30(ILE50)Li(Al2)	27.01	11.59	50	9.4	-	2

Procedure A (for samples without nanoparticles): In a flat bottom jar the lithium salt was dissolved in the ionic liquid. Then, both resins L and P were weighed in the same jar and the mixture was degassed under vacuum at 80°C for 15 min. Next, the hardeners XB3473 and DDS were added, and the mixture was stirred under vacuum for 10 more minutes. The mixture was placed on a metallic mold and left to cure at 140°C for 8 h, as previously described.

Procedure B (for samples containing Al_2O_3 nanoparticles): In a flat bottom jar the lithium salt was dissolved in the ionic liquid. Alumina was added to the previous solution and the nanoparticles were dispersed by ultrasonication in a Hielscher ultrasonic processor UP400 St at 0.5 pulse cycles and 50% amplitude for 1.5 h. Then, both resins L and P were weighted in the same jar and the mixture was degassed under vacuum at 80 °C for 15 min. Next, the hardeners DDS and XB3473 were added, and the mixture was stirred under vacuum for 10 more minutes. The mixture was placed on a metallic mold and left to cure at 140 °C for 8 h, as previously described.

2.4. Supercapacitor Fabrication

The electrodes of CuO NPs on woven carbon fiber were fabricated following the procedure previously reported in the literature [28]. For this, 1 wt.% of PVP K90 was added to a solution containing $CuCl_2$/NaCl/L-Ascorbic acid (molar ratio of 1:3.5:15) in deionized water (DIW, 375 mL). The pH was adjusted to 3.9 using NaOH, and this solution was used as a growth solution for the hydrothermal modification of CuO on WCF. The reaction was left in the autoclave at 80 °C for 2 h. The fibers obtained were washed with DIW and dried overnight in the oven at 60 °C to use it as electrodes. The structural supercapacitor was fabricated by vacuum-assisted resin infusion molding (VARIM) as reported [24] and using the CPE L70P30(ILE40)Li(Al2). The composite was cured in an oven at 140 °C for 8 h. The supercapacitors were composed of two layers of WCF electrodes with a two-layer GF separator between them. Copper sheets were also attached to each CFRP electrode by using silver ink to ensure a good electrical contact. The EIS test was carried out on the composites in a frequency range of 10×10^6–0.1 Hz. The tests were also performed in an AUTO- LAB PGSTAT302N module with Nova 2.1 software.

3. Results and Discussion

3.1. Thermomechanical and Morphologycal Characterization of the Electrolytes

In this study, samples of electrolytes were prepared keeping in mind the ability to enhance the electrochemical and mechanical properties simultaneously. We studied the effect of increasing the amount of ionic liquid (40, 45, and 50 wt. % of ILE) and the structural resin content (60, 65, and 70 wt. % of L resin). The effect of the addition of PC (5 wt.%) and ceramic nanoparticles (2 wt.% of alumina, Al2) was also studied. Table 2 shows the mean values of the storage modulus (E′), loss modulus (E″), and glass transition temperature (T_g), as well as their standard deviations obtained through the DMTA analysis.

Concerning the amount of ionic liquid included (40, 45, or 50 wt.%), it was possible to observe a higher storage modulus (E′) at lower ionic liquid content for all the formulations. The highest values were obtained for those samples containing 40% of ILE (from 539 to 1032 MPa at 30 °C, entries 1–3). Surprisingly, the T_g values do not vary in this regard, being around 70 °C for the samples L70P30 containing 40, 45, and 50 wt.% of ILE (entries 1, 4, and 7), meaning that these values are more susceptible to the resin's ratio (L/P) than to the ionic liquid content. The same behavior was observed for the T_g values of samples L65P35 and L60P40 (entries 2, 5, and 8 and 3, 6, and 9, respectively). Regarding the amount of the more structural resin (L, from 60 to 70 wt.%), when increases up to 70% the T_g and the storage modulus (E′) were also found to increase (entries 1–9), as was expected and observed in our previous study [25]. From a mechanical perspective, the samples containing 70% L resin appear to be the best suited for structural applications, since as soon as the L content is below 70 wt.%, the storage modulus drops irremediably.

Table 2. Thermomechanical properties obtained by DMTA analysis.

Entry	Sample	T_g (°C)	E′ (MPa) (T = 30 °C)	E″ (MPa) (T = 30 °C)
1	L70P30(ILE40)Li	70.7 ± 0.9	1032 ± 41	94 ± 6
2	L65P35(ILE40)Li	57 ± 2	659 ± 27	108 ± 8
3	L60P40(ILE40)Li	55 ± 2	539 ± 87	104 ± 17
4	L70P30(ILE45)Li	69 ± 2	773 ± 127	71 ± 2
5	L65P35(ILE45)Li	56 ± 3	506 ± 93	82 ± 4
6	L60P40(ILE45)Li	44.8 ± 0.5	222 ± 19	72 ± 3
7	L70P30(ILE50)Li	73 ± 1	675 ± 81	61.8 ± 0.9
8	L65P35(ILE50)Li	57 ± 4	530 ± 145	73 ± 14
9	L60P40(ILE50)Li	50 ± 2	328 ± 79	74 ± 12
10	L70P30(ILE40)(PC5)Li	63 ± 2	603 ± 122	76 ± 7
11	L65P35(ILE40)(PC5)Li	50 ± 1	504 ± 26	101 ± 5
12	L70P30(ILE40)Li(Al2)	72 ± 2	1231 ± 84	112 ± 5
13	L70P30(ILE45)Li(Al2)	72 ± 2	781 ± 77	74 ± 3
14	L70P30(ILE50)Li(Al2)	76.6 ± 0.8	696 ± 18	69 ± 1

The incorporation of PC was then studied in two formulations, L70P30(ILE40)Li and L65P35(ILE40)Li (entries 10 and 11), and it was observed that there was a more remarkable drop in the storage modulus for the resin L70P30(ILE40) (from 1032 to 603 MPa) than for the resin with a lesser L content (from 659 to 504 MPa).

This additive was only added to samples containing 40% ILE to avoid an extreme loss of storage modulus. The T_g were lower, decreasing 7 °C in both samples. Alumina nanoparticles were then added to samples with formula L70P30(ILEX)Li (entries 12–14), to improve and recover mechanical strength. The sample with 40 wt.% of ILE (entry 12), exhibits the best improvement with a 1.23 GPa of storage modulus. For the other samples (45 and 50 wt.% ILE), this effect was not so remarkable, as the values were practically the same as without nanoparticles. The T_g values with the addition of nanoparticles slightly increased in the three L/P resin's ratio studied.

Figure 1 shows the results of storage modulus E′ and T_g for all of the samples studied. It can be observed that when alumina (2 wt.%) is included in the electrolytes, L70P30, for 40, 45, and 50% ILE, slightly increases the T_g of the samples regarding the initial values. The best sample performance was found for the sample containing 40% ILE and alumina, L70P30(ILE40)Li(Al2) (entry 12).

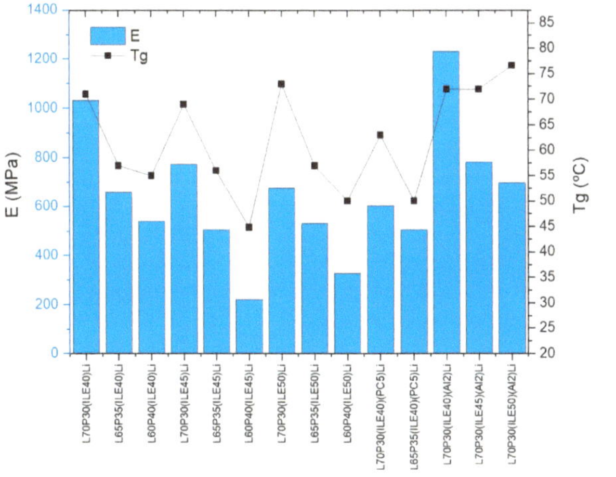

Figure 1. Storage modulus (E′, blue bars) and glass transition temperature (T_g, black squares) of the samples.

3.2. *Electrochemical Characterization of Electrolytes*

All the samples were analyzed by EIS to calculate the ionic conductivity. The values are listed in Table 3 (the Nyquist plots equivalent circuit fittings are shown in Supplementary Materials, Figures S1–S5).

Table 3. Electrochemical data for the solid electrolytes.

Entry	Sample	σ_0 ($S \cdot cm^{-1}$)	σ_1 ($S \cdot cm^{-1}$)	C_{sp} ($\mu F/cm^2$)	Stability Range (V)
1	L70P30(ILE40)Li		2.6×10^{-7}	0.30	0.3
2	L65P35(ILE40)Li		2.9×10^{-7}	1.74	0.3
3	L60P40(ILE40)Li		3.5×10^{-7}	3.76	0.8
4	L70P30(ILE45)Li		1.6×10^{-6}	21.22	1.2
5	L65P35(ILE45)Li		1.5×10^{-6}	44.03	1.0
6	L60P40(ILE45)Li		2.6×10^{-6}	109.39	1.7
7	L70P30(ILE50)Li		7.2×10^{-7}	29.64	2.2
8	L65P35(ILE50)Li		8.8×10^{-7}	18.44	1.1
9	L60P40(ILE50)Li		6.1×10^{-6}	67.57	1.8
10	L70P30(ILE40)(PC5)Li		4.8×10^{-7}	7.50	1.4
11	L65P35(ILE40)(PC5)Li		1.2×10^{-6}	30.44	1.2
12	L70P30(ILE40)Li(Al2)	2.0×10^{-5}	3.6×10^{-7}	4.61	2.4
13	L70P30(ILE45)Li(Al2)	2.5×10^{-5}	4.2×10^{-6}	9.11	2.6
14	L70P30(ILE50)Li(Al2)	2.7×10^{-5}	5.7×10^{-6}	2.95	2.3

Before discussing ionic conductivity, it is necessary to understand the ionic movement in hybrid electrolytes with non-active fillers [29]. The ionic movements in a HE can occur via a hopping mechanism through the polymer chain, where the Li^+ and other cations hop though the oxygen atoms in ethylene glycol polymer chains ($-O-CH_2-CH_2-O-$); this is usually known as bulk polymer conductivity (σ_1). The other way the ions can move is by percolation through the interfacial layer, that can be defined as that region were the conducting dispersing medium (Li in ILE) is immobilized or confined on the surface of the nanofiller [30]. In this article, this interfacial ionic conductivity will be referred to as interfacial conductivity (σ_0), and it will be calculated only for those electrolyte formulations including alumina.

The ionic conductivities of all the polymer electrolytes prepared in this work are between 2.6×10^{-7} and 5.7×10^{-6} S/cm. In general, as we decrease the L content in the resin blends, the ionic conductivities increase due to the more elastomeric nature of resin P, which favors the ionic movements through the polymer chains. This behavior has been observed in other blends where there are two or more resins with a different nature [20].

The addition of ionic liquid also affects the ionic conductivities; for 40 wt.%, the samples exhibit the lowest ionic conductivity values ($\sim \times 10^{-7}$ $S \cdot cm^{-1}$, entries 1–3). These values can be improved by at least one order of magnitude when a plasticizer such as propylene carbonate (PC) is added, rising to 1.2×10^{-6} $S \cdot cm^{-1}$ (entry 11). On the contrary, the samples containing 50 wt.% of ionic liquid did not show the best ionic conductivity values, being between 7.2×10^{-7} and 6.1×10^{-6} $S \cdot cm^{-1}$ (entries 7–9). For these samples, an exudate of ionic liquid appears after a few days, which means that so much ionic liquid could reduce the stability of the electrolyte in operation.

According to the electrochemical stability range, values obtained from the LSV tests (Table 3) were found to increase with greater ionic liquid and lithium salt content. Thus, the samples with 40 wt.% of ionic liquid (7.6 wt.% Li) showed the lowest stability, followed by those with 45% and 50 wt. %, containing (8.5 and 9.4 wt.% of Li, respectively). Similar behavior has been reported for polymer electrolytes containing ionic liquid and lithium salts [31,32]. This improvement can be attributed to the synergic effect of lithium salt in ionic liquids.

The introduction of alumina nanoparticles improves the ion mobility through the bulk, in all cases, and the interfacial ionic conductivity obtained shows much higher values (Table 3, entries 12–14, and Figure 2). Moreover, the electrochemical window is greater, meaning the addition of alumina nanoparticles also enhances the electrochemical stability of the electrolytes.

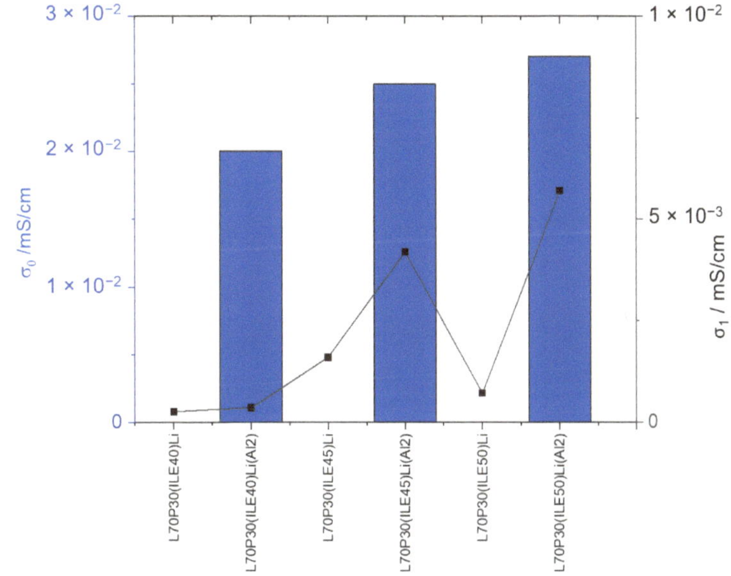

Figure 2. Ionic conductivities of the electrolytes with and without alumina nanoparticles.

In general terms, the specific capacitance values (shown in Table 3) increase within each percentage of ionic liquid (40, 45, and 50 wt.%), as the percentage in P resin is greater. This result is justified since the mobility of the ions through the polymer chains is greater as the number of groups of ethylene oxide units (EO), present in the P resin, increases. This favors ionic transport, improving the electrochemical properties of the electrolyte.

To discuss the best systems found with the addition of alumina nanoparticles, regarding not only the electrochemical behavior but the mechanical properties, a representation of the storage modulus and the T_g for samples with and without the reinforcement is shown in Figure 3. The only sample that reaches $E' > 1$ GPa is that containing 40 wt. % of ionic liquid and 7.6 wt.% of lithium content; then, at higher ionic liquid content, the storage modulus drops dramatically.

Concerning the T_g, the addition of alumina nanoparticles increases the values, which is more remarkable for the samples with 40% ionic liquid. From a structural point of view, the best equilibrium was obtained for the sample L70P30(ILE40)LiAl2, where the modulus and T_g are very high with a modest ionic conductivity.

Some samples were analyzed using FEGSEM to evaluate the differences associated with the ionic liquid content and the addition of nanoparticles. Figures 4a–c, 4d–f, and 4g–i show images at different magnifications of cryofractured samples of L70P30(ILE40)Li, L70P30(ILE50)Li and L70P30(ILE50)LiAl2, respectively. The microstructure of the crosslinked resins with different ionic liquid contents (40 and 50 wt.%) is entirely amorphous and does not exhibit either a biphasic or porous structure, even at the highest ionic liquid content. We have observed that the sample with a lower ionic liquid content (Figure 4a–c) has a smoother and less rough surface at the micrometric level compared to samples containing 50% ionic liquid (Figure 4d–f). The ionic liquid facilitates plastification mechanisms, activating mechanisms of crack deviation and fibrillation. The incorporation of nanoparticles reduces the size of the granular structures observed on the fracture surfaces (Figure 4c,f,i).

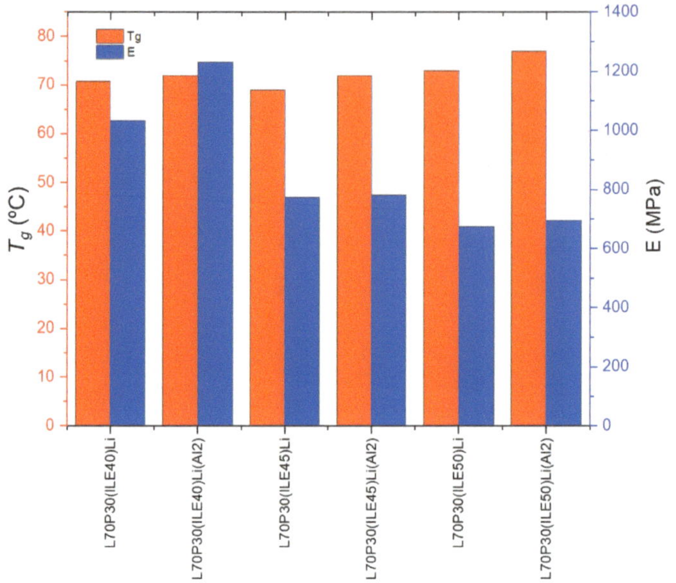

Figure 3. Thermomechanical parameters of the electrolytes with and without alumina nanoparticles.

When alumina nanoparticles were introduced in the formulation (Figure 4g–i), an even rougher surface was obtained. For this sample, at higher magnification (Figure 4i) the nanoparticles seem to be very well dispersed and there is no evidence of big agglomeration or nanoparticle clusters on the surface. More stress concentration points were created, and the cracks interacted with the nanoparticles interface, increasing the number of cavities on the fractured surface. These stress points or defects are used to improve the ionic conductivity of the solid as it has been previously observed (Figure 2). For similar epoxy resin systems, a better nanoparticles dispersion improves the interfacial interaction between the reinforcement and the matrix [14], which explains the enhancement of the T_g in all the cases studied [25].

In a previous study performed using the same epoxy resin system and ionic liquid, the sample L65P35(ILE30)Al2 exhibited the best balance between electrochemical and mechanical properties [25]. To compare the effect of the addition of lithium salt and alumina simultaneously in this formulation, another set of samples was prepared (Table 4). As it has been previously observed, the addition of lithium salt can have a positive effect on the stability of the electrolyte [20], and it was also observed in our system where the storage modulus had increased when the lithium salt was added (Table 1, entries 2 and 5 vs. Table 4, entries 2 and 3). However, in those cases the T_g was slightly lower when lithium was included. The addition of alumina increases the T_g of the electrolytes, which is more remarkable for those samples with a lower ionic liquid content (30 wt.%, Table 4 entries 1 and 4). The best ionic conductivity values were obtained for those samples containing alumina (Table 4, entries 4–6) compared to the plain samples without either lithium salt or nanoparticles (Table 4, entries 1–3). The simultaneous addition of lithium salt and alumina improves the storage modulus and the ionic conductivity in comparison to the samples modified only with lithium salt (Table 4, entries 7 and 8), values only surpassed by the sample containing alumina (Table 4, entries 4 and 8). A detrimental effect on the ionic conductivity with the addition of lithium salt has been previously observed [18,27], in some cases where the lithium concentration in an imidazole-based ionic liquid solution is low (<0.3 M) or high (>1.0 M) [33]. Some authors associate this effect with an increase in viscosity and a reduction in the mobility of ions in the electrolyte [19,33–35]. Choi and coworkers have found that the lithium salt (LiTFSI) concentration within the electrolyte mixture plays a crucial role in the formation of a bicontinuous nanoscale ion channel confined in the epoxy matrix, changing the ionic and epoxy domains [24].

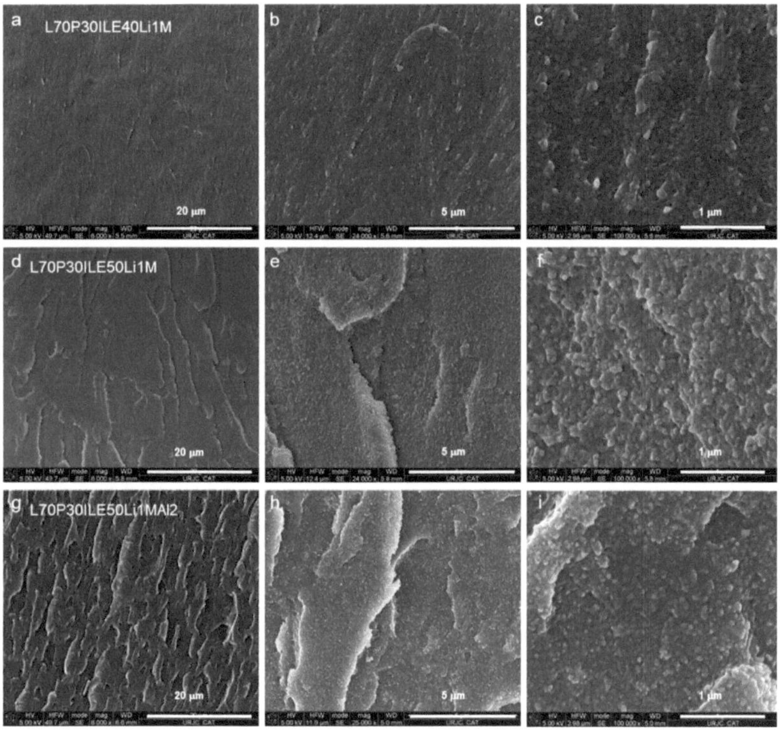

Figure 4. FEGSEM images at different magnifications of the epoxy resins (**a**–**c**) L70P30(ILE40)Li, (**d**–**f**) L70P30(ILE50)Li, and (**g**–**i**) L70P30(ILE50)LiAl2.

Table 4. Electrochemical and thermomechanical data for the polymer electrolytes.

Entry	Sample	T_g (°C)	E′ (MPa) (T = 30 °C)	σ_0 (S·cm^{-1})	σ_1 (S·cm^{-1})
1 [1,2]	L65P35(ILE30)	65 ± 2	1202 ± 10	-	1.3×10^{-7}
2 [1]	L65P35(ILE40)	68 ± 2	495 ± 60	-	1.4×10^{-5}
3 [1]	L65P35(ILE45)	63 ± 3	401 ± 80	-	4.5×10^{-5}
4 [1,2]	L65P35(ILE30)Al_2	83 ± 1	1213 ± 164	7.0×10^{-4}	1.6×10^{-6}
5 [1]	L65P35(ILE40)Al_2	70 ± 2	461 ± 62	6.8×10^{-5}	9.6×10^{-6}
6 [1]	L65P35(ILE45)Al_2	65 ± 3	469 ± 89	1.8×10^{-4}	1.8×10^{-7}
7	L65P35(ILE30)Li	68 ± 1	1235 ± 20	-	7.6×10^{-8}
8	L65P35(ILE30)Li(Al_2)	85 ± 1	1224 ± 103	1.4×10^{-5}	5.5×10^{-7}

[1] No lithium [2] Ref. [25].

A comparison of the elastic modulus and ionic conductivities of the SPEs/CPEs developed in this work with some of the best representative systems found in the literature [21,24,36–39] is depicted in Figure 5.

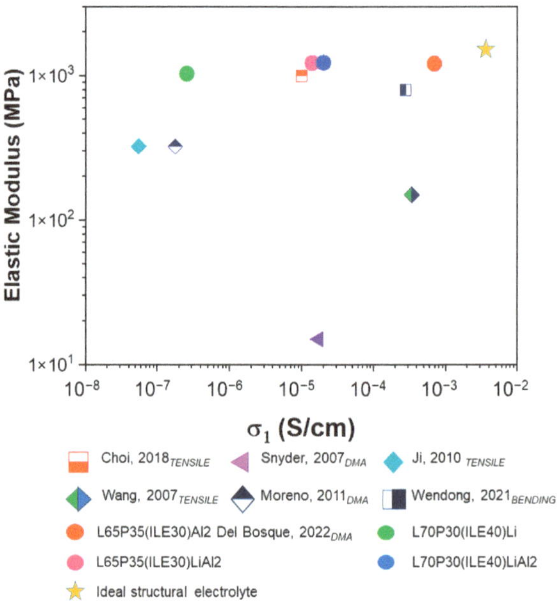

Figure 5. Mechanical and electrochemical comparison of the electrolytes. The subscript shows the mechanical test performed to establish the elastic modulus. Choi, 2018 [24]; Snyder, 2007 [24], Ji, 2010 [36], Wang, 2007 [37], Moreno, 2011 [38], Wendong, 2021 [21]; Del Bosque, 2022 [25].

The yellow star indicates ideal values of ionic conductivity for a neat ionic liquid and the Young modulus for a structural matrix based on epoxy resins. The colored circles are the electrolytes developed in this work. The green circle corresponds to our system L70P30(ILE40)Li without alumina, which is very robust. In the samples containing alumina (pink and blue circles), the ionic conductivity exhibits a very competitive value. The best CPE previously developed using the same resin blend but without lithium is L65P35(ILE30)Al2 (red circle) [25]. This electrolyte was very close to the ideal performance, but it does not contain lithium ions, so it cannot be used for structural batteries. Thus, the idea to reproduce an electrolyte for LIBs with this outstanding electrochemical and mechanical behavior was not possible only with the incorporation of lithium salt. In general, these electrolytes constitute a promising alternative considering the ionic conductivities obtained using less than 10% lithium as well as the competitive values in storage modulus.

The effect of temperature on the ionic conductivities of the best systems found in this study, L65P35(ILE30)Al2 and L70P30(ILE40)Al2, was checked. The samples were heated in an oven and connected to the potentiostat, and after reaching the appropriate temperature, the samples were led to equilibrate for a certain time (30 min). EIS analyses were measured to calculate the ionic conductivities through the impedances.

The Ln of ionic conductivity values (those corresponding to the bulk and interfacial regions) were plotted against the inverse of temperature ($1000/T(K^{-1})$). The plots are shown in Figure 6a,b. Both composite electrolytes were fitted to the Arrhenius model, and the activation energies were calculated and shown in Figure 6c.

For the sample L65P35(ILE30)LiAl2, the two activation energy values are higher than those obtained for the sample L70P30(ILE40)LiAl2; for the two ion-conduction paths, the values obtained for the sample L65P35(ILE30)LiAl2 were the same. Regarding the system L70P30(ILE40)LiAl2, two different values for the activation energy were obtained, and the activated energy associated with the interfacial layer was smaller, which means that the ionic movements associated with the Li/ions in the bulk are favored. These results can be explained if it is considered that the lithium salt is dissolved in an ionic liquid and this mixture is able to form a thicker layer on the nanoparticles surface (Figure 6d,e). In Figure 6e, the blue line path is the lithium ionic conduction through the bulk, as in typical SPEs. The ochre line path corresponds to the lithium ionic conduction way by a percolation through the interfacial layer, as it has been previously proposed for hybrid electrolytes [29,30]. For L70P30(ILE40)LiAl2, as the content of the ionic liquid is higher, the ions can move more easily compared to the ions hopping through the bulk.

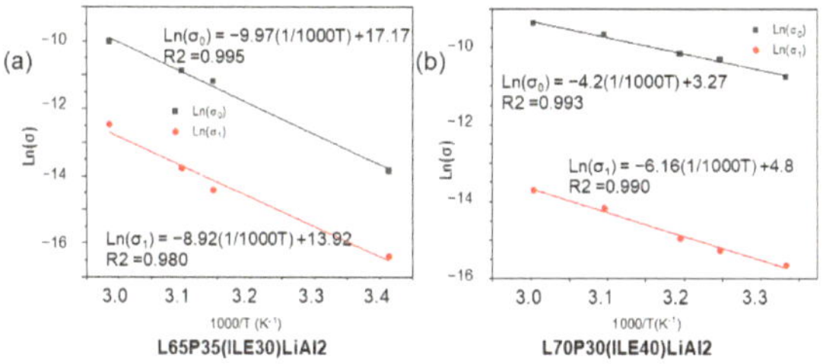

(c)

Ea (eV)	L65P35(ILE30)LiAl2	L70P30(ILE40)LiAl2
Interfacial	0.78	0.34
Bulk	0.77	0.53

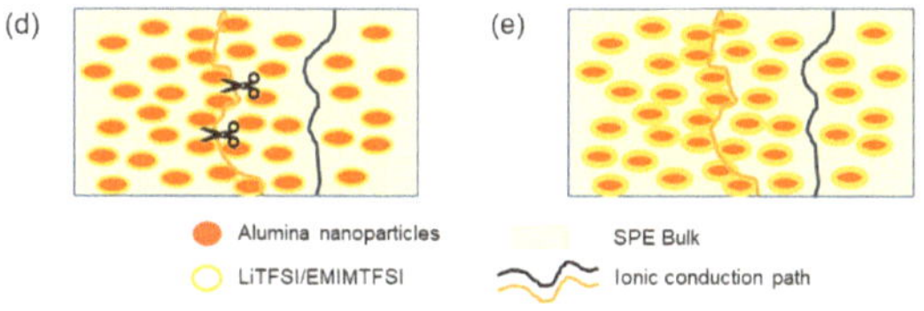

Figure 6. (**a**) Arrhenius type ionic conduction mechanism for L65P35(ILE)30LiAl2 and (**b**) for L70P30(ILE)40LiAl2 ionic conductivities in the electrolytes, including (**c**) activation energy in eV for the two ionic conductivities, (**d,e**) conduction paths in the CPEs through the bulk (black line), and percolation through the interfacial region (ochre line). Scissors represent the lost in the ionic conduction path due to the poor interfacial region in lower ionic liquid content electrolytes.

The lithium transference number of electrolyte L70P30(ILE40)LiAl2 was calculated, and it was found to be $t_{Li}{}^{+} = 0.13$. This number is usually related to the availability of free lithium in the mixture [40,41]. For systems based on PEO and LiTFSI, when the ratio of EO:Li is over three the $t_{Li}+$ decreased dramatically [42]. In our case, the EO units (from PEGDGE) gave an EO:Li ratio of 14:1, meaning that the whole network is more hindered and there are countless interactions between lithium and its environment (anions, alumina NPs, etc.) that can explain the low value obtained. Nevertheless, the ionic conductivity that can be obtained for this system at 60 °C is 0.085 mS/cm, closer to the optimum value expected for a solid electrolyte.

3.3. Supercapacitors Performance of Composite Polymer Electrolytes

To test the applicability of the developed CPEs, we fabricated a structural supercapacitor using carbon fiber electrodes modified with CuO nanoparticles, following a procedure previously reported in our group [28]. In this procedure, woven carbon fiber (10 × 10 cm, approximately) was placed in a reactor containing a copper precursor ($CuCl_2$) and some additives (NaCl, PVP K90, and L-AA) to favor the nanoparticles growing process. The hydrothermal process was left to react at 80 °C for 2 h.

After this time, the woven carbon fiber was removed, washed with distilled water, and dried in an oven at 60 °C overnight. These modified carbon fibers have been successfully used as electrodes for flexible supercapacitor using a PVA/H_3PO_4 matrix as gel electrolyte [28]. The symmetric supercapacitor was assembled by piling up layers of modified carbon fiber, glass fiber, glass fiber, and modified carbon fiber, in that order, and introduced in a vacuum bag for the VARIM process (Figure 7a). The electrolyte L70P30(ILE40)LiAl2, chosen as the best option for the supercapacitor fabrication, was prepared following procedure B, described in Section 2.3. The quantities of resins weighted were recalculated for a total amount of 60 g.

Figure 7b,c shows the vacuum bag after the curing process, and the demolded structural composite after copper current collectors were glued using silver ink. The device was characterized by EIS, which shows an incomplete, depressed semicircle at high frequencies followed by an irregular spike at low frequencies. The data were fitted to a simple modified Randles circuit (typical for most EDLC). The equivalent circuit fitted is composed by a constant phase element (CPE) and in parallel a resistance and a Warburg impedance connected in series (Figure 7d, Table S1). Even when the fitted is very good, the interpretation of these values can be quite confusing, and they sometimes do not represent a real physical meaning. Mei and coworkers have established a physical interpretation of Nyquist plots without relating them to an electric circuit [43]. According to their findings, they could identify regions of electrode–electrolyte resistances (orange region in Figure 7d), the equilibrium differential capacitance (pink region in Figure 7d), and a diffusion layer resistance (gray region in Figure 7d), where the slope can be used to indicate if the charging process can be controlled by EDL formation or by ion diffusion in the electrolyte. Our device has a slope of 0.64 (R^2 = 0.997), suggesting that the process is limited by ion diffusion, as it was expected, considering the nature of the CPE used. In the equilibrium differential capacitance region (pink in Figure 6d), a pure capacitive behavior exhibits a completely vertical line. Relating these results to the fitted equivalent circuit, the constant phase element (Q), as a non-ideal capacitor, denoted a mostly capacitive behavior (n~0.6) but was still far from an ideal capacitor. This interpretation agrees with the system studied, since the lithium transference number was found to be low enough to exhibit outstanding performance.

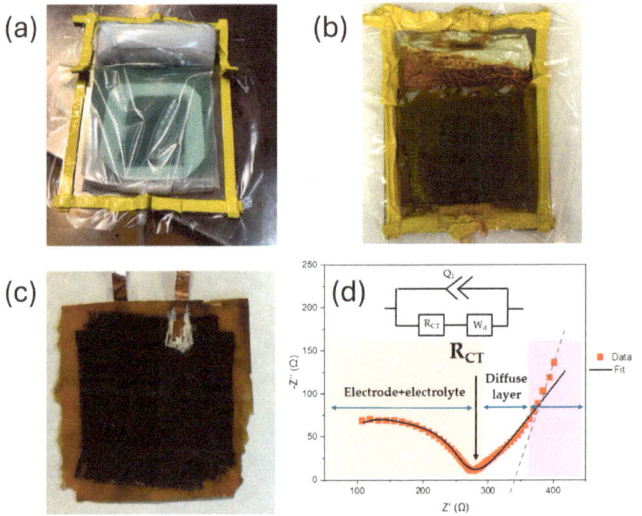

Charge Transfer R_{CT}(Ω)	Charge Transfer Q_2 (μΩ·s$^{(n)}$/n)	Warburg mΩ·s$^{(0.5)}$
282.5	2.23/0.579	6.59

Figure 7. Sequence for the fabrication of the structural supercapacitor. (**a**) VARIM set-up before resin infusion. (**b**) VARIM after cured in an oven 8 h at 140 °C. (**c**) Structural supercapacitor demolded. (**d**) EIS Nyquist plot and equivalent circuit showed inset.

Finally, to carry out the proof-of-concept, the device, CuO-WCF/WGF/CuO-WCF embedded in the L70P30(ILE40)LiAl2 matrix, was connected to a power source for charging

(Figure 8a) at 12 V for 2 h. Then, the device was disconnected from the power source (Figure 8b) and successfully powered a red LED light, but only for a couple of minutes (Figure 8c).

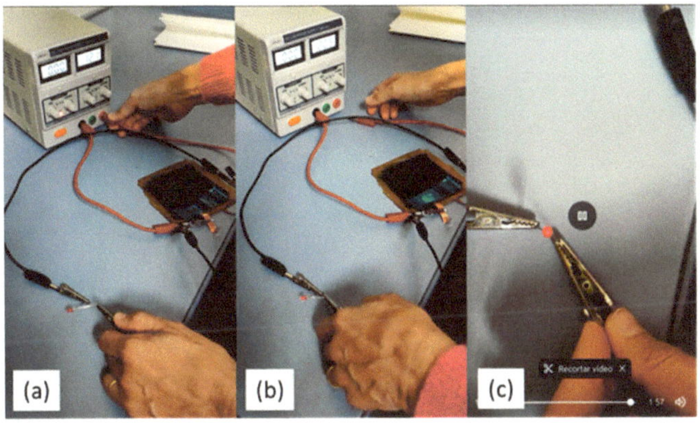

Figure 8. Sequence of the proof of concept by switching on a red LED (**a**) charging the device. (**b**) Switching on the red LED. (**c**) Red LED on after almost 2 min.

Attempts to manufacture a structural supercapacitor and battery completely inside the glove box in a pouch cell are currently in progress.

4. Conclusions

New SPEs and CPEs based on epoxy resin blends using 1-ethyl-3-methyl-imidazolium bis(trifluoromethylsulfonyl)imide ionic liquid, lithium bis(trifluoromethylsulfonyl)imide salt, and alumina nanoparticles (13 nm) were prepared, and the effect of resin composition (60–70 wt.% of structural based on DGEBA) and the ionic liquid content (from 30 to 50 wt.%) were studied. The resin blends containing 40 wt.% of ionic liquid were found to be optimal without observing any trace of exudate. The mechanical properties were improved by increasing the amount of DGEBA-based epoxy resin (**L**) in their final composition. Regarding the electrochemical properties, the addition of alumina nanoparticles was found to increase the ionic conductivities, and this effect was attributed to the interfacial alumina–electrolyte interaction in the polymer matrix. We have reached good thermal and mechanical properties for systems L70P30, higher than other similar systems previously reported ($T_g > 70$ °C, E′ > 1.2 GPa). The new formulation found fulfills not only the requirements of a SPE, but it is also a cheaper alternative, considering it is fabricated using commercially available epoxy resins, and the ionic conductivity reported here is higher than other electrolytes with less than 10 wt.% lithium salt.

Supplementary Materials: The following supporting information can be downloaded at: https://www.mdpi.com/article/10.3390/polym16142048/s1. Figure S1. Nyquist Plots for electrolytes containing 40 wt % ILE; Figure S2. Nyquist Plots for electrolytes containing 45 wt % ILE; Figure S3. Nyquist Plots for electrolytes containing 50 wt % ILE; Figure S4. Nyquist Plots for electrolytes containing 5 wt %PC; Figure S5. Nyquist Plots for electrolytes containing alumina. (a) Electrolytes 45 wt %ILE and comparison with its analogous without alumina. (b) Electrolytes 50 wt %ILE and comparison with its analogous without alumina. Table S1. Fitting for the equivalent circuit of Nyquist Plot for structural supercapacitor fabricated using L70P30(ILE40)LiAl2 electrolyte.

Author Contributions: B.K.M.: investigation, methodology, conceptualization, formal analysis, visualization, and writing—original draft; J.L.: investigation, methodology, and formal analysis; M.S.: writing—review and editing, and funding acquisition; A.U.: writing—review and editing, and funding acquisition. All authors have read and agreed to the published version of the manuscript.

Funding: This work was supported by the Agencia Estatal de Investigación of the Spanish Government [Project MULTISENS PID2022-136636OB-I00].

Institutional Review Board Statement: Not applicable.

Data Availability Statement: The original contributions presented in the study are included in the article/Supplementary Materials, further inquiries can be directed to the corresponding author/s.

Conflicts of Interest: The authors declare no conflicts of interest.

References

1. Wu, K.J.; Young, W.B.; Young, C. Structural supercapacitors: A mini-review of their fabrication, mechanical & electrochemical properties. *J. Ener. Storage Part B* **2023**, *72*, 108358.
2. Shi, C.; Yu, M. Flexible solid-state lithium-sulfur batteries based on structural designs. *Ener. Storage Mater.* **2023**, *57*, 429–459. [CrossRef]
3. Wazeer, A.; Das, A.; Abeykoon, C.; Sinha, A.; Karmakar, A. Composites for electric vehicles and automotive sector: A review. *Green Energy Intell. Transp.* **2023**, *2*, 100043. [CrossRef]
4. Cheah, L. Cars on a Diet: The Material and Energy Impacts of Passenger Vehicle weight Reduction in the US. Ph.D. Thesis, Massashussets Institute of Technology, Cambridge, MA, USA, 2010. Available online: https://dspace.mit.edu/handle/1721.1/62760 (accessed on 5 June 2024).
5. Jin, T.; Singer, G.; Liang, K.; Yang, Y. Structural batteries: Advances, challenges and perspectives. *Mater. Today* **2023**, *62*, 151–167. [CrossRef]
6. Greenhalgh, E.S.; Nguyen, S.; Valkova, M.; Shirshova, N.; Shaffer, M.S.P.; Kucernak, A.R.J. A critical review of structural supercapacitors and outlook on future research challenges. *Compos. Sci. Technol.* **2023**, *235*, 109968. [CrossRef]
7. Klongkan, S.; Pumchusak, J. Effects of nano alumina and plasticizers on morphology, ionic conductivity, thermal and mechanical properties of PEO-$LiCF_3SO_3$ solid polymer electrolyte. *Electrochim. Acta* **2015**, *161*, 171–176. [CrossRef]
8. Lin, C.; Hung, C.; Venkateswarlu, M.; Hwang, B. Influence of TiO_2 nano-particles on the transport properties of composite polymer electrolyte for lithium-ion batteries. *J. Power Sources* **2005**, *146*, 397–401. [CrossRef]
9. Jiang, G.; Maeda, S.; Yang, H.; Saito, Y.; Tanase, S.; Sakai, T. All solid-state lithiumpolymer battery using poly (urethane acrylate)/nano-SiO_2 composite electrolytes. *J. Power Sources* **2005**, *141*, 143–148. [CrossRef]
10. Feng, Q.; Yang, J.; Yu, Y.; Tian, F.; Zhang, B.; Feng, M.; Wang, S. The ionic conductivity, mechanical performance and morphology of two-phase structural electrolytes based on polyethylene glycol, epoxy resin and nano-silica. *Mater. Sci. Eng. B Solid-State Mater. Adv. Technol.* **2017**, *219*, 37–44. [CrossRef]
11. Zhang, Y.; Zhao, Y.; Gosselink, D.; Chen, P. Synthesis of poly (ethylene-oxide)/nanoclay solid polymer electrolyte for all solid-state lithium/sulfur battery. *Ionics* **2015**, *21*, 381–385. [CrossRef]
12. Chand, N.; Rai, N.; Agrawal, S.L.; Patel, S.K. Morphology, thermal, electrical and electrochemical stability of nano aluminium-oxide-filled polyvinyl alcohol composite gel electrolyte. *Bull. Mater. Sci.* **2011**, *34*, 1297–1304. [CrossRef]
13. Song, Z.; Liu, X.; Ding, J.; Liu, J.; Han, X.; Deng, Y.; Zhong, C.; Hu, W. Poly(Acrylic Acid)-Based Nanocomposite Gel Polymer Electrolyte with High Mechanical Strength and Ionic Conductivity Towards Long-Cycle-Life Flexible Zinc–Air Battery. *SSRN Electron. J.* **2022**. [CrossRef]
14. Kwon, S.J.; Jung, B.M.; Kim, T.; Byun, J.; Lee, J.; Lee, S.B.; Choi, U.H. Influence of Al_2O_3 nanowires on ion transport in nanocomposite solid polymer electrolytes. *Macromolecules* **2018**, *51*, 10194–10201. [CrossRef]
15. Wang, J.; Fan, L.; Du, Q.; Jiao, K. Lithium ion transport in solid polymer electrolyte filled with alumina nanoparticles. *Energy Adv.* **2022**, *1*, 269. [CrossRef]
16. Croce, F.; Persi, L.; Scrosati, B.; Serraino-Fiory, F.; Plichta, E.; Hendrickson, M.A. Role of the Ceramic Fillers in Enhancing the Transport Properties of Composite Polymer Electrolytes. *Electrochim. Acta* **2001**, *46*, 2457–2461. [CrossRef]
17. Dissanayake, M.A.K.L.; Jayathilaka, P.A.R.D.; Bokalawala, R.S.P.; Albinsson, I.; Mellander, B.-E. Effect of Concentration and Grain Size of Alumina Filler on the Ionic Conductivity Enhancement of the $(PEO)_9LiCF_3SO_3$:Al_2O_3 Composite Polymer Electrolyte. *J. Power Sources* **2003**, *119–121*, 409–414.
18. Chandra, A.N.; Sungmook, L.; Guk-Hwan, L.; Wonoh, L. Epoxy-based multifunctional solid polymer electrolytes for structural batteries and supercapacitors. a short review. *Front. Chem.* **2024**, *12*, 1330655.
19. Shirshova, N.; Bismarck, A.; Greenhalgh, E.S.; Johansson, P.; Kalinka, G.; Marczewski, M.J.; Shaffer, M.S.P.; Wienrich, M. Composition as a Means to Control Morphology and Properties of Epoxy Based Dual-Phase Structural Electrolytes. *J. Phys. Chem. C* **2014**, *118*, 28377–28387. [CrossRef]
20. Muñoz, B.K.; del Bosque, A.; Sánchez, M.; Utrilla, V.; Prolongo, S.G.; Prolongo, M.G.; Ureña, A. Epoxy Resin Systems Modified with Ionic Liquids and Ceramic Nanoparticles as Structural Composites for Multifunctional Applications. *Polymer* **2021**, *214*, 123233. [CrossRef]
21. Wendong, Q.; Dent, J.; Arrighi, V.; Cavalcanti, L.; Shaffer, M.S.P.; Shirshova, N. Biphasic Epoxy-Ionic Liquid Structural Electrolytes: Minimising Feature Size through Cure Cycle and Multifunctional Block-Copolymer Addition. *Multifunct. Mater.* **2021**, *4*, 035003. [CrossRef]

22. Shirshova, N.; Bismarck, A.; Carreyette, S.; Fontana, Q.P.V.; Greenhalgh, E.S.; Jacobsson, P.; Johansson, P.; Marczewski, M.J.; Kalinka, G.; Kucernak, A.R.J.; et al. Structural supercapacitor electrolytes based on bicontinuous ionic liquid–epoxy resin systems. *J. Mater. Chem. A* **2013**, *1*, 15300–15309. [CrossRef]
23. Tu, V.; Asp, L.E.; Shirshova, N.; Larsson, F.; Runesson, K.; Jänicke, R. Performance of bicontinuous structural electrolytes. *Multifunct. Mater.* **2020**, *3*, 025001. [CrossRef]
24. Choi, U.H.; Jung, B.M. Ion Conduction, Dielectric and Mechanical Properties of Epoxy-Based Solid Polymer Electrolytes Containing Succinonitrile. *Macromol. Res.* **2018**, *26*, 459–465. [CrossRef]
25. Del Bosque, A.; Muñoz, B.K.; Sánchez, M.; Ureña, A. Thermomechanically Robust Ceramic/Polymer Nanocomposites Modified with Ionic Liquid for Hybrid Polymer Electrolyte Applications. *ACS Appl. Energy Mater.* **2022**, *5*, 4247–4258. [CrossRef]
26. Sánchez-Romate, X.F.; Del Bosque, A.; Artigas-Arnaudas, J.; Muñoz, B.K.; Sánchez, M.; Ureña, A. A proof of concept of a structural supercapacitor made of graphene coated woven carbon fibers: EIS study and mechanical performance. *Electrochim. Acta* **2021**, *370*, 137746. [CrossRef]
27. *ASTM D5418*; Standrad Test Method for Plastics. Dynamic Mechanical Properties: In Flexure (Dual Cantilever Beam). ASTM: West Conshohocken, PA, USA, 2023.
28. Muñoz, B.K.; González-Banciella, A.; Ureña, D.; Sánchez, M.; Ureña, A. Electrochemical Comparison of 2D-Flexible Solid-State Supercapacitors Based on a Matrix of PVA/H3PO4. *Polymers* **2023**, *15*, 4036. [CrossRef] [PubMed]
29. Zaman, W.; Hortance, N.; Dixit, M.B.; De Andrade, V.; Hatzell, K.B. Visualizing percolation and ion transport in hybrid solid electrolytes for Li–metal batteries. *J. Mater. Chem. A* **2019**, *7*, 23914–23921. [CrossRef]
30. Dudney, N.J. Effect of Interfacial Space-Charge Polarization on the Ionic Conductivity of Composite Electrolytes. *J. Amer. Cer. Soc.* **1985**, *68*, 538–545. [CrossRef]
31. Chen, L.; Fu, J.; Lu, Q.; Shi, L.; Li, M.; Dong, L.; Xu, Y.; Jia, R. Cross-linked polymeric ionic liquids ion gel electrolytes by in situ radical polymerization. *Chem. Eng. J.* **2019**, *378*, 122245. [CrossRef]
32. Dong, L.; Zeng, X.; Fu, J.; Chen, L.; Zhou, J.; Dai, S.; Shi, L. Cross-linked ionic copolymer solid electrolytes with loose Coordination-assisted lithium transport for lithium batteries. *Chem. Eng. J.* **2021**, *423*, 130209. [CrossRef]
33. Seki, S.; Ohno, Y.; Kobayashi, Y.; Miyashiro, H.; Usami, A.; Mita, Y.; Tokuda, H.; Watanabe, M.; Hayamizu, K.; Tsuzuki, S.; et al. Imidazolium-based room-temperature ionic liquid for lithium secondary batteries. *J. Electrochem. Soc.* **2007**, *154*, A173–A177. [CrossRef]
34. Rosol, Z.P.; German, N.J.; Gross, S.M. Solubility, ionic conductivity and viscosity of lithium salts in room temperature ionic liquids. *Green Chem.* **2009**, *11*, 1453–1457. [CrossRef]
35. Song, Y.H.; Kim, T.; Choi, U.H. Tuning morphology and properties of epoxy-based solid-state polymer electrolytes by molecular interaction for flexible all-solid-state supercapacitors. *Chem. Mater.* **2020**, *32*, 3879–3892. [CrossRef]
36. Snyder, J.F.; Carter, R.H.; Wetzel, E.D. Electrochemical and mechanical behavior in mechanically robust solid polymer electrolytes for use in multifunctional structural batteries. *Chem. Mater.* **2007**, *19*, 3793–3801. [CrossRef]
37. Ji, J.; Li, B.; Zhong, W.-H. Simultaneously enhancing ionic conductivity and mechanical properties of solid polymer electrolytes via a copolymer multi-functional filler. *Electrochim. Acta* **2010**, *55*, 9075–9082. [CrossRef]
38. Wang, Y.J.; Kim, D. Crystallinity, morphology, mechanical properties and conductivity study of in situ formed PVdF/LiClO4/TiO2 nanocomposite polymer electrolytes. *Electrochim. Acta* **2007**, *52*, 3181–3189. [CrossRef]
39. Moreno, M.; Quijada, R.; Ana, M.A.S.; Benavente, E.; Gomez-Romero, P.; González, G. Electrical and mechanical properties of poly(ethylene oxide)/intercalated clay polymer electrolyte. *Electrochim. Acta* **2011**, *58*, 112–118. [CrossRef]
40. Tan, S.J.; Zeng, X.X.; Ma, Q.; Wu, X.W.; Guo, Y.G. Recent Advancements in Polymer-Based Composite Electrolytes for Rechargeable Lithium Batteries. *Electrochem. Energ. Rev.* **2018**, *1*, 113–138. [CrossRef]
41. Jia, M.; Tufail, M.K.; Guo, X. Insight into the Key Factors in High Li+ Transference Number Composite Electrolytes for Solid Lithium Batteries. *ChemSusChem* **2023**, *16*, e2022018. [CrossRef] [PubMed]
42. Pożyczka, K.; Marzantowicz, M.; Dygas, J.R.; Krok, F. Ionic conductivity and lithium transference number of poly(ethylene oxide):litfsi system. *Electrochimica Acta* **2017**, *227*, 127–135. [CrossRef]
43. Mei, B.-A.; Munteshari, O.; Lau, J.; Dunn, B.S.; Pilon, L. Physical Interpretations of Nyquist Plots for EDLC Electrodes and Devices. *J. Phys. Chem. C* **2018**, *122*, 194–206. [CrossRef]

polymers

MDPI

Article

Polyethylene Film Surface Modification via Benzoic Acid Grafting

Ana Luisa Grafia [1,2] and Silvia Elena Barbosa [1,2,*]

1 Planta Piloto de Ingeniería Química, PLAPIQUI (UNS-CONICET), Bahía Blanca 8000, Argentina; agrafia@plapiqui.edu.ar
2 Departamento de Ingeniería Química, Universidad Nacional del Sur (UNS), Bahía Blanca 8000, Argentina
* Correspondence: sbarbosa@plapiqui.edu.ar

Abstract: A polyethylene (PE) film surface modification method is proposed via benzoic acid (BA) alkylation grafting to improve the surface affinity to polar substances. The procedure involves sequentially spraying $AlCl_3$ and BA onto the heat-softened PE surface. The occurrence of the alkylation reaction was evaluated through comparative chemical, morphological, and thermal analyses. It was demonstrated that the grafting reaction of BA onto the PE film surface took place, limited to the surface layer, while preserving the bulk properties of PE. The reaction resulted in the formation of aluminum benzoate complexes, which improved the surface affinity to polar compounds. The impact of grafting on the surface properties of PE was further assessed by comparing the behavior of PE films treated with BA and untreated PE films when painted with watercolors. The PE film grafted with BA exhibited increased affinity towards watercolors, providing strong evidence of a change in surface polarity from hydrophobic to hydrophilic. These findings indicate that the proposed methodology effectively renders the PE surface paintable, even with non-toxic water-based inks, making it suitable for applications such as packaging.

Keywords: polyethylene surface modification; grafting; aluminum benzoate complexes; hydrophobic to hydrophilic

Citation: Grafia, A.L.; Barbosa, S.E. Polyethylene Film Surface Modification via Benzoic Acid Grafting. *Polymers* **2024**, *16*, 1291. https://doi.org/10.3390/polym16091291

Academic Editor: Markus Gahleitner

Received: 22 February 2024
Revised: 15 April 2024
Accepted: 23 April 2024
Published: 5 May 2024

1. Introduction

Polyethylene (PE) stands as one of the most widely utilized thermoplastics globally due to its outstanding cost/performance/sustainability ratio. Nevertheless, despite its excellent bulk properties, the inert, non-polar, and hydrophobic PE surface imposes limitations on its applications. Properties such as printability, paintability, and adhesion, which are determined by the surface affinity with polar substances, require enhancement [1,2].

To address this demand, strategies involve changing the polymer surface through chemical or physical surface modification methods. These methods offer the advantage of preserving the material bulk properties by confining the modification to a thin surface layer. Among the most industrially used methods for PE are dielectric barrier discharges, owing to their suitability for continuous processes. Consequently, oxidized products, predominantly ozonides, hydroperoxides, and peroxides, form on the PE surface, increasing its polarity and surface free-energy. However, many of these methods suffer from instability, lack of control, nonspecificity, surface degradation, susceptibility to polar liquids, heat, or friction, and diminish over time due to aging effects, which constitute their main limitation [1–3].

An alternative, versatile, and tailored modification method involves surface grafting of specific molecules via wet or melting processing [1,2]. These methods entail introducing a new functional group into a polymer to modify its properties or serve as an anchor point for other molecules. By employing these approaches, stable reaction products with robust anchoring bonds to the polymer can be generated. Grafting voluminous molecules, such as chains or aromatics ones, hinders their diffusion and then probably mitigates the aging effects.

Given PE's inherent inertness, different specific molecules, mainly organic acids and vinyl compounds, have been selected as grafting compounds [4]. For instance, Zhu et al. [5] grafted caffeic acid functionalized with ε-polylysine onto PE via reactive co-extrusion to yield antimicrobial and antioxidant active films. El-Wakil et al. [6] grafted acrylic acid onto low-density PE through melt blending to enhance the compatibility with a composite containing neat PE and rice bran impregnated with tea tree oil. Similarly, Versteeg et al. [7] improved the mechanical properties of recycled PE by grafting glycidyl methacrylate via reactive extrusion with supercritical carbon dioxide. Additionally, Wei et al. [8] enhanced the insulation properties of low-density PE and high-density PE blends by grafting 3-aminobenzoic acid via free radical grafting using solution grafting. All the above mention methods not only modify the surface of the PE but also change its bulk.

A promising compound for modifying only the PE surface is benzoic acid (BA). BA possesses advantageous characteristics of polarity from its carboxylic group and aromaticity from its benzene ring. Thus, enhancing the PE film's surface affinity for polar and aromatic substances could render it suitable as a carrier for polar active or functional agents. Given that organic pigments, the preferred color components in printing inks used in flexible polyethylene-based packaging, are aromatic and polar in nature, they should exhibit good adhesion and compatibility with surface rich in BA-derived groups [9].

Friedel–Crafts alkylation emerges as an effective grafting reaction to obtain a copolymer from an aromatic monomer and a hydrocarbon chain [10]. This reaction entails the alkylation of an aromatic ring and an alkyl halide using a strong Lewis acid catalyst such as aluminum chloride ($AlCl_3$). Previous studies have demonstrated the successful grafting of a hydrocarbon chain, like PE, chemically bonded to the styrene benzene ring via aromatic electrophilic substitution [11–14]. Notably, in our prior work [13], we styrene-grafted a PE film by subsequently spraying catalysts and reagents directly onto its softened surface. This PE surface modification method could be seamlessly integrated into continuous film production as a direct and expeditious post-processing treatment. Taking into account these results, this alkylation reaction seems promising for use for BA grafting onto PE.

The current study aims to enhance PE surface affinity to polar substances through BA grafting using a direct and rapid method that preserves PE bulk properties. The modification procedure involves a spraying methodology where the reagents and catalyst are sprayed onto a heat-softened PE film surface. The occurrence of the reaction was determined through comparative chemical, morphological, and thermal analyses, employing five different and independent analytical techniques. Surface compatibility with polar compounds was verified through contact angle measurements, surface free energy calculations, testing with a dyna pen, painting with water-based paints, and testing paint adhesion.

2. Experimental

2.1. Materials

PE blown films, 80 μm thick, made from LDPE 203, were provided by Dow Polisur, Bahía Blanca, Argentina (Mw: 229,300 g/mol, Mn: 22,500 g/mol). BA (ACS) and anhydrous $AlCl_3$ ($\geq$98% purity), both obtained from Merck, Darmstadt, Germany, were used as reagents. N-heptane and absolute ethanol, both obtained from Cicarelli, Buenos Aires, Argentina, with a purity higher than 99.5%, were used as solvents. Distilled water and diiodomethane, 99% purity, from Cicarelli were used for contact angle measurements. A red watercolor (0.37 g of dry paint per water mL) with permanent red tonality, pigment: PR170 $C_{26}H_{22}N_4O_4$ CAS: 2786-76-7 and blue tempera (5 g paint per water mL) with ultramarine blue tonality, pigment: PB29 $Na_6Al_4Si_6S_4O_{20}$ CAS: 57455-37-3, both from Alba (professional artistic line), were used for paintability tests.

2.2. Grafting Reaction Procedure

PE film squares (15 × 15 cm) were surface-softened by heating at circa 95 °C using an infrared system. It was previously demonstrated that PE is softened but not melted at this temperature [13]. The softened PE surface was consecutively sprayed with a suspension of 155 mg of $AlCl_3$ in 25 mL of n-heptane and with a solution of 500 mg of BA in 25 mL of absolute ethanol. The pH of the ethanol solution was adjusted to 3 by addition of HCl solution. The spray equipment was placed 15 cm away from the film surface. It had a conventional 120° fan nozzle connected to a dry air stream at 3 bars. Experiments were performed at room temperature with a flow rate of 6 mL/min. It is important to note that the solvent is evaporated during spraying. In order to remove residual unreacted reagents and not grafted compounds, the modified films were sonicated for 2 h in absolute ethanol and then dried for 48 h under an extractor hood. Modified sonicated films were named PE-BA.

2.3. *Characterization*

The first evidence of a reaction occurrence was assessed by determining the aromatic presence in PE-BA with a colorimetric assay. This test is based on the aromatic identification of chloroform and $AlCl_3$ proposed by Shriner et al. [15]. PE and PE-BA film samples were placed in a beaker with 20 mL of chloroform and 0.2 g of $AlCl_3$ under agitation for 5 min. Then, the samples were removed from the beaker and the change in its surface color was observed with the naked eye and photographed. The chemical structure of PE, pure reagents, and PE-BA were analyzed by Fourier transform infrared spectroscopy (FTIR) in a Nicolet 520-FTIR (Waltham, MA, USA) spectrometer. Spectra were performed in two modes: attenuated total reflection (ATR) using zinc selenide crystal, and a transmission one. The effect of the reaction products on the thermal decomposition of modified films was studied with thermogravimetric analysis (TGA) in a temperature range of 30–800 °C at a heating rate of 10 °C/min under a nitrogen atmosphere. Thermogravimetric (TG) curves were recorded using Discovery TGA equipment from TA Instruments, New Castle, DE, USA.

Surface morphology and elemental composition of films were assessed by scanning electron microscopy with energy disperse X-ray microanalysis (SEM-EDX) in a JEOL-35CFelectron scanning microscope equipped with a EDAX DX4 microanalyzer, all from Tokyo, Japan. Samples were previously coated with Au in a sputter coater PELCO 91000, New York, NY, USA. EDAX DX4 allows elements to be detected from B to U with a surface penetration of 1 μm.

2.4. *Wettability and Paintability Tests*

The effect of the grafting reaction on the film surface compatibility with polar compounds was tested using qualitative and quantitative wettability and paintability tests.

Contact angle measurements were performed in static mode according to the sessile drop method using a goniometer OCA-15 (Data Physics, Filderstadt, Germany). More than 50 drops of 1 μm of water and diiodomethane were placed onto different locations of PE and PE-BA surfaces, in a 25 °C and 50% RH room.

The measured contact angle values of PE and PE-BA were used to calculate the respective total surface energy and its dispersive and polar component. Table 1 shows the surface energy parameters of distilled water and diiodomethane. The calculation of the surface energy was carried out using the Owens and Wendt equation [16].

$$\gamma_L(1 + cos\varnothing) = 2\left[\left(\gamma_L^d\gamma_S^d\right)^{0.5} + \left(\gamma_L^p\gamma_S^p\right)^{0.5}\right] \quad (1)$$

where θ is the contact angle; γ_L is the surface tension; and γ_L^p and γ_L^d are the polar and dispersive components of the testing liquid. Meanwhile, γ_S^p and γ_S^d are the polar and dispersive components of the tested films. The sum of the polar and dispersive components is the total surface energy or surface energy of the film γ_S.

Table 1. Surface tensions (γ_L) of testing liquids and their dispersive (γ_L^d) and polar (γ_L^p) components [17].

Liquid	γ_L^d [mJ. m^{-2}]	γ_L^p [mJ. m^{-2}]	γ_L [mJ. m^{-2}]
water	21.8	51.0	72.8
diiodomethane	50.42	0.38	35.00

Surface energies were also tested with a quick method, widely used industrially, using dyne pens. A line was written onto the surface of each film (PE and PE-BA) using dyne pens of 38 and 44 dynes/cm (Lemdisa S. A., Buenos Aires, Argentina). If the ink remains as a line without forming individual droplets, then the surface has a dyne level at least as high as the ink used. If the line is continuous, without the presence of individual drops, the surface has a dyne level at least as high as the dyne pen used.

The effect of the grafting on the paintability of the PE was comparatively and qualitatively tested by painting the surface of PE and PE-BA films with red watercolor and blue tempera. Paint coverage was observed with the naked eye and photographed for later analysis. Furthermore, only on the films painted with tempera, the paint adhesion was evaluated by a tape test according to ASTM 3359-02 [18]. The classification of the results is based on the percentage of removed area (r.a.) and divided into six categories, from 5B (none r.a.) to 0B (>65% r. a.). The r.a. was calculated using the image analysis software Image J Java 8.

3. Results and Discussion

Reaction occurrence and effectiveness were evaluated with a comparative chemical, thermal, and morphological analysis of the PE and PE-BA films.

First, we analyzed the chemical reaction occurrence with a colorimetric assay. Figure 1 shows the surface photographs of the PE and PE-BA films after testing. The PE-BA film shows an orange appearance giving a positive test result, which evidences the aromatic presence from grafted BA; in contrast, no color changes are detected on neat PE.

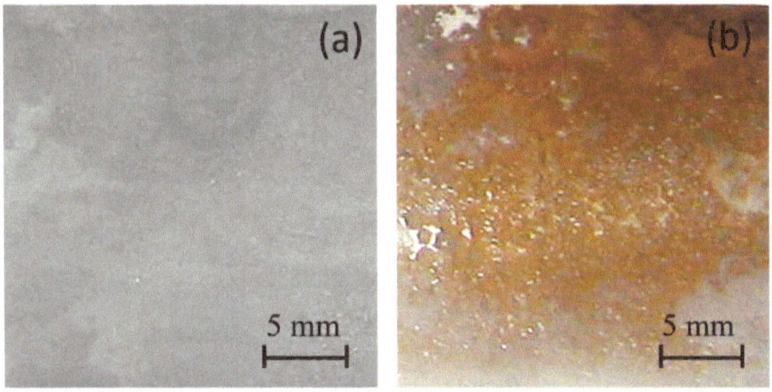

Figure 1. Photographs after chloroform and $AlCl_3$ test of films from (**a**) PE and (**b**) PE-BA.

Second, in order to elucidate the chemistry of the reaction product, FTIR-ATR spectroscopy was performed. The FTIR spectra of PE-BA compared with those of PE and BA are shown in Figure 2. The main spectrum peak assignments are summarized in Table 2. In addition to the typical PE bands [19], the spectra of PE-BA exhibit new bands attributed to the reaction product, particularly in the range of 1700–700 cm^{-1}. The aromatic nature of the graft product is revealed by the band at 1496 cm^{-1} corresponding to the CC stretching vibration in the plane of the aromatic ring [20–22], while the rest of the BA characteristic bands undergo notable changes and shifts. One of these is the absence of the bands assigned to the carboxyl group (COOH), at 1685 cm^{-1}, 1326 cm^{-1}, and 1292 cm^{-1} [20,21] of BA, which allows the inference that the graft occurred through the acid group.

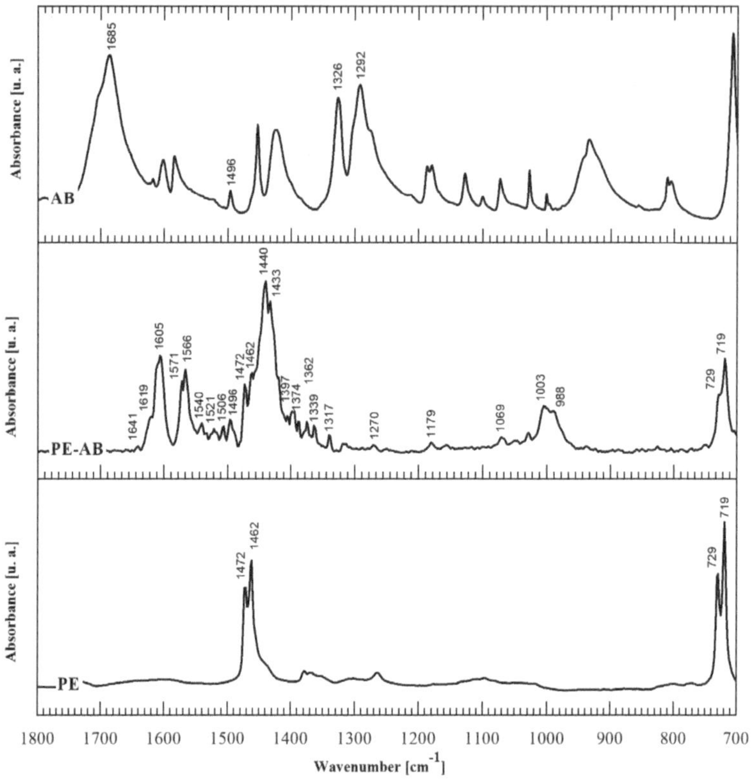

Figure 2. FTIR ATR spectra of BA, PE-BA, and PE.

Table 2. FTIR identification and assignment of the main absorption bands for PE-BA, PE, and BA *.

Frequency [cm^{-1}]			Assignment	References
PE-BA	PE	BA		
		1685	C=O free carboxylic acid stretching mode	[20,21]
1641			C=O stretching vibration, in monodentate aluminum benzoate complexes (1645 cm^{-1}).	[23]
1619 **1571** **1440** **1317**			Vibrations related to monodentate aluminum benzoates complexes (1622 cm^{-1}, 1571 cm^{-1}, 1440 cm^{-1}, 1317 cm^{-1})	[24–26]
1605 **1566** **1433**			C-O in bridge or bidentate complexes	[23,27]
1506 **1540**			-COO^{-} benzoate ion (1550 cm^{-1}) and oriented benzoate on aluminum oxide (1500 cm^{-1})	[28]
1521 **1405**			-COO^{-} Al^{+} aluminum carboxylate (1523–1410 cm^{-1})	[29]

Table 2. *Cont.*

Frequency [cm^{-1}]			Assignment	References
PE-BA	PE	BA		
1496		1496	C-C of aromatic ring	[20–22]
1472 1462	1472 1462		CH bending deformation	[19]
1397			$-COO^{-}$ carboxylate ion on aluminum surface	[30,31]
1378	1378		CH_3 symmetric deformation	[19]
1306	1306		CH_3 twisting deformation	[19]
		1326 1292	C-O of COOH-coupled stretching vibrations	[21]
1270			CH_2-Al	[32]
1179 **1069**			Al-O-C	[33]
1003 **990**			Al-O	[30,34]
721 719	721 719		CH_2 rocking deformation	[19]

* The bands assigned to the grafting product are highlighted in bold.

A detailed analysis of the new spectrum bands of PE-BA revealed that the aluminum incorporated through $AlCl_3$ was part of the reaction product. The bands assigned to the carboxylate ion (COO-) associated with Al were identified. Olafsson et al. [30], Olafsson and Hildingsson [31], and Olafsson et al. [35] studied the organic acid migration in laminated packaging and its adhesion effect between sheets of PE and aluminum, by analyzing the FTIR spectra of this system. Based on these studies, the band at 1397 cm^{-1} in PE-BA can be assigned to the carboxylate ion (COO-) in contact with aluminum surfaces. The pair of bands at 1405 and 1521 cm^{-1} have also been identified as aluminum carboxylates [29]. Consistently, the bands at 1540 and 1506 cm^{-1} are attributed to benzoate ion complexed with Al [28]. Other absorption bands assigned to aluminum chemical bonds in PE-BA are as follows: the bands at 1179 and 1069 cm^{-1} are attributed to link Al-O-C [33]; the bands at 1003 and 990 cm^{-1} are specific to Al-O [30,34]; and at 1270 cm^{-1}, the band is assigned to the CH_2-Al bond [32]. These results allow us to corroborate that grafting occurs through the acidic group of BA. In turn, $AlCl_3$ not only acts as a catalyst, but also participates in the formation of the reaction product. Furthermore, Al is involved in BA anchoring to the PE matrix. Al bonds with PE chains through the methylene group (CH_2-Al) and the carboxylic group of BA.

Moreover, $AlCl_3$ typical infrared absorption bands were not detected in any of the modified film spectra. In particular, neither the characteristic band of $AlCl_3$ at 1650 cm^{-1} nor the less intense band at 1158 cm^{-1} assigned to the ClO^{4-} group were identified [36,37]. The absorption bands in the modified spectra films assigned to aluminum bindings were attributed to the reaction product. Furthermore, no Cl infrared absorption bands were identified on modified film spectra; this element was released during the reaction experiment, not forming part of the product [38]. This was expected because $AlCl_3$, as demonstrated above, acts both as a catalyst and reagent and the surplus is dissolved in ethanol [39], the solvent used for the film-washing process.

It is important to note that the PE-BA spectrum includes several bands that can be assigned to aluminum aromatic complexes. In the literature, it was demonstrated that aluminum with BA forms different coordination complexes of aluminum benzoates [27,40]. In particular, $AlCl_3$ was previously used as an Al source to obtain different aluminum carboxylate complexes, as in the present study. Pękal and Pyrzynska [41] determined the flavonoid content through the aluminum-flavonoid complexes formed from the reaction of flavonoid, an aromatic compound with carbonyl groups, with a solution of $AlCl_3$. Yokel et al. [42] prepared Al citrate from $AlCl_3{\cdot}6H_2O$ and citric acid trisodium salt dehydrate. Karlik et al. [43] studied the interactions between Al(III) and carboxylate ligands from salts of lactate, citrate, and EDTA prepared by mixing $AlCl_3$ and the corresponding sodium salt. Motekaitis and Martell [44] studied the complexation of Al(III) with hydroxyl carboxylic acids using potentiometric methods. They used $AlCl_3{\cdot}6H_2O$ and crystalline monopotassium salts of citric acid, tartaric acid, gluconic acid, saccharic acid, glyceric acid, bis(hydroxyethyl) glycine, and catechol as reagents.

Table 3 summarizes the benzoate complex configurations depending on the amount and type of associations generated between the metal and the carboxylate group (COO-) of the acid: mononuclear monodentates and mono- or binuclear bidentates. The absorption spectra corresponding to these complexes have been analyzed in detail in reports focused on the absorption of BA in minerals containing aluminum [25,45,46], aluminum hydroxides [23,24,40], and alumina [47]. In the PE-BA spectrum (Figure 2, Table 2), the bands at 1619, 1641, 1571, 1440, and 1317 cm^{-1} are attributed to monodentate complexes [23–26]. The bands at 1433, 1605, and 1566 cm^{-1} in the PE-BA spectrum are attributed to bidentate complexes, chelates, or bridge [23,27]. A higher proportion of bands assigned to the monodentate complexes is observed, suggesting a reaction product with a predominance of this complex type.

Table 3. Aluminum benzoate coordination complexes [24].

Salt	Monodentate Complexes	Bidentate Complexes	
		Mononuclear	**Binuclear**
		Chelate	Bridge
Aluminum benzoates	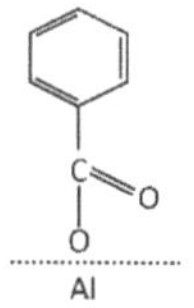	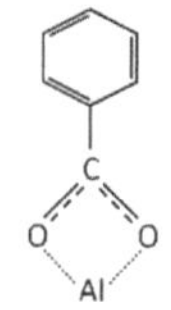	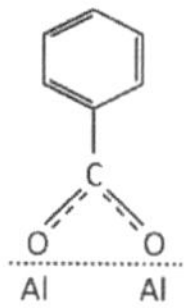

It is important to note that the reaction occurs only in the film surface. To prove this point, an FTIR was performed in transmission mode on PE-BA and PE films. Contrary to the FTIR-ATR spectra, in transmission spectra it is not possible to detect bands corresponding to the graft reaction products, as is shown in Figure 3. ATR-FTIR spectroscopy is a surface technique with a penetration depth that depends on IR wavelength, the refractive index of the sample, the refractive index of used crystal, and angle of incident light. The depth of penetration is up to 2 μm when organic materials are tested in a standard setup ATR-FTIR accessories using zinc selenide crystal and a 45° incident angle [48]. Then, considering that the PE film used as a substrate for the reaction is 100 microns thick, the reaction is circumscribed to its surface zone, penetrating less than 2% of the whole film thickness; then, the major volume of the film remains unaltered.

Third, we compared the thermal decomposition behavior of PE-BA with that of pure PE. Figure 4 shows complete (Figure 4a) and expanded (Figure 4b) thermogravimetric curves of PE and PE-BA. Both weight loss and the derivative of weight loss are depicted in order to assess thermal events related to reaction products resulting from surface modification. Both thermograms exhibited an overall similar behavior (Figure 4a), typical of PE, with a significant drop at around 450–460 °C associated with hydrocarbon degradation [49,50]. However, the thermogram of PE-BA shows a slight shift towards higher temperatures compared to that of PE. This shift is clearer in the derivative curves. This is the first thermal evidence of the presence of reaction products on PE-BA because aluminum compounds act as PE stabilizers, causing the loss weight at higher temperature values [51,52].

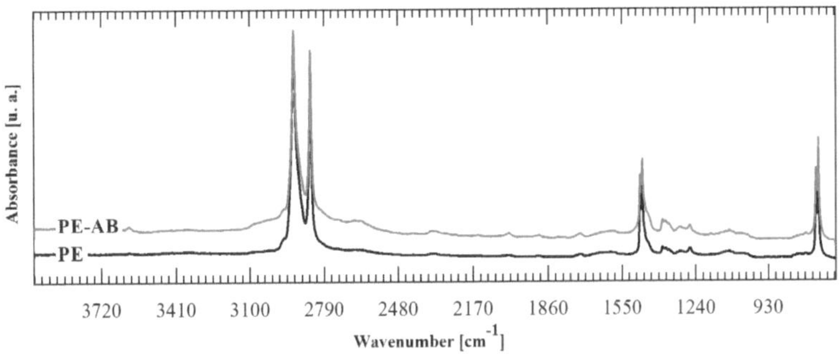

Figure 3. FTIR transmission spectra of PE-BA and PE.

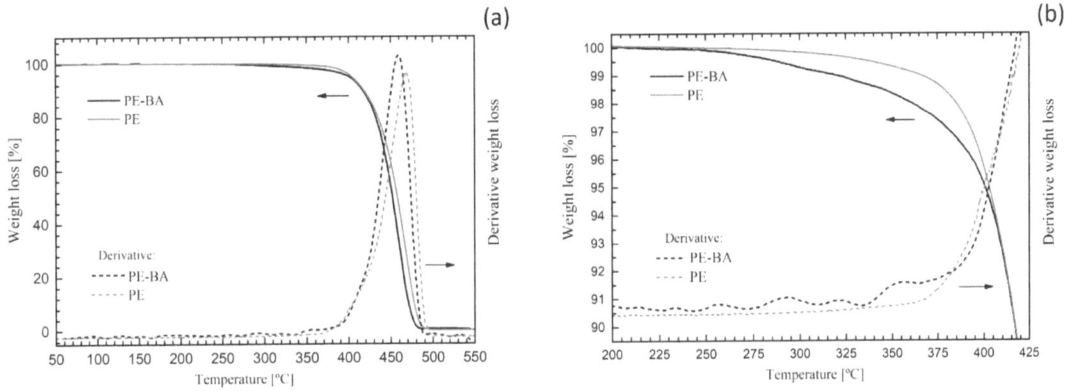

Figure 4. PE and PE-BA thermogravimetric curves: complete (**a**) and expanded area (**b**).

The main differences in the thermal behavior of PE and PE-BA were detected in the temperature range from 200 °C to 450 °C (Figure 4b). PE-BA begins to lose weight at around 225 °C, earlier than PE, which starts at around 290 °C. At 350 °C, PE-BA has lost 1% more mass than in PE. In the derivative curve of PE-BA, several small peaks associated with weight loss between 250 °C and 380 °C can be identified, with the largest one around 350 °C. This is attributed to the degradation of aluminum benzoate complexes formed during the grafting reaction, as identified with FTIR analysis. Furthermore, thermograms of this type of complex formed on clay surfaces [46] and of Ni-Al-benzoate obtained for amoxicillin absorbent [53] have shown degradation steps in the range of 200 °C to 400 °C, consistent with the assessed temperature range. Note that the degradation temperatures of $AlCl_3$ and BA are lower than the events detected in PE-BA [54]; thus, the presence of these reagents was not detected, as expected.

Finally, we performed a comparative and combined morphological–elemental analysis of PE-BA and pure PE surfaces. SEM micrographs of both film surfaces with their corresponding EDX spectra are shown in Figure 5. PE films (Figure 5a) displayed a smooth homogeneous surface containing only C according to the chemical composition of this polymer. In contrast, PE-BA films (Figure 5b) exhibited a wrinkled and irregular surface containing C, O, and Al, according to the appearance of a growing specimen in the form of a surface layer and with the chemical composition of the previously proposed aluminum benzoate complexes. It should be noted that Cl was not distinguished during analysis [54], so it was ruled out that it was part of the graft, as predicted for the reaction and as suggested by the FTIR analysis.

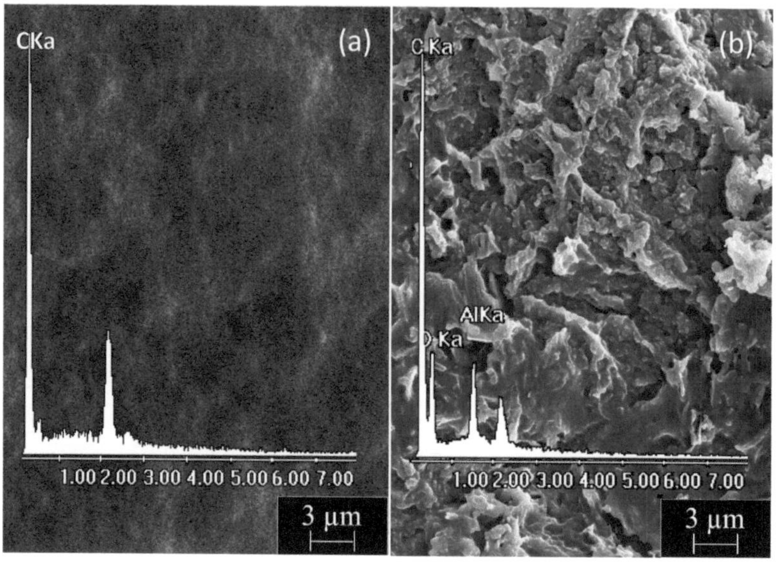

Figure 5. SEM micrograph (5000×) with corresponding EDX spectrum from film surfaces of (**a**) PE and (**b**) PE-BA.

Figure 6 includes a SEM micrograph with corresponding EDX mapping of C, O, and Al of the PE-BA surface. It was shown that both C, main component of PE, and O, as well as Al, components of the graft, were uniformly distributed throughout the analyzed surface. Therefore, this result revealed that the reaction on PE surface was uniform and massive. In fact, the PE-BA surface was differentiable from the PE surface even at 100×, the smallest magnification tested by SEM. Furthermore, from visual examination, PE-BA and PE films did not seem to be exactly equal. Even the texture to the touch of both films differed, that of PE-BA being rougher. These observations demonstrated a good reaction extent covering the complete, treated PE film surface.

In order to validate that the grafting reaction changes the surface polarity of PE films, the wettability was evaluated using contact angle measurements and surface energy calculations. Table 4 shows the measured values of the contact angle using water and diiodomethane and the calculated values of the dispersive (γ_s^d) and polar (γ_s^p) components and the total (γ_s) of the surface energy for PE-BA and PE. The water contact angle for PE is 5° greater than 90°, meaning that its surface is strongly hydrophobic, as is expected. The measured angles are comparable to those reported in the literature, ranging from 90 to 100° for PE films [55–59]. In contrast, the value of the water contact angle for PE-BA is more than 10° lower than that of PE, suggesting a hydrophilic nature as it falls below 90°. This clearly indicates the transformation of the surface character of PE from hydrophobic to hydrophilic upon grafting reaction. Furthermore, compared to aromatic polymers such as polystyrene, which typically exhibit a water contact angle of around 94° [60,61], the water contact angle of PE-BA is even lower, indicating a more polar surface. PE-BA water contact angle values are similar to an oxidized polystyrene (PS) with an oxygen atom percentage slightly higher than 3%, that is equivalent to a 0.013 oxygen atom/Å^2 [60]. These results are in concordance with the chemistry structure of the grafting products suggested by FTIR analysis. Benzoate complexes linked to PE by bridges between the carboxylic group and Al, but also free carboxylic acid groups, were identified, which together give greater polarity compared to the grafting of neutral aromatic groups. Meanwhile, the values of the contact angle with diiodomethane did not change significantly. Diiodomethane is a slightly polar liquid with a surface energy less than half that of water, but sufficient to produce easily measurable contact angles. This characteristic makes it ideal for complementing water contact angle measurements in surface energy analysis. However, due to its slight polarity and primarily dispersive forces, it is less sensitive to changes in the substrate polarity [62]. The measured values are comparable to those reported in the literature for untreated PE, which are 51° [63], 55.6° [56], and 52.8° [57], among other similar values.

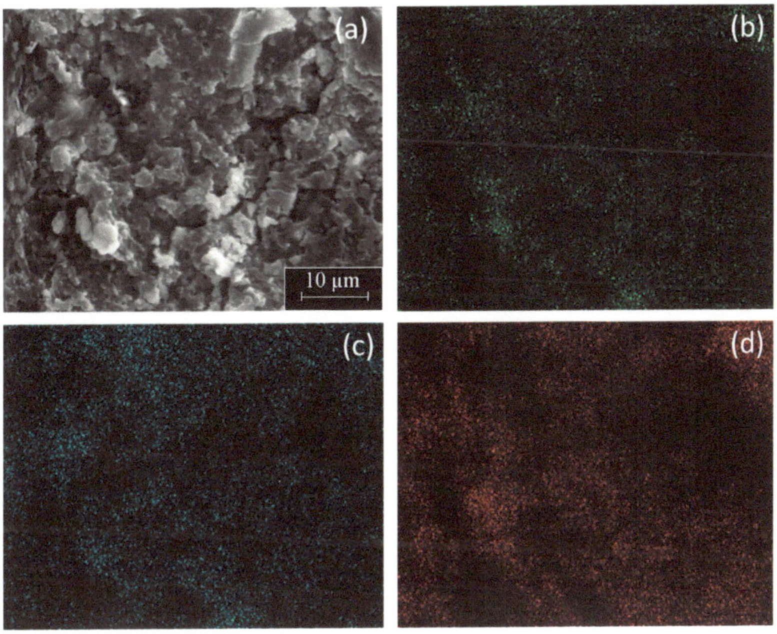

Figure 6. SEM micrograph (2000×) of PE-BA film surface (**a**) with corresponding elemental EDX mapping of (**b**) O, (**c**) Al, and (**d**) C.

Table 4. Water and diiodomethane contact angle and surface energy of PE-BA and PE.

Sample	Water Contact Angle [°]	Diiodomethane Contact Angle [°]	Surface Energy [mJ. m^{-2}]		
			γ_s^d	γ_s^p	γ_s
PE	95 ± 2	52 ± 1	31.71	0.95	32.71
PE-BA	84 ± 3	51 ± 2	30.97	4.04	35.00

Surface energy parameters, calculated from contact angle measurements using Equation (1), show that the values of γ_s and γ_s^p for PE-BA are higher than for PE. The calculated values for PE agree with the literature for LDPE films, which reports values around 30 mJ. m^{-2} [55–57,59]. Although the values of total surface energy are quite close between them, the values corresponding to PE-BA remained slightly higher, denoting the change in the total polarity of the surface. Meanwhile, the greatest changes are observed in the values of the polar component of PE-BA, which increases more than 4 times compared to PE. The percentage of the polar component, with respect to the total surface energy, increased from 3% to 11%. This change can be attributed to a polarity increase by the surface grafting products. Similar delta and total surface energy values were reported by

Rytlewski and Żenkiewicz, [64] for PS and PS oxidized by laser treatment, using a pulse number 10, with an atomic oxygen concentration close to 6.4%. This oxygen concentration is at least 2 times higher than the value informed by the work of Bekele and Tsige [60], discussed previously for similar water contact angle of an oxidized PS.

A commonly used semiquantitative analysis to determine surface energy in the industry is the dyne pen test. In Figure 7, the results of this test for both PE and PE-BA samples are presented. No continuous lines are observed on the PE surface; only drops are visible with the 38 dynes/cm pen, and even smaller individual drops with the 44 dynes/cm pen. Conversely, continuous lines are observed on the PE-BA samples drawn with both the 38 and 44 dynes/cm pens. These results suggest that the surface energy of PE is below 38 dynes/cm, while that of PE-BA is even higher than 44 dynes/cm. While the behavior of PE aligns with expectations, that of PE-BA surpasses what was calculated from the contact angle (refer to Table 4). These disparities may stem from the fact that the contact angle is influenced not only by surface chemistry but also by surface topography. As shown in SEM images, PE-BA is rougher than PE.

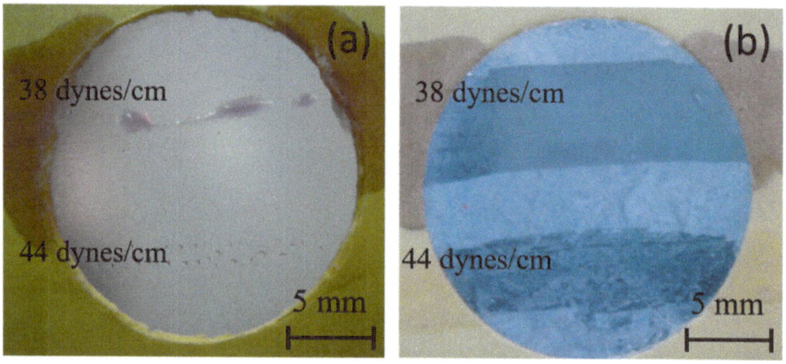

Figure 7. Photographs of dyne pen testing, using 38 and 44 dynes/cm pens, on the surfaces of (**a**) PE and (**b**) PE-BA.

Generally, industrial printing requirements are approximately 38–40 dynes/cm for solvent-based inks and around 42–22 dynes/cm for water-based inks [9]. Accordingly, PE-BA films would be suitable for painting with water-based inks. In order to demonstrate this fact, several experiments using watercolor and tempera paints were performed on both films.

Two different water-based paints, red watercolor and blue tempera, were utilized to conduct a comparative paintability test on PE and PE-BA films. In Figure 8, the surfaces painted with red watercolor are depicted for both PE (a) and PE-BA (b) films. While the watercolor evenly covers the entire surface of the PE-BA film, it is repelled from the PE surface, forming independent droplets. The increased hydrophilicity of PE-BA is evident, confirming the alteration in surface activity resulting from the chemical bonding of polar and aromatic groups onto the PE surface. Figure 9 displays the surfaces painted with blue tempera for both PE and PE-BA. Although both surfaces are painted, the coverage is notably better on PE-BA. Brush strokes fail to completely cover the PE background, leaving white spaces, a phenomenon absent on PE-BA. This change in behavior is more pronounced when using watercolor, as it has a higher water proportion and lower viscosity compared to tempera. Additionally, the watercolor pigment, naphthol red ($C_{26}H_{22}N_4O_4$), contains aromatic rings in its structure, rendering it more compatible with other aromatic compounds such as benzoates present on the PE-BA surface.

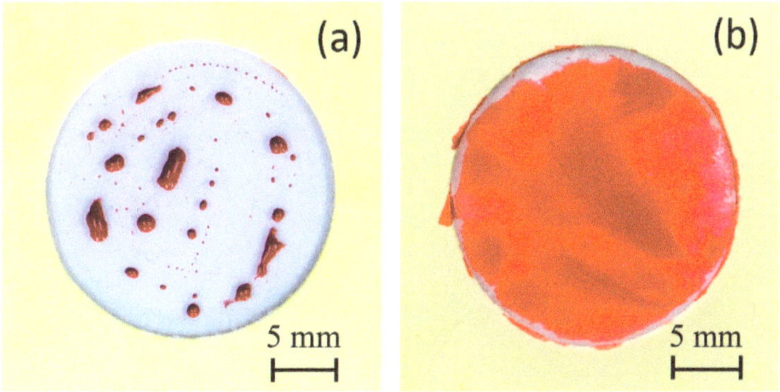

Figure 8. Photographs of painting experiments with red watercolor on the surfaces of (**a**) PE and (**b**) PE-BA.

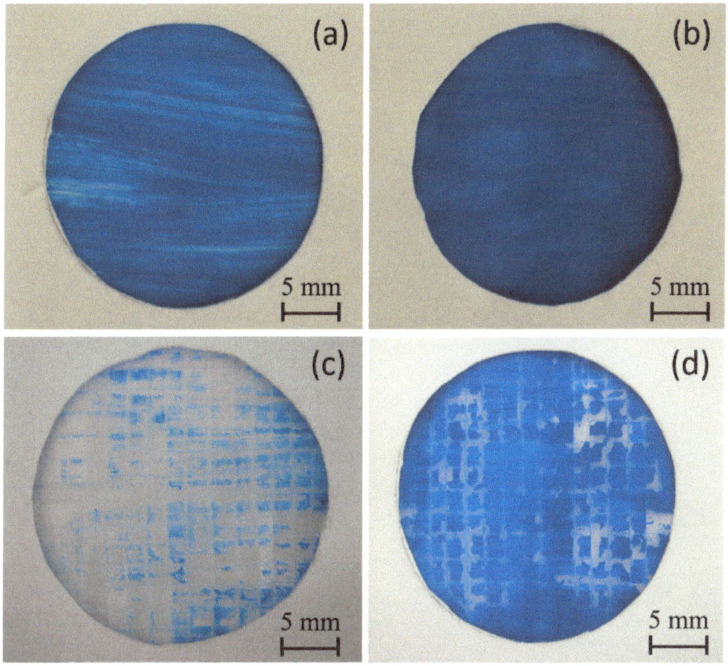

Figure 9. Photographs of painting experiments with blue tempera on the surfaces of (**a**) PE and (**b**) PE-BA and after adhesion test (**c**) PE and (**d**) PE-BA.

Contact angle measurements, dyne pen testing and painting with watercolor were also performed with PE-BA samples that were stored for more than nine months. No changes in the performance of such samples compared with new ones were detected. This behavior could be attributed to the thermodynamic stability of the benzozed aluminum complexes generated on the PE-BA surface and to their voluminous molecules, since, in addition, the complexes tend to associate with each other, which makes their diffusion through the PE matrix difficult [65]. According to these results, the aging effect of grafting could be not considered as an important detrimental factor of the surface properties.

In Figure 9, photographs c and d display the surfaces of the PE and PE-BA films, respectively, after undergoing an adhesion test [18]. The improvement in paint adhesion is evident, as the surface of the PE film is nearly devoid of paint, while the PE-BA surface retains good coverage. The estimated values of r.a. for PE are approximately 91%, whereas for PE-BA, they are around 17%, representing an adhesion enhancement of more than 5 times. According to the classification of ASTM 3359-02, these results correspond to 2B (15–35% r.a.) for PE-BA and 0B (>65% r.a.) for PE. Similar findings for neat PE have been previously reported by Ataeefard [66] and Grafia et al. [13].

These results demonstrate that the surface modification effectively increases the affinity of polar substances with the PE surface, owing to the grafting of oxygenated and aromatic groups. In particular, benzoates and derivatives enhance polyester dyeing by serving as dye carriers [67,68]. Polystyrenes recovered from waste are converted into dye absorbents by grafting 1,2,4-benzenetricarboxylic anhydride [4] or maleic anhydride [69] through Friedel–Crafts reactions, employing $AlCl_3$ as a catalyst. These graft molecules, which share chemical structures in common with those generated on PE-BA, increase the dye-receptive sites of the polymer. The dye–polymer interactions proposed in these works, such as electrostatic interaction, hydrogen bonding, n-π interaction, π-π interaction, and complexation, could be similar to those occurring with PE-BA. All of these interactions create different dye-anchoring pathways that enhance the polymer–dye affinity. In contrast, interactions of dyes with oxidized polymers, such as through corona treatment, primarily occur with the hydroxyl, carbonyl, and other oxygen-containing groups on their polarized surface [70].

4. Conclusions

A method for introducing hydrophilic characteristics to PE film through grafting reactions is proposed. This method involves an alkylation reaction on the molten film surface using $AlCl_3$ as a catalyst, enabling the grafting of BA. Surface reaction occurrence and identification of reaction products were assessed through a comparative chemical, thermal, and morphological analysis of pure PE and reacted films, previously extracted with ethanol.

The presence of aromatic compounds on the modified film surface was initially confirmed using the chloroform and $AlCl_3$ aromatic identification test. The chemical nature of the reaction product was determined through FTIR-ATR spectroscopy analysis, revealing that BA is indeed grafted to PE and that aluminum is part of the reaction product, resulting in a mixture of mono and bidentate benzoate complexes. The existence of such complexes on PE-BA was further confirmed by thermogravimetric analysis. Additionally, it was demonstrated that the chlorine incorporated through $AlCl_3$ is not present in the modified films. Furthermore, morphological-elementary analysis provided additional evidence of reaction occurrence.

It was also demonstrated that the grafting reaction of BA onto PE occurs exclusively on the film surface, transforming PE films from hydrophobic to hydrophilic and subsequently improving surface affinity towards polar compounds. This change in surface polarity was assessed using diverse qualitative and quantitative methods: contact angle measurements, surface free energy calculations, dyne pen test, paintability with watercolors and tempera, and adhesion tests for these water-based paints. These results can be explained by a synergistic effect of included polarity and aromaticity PE surface from BA-grafted molecules. Thus, it not only increases compatibility with polar compounds but also with aromatic compounds. Considering that many inks and pigments are aromatic in nature, this is an added advantage. Furthermore, no aging effects were detected.

This method could be scalable for direct application in blown film processes. It can be applied immediately above the freeze line, using two successive spraying annular systems. At this location of the PE bubble, the surface film is molten, allowing reagents to be sprayed successively and reacts at the rates of the process. Furthermore, the modification proposed in this work could be extended to other thermoplastics polymeric substrates such as polypropylene and with other grafting molecules such as salicylic acid. In this way, the surface of the polymers could be tailored according to each specific application required.

Author Contributions: Methodology, A.L.G.; Formal analysis, A.L.G. and S.E.B.; Investigation, A.L.G. and S.E.B.; Writing—original draft, A.L.G.; Writing—review & editing, S.E.B.; Supervision, S.E.B. All authors have read and agreed to the published version of the manuscript.

Funding: The authors acknowledge the financial support of CONICET (National Research Council of Argentina), ANPCyT (National Agency for Promotion of Science and Technology of Argentina), and UNS (Universidad Nacional del Sur). This research was funded by ANPCyT grant number PICT-2018-04606, and UNS grant number PGI 24/M174.

Institutional Review Board Statement: Not applicable.

Data Availability Statement: The original contributions presented in the study are included in the article, further inquiries can be directed to the corresponding authors.

Conflicts of Interest: The authors declare no conflict of interest.

References

1. Narimisa, M.; Ghobeira, R.; Onyshchenko, Y.; Geyter, N.D.; Egghe, T.; Morent, R. Different Techniques Used for Plasma Modification of Polyolefin Surfaces. In *Plasma Modification of Polyolefins*; Springer: Cham, Switzerland, 2022; pp. 15–56.
2. Nemani, S.K.; Annavarapu, R.K.; Mohammadian, B.; Raiyan, A.; Heil, J.; Haque, M.A.; Sojoudi, H. Surface modification of polymers: Methods and applications. *Adv. Mater. Interfaces* **2018**, *5*, 1801247. [CrossRef]
3. Mozetič, M. Aging of Plasma-Activated Polyethylene and Hydrophobic Recovery of Polyethylene Polymers. *Polymers* **2023**, *15*, 4668. [CrossRef] [PubMed]
4. Li, T.T.; Cheng, S.B.; Feng, L.F.; Gu, X.P.; Duan, J.T.; Jiang, M.Z.; Zhang, C.L. A review on the free radical grafting of vinyl monomers onto polyethylene and polypropylene by reactive extrusion. *Chem. Eng. Sci.* **2023**, *278*, 118916. [CrossRef]
5. Zhu, J.; Tang, Y.; Lu, L.; Qiu, X.; Pan, L. Reactive extrusion of caffeic acid functionalized ε-polylysine with low-density polyethylene as an antimicrobial and antioxidant film. *React. Funct. Polym.* **2024**, *197*, 105861. [CrossRef]
6. El-Wakil, A.E.A.A.; Moustafa, H.; Youssef, A.M. Antimicrobial low-density polyethylene/low-density polyethylene-grafted acrylic acid biocomposites based on rice bran with tea tree oil for food packaging applications. *J. Thermoplast. Compos. Mater.* **2022**, *35*, 938–956. [CrossRef]
7. Versteeg, F.A.; Benita, B.B.; Jongstra, J.A.; Picchioni, F. Reactive Extrusion Grafting of Glycidyl Methacrylate onto Low-Density and Recycled Polyethylene Using Supercritical Carbon Dioxide. *Appl. Sci.* **2022**, *12*, 3022. [CrossRef]
8. Wei, Z.; Liu, H.; Yu, L.; Xiao, S.; Hou, Y.; Chen, X. Delocalized aromatic molecules with matched electron-donating and electron-withdrawing groups enhancing insulating performance of polyethylene blends. *J. Appl. Polym. Sci.* **2020**, *137*, 49185. [CrossRef]
9. Izdebska-Podsiadły, J.; Thomas, S. (Eds.) *Printing on Polymers: Fundamentals and Applications*; William Andrew: Chadds Ford, PA, USA, 2015.
10. Kennedy, J.P.; Maréchal, E. *Carbocationic Polymerization*; J. Wiley & Sons: New York, NY, USA, 1982.
11. Carrick, W.L. Reactions of polyolefins with strong lewis acids. *J. Polym. Sci. Part A-1 Polym. Chem.* **1970**, *8*, 215–223. [CrossRef]
12. Díaz, M.F.; Barbosa, S.E.; Capiati, N. J Reactive compatibilization of PE/PS blends. Effect of copolymer chain length on interfacial adhesion and mechanical behavior. *Polymer* **2007**, *48*, 1058–1065. [CrossRef]
13. Grafia, A.L.; Martini, R.E.; Barbosa, S.E. Spray process to styrene grafting onto polyethylene film surface for paintability enhancement. *Prog. Org. Coat.* **2018**, *117*, 91–101. [CrossRef]
14. Martini, R.E.; Brignole, E.A.; Barbosa, S.E. Grafting of styrene onto polyethylene in near critical media. *J. Appl. Polym. Sci.* **2012**, *123*, 2787–2799. [CrossRef]
15. Shriner, R.L.; Hermann, C.K.; Morrill, T.C.; Curtin, D.Y.; Fuson, R.C. *The Systematic Identification of Organic Compounds*; John Wiley and Sons: New York, NY, USA, 1980.
16. Żenkiewicz, M. Methods for the calculation of surface free energy of solids. *J. Achiev. Mater. Manuf. Eng.* **2007**, *24*, 137–145.
17. Park, S.J.; Jin, J.S. Effect of corona discharge treatment on the dyeability of low-density polyethylene film. *J. Colloid Interface Sci.* **2001**, *236*, 155–160. [CrossRef] [PubMed]
18. *ASTM D3359-02*; Standard Test Methods for Measuring Adhesion by Tape Test. ASTM International: West Conshohocken, PA, USA, 2002.
19. Gulmine, P.R.; Janissek, P.R.; Heise, H.M.; Akcelrud, L. Polyethylene characterization by FTIR. *Polym. Test.* **2002**, *21*, 557–563. [CrossRef]
20. Abounassif, M.A.; Mian, M.S.; Mian, N.A. Salicylic acid. *Anal. Profiles Drug Subst. Excip.* **1994**, *23*, 421–470.
21. Hayashi, S.; Kimura, N. Infrared Spectra and Molecular Configuration of Benzoic acid. *Bull. Inst. Chem. Res.* **1966**, *44*, 335–340.
22. Indrayanto, G.; Syahrani, A.; Rahman, A.; Tanudjojo, W.; Susanti, S.; Yuwono, M.; Ebel, S. Benzoic acid. In *Analytical Profiles of Drug Substances and Excipients*; Academic Press: Cambridge, MA, USA, 1999; pp. 1–46.
23. Guan, X.; Chen, G.; Shang, C. ATR-FTIR and XPS study on the structure of complexes formed upon the adsorption of simple organic acids on aluminum hydroxide. *J. Environ. Sci.* **2007**, *19*, 438–443. [CrossRef] [PubMed]

24. Biber, M.V.; Stumm, W. An In-Situ ATR-FTIR Study: The surface Coordination of Salicylic Acid on Aluminum and Iron (III) Oxides. *Environ. Sci. Technol.* **1994**, *28*, 763–768. [CrossRef]
25. Kubicki, J.D.; Schroeter, L.M.; Itoh, M.J.; Nguyen, B.N.; Apitz, S.E. Attenuated total reflectance Fourier-transform infrared spectroscopy of carboxylic acids adsorbed onto mineral surfaces. *Geochem. Cosmochim. Acta* **1999**, *63*, 2709–2725. [CrossRef]
26. Schultz, J.; Carré, A.; Mazeau, C. Formation and rupture of grafted polyethylene/aluminium interfaces. *Int. J. Adhes. Adhes.* **1984**, *4*, 163–168. [CrossRef]
27. Lewandowski, W.; Baranska, H. The influence of Selected Metals on the Aromatic System of Salicylic Acid. *Appl. Spectrosc.* **1987**, *41*, 976–980. [CrossRef]
28. Groff, R.P. Adsorption and Orientation of Benzoic Acid on Aluminum Oxide: An Infrared Study. *J. Od Catal.* **1983**, *79*, 259–263. [CrossRef]
29. Strälin, A.; Hjertberg, T. Hydrated Aluminum and Polar Groups in Ethylene Copolymers. *Surf. Interface Anal.* **1993**, *20*, 337–340. [CrossRef]
30. Olafsson, G.; Hildingsson, I. Sorption of Fatty Acids into Low-Density Polyethylene and Its Effect on Adhesion with Aluminum Foil in Laminated Packaging Material. *J. Agric. Food Chem.* **1995**, *43*, 306–312. [CrossRef]
31. Olafsson, G.; Hildingsson, I.; Bergnstahl, B. Transport of Oleic and Acetic Acids from Emulsions into Low Density Polyethylene; Effects on Adhesion with Aluminum Foil in Laminated Packaging. *J. Food Sci.* **1995**, *60*, 420–425. [CrossRef]
32. Aoyagi, T.; Araki, T.; Oguni, N.; Mikumo, M.; Tani, H. Synthesis of inuclear organoaluminum compounds containing oxygen, nitrogen, and sulfur atoms. *Inorg. Chem.* **1973**, *12*, 2702–2707. [CrossRef]
33. Guertin, D.L.; Wiberley, S.E.; Bauer, W.H.; Goldenson, J. The infrared spectra of three Aluminum Alkoxides. *J. Phys. Chem.* **1956**, *60*, 1018–1019. [CrossRef]
34. Scott, F.A.; Goldenson, J.; Wiberley, S.E.; Bauer, W.H. Infrared spectra of Aluminum soaps and soaps-hydrocarbon gels. *J. Phys. Chem.* **1953**, *58*, 61–64. [CrossRef]
35. Olafsson, G.; Jägerstad, M.; Öste, R.; Wesslén, B. Effects of different organic acids on the adhesion between polyethylene film and aluminium foil. *Food Chem.* **1993**, *47*, 227–233. [CrossRef]
36. McIntyre, J.F.; Foley, R.T.; Brown, B.F. Infrared spectra of aluminum salt solutions. *Appl. Spectrosc.* **1982**, *36*, 128–136. [CrossRef]
37. Rabek, J.F.; Lucki, J. Crosslinking of polystyrene under friedel–crafts conditions in dichloroethane and carbon tetrachloride solvents through the formation of strongly colored polymer–$AlCl_3$–solvent complexes. *J. Polym. Sci. Part A Polym. Chem.* **1988**, *26*, 2537–2551. [CrossRef]
38. Pretsch, E.; Clerc, T.; Seibl, J.; Simon, W. *Tables of Spectral Data for Structure Determination of Organic Compounds*; Springer Science & Business Media: New York, NY, USA, 2013.
39. Thomas, C.A. *Anhydrous $AlCl_3$ in Organic Chemistry*; Reinhold: New York, NY, USA, 1941; p. 716.
40. Phambu, N. Adsorption of carboxylic acids on submicrocrystalline aluminum hydroxides in aqueous solution. Part I: Qualitative study by infrared and Raman spectroscopy. *Appl. Spectrosc.* **2002**, *56*, 756–761. [CrossRef]
41. Pękal, A.; Pyrzynska, K. Evaluation of aluminium complexation reaction for flavonoid content assay. *Food Anal. Methods* **2014**, *7*, 1776–1782. [CrossRef]
42. Yokel, R.A.; Wilson, M.; Harris, W.R.; Halestrap, A.P. Aluminum citrate uptake by immortalized brain endothelial cells: Implications for its blood–brain barrier transport. *Brain Res.* **2002**, *930*, 101–110. [CrossRef] [PubMed]
43. Karlik, S.J.; Tarien, E.; Elgavish, G.A.; Eichhorn, G.L. Aluminum-27 nuclear magnetic resonance study of aluminum (III) interactions with carboxylate ligands. *Inorg. Chem.* **1983**, *22*, 525–529. [CrossRef]
44. Motekaitis, R.J.; Martell, A.E. Complexes of aluminum (III) with hydroxy carboxylic acids. *Inorg. Chem.* **1984**, *23*, 18–23. [CrossRef]
45. Xin, X.; Si, W.; Yao, Z.; Feng, R.; Du, B.; Yan, L.; Wei, Q. Adsorption of benzoic acid from aqueous solution by three kinds of modified bentonites. *J. Colloid Interface Sci.* **2011**, *359*, 499–504. [CrossRef] [PubMed]
46. Yariv, S.; Cross, H. *Organo-Clay Complexes and Interactions*; The Clay Minerals Society: New York, NY, USA, 2002.
47. Niwa, M.; Inagaki, S.; Murakami, Y. Alumina: Sites and mechanism for benzaldehyde and ammonia reaction. *J. Phys. Chem.* **1985**, *89*, 2550–2555. [CrossRef]
48. Almond, J.; Sugumaar, P.; Wenzel, M.N.; Hill, G.; Wallis, C. Determination of the carbonyl index of polyethylene and polypropylene using specified area under band methodology with ATR-FTIR spectroscopy. *e-Polymers* **2020**, *20*, 369–381. [CrossRef]
49. Anderson, D.A.; Freeman, E.S. The kinetics of the thermal degradation of polystyrene and polyethylene. *J. Polym. Sci.* **1961**, *54*, 253–260. [CrossRef]
50. Peterson, J.D.; Vyazovkin, S.; Wight, C.A. Kinetics of the thermal and thermo-oxidative degradation of polystyrene, polyethylene and poly (propylene). *Macromol. Chem. Phys.* **2001**, *202*, 775–784. [CrossRef]
51. Fuzail, M.; Shah, G.; Anwar, J. Modification of polyethylene and incorporation of Al (OH) 3 for improvement of mechanical properties, burning behaviour and thermal stability. *Iran. Polym. J.* **2010**, *19*, 47–56.
52. Wilkie, C.A.; Morgan, A.B. *Fire Retardancy of Polymeric Materials*; CRC Press: New York, NY, USA, 2009.
53. Dwiasi, D.W.; Mudasir, M.; Roto, R. Ion exchange of benzoate in Ni-Al-benzoate layered double hydroxide by amoxicillin. *Open Chem. J.* **2019**, *17*, 1043–1049. [CrossRef]
54. Hartman, M.; Trnka, O.; Šolcová, O. Thermal decomposition of aluminum chloride hexahydrate. *Ind. Eng. Chem. Res.* **2005**, *44*, 6591–6598. [CrossRef]

55. Aliev, R. Hydrophilicity and surface energy of polyethylene modified by radiation grafting of acrylamide. *Polym. Bull.* **2001**, *47*, 99–104. [CrossRef]
56. De Geyter, N.; Morent, R.; Leys, C. Surface characterization of plasma-modified polyethylene by contact angle experiments and ATR-FTIR spectroscopy. *Surf. Interface Anal. Int. J. Devoted Dev. Appl. Tech. Anal. Surf. Interfaces Thin Film.* **2008**, *40*, 608–611. [CrossRef]
57. Jańczuk, B.; Białopiotrowicz, T. Surface free-energy components of liquids and low energy solids and contact angles. *J. Colloid Interface Sci.* **1989**, *127*, 189–204. [CrossRef]
58. Seyhan, A.; Gunaydin, B.N.; Polat, Y.; Kilic, A.; Demir, A.; Avci, H. Improvement of polyethylene fiber wettability and mechanical properties through an environmentally sustainable spinning process. *Int. J. Adhes. Adhes.* **2022**, *119*, 103250. [CrossRef]
59. Subedi, D.P. Contact angle measurement for the surface characterization of solids. *Himal. Phys.* **2011**, *2*, 1–4. [CrossRef]
60. Bekele, S.; Tsige, M. Interfacial properties of oxidized polystyrene and its interaction with water. *Langmuir* **2013**, *29*, 13230–13238. [CrossRef]
61. Brown, P.S.; Bhushan, B. Mechanically durable liquid-impregnated honeycomb surfaces. *Sci. Rep.* **2017**, *7*, 6083. [CrossRef]
62. Strobel, M.; Kirk, S.M.; Heinzen, L.; Mischke, E.; Lyons, C.S.; Endle, J.; Dillingham, G. Contact angle measurements on oxidized polymer surfaces containing water-soluble species. *J. Adhes. Sci. Technol.* **2015**, *29*, 1483–1507. [CrossRef]
63. Zdziennicka, A.; Szymczyk, K.; Krawczyk, J.; Jańczuk, B. Some remarks on the solid surface tension determination from contact angle measurements. *Appl. Surf. Sci.* **2017**, *405*, 88–101. [CrossRef]
64. Rytlewski, P.; Żenkiewicz, M. Laser-induced surface modification of polystyrene. *Appl. Surf. Sci.* **2009**, *256*, 857–861. [CrossRef]
65. Florjanczyk, Z.; Bury, W.; Zygadło-Monikowska, E.; Justyniak, I.; Balawender, R.; Lewinski, J. Structure Investigations of Dichloroaluminum Benzoates: An Unprecedented Example of a Monomeric Aluminum Complex with a Chelating Carboxylate Ligand. *Inorg. Chem.* **2009**, *48*, 10892–10894. [CrossRef] [PubMed]
66. Ataeefard, M. Study of PLA printability with flexography ink: Comparison with common packaging polymer. *Prog. Color Color. Coat.* **2019**, *12*, 101–105.
67. Derbyshire, A.N.; Mills, W.P.; Shore, J. The role of auxiliary products in the high–temperature dyeing of polyester. *J. Soc. Dye. Colour.* **1972**, *88*, 389–394. [CrossRef]
68. Roberts, G.A.F.; Solanki, R.K. Carrier Dyeing of Polyester Fibre: Part I–Studies of Carrier Diffusion. *J. Soc. Dye. Colour.* **1979**, *95*, 226–229. [CrossRef]
69. Ye, H.; Xie, Z.; Li, W.; Pu, Y.; Liu, M.; Wen, Y.; Liu, Y. Converting waste polystyrene foam into new value-added materials: A large-capacity scavenger to remove cationic dyes and heavy metals. *J. Appl. Polym. Sci.* **2022**, *139*, 51868. [CrossRef]
70. Aydemir, C.; Altay, B.N.; Akyol, M. Surface analysis of polymer films for wettability and ink adhesion. *Color Res. Appl.* **2021**, *46*, 489–499. [CrossRef]

polymers

MDPI

Article

Critical Cooling Rate of Fast-Crystallizing Polyesters: The Example of Poly(alkylene *trans*-1,4-cyclohexanedicarboxylate)

Kylian Hallavant [1], Michelina Soccio [2,3,4], Giulia Guidotti [2], Nadia Lotti [2,3,5], Antonella Esposito [1,*] and Allisson Saiter-Fourcin [1]

[1] INSA Rouen Normandie, CNRS, Groupe de Physique des Matériaux UMR 6634, University Rouen Normandie, F-76000 Rouen, France; kylian.hallavant@univ-rouen.fr (K.H.); allison.saiter@univ-rouen.fr (A.S.-F.)
[2] Department of Civil, Chemical, Environmental and Materials Engineering, University of Bologna, Via Terracini 28, 40131 Bologna, Italy; m.soccio@unibo.it (M.S.); giulia.guidotti9@unibo.it (G.G.); nadia.lotti@unibo.it (N.L.)
[3] Interdepartmental Center for Industrial Research on Advanced Applications in Mechanical Engineering and Materials Technology, CIRI-MAM, Viale del Risorgimento 2, 40136 Bologna, Italy
[4] Interdepartmental Center for Industrial Research on Buildings and Construction, CIRI-EC, Via del Lazzaretto 15/5, 40131 Bologna, Italy
[5] Interdepartmental Center for Agro-Food Research, CIRI-AGRO, Via Quinto Bucci 336, 47521 Cesena, Italy
* Correspondence: antonella.esposito@univ-rouen.fr

Abstract: Controlling the cooling rate experienced by a material during a manufacturing process is a challenge and a major issue. Industrial processing techniques are very diverse and may involve a whole range of cooling rates, which are sometimes extremely high for small and/or thin manufactured parts. For polymers, the cooling rate has consequences on both the microstructure and the time-dependent properties. The common cooling rates associated with conventional calorimetric measurements are generally limited to a few tens of degrees per minute. This work combines several calorimetric techniques (DSC, modulated-temperature DSC, stochastically-modulated DSC and Fast Scanning Calorimetry) to estimate the critical cooling rate required to melt-quench fast-crystallizing polyesters to their fully amorphous state, based on the example of a series of poly(alkylene *trans*-1,4-cyclohexanedicarboxylate) (PCHs) with a number of methylene groups in the main structure of the repeating unit n_{CH_2} varying from 3 to 6. The even-numbered ones require faster cooling rates (about 3000 K s^{-1} for n_{CH_2} = 4, between 500 and 1000 K s^{-1} for n_{CH_2} = 6) compared to the odd-numbered ones (between 50 K min^{-1} and 100 K s^{-1} for n_{CH_2} = 3, between 10 and 30 K min^{-1} for n_{CH_2} = 5).

Keywords: crystallization kinetics; odd-even effect; microstructure control; quenching

Citation: Hallavant, K.; Soccio, M.; Guidotti, G.; Lotti, N.; Esposito, A.; Saiter-Fourcin, A. Critical Cooling Rate of Fast-Crystallizing Polyesters: The Example of Poly(alkylene *trans*-1,4-cyclohexanedicarboxylate). *Polymers* **2024**, *16*, 2792. https://doi.org/10.3390/polym16192792

Academic Editor: Nicolas Sbirrazzuoli

Received: 8 September 2024
Revised: 28 September 2024
Accepted: 29 September 2024
Published: 1 October 2024

1. Introduction

Polymers play an increasing role in manufacturing processes and their production keeps on growing, especially in the packaging industry [1], which is the biggest consumer and producer of plastic waste. However, the decline of fossil resources, and the raise of collective awareness about the impact that waste has on the environment, impose to look for possible alternatives to petroleum-based polymers with reduced carbon footprint and environmental risks. Plastic waste started accumulating in increasingly large amounts in the fifties, and nowadays the environmental pollution due to plastic waste is becoming severe. For this reason, bio-sourced and/or biodegradable polymers, in particular polyesters, have attracted much attention from both academic researchers and industrials—because they appear as one of the most promising solutions for next-generation sustainable packaging, with improved barrier properties against gas and water vapor, and balanced mechanical properties [2,3].

Industrial processing techniques, such as injection molding or additive manufacturing (laser bed fusion, 3D-printing...), involve a whole range of cooling rates (Figure 1), which depends on the surface-to-volume ratio of the manufactured item, the cooling technology and the temperature of the cooling medium, the covered temperature range, and the thermal conductivity of the material. Sometimes the cooling rates experienced by the material are extremely high, especially when the manufactured parts are small and/or thin. Of course, increasing the cooling rates used for processing helps increasing cost-effectiveness, however it is appropriate to recall that cooling may also have major consequences on polymer microstructure and molecular mobility [4]. In any case, controlling the local cooling rate and understanding its consequences on polymer properties remain a major challenge.

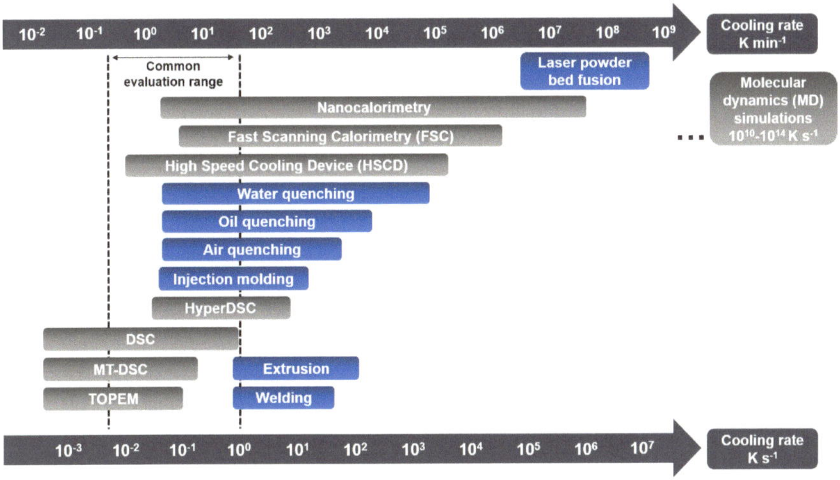

Figure 1. Typical range of cooling rates achievable with different processing and characterization techniques for different materials including polymers [5–24].

A series of biodegradable and potentially bio-based thermoplastic polyesters has been recently synthesized from *trans*-1,4-cyclohexanedicarboxylic acid and linear diols of different lengths. These materials showed interesting barrier and mechanical properties for food packaging applications [25–29], but also a very different aptitude to crystallize depending on the number of methylene groups $-CH_2-$ in their repeating unit (odd-even effect) [30–35]. The cooling rate required for full quenching is expected to be very different for each one of these materials, with different consequences for the ones that are able to crystallize compared to the ones that cannot [36]. It has already been shown that, besides making a difference in terms of nucleus density [37], crystal fraction [38], crystallite size, lamellar thickness and spherulite morphology [39], playing on the cooling rate may significantly affect physical ageing [40] and consequently any other time-dependent property [41,42], even when the polymers are melt-quenched to their fully amorphous state and kept at temperatures below their glass transition.

Schawe and Löffler [43] took the example of Au-based glasses to discuss the existence of multiple critical cooling rates for the generation of different types of glasses. They identified three thresholds, which they called $\beta_{c,SCG}$ (critical cooling rate to obtain a semi-crystalline glass), $\beta_{c,SDG}$ (critical cooling rate to obtain a self-doped glass), and $\beta_{c,CHG}$ (critical cooling rate to obtain a chemically-homogeneous glass). Transposing and adapting this terminology to macromolecular glasses, one may expect that (1) increasing the cooling rate is an efficient way to melt-quench polymers that are intrinsically able to crystallize, (2) crystallization is observed only if the cooling rate is kept below a critical value $\beta_{c,SDG}$, (3) crystallinity reaches different extents depending on the applied cooling rate $\beta_{c,SCG}$ (faster cooling reduces it since it shortens the time allowed for chain folding and vice-versa), (4) because of chain entanglements, crystallinity increases as the cooling rate decreases, until hitting a plateau (100% crystallinity is never reached, not even at the lowest values of

$\beta_{c,SCG}$ or in isothermal conditions), (5) when the cooling rate reaches the threshold $\beta_{c,SDG}$, a self-doped glass (SDG) is obtained, and (6) it is necessary to reach values of cooling rate at least equal to a further threshold $\beta_{c,CHG}$ to obtain a so-called chemically-homogeneous glass (CHG). Cooling down crystallizable polymers at different cooling rates $\beta_c < \beta_{c,SDG}$ may affect their crystalline phase not only quantitatively, but also qualitatively, worsening the intrinsic complexity due to chemistry (polymorphism [44–47], imperfect chain folding with extended coupling between the crystalline and amorphous domains [48–50], development of mesophases [51,52]. . .). In general, reducing the cooling rate favors the development of larger amounts of more perfect crystals, with positive consequences on properties such as the Young's modulus [53], whereas increasing the cooling rate improves transparency, toughness and elongation at break [54].

Conventional calorimetric measurements are performed at scanning rates generally limited to a few tens of degrees per minute. Fast Scanning Calorimetry (FSC) allows the extension of the experimental window of several orders of magnitude in terms of scanning rate, giving access to cooling rates up to 40,000 K s^{-1} [41,55], which is ideal to reproduce industrial processing conditions (Figure 1). This work illustrates how to estimate the critical cooling rate $\beta_{c,SDG}$ (threshold between semi-crystalline and fully amorphous samples obtained during melt-quenching) for fast-crystallizing polyesters, taking the example of a series of poly(alkylene *trans*-1,4-cyclohexanedicarboxylate) (PCHs) with a number of methylene groups in the main structure of the repeating unit n_{CH_2} varying from 3 to 6, and combining several calorimetric techniques, i.e., Differential Scanning Calorimetry (DSC), stochastically-modulated DSC (TOPEM) and Fast Scanning Calorimetry (FSC).

2. Materials and Methods

2.1. Reagents

Trans-1,4-cyclohexanedicarboxylic acid (95%, 5 mol% *cis*-isomers) (CHDA) was purchased from Fluorochem (Hadfield, UK). 1,3-propanediol (PD) (98%) was purchased from Carbosynth. 1,4-butanediol (BD) (99%) was purchased from Sigma-Aldrich. 1,5-pentanediol (PeD) (97%) was purchased from Fluka Chemika. 1,6-hexanediol (HD) (>97%) was purchased from TCI. Titanium tetrabutoxide $Ti(OBu)_4$ (TBT) was purchased from Sigma-Aldrich. All reagents were used as received.

2.2. Polymer Synthesis

Four poly(alkylene *trans*-1,4-cyclohexanecarboxylate)s (PCHs) were synthesized by a two-step solvent-free melt polycondensation procedure consisting in the esterification reaction between a bifunctional alicyclic acid (CHDA) and a bifunctional linear aliphatic glycol. Depending on the selected glycol, four homopolymers were obtained with a variable number of methylene groups n_{CH_2} in the glycolic fraction of their repeating unit. Poly(propylene *trans*-1,4-cyclohexanedicarboxylate) (PPCE) with n_{CH_2} = 3 was obtained with PD. Poly (butylene *trans*-1,4-cyclohexane dicarboxylate) (PBCE) with n_{CH_2} = 4 was obtained with BD. Poly(pentamethylene *trans*-1,4-cyclohexanedicarboxylate) (PPeCE) with n_{CH_2} = 5 was obtained with PeD. Poly (hexamethylene *trans*-1,4-cyclohexanedicarboxylate) (PHCE) with n_{CH_2} = 6 was obtained with HD. Each synthesis was carried out starting from CHDA, the selected diol (100% molar excess with respect to the diacid), and TBT (200 ppm) as catalyst, within a 250-mL glass reactor stirred at about 50 rpm in a thermostated oil bath connected to a six-bulb condenser to prevent the evaporation of the reagents. The molar excess of diol promoted the dissolution of the diacid. In the first step, the temperature was set to 190 °C and kept constant for about 1.5 h under a continuous flow of pure nitrogen (50 mL min^{-1}) and a pressure of 1 atm, allowing to distill off up to 90% of the theoretical amount of water. At the beginning of the second step, stirring was increased to 100 rpm and the temperature was raised to 200 °C to remove the excess of diol, then the temperature was further raised to 210 °C and the pressure was gradually reduced to 0.06 mbar to promote the transesterification reactions. The synthesis was carried out for 2 h, until a high and constant torque value was measured indicating that a high molecular weight was reached, and no more distillation was observed. After the synthesis, all the PCHs were purified by dissolution in chloroform and precipitation in a beaker filled with methanol in large excess (chloroform:methanol 1:10). After purification by dissolution/precipitation, the samples were dried under vacuum (1 h at 0.1 mbar). The samples were continuously stored under vacuum at room temperature, in a dessicator with phosphorus pentoxide P_2O_5 to reduce exposure to humidity until characterization.

2.3. Chemical Characterizations

^{1}H-NMR analyses were carried out to confirm the chemical structure and the *cis*/*trans* isomer ratio of the synthesized PCHs. The spectra were acquired using a Varian XL-400 NMR spectrometer (Palo Alto, CA, USA) at room temperature (relaxation time = 0 s, acquisition time = 1 s, 100 repetitions). The polymers were dissolved by introducing about 15 mg of sample in 1 mL of deuterated chloroform (containing 0.03 % tetramethylsilane, TMS, as internal reference). The number-average molecular weight ($\overline{M}_n$), the weight-average molecular weight ($\overline{M}_w$), the dispersity ($Đ$) and the degree of polymerization (DP_n) of the samples were estimated thanks to the data obtained by Gel Permeation Chromatography (GPC) analysis performed at 30 °C using an HPLC 1100 chromatograph (Agilent Technologies, Santa Clara, CA, USA) equipped with a PLgel 5 mm MiniMIX-C column. A chloroform solution was used as eluent with a flow of 0.3 mL min^{-1}, and sample concentrations of about 2 mg mL^{-1} were adopted. The calibration curve was obtained using polystyrene standards with a molecular range of 800–100,000 g mol^{-1}. ^{1}H-NMR spectra of the four polyesters can be found in the Supporting Information (Figures S1–S4). A summary of the chemical features of the investigated PCHs is given in Table 1.

Table 1. List of poly(alkylene *trans*-1,4-cyclohexanedicarboxylate)s (PCHs) investigated in this study, along with their repeating unit, the molar mass of their repeating unit (M_0), the number-average molecular weight ($\overline{M}_n$), the weight-average molecular weight ($\overline{M}_w$), the dispersity ($Đ$), the degree of polymerization (DP_n) and the percentage of *cis*-isomers in 1,4-cyclohexanedicarboxylic acid.

Sample	Repeating Unit	M_0 (g mol^{-1})	$\overline{M}_n$ (g mol^{-1})	$\overline{M}_w$ (g mol^{-1})	$Đ$	DP_n	*cis*-Isomers (%)
PPCE	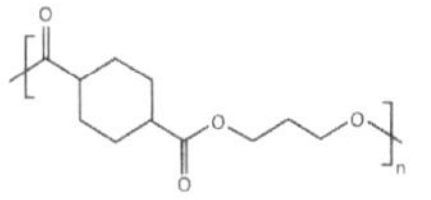	212	62,462	96,657	1.5	295	6.6
PBCE	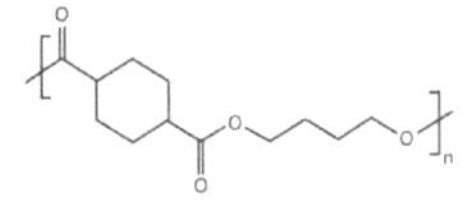	226	68,703	93,382	1.4	304	5.5
PPeCE	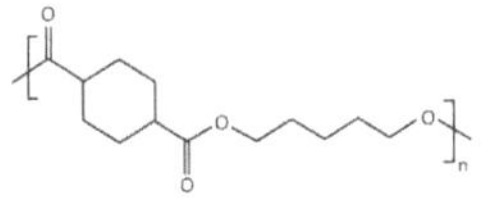	240	57,855	83,592	1.4	241	5.4
PHCE	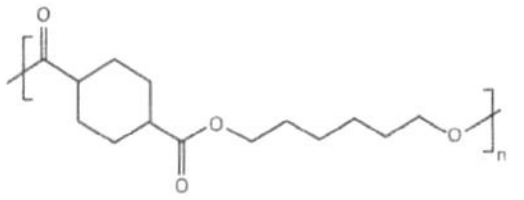	254	38,666	58,734	1.5	152	9.4

PPCE = poly(propylene *trans*-1,4-cyclohexanedicarboxylate); PBCE = poly(butylene *trans*-1,4-cyclohexanedicarboxylate); PPeCE = poly(pentamethylene *trans*-1,4-cyclohexanedicarboxylate); PHCE = poly(hexamethylene *trans*-1,4-cyclohexanedicarboxylate).

2.4. Thermal Characterizations

Given the aim of the study, the thermal characterizations mostly consisted in calorimetric investigations. Different equipments were used to cover the largest possible range of cooling rates. The lowest cooling rates were covered by Differential Scanning Calorimetry (DSC) and stochastically-modulated DSC (TOPEM), whereas the highest cooling rates were covered by Fast Scanning Calorimetry (FSC) with either UFS1 (lower range) or UFH1 (upper range) MultiSTAR sensors.

DSC and TOPEM experiments were performed using a DSC 3+ (Mettler Toledo) equipped with a FRS 6+ sensor and piloted by the STARe software. Calibrations for temperature, enthalpy, and thermal lag (tau lag) were achieved using zinc, indium, and water standards. The samples for both DSC and TOPEM (masses comprised between 2 and 10 mg) were placed in 40 µL sealed aluminum pans. The measurements were conducted under a constant nitrogen flow (50 mL min^{-1}). TOPEM measurements were performed with a pulse height of 0.1 K and pulse widths stochastically varying between 15 and 30 s, then an extrapolation to zero frequency was performed to obtain the quasi-static heat capacity $c_{p,0}$ as a function of temperature. More details about TOPEM can be found elsewhere [56].

FSC measurements were performed using a Flash DSC 2+ (Mettler Toledo) equipped with a Huber intracooler TC100 and also piloted by the STARe software. The samples for *FSC* (masses comprised between 10 and 200 ng) were obtained by successively cutting bulk samples into smaller pieces under a microscope, and then transferred by a hair of a paint brush onto the center of the active zone of a conditioned and temperature-corrected MultiSTAR sensor. The measurements were conducted under a constant argon flow (60 mL min^{-1}). The sensor support temperature was set at −95 °C.

For samples that could be successfully melt-quenched by conventional *DSC*, the mass deposited onto the *FSC* sensor was estimated according to Equation (1), i.e., by comparing the change in heat capacity at the glass transition temperature ΔC_p obtained with *FSC* and *DSC*, as previously reported in the literature [40].

$$m = \frac{\Delta C_p^{FSC}}{\Delta c_p^{DSC}} \tag{1}$$

where ΔC_p^{FSC} [J K^{-1}] is the heat capacity step estimated by *FSC* and Δc_p^{DSC} [J g^{-1} K^{-1}] is the heat capacity step measured by *DSC*, both at the glass transition temperature.

For samples that could not be obtained in their fully amorphous state within the cooling range provided by conventional *DSC*, the mass deposited onto the *FSC* sensor was estimated by comparing the values of heat capacity measured by *FSC* and specific heat capacity measured by TOPEM at the same temperature, according to Equation (2).

$$m = \frac{C_p^{FSC}(T)}{c_p^{TOPEM}(T)} \tag{2}$$

where C_p^{FSC} [J K^{-1}] is the heat capacity measured by *FSC* and c_p^{TOPEM} [J g^{-1} K^{-1}] is the specific heat capacity measured by TOPEM at the same temperature T. At least two temperatures were selected, one in the glassy and one in the liquid state, and the results were averaged.

Preliminary characterizations of the thermal behavior of each PCH were done by DSC through typical heating-cooling-heating ramps, with heating and cooling rates $\beta_h = |\beta_c|$ = 10 K min^{-1}. The first heating ramp brought the samples up to the melt, thus ensuring the best thermal contact with the bottom of the aluminum pan and erasing any previous thermal history. The subsequent cooling ramp allowed to compare the relative aptitude to crystallize when the four samples were cooled from the melt at a constant arbitrary cooling rate, providing the temperature range over which crystallization is expected to occur. The second heating ramp provided a glimpse on the microstructural differences induced on each sample due to their different chemical composition, despite the common cooling conditions.

Based on these preliminary results, a thermal protocol consisting in a series of successive cooling and heating ramps was designed for each PCH, with the purpose of evaluating its critical cooling rate $\beta_{c,SDG}$, i.e., the minimum cooling rate at which no crystallization is observed when the polymer is cooled down from the melt. The protocol consisted in heating up each sample with a constant heating rate β_h to a temperature slightly above its melting, holding it for 0.1 s to ensure that melting is complete, cooling it down through the glass transition to −90 °C at a constant cooling rate β_c, then heating it up again at 1000 K s^{-1} to check for any possible sign of crystals through melting. The protocol, whose temperature ranges and cooling rates were adjusted to each sample's thermal behavior (crystallization temperature T_c, glass transition temperature T_g, melting temperature range ΔT_m), is schematically represented in Figure 2. Five decades of cooling rates (from 2 K min^{-1} to 5000 K s^{-1}) were investigated to assess the critical cooling rates of PPCE, PBCE, PPeCE and PHCE.

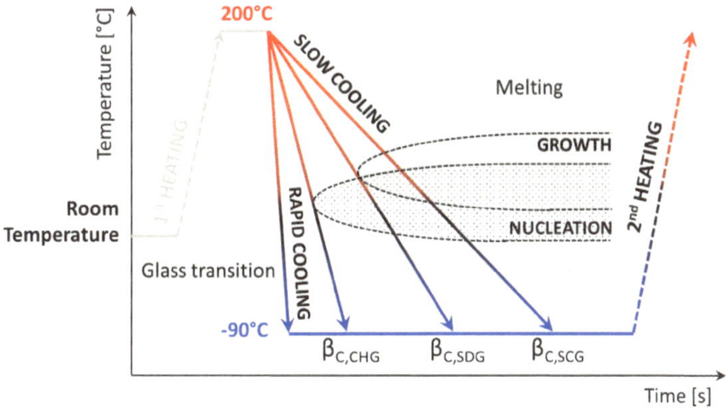

Figure 2. Schematic representation of the thermal protocol used to evaluate the critical cooling rate $\beta_{c,SDG}$, i.e., the minimum cooling rate allowing to obtain a fully amorphous polymer upon quenching from the molten state. The scans in terms of cooling rate covered five decades (from 2 K min^{-1} to 5000 K s^{-1}) thanks to a combination of different calorimetric techniques, i.e., Differential Scanning Calorimetry (DSC), stochastically-modulated DSC (TOPEM) and Fast Scanning Calorimetry (FSC).

3. Results and Discussion

Whenever an alicyclic moiety is introduced in a polymer backbone, the *cis*/*trans* isomer ratio is expected to potentially influence its crystallization behavior. In particular, the crystallization temperature T_c, glass transition temperature T_g and melting temperature T_m are known to depend on the *trans*-isomer content [57]. Even though the diacid selected for this work mostly contained *trans*-isomers (with only 5% *cis*-isomers), the *cis*/*trans* isomer ratio was evaluated for all the synthesized PCHs prior to any other thermal characterization (results reported in Table 1). Indeed, it cannot be excluded that the configuration of some molecules changes from *trans* to *cis* during polymerization, the process being generally favored by long exposures to relatively high temperatures in the presence of a catalyst. To minimize the occurrence of undesired *trans*/*cis* isomerization, i.e., to maintain the highest possible content of *trans*-isomers and consequently ensure high values of T_c, the temperature, time and catalyst content used for the synthesis should be minimized. On the other hand, these very same variables also control the increase in molecular weight during polymerization. Therefore, a compromise had to be found between the best conditions to get high values of T_c, and the best conditions to get a high molecular weight [58]. The synthesis performed in this work was optimized with respect to previous batches [36], for which slightly higher *cis*-isomer contents were obtained. The highest efficiency in maintaining a fixed *trans*-isomer content was observed when CHDA was combined with BD and PeD, i.e., for the synthesis of PBCE

(n_{CH_2} = 4) and PPeCE (n_{CH_2} = 5). The largest extent of undesired *trans/cis* isomerization was observed during the synthesis of PHCE (n_{CH_2} = 6), with a *cis*-isomer content however limited to less than 10 %. Both the molecular weight and the degree of polymerization decreased as n_{CH_2} increased, to the point that PHCE ended with a two-fold smaller molecular weight and DP$_n$ compared to PPCE. However, all PCHs were synthesized with satisfactory molecular weights and the same dispersity $Đ \approx 1.5$ (Table 1).

Figure 3 shows the preliminary results obtained by DSC on all the considered PCHs. The initial melting (first heating ramps) are not shown; the cooling ramps from the melt down to negative temperatures at $|\beta_c|$ = 10 K min^{-1} are represented by solid lines; the subsequent heating ramps at β_h = 10 K min^{-1} are represented by dashed lines. A zoom into the glass transition temperature range is reported in the inset to each graph. The thermal characteristics evidenced by the preliminary results in Figure 3 are summarized in Table 2. The most striking result is that PBCE and PHCE, i.e., the PCHs with an even number of methylene groups (n_{CH_2} = 4 and 6), are prone to a very rapid crystallization process (ΔT_c < 10 °C), whereas PPCE (n_{CH_2} = 3) crystallizes much slower and PPeCE (n_{CH_2} = 5) barely has the time to start the process and shows a wide crystallization peak (large ΔT_c) with a low intensity (Δh_c of 0.5 J g^{-1}). According to these results, the PCHs with an even value of n_{CH_2} are expected to have the highest critical cooling rates $|\beta_{c,SDG}|$, whereas the ones with an odd value of n_{CH_2} should be the easiest to melt-quench. In both cases, an increase in the number of methylene groups appears to profoundly change the crystallization behavior. When the number of methylene groups is even, the change is mostly seen in terms of T_c (with a decrease of about 45 °C as n_{CH_2} increases from 4 to 6) with no significant changes in ΔT_c. When the number of methylene groups is odd, the change is more dramatic. As n_{CH_2} increases from 3 to 5, the temperature range for crystallization spreads from about 20 °C to more than 40 °C, and the crystal growth is almost entirely suppressed. For purposes such as minimizing the cycle time for injection molding and yet developing a semi-crystalline microstructure, PCHs with an even number of methylene groups would be the best choice, because crystallization starts at high temperatures and proceeds very fast; the subsequent heating ramp, though, suggests that the crystalline phase grown upon cooling is complex, with a marked trend to reorganization for improved crystal perfection (the melting temperature range ΔT_m spreads over ≈80 °C, and the melting peak has a shape that recalls the double melting peak of polyhydroxyalkanoates [59,60], which are known to be prone to extensive crystalline reorganization upon heating). Crystalline reorganization occurs also in PCHs with an odd number of methylene groups, but apparently to a less extent, with a much narrower melting temperature range (ΔT_m < 25 °C) and just a small exotherm preceding the melting endotherm.

Irrespective of the odd or even character of n_{CH_2}, the melting temperature generally decreases as the number of methylene groups increases. However, the lowest melting (and crystallization) temperature is observed for PPeCE (n_{CH_2} = 5). This particular behavior has been reported for other polymers containing five methylene groups in their backbone [31,35]. This phenomenon could be due to some local polarization that builds up in odd-numbered polyesters, hindering the crystallization process, whereas in even-numbered polyesters the dipoles are aligned in opposite directions, allowing a more efficient chain folding and packing [34,61]. NMR experiments conducted on flexible alkyl chains with mesogenic groups at either end showed that, for n_{CH_2} = 5, the alkyl chain is in an all-*trans* conformation, except that the conformation around the C-O single bonds was found to be approximately *gauche* [61]. This *gauche* conformation forces the molecule to bend more with respect to an all-*trans* conformation, causing less efficient packing and therefore improving its glass-forming ability. A recent investigation on the crystal structure of PBCE confirms the all-*trans* conformation of its alkyl segments [62]. The values of glass transition temperature T_g consistenly decrease with the number of methylene groups introduced in the polymer backbone, from about 8 °C for PPCE (n_{CH_2} = 3) to about −21 °C for PHCE (n_{CH_2} = 6).

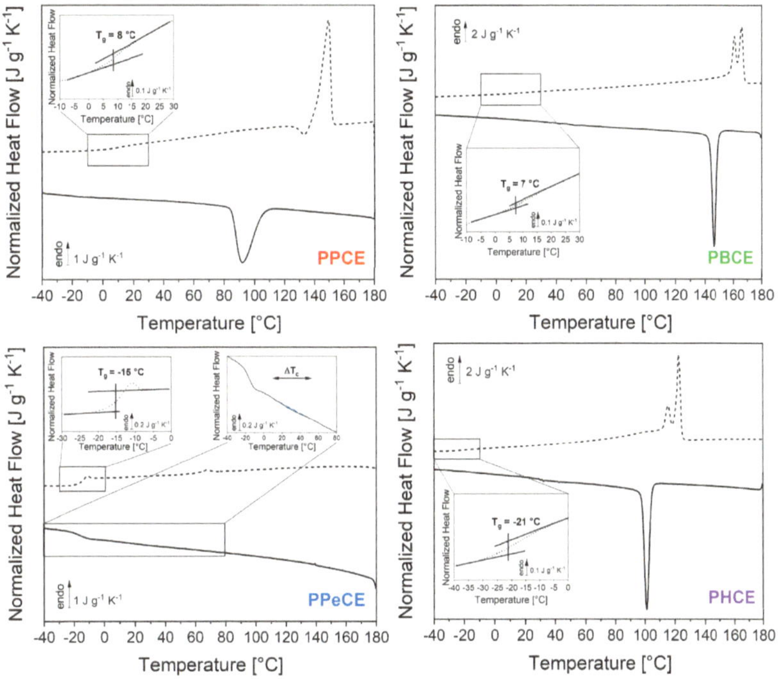

Figure 3. Thermal behavior recorded by DSC with cooling and heating rates $|\beta_c| = \beta_h = 10\ \text{K min}^{-1}$ after melting (the first heating ramp is not shown). A zoom into the glass transition temperature range is reported for each sample in the corresponding inset. For PPeCE, an additional inset zooms into the crystallization temperature range (dashed blue area). PPCE = poly(propylene *trans*-1,4-cyclohexanedicarboxylate). PBCE = poly(butylene *trans*-1,4-cyclohexanedicarboxylate) (PBCE). PPeCE = poly(pentamethylene *trans*-1,4-cyclohexanedicarboxylate). PHCE = poly(hexamethylene *trans*-1,4-cyclohexanedicarboxylate).

Table 2. Thermal characteristics of the investigated PCHs extracted from the thermograms in Figure 3. T_c is the crystallization temperature measured at the maximum of the exothermic peak observed upon cooling, ΔT_c is the temperature range over which crystallization occurs, Δh_c is the enthalpy of crystallization, T_g is the mid-point glass transition recorded upon heating (insets in Figure 3), ΔT_m is the melting temperature range, Δh_m is the enthalpy of melting calculated as the algebraic area under the curve in the melting temperature range.

SAMPLE	T_c (°C)	ΔT_c (°C)	Δh_c (J g^{-1})	T_g (°C)	ΔT_m (°C)	Δh_m (J g^{-1})
PPCE	92.2 ± 0.5	85–106	33 ± 1	8 ± 2	136–160	32 ± 2
PBCE	146.5 ± 0.5	144–150	39 ± 2	7 ± 2	92–171	55 ± 5
PPeCE	25 ± 5	6–51	0.5 ± 0.1	−15 ± 1	60–76	0.6 ± 0.1
PHCE	101.3 ± 0.5	98–105	43 ± 3	−21 ± 2	50–129	57 ± 5

PPCE = poly(propylene *trans*-1,4-cyclohexanedicarboxylate). PBCE = poly(butylene *trans*-1,4-cyclohexanedicarboxylate). PPeCE = poly(pentamethylene *trans*-1,4-cyclohexanedicarboxylate). PHCE = poly(hexamethylene *trans*-1,4-cyclohexanedicarboxylate).

Figure 4 (left column) reports the cooling ramps obtained for each PCH according to the thermal protocol sketched in Figure 2. The cooling ramps recorded at the lowest cooling rate $|\beta_c| = 0.033\ \text{K s}^{-1}$ were obtained by TOPEM (dashed-dotted lines). The cooling ramps plotted with dashed lines were obtained by conventional DSC. The cooling ramps recorded at the highest cooling rates were obtained by FSC (solid lines). A quick glance at the cooling curves confirms that PPeCE is the easiest to melt-quench, whereas PBCE is the most challenging. It is indeed necessary, at first, to check that no exothermal signals associated with crystallization are recorded during the cooling ramp. Based on this criterion, a threshold cooling rate could be identified for each PCH (left column, curves in colors). The corresponding critical cooling rates would be somewhere between 0.833 K s^{-1} (50 K min^{-1}) and 100 K s^{-1} for PPCE, between 400 and 500 K s^{-1} for PBCE, between 0.167 K s^{-1} (10 K min^{-1}) and 0.5 K s^{-1} (30 K min^{-1}) for PPeCE, and between 100 and 200 K s^{-1} for PHCE. It is however necessary to double-check these values by considering the heating ramps recorded right after cooling.

In Figure 4 (right column), the heating ramps recorded by TOPEM and DSC were obtained with $\beta_h = |\beta_c|$, and all the heating ramps obtained by FSC were recorded at β_h = 1000 K s^{-1}. The double-check consists in verifying that no endothermal signals associated with melting are recorded upon heating, or that the recorded enthalpy of melting is perfectly balanced by the enthalpy of cold crystallization (if cold crystallization occurs). Based on this additional criterion, the threshold cooling rate for melt-quenching can in some cases be readjusted (right column, curves in colors). About PPCE, for instance, Figure 4 shows that (1) the slowest cooling rate leads to a fully crystallized sample (TOPEM), (2) a progressive increase in the cooling rate hinders the crystallization process and let appears cold crystallization (DSC curves), and (3) FSC allows to melt-quench the sample (no cold crystallization is observed during the subsequent heating ramp because of the higher heating rate with respect to DSC measurements). In the case of PBCE, it is worth noting that (1) only FSC is able to provide sufficiently high cooling rates for an efficient melt-quenching, (2) the range of cooling rates previously identified as critical (400–500 K s^{-1}) is sufficient to suppress the crystallization from the melt and cold crystallization occurs during the subsequent heating ramps, (3) a heating rate faster than 1000 K s^{-1} is required to suppress cold crystallization and any other possible melting-recrystallization process. Focusing on the heating ramps recorded after cooling at 300, 400, 500 and 600 K s^{-1}, one may notice that the peak of cold crystallization keeps on evolving, which confirms that vitrification is more and more efficient; the shape of the peak stabilizes between 500 and 600 K s^{-1}, which is therefore considered as a better estimation of the range within which the critical cooling rate is supposed to fall. PPeCE is easily quenched with conventional cooling rates. As for PHCE, based solely on the cooling ramps one may guess that the critical cooling rate is between 100 and 200 K s^{-1} (left column, curves in colors), however the heating ramps clearly show that a crystalline phase is formed for cooling rates up to 500–1000 K s^{-1} (right column, curves in colors).

Of all the considered samples, PBCE is the only one being able to cold-crystallize despite the relatively high heating rates used in FSC experiments (1000 K s^{-1}) (Figure 4, right column). Figure 5 illustrates the additional criterion to meet to make sure that a polymer able to cold-crystallize is completely vitrified during melt-quenching. Figure 5 (left) shows a selection of heating ramps recorded after cooling from the melt at different $|\beta_c|$. The additional criterion consists in calculating the enthalpy of cold crystallization Δh_{cc} and the enthalpy of melting Δh_m, and then verifying if they are perfectly balanced. For cooling rates below 100 K s^{-1}, no cold crystallization is observed during the subsequent heating. When faster cooling rates are used, cold crystallization occurs, and the associated enthalpy Δh_{cc} gradually increases until reaching a plateau (18 J $g^{-}1$ for $|\beta_c|$ = 1000 K s^{-1}). On the other hand, the enthalpy of melting Δh_m decreases from 40 to 18 J g^{-1} as the cooling rate increases up to 100 K s^{-1}, and then stabilizes at 18 J g^{-1} as well. Figure 5 (right) reports the values of Δh_{cc}, Δh_m, and their algebraic difference $\Delta h_m - \Delta h_{cc}$, plotted against the cooling rate $|\beta_c|$ previously used for melt-quenching. With this additional criterion, the critical cooling rate $|\beta_{c,SDG}|$ for PBCE is rather estimated at about 3000 K s^{-1} (when $\Delta h_m - \Delta h_{cc}$ = 0).

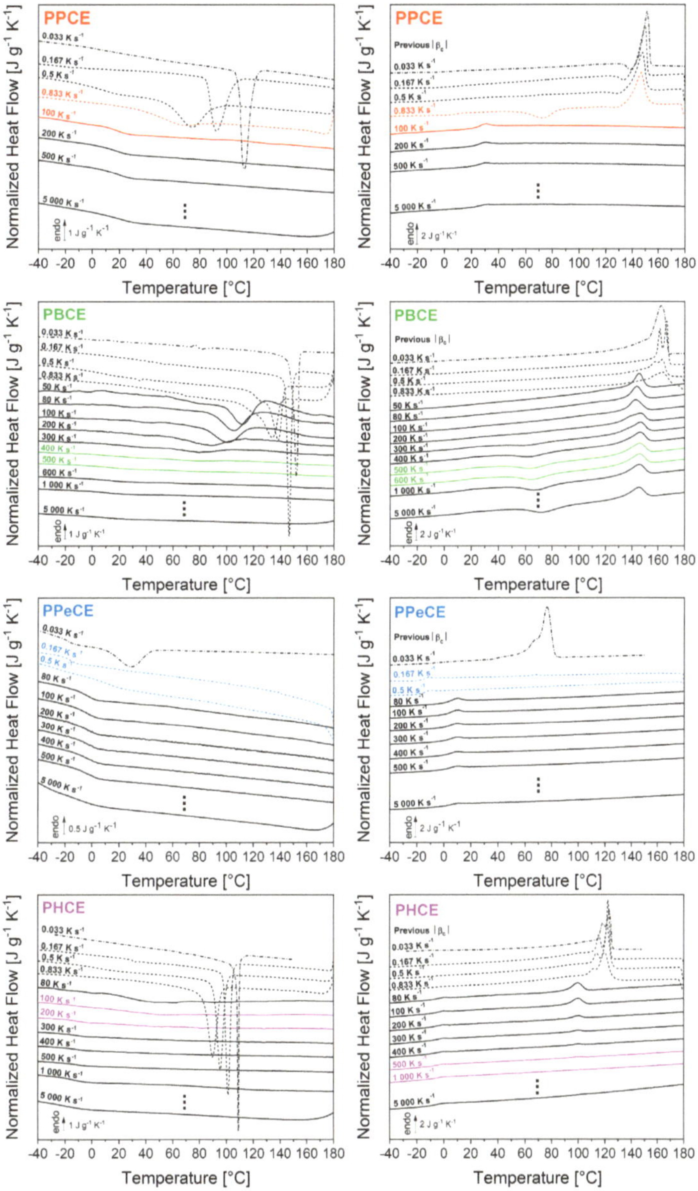

Figure 4. Normalized heat flow recorded upon cooling from the melt (left column) and during the subsequent heating ramp (right column). The curves at the lowest cooling rate $|\beta_c|$ = 0.033 K s^{-1} (dashed-dotted lines) were obtained by TOPEM. The curves plotted with dashed lines were obtained by conventional DSC. The curves at the highest cooling rates were obtained by FSC (solid lines). The heating ramps were obtained with $\beta_h = |\beta_c|$ (TOPEM and DSC) or β_h = 1000 K s^{-1} (FSC). The curves in colors highlight the range within which the critical cooling rate $|\beta_{c,SDG}|$ is supposed to fall based on either crystallization (left column) or subsequent melting (right column). PPCE = poly(propylene *trans*-1,4-cyclohexanedicarboxylate). PBCE = poly(butylene *trans*-1,4-cyclohexanedicarboxylate). PPeCE = poly(pentamethylene *trans*-1,4-cyclohexanedicarboxylate). PHCE = poly(hexamethylene *trans*-1,4-cyclohexanedicarboxylate).

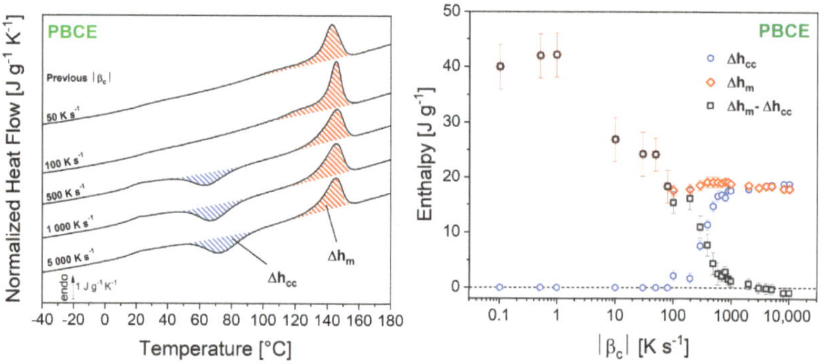

Figure 5. (**Left**) Calculation of the enthalpies of cold crystallization Δh_{cc} (blue-hatched areas) and melting Δh_m (red-hatched areas) for poly (butylene *trans*-1,4-cyclohexanedicarboxylate) (PBCE) previously cooled down from the melt at different cooling rates $|\beta_c|$. (**Right**) The enthalpies of cold crystallization Δh_{cc} (blue circles) and melting Δh_m (red diamonds), along with their algebraic difference $\Delta h_m - \Delta h_{cc}$ (black squares) are then plotted against the cooling rate β_c previously used to attempt melt-quenching.

Figure 6 shows the influence of the cooling rate on the crystallization temperature T_c (symbols), and more generally on the temperature range at which crystallization occurs (bars). Two phenomena are evidenced. The first observation is that, irrespective of n_{CH_2}, an increase in the cooling rate $|\beta_c|$ leads to a decrease in the crystallization temperature T_c and a broadening of the crystallization peak (increase in ΔT_{cc}), from less than 10 °C at the slowest cooling rates to more than 30 °C at the fastest cooling rates. The shift of T_c and the increase in ΔT_{cc} both prove that the observed transformation is controlled by nucleation; indeed, slow cooling enables the activation of the nuclei at higher temperature, whereas fast cooling retards and slows down the nucleation process [63].

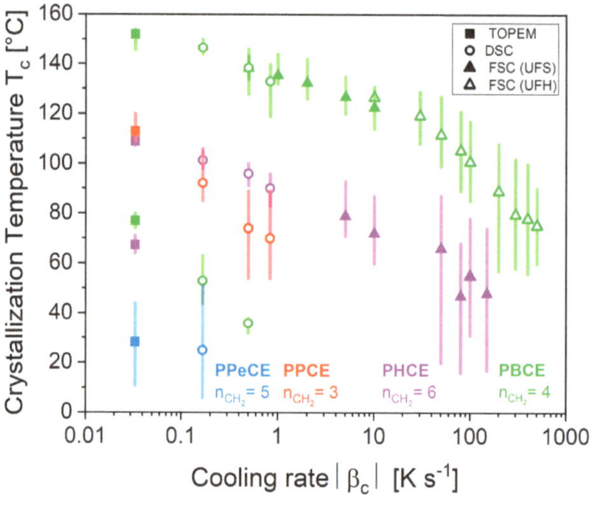

Figure 6. Crystallization temperatures T_c (symbols) and the temperature ranges at which crystallization occurs ΔT_{cc} (bars) measured upon cooling from the melt at different cooling rates $|\beta_c|$. The bars represent the temperature range between the onset ($T_{c,on}$) and endset ($T_{c,end}$) of the crystallization peak. PPCE = poly(propylene *trans*-1,4-cyclohexanedicarboxylate). PBCE = poly(butylene *trans*-1,4-cyclohexanedicarboxylate). PPeCE = poly(pentamethylene *trans*-1,4-cyclohexanedicarboxylate). PHCE = poly(hexamethylene *trans*-1,4-cyclohexanedicarboxylate).

The second observation is that the odd-numbered polyesters are associated with larger crystallization peaks and lower crystallization temperatures in comparison with their even-numbered counterparts at the same cooling rate. It should also be mentioned that PBCE has recently been shown to crystallize in at least two polymorphic forms, α and β, with the α-form observed upon slow cooling, and the β-form generated with sufficiently fast cooling from the melt (however the explored range of cooling rates did not exceed 50 K min^{-1}) [62]. With this in mind, the shift of T_c from 80 to 150 °C could be interpreted as the progressive transformation of the metastable β-form into the more stable α-form as the undercooling is reduced. One may also notice that both PBCE and PHCE have a double value of T_c at the lowest cooling rates. This could be due to either polymorphism (as previously mentioned for PBCE) or just to crystalline perfection through melting-recrystallization upon heating. So far, the literature has reported no evidence of polymorphism for PHCE.

Figure 7 shows a summary of the critical cooling rates $|\beta_{c,crit}|$ reported in the literature for common polymers and for a few other materials (silica, benzocaine, water). The critical cooling rates estimated in this work for PPCE, PBCE, PPeCE and PHCE are also reported for comparison purposes. As expected, the even-numbered PCHs (PBCE with n_{CH_2} = 4 and PHCE with n_{CH_2} = 6) require faster cooling rates to be effectively melt-quenched as compared to the odd-numbered PCHs (PPCE with n_{CH_2} = 3 and PPeCE with n_{CH_2} = 5). It also appears that, irrespective of the odd or even character of n_{CH_2}, the value of critical cooling rate decreases as the length of the alkyl chain within the repeating unit increases, suggesting that the methylene groups act as defects for crystal formation. This observation cannot be extended to other systems, such as poly (ethylene terephthalate) (PET) and poly (butylene terephthalate) (PBT), for which an increase of n_{CH_2} from 2 to 4 leads to a four-decade decrease in $|\beta_{c,crit}|$.

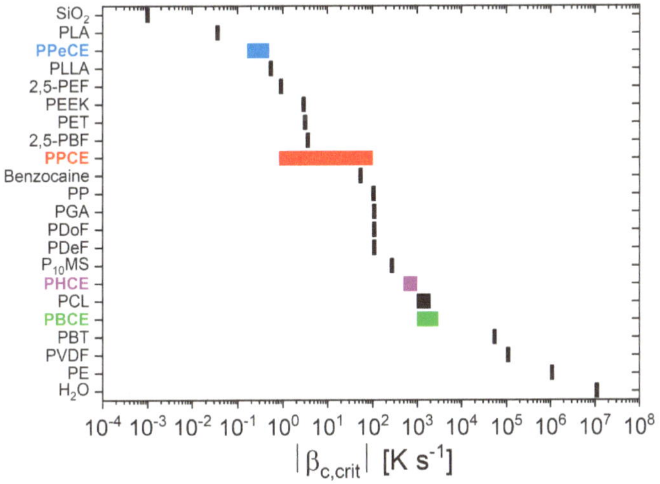

Figure 7. Values of critical cooling rates $|\beta_{c,crit}|$ issued from the literature [9,37,64–71] for common polymers as well as for a few other materials (silica, benzocaine, water). PPCE = poly(propylene *trans*-1,4-cyclohexanedicarboxylate). PBCE = poly(butylene *trans*-1,4-cyclohexanedicarboxylate). PPeCE = poly(pentamethylene *trans*-1,4-cyclohexanedicarboxylate). PHCE = poly(hexamethylene *trans*-1,4-cyclohexanedicarboxylate).

4. Conclusions

This work combines several calorimetric techniques, i.e., Differential Scanning Calorimetry (DSC), modulate-temperature DSC (MT-DSC), stochastically-modulated DSC (TOPEM) and Fast Scanning Calorimetry (FSC), to estimate the critical cooling rate necessary to melt-quench fast-crystallizing polyesters to their fully amorphous state. The method is applied to a series of poly(alkylene *trans*-1,4-cyclohexanedicarboxylate) (PCHs) with a number of methylene groups n_{CH_2} varying from 3 to 6, i.e., poly(propylene *trans*-1,4-cyclohexanedicarboxylate) (PPCE), poly(butylene *trans*-1,4-cyclohexanedicarboxylate) (PBCE), poly(pentamethylene *trans*-1,4-cyclohexanedicarboxylate) (PPeCE), and poly(hexamethylene *trans*-1,4-cyclohexanedicarboxylate) (PHCE).

MT-DSC and TOPEM (0.02–5 K min^{-1}) cannot give an insight of what happens during quenching, and DSC (5–50 K min^{-1}) is limited to slow-crystallizing polymers. FSC is sometimes necessary to emulate the cooling rates experienced by a material during manufacturing, even in the most extreme industrial processing conditions. Since polyesters can crystallize both during the cooling ramp from the melt and during the subsequent heating ramp (cold crystallization), three criteria should be considered: (1) the absence of exothermic events recorded during the cooling ramp (crystallization), (2) the absence of endothermic events recorded during the heating ramp (melting), and (3) the balance of exothermic and endothermic events recorded during the heating ramp (cold crystallization followed by melting).

The results show that, among the four PCHs investigated in this work, the even-numbered ones (PBCE and PHCE) require faster cooling rates compared to the odd-numbered ones (PPCE and PPeCE). Irrespective of the odd or even number of methylene groups introduced in the main structure of the repeating unit, the critical cooling rate decreases as n_{CH_2} increases, suggesting that the methylene groups act as defects during crystal formation. This behaviour seems to be specific to alicyclic polyesters, since the opposite is observed in their aromatic counterparts (poly (butylene terephthalate) (PBT) has a critical cooling rate four-decade smaller compared to poly (ethylene terephthalate) (PET)). PCHs with an even number of methylene groups would be the best choice to minimize the cycle time for injection molding without renouncing to developing a semi-crystalline microstructure, because crystallization starts at high temperatures; however, controlling crystal perfection could be challenging, because their crystallization proceeds very fast and the risk of crystalline reorganization upon further heating is high. For vitrification purposes, PCHs with an odd number of methylene groups are recommended, in particular PPeCE, which is the easiest to melt-quench to its fully amorphous state.

Supplementary Materials: The following supporting information can be downloaded at: https://www.mdpi.com/article/10.3390/polym16192792/s1, Figure S1: ^{1}H-NMR spectrum of poly(propylene *trans*-1,4-cyclohexanedicarboxylate) (PPCE). Figure S2: ^{1}H-NMR spectrum of poly(butylene *trans*-1,4-cyclohexanedicarboxylate) (PBCE). Figure S3: ^{1}H-NMR spectrum of poly(pentamethylene *trans*-1,4-cyclohexanedicarboxylate) (PPeCE). Figure S4: ^{1}H-NMR spectrum of poly(hexamethylene *trans*-1,4-cyclohexanedicarboxylate) (PHCE).

Author Contributions: Conceptualization, A.E. and A.S.-F.; data curation, K.H. and G.G.; formal analysis, K.H.; funding acquisition, A.S.-F. and N.L.; investigation, K.H. and G.G.; methodology, A.E. and A.S.-F.; project administration, A.E. and A.S.-F.; resources, A.E., A.S.-F., M.S. and N.L.; supervision, A.E., A.S.-F., M.S. and N.L.; validation, A.E., A.S.-F., M.S. and N.L.; visualization, K.H.; writing—original draft preparation, A.E. and K.H.; writing—review and editing, A.E., K.H., A.S.-F., G.G., M.S. and N.L. All authors have read and agreed to the published version of the manuscript.

Funding: This research received no external funding besides the Erasmus+ mobility grant and the PhD financial support for Kylian Hallavant.

Institutional Review Board Statement: Not applicable.

Informed Consent Statement: Not applicable.

Data Availability Statement: Dataset available on request from the authors.

Acknowledgments: The authors acknowledge the Erasmus+ Agency for providing Kylian Hallavant a mobility grant that allowed him to go to Bologna and participate in polymer synthesis. The authors acknowledge the Normandy Region for providing the PhD financial support for Kylian Hallavant. Mettler-Toledo is acknowledged for lending the DSC 3+ apparatus. The LabEx EMC^3 (Energy Materials and Clean Combustion Center) is acknowledged for financing the acquisition of the Flash DSC 2+

apparatus. Antonella Esposito is grateful to José Maria Kenny for having transmitted her the passion for scientific research about materials. It all started at Università degli Studi di Perugia, Terni, Italy.

Conflicts of Interest: The authors declare no conflicts of interest.

References

1. Geyer, R.; Jambeck, J.R.; Law, K.L. Production, use, and fate of all plastics ever made. *Sci. Adv.* **2017**, *3*, e1700782. [CrossRef] [PubMed]
2. Zia, K.M.; Noreen, A.; Zuber, M.; Tabasum, S.; Mujahid, M. Recent developments and future prospects on bio-based polyesters derived from renewable resources: A review. *Int. J. Biol. Macromol.* **2016**, *82*, 1028–1040. [CrossRef] [PubMed]
3. Wu, F.; Misra, M.; Mohanty, A.K. Challenges and new opportunities on barrier performance of biodegradable polymers for sustainable packaging. *Prog. Polym. Sci.* **2021**, *117*, 101395. [CrossRef]
4. Herrera, N.; Salaberria, A.M.; Mathew, A.P.; Oksman, K. Plasticized polylactic acid nanocomposite films with cellulose and chitin nanocrystals prepared using extrusion and compression molding with two cooling rates: Effects on mechanical, thermal and optical properties. *Compos. Part A Appl. Sci. Manuf.* **2016**, *83*, 89–97. [CrossRef]
5. Adamovsky, S.; Minakov, A.; Schick, C. Scanning microcalorimetry at high cooling rate. *Thermochim. Acta* **2003**, *403*, 55–63. [CrossRef]
6. Bardelcik, A.; Salisbury, C.P.; Winkler, S.; Wells, M.A.; Worswick, M.J. Effect of cooling rate on the high strain rate properties of boron steel. *Int. J. Impact Eng.* **2010**, *37*, 694–702. [CrossRef]
7. Buczek, A.; Telejko, T. Investigation of heat transfer coefficient during quenching in various cooling agents. *Int. J. Heat Fluid Flow* **2013**, *44*, 358–364. [CrossRef]
8. Carmeli, E.; Cavallo, D.; Tranchida, D. Instrument for mimicking fast cooling conditions of polymers: Design and case studies on polypropylene. *Polym. Test.* **2021**, *97*, 107164. [CrossRef]
9. Gradys, A.; Sajkiewicz, P.; Zhuravlev, E.; Schick, C. Kinetics of isothermal and non-isothermal crystallization of poly(vinylidene fluoride) by fast scanning calorimetry. *Polymer* **2016**, *82*, 40–48. [CrossRef]
10. Han, T.; Chen, J.; Wei, Z.; Qu, N.; Liu, Y.; Yang, D.; Zhao, S.; Lai, Z.; Jiang, M.; Zhu, J. Effect of cooling rate on microstructure and mechanical properties of AlCrFe2Ni2 medium entropy alloy fabricated by laser powder bed fusion. *J. Mater. Res. Technol.* **2023**, *25*, 4063–4073. [CrossRef]
11. Hyer, H.; Zhou, L.; Park, S.; Gottsfritz, G.; Benson, G.; Tolentino, B.; McWilliams, B.; Cho, K.; Sohn, Y. Understanding the Laser Powder Bed Fusion of AlSi10Mg Alloy. *Metallogr. Microstruct. Anal.* **2020**, *9*, 484–502. [CrossRef]
12. Kamal, M.R.; Kalyon, D. Heat transfer and microstructure in extrusion blowmolding. *Polym. Eng. Sci.* **1983**, *23*, 503–509. [CrossRef]
13. Li, X.; Song, W.; Yang, K.; Krishnan, N.A.; Wang, B.; Smedskjaer, M.M.; Mauro, J.C.; Sant, G.; Balonis, M.; Bauchy, M. Cooling rate effects in sodium silicate glasses: Bridging the gap between molecular dynamics simulations and experiments. *J. Chem. Phys.* **2017**, *147*, 074501. [CrossRef] [PubMed]
14. Minakov, A.A.; Schick, C. Dynamics of the temperature distribution in ultra-fast thin-film calorimeter sensors. *Thermochim. Acta* **2015**, *603*, 205–217. [CrossRef]
15. Pijpers, T.F.; Mathot, V.B.; Goderis, B.; Scherrenberg, R.L.; van der Vegte, E.W. High-Speed Calorimetry for the Study of the Kinetics of (De)vitrification, Crystallization, and Melting of Macromolecules. *Macromolecules* **2002**, *35*, 3601–3613. [CrossRef]
16. Poorhaydari, K.; Patchett, B.M.; Ivey, D.G. Estimation of cooling rate in the welding of plates with intermediate thickness. *Weld. J.* **2005**, *84*, 149s–155s.
17. Seppala, J.E.; Migler, K.D. Infrared thermography of welding zones produced by polymer extrusion additive manufacturing. *Addit. Manuf.* **2016**, *12*, 71–76. [CrossRef]
18. Shen, Y.; Li, Y.; Chen, C.; Tsai, H.L. 3D printing of large, complex metallic glass structures. *Mater. Des.* **2017**, *117*, 213–222. [CrossRef]
19. Suplicz, A.; Szabo, F.; Kovacs, J. Injection molding of ceramic filled polypropylene: The effect of thermal conductivity and cooling rate on crystallinity. *Thermochim. Acta* **2013**, *574*, 145–150. [CrossRef]
20. Vanden Poel, G.; Mathot, V.B. High performance differential scanning calorimetry (HPer DSC): A powerful analytical tool for the study of the metastability of polymers. *Thermochim. Acta* **2007**, *461*, 107–121. [CrossRef]
21. Takata, N.; Liu, M.; Li, H.; Suzuki, A.; Kobashi, M. Fast scanning calorimetry study of Al alloy powder for understanding microstructural development in laser powder bed fusion. *Mater. Des.* **2022**, *219*, 110830. [CrossRef]
22. Zhuravlev, E.; Schick, C. Fast scanning power compensated differential scanning nano-calorimeter: 1. The device. *Thermochim. Acta* **2010**, *505*, 1–13. [CrossRef]
23. Zuidema, H.; Peters, G.W.M.; Meijer, H.E.H. Influence of cooling rate on pVT-data of semicrystalline polymers. *J. Appl. Polym. Sci.* **2001**, *82*, 1170–1186. [CrossRef]
24. Monnier, X.; Maigret, J.E.; Lourdin, D.; Saiter, A. Glass transition of anhydrous starch by fast scanning calorimetry. *Carbohydr. Polym.* **2017**, *173*, 77–83. [CrossRef]
25. Genovese, L.; Lotti, N.; Gazzano, M.; Finelli, L.; Munari, A. New eco-friendly random copolyesters based on poly(propylene cyclohexanedicarboxylate): Structure-properties relationships. *EXPRESS Polym. Lett.* **2015**, *9*, 972–983. [CrossRef]
26. Gigli, M.; Lotti, N.; Gazzano, M.; Siracusa, V.; Finelli, L.; Munari, A.; Dalla Rosa, M. Fully Aliphatic Copolyesters Based on Poly(butylene 1,4-cyclohexanedicarboxylate) with Promising Mechanical and Barrier Properties for Food Packaging Applications. *Ind. Eng. Chem. Res.* **2013**, *52*, 12876–12886. [CrossRef]

27. Guidotti, G.; Soccio, M.; Siracusa, V.; Gazzano, M.; Munari, A.; Lotti, N. Novel Random Copolymers of Poly(butylene 1,4-cyclohexane dicarboxylate) with Outstanding Barrier Properties for Green and Sustainable Packaging: Content and Length of Aliphatic Side Chains as Efficient Tools to Tailor the Material's Final Performance. *Polymers* **2018**, *10*, 866. [CrossRef]
28. Soccio, M.; Lotti, N.; Finelli, L.; Gazzano, M.; Munari, A. Aliphatic poly(propylene dicarboxylate)s: Effect of chain length on thermal properties and crystallization kinetics. *Polymer* **2007**, *48*, 3125–3136. [CrossRef]
29. Siracusa, V.; Genovese, L.; Ingrao, C.; Munari, A.; Lotti, N. Barrier Properties of Poly(Propylene Cyclohexanedicarboxylate) Random Eco-Friendly Copolyesters. *Polymers* **2018**, *10*, 502. [CrossRef]
30. Kumar, N.; Chaudhary, S.; Singh, P.; Thapa, K.B.; Kumar, D. Electro-optical odd-even effect of APAPA liquid crystal molecules studied under the influence of an extraneous electric field (THz): A theoretical approach. *J. Mol. Liq.* **2020**, *318*, 114254. [CrossRef]
31. Zhou, C.; Wei, Z.; Yu, Y.; Shao, S.; Leng, X.; Wang, Y.; Li, Y. Biobased long-chain aliphatic polyesters of 1,12-dodecanedioic acid with a variety of diols: Odd-even effect and mechanical properties. *Mater. Today Commun.* **2019**, *19*, 450–458. [CrossRef]
32. Zhang, X.; Zuo, X.; Ortmann, P.; Mecking, S.; Alamo, R.G. Crystallization of Long-Spaced Precision Polyacetals I: Melting and Recrystallization of Rapidly Formed Crystallites. *Macromolecules* **2019**, *52*, 4934–4948. [CrossRef]
33. Pérez-Camargo, R.A.; Meabe, L.; Liu, G.; Sardon, H.; Zhao, Y.; Wang, D.; Müller, A.J. Even–Odd Effect in Aliphatic Polycarbonates with Different Chain Lengths: From Poly (Hexamethylene Carbonate) to Poly (Dodecamethylene Carbonate). *Macromolecules* **2021**, *54*, 259–271. [CrossRef]
34. Flores, I.; Pérez-Camargo, R.A.; Gabirondo, E.; Caputo, M.R.; Liu, G.; Wang, D.; Sardon, H.; Müller, A.J. Unexpected Structural Properties in the Saturation Region of the Odd–Even Effects in Aliphatic Polyethers: Influence of Crystallization Conditions. *Macromolecules* **2022**, *55*, 584–594. [CrossRef]
35. Masubuchi, T.; Sakai, M.; Kojio, K.; Furukawa, M.; Aoyagi, T. Structure and Properties of Aliphatic Poly(carbonate) glycols with Different Methylene Unit Length. *e-J. Soft Mater.* **2007**, *3*, 55–63. [CrossRef]
36. Guidotti, G.; Fosse, C.; Soccio, M.; Gazzano, M.; Siracusa, V.; Delbreilh, L.; Esposito, A.; Lotti, N. Synthesis and properties of poly(alkylene trans-1,4-cyclohexanedicarboxylate)s with different glycolic subunits. *Polym. Degrad. Stab.* 2024, *under review*.
37. Salmerón Sánchez, M.; Mathot, V.B.F.; Vanden Poel, G.; Gómez Ribelles, J.L. Effect of the Cooling Rate on the Nucleation Kinetics of Poly(L-Lactic Acid) and Its Influence on Morphology. *Macromolecules* **2007**, *40*, 7989–7997. [CrossRef]
38. Hu, Y.; Liao, Y.; Zheng, Y.; Ikeda, K.; Okabe, R.; Wu, R.; Ozaki, R.; Xu, J.; Xu, Q. Influence of Cooling Rate on Crystallization Behavior of Semi-Crystalline Polypropylene: Experiments and Mathematical Modeling. *Polymers* **2022**, *14*, 3646. [CrossRef]
39. Boyer, S.; Haudin, J.M. Crystallization of polymers at constant and high cooling rates: A new hot-stage microscopy set-up. *Polym. Test.* **2010**, *29*, 445–452. [CrossRef]
40. Mejres, M.; Hallavant, K.; Guidotti, G.; Soccio, M.; Lotti, N.; Esposito, A.; Saiter-Fourcin, A. Physical aging of a biodegradable alicyclic polymer: Poly (pentamethylene trans-1,4-cyclohexanedicarboxylate). *J. Non-Cryst. Solids* **2024**, *629*, 122874. [CrossRef]
41. Hallavant, K.; Mejres, M.; Schawe, J.E.K.; Esposito, A.; Saiter-Fourcin, A. Influence of Chemical Composition and Structure on the Cooperative Fluctuation in Supercooled Glass-Forming Liquids. *J. Phys. Chem. Lett.* **2024**, *15*, 4508–4514. [CrossRef] [PubMed]
42. Monnier, X.; Saiter, A.; Dargent, E. Vitrification of PLA by fast scanning calorimetry: Towards unique glass above critical cooling rate? *Thermochim. Acta* **2017**, *658*, 47–54. [CrossRef]
43. Schawe, J.E.; Löffler, J. Existence of multiple critical cooling rates which generate different types of monolithic metallic glass. *Nat. Commun.* **2019**, *10*, 1337. [CrossRef] [PubMed]
44. Zheng, Y.; Pan, P. Crystallization of biodegradable and biobased polyesters: Polymorphism, cocrystallization, and structure-property relationship. *Prog. Polym. Sci.* **2020**, *109*, 101291. [CrossRef]
45. Corradini, P.; Guerra, G. Polymorphism in polymers. In *Macromolecules: Synthesis, Order And Advanced Properties*; Springer: Berlin/Heidelberg, Germany, 1992; pp. 183–217.
46. Di Lorenzo, M.L.; Cocca, M.; Malinconico, M. Crystal polymorphism of poly(l-lactic acid) and its influence on thermal properties. *Thermochim. Acta* **2011**, *522*, 110–117. [CrossRef]
47. Cocca, M.; Lorenzo, M.L.D.; Malinconico, M.; Frezza, V. Influence of crystal polymorphism on mechanical and barrier properties of poly(l-lactic acid). *Eur. Polym. J.* **2011**, *47*, 1073–1080. [CrossRef]
48. Esposito, A.; Delpouve, N.; Causin, V.; Dhotel, A.; Delbreilh, L.; Dargent, E. From a Three-Phase Model to a Continuous Description of Molecular Mobility in Semicrystalline Poly(hydroxybutyrate-co-hydroxyvalerate). *Macromolecules* **2016**, *49*, 4850–4861. [CrossRef]
49. Fosse, C.; Esposito, A.; Lemechko, P.; Salim, Y.S.; Causin, V.; Gaucher, V.; Bruzaud, S.; Sudesh, K.; Delbreilh, L.; Dargent, E. Effect of Chemical Composition on Molecular Mobility and Phase Coupling in Poly(3-hydroxybutyrate-co-3-hydroxyvalerate) and Poly(3-hydroxybutyrate-co-3-hydroxyhexanoate) with Different Comonomer Contents. *J. Polym. Environ.* **2023**, *31*, 4430–4447. [CrossRef]
50. Bourdet, A.; Fosse, C.; Garda, M.R.; Thiyagarajan, S.; Delbreilh, L.; Esposito, A.; Dargent, E. Microstructural consequences of isothermal crystallization in homo- and co-polyesters based on 2,5- and 2,4-furandicarboxylic acid. *Polymer* **2023**, *272*, 125835. [CrossRef]
51. Gradys, A.; Sajkiewicz, P.; Minakov, A.; Adamovsky, S.; Schick, C.; Hashimoto, T.; Saijo, K. Crystallization of polypropylene at various cooling rates. *Mater. Sci. Eng. A* **2005**, *413–414*, 442–446. [CrossRef]
52. Androsch, R.; Di Lorenzo, M.L.; Schick, C.; Wunderlich, B. Mesophases in polyethylene, polypropylene, and poly(1-butene). *Polymer* **2010**, *51*, 4639–4662. [CrossRef]
53. Li, J.; Zhu, Z.; Li, T.; Peng, X.; Jiang, S.; Turng, L.S. Quantification of the Young's modulus for polypropylene: Influence of initial crystallinity and service temperature. *J. Appl. Polym. Sci.* **2020**, *137*, 48581. [CrossRef]

54. Wright, D.; Dunk, R.; Bouvart, D.; Autran, M. The effect of crystallinity on the properties of injection moulded polypropylene and polyacetal. *Polymer* **1988**, *29*, 793–796. [CrossRef]
55. Schick, C.; Mathot, V. (Eds.) *Fast Scanning Calorimetry*; Springer: Cham, Switzerland, 2016.
56. Schawe, J.; Hütter, T.; Heitz, C.; Alig, I.; Lellinger, D. Stochastic temperature modulation: A new technique in temperature-modulated DSC. *Thermochim. Acta* **2006**, *446*, 147–155. [CrossRef]
57. Klein, J.P.; Gdowski, Z.M.; Register, R.A. Controlling thermomechanical behavior of semicrystalline hydrogenated polynorbornene through the *cis*- to *trans*-cyclopentylene ratio. *J. Polym. Sci.* **2022**, *60*, 266–275. [CrossRef]
58. Brunelle, D.J.; Jang, T. Optimization of poly(1,4-cyclohexylidene cyclohexane-1,4-dicarboxylate) (PCCD) preparation for increased crystallinity. *Polymer* **2006**, *47*, 4094–4104. [CrossRef]
59. Puente, J.A.S.; Esposito, A.; Chivrac, F.; Dargent, E. Effects of Size and Specific Surface Area of Boron Nitride Particles on the Crystallization of Bacterial Poly(3-hydroxybutyrate-co-3-hydroxyvalerate). *Macromol. Symp.* **2013**, *328*, 8–19. [CrossRef]
60. Puente, J.A.S.; Esposito, A.; Chivrac, F.; Dargent, E. Effect of boron nitride as a nucleating agent on the crystallization of bacterial poly(3-hydroxybutyrate). *J. Appl. Polym. Sci.* **2013**, *128*, 2586–2594. [CrossRef]
61. Duer, M.J.; Roper, C. A solid-state NMR investigation of the odd–even effect in a series of liquid-crystal dimers. *Phys. Chem. Chem. Phys.* **2003**, *5*, 3034–3041. [CrossRef]
62. Hu, W.; Ma, T.; Zhou, Y.; Soccio, M.; Lotti, N.; Cavallo, D.; Wang, D.; Liu, G. Crystal Structure and Polymorphism of Poly(butylene-trans-1,4-cyclohexane dicarboxylate). *Macromolecules* **2024**, *57*, 4374–4384. [CrossRef]
63. Martins, J.C.; Novack, K.M.; Gomes, A.S. Non-isothermal crystallization kinetics of thermotropic polyesters with flexible spacers in the main chain. *Polymer* **1998**, *39*, 6941–6944. [CrossRef]
64. Barandiarán, J.; Colmenero, J. Continuous cooling approximation for the formation of a glass. *J. Non-Cryst. Solids* **1981**, *46*, 277–287. [CrossRef]
65. Chen, Y.; Xie, K.; He, Y.; Hu, W. Fast-Scanning Chip-Calorimetry Measurement of Crystallization Kinetics of Poly(Glycolic Acid). *Polymers* **2021**, *13*, 891. [CrossRef]
66. Furushima, Y.; Kumazawa, S.; Umetsu, H.; Toda, A.; Zhuravlev, E.; Schick, C. Melting and recrystallization kinetics of poly(butylene terephthalate). *Polymer* **2017**, *109*, 307–314. [CrossRef]
67. Kagawa, F.; Oike, H. Quenching of Charge and Spin Degrees of Freedom in Condensed Matter. *Adv. Mater.* **2017**, *29*, 1601979. [CrossRef]
68. Mukhametzyanov, T.A.; Andrianov, R.A.; Bolmatenkov, D.N.; Yagofarov, M.I.; Solomonov, B.N.; Schick, C. Nucleation and crystallization of deeply supercooled benzocaine, a rapidly crystallizing organic compound: A Fast scanning calorimetry investigation. *Thermochim. Acta* **2023**, *730*, 179613. [CrossRef]
69. Papageorgiou, G.Z.; Papageorgiou, D.G.; Tsanaktsis, V.; Bikiaris, D.N. Synthesis of the bio-based polyester poly(propylene 2,5-furan dicarboxylate). Comparison of thermal behavior and solid state structure with its terephthalate and naphthalate homologues. *Polymer* **2015**, *62*, 28–38. [CrossRef]
70. Papageorgiou, D.G.; Guigo, N.; Tsanaktsis, V.; Exarhopoulos, S.; Bikiaris, D.N.; Sbirrazzuoli, N.; Papageorgiou, G.Z. Fast Crystallization and Melting Behavior of a Long-Spaced Aliphatic Furandicarboxylate Biobased Polyester, Poly(dodecylene 2,5-furanoate). *Ind. Eng. Chem. Res.* **2016**, *55*, 5315–5326. [CrossRef]
71. Stoclet, G.; Gobius du Sart, G.; Yeniad, B.; de Vos, S.; Lefebvre, J. Isothermal crystallization and structural characterization of poly(ethylene-2,5-furanoate). *Polymer* **2015**, *72*, 165–176. [CrossRef]

MDPI AG
Grosspeteranlage 5
4052 Basel
Switzerland
Tel.: +41 61 683 77 34

Polymers Editorial Office
E-mail: polymers@mdpi.com
www.mdpi.com/journal/polymers

www.ingramcontent.com/pod-product-compliance
Ingram Content Group UK Ltd.
Pitfield, Milton Keynes, MK11 3LW, UK
UKHW061544051026
15625UKWH00012B/607
9783725840236